Robert Riding (Ed.)

Calcareous Algae and Stromatolites

With 16 Plates and 202 Figures

Springer-Verlag
Berlin Heidelberg New York London
Paris Tokyo Hong Kong Barcelona

Dr. ROBERT RIDING
University of Wales College of Cardiff
Cardiff CF1 3YE
United Kingdom

ISBN 3-540-52373-1 Springer-Verlag Berlin Heidelberg New York
ISBN 0-387-52373-1 Springer-Verlag New York Berlin Heidelberg

Library of Congress Cataloging in Publication Data. Calcareous algae and stromatolites / R. Riding [editor]. p. cm. Includes articles originally presented at the 4th International Symposium on Fossil Algae, held in Cardiff, July 1987. Includes index. ISBN 3-540-52373-1 (Berlin: alk. paper). – ISBN 0-387-52373-1 (New York: alk. paper) 1. Algae, Fossil – Congresses. 2. Stromatolites – Congresses, I. Riding, Robert. II. International Symposium on Fossil Algae (4th: 1987: Cardiff, Wales) QE955.C338 1990 561'.93 – dc20 90-10112

Typesetting: International Typesetters Inc., Makati, Philippines
32/3145(3011)-543210 – Printed on acid-free paper

Preface

"Calcareous algae and stromatolites" is shorthand for a wider array of organisms and fabrics that also includes calcified cyanobacteria, plus thrombolites and other microbial carbonates. Composition is the link: these are all important components of $CaCO_3$ sediments, from Archaean to present and from the ocean floor to streams and lakes. It is hardly possible to examine limestones of any age without encountering them. Simultaneously they are fossils, sediments, and environmental indicators. It is the range of significance, coupled with the breadth of their distribution in time and space, which compels their study.

Modern calcareous marine algae mainly include reds (corallines, squamariaceans, and the nemalialean *Galaxaura*) and greens (dasycladaleans, udoteaceans, halimedaceans). Blue-greens, of course, are cyanobacteria and not algae, and significantly, although they are largely responsible for Recent tidal flat stromatolites, they are not calcified in the same way that pre-Cenozoic marine blue-greens are. It is in the freshwater environment of calcareous streams and lakes that we find modern calcified cyanobacteria, and they are commonly associated with the only major group of non-marine calcareous algae, the charophytes.

However, in the past, and especially in the Palaeozoic and Mesozoic, things look radically different. Mingling with the ancestors of the modern flora are distinct, and often problematic, organisms. Their organization could be consistent with an algal or cyanobacterial affinity, and they have been treated as such; but sometimes they hardly seem to fit conceptions founded in the Recent. This is another world, and it informs and confuses at the same time.

Most reported calcareous "algae" in the Cambrian are actually more likely to be cyanobacteria, and are especially important in reefs. Never again, except briefly in the Upper Devonian, do these forms assume such importance in marine environments. Dasycladaleans and halimedacean-like greens, strikingly similar to Recent forms, appear in the Ordovician. But any semblance of modernity is destroyed by other newcomers which raise possibilities and problems for algal evolution which are still unresolved. For the greens there are Ordovician receptaculitids and cyclocrinitids to be considered, together with Late Palaeozoic "green" phylloids, and also "erect udoteaceans", which so often resemble cyanobacteria. For the reds there are Solenoporaceae and Moniliporellaceae to be taken into account in the Middle Palaeozoic, and these are augmented by "ancestral coralline" and squamariacean-like phylloid algae, as well as by Gymnocodiaceae, in the Late Palaeozoic. Along with these a host of more minor problematica, which for better or worse have been, entrusted to palaeolagologists for study, create a systematists jungle in the Carboniferous and Permian.

End-Palaeozoic extinctions failed to eliminate this disorder. During the Mesozoic, despite dasycladalean diversification and the appearance of coralline red algae, the undercurrent of problematic forms is still evident, with bizarre newcomers mingling with survivors from the Late Palaeozoic.

No less do stromatolites present problems and show dramatic changes with time. The striking differences between Recent forms and those of the Precambrian are often glossed over. More than most other fossils, stromatolites exist at two levels in the minds of those who study them. On the one hand, there are the tangible deposits: laminated layers, domes, and columns which are fairly easy to discern in isolation but whose clotted, spongy, sometimes filamentous, microfabrics also often, since the Early Cambrian, lurk obscurely in skeletal reefs. On the other hand, there are the shifting dogmas of explanation: environment of origin (freshwater, intertidal, subtidal); microbial community (cyanobacterial, bacterial, algal); process of formation (trapping and binding, calcification); reason for decline (oxygen levels, invertebrate grazing, coarse bioclastic sediment); reason for survival (saline refuges and desiccated tidal flats). Descriptive emphases have also varied: geometric or Linnean classifications, macrostructures or microfabrics.

The problem with stromatolites is not their complexity, but their simplicity. They offer so little to grasp. But, like Plautus's Pseudolus, we so often seem to have overlooked the certainties while seeking uncertainties. Stromatolites are just the tip of a microbial carbonate iceberg: lamination varies and may disappear; there are gradations between clotted and peloidal spongiostrome fabrics, porostromate fabrics, and macroscopically clotted thrombolite fabrics; calcification or cement encrustation of the microbes may radically alter the appearance of the deposit. The clues have always been there in the carbonate fabrics. The challenge is not only to correctly interpret them but also to objectively and accurately describe them.

The incentive for this book came about because of the 4th International Symposium on Fossil Algae, held at Cardiff in July 1987. But I decided early on that there was much more need for an overview of the subject as a whole, than for a compilation of conference papers. Only just over half of the articles here were originally presented as talks at the meeting, so this is not strictly a symposium volume.

Nevertheless, were it not for the stimulus of the Symposium this book would not have appeared.

The title of the Fossil Algae Symposium is an overstatement. The four meetings so far (Erlangen 1975, Paris 1979, Denver 1983, Cardiff 1987) have mainly attracted researchers interested in benthic calcified algae, cyanobacteria, stromatolites, and other microbial carbonates. But even within this much more limited area there is an enormous variety of information and it is very scattered. Calcareous algae and prokaryotes have yet to benefit from the kind of general synthetic treatment applied to invertebrate fossils, on the one hand, and to vascular plant fossils, on the other.

Precisely because of this, it is worth trying to draw together a review volume. I cannot claim to have achieved anything close to an even, comprehensive coverage. What has emerged, however, does demonstrate the breadth of the subject, the variety of approach, and the state of the science at present. Despite omissions due to constraints on length or non-availability of authors, this volume does include discussions of major groups of algae and cyanobacteria, stromatolites, and other microbial carbonates, of all ages, both extant and extinct, marine and non-marine and, where appropriate, benthic and planktonic.

I am very grateful indeed to the following colleagues who generously helped by giving their time and expertise to referee manuscripts: Stanley M. Awramik, Jack A. Babcock, Michael A. Borowitzka, Candace M. Brooke, Yvonne M. Chamberlain, Barrie Dale, Graham F. Elliott, Monique Feist, Erik Flügel, Richard A. Fortey, Patrick Génot, Nicole Grambast-Fessard, John R. Groves, Llewellya Hillis-Colinvaux, Hans J. Hofmann, David M. John, Andrew R. Leitch, Bernard L. Mamet, Xi-nan Mu, Ernst Ott, Allan Pentecost, Katharina Perch-Nielsen, Brian R. Pratt, Alain Roux, Daniel Vachard, Malcolm R. Walter, V. Paul Wright.

It is a pleasure to thank the authors for their willingness to contribute to this volume, and for their patience with my suggestions. The skill and care of the staff of the Springer Geosciences editorial office in Heidelberg brought the whole to fruition.

Cardiff, August 1990 ROBERT RIDING

Contents

List of Contributors

1 Introduction

Chapter 1

Calcification Processes in Algae and Cyanobacteria

A. PENTECOST[1]

Abstract

In algae and cyanobacteria, calcification is always associated with biosynthetic products and, where analyses have been made, these products have been found to consist predominantly of polysaccharide, although no specific polymer seems to be associated with calcification.

Two aspects of the mineralization merit special consideration: photosynthesis has long been considered the main driving force causing a depletion in aqueous carbon dioxide and an increase in CO_3^{2-}. Photosynthesis will always favour calcification and the magnitude of the effect will depend upon photosynthetic rates and the lengths of diffusion paths for the ions and molecules involved in calcification. Second, at the ocean surface the precipitation of calcium carbonate is thermodynamically favoured even in the absence of photosynthesis, which may indicate either a widespread evolution of calcification inhibitors or absence of suitable catalysts. Today, the majority of aquatic plants remain uncalcified.

1 Introduction

During the past 15 years the topic of algal calcification has been reviewed many times. A detailed overview of the subject is not the purpose of this essay. Instead, a brief description of calcification processes in the algal phyla will be presented with emphasis on recent progress in the field. The earlier work has been discussed in detail in a number of papers, a brief outline of which is given below. Reviews of algal calcification with an emphasis on mechanisms have been presented by Borowitzka (1977, 1982a), Krumbein (1979), Pentecost (1980), Okazaki and Furuya (1985) and Watabe (1989). Information on carbon sources and anatomy can be obtained from Borowitzka (1982b) and Pentecost (1985). A few earlier reviews, notably Lewin (1962) and Arnott and Pautard (1970) are also useful sources of information. In addition there are several specialized reviews dealing with individual phyla or classes plus a large body of information on calcification processes in animals. Some important principles have now been established by combining theory with some experimentation. Our knowledge is sufficient to state that the evolution of calcification within the algae is polyphyletic and even within genera the process can be remarkably diverse.

Much has been written about the degree of control exercised by organisms over calcification. Lowenstam (1981, 1986) distinguished 'organic matrix-mediated' biomineralization, which is biologically controlled by means of an

[1] Division of Biosphere Sciences, King's College London, Campden Hill Road, London W8 7 AH, UK

organically constructed template or framework, from 'biologically induced' mineralization, where there is apparently no biological control over the process. This allows mineralizing organisms to be classified according to the degree of control they exercise over the process. The importance of organic frameworks for calcification has been stressed by several researchers, stimulated in part by the close association between mineral and organic phases in bone formation and many types of invertebrate calcification (Lowenstam and Wiener 1983; Simkiss 1986). Research into algal calcification proceeds at a slow rate and we owe much to research in the medical field despite the hiatus resulting from photosynthesis which separates algal calcification from any other mineralization process. In some ways, photosynthesis is a *bête noire*, as it adds the complicating factor of a diurnal, light-dependent sink for carbon dioxide. Nevertheless, as the following discussion will show, photosynthesis plays some part, large or small, in all of the calcification processes so far investigated in the cyanobacteria and algae.

Figure 1 illustrates the approximate relative abundance of modern calcified genera within the algal phyla divided between freshwater and the sea. Most phyla are seen to possess only small numbers of genera with calcified species. Considerably more cyanobacteria have been observed calcified in freshwater compared with the sea, but in the Chlorophyta the situation is reversed. Some species of all six charophyte genera calcify, although the degree of mineralization is often slight and sometimes confined to the reproductive organs. This phylum is barely represented in salt water but is probably the major contributor towards modern freshwater algal carbonate.

One phylum, the Prymnesiophyta, holds a pre-eminent position, with more than 50% of the genera having calcified species (the coccolithophorids). These minute algae are major primary producers in the oceans and must form more biogenic carbonate than any other group of plants. A small but significant contribution is made by another planktonic group, the Dinophyta, though unfortunately nothing is known of the calcification processes in this great phylum. The two remaining groups are benthic, mainly neritic algae. In the Phaeophyta there is one genus, *Padina*, whose calcified species are often abundant in warmer waters. The Rhodophyta is a large phylum containing algae of great diversity with over 500 genera currently recognised. The comparatively small number of calcified genera probably gives a false impression of the actual quantity of calcium carbonate produced, as several genera are found in all of the world's oceans.

2 Calcification Processes in Cyanobacteria

In common with the eukaryotic algae, the cyanobacteria have left an extensive fossil record primarily as a result of their intimate association with calcium carbonate. Unlike many algae, these microbes have a simple organization and lack the complex anatomy which facilitates calcification in some chlorophycean genera such as *Halimeda*. Pentecost and Riding (1986) defined calcification in

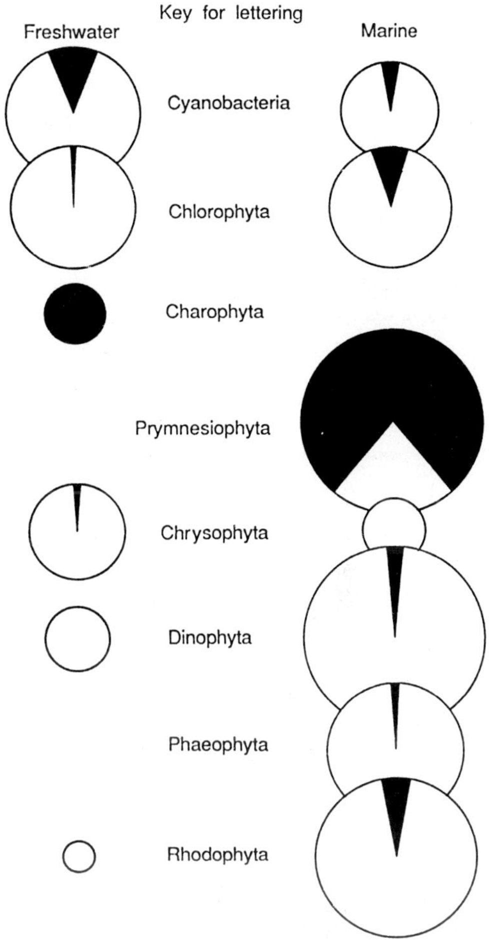

Fig. 1. Pie charts illustrating the approximate relative proportion of calcified genera to uncalcified genera in the major algal phyla. Circle size is approximately proportional to the known abundance of the groups in freshwater and the oceans respectively. Minor groups are omitted. Calcified genera *black*, uncalcified *white*

cyanobacteria as "the nucleation of calcium carbonate upon or within the mucilaginous sheath". It is necessary to make this distinction so that trapping and binding of carbonate can be excluded as a calcification process. The problem arises because the sheaths surrounding the cells of most cyanobacteria promote the adhesion of particles suspended in the surrounding water, irrespective of their origin. Larger particles can also become trapped between the open framework produced by filamentous genera. This is assisted by the motility of the filaments which secrete a sticky mucilage whilst gliding phototactically to the mat surface. Calcification is unlikely to occur with the complete exclusion of trapping and binding. The problem is illustrated in Fig. 2 where three calcification processes are shown in addition to binding. It is assumed, in this model, that the surrounding water is supersaturated with respect to a $CaCO_3$ mineral and that nucleation begins both upon and within the sheath. In Fig. 2a,

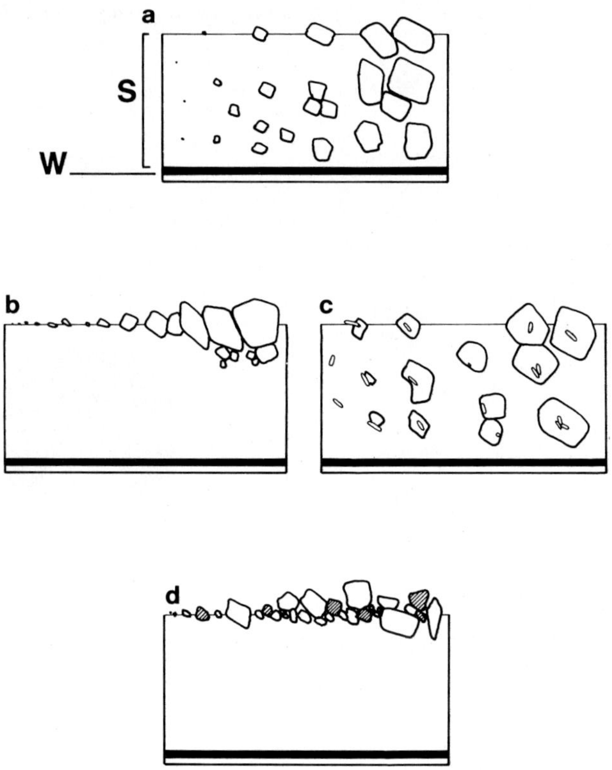

Fig. 2a-d. Calcification and carbonate-binding by the cyanobacterium sheath. **a** Calcification resulting from biogenic nucleation within and upon the sheath. **b** Calcification at the sheath surface resulting from the adhesion of seed crystals or heteronuclei originally present in the surrounding water. Note secondary nucleation at right caused by the fragmentation of larger crystals. **c** Calcification associated with bacteria upon and within the sheath. **d** Carbonate binding at the sheath surface. This process is non-selective and includes non-carbonate grains. Time increases from *left* to *right* along the abscissa. *S* sheath, *W* cell wall

heterogeneous nucleation occurs at favourable sites in a sheath composed of diverse polysaccharides and some protein (Schrader et al. 1982). Sheath chemistry is known to be complex, but no detailed studies have so far been made on calcified species. In Fig. 2b, abiogenic nucleation results from the attachment of crystal seeds originating from the surrounding milieu. Calcification would begin superficially but could proceed into the sheath by crystal growth or secondary nucleation. Finally, precipitation could be initiated by bacteria which are invariably present in natural mats. The sheath either provides a substrate or a suitable attachment point for epiphytic bacteria. The association between certain bacteria and cyanobacteria is well known (Paerl 1982). Calcification could be induced by the presence of a low-energy bacterium surface reducing the energy barrier to nucleation, or result from metabolic activities, e.g. base formation by ammonification or nitrate-reduction (Krumbein 1979). The binding of particles (Fig. 2d) is by contrast, non-selective but difficult to distinguish from 2b. In fact, there is no certain way of distinguishing the processes if binding is accompanied by the growth of mineral grains in a supersaturated solution.

Recent studies of the freshwater cyanobacterium *Rivularia haematites* (D.C.) Ag. have demonstrated a pronounced seasonal pattern of growth and calcification (Pentecost 1987). Calcification results from two separate processes. During winter, when growth was less than 2 μm day^{-1}, a surface precipitate of calcite formed on the hemispherical colonies with crystallization proceeding inwards. The process involved could be any of those described above. During spring, growth was more rapid, reaching a maximum in midsummer when the trichomes overgrew the winter layer but were initially less heavily calcified (Fig. 3). As growth continued, narrow calcified layers, termed 'summer bands' appeared (Figs. 3, 4a). These bands could result from a pH rise induced by photosynthesis in the region of active growth, 1–2 mm below the colony surface. Since the entire colony is invested in a polysaccharide gel, the flow of ions is diffusion-controlled. Counter-diffusion of Ca^{2+} and CO_3^{2-} ions is possible and periodic precipitation (Keller and Rubinow 1981) might occur. In any case, calcification during this period would seem to be induced by photosynthesis. The water at these sites was supersaturated with respect to calcite. During

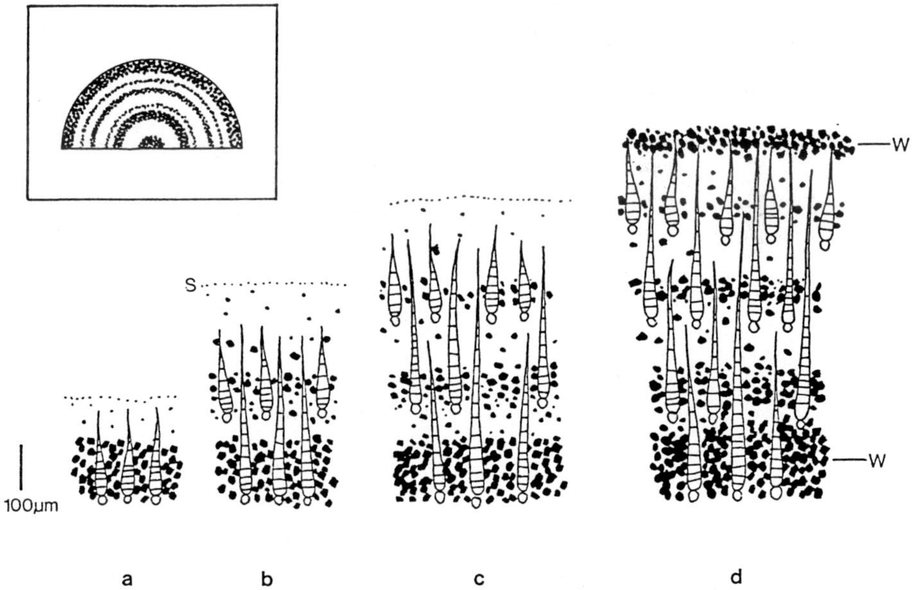

Fig. 3a-d. Diagram illustrating the progress of growth and calcification in *Rivularia* colonies. Calcification shown by *stippling* consists of a thick 'winter' band and a number of narrower 'summer' bands. *Inset* shows a section through a *Rivularia* colony. *W* winter bands; *S* surface of mucilaginous sheath. **a** Spring; **b** early summer; **c** mid-summer; **d** winter

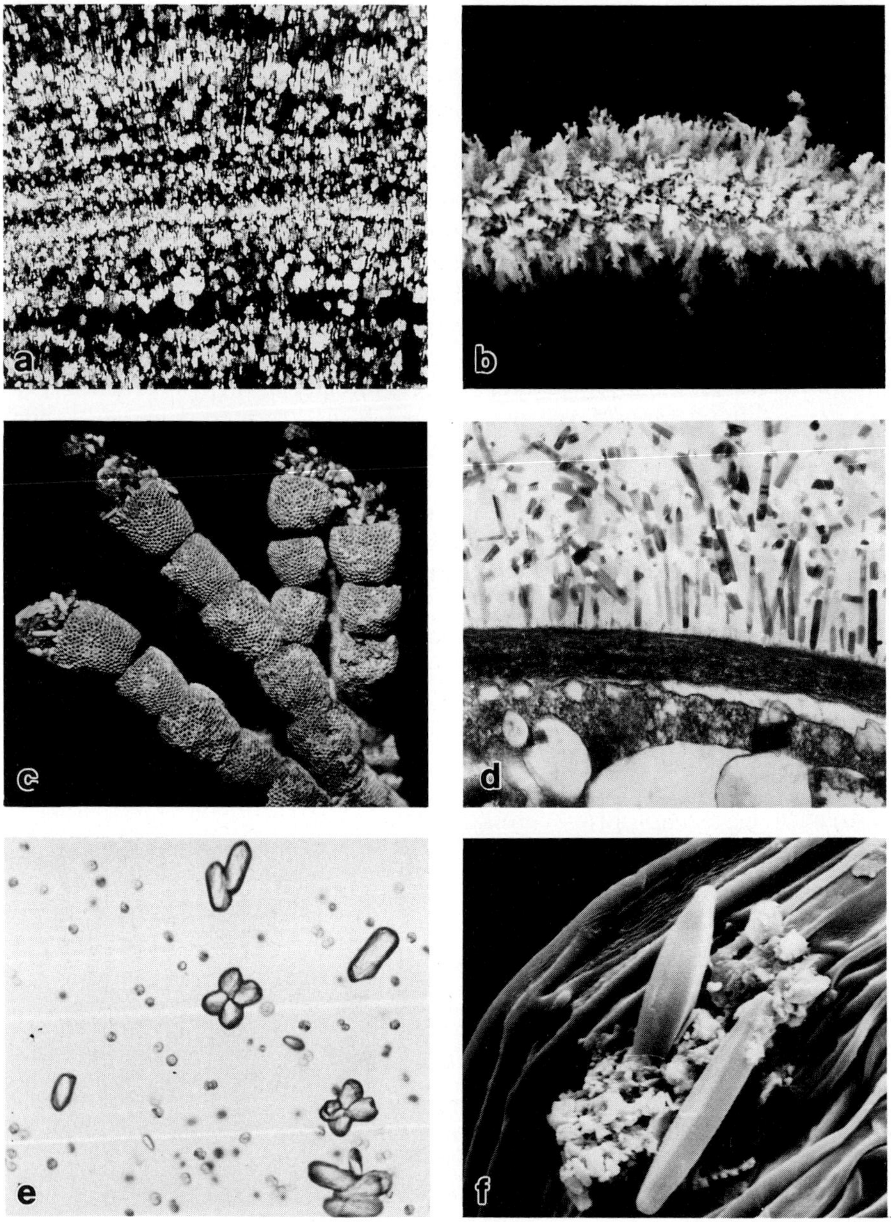

Fig. 4. **a** Section through calcified *Rivularia* colony showing a series of calcified 'summer' bands. Crossed nicols, x250, Mastiles Stream, N. Yorks. **b** Calcified *Scytonema* (cyanobacterium) trichome from a subaerial site. Note the dendritic crystals radiating out from the sheath surface. SEM photograph, x1250, Cowside Beck, N. Yorks. **c** Part of the articulated thallus of *Cymopolia barbata* (L.) Harv. (Chlorophyta, Dasycladales) showing intense ICS calcification adjacent to the outer cortex. Tufts of conglutinated uncalcified hairs appear at the branch apices. x5, Jamaica. **d** Section through the infolded margin of *Padina japonica* showing crystals of aragonite perpendicular to the

winter, the calcification is abiogenic, whilst in summer, it results from a combination of abiogenic and biogenic (photosynthetic) processes. In contrast, regular banding is not apparent in marine species of *Rivularia* which are occasionally found calcified with aragonite and amorphous carbonate (Golubic and Campbell 1981).

A second colonial, calcified cyanobacterium, *Homoeothrix crustacea* Woronichin, is often associated with *Rivularia*. Despite the close association, the progress of calcification is quite different in this species. During winter a soft spongy deposit is the norm, whilst in summer the deposit is more dense, and forms 100 μm below a surface layer of radiating trichomes (Pentecost 1988a). Summer bands also occur, and probably relate to periods of high photosynthetic activity, as in *Rivularia*.

In *Calothrix*, a closely related genus, calcification follows neither of these patterns. Growth rates of *Calothrix* oncoids have been found to be an order of magnitude lower than *Rivularia* and *Homoeothrix* even where the genera coexist (Pentecost 1989). *Calothrix* does not form mucilaginous colonies and calcification is confined to the sheath which adheres closely to the trichomes. Irregular carbonate grains permeate the sheath and extend little, if at all, beyond the sheath boundary. There are no narrow 'summer' bands and inorganic nucleation appears to predominate throughout the year. Photosynthesis must play a subordinate role, considering the low rates of summer growth, amounting to just 1 μm day^{-1}.

Other calcified cyanobacteria include *Geitleria*, *Lyngbya*, *Phormidium*, *Plectonema*, *Schizothrix* and *Scytonema* (Golubic 1973; Pentecost and Riding 1986). In *Geitleria* the discovery of regular arrays of calcite crystals within and external to the sheath has led to the idea that the sheath may regulate this orderly form of mineralization (Lowenstam 1986), although fine structure studies in other genera have not revealed pronounced periodicity in the sheath. The conditions under which such calcification occurs are uncertain. Similar extracellular calcification is found in *Scytonema* in cave entrances (Fig. 4b) where condensation water may be the only significant vehicle for the reacting ions. Dendritic calcite associated with algal (probably cyanobacterial) sheaths has been described by Jones and Kahle (1986).

The problem of species-specificity of calcification remains, but some interesting patterns are emerging. For example, clear differences in the calcification of *Calothrix*, *Homoeothrix* and *Rivularia* are apparent. By placing cyanobacteria into a gel containing counterdiffusing calcium and carbonate ions, Pentecost and Bauld (1988) demonstrated sheath nucleation of calcite in

◄

cell wall surface. Note the association between the fibrous wall and the crystals. Electron micrograph Pb stained, x22000 (Okazaki et al. 1986). **e** Calcified colonies of *Chrysonebula holmesii* Lund. Prismatic calcite crystals, growing along their c-axes develop in diffluent mucilage surrounding the cells. Note penetration twins (?on 0112). x140, Waterfall Beck, N. Yorks. **f** Calcification associated with freshwater pennate diatoms attached to a moss leaf. The carbonate crystals (identified with EPMA) are 2–3 μm in diameter. SEM photograph x1150, Matlock, Derbyshire

Lyngbya and *Scytonema*. Under the same conditions, *Entophysalis* and *Pseudanabaena* failed to nucleate the mineral, providing further evidence for 'species' specificity.

3 Calcification Processes in Algae

3.1 Chlorophyta

Most calcified genera are marine macrophytes belonging to the Dasycladales and Caulerpales. Whilst the former have an extensive fossil record, it is only in the latter that there has been much experimental work (Borowitzka 1986). The process of calcification is well understood in *Halimeda* and is dependent upon thallus anatomy. The outermost cortex is composed of apressed clavate cells enclosing a series of aqueous spaces adjacent to the photosynthetic cortex. These intercellular spaces (ICS) are believed to contain seawater which exchanges carbon dioxide rapidly with the adjacent cells as a result of photosynthesis and respiration. The resulting alkalinization during CO_2 removal favours the precipitation of aragonite. Although no direct measurements have been made on these spaces the model is supported by several lines of evidence using radioactive tracers and metabolic inhibitors (Borowitzka and Larkum 1976a,b; Borowitzka 1976, 1986). This model can be readily extended to green algae possessing a similar anatomy, e.g. *Cymopolia* (Fig. 4c), *Dasycladus*, *Neomeris* and some species of *Udotea* (Borowitzka 1982b; Pentecost 1985) but the mechanism cannot explain calcification in *Acetabularia*, *Rhipocephalus* and *Penicillus*. Here, calcification is generally superficial but associated with a 'sheath-like' layer of the cell wall. In *Acetabularia* aragonite precipitation is initiated within the wall and is possibly augmented by the secretion of Ca-rich material from the cytosol (Okazaki and Furuya 1985). Wall material is also associated with the initial calcification sites of *Halimeda* but the composition of the material has yet to be determined in any of the calcified genera. Calcium-binding polysaccharides have been reported from *Halimeda* (Böhm and Goreau 1973; Misonou et al. 1980a,b) but polysaccharides with similar properties are also known in non-calcifying genera. Further studies are therefore needed in some of these genera, particularly *Acetabularia*, *Penicillus* and *Udotea*.

A small number of calcified freshwater species occasionally produce significant amounts of calcareous sediment (Müller and Oti 1981). The most interesting of these are the planktonic species *Phacotus* and *Coccomonas* often seen in hard-water lakes. In these unicellular forms anatomy plays no part in calcification and it is doubtful whether these organisms alone could always remove sufficient CO_2 from the water to cause precipitation. *Phacotus* possesses a thickened, calcified mucilage envelope. The breadth of the envelope undoubtedly increases the diffusion path of dissolved gases entering the cell but there are similar uncalcified genera with thickened envelopes e.g. *Sphaerellopsis*. In common with some other algae, calcification of *Phacotus* is not always observed in culture but a non-motile colonial phytoplankter, *Gloeotaenium*, calcifies in vitro at high pH, possibly stimulated by photosynthesis (Devi-Prasad and Chowdary 1979). Calcification also occurs in three epilithic freshwater Chlorophytes, *Cladophora* (Wood 1975), *Gongrosira* (Golubic and Fischer 1975; Pentecost 1988b) and *Oocardium* (Golubic and Marcenko 1965). These algae are often associated with calcified cyanobacteria in highly calcareous streams, and like them, the precipitation is thought to be partly the result of photosynthesis and partly by CO_2 evasion to the atmosphere.

3.2 Charophyta

Extracellular deposits of calcium carbonate occur on the internodal and some reproductive cells of charophytes. Calcification has been reported in all genera but is most apparent in *Chara*. Most of our knowledge of calcification comes from studies of the internodal cells which have long been favourite experimental objects for plant physiologists. Regular bands of calcite are often a conspicuous feature of certain ecorticate species and these are known to be associated with alkaline regions of the cell wall where the pH attains values of 9 or more. This high pH is thought to result primarily from a passive influx of H^+ through the plasmalemma (Raven et al. 1986). An active efflux of H^+ occurs in adjacent uncalcified acid bands resulting in the circulation of electric current. This efflux is powered by ATP, the ultimate energy source being provided by photosynthetic reactions. This circulatory behaviour is unusual and has not been detected in any other algae and is probably linked to photosynthetic inorganic carbon uptake.

In the acid bands, the extra H^+ may combine with bicarbonate, which is the predominant anion in most freshwaters, to form unionized carbon dioxide. The plasma membrane is known to be permeable to CO_2, allowing the gas to diffuse into the cells, ultimately forming photosynthetic products within the chloroplast stroma. This hypothetical mechanism is called 'external conversion' (Walker 1983; Price et al. 1985). An alternative hypothesis involves the combined active uptake of H^+ and HCO_3^- via an electrochemical pump followed by intracellular combination perhaps catalyzed by carbonic anhydrase. Whichever hypothesis is correct, quantitatively the effect is the same and the precipitation of $CaCO_3$ in the alkaline bands is seen to be indirectly the result of photosynthetically driven processes. Whether such a mechanism operates in the oogonia of charophytes (Daily 1975) remains to be seen, but this appears unlikely, and calcification may instead result from a favourable anatomy.

3.3 Prymnesiophyta

The physiology and fine structure of coccolith formation has been studied in depth in three coccolithophorid species; *Emiliania huxleyi* (Lohmann) Kamptner, *Pleurochrysis carterae* Braarud and Fagerl. and *Coccolithus* (*Crystallolithus*) *pelagicus* (Wallich) Schiller. The calcification process differs in each case but there are similarities between *Emiliania* and *Pleurochrysis. E. huxleyi* is the best known species. Although a common phytoplankter in the oceans, the organic scale upon which all coccoliths form is not as apparent in this species as it is in many others. There has been little work to date on scale analysis (Klaveness and Paasche 1979) and many coccolithophorids possess uncalcified scales in addition to coccoliths. Sonication of the coccolith-bearing scales of *Emiliania* (the 'c' cells, see Westbroek et al. 1986) followed by decalcification has yielded a water-soluble polysaccharide intimately associated with the carbonate (de Vrind de Jong et al. 1986). The polysaccharide is complex, with a molecular weight ca. 60 000, and contains carboxyl and ester-sulphate groups.

In *Emiliania* c-cells, calcification is intracellular and begins in small membrane-bound vesicles. These bodies appear to develop into more obvious 'coccolith' vesicles within which an organic scale is formed (Klaveness 1976). Calcite is observed forming along the rim of the scale until the fully developed coccolith, consisting of a mosaic of calcite crystals (Parker et al. 1983), is released onto the cell surface. Staining has shown that polysaccharide is in close contact with the vesicle membrane and the mineral surface and appears to penetrate the crystalline areas (van der Wal et al. 1983a; Westbroek et al. 1986). Intracellular calcification resembling that found in *Emiliania* has also been reported in the genus *Cruciplacolithus* (Fresnel 1986), and is probably widespread in the group.

In *Pleurochrysis carterae* the vesicles are associated with the Golgi apparatus and contain coccolithosomes (granular particles about 25 nm in diameter) which, by vesicle fusion with scale-containing cisternae, appear to initiate crystallization. Two polysaccharides have been isolated from coccolith-rich fractions of *Pleurochrysis* and they show some similarity to the polysaccharide of *Emiliania* (van der Wal et al. 1983b; Westbroek et al. 1986).

It has been known for some time, either through direct observation (Manton and Leedale 1963; Green 1986; Manton 1986) or by inference (Pentecost 1980), that scales of certain species can undergo extracellular calcification. In a study of the '*Crystallolithus*' stage of *Coccolithus hyalinus* Gaarder and Markali, recalcification of decalcified scales has been observed in vitro (Green 1986; Rowson et al. 1986). Organic matter is again closely associated with the calcite rhombs and the entire assemblage is covered with an organic 'skin' marking the periplast surface. An important observation here is the recognition of two apparently disparate mechanisms operating in the same genotype, namely intracellular calcification in the *Coccolithus* stage and extracellular calcification in the *Crystallolithus* stage (Green 1986).

Elaborate intracellular construction seems to be required for heterococcoliths, whilst the simpler holococcoliths seem able to calcify at the cell surface. Other forms of calcification are known in the Prymnesiophyta. These include the 'vacuolar clusters' found in *Pleurochrysis* (Westbroek et al. 1986) and the sheath calcification of *Chrysotila* (Green and Course 1983; Green 1986). In the latter, calcification is initiated within mucilage surrounding the cells in vitro. Minor differences in deposition have been observed between strains and may be attributable to different polysaccharide compositions, though none has yet been analyzed. Within the Prymnesiophyta as a whole, the link between specific polysaccharides and calcification is strong, but mechanisms supported by a rigid theoretical background are still lacking.

3.4 Phaeophyta

Calcification has only been reported in one genus, *Padina* (Dictyotales). *Padina* species are widely distributed in warmer waters but only about half of them calcify. The calcification is superficial and consists of band-like deposits of aragonite on the thallus surface. In some species, calcification occurs on both dorsal (outward-facing) and ventral surfaces (*P. japonica* Yamada), whilst in others it occurs only on the ventral surface (*P. pavonica* Gaillon). In both species, calcification has been shown to be initiated in the enrolled thallus margin, which in section, resembles a fern crozier and contains meristematic cells (Okazaki et al. 1986). This circinate margin behaves as a semi-enclosed space, containing a film of sea water a few micrometres in thickness. The surrounding cells contain chloroplasts, and calcification is probably initiated within this space as a result of photosynthesis. Initially, crystals are found within an outer fibrous wall layer (Fig. 4d) and remain associated with the surface after thallus expansion. Further calcification occurs on the expanded thallus, perhaps seeded by crystals already present on the surface. Most species of *Padina* calcify only on their ventral surface suggesting that the surfaces differ in composition. Of greater interest, however, is the observation that a non-calcifying species, *P. arborescens* Holmes, lacks the fibrous wall layer previously referred to, but possesses a smooth 'cuticular' layer instead. This suggests that the wall plays an important part in the calcification either by inhibiting or directly promoting deposition. The origins of the concentric calcified bands are still uncertain but they are sometimes associated with zones of developing tetrasporangia showing enhanced photosynthetic activity.

3.5 Rhodophyta

Most intensively studied are the coralline algae, the most widely distributed calcified benthic algae in the seas today. Most vegetative cells of the crustose and non-articulated branched species are calcified to some extent. Calcification is observed to begin immediately after segmentation of the tetraspores and is entirely extracellular (Vesk and Borowitzka 1984). Two phases of mineralization are apparent (Cabioch and Giraud 1986). First, acicular crystals, probably of calcite, form in the 'middle lamella' and outer cell walls and in some cases these grow into tabular crystals tangential to the wall surface. More intense calcification follows within the inner cell walls and perpendicular to the wall surface. The growth of crystals follows the direction of polysaccharide fibrils in the outer wall and middle lamella, but this is less apparent in the inner wall.

Nevertheless there does seem to be a close association between the wall material and deposition. The coralline algae have a complex organization and compact structure but there are no obvious differences between the cellular organization of the corraline algae and many other Florideophyceae which might account for the mineralization. A calcification mechanism has been proposed by Digby (1977) which involves the photosynthetic oxidation of water followed by export of the hydrogen ions to sea water. Intracellular hydroxyl ions react with bicarbonate and the resulting carbonate ion is exported to the cell wall where it reacts with calcium to form calcite. An alternative model involving localized export of H^+ and OH^- has also been proposed (Pentecost 1985). There is some evidence for H^+ efflux to seawater (Digby 1977) and, if this is a local effect, parallels can be seen with the Charophyta.

Rates of calcification and photosynthesis are significantly correlated in *Corallina* (Pentecost 1978; Borowitzka 1981) but correlation is not evidence for direct participation. Lavelle (1979) has demonstrated the transport of calcium to the calcifying tips of *Calliarthron* which considerably complicates the picture. Intercalary growth occurs in the Corallinaceae but it is not known whether the meristematic cells are always photosynthetically active. The link between calcification and photosynthesis might be further explored through a careful examination of these cells.

Calcification also occurs in a small number of genera belonging to other groups. For example, in *Galaxaura fastigiata* Decaisne, ICS calcification occurs in the open spaces of the young cortex (Okazaki et al. 1982) and the same mechanism might account for calcification in a related genus, *Liagora* (Okazaki and Furuya 1985).

3.6 Dinophyta, Chrysophyta, and Bacillariophyta

Calcification is rarely reported in these groups and deposition is varied, unusual, and in the Dinophyta, may be on a sufficient scale to produce substantial amounts of sediment. The best documented case of dinophyte calcification is afforded by *Thoracosphaera*, which until recently was classified as a coccolithophorid (Tangen et al. 1982). The oceanic species *T. heimii* Kamptner has a calcified cell wall in the vegetative coccoid stage. Calcite deposition begins, in vitro, at numerous crystallization centres on the wall surface and eventually covers the entire cell. The mineralization is extracellular and no evidence of intracellular calcification has been obtained. Cyst calcification also occurs (Keupp, this Vol.).

In the other phyla, calcification has been observed only in freshwater and is usually associated with mucilage coatings. Calcite deposition within a diffluent gel surrounding unicells has been observed in *Tetrasporopsis* (Tschermak-Woess 1980) and *Chrysonebula* (Fig. 4e). In *Chrysonebula* nucleation is confined to the gel rather than the cell walls and deposition only occurs in waters supersaturated with respect to calcium carbonate. An unusual case of calcification has been reported in the lorica of some *Pseudokephyrion* species (Tappan 1980), a genus normally characteristic of aggressive, oligotrophic waters.

Diatom calcification is rarely observed but a close association between carbonate deposition and diatom mucilage is sometimes found (Winsborough and Golubic 1987), again, in calcite-supersaturated water (Fig. 4f). However, in both phyla, bacteria can be observed in the mucilage and may provide nucleation centres, although Green (1986) found no evidence for bacterial nucleation in *Chrysotila* (Prymnesiophyta) which also produces copious mucilage.

4 Discussion and Conclusions

Most calcifying algal regimes are characterized by a water pH of 7.8–9.5. Logic dictates that photosynthesising cells will utilize HCO_3^- rather than CO_2 under these conditions because the concentration of CO_2 will be low, typically 5–10 μM. Uptake of HCO_3^- may be either active or passive, depending upon the sign of the free energy change resulting from the transport of the ion across the diffusion gradient of an electrically charged membrane. Calculations made by Walker (1983) and others employing estimated cytosol pH, membrane potential, and the K_m of ribulose biphosphate carboxylase-oxidase indicate that HCO_3^- transport is almost certainly an active process. To maintain electroneutrality in the external medium, H^+ must also be removed resulting in an effective net loss of H_2CO_3. In an unbounded medium this will be replaced by CO_2 arriving through laminar and turbulent diffusion, with the length of the diffusion path, depending upon relative water movement. Within the diffusion zone, depletion of H^+ and HCO_3^- results in a pH rise and CO_3^{2-} formation, assuming instantaneous equilibration. Such conditions favour $CaCO_3$ precipitation and increasing the photosynthesis rate will therefore favour calcification.

As the volume of the medium decreases relative to that of the cell, the CO_2 exchange will be more rapid, resulting in a greater pH rise. In a semi-enclosed space, high pH may be maintained between two diffusion gradients, namely that across the cell wall plus plasma membrane and that between the intercellular space and the external medium, separated by a layer of wall material (Borowitzka 1982a; 1986).

The medium surrounding calcifying algae is normally supersaturated with respect to mineral phases of calcium carbonate. Although the degree of supersaturation at the ocean surface is not as great as originally supposed (Cooke and Kepkay 1984) the formation of (magnesian) calcite is still accompanied by a favourable free energy change of about -1 Kcal $Mole^{-1}$. The supersaturation comes from five sources whose relative significance depends upon the local environment. These are photosynthesis, CO_2 evasion to the atmosphere, evaporation and changes in water temperature and pressure. In freshwater the same factors operate, apart from pressure, and high supersaturations are sometimes achieved. Calcified algae have not been found in waters undersaturated with $CaCO_3$. This is probably due to the lack of suitable protective coatings preventing dissolution of the mineral phase, which is often extracellular, and perhaps the extra energy required for mineralization results in no benefits for the species involved.

Calcification should occur spontaneously at the ocean surface and in many calcium-rich freshwaters. This is sometimes observed but it is by no means universal and the majority of algae are uncalcified (Fig. 1), even where there is additional CO_2 depletion caused by photosynthesis. Two principles are at work here. First, the kinetics of calcite nucleation are slow, especially in seawater (Pytkowicz 1965). Furthermore, nucleation and subsequent crystal growth can be inhibited by a variety of substances. Inhibitors of carbonate precipitation include a wide range of compounds. Particularly effective are poly (meta)phosphates, phosphonates (Gill and Varsanik 1986), phosphate esters and orthophosphate (House and Donaldson 1986). Other substances include polyphenols and a range of polysaccharides. Of particular interest are polymers isolated from calcified species. A polysaccharide associated with *Emiliania* was found to be an effective inhibitor of calcite nucleation and uronic acid groups are thought to be involved (Borman et al. 1982). The polysaccharide binds several ions including calcium and the degree of inhibition varies with the ion bound. Conformational changes seem able to alter the inhibitory effect which suggests that such substances might control mineralization. A polysaccharide purified from *Rivularia* also completely inhibited crystal growth at 25 ppm concentration (unpublished observations) and the sheaths of some cyanobacteria have Ca-binding properties (Somers and Brown 1978). Further evidence for the role of inhibitors has been obtained from fine structure studies. In *Acetabularia*, calcification is confined to the inner regions of the cell wall (Okazaki and Furuya 1985) but in the coralline algae, mineralization is initially only present in the outer wall. Deposition would normally be expected close to the cell surface where large concentration gradients would occur through photosynthetic activity or ion pumping.

In the absence of inhibitors, calcification may proceed, although rates could be low. No catalysts are known to facilitate the direct combination of Ca^{2+} and CO_3^{2-}/HCO_3^- but they may exist. Such activity might occur in the extracellular calcification of the crystalloliths of *Coccolithus pelagicus*. Here there is a paradox because in one stage of the life cycle, calcification is intracellular and must involve the transport of ions across a minimum of two membranes with the expenditure of metabolic energy, whilst in the other, mineralization seems to proceed in the complete absence of membrane transport. Rowson et al. (1986) noted calcification in the dark, in which case a local photosynthetic depletion of CO_2 cannot be involved.

In all of the algae studied so far, calcification occurs associated with organic matter, primarily polysaccharide. In some groups, particularly the Corallinaceae and the coccolithophorids, the association is intimate and mechanisms involving epitaxial growth are often invoked. It is possible for an organic matrix to stimulate nucleation and control the subsequent growth of crystals but the mechanisms are far from clear. It is too early to state how widespread matrix control is in the algae, and a continuum of organization levels seems to exist from the synthesis of intracellular coccoliths down to the random deposition of calcite in a *Chrysotila* colony. If organic matrices do play a major role, then we would not expect to see them at these lower levels. Calcite is often grown in artificial gels and this mineral appears to be unusual in maintaining growth whilst incorporating the gel matrix (Henisch 1970), so this intimate association in nature may be to some extent accidental. If calcification occurred in an initially undersaturated medium the problem would be more satisfactorily defined and ironically, easier to tackle.

Acknowledgements. Thanks are extended to Jim Price, John C. Green, and John Dodge for useful discussion and suggestions.

References

Arnott HJ, Pautard FGE (1970) Calcification in Plants. In: Schraer H (ed) Biological calcification. Elsevier/North Holland Biomedical Press, Amsterdam New York, pp 375–446

Böhm EL, Goreau TF (1973) Composition and calcium binding properties of the water soluble polysaccharides in the calcareous alga *Halimeda opuntia* (L.). Int Rev Ges Hydrobiol 58:117–126

Borman AH, de Jong EW, Huizinga M, Kok DJ, Westbroek P, Bosch L (1982) The role in $CaCO_3$ crystallization of an acid Ca^{2+}-polysaccharide associated with coccoliths of *Emiliania huxleyi*. Eur J Biochem 129:179–183

Borowitzka MA (1976) Calcification in the green alga *Halimeda*. II. The exchange of Ca^{2+} and the occurrence of age gradients in calcification and photosynthesis. J Exp Bot 27:864–878

Borowitzka MA (1977) Algal calcification. Oceanogr Mar Biol 15:189–223

Borowitzka MA (1981) Photosynthesis and calcification in the articulated coralline red algae *Amphiroa anceps* and *A. foliacea*. Mar Biol 62:17–23

Borowitzka MA (1982a) Mechanisms in algal calcification. In: Round FE, Chapman CH (eds) Progress in phycological research, vol 1. Elsevier/North Holland Biomedical Press, Amsterdam New York, pp 137–177

Borowitzka MA (1982b) Morphological and cytological aspects of algal calcification. Int Rev Cytol 74:127–162

Borowitzka MA (1986) Physiology and biochemistry of calcification in the Chlorophyceae. In: Leadbeater BSC, Riding R (eds) Biomineralization in lower plants and animals. Clarendon, Oxford, pp 107–124

Borowitzka MA, Larkum AWD (1976a) Calcification in the green alga *Halimeda*. IV. The action of metabolic inhibitors on photosynthesis and calcification. J Exp Bot 27:894–907

Borowitzka MA, Larkum AWD (1976b) Calcification in the green alga *Halimeda*. III. The sources of inorganic carbon for photosynthesis and calcification and a model of the mechanism of calcification. J Exp Bot 27:879–893

Cabioch J, Giraud G (1986) Structural aspects of biomineralization in the coralline algae (calcified Rhodophyceae). In: Leadbeater BSC, Riding R (eds) Biomineralization in lower plants and animals. Clarendon, Oxford, pp 141–156

Cooke RC, Kepkay PE (1984) Apparent calcite supersaturation at the ocean surface. Why the present solubility product of pure calcite in seawater does not predict the correct solubility of the salt in nature. Mar Chem 15:59–70

Daily FK (1975) A note concerning calcium carbonate deposition in Charophytes. Phycologia 14:331–332

Devi-Prasad PV, Chowdary YBY (1979) Factors influencing calcification of *Gloeotaenium loitlesbergianum* Hansgirg. Indian J Exp Biol 17:1402–1403

Digby PSB (1977) Growth and calcification in the coralline algae. *Clathromorphum circumscriptum* and *Corallina officinalis* and the significance of pH in relation to precipitation. J Mar Biol Assoc UK 57:1095–1109

Fresnel J (1986) Nouvelles observations sur une coccolithacée rare: *Cruciplacolithus neohelis* (McIntyre et Bé) Reinhardt (Prymnesionphyceae). Protistologica: 22:193–204
Gill JS, Varsanik RG (1986) Computer modelling of the specific matching between scale inhibitors and crystal structure of scale forming minerals. J Crystal Growth 76:57–62
Golubic S (1973) The relationship between blue-green algae and carbonate deposits. In: Carr NG, Whitton BA (eds) The biology of blue-green algae. Blackwell, London, pp 434–472
Golubic S, Campbell SE (1981) Biogenically formed aragonite in marine *Rivularia*. In: Monty CLV (ed) Phanerozoic stromatolites. Springer, Berlin Heidelberg New York, pp 209–229
Golubic S, Fischer AG (1975) Ecology of calcareous nodules forming in Little Conestoga Creek near Lancaster, Pennsylvania. Verh Int Verein Limnol 19:2315–2323
Golubic S, Marcenko E (1965) Über Konvergenszerscheinungen bei Standortsformen der Blaualgen unter extremem Lebensbedingungen. Schweiz Z Hydrol 27:207–217
Green JC (1986) Biomineralization in the algal class Prymnesiophyceae. In: Leadbeater BSC, Riding R (eds) Biomineralization in lower plants and animals. Clarendon, Oxford, pp 173–188
Green JC, Course PA (1983) Extracellular calcification in *Chrysotila lamellosa* (Prymnesiophyceae). Br Phycol J 18:367–382
Henisch HK (1970) Crystal growth in gels. Pennsylvania Univ Press, Pennsylvania, USA
House WA, Donaldson L (1986) Adsorption and coprecipitation of phosphate on calcite. J Colloid Interf Sci 112:309–324
Jones B, Kahle CF (1986) Dendritic calcite crystals formed by calcification of algal filaments in a vadose environment. J Sediment Petrol 56:217–227
Keller JB, Rubinow SI (1981) Recurrent precipitation and Liesegang rings. J Chem Phys 74:5000–5007
Klaveness D (1976) *Emiliania huxleyi* (Lohmann) Hay and Mohler. III. Mineral deposition and the origin of the matrix during coccolith formation. Protistologica 12:217–224
Klaveness D, Paasche E (1979) Physiology of coccolithophorids. In: Levandowsky M, Hunter SH (eds) Biochemistry and physiology of protozoa, vol 1, 2nd edn. Academic Press, New York London, pp 191–213
Krumbein WF (1979) Calcification by bacteria and algae. In: Trudinger PA, Swaine DJ (eds) Biogeochemical cycling of mineral-forming elements, vol 3. Elsevier, Amsterdam, pp 47–68
Lavelle JM (1979) Translocation in *Calliarthron* and its role in the light-enhancement of calcification. Mar Biol 55:37–44
Lewin JC (1962) Calcification. In: Lewin RA (ed) Physiology and biochemistry of algae. Academic Press, New York London, pp 457–464
Lowenstam HA (1981) Minerals formed by organisms. Science 211:1126–1131
Lowenstam HA (1986) Mineralization processes in monerans and protoctists. In: Leadbeater BSC, Riding R (eds) Biomineralization in lower plants and animals. Clarendon, Oxford, pp 1–17
Lowenstam HA, Wiener S (1983) Mineralization by microorganisms and the evolution of biomineralization. In: Westbroek P, de Jong EW (eds) Biomineralization and biological metal accumulation. Elsevier, Amsterdam, pp 191–204
Manton I (1986) Functional parallels between calcified and uncalcified periplasts. In: Leadbeater BSC, Riding R (eds) Biomineralization of lower plants and animals. Clarendon, Oxford, pp 157–172
Manton I, Leedale GF (1963) Observations on the microanatomy of *Crystallolithus hyalinus* Gaarder and Markali. Arch Mikrobiol 47:115–136
Misonou T, Furuya K, Nisizawa K (1980a) Particular Ca-binding substances from marine macro-algae. II. Incorporation of ^{45}Ca into acid-insoluble residues from various algae and the solubilization of Ca-binding substances from the residues. Jpn J Phycol 28:105–112
Misonou T, Okazaki M, Nisizawa K (1980b) Particular Ca-binding substances in marine macro-algae. I. Uptake of ^{45}Ca by various algae and extraction of the Ca-binding substances. Jpn J Phycol 28:31–36
Müller G, Oti M (1981) The occurrence of calcified planktonic green algae in freshwater carbonates. Sedimentology 28:897–907
Okazaki M, Furuya K (1982) Studies on calcium carbonate deposition of algae-IV. Initial calcification sites of calcareous red alga *Galaxaura fastigiata* Decaisne. Bot Mar 25:511–517
Okazaki M, Furuya K (1985) Mechanisms in algal calcification. Jpn J Phycol 33:328–344

Okazaki M, Ichikawa K, Furuya K (1982) Studies on the calcium carbonate deposition of algae: 4. Initial calcification site of *Galaxaura fastigiata*. Bot Mar 25:511–519

Okazaki M, Pentecost A, Tanaka Y, Miyata M (1986) A study of calcium carbonate deposition in the genus *Padina* (Phaeophyceae, Dictyotales). Br Phycol J 21:217–224

Paerl HW (1982) Interactions with bacteria. In: Carr NG, Whitton BA (eds) The biology of cyanobacteria. Blackwell, Oxford, pp 441–461

Parker SB, Skarnulis AJ, Westbroek P, Williams RJP (1983) The ultrastructure of coccoliths from the marine alga *Emiliania huxleyi* (Lohmann) Hay and Mohler: an ultra-high resolution electron microscope study. Proc R Soc London Ser B 219:111–117

Pentecost A (1978) Calcification and photosynthesis in *Corallina officinalis* L. using the $^{14}CO_2$ method. Br Phycol J 13:383–390

Pentecost A (1980) Calcification in plants. Int Rev Cytol 62:1–27

Pentecost A (1985) Photosynthetic plants as intermediary agents between environmental bicarbonate and carbonate deposition. In: Lucas WJ, Berry JA (eds) Inorganic carbon uptake by aquatic photosynthetic organisms. Am Soc Plant Physiol, Rockville, Maryland, pp 459–476

Pentecost A (1987) Growth and calcification of the freshwater cyanobacterium *Rivularia haematites*. Proc R Soc London Ser B 232:125–136

Pentecost A (1988a) Growth and calcification of the cyanobacterium *Homoeothrix crustacea*. J Gen Microbiol 134:2665–2671

Pentecost A (1988b) Observations on growth rates and calcium carbonate deposition in the green alga *Gongrosira*. New Phytol 110:249–253

Pentecost A (1989) Growth and calcification of *Calothrix*-dominated oncolites from Northern England. In: Crick RE (ed) 5th Int Conf Biomineralization: Origin, Evolution and Modern Aspects of Biomineralization in Plants and Animals. Plenum, New York, pp 443–454

Pentecost A, Bauld J (1988) Nucleation of calcite on the sheaths of cyanobacteria using a simple diffusion cell. Geomicrobiol J 6:129–135

Pentecost A, Riding R (1986) Calcification in Cyanobacteria. In: Leadbeater BSC, Riding R (eds) Biomineralization of lower plants and animals. Clarendon, Oxford, pp 73–90

Price GD, Badger MR, Bassett ME, Whitecross MI (1985) Involvement of Plasmalemmasomes and carbonic anhydrase in photosynthetic utilization of bicarbonate in *Chara corallina*. Aust J Plant Physiol 12:241–256

Pytkowicz RM (1965) Rates of inorganic calcium carbonate nucleation. J Geol 73:196–199

Raven JA, Smith FA, Walker NA (1986) Biomineralization in the Charophyceae *sensu lato*. In: Leadbeater BSC, Riding R (eds) Biomineralization of lower plants and animals. Clarendon, Oxford, pp 125–139

Rowson JD, Leadbeater BSC, Green JC (1986) Calcium carbonate deposition in the motile (*Crystallolithus*) phase of *Coccolithus pelagicus*. Br Phycol J 21:359–370

Schrader M, Drews G, Golecki JR, Weckesser J (1982) Isolation and characterization of the sheath from the cyanobacterium *Chlorogloeopsis* PCC 6912. J Gen Microbiol 128:267–272

Simkiss K (1986) The processes of biomineralization in lower plants and animals—an overview. In: Leadbeater BSC, Riding R (eds) Biomineralization of lower plants and animals. Clarendon, Oxford, pp 19–37

Somers GF, Brown M (1978) The affinity of trichomes of blue-green algae for calcium ions. Estuaries 1:17–28

Tangen K, Brand LE, Blackwelder PL, Guillard RRL (1982) *Thoracosphaera heimii* is a dinophyte. Observations on its morphology and life cycle. Mar Micropaleontol 7:193–212

Tappan H (1980) The paleobiology of plant protists. Freeman, San Francisco

Tschermak-Woess E (1980) Zur Kenntnis von *Tetrasporopsis fuscescens*. Plant Syst Evol 133:121–133

Vesk M, Borowitzka MA (1984) Ultrastructure of tetrasporogenesis in the coralline alga *Haliptilon cuvieri* (Rhodophyta). J Phycol 20:501–515

de Vrind de Jong EW, Borman AH, Thierry R, Westbroek P, Gruter M, Kamerling JP (1986) Calcification in the coccolithophorids *Emiliania huxleyi* and *Pleurochrysis carterae*. II. Biochemical aspects. In: Leadbeater BSC, Riding R (eds) Biomineralization in lower plants and animals. Clarendon, Oxford, pp 205–218

van der Wal P, de Jong L, Westbroek P, de Bruijn WC (1983a) Ultrastructural polysaccharide localization in calcifying and naked cells of the coccolithophorid *Emiliania huxleyi*. Protoplasma 118:157–168

van der Wal P, de Jong L, Westbroek P, de Bruijn WC, Mulder-Stapel AA (1983b) Polysaccharide localization, coccolith formation and Golgi dynamics in the coccolithophorid *Hymenomonas carterae*. J Ultrastruct Res 85:1139–1158

Walker NA (1983) The uptake of inorganic carbon by freshwater plants. Plant Cell Environ 6:323–328

Watabe N (1989) Intracellular mineralization of invertebrates and algae. In: Crick RE (ed) 5th Int Conf Biomineralization: evolution of ocean chemistry and its significance to biomineralization. (in press)

Westbroek P, van der Wal P, van Emburg PR, de Vrind de Jong EW, de Bruijn WC (1986) Calcification in the coccolithophorids *Emiliania huxleyi* and *Pleurochrysis carterae*. I. Ultrastructural aspects. In: Leadbeater BSC, Riding R (eds) Biomineralization in lower plants and animals. Clarendon, Oxford, pp 189–204

Winsborough BM, Golubic S (1987) The role of diatoms in stromatolite growth: two examples from modern freshwater settings. J Phycol 23:195–201

Wood KG (1975) Photosynthesis of *Cladophora* in relation to light, and CO_2 limitation; $CaCO_3$ precipitation. Ecology 56:479–484

Chapter 2

Classification of Microbial Carbonates

R. RIDING[1]

Abstract

The main groups of organisms participating in the formation of benthic microbial carbonates are bacteria, cyanobacteria and algae. Three main processes are involved in the formation of microbial carbonates: (1) trapping of sedimentary particles; (2) biomineralization of organic tissue; (3) surficial precipitation of minerals on organisms and sediments. Of these, biomineralization preserves the most information concerning the organisms involved, and trapping preserves the least. Surface mineralization is intermediate, in that it preserves information concerning the exterior size and shape of the organisms.

The principal categories of benthic microbial carbonate deposits are:

1. stromatolite, laminated microbial deposit:
   a) agglutinated stromatolite
      (i) fine-grained, well-laminated
      (ii) coarse-grained, crudely laminated
   b) tufa stromatolite
   c) skeletal stromatolite
   d) subaerial stromatolite
2. dendrolite, dendritic microbial deposit
3. thrombolite, clotted microbial deposit
4. travertine, layered microbial deposit with a dendritic macrofabric
5. cryptic microbial carbonates: microbial deposits with micritic, clotted, peloidal or sparitic microfabrics, but lacking distinctive macrofabrics.

Only agglutinated stromatolites are dominated by trapping. Skeletal stromatolites, dendrolites, and thrombolites are dominated by biomineralization. Tufa stromatolites, subaerial stromatolites and travertine are dominated by surficial mineralization.

Tufa stromatolites, subaerial stromatolites, and travertine are essentially non-marine, whereas skeletal stromatolites, dendrolites, and thrombolites are essentially marine and restricted to the Palaeozoic and Mesozoic. Agglutinated stromatolites occur in both marine and non-marine environments. Travertine, and especially tidal flat and subaerial stromatolites, are strongly influenced by desiccation.

1 Introduction

The history of the development of ideas concerning stromatolites has often been reviewed (Maslov 1960; Hoffman 1973; Hofmann 1973; Monty 1977), and is instructive in understanding these and other benthic microbial deposits. But it is also true that the intricacies of the uncertainties and confusions along this path readily obscure the results of this research during the past century. The fact is that we can now recognize a variety of benthic calcareous deposits formed

[1] Department of Geology, University of Wales College of Cardiff, Cardiff CF1 3YE, UK

primarily by microbial organisms such as bacteria, cyanobacteria, and algae, and I propose to concentrate here on describing in a straightforward way what I consider to be the main processes, components, and diagnostic features of the principal types of benthic calcareous microbial deposits, together with an outline of the organisms most likely to be involved in their formation.

The term microbolite (adapted from microbialite of Burne and Moore 1987) is useful because it encourages us to view benthic microbial carbonate deposits as a whole, rather than concentrating on one type, such as stromatolites, alone. However, this broad view requires both comparative terminology and a system of classification which can accurately, but without unnecessary complication, provide basic information linking the organisms, processes and products involved. In particular, terms are required for non-calcified, non-mineralized "classic" stromatolites formed by trapping of particulate sediment which I here propose to call agglutinated stromatolites; and a term is required for calcified but non-laminated microbial deposits with a dendritic fabric which I propose to call dendrolites (Riding 1988). Travertine and some types of tufa need to be included in microbial carbonates, although much still remains to be learned about these non-marine deposits.

Microbial carbonates are defined here as the carbonate deposits produced or localized by benthic (to exclude coccolith and foraminiferal oozes) microbial communities. The term microbe is used here for any microscopic organism. The principal benthic microbes involved in these deposits are bacteria, cyanobacteria (both of which are prokaryotic) and algae such as diatoms and various chlorophytes and rhodophytes. The principal invertebrate microbes locally involved in construction of benthic microbial carbonates are foraminifers. Cyanobacteria (cyanophytes, blue-greens) are especially important, and reviews of the groups prominent in carbonate deposition have been made by Golubic (1973) and Pentecost and Riding (1986).

The communities of organisms creating these deposits are often termed "microbial mats" in the current literature (Gerdes and Krumbein 1987, p. 5), and "algal mats" in earlier literature, a reference to the densely interlayered, intertwined and variously prostrate or erect orientations of the coccoid and filamentous cells involved. The composition of these communities will have changed through time (Fig. 1).

2 Processes

The main processes of microbial carbonate formation are the same as for stromatolite formation: (1) trapping (agglutination) of sedimentary particles; (2) biomineralization of organic tissues; (3) mineralization, i.e. surficial precipitation of minerals on organisms and/or sediment (Fig. 2). The most important distinction to be made here is between biomineralization (2 above) and mineralization (3 above). Biomineralization here signifies precipitation within organisms or organic tissue during life and implies a significant degree of biological control over precipitation. Mineralization, in contrast, is here regarded as precipitation outside organisms.

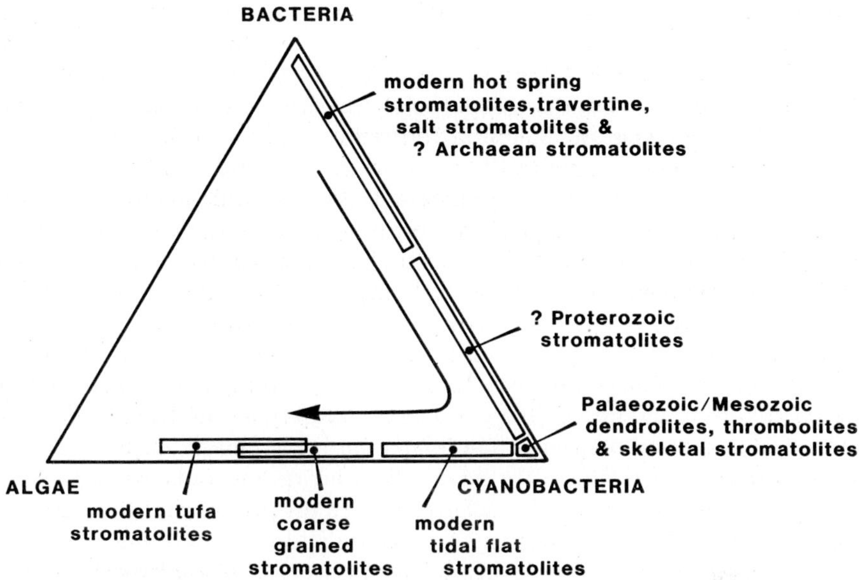

Fig. 1. ABC (algae, bacteria, cyanobacteria) diagram showing possible variations in time and space of the principal constructional organisms in benthic microbial carbonate deposits. *Arrow* indicates likely time trend for marine forms. Subaerial stromatolites (not shown) contain fungi and lichens in addition to prokaryotes and algae

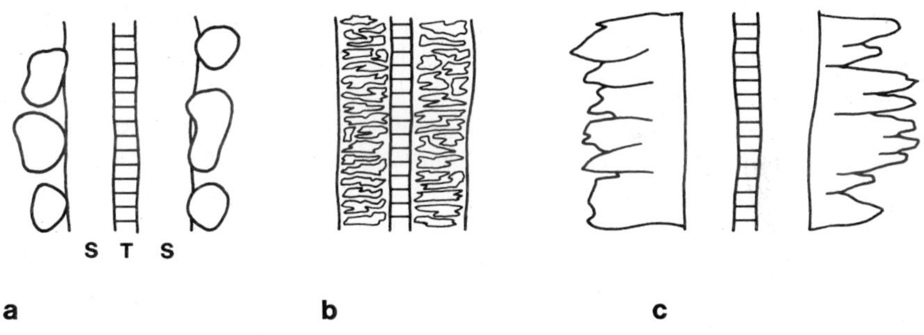

Fig. 2a-c. Processes of benthic microbial carbonate deposition exemplified by a filamentous cyanobacterium; *T* trichome; *S* sheath. **a** Trapping of sediment particles on the sheath exterior; **b** calcification (biomineralization) during life of the sheath interior; **c** precipitation of a surficial crust of $CaCO_3$ upon the sheath exterior (mineralization). None of these processes is mutually exclusive, although trapping will be by baffling rather than by adhesion if either calcification or mineralization of the sheath has occurred

2.1 Trapping

Stickiness and relief (even if the relief is only microscopic) on a surface can both result in retention of sedimentary particles. These two sub-processes of trapping are theoretically separate: adhesion due to the sticky or soft nature of mucilage produced by microbes, and baffling due to organisms impeding movement of particles by physical obstruction. In practice, adhesion and baffling may not be readily distinguishable. For example, erect cyanobacterial filaments will act as baffles and may also be encased in a sticky mucilaginous sheath. On a larger scale, pustular colonies of cyanobacteria and/or diatoms can create microrelief and may also be sufficiently soft for particles to become imbedded within the sticky gel which they create.

Sedimentary particles will be transported by currents onto the microbolite surface. They may then be removed again by currents which are sufficiently strong, or they may be trapped. The longer the grain remains on the microbolite surface, the greater is the chance that it will be incorporated into the deposit. This is due to the vertical growth of the organisms, a process generally referred to as binding. "Trapping and binding" is a phrase widely used in the stromatolite literature to indicate the initial retention of particles on a microbial substrate and their subsequent incorporation.

The sizes of particles trapped will depend mainly on: (1) size and abundance of particles introduced by currents; (2) strength and frequency of currents; (3) stickiness of the surface; and (4) relief on the surface. Smaller, micrite-grade, particles will be preferentially trapped. The stickiness and relief of most microbial surfaces are generally insufficient to trap sand-size particles (Fig. 3). An exception to this is the more complex, more sticky, and more irregular surfaces of algal-cyanobacterial communities (Awramik and Riding 1988).

This relationship between particle size and trapping potential of the microbolite community is probably important in permitting long-term accretion. If large particles were deposited relatively rapidly on a surface occupied by small, relatively slow-growing organisms, then the survival of the community would be threatened. Pratt (1982) is probably correct to suggest that the widespread production of coarse bioclastic sediment due to metazoan biomineralization has been a factor of equal significance to the grazing effects of herbivorous animals in the Phanerozoic decline of stromatolites formed by trapping of sediment.

Trapping appears to operate in all types of microbial carbonate deposition, but it is the pre-eminent process only where calcification, whether by biomineralization or by inorganic mineralization, is relatively unimportant. It is therefore most conspicuous in environments, or at times, where calcification is reduced. In non-marine environments this is more likely to be in lakes than in rivers. In shallow marine environments it has mainly been during the Precambrian and Cenozoic (see Distribution, below).

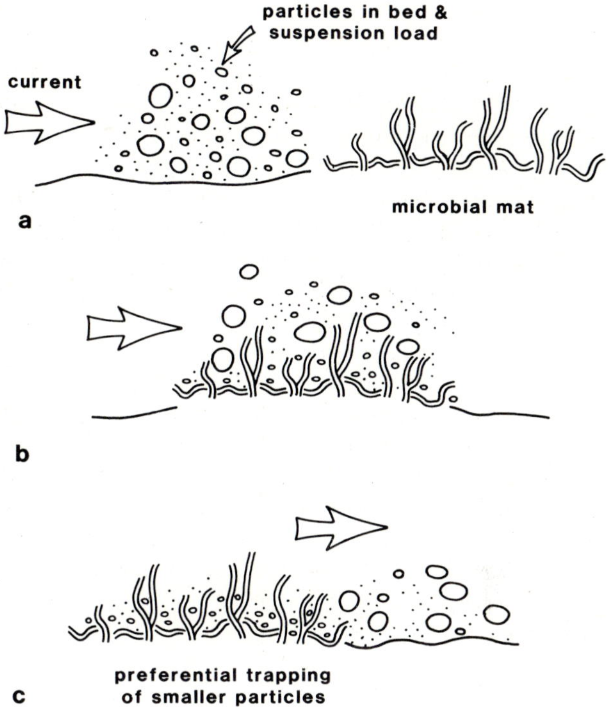

Fig. 3a-c. Selective trapping of smaller particles by a benthic microbial mat community. Larger particles are not easily retained by small organisms, whereas small particles are

2.2 Biomineralization

The degree of control over mineralogy and crystal growth in biomineralization varies with the organism. Attempts to clarify these processes, and the terminology used to describe them, are still being debated. Lowenstam (1981) and Lowenstam and Weiner (1983) recognized biologically-induced mineralization (resulting in random crystal aggregates) and organic matrix-mediated mineralization (where minerals adopt unique crystal habits with well-defined orientation) as end-members of an intergrading spectrum. Simkiss (1986) emphasized four factors controlling biomineralization: 1) modification of ion-activity, 2) presence of a diffusion limited site, 3) presence of a nucleating surface, 4) lattice modifiers and crystal inhibitors.

In general, microbial calcification appears to be induced rather than mediated, although details of the operation of Simkiss' factors still need to be elucidated in most of the groups involved. Bacterial calcification (Krumbein 1986) is less well understood than cyanobacterial calcification but principal products are spherulites (Danielli and Edington 1983) and bushy growths (Chafetz and Folk 1984). Cyanobacterial calcification (Pentecost and Riding

1986) produces a wide range of bushy, dendritic and tubular fabrics. Macroscale deposits of these, such as skeletal stromatolites, dendrolites, etc. are discussed here, but discrete microfossils, such as *Girvanella*, *Cayeuxia*, and so on, are not included (for details see Riding, calcified cyanobacteria, this Vol.). Calcification of eukaryotic algae is less important in microbial carbonate formation, not because in place deposits produced by benthic algae are uncommon, but because the large size of the algae generally involved precludes them from being regarded as microbes.

Calcification during life of the organisms enhances the accretion rate of the microbial deposit and, by preserving organisms in growth position, it commonly increases micro-relief of the surface and thereby enhances trapping of particles. At the same time, it may disrupt lamination if the calcified organisms have erect orientation. Consequently, calcification often results in a coarse, unlaminated or at best crudely laminated fabric in which the original organisms are preserved to some degree; usually their size distribution and general morphology are retained. Closely spaced erect calcified tufts and bushes may create intervening shelter cavities. Where these remain unfilled by particulate sediment, coarse fenestrae then develop which differ in size and orientation from those associated with desiccation in, for example, tidal flat stromatolites. Another product of calcification is post-mortem nucleation of cements upon the microbes. These are often conspicuous, for example, in Early Palaeozoic dendrolites formed by *Angulocellularia* Vologdin, *Renalcis* Vologdin and *Epiphyton* Bornemann (Fig. 4; see Dendrolites below). Thus, microbial carbonates formed by biomineralization basically consist of:

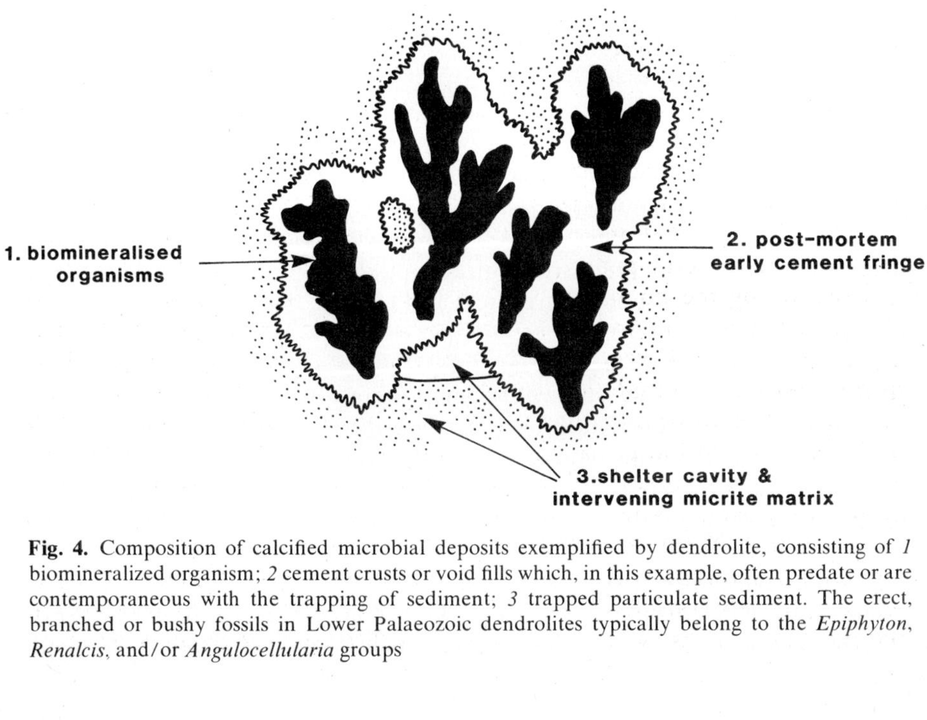

Fig. 4. Composition of calcified microbial deposits exemplified by dendrolite, consisting of *1* biomineralized organism; *2* cement crusts or void fills which, in this example, often predate or are contemporaneous with the trapping of sediment; *3* trapped particulate sediment. The erect, branched or bushy fossils in Lower Palaeozoic dendrolites typically belong to the *Epiphyton*, *Renalcis*, and/or *Angulocellularia* groups

1. Calcified organisms;
2. Trapped particulate sediment;
3. Cement crusts or void fills.

In terms of the fabrics and degree of lamination they possess, these microbial carbonates may be subdivided into skeletal stromatolites (crudely laminated, with prostrate or erect calcified organisms), dendrolites (unlaminated, with erect, or pendent dendritic calcified organisms), and thrombolites (unlaminated, with a macroscopic clotted fabric which may have a variety of origins, including the presence of rounded, or irregular, calcified organisms). Composition and fabrics of microbial carbonates are discussed further below (see Composition and fabrics).

2.3 External Surficial Precipitation of Minerals

This process, here termed mineralization, is, in effect, early cementation. In microbial carbonates it is important in the formation of subaerial and tufa stromatolites and travertine. It results in crystalline veneers on the surfaces of organisms, irrespective of whether or not they are biomineralized, and on associated particulate sediment. Mineralization in marine microbial environments is most conspicuous in reefal and intertidal situations. In non-marine environments it is most prominent in areas of rapid CO_2 evasion, such as springs and waterfalls, or high evaporation.

3 Types of Benthic Microbial Carbonates

3.1 Microbolites

The term microbialite was proposed by Burne and Moore (1987), and defined as "organosedimentary deposits that have accreted as a result of a benthic microbial community trapping and binding detrital sediment and/or forming the locus of mineral precipitation". Here I replace the adjectival prefix "microbial" by the noun-based prefix "microbo-" to conform with similar established usage, as in stromatolite and thrombolite. Burne and Moore's (1987) definition is, in fact, very similar to Awramik and Margulis' definition of stromatolite (see Stromatolites, below).

The processes of microbial carbonate formation outlined above can be related directly to the principal calcareous microbolite types: stromatolites, dendrolites, thrombolites, and travertine (Table 1). Sediment trapping results in agglutinated stromatolites, characterized by a particulate microstructure which preserves few, if any, details of original size, shape or orientation of the organisms responsible for localizing deposition. Biomineralization, in contrast, retains relatively considerable detail of the organisms, which are preserved as skeletal microfossils. These can often be recognized as cyanobacteria or,

Table 1. Possible main relationships between the processes and products of calcareous microbial deposits[a]

Process	Product		
Mineralization	Subaerial stromatolite		
	Tufa stromatolite	Travertine	
Biomineralization	Skeletal stromatolite	Dendrolite	Thrombolite
Trapping	Agglutinated stromatolite		
Microbial growth	Prostrate	Erect	Coccoid
	Microbial mat		

[a] Note that (1) all thrombolites are not necessarily produced by biomineralization of coccoid microorganisms; (2) biomineralization influences subaerial stromatolites, but they are dominated by mineralization processes; (3) prostrate, erect and coccoid growth styles strongly influence the fabrics of biomineralized microbial deposits (skeletal stromatolites, dendrolites, thrombolites), but are less well reflected in agglutinated and subaerial stromatolites; (4) lamination (the characteristic feature uniting stromatolites) is due to episodic sedimentation, episodic growth, or prostrate habit, or a combination of these.

especially in the Early Palaeozoic, as probable cyanobacteria. Consequently, skeletal stromatolites, dendrolites and thrombolites generally reveal much more information concerning the microbes responsible for their formation than do agglutinated stromatolites (non-skeletal stromatolites of Riding 1977).

Mineralization is especially important in non-marine microbolites. Tufa stromatolites form in freshwater environments where inorganic $CaCO_3$ precipitation is enhanced by CO_2 evasion. This results in surficial veneering of microbes, mainly cyanobacteria, and sediment by $CaCO_3$ which would be termed cement if it were interstitial. There is no general term for this surface crystalline veneer; I term it here "surface cement". Processes of travertine formation are as yet poorly understood, but it is likely that mineralization is also a major factor for these non-marine, warm spring, deposits and for subaerial stromatolites. Exposure results in desiccation, which promotes mineralization and early lithification. There is thus an important relationship between the environment and the processes of formation of microbial carbonates. Dendrolites, skeletal stromatolites, subtidal agglutinated stromatolites and, probably, thrombolites are subaqueous in origin. Tufa stromatolites are essentially subaqueous, but may be occasionally exposed by falls in lake or river levels. On the other hand, tidal flat agglutinated stromatolites, travertine and subaerial stromatolites represent a trend towards increasing exposure which will be reflected in their processes of formation and the fabrics they contain.

The important distinction between stromatolites on the one hand and dendrolites and thrombolites on the other is based upon the presence or absence of lamination. Any of the three principal processes (trapping, biomineralization, mineralization) may produce stromatolites, provided that lamination (due to episodic sedimentation and/or growth, or to prostrate habit of the organisms) is also developed, even though the quality of the lamination will vary.

The principal categories of benthic microbial carbonate deposits recognized here are as follows:

1. Stromatolite: macroscopically laminated benthic microbial deposit.
   a) Agglutinated stromatolite, produced by trapping/binding of particulate sediment:
      (i) fine-grained, well-laminated;
      (ii) coarse-grained, crudely laminated.
   b) Tufa stromatolite, produced by precipitation of minerals on (rather than within) organic tissue.
   c) Skeletal stromatolite, produced by calcified in place organisms.
   d) Subaerial stromatolite, dominated by mineralization in a desiccating environment.
2. Dendrolite: macroscopically dendritic benthic microbial deposit.
3. Thrombolite: macroscopically clotted benthic microbial deposit.
4. Travertine: layered benthic microbial deposit with a dendritic macrofabric.
5. Cryptic microbial carbonate: benthic microbial carbonates which have micritic, clotted, peloidal or sparitic microfabrics, but which lack distinctive macrofabrics.

3.2 Stromatolites

Kalkowsky (1908) introduced the term stromatolith (for details see Monty 1977), but research on stromatolites preceded his paper by at least twenty-five years. Hall's (1883) description of *Cryptozoon* from the Cambrian of New York marks the beginning of close attention to stromatolites. Definition of the term has oscillated between genetic and descriptive approaches. Genetically, stromatolites may be defined as: "organosedimentary structures produced by sediment trapping, binding and/or precipitation as a result of the growth and metabolic activity of microorganisms, principally cyanophytes". This definition was presented by S.M. Awramik and L. Margulis in the Stromatolite Newsletter in 1974 (see Walter 1976a, p. 1). It is noteworthy that it makes no reference to lamination and is thus applicable to benthic microbial deposits generally. In fact, Burne and Moore (1987) use virtually the same definition for microbolites. Descriptively, stromatolites may be defined as: "attached, laminated, lithified, sedimentary growth structures, accretionary away from a point or limited surface of initiation" (Semikhatov et al. 1979, p. 993), an all-embracing view which would readily include many oöids and other coated grains, as well as sinter and some calcrete fabrics. A more simple, but combined, definition proposed here is "stromatolites are laminated benthic microbial deposits".

In Recent marine stromatolites trapping of sediment is by far the most important process involved. In non-marine stromatolites at the present day, in contrast, biomineralization and mineralization are usually dominant. Walcott (1914) compared Proterozoic stromatolites with modern tufa, and this freshwater analogue was popular until Black's (1933) recognition of sediment trapping in intertidal and supratidal stromatolites led to the widely held "trapping and binding" view of stromatolite which persists to the present. It is not difficult, in hindsight, to see that conflicting views of stromatolite formation are largely due to failure to discriminate between space and time variations in stromatolite development. As has already been stated, stromatolites may be formed by one of several processes: trapping, biomineralization, or mineralization. Furthermore, these processes have in the past operated in environments in which they are now rare. For example, cyanobacterial biomineralization has clearly taken place in marine environments in the Palaeozoic and Mesozoic, but has been rare during the Cenozoic (Riding 1982).

3.2.1 Agglutinated Stromatolites

This term is introduced here for stromatolites produced by trapping of particulate sediment. Two types can be distinguished: (1) fine-grained, well-laminated and (2) coarse-grained, crudely laminated (Awramik and Riding 1988).

Fine-Grained, Well-Laminated. This is the predominant Precambrian stromatolite (Fig. 5), but it persists in marine and non-marine environments until the present-day. Fine grain size is often due to a combination of supply and selection: the microbes involved (probably bacteria and/or cyanobacteria) will have difficulty in coping with coarse sediment and fine-grained stromatolites can be observed to develop even in environments where coarse sediment is available. Coarse, sand-sized material will not be retained by these small organisms, just as a football is unlikely to adhere to glue-covered strands if they are only a few millimetres in diameter. And if a cyanobacterial layer should be covered by coarse sediment, it is unlikely to successfully colonize such large grains, whereas micritic particles may be readily trapped and bound as the organic filaments move up and over them.

Episodic sedimentation, linked to periodic microbial growth, will result in good lamination, and this is a characteristic feature of many fine-grained stromatolites. Lamination, even when well-developed, is however not necessarily smooth, and we can distinguish:

1. Smooth lamination, of subaqueous origin;
2. Crinkly lamination, with fenestrae, due to exposure and dessication on tidal flats or lake margins.

Fig. 5. Precambrian columnar stromatolites, ca. 1100 million years; Tieling Formation, near Tielingzi, Jixian County, Tianjian Province, northern China. Note the good lamination. Scale *bar* is 5 cm long

Coarse-Grained, Crudely Laminated. Awramik and Riding (1988) recognized a probable link between an algal (as opposed to only cyanobacterial) floral component and coarse grain size in Shark Bay subtidal columnar stromatolites. They contrasted coarse-grained, crudely laminated, algal-cyanobacterial stromatolites with fine, well-laminated cyanobacterial stromatolites and cast doubt on the previously widely established view that Shark Bay columnar stromatolites are analogues for most Precambrian stromatolites. Subsequently, this link between coarse grain size and the presence of eukaryotic algae has been supported by study of "giant" Bahamian stromatolites at Lee Stocking Island, which have sand-size grains and a flora showing an approximately equal mix of algae and cyanobacteria (Riding et al., in preparation).

No ancient examples of similar coarse-grained stromatolites have yet been reported, and it remains to be seen whether these may date from the evolution of algae in general (probably 1000 Ma) or from the appearance of diatoms during the Cretaceous.

3.2.2 Tufa Stromatolites

This term is introduced here for stromatolites dominated by precipitation of minerals on (as opposed to within) organic substrates. These deposits are characterized by crystalline precipitated veneers of calcium carbonate upon the external surfaces of the organisms. These crusts interfere with organic growth and result in rapid sedimentary accretion. Cyanobacteria are the principal organisms involved in tufa stromatolite formation. Other types of tufa, not considered here because they are not laminated, include algal and moss tufa.

Tufa stromatolites are not known from marine environments. They are common in modern calcareous freshwater lakes and streams (e.g. Eggleston and Dean 1976; Winsborough 1987; Winsborough et al., in prep.). In particular they are prominent here as oncoids (e.g. Clarke 1900; Roddy 1915) which in freshwater may combine features of tufa stromatolites and skeletal stromatolites. These are produced by both biomineralization and inorganic mineralization. The former process is likely to be dominant in lakes, the latter in streams. These modern freshwater tufa oncoids have been compared with ancient marine oncoids, but there are often pronounced differences: ancient marine oncoids are usually either agglutinated oncoids or skeletal oncoids, and the latter normally lack the component of surficial inorganic mineralization which is characteristic of tufa stromatolites. Nevertheless, there is intergradation between tufa stromatolites and skeletal stromatolites.

3.2.3 Skeletal stromatolites

Riding (1977) introduced this term to describe "stromatolites in which the organisms responsible for their formation are commonly preserved as calcified fossils" (Riding 1977, p. 57). It is important to distinguish now between physicochemical calcification (here termed mineralization) and biochemical calcification (here termed biomineralization). Skeletal stromatolites refer to stromatolites formed by biochemical calcification, as was shown on the original triangular diagram (Fig. 6), and this distinguishes them from tufa stromatolites which are dominated by mineralization (surficial inorganic precipitation).

Skeletal stromatolites occur in marine environments in Palaeozoic and Mesozoic rocks. They are typically constructed by calcified cyanobacteria such as *Girvanella* Nicholson and Etheridge and *Ortonella* Garwood. These fossils can be classified as Porostromata (Pia 1927, p. 37) and skeletal stromatolites have also been termed porostromate stromatolites (Monty 1981). However, the Porostromata is a grouping of heterogeneous fossils and the term skeletal is preferred here. Some Recent lacustrine stromatolites resemble skeletal stromatolites. However, freshwater microbial carbonates generally possess an element of mineralization which places them in the category of tufa stromatolites. Nevertheless, this physicochemical precipitation effect is generally weaker in lacustrine than in fluviatile freshwater environments because CO_2

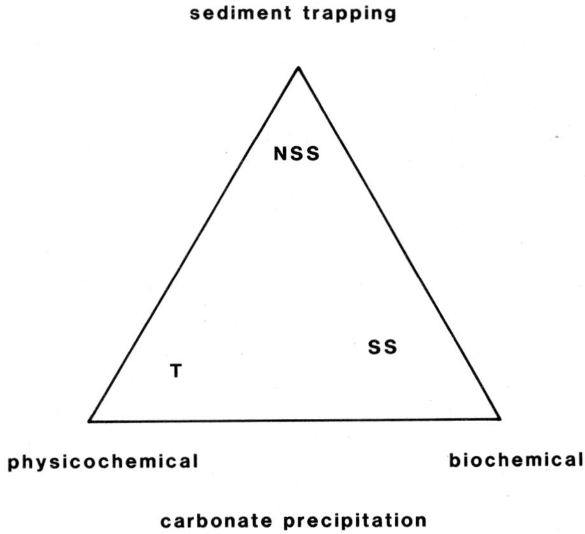

Fig. 6. Relationships between processes of carbonate deposition in the formation of nonskeletal stromatolites (*NSS*), tufa (*T*), and skeletal stromatolites (*SS*) (Riding 1977, Fig. 1)

evasion is enhanced by water movement. As a result, it is possible for lacustrine stromatolites to approach skeletal stromatolites in appearance.

The term cyanoid, describing oncoids constructed by calcified cyanobacteria (Riding 1979, 1983), is applicable to both tufa oncoids and skeletal oncoids.

The time distribution of skeletal stromatolites in marine environments parallels that of other biomineralized microbial carbonate deposits such as dendrolites (and also thrombolites containing calcified microfossils). These appear to be restricted to the Palaeozoic and Mesozoic, probably in response to an environmental control such as sea-water chemistry (Riding 1982).

3.2.4 Subaerial Stromatolites

Research on highly calcareous soils and subaerially exposed limestone surfaces has resulted in progressive refinement of understanding of calcrete and caliche deposits. The importance of micro-organisms, particularly prokaryotes, fungi, lichens and green algae, in subaerial laminated crusts has been increasingly recognized (Krumbein 1968; Kahle 1977; Klappa 1979a) with the result that the distinction between calcretes and stromatolites (Read 1976) has become blurred. As a result, terms such as lichen stromatolite (Klappa 1979b) and desert stromatolite (Krumbein and Giele 1979) have been applied, and it is now necessary to recognize laminated microbial calcretes as varieties of stromatolite (Wright 1981).

Subaerial stromatolites form in environments dominated by desiccation and vadose diagenesis. They are highly variable, but features similar to those of subaqueous and tidal flat stromatolites are conspicuous. These include well-developed lamination, fenestrae and also coated grains resembling oncoids (but usually referred to as vadose pisoids, e.g. Dunham 1969; Esteban and Pray 1983). In addition, pedogenic fabrics, such as glaebules, alveolar texture, circumgranular cracks and micrite bridges are usually present (see Esteban and Klappa 1983). Intense desiccation, as well as biogenic processes, result in early lithification of subaerial stromatolites, and carbonate precipitation and replacement dominate. Trapping of particles appears to be of relatively minor importance. There are similarities between subaerial stromatolites and travertines: both form in environments strongly influenced by desiccation and carbonate precipitation. Thus, subaerial stromatolites combine features of subaqueous stromatolites with distinctive pedogenic fabrics. Their recognition is further assisted by their intimate association with weathering profiles.

3.3 Dendrolites

This term was introduced by Riding (1988) for non-laminated biomineralized microbial deposits with a dominant dendritic macrofabric. Dendrolites formed by genera such as *Epiphyton, Renalcis* and *Angulocellularia* are conspicuous in Cambrian and Lower Ordovician reefs (Fig. 7), commonly in association with

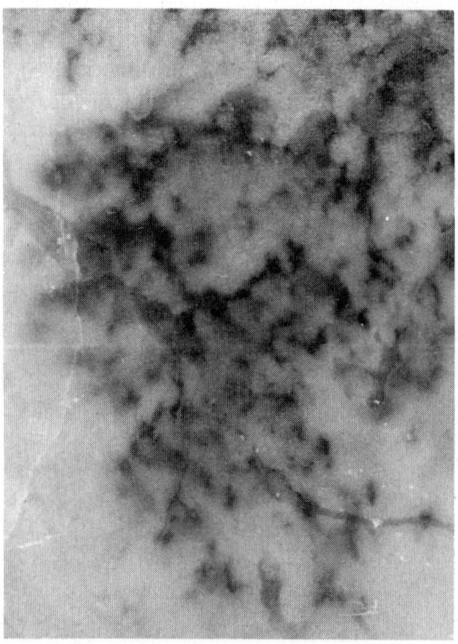

Fig. 7. Dendrolite. Dendritic fabric created by ERA (*Epiphyton, Renalcis* and/or *Angulocellularia*) microfossils, apparently pendant in reef cavity. Serie Schisteuse, Lower Cambrian, Assads, Anti-Atlas Mountains, Morocco. Specimen provided by Andrew Latham. Width of view 38 mm

sponge-like organisms such as archaeocyaths. They reappear in Upper Devonian reefs where the dendritic fossils are again *Epiphyton* and *Renalcis* Vologdin, or close relatives of these genera, and the principal metazoans with which they are associated are stromatoporoids. The "bacterial shrubs" which occur in some travertines (see Chafetz and Folk 1984, Fig. 1) have dendrolitic fabrics, but are arranged in layers. There is also a link with skeletal stromatolites, which differ from dendrolites principally in the presence of lamination, and with thrombolites which may show intergradation with dendrolites.

Epiphyton-Renalcis-Angulocellularia (ERA) microfossils are common constituents of Cambrian reefs, often completely dominating the rock fabric. They are frequently associated with early cementation localized between their filamentous and bushy masses (Fig. 4). Although generally individually microscopic, these fossils amalgamate to form dendritic fabrics, either vertically erect or pendant, which are readily visible in hand-specimen. The result of this growth, calcification, and associated early cementation, was the formation of microframe reef structures which provided hard substrates for attached metazoans, notably archaeocyaths. This substrate-attachment relationship could be reversed, with ERA microfossils clustered upon archaeocyaths. Thus, Cambrian reefs typically show two basic skeletal end-members: ERA dendrolite and archaeocyaths.

Although the precise affinities of most ERA microfossils are uncertain; their probable cyanobacterial affinities (see Riding, Calcareous cyanobacteria and algae, this Vol.) would ensure their placement in microbolites. Their lack of lamination, combined with biomineralization, ally them with thrombolites, although more remains to be understood concerning the origins of the latter. The diversification of metazoan reef-builders in the Middle Ordovician modified reef composition considerably, and substantially reduced the importance of ERA micro-organisms in reef ecosystems. Consequently, ERA dendrolites are most conspicuous at times when the importance of metazoans was reduced: (1) Cambrian to Lower Ordovician, prior to the advent of the stromatoporoid-tabulate-bryozoan association; (2) Upper Devonian, during the decline of stromatoporoids; (3) Lower Carboniferous, following the extinction of most Palaeozoic stromatoporoids.

Further documentation of ERA and other dendrolites will clarify their relationships to both thrombolite, skeletal stromatolite, and travertine.

3.4 Thrombolites

Aitken (1967) introduced the term thrombolite for non-laminated cryptalgal (i.e. microbial) carbonates with a clotted fabric. In the field thrombolites commonly possess domical morphologies resembling stromatolites, but they lack lamination. The clotted fabrics of thrombolites are quite variable and may alternate with stromatolitic lamination within a single dome. Aitken (1967) noted that thrombolites are common in the Cambrian and Ordovician and subsequent studies have confirmed this: they have been reported from the

Cambrian, Ordovician, Silurian, Carboniferous and Permian, but are especially abundant in the Upper Cambrian and Lower Ordovician (Kennard and James 1986, Fig. 8). Despite recognition for more than twenty years, the origins of thrombolites remain uncertain. Walter and Heys (1985, p. 164) considered that thrombolites are disturbed stromatolites in which the activity of animals and algae has disrupted and destroyed the lamination. However, Kennard and James (1986, p. 494) attributed the clotted fabrics of thrombolites to calcification of coccoid microbial communities. Latham and Riding (1988) recognized *Renalcis* Vologdin and *Tarthinia* Drosdova in thrombolites in the Lie de Vin Series of Morocco. These calcareous microfossils create the irregular clotted fabrics of the thrombolites (Fig. 8). Their precise origins are uncertain but they probably represent cyanobacteria. However, it remains quite likely that there is a variety of origins for the deposits which have been termed thrombolites. For example, digitate plumose algal-foraminiferal colonies with a blotchy appearance, termed "leopard rock" in the Lower Permian of New Mexico (Toomey and Cys 1979, p. 66, Figs. 3b and 4c) resemble thrombolites. At present any deposit that appears to be microbial, and which is unlaminated and clotted, can be termed a thrombolite. Future investigations should aim to discriminate between different types of thrombolites and their various origins.

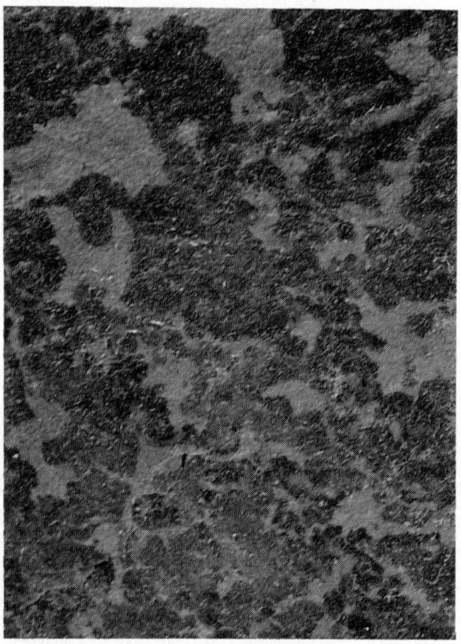

Fig. 8. Thrombolite. Clotted macrofabric in a domical thrombolite created by clusters of botryoidal microfossils (*Renalcis* and *Tarthinia*) (Latham and Riding 1988) and enhanced by partial dolomitization of intervening matrix. Upper part of Serie Lie de Vin, Lower Cambrian, Tiout, Anti-Atlas Mountains, Morocco. Width of view 30 mm

3.5 Travertine

Travertine is named from the old name for Tivoli, east of Rome, where carbonates deposited from hot springs have been mined for thousands of years. Travertine is used here to describe a crudely layered carbonate rock, commonly with a bushy fabric. The latter has been attributed to bacterial activity (see Chafetz and Folk 1984). The terminology of travertine, tufa and sinter has not been standardized. Here I regard travertine as a product of warm carbonate springs where the elevated temperatures, together with the dissolved materials present in these warm waters, exclude most eukaryotic organisms. The result is that any mediation of carbonate precipitation in these environments is by prokaryotes. Consequently the bushy fabrics which are conspicuous in travertines may be bacterial and/or cyanobacterial in origin, and the shrubs described as bacterial by Chafetz and Folk (1984, Fig. 1) also resemble *Angulocellularia*, although further work is required to determine precisely what organisms are present and exactly which processes result in the formation of travertine.

In fact, at least two types of shrub appear to be involved in travertine formation: (1) the micritic shrubs, such as those described by Chafetz and Folk (1984) which may be bacterial or cyanobacterial, and (2) needle shrubs described by Casanova (1986, p. 204 and Fig. 4) as bacterial from pool rims of hot springs in Kenya (see also Tiercelin et al. 1987, plates 32 and 33). Despite assertion of the microbial nature of these latter needle shrubs, they resemble inorganic precipitates.

A parallel can be drawn between crudely laminated travertines with bushy microfabrics and skeletal stromatolites. I prefer to maintain a distinction between travertine and skeletal stromatolites, until further work can show how similar (or distinct) the processes of formation of travertine are. In the absence of banding and lamination, the bushy fabrics of travertines closely resemble dendrolite.

3.6 Cryptic Microbial Carbonates

Stromatolites, and the other types of microbial carbonates described above, generally have quite distinctive macroscale fabrics and features which assist in their recognition. However, there is a large volume of carbonate rock, especially in reefal facies, which may be microbial in origin but whose origins are nevertheless obscure and for which descriptive criteria are also difficult to establish. These are here termed *cryptic microbial carbonates*. They lack distinctive macrofabrics, but typically possess micritic, clotted or peloidal microfabrics, in which there may also be traces of filaments. Some cement-like spar fabrics and dolomite crystals have also been interpreted as microbial.

Many of these microfabrics are also present in stromatolites, dendrolites, thrombolites and travertines. However, these latter also possess characteristic macrofabrics which are lacking in the case of cryptic microbial carbonates.

Consequently, cryptic microbial carbonates have to be interpreted purely on the basis of microfabrics, and it is the uncertainties inherent in this which create the particular difficulties which surround them. It is difficult to establish criteria which allow confident interpretation of a microbial origin for these micritic, peloidal and other microfabrics. It is sometimes just as difficult to decide whether they are in place or allochthonous.

Nevertheless, in recent years a number of workers appear to have had an innate sense of the presence and importance of these deposits and, despite the obstacles, there has been a growing attempt to deal with the problem. Much current work on "microbial carbonates" is, in fact, aimed at this cryptic side of the subject (Maurin 1987). The origins of this research can be seen in early work such as that by Drew (1911). Part of its revival can be attributed to the need to understand "cryptalgal" fabrics (e.g. Monty 1976) and also to detailed scrutiny of the organisms and the deposits within modern microbial mats (e.g. Krumbein and Cohen 1977); these were developments of the stromatolite work of the 1960s and early 1970s. At the same time, a few workers were starting to report bacterium-like objects in ancient limestones (e.g. Maurin and Noël 1977). Subsequent work has involved spars (Monty 1982), Devonian reefs (Monty et al. 1982), mud mounds (Maurin et al. 1981; Tsien 1985b), *Stromatactis* (Tsien 1985a), oöids (Monty and van Laer 1984; Castanier et al. 1989), peloids (Chafetz 1986), and dolomites (Monty 1986; Castanier et al. 1988).

Yet problems of establishing organic versus inorganic, and in place versus allochthonous, origins remain. It is particularly difficult to discriminate between cause and effect: are microbes associated with a crystal of calcium carbonate because they have caused its precipitation, or because it is a convenient substrate to live on, or both, or neither? The way forward most likely lies in laboratory studies (e.g. Castanier 1987). But apart from these fundamental problems of microbial biomineralization, a descriptive vocabulary for the microfabrics (and nanofabrics) of these cryptic microbial carbonates also remains to be established. This should allow improved comparison to be made between ancient examples and modern analogues which, in combination with experimental work, will probably provide the key to understanding these enigmatic, but potentially very important, deposits.

4 Composition and Fabrics

Calcareous benthic microbial deposits consist, basically, of one or more of the following: trapped particulate sediment, biomineralized skeletons, and mineralized crust, either surficial or as a void-fill. These may be laminated and/or dendritic, or clotted in appearance (Fig. 9), and they may possess fenestrae. These variations in composition and fabric are controlled both by the organisms involved and by the environment of formation of the microbial carbonate. Particulate sediment is predominant in agglutinated stromatolites, calcareous skeletons in skeletal stromatolites, dendrolites and thrombolites, and mineralized crusts in tufa stromatolites and travertine. Lamination is good in

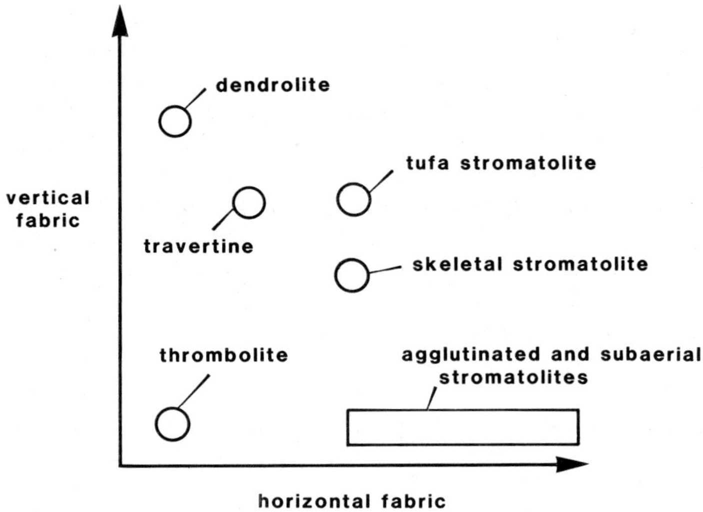

Fig. 9. Preponderance of vertical (and radial) versus horizontal (laminated/layered) fabrics in benthic microbial carbonates

fine-grained agglutinated stromatolites, and crude in coarse-grained agglutinated, skeletal and tufa stromatolites, and also in travertine. Dendritic and clotted fabrics are characteristic of dendrolites and thrombolites respectively, both of which are non-laminated. Fenestrae are especially present in tufa stromatolites, skeletal stromatolites and coarse-grained agglutinated stromatolites. In all of these they can be produced in subaqueous environments. They also form in tidal flat stromatolites.

The interrelationships of lamination, mineralized crusts and biomineralized skeletons are shown in Fig. 10 and Table 2. Certain microbial carbonate types are closely related and therefore may show intergradation. This is particularly true of (1) skeletal stromatolites and travertine; (2) skeletal stromatolites and dendrolites; (3) skeletal stromatolites and tufa stromatolites; (4) dendrolites and thrombolite; and (5) tufa stromatolites and travertine.

5 Distribution

There are six main stages in the geological history of stromatolites, dendrolites, and thrombolites. The following outline applies principally to marine microbial carbonates. The time distribution of non-marine benthic microbial carbonates is less-well recorded, although it must be pointed out that it is often not clear whether Precambrian stromatolites occur in marine or lacustrine environments. It is likely that there is a direct relationship between the organisms and processes available for the formation of benthic microbial deposits, and the overall temporal sequence that they show (Fig.1).

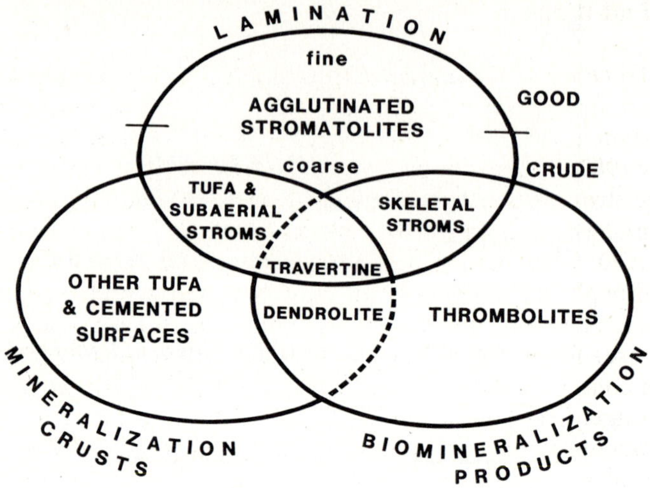

Fig. 10. Types of microbial carbonate deposits plotted according to development of lamination, mineralization and biomineralization. Note that potentially close intergradations occur between skeletal stromatolites and travertine, skeletal stromatolites and dendrolites, dendrolites and thrombolites, tufa stromatolites and travertine, subaerial stromatolites and travertine, and tufa stromatolites and skeletal stromatolites. Note that it is likely that not all thrombolites are dominated by biomineralization

Table 2. Fabrics and principal components typical of major microbial carbonate types[a]

		Principal components		
		Calcified organism		Particulate sediment
Lamination	good / poor	Skeletal stromatolites Tufa stromatolites Travertine	fine / coarse	Agglutinated stromatolites
Dendritic		Dendrolites		
Clotted		Thrombolites		Thrombolites (?)

[a] Subaerial stromatolites have good lamination but appear to be dominated by mineralization products rather than calcified organisms.

5.1 Archaean

The oldest definite record of stromatolites is from the Warrawoona Group (3.5 Ga) of Western Australia (Dunlop et al. 1978). These occur in nearshore environments dominated by volcaniclastic sediments. Silicified filamentous microfossils in cherts from these stromatolites could be either bacterial or cyanobacterial (Awramik et al. 1983; Schopf and Walter 1983, pp. 232–237). The Archaean stromatolite record is reviewed by Walter (1983).

5.2 Proterozoic to End-Riphean

Stromatolite abundance increases rapidly during the Early Proterozoic, possibly in response to both the evolution of cyanobacteria, whose definite fossil record appears to begin in the Fortescue Group (2.7 Ga) of Western Australia (Schopf and Walter 1982), and to the progressive development of continental cratons. Proterozoic stromatolites are characteristically, although not exclusively, fine-grained and well-laminated (Awramik and Riding 1986). It has often been stated that $CaCO_3$ precipitation was significant in their formation (e.g. Gebelein 1976), although this is not proven and often they resemble agglutinated stromatolites. Nevertheless, microdigitate ministromatolites with a radial fibrous fabric are a feature of the Early and Middle Proterozoic (900–2500 Ma) and have been compared with calcified *Rivularia* and *Phormidium* mats (Walter 1988; Grotzinger 1989). The Riphean (800–1650 Ma) was the climax of development of branching and conical stromatolites, and these have been applied to stratigraphical correlation following the lead of Soviet geologists (Semikhatov 1976).

5.3 Sturtian and Vendian

Awramik (1971) recognized a major decline in both the abundance and diversity of stromatolites following the Riphean, which he attributed to the appearance of metazoans. Subsequent studies have confirmed this decline (Walter and Heys 1985). The deleterious effects of metazoans are likely to have been various: grazing, burrowing, competition for space. To these were added the production of coarse bioclastic sediment following the onset of biomineralization close to the base of the Cambrian (Pratt 1982).

5.4 Cambrian-Lower Ordovician

Fine-grained agglutinated stromatolites never recovered from the impact of metazoan evolution. However, cyanobacterial biomineralization, which appeared as dramatically as that of metazoans near the base of the Tommotian (Riding and Voronova 1984), resulted in the formation of dendrolites and thrombolites which are both particularly common in the Cambrian and Lower Ordovician. Skeletal stromatolites also make their appearance at this time.

5.5 Middle Ordovician to Middle Cretaceous

Calcified cyanobacterial dendrolites, thrombolites and skeletal stromatolites are sporadically prominent during this long period of overall decline in calcareous microbial deposits in marine environments. Dendrolites and thrombolites appear to be virtually restricted to the Palaeozoic. Decline is likely to be due to the increasing success, and therefore, competition, predation and disturbance, of metazoans. And, particularly in the case of dendrolites and skeletal stromatolites, it may also be due to a general long-term reduction in $CaCO_3$ precipitation in the world oceans. Only in tidal flat environments did stromatolites (represented by fine-grained agglutinated forms) retain their overall success; and this appears to clearly relate to the inhibition of metazoans by strongly fluctuating physical and chemical parameters in these environments (Garrett 1970).

5.6 Mid-Cretaceous to Recent

Calcified cyanobacteria disappear, or virtually disappear, from normal subtidal marine environments sometime during the Mesozoic-Cenozoic transition (Riding 1982). Probably the final occurrence of fairly common marine calcified cyanobacteria was during the Middle Cretaceous, but this remains to be confirmed. During the Cenozoic stromatolites have a fine record in lacustrine environments, for example in the Middle Eocene Green River Formation of Wyoming (see Surdam and Wray 1976), but the principal marine occurrence of stromatolites appears to have been the tidal-flat environment, and this can be seen as a refuge for fine-grained cyanobacterial-dominated stromatolites (Garrett 1970; Awramik and Riding 1988). Undoubtedly, the conspicuous survival of this type of stromatolite in this environment led to the "intertidal dogma" of stromatolitology (Hoffman 1973, p. 181) following Black's (1933) work in Andros. However, a significant new development has been the origin of the eualgal-cyanobacterial stromatolite-forming community which characteristically produces coarse-grained stromatolites (Awramik and Riding 1988). These are exemplified by the Recent well-known subtidal Shark Bay and Lee Stocking columnar and domical forms. Both diatoms and filamentous green algae are conspicuous in the formation of these distinctive stromatolites which have widely, but probably mistakenly, been regarded as analogues of Precambrian stromatolites. It remains to be seen whether eualgal-cyanobacterial stromatolites date from the time of appearance of diatoms, in the Cretaceous, or earlier. So far no closely similar fossil stromatolites have been identified.

6 Outline History of Research

The history of work on microbial carbonate deposits is dominated by research on stromatolites. This has been reviewed a number of times (Maslov 1960; Hoffman 1973; Hofmann 1973; Monty 1977; Krumbein 1983).

Twentieth-century progress in stromatolite studies can be characterized as a succession of dogmas (Monty 1977). Walcott's (1914) views concerning Proterozoic Belt Supergroup stromatolites in Montana led to a general freshwater interpretation until Black's (1933) study of modern Bahamian stromatolites redirected attention to tidal flats and also introduced the important trapping and binding concept of formation. During the 1960s and 1970s, interest in morphological classifications of stromatolites was coupled with the development of Proterozoic stromatolite biostratigraphy. Two results of this have been the production of a fairly complex Linnean-style nomenclatural system, especially for branching stromatolites (for discussion see Cloud and Semikhatov 1969; Walter 1972, pp. 14–27; Semikhatov 1976), which still needs to be revised and consolidated into a readily applicable scheme and, on the other hand, a rather over-simple morphological descriptive terminology (Logan et al. 1964) which awaits refinement.

Simultaneously, the search has gone on for modern analogues of ancient stromatolites. Much has been learned concerning the biology of modern microbial mats, particularly the complex physiological adaptations which are linked to the vertical stratification of various bacterial and cyanobacterial communities within the mats (see papers in Cohen et al. 1984). However, the lack of obvious information preserved in most ancient stromatolites has hindered progress in extrapolating results from the modern to the ancient. Nevertheless, it does seem likely that petrographic evidence in ancient stromatolites still awaits interpretation. Most studies of Precambrian stromatolites, for example, have been made by palaeobiologists and stratigraphers who may have overlooked useful, if somewhat cryptic, information within the carbonate microfabrics. Recognition that the modern coarse-grained, crudely laminated subtidal Shark Bay stromatolites are really not closely comparable with the ancient fine-grained well-laminated fossils with which they have generally been compared (Awramik and Riding 1988) illustrates the type of comparative approach which may lead to significant new insights into stromatolites.

Stromatolites can form in hot springs (Doemel and Brock 1974) and work on hot spring sinters (geyserites) has been related to stromatolite research by Walter (1976b). However, in general, travertine and geyserite studies have developed separately, and more slowly, than those on stromatolites, and have been more closely linked with work on tufa. Travertines and tufas are both non-marine deposits, and there has been terminological confusion between them, usually with fluviatile tufas being termed travertine (e.g. Love and Chafetz 1988), rather than the other way round. Work on tufa, especially, has often been concentrated on attempting to discriminate between organic and inorganic origins (e.g. Emeis et al. 1987). Tufa dams in rivers have been termed phytoherms (Pedley 1987).

Fluviatile and lacustine oncoids have long attracted attention (see Fritsch 1959, p. 868, and references therein), but their in-place counterparts, tufa stromatolites, are probably among the most neglected of all the microbial deposits considered here, possibly because of their intergradation to, and incorporation with, more diverse tufa deposits.

Stromatolites and other microbial carbonates reviewed here possess macroscopically distinctive fabrics. However, research has also been moving towards studies of less distinctive, and often obscure, fabrics which may be microbial in origin and are here termed cryptic microbial carbonates. These include an array of micritic, clotted, peloidal and spongiostrome microfabrics (see Maurin and Noël 1977; Monty et al. 1982; Chafetz 1986). In hindsight, we can see that research on some of these was begun by earlier workers (Gürich 1906; Pia 1927; Maslov 1960).

Introduction of a microbolite, rather than specifically stromatolite, orientated approach (Burne and Moore 1987) to this field will no doubt assist comparative studies. Kennard and James (1986) emphasized that thrombolites and stromatolites are two classes of microbial structure. This present paper emphasizes dendrolite as yet another type. Dendrolites, although not hitherto dignified by a specific name, have been obvious fabrics within Early Palaeozoic reefs for many years. Thrombolites have themselves been named for more than twenty years. By drawing these deposits, along with travertine, into a broad category of calcareous benthic microbial deposits we can now begin to see the similarities which they share, alongside their differences, and thereby more precisely interpret their origins, distribution and development.

7 Conclusions

Riding (1977, Fig. 1) used a triangular diagram (Fig. 6) to show the relationships between the processes of carbonate deposition (sediment trapping, physico-chemical precipitation, biochemical trapping) and the products: non-skeletal stromatolites (which here are termed agglutinated stromatolites), tufa, and skeletal stromatolites. The same diagram can be used to express the array of microbial carbonates described here (Fig. 11).

Agglutinated stromatolites are the classic stromatolites of the stromatolite literature, referred to as 'normal' or nonskeletal stromatolites by Riding (1977, p. 57). It is precisely because they have been recognized as the most common type of stromatolite that they have not been given a more specific name until introduction of the term agglutinated stromatolite here. Fine-grained agglutinated stromatolites are probably dominantly prokaryotic in origin and range from the Archaean to Recent. However, there appear to be no good modern analogues for the complex branching forms which they produced during the climax of their development during the Riphean. The Shark Bay and Lee Stocking subtidal stromatolites not only differ in morphology, but are coarse grained and crudely laminated: features which can be attributed to the prominent eualgal component of their mat community. The marine refuge of fine-grained agglutinated stromatolites is the tidal flat environment, which is somewhat similar to the Precambrian in exhibiting reduced metazoan competition and abundant fine-grained sediment, but in which the stromatolites characteristically are flat and crinkly laminated due to the exposure characteristic of this environment. It is likely that modern lacustrine environments possess better, although nevertheless not precise, analogues for subtidal Precambrian stromatolites.

Although it is probable that the carbonate microfabrics of fossil agglutinated stromatolites possess valuable information concerning their origins which

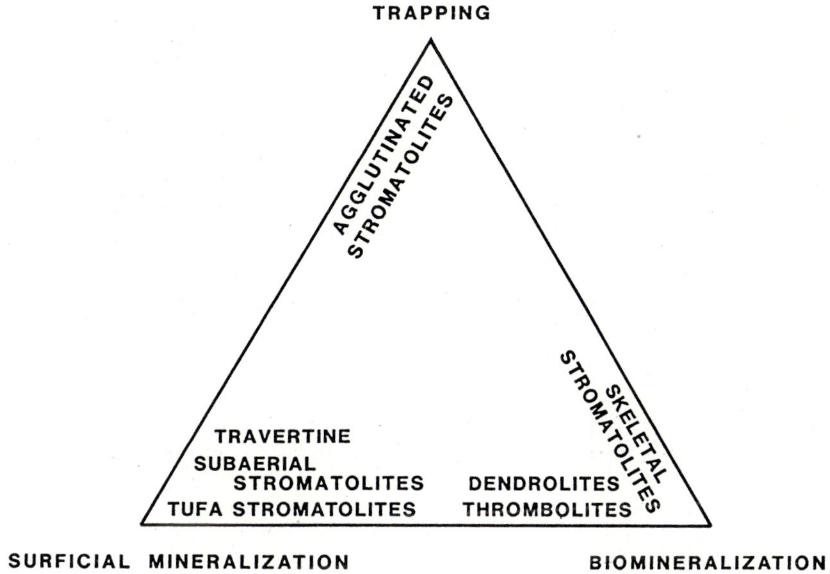

Fig. 11. Triangular diagram adapted from Riding (1977, Fig. 1; see Fig. 6) showing major microbial carbonate types according to processes of formation. Note that it is likely that not all thrombolites are dominated by biomineralization. Skeletal stromatolites overlap with thrombolites and dendrolites on this diagram. The lamination which distinguishes stromatolites is due to episodic accretion and/or prostrate orientation of organisms. Biomineralization here is regarded as precipitation within organisms, or within organic tissue, during life. Surficial mineralization is external to organisms and their tissues, but does not here preclude the possibility of organic influence on this precipitation

remains to be exploited, they nevertheless retain relatively little information concerning the organisms responsible for their formation. This is not the case with biomineralized and mineralized microbial deposits. Skeletal stromatolites, tufa stromatolites, dendrolites, some thrombolites and travertine all possess fabrics which reflect the size, shape and orientation of individuals, or clusters of individuals, of mat-forming organisms to varying extents. Much, however, remains to be learned about all these categories of calcareous microbolite. In the case of mineralized varieties (tufa stromatolites, subaerial stromatolites, travertine) superficial crusts preserve, but also complicate the morphology of the organisms, whereas biomineralized fabrics within skeletal stromatolites, dendrolites and some thrombolites probably retain clearer details. Of these latter three deposits, skeletal stromatolites possess at least poor lamination, and this distinguishes them from dendrolites and thrombolites. If they are mineralized, then skeletal stromatolites equal tufa stromatolites. If they are unlaminated, then they may equal dendrolites. Whether or not all varieties of thrombolite contain an important component of calcified microfossils remains to be elucidated; it is still likely that non-skeletal thrombolites will be distinguished.

If unlaminated, travertine may equal dendrolite. But travertine can also be viewed as intermediate between tufa and calcareous sinter, both with respect to lamination (tufa unlaminated, travertine crudely laminated, sinter well laminated) and porosity (tufa, high porosity; travertine, medium to low porosity; sinter, zero porosity). As a rough generalization, tufa is formed by precipitation of $CaCO_3$ from cold water onto cyanobacteria, algae, moss, grass, etc. Travertine is formed by precipitation from warm water mainly onto cyanobacteria and bacteria, although diatoms may also be present; sinter is either precipitated from hot water (at the Earth's surface) in the presence of bacteria and subsidiary cyanobacteria, or from cold water in caves where bacteria may be present. Elevated temperatures act to progressively exclude eukaryotes.

Cryptic microbial carbonates possess micritic clotted, peloidal or sparitic microfabrics, similar to those of evident benthic microbial carbonates such as stromatolites, etc., but they lack distinctive macrofabrics. They are probably widespread and important in Phanerozoic reef facies. Criteria are still required to distinguish them from particulate or cement fabrics of inorganic origin, and to elucidate the particular micro-organisms and processes responsible for their formation.

Since benthic microbial carbonates vary in time and space, and also according to the organisms and precipitating fluids involved, there are notable contrasts between the different types of deposits which have been reviewed here. In hindsight, therefore, it is not surprising that difficulties have been encountered in the past in attempting to draw too close analogies between, for example, freshwater and marine, and modern and ancient microbial carbonates. The present is only a key to the past if the locks have not been changed.

Appendix

Summary of terms used here

Agglutinated stromatolite (this paper), produced by trapping/binding of particulate sediment.

Biomineralization, mineral precipitation within organisms, or within organic tissue, during life.

Cryptic microbial carbonate (this paper), benthic microbial deposit with micritic, clotted, peloidal or sparitic microfabrics, but lacking distinctive macrofabrics.

Cyanoid (Riding 1979, 1982), oncoid constructed by calcified cyanobacteria.

Dendrolite (Riding 1988), non-laminated biomineralized microbial deposit with a dominant dendritic macrofabric.

Microbial carbonate, carbonate deposits produced or localized by benthic microbial communities.

Microbolite (Burne and Moore 1987, name emended here), benthic microbial deposit.

Mineralization, mineral precipitation on inorganic surfaces and on external surfaces of organic tissue.

Nonskeletal stromatolite (Riding 1977), stromatolite in which the microorganisms responsible for its formation are not calcified. Equivalent to agglutinated stromatolite.

Oncoid (from oncolith, Pia 1927), nodular, originally loose, stromatolite.

Oncolite (from oncolith, Pia 1927), lithified deposit of oncoids.

Sinter, dense, laminated deposit, precipitated often as an incrustation upon rock.

Skeletal stromatolite (Riding 1977), stromatolite in which the organisms responsible for its formation are commonly preserved as calcified fossils.

Stromatolite (from stromatolith, Kalkowsky 1908), laminated benthic microbial deposit.

Subaerial stromatolite, microbial calcrete, formed at subaerial surfaces and dominated by desiccation which promotes rapid lithification; characterized by pedogenic fabrics.

Thrombolite (Aitken 1967), non-laminated benthic microbial deposit with a dominant clotted macrofabric.

Travertine (from the ancient name for Tivoli, Italy; tivertino, tiburtinus), layered deposit of $CaCO_3$ with moderate to low primary porosity, commonly with a dendritic macrofabric, produced by precipitation associated with warm springs.

Tufa (an old Italian word from the Latin name *tofus* for a porous rock), originally porous, laminated to massive deposit mainly produced by precipitation of minerals on organic substrates in freshwater.

Tufa stromatolite (this paper), stromatolite produced by precipitation of minerals on (rather than within) organic tissue.

Acknowledgements. My interest in tufas was stimulated by Ernst Ott and supported by an Alexander von Humboldt-Stiftung research fellowship at the Technical University, Munich. Anna Gandin introduced me to travertines in Tuscany and TOTAL, through the encouragement of André Maurin, supported their study. Discussions with Stanley Awramik, facilitated by NATO Research Grant 84/0176, were a fertile source of insights into stromatolites past and present; not a few of these views are included here, and I am indebted to Stan for the stimulus, and the fun, which our debates created. A visit to Lee Stocking Island was supported by The Royal Society. Constructive comments by Stanley Awramik and V. Paul Wright improved the manuscript.

References

Aitken JD (1967) Classification and environmental significance of cryptalgal limestones and dolomites, with illustrations from the Cambrian and Ordovician of southwestern Alberta. J Sediment Petrol 37:1163–1178

Awramik SM (1971) Precambrian columnar stromatolite diversity: reflection of metazoan appearance. Science 174:825–827

Awramik SM, Riding R (1986) "Golden age" stromatolites and modern analogs. SEPM Annu Midyear Meet, Raleigh, NC, Abstr, pp 3–4

Awramik SM, Riding R (1988) Role of algal eukaryotes in subtidal columnar stromatolite formation. Proc Natl Acad Sci USA 85:1327–1329

Awramik SM, Schopf JW, Walter MR (1983) Filamentous fossil bacteria from the Archean of Western Australia. Precambrian Res 20:357–374
Black M (1933) The algal sediments of Andros Island, Bahamas. Philos Trans R Soc London Ser B 222:165–192
Burne RV, Moore LS (1987) Microbialites: organosedimentary deposits of benthic microbial communities. Palaios 2:241–254
Casanova J (1986) East African Rift stromatolites. In: Frostick LE, Renaut RW, Reid I, Tiercelin JJ (eds) Sedimentation in the African rifts. Geol Soc London Spec Publ 25:201–210
Castanier S (1987) Microbiogéologie: Processus et modalités de la carbonatogenèse bactérienne. Thèse Doctorat Etat Univ Nantes, 541 pp
Castanier S, Maurin A, Perthuisot J-P (1988) Les cugnites: carbonates amorphes de Ca et Mg, précurseurs possibles de la dolomite. CR Acad Sci Paris 306 Ser II:1231–1235
Castanier S, Maurin A, Perthuisot J-P (1989) Production bactérienne expérimentale de corpuscles carbonatés, sphéroidaux à structure fibro-radiaire. Bull Soc Geol Fr 8, 3:589–595
Chafetz HS (1986) Marine peloids: a product of bacterially induced precipitation of calcite. J Sediment Petrol 56:812–817
Chafetz HS, Folk RL (1984) Travertines: depositional morphology and the bacterially constructed constituents. J Sediment Petrol 54:289–316
Clarke JM (1900) The water biscuit of Squaw Island, Canandaigua Lake, N.Y. Bull NY State Mus 8(39):195–198
Cloud Jr PE, Semikhatov MA (1969) Proterozoic stromatolite zonation. Am J Sci 267:1017–1061
Cohen Y, Castenholz RW, Halvorson HO (eds) (1984) Microbial mats: stromatolites. MBL Lectures in biology, vol 3. Liss, New York, 498 pp
Danielli HMC, Edington MA (1983) Bacterial calcification in limestone caves. Geomicrobiol J 3:1–16
Doemel WN, Brock TD (1974) Bacterial stromatolites: origin of laminations. Science 184:1083–1085
Drew GH (1911) The action of some marine denitrifying bacteria in tropical and temperate seas, and the bacterial precipitation of calcium carbonate in the sea. J Mar Biol Assoc UK NS 9:142–155
Dunlop JSR, Muir MD, Milne VA, Groves DI (1978) A new microfossil assemblage from the Archean of Western Australia. Nature (London) 274:676–678
Dunham RJ (1969) Vadose pisolite in the Capitan reef (Permian), New Mexico and Texas. In: Symp Depositional environments in carbonate rocks. Soc Econ Paleontol Mineral Permian Basin Sect Publ 14:182–191
Eggleston JR, Dean WE (1976) Freshwater stromatolitic bioherms in Green Lake, New York. In: Walter MR (ed) Stromatolites. Elsevier, Amsterdam, pp 479–488
Emeis K-C, Richnow H-H, Kempe S (1987) Travertine formation in Plitvice National Park, Yugoslavia: chemical versus biological control. Sedimentology 34:595–609
Esteban M, Klappa CF (1983) Subaerial exposure environment. In: Scholle PA, Bebout DG, Moore CH (eds) Carbonate depositional environments. Am Assoc Petrol Geol, Tulsa, pp 1–54
Esteban M, Pray LC (1983) Pisoids and pisolite facies (Permian), Guadalupe Mountains, New Mexico and West Texas. In: Peryt TM (ed) Coated grains. Springer, Berlin Heidelberg New York, pp 503–537
Fritsch FE (1959) The structure and reproduction of the algae, vol 2. Univ Press, Cambridge, 939 pp
Garrett P (1970) Deposit feeders limit development of stromatolites. Am Assoc Petrol Geol Bull 54:848
Gebelein CD (1976) The effects of the physical, chemical and biological evolution of the earth. In: Walter MR (ed) Stromatolites. Developments in sedimentology, vol 20. Elsevier, Amsterdam, pp 499–515
Gerdes G, Krumbein WE (1987) Biolaminated deposits. Lecture Notes in Earth Sciences 9, Springer, Berlin Heidelberg New York London Paris Tokyo, 183 pp
Golubic S (1973) The relationship between blue-green algae and carbonate deposits. In: Carr NG, Whitton BA (eds) The biology of blue-green algae. Blackwell, London, pp 434–472
Grotzinger JP (1989) Facies and evolution of Precambrian carbonate depositional systems:

emergence of the modern platform archetype. In: Crevello P, Sarg R, Read JF, Wilson JL (eds) Controls on carbonate platforms and basin development. SEPM Spec Publ 44:79–109
Gürich G (1906) Les spongiostromides du viséen de la Province de Namur. Polleunis & Ceuterick, Bruxelles, 55 pp
Hall JD (1883) Cryptozoon (proliferum) NG (and sp). NY State Mus Annu Rep 36
Hoffman P (1973) Recent and ancient algal stromatolites: seventy years of pedagogic cross-pollination. In: Ginsburg RN (ed) Evolving concepts in sedimentology. Johns Hopkins Univ Stud Geol 21:178–191
Hofmann HJ (1973) Stromatolites: characteristics and utility. Earth Sci Rev 9: 339–373
Kahle CF (1977) Origin of subaerial Holocene calcareous crusts: role of algae, fungi and sparmicritisation. Sedimentology 24:413–435
Kalkowsky E (1908) Oolith und Stromatolith im norddeutschen Buntsandstein. Z Dtsch Geol Ges 60:68–125
Kennard JM, James NP (1986) Thrombolites and stromatolites: two distinct types of microbial structures. Palaios 1: 492–503
Klappa CF (1979a) Calcified filaments in Quaternary calcretes: organomineral interactions in the subaerial vadose environment. J Sediment Petrol 49:955–968
Klappa CF (1979b) Lichen stromatolites: criterion for subaerial exposure and a mechanism for the formation of laminar calcretes (caliche). J Sediment Petrol 49:387–400
Krumbein WE (1968) Geomicrobiology and geochemistry of the "Nari-Lime-Crust" (Israel). In: Müller G, Friedman GM (eds) Recent developments in carbonate sedimentology in central Europe. Springer, Berlin Heidelberg New York, pp 138–147
Krumbein WE (1983) Stromatolites – the challenge of a term in space and time. Precambrian Res 20:493–531
Krumbein WE (1986) Biotransfer of minerals by microbes and microbial mats. In: Leadbeater BSC, Riding R (eds) Biomineralization in lower plants and animals. Syst Assoc Spec Vol 30:55–72
Krumbein WE, Cohen Y (1977) Primary production, mat formation and lithification: contribution of oxygenic and facultative anoxygenic cyanobacteria. In: Flügel E (ed) Fossil algae, recent results and developments. Springer, Berlin Heidelberg New York, pp 37–56
Krumbein WE, Giele C (1979) Calcification in a coccoid cyanobacterium associated with the formation of desert stromatolites. Sedimentology 26:593–604
Latham A, Riding R (1988) Thrombolites formed by calcified cyanobacteria in the Lie de Vin Series, Morocco. In: Int Assoc Sediment 9th Eur Region Meet Abstr, Leuven, Belgium, pp 127–128
Logan BW, Rezak R, Ginsburg RN (1964) Classification and environmental significance of algal stromatolites. J Geol 72:68–83
Love KM, Chafetz HS (1988) Diagenesis of laminated travertine crusts, Arbuckle Mountains, Oklahoma. J Sediment Petrol 58:441–445
Lowenstam HA (1981) Minerals formed by organisms. Science 211:1126-1231
Lowenstam HA, Weiner S (1983) Mineralization by organisms and the evolution of biomineralization. In: Westbrook P, de Jong EW (eds) Biomineralization and biological metal accumulation. Reidel, Dordrecht, pp 191–203
Maslov VP (1960) Stromatolites. USSR Acad Sci Trans Geol Inst 41:188 pp (in Russian)
Maurin AF (1987) Microbial micrites. In: 4th Int Symp Fossil algae, Cardiff, Abstr, pp 24–25
Maurin AF, Noël D (1977) A possible bacterial origin for Famennian micrites. In: Flügel E (ed) Fossil algae, recent results and developments. Springer, Berlin Heidelberg New York, pp 136–142
Maurin AF, Philip J, Brunel P (1981) Possible microbial accretions in Cenomanian mounds, SE France. In: Monty C (ed) Phanerozoic stromatolites, case histories. Springer, Berlin Heidelberg New York, pp 121–133
Monty CLV (1976) The origin and development of cryptalgal fabrics. In: Walter MR (ed) Stromatolites. Developments in sedimentology, vol 20. Elsevier, Amsterdam, pp 193–249
Monty CLV (1977) Evolving concepts on the nature and ecological significance of stromatolites. In: Flügel E (ed) Fossil algae. Springer, Berlin Heidelberg New York, pp 15–35
Monty CLV (1981) Spongiostromate vs porostromate stromatolites and oncolites. In: Monty CLV (ed) Phanerozoic stromatolites. Springer, Berlin Heidelberg New York, pp 1–4

Monty CLV (1982) Microbial spars. In: Int Assoc Sediment (ed) 11th Int Congr Sedimentology, Hamilton, Abstr

Monty CLV (1986) Microbial dolomites. In: 12th Int Sedimentology Congr, Canberra, Abstr

Monty CLV, van Laer P (1984) Experimental radial calcite oöids of microbial origin and fossil counterparts. In: Int Assoc Sediment (ed) 5th Eur Reg Meet Sedimentology, Marseille, Abstr, pp 296–297

Monty CLV, Bernet-Rollande MC, Maurin AF (1982) Reinterpretation of the Frasnian classical "reefs" of the southern Ardennes, Belgium. Ann Soc Geol Belg 105:339–341

Pedley HM (1987) The Flandrian (Quaternary) Caerwys tufa, North Wales: an ancient barrage tufa deposit. Proc Yorks Geol Soc 46:141–152

Pentecost A, Riding R (1986) Calcification in cyanobacteria. In: Leadbeater BSC, Riding R (eds) Biomineralization in lower plants and animals. Syst Assoc Spec Vol 30:73–90

Pia J (1927) Thallophyta. In: Hirmer M (ed) Handbuch der Paläobotanik, vol 1. Oldenbourg, Munich, pp 31–136

Pratt BR (1982) Stromatolite decline – a reconsideration. Geology 10:512–515

Read JF (1976) Calcretes and their distinction from stromatolites. In: Walter MR (ed) Stromatolites. Developments in sedimentology, vol 20. Elsevier, Amsterdam, pp 55–71

Riding R (1977) Skeletal stromatolites. In: Flügel E (ed) Fossil algae, recent results and developments. Springer, Berlin Heidelberg New York, pp 57–60

Riding R (1979) Cyanoliths: skeletal oncolites formed by calcified blue-green algae. In: 2nd Symp Int Algues fossiles, Abstr

Riding R (1982) Cyanophyte calcification and changes in ocean chemistry. Nature (London) 299:814–815

Riding R (1983) Cyanoliths (cyanoids): oncoids formed by calcified cyanophytes. In: Peryt TM (ed) Coated grains. Springer, Berlin Heidelberg New York, pp 276–283

Riding R (1988) Classification of microbial carbonates. In: 6th Int Coral Reef Symp Benthic microbes and reefs, Abstr, Townsville, Australia, p 5

Riding R, Voronova L (1984) Assemblages of calcareous algae near the Precambrian/Cambrian boundary in Siberia and Mongolia. Geol Mag 121:205–210

Riding R, Awramik SM, Winsborough BM, Griffin KM, Dill RF (in preparation) Giant bahamian stromatolites: composition of surface mats.

Roddy HJ (1915) Concretions in streams formed by the agency of blue-green algae and related plants. Proc Am Philos Soc 59:246–258

Schopf JW, Walter MR (1982) Origin and early evolution of cyanobacteria. In: Carr NG, Whitton BA (eds) The biology of cyanobacteria. Blackwell, Oxford London, pp 543–564

Schopf JW, Walter MR (1983) Archean macrofossils: new evidence of ancient microbes. In: Schopf JW (ed) Earth's earliest biosphere, its origin and evolution. Univ Press, Princeton, pp 214–239

Semikhatov MA (1976) Experience in stromatolite studies in the U.S.S.R. In: Walter MR (ed) Stromatolites. Elsevier, Amsterdam, pp 337–357

Semikhatov MA, Gebelein CD, Cloud P, Awramik SM, Benmore WC (1979) Stromatolite morphogenesis – progress and problems. Can J Earth Sci 16(5):992–1015

Simkiss K (1986) The processes of biomineralization in lower plants and animals – an overview. In: Leadbeater BSC, Riding R (eds) Biomineralization in lower plants and animals. Syst Assoc Spec Vol 30:19–37

Surdam RC, Wray JL (1976) Lacustrine stromatolites, Eocene Green River Formation, Wyoming. In: Walter MR (ed) Stromatolites. Developments in sedimentology, vol 20. Elsevier, Amsterdam, pp 535–541

Tiercelin J-J, Vincens A, Barton CE, Carbonel P et al. (1987) Le demi-graben de Baringo-Bogoria, Rift Gregory, Kenya. 30 000 ans d'histoire hydrologique et sédimentaire. Bull Centres Rech Explor Prod Elf-Aquitaine 11(2):249–540

Toomey DF, Cys JM (1979) Community succession in small bioherms of algae and sponges in the Lower Permian of New Mexico. Lethaia 12:65–74

Tsien HH (1985a) Origin of Stromatactis – a replacement of colonial microbial accretions. In: Toomey DF, Nitecki MH (eds) Paleoalgology, contemporary research and applications. Springer, Berlin Heidelberg New York Tokyo, pp 274–289

Tsien HH (1985b) Algal-bacterial origin of micrites in mud mounds. In: Toomey DF, Nitecki MH (eds) Paleoalgology, contemporary research and applications. Springer, Berlin Heidelberg New York Tokyo, pp 290–296

Walcott CD (1914) Precambrian algonkian flora. Smithsonian Misc Coll 64:77–156

Walter MR (1972) Stromatolites and the biostratigraphy of the Australian Precambrian and Cambrian. Palaeontology Spec Pap 11:190 pp, 33 pls

Walter MR (1976a) Introduction: In: Walter MR (ed) Stromatolites. Developments in sedimentology, vol 20. Elsevier, Amsterdam, pp 1–3

Walter MR (1976b) Geyserites of Yellowstone National Park: an example of abiogenic "stromatolites". In: Walter MR (ed) Stromatolites. Developments in sedimentology, vol 20. Elsevier, Amsterdam, pp 87–112

Walter MR (1983) Archean stromatolites: evidence of the Earth's earliest benthos. In: Schopf JW (ed) Earth's earliest biosphere, its origin and evolution. Univ Press, Princeton, pp 187–213

Walter MR (1988) Major features in the record of Proterozoic stromatolites. In: Symp Proterozoic biosphere, Los Angeles, Abstr, pp 27–28

Walter MR, Heys GR (1985) Links between the rise of the Metazoa and the decline of stromatolites. Precambrian Res 29:149–174

Winsborough B (1987) Fossil and living stromatolites from a desert basin in northeastern Mexico. In: 4th Int Symp Fossil algae, Cardiff, Abstr, p 33

Winsborough BM, Seeler J-S, Golubic S, Folk RL, Maguire Jr B (in prep). Freshwater stromatolites, oncoids and coated pebbles from a spring-fed lake in northeastern Mexico

Wright VP (1981) A subaerial stromatolite from the Lower Carboniferous of South Wales. Geol Mag 118:97–100

II Major Groups

Chapter 3

Calcified Cyanobacteria

R. Riding[1]

Abstract

Calcification affects the sheath of some coccoid and filamentous cyanobacteria during their life, under suitable environmental conditions. There is probably a link between sheath character, the environment and nucleation of $CaCO_3$ within or upon the sheath. Calcification is thus extracellular and is influenced, but not fully controlled, by the cyanobacterium. It is promoted by thick sheaths, which provide favourable sites for $CaCO_3$ nucleation, and by conditions favouring the physico-chemical precipitation of $CaCO_3$.

In the Recent, calcification is widespread in freshwater lakes and streams in limestone areas, but is virtually unknown in marine environments. However, calcification occurred quite widely in marine cyanobacteria during the Palaeozoic and Mesozoic. This could reflect temporal changes in oceanic carbonate chemistry.

Calcification results in (1) discrete skeletons; (2) skeletal stromatolites; (3) dendrolites and thrombolites; and (4) tufa stromatolites. Where calcification is uncommon the mucilaginous sheaths and felted masses of cyanobacterial mats trap particulate sediment to form stromatolites which may be calcareous in composition, but which do not consist of calcified cyanobacteria in the sense of precipitation of $CaCO_3$ on or within the sheath.

The main groups of fossil calcified cyanobacteria may be exemplified by the following genera: *Angulocellularia*, *Epiphyton*, *Girvanella*, *Hedstroemia*, *Renalcis* and *Garwoodia*/*Mitcheldeania*, although some of these are not definitely known to by cyanobacteria. In addition there is a variety of fossil calcareous microproblematica which have been widely attributed to the cyanobacteria.

Well-defined and diverse calcified cyanobacteria first appear near the Precambrian-Cambrian boundary. Dendrolites and thrombolites composed of probable cyanobacteria are common in the Cambrian and Lower Ordovician. Subsequently, calcified cyanobacteria are generally less common in marine environments, although they reappear in relative abundance in the Upper Devonian and Lower Carboniferous, and in the Middle Triassic to mid-Cretaceous. A number of these fossils may be compared with modern analogues in freshwater lakes and streams, but the precise affinities of others are still uncertain.

1 Introduction

Cyanobacteria, also known as blue-green algae and cyanophytes, are prokaryotes. They are thus more closely related to bacteria than to eukaryotic algae. Cyanobacteria are responsible for a variety of calcified microfossils, such as *Girvanella* Nicholson and Etheridge and *Ortonella* Garwood, and are also significantly involved in stromatolites formed by trapping of sedimentary particles.

Calcified cyanobacteria are individually normally microscopic, but the deposits which they construct, such as oncoids, stromatolites, tufas and reefal

[1] Department of Geology, University of Wales College of Cardiff, Cardiff CF1 3YE, UK

fabrics, are macroscopic and conspicuous. Thus, they may occur either as very small individual skeletal fossils, or be amalgamated to form stromatolitic and reefal deposits. In both cases they have a range in marine environments from the earliest Cambrian to late Cretaceous or early Cenozoic. Oncoids and stromatolites are also formed by non-calcified cyanobacteria. Consequently, it is necessary to check, by microscopic examination, whether stromatolite fabrics are composed mainly by agglutinated sediment particles or by calcified cyanobacteria. In the Recent, calcified cyanobacteria are often conspicuous in freshwater calcareous rivers and lakes where analogues for marine fossils exist and provide information concerning these simple, but important, fossils.

Calcified cyanobacterial sheaths are commonly irregular bushy masses, or tubes arranged in radial fans or tangled masses. This simplicity, coupled with their small size, creates rather than reduces systematic problems. Gradation between taxa makes it difficult to establish diagnoses, due to lack of distinguishing features and variation in sheath shape. The tubular construction of filamentous sheaths causes some of them to be confused with some small green and red algae and invertebrates. A wide range of fossils of uncertain affinities, which should properly be regarded as microproblematica until their relationships are better understood, have been referred to the cyanobacteria. Some calcified cyanobacteria, particularly forms such as *Cayeuxia* Frollo and *Ortonella*, have also been confused with udoteacean green algae.

Calcified cyanobacteria have a distinctive distribution as fossils in marine environments. They are relatively common in the Palaeozoic and Mesozoic, but are rare or absent in the Precambrian and Cenozoic. Recent marine calcified cyanobacteria are virtually unknown, although freshwater tufas and oncoids commonly contain them in abundance. This long-term pattern: Precambrian scarcity, Palaeozoic-Mesozoic presence, followed again by Cenozoic scarcity of marine calcified cyanobacteria contrasts with the geological range of non-mineralized cyanobacteria. The group as a whole possesses the longest range of any class of organism, apart from bacteria. Cyanobacteria are known from the late Archaean onwards. An explanation for the appearance and disappearance of calcification in marine cyanobacteria may lie in a change, probably subtle but nonetheless significant, in sea-water chemistry (Riding 1982), although it is also likely that evolutionary events, possibly involving specificity for calcification, have interacted with this environmental control (Riding 1989). The calcification site in living cyanobacteria appears to be essentially extracellular and lies in the mucilagenous protective sheath. This accounts for both the environmental dependence of cyanobacterial calcification, and also for the morphological simplicity of the resulting fossils.

2 Calcification

Cyanobacterial calcification is reviewed by Pentecost and Riding (1986) and has been described in detail from freshwater environments by Pentecost (1987, 1988). Two key features are: (1) the site of calcification is the enveloping sheath, rather than the cells, of the organisms, and (2) calcification appears to proceed only under conditions where $CaCO_3$ precipitation is thermodynamically favoured. The result of calcification is the formation of discrete fossils whose biomineralized sheaths rank as skeletons (see Discussion), albeit very small ones, if they function for support and protection of the organism. Numerous individual calcified cyanobacteria may be amalgamated into dendrolites, thrombolites and other large reefal masses or, if they possess lamination, into skeletal stromatolites and skeletal oncoids (cyanoids, Riding 1983).

2.1 Processes

Cyanobacterial calcification is sited in or on the mucilaginous sheath. The sheath itself is secreted by the cell for which it forms a protective envelope (Fig. 1). Coccoid cells are rounded and form clusters within a mass of sheath material. Filamentous forms consist of hairlike trichomes which are strands of cells attached end-to-end. One, several, or a number of trichomes may share a common tubular sheath. Trichomes and sheath together are then termed a filament. If the trichomes branch, or are arrayed in a fan-shape, then the resulting sheath may have a bush-like or radial form.

Sheath production is a prerequisite for calcification, but under conditions where there is little need for protection from water-movement, sediment particles, sunlight, desiccation or grazing organisms the sheath may develop only weakly or not at all. Cyanobacteria under laboratory culture commonly produce abundant soluble polysaccharides, but do not usually develop sheaths (Rippka et al. 1979, pp. 23–24). They consequently differ from specimens in natural environments, not only in the absence of the sheath but also in features which may be sheath-related, such as branching-style and colony form.

The presence of a sheath, and also its consistency (whether dense or diffluent), influence calcification because the crystals of $CaCO_3$ nucleate within

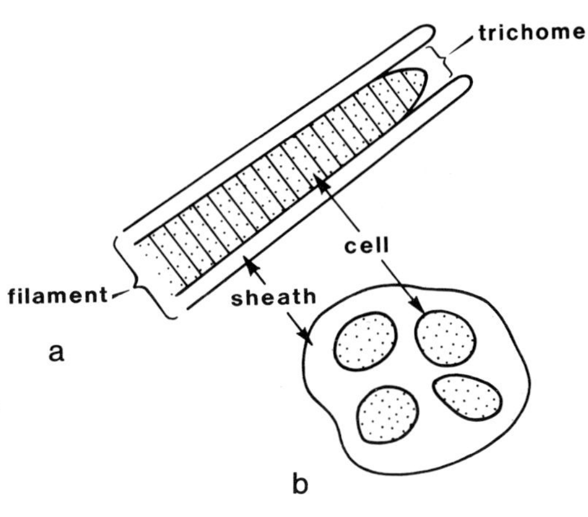

Fig. 1a,b. Morphology of **a** filamentous, and **b** coccoid, cyanobacteria. The sheath is the normal site of calcification

or upon it, possibly as a result of the attraction of calcium ions by uronic acids. Trapping of sediment grains by cyanobacteria, forming agglutinated stromatolites, is also primarily effected by the sheath; the grains may then be bound into a felted mass of filaments by subsequent cyanobacterial growth and movement. Calcification results in either impregnation or encrustation of the sheath, or a combination of these (Fig. 2). In impregnation, $CaCO_3$ crystals nucleate within the sheath. They are generally small, bladed, and may show orientation in response to the sheath structure. In encrustation, $CaCO_3$ crystals nucleate upon the sheath and generally form larger crystals. They range in form from simple rhombs to complex dendritic crystallites (Pentecost and Riding 1986).

Of equal importance to the presence of a sheath for calcification, is a calcareous environment. Cyanobacterial metabolism may stimulate $CaCO_3$ precipitation by uptake of CO_2 or HCO_3^- but, so far as is presently known, the physico-chemical conditions must also favour calcification. This is readily achieved in calcifying freshwater streams where turbulence promotes CO_2 evasion and leads to tufa deposition. However, in marine environments at the present-day, even where early cementation is occurring, cyanobacterial calcification appears at best to be rare, although it was common during the Palaeozoic and Mesozoic.

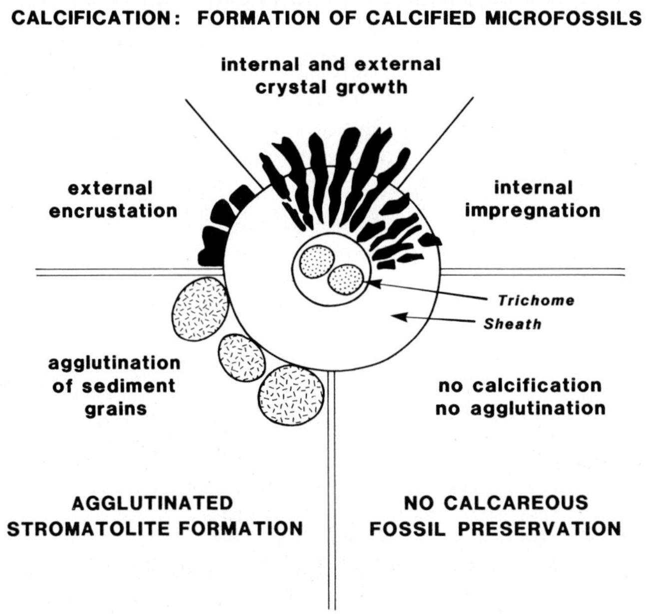

Fig. 2. Filamentous cyanobacterium in transverse section to show sites and styles of calcification (*upper part* of diagram) and sediment agglutination (*lower left part* of diagram). Calcification may be either by external encrustation (*upper left*), or internal impregnation (*upper right*) of the sheath, or by a combination of these (*upper centre*). Calcification will result in the formation of calcified microfossils. Agglutination of particulate sediment will result in the formation of agglutinated stromatolites

2.2 Products

Individuals or colonies of calcified cyanobacteria can occur as discrete skeletons, or they may be arrayed in a variety of massive deposits, which can be divided on the basis of macroscopic fabrics into skeletal stromatolites and oncoids (laminated), thrombolites (clotted), dendrolites (dendritic), and closely spaced microbush fabrics, as well as less distinctive reefal masses (Fig. 3). Where these deposits are marine, they are often associated with skeletal invertebrates. In freshwater they are often associated with enhanced physicochemical precipitation which promotes external encrustation, and then form tufa. Travertine deposits formed by warm springs may contain cyanobacteria (Copeland 1936), but they have been reported to be dominated by bacteria (Chafetz and Folk 1984). They require further study.

2.2.1 Discrete Skeletons

Calcification of the sheaths of individual filaments or of filamentous colonies results in morphologies based on tubes or bush-shaped masses (Fig. 4). The

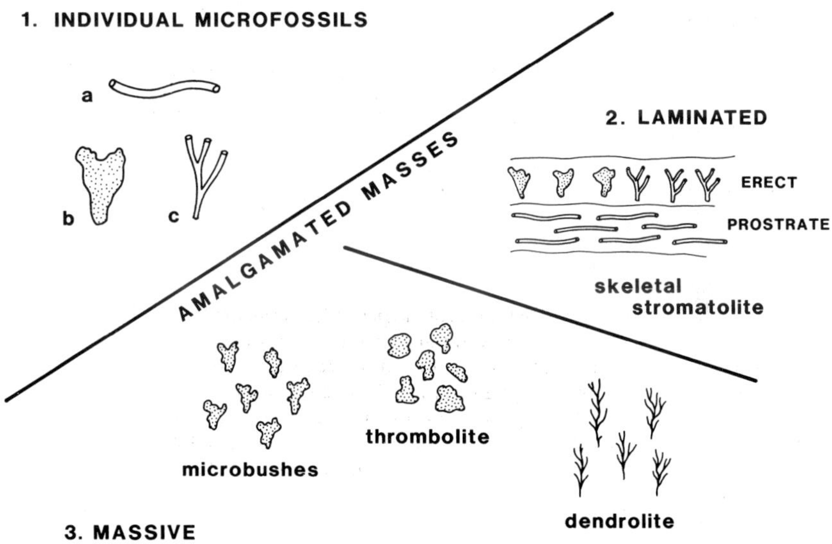

Fig. 3. Modes of occurrence of calcified cyanobacteria. *1* Discrete individual microfossils (e.g. *a Girvanella*; *b Angulocellularia*; *c Ortonella*). *2* and *3* Amalgamated masses, which may be either *2*, laminated (skeletal stromatolites) or *3*, massive (microbushes, thrombolite, dendrolite). Skeletal stromatolites may, within their layers, have either erect or prostrate fabrics, depending upon the orientation of the calcified microfossils. Examples of massive amalgamated masses include, for microbushes, Lower Ordovician *Angulocellularia* ("*Epiphyton*, *Renalcis*", Riding and Toomey 1972), and Upper Devonian "fenestral renalcid micrite" (Mountjoy and Riding 1981); for thrombolite, Lower Cambrian *Renalcis* and *Tarthinia* thrombolites (Latham and Riding 1988); and for dendrolite, Lower Cambrian *Epiphyton-Renalcis-Angulocellularia* fabrics (Riding, this Vol.)

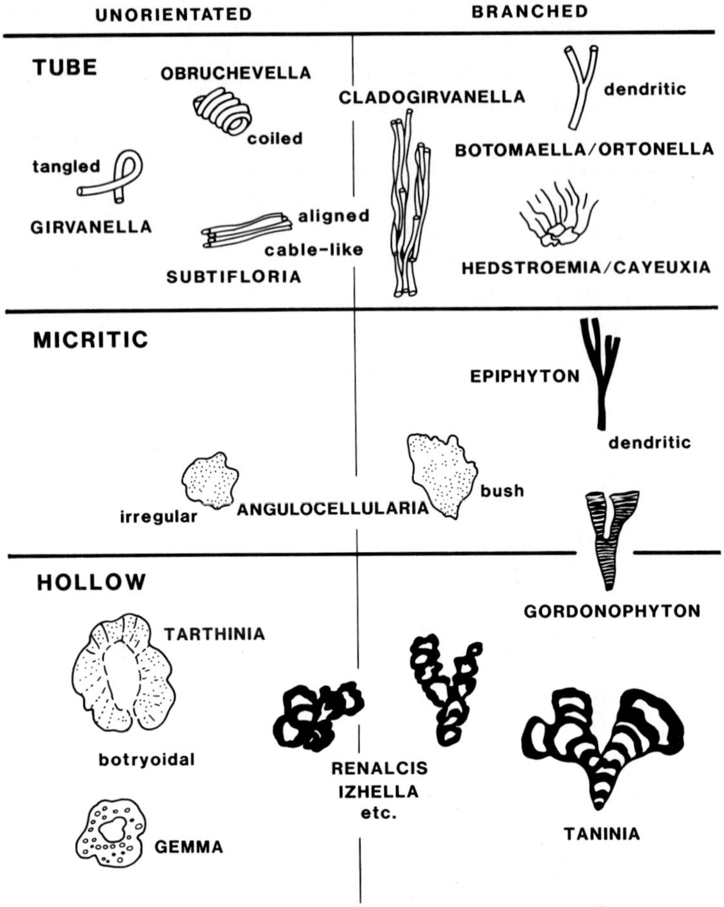

Fig. 4. Simplified outline of the main morphologies of fossil calcified cyanobacteria together with possible cyanobacteria of problematic affinity. Examples shown are believed to be impregnated sheaths (rather than encrusted sheaths, see Fig. 2). Three principal forms (tubes solid micritic, and hollow) exhibit organizations ranging from unorientated to erect and branched. Two of the tube forms (*Obruchevella*, *Subtifloria*) show coiling or alignment without obvious substrate attachment. Named genera are a selection of principal examples. See also Riding and Voronova (1985) for details of this type of morphological approach to Cambrian calcareous "algae", many of which may be cyanobacterial

tubes are generally either erect radial clusters (e.g. *Ortonella*, *Cayeuxia*) or are prostrate felts (e.g. *Girvanella*) or bundles (e.g. *Subtifloria* Maslov). The bush-shaped masses, such as *Angulocellularia* Vologdin, usually appear to be densely micritic, but Recent examples formed by *Schizothrix calcicola* (Agardh) Gomont show tubular traces of trichomes within the bush (Riding and Voronova 1982a). A dendritic appearance, but with a dense rod-like rather than tubiform structure, is also shown in members of the *Epiphyton*-group. It is possible that *Epiphyton* Bornemann and its allies represent calcified filamentous cyanobacteria: their size and morphology are consistent with this, and *Tubomorphophyton* Korde, which appears to be related to *Epiphyton*, has been interpreted as a cyanobacterium (Riding and Voronova 1982b). Thus, filamentous forms are known in the Recent to produce tubiform and bush-like calcified sheaths, and it is possible that fossil rod-like dendritic forms are also produced by filamentous cyanobacteria. Members of the *Epiphyton*-group are only known for certain from the Palaeozoic and, in the absence of Recent analogues, their exact affinities remain uncertain.

Calcified coccoid forms are less well-known. They can be expected to occur as either micro-spherulites, where small clusters of individual cells are calcified, or as botroyoidal and reniform masses (which may show vertical elongation) where the mucilaginous sheath of larger coccoid colonies are peripherally calcified. Possible examples of these are the spherulites in Pleistocene tufas of the Dead Sea area (Buchbinder 1981), and the *Renalcis*-group fossils which are common in the Cambrian and range into the Carboniferous (Hofmann 1975; Pratt 1984; Fig. 4). Pratt (1984) has also suggested that *Epiphyton* is a coccoid cyanobacterium. But the fact remains that because Recent calcified coccoid forms are poorly known, their ancient counterparts are difficult to interpret.

2.2.2 Stromatolites

Stromatolites formed by calcified cyanobacteria have been referred to as skeletal stromatolites (Riding 1977a) because they are distinguished from agglutinated stromatolites. by possessing calcified microfossils which both construct them and show how they originated. Agglutinated stromatolites (Riding, this Vol.), in contrast, are essentially sediment grains trapped by a microbial mat which has since been destroyed and whose precise original component organisms are usually unclear.

Calcification of cyanobacteria hardens the otherwise sticky sheath and inhibits agglutination of particles. However, erect filaments are preserved in growth position by calcification and this promotes trapping of particles between the calcified tufts. Thus, growth rates may be faster for skeletal than for agglutinated stromatolites, and the erect growth form of many filamentous cyanobacteria results in the formation of relatively large irregular fenestrae between the filament clusters which are constructional in origin and not related to desiccation or gas bubbling. At the same time, lamination is cruder than in most agglutinated stromatolites and may disappear entirely. When lamination is altogether absent then the deposit must be regarded as non-stromatolitic (see Reefs, below).

Nodular, unattached stromatolites are termed oncoids, and skeletal oncoids formed by calcified cyanobacteria have been termed cyanoids (Riding 1983). As in skeletal stromatolites, the lamination is made cruder by the irregularities created by calcification of the cyanobacteria, and cyanoids may acquire an additional radial fabric due to the preservation of erect filaments (Fig. 5). Cyanoids are a type of coated grain; they possess a nucleus, usually a rock or

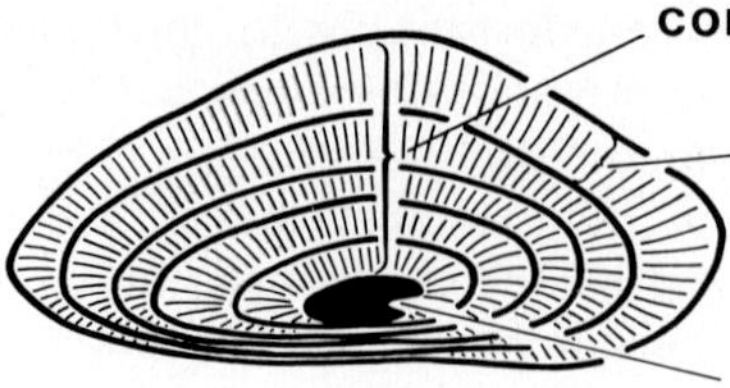

Fig. 5. Typical skeletal oncoid produced by cyanobacterial calcification. Note that the laminae thicken towards the upper surface. Episodic, usually seasonal, growth results in lamination, whereas the erect growth of individual filaments may impart an additional radial fabric

shell fragment (or a piece of grass or reed in freshwater environments), and a layered cortex composed of calcified sheaths. Recent cyanoids are best developed in calcareous freshwater lakes and streams where the most common constituent cyanobacteria are species of *Calothrix* Agardh ex Bornet and Flahault, *Phormidium* Kützing and *Schizothrix* Kützing. In lakes cyanoids are generally small, typically a few centimetres in size, but they may form spheroids up to several decimetres in size in fast-flowing rivers where turbulence promotes $CaCO_3$ precipitation. However, in addition to sheath impregnation, fluviatile freshwater oncoids and stromatolites both commonly possess surficial $CaCO_3$ sheath encrustation, similar to the crusts which form tufa deposits, and they grade into tufa stromatolites (see Tufa, below). Consequently, freshwater examples of cyanoids and skeletal stromatolites are only partial analogues of fossil marine examples in which calcification during life is more strictly limited to the sheath. Neither skeletal stromatolites nor cyanoids are known in Recent marine environments.

2.2.3 Reefs

The hard substrates created by skeletal stromatolites and cyanoids (and dendrolites and thrombolites, see below) are attractive to both encrusting organisms and to endoliths. In restricted marine environments, and in freshwater, encrusters will generally be limited to molluscs and worms, but in ancient normal marine waters the diversity of encrusters may result in complex fabrics whose stromatolitic, dendrolitic, or thrombolitic origin is obscured by an invertebrate-dominated reef structure. Calcified cyanobacteria are particularly conspicuous in Cambrian, Lower Ordovician, Upper Devonian, and Middle to Upper Triassic reefs (see Geological History below).

2.2.4 Dendrolites, Thrombolites, and Microbushes

Calcified erect bushes and radial tufts of filamentous cyanobacteria, especially such as members of the *Epiphyton*, *Angulocellularia* and other dendritic groups (see Table 1 in Riding and Voronova 1985), if arranged en echelon in vertical section, rather than being in layers, form unlaminated masses which are not stromatolitic. These may form *dendrolites*, defined as: microbial deposits with a dominant dendritic macrofabric (Riding 1988), or closely spaced microbush fabrics. No Recent marine examples of these are known but they are locally important components of Palaeozoic reefs, particularly in association with archaeocyaths and stromatoporoids. The absence of lamination, presumably due to lack of well-defined periodicity in their growth, is important in distinguishing dendrolites, thrombolites and microbush fabrics from skeletal stromatolites (Fig. 3).

Dendrolites are especially prominent in Cambrian reefs, but following the diversification of middle Palaeozoic reef-builders in the Ordovician they may be obscured, as are skeletal stromatolites, by associated invertebrates. Microbush fabrics are nevertheless well-seen in many Lower Ordovician (Riding and Toomey 1972) and Upper Devonian (e.g. Mountjoy and Riding 1981) reefs.

Thrombolites (Aitken 1967) are non-laminated microbial carbonates with a clotted fabric. They appear to be preferentially developed in the Cambrian and Ordovician (Kennard and James 1986). Thrombolites are similar to dendrolites in that they may be formed by calcified microbes, but thrombolites may form domes which resemble stromatolites, whereas dendrolites typically form reefal masses. Uncertainty exists concerning the range of structures which may be included within thrombolites. The nature of the thrombolites originally described by Aitken still needs to be established in detail. Walter and Heys (1985; p. 164) have suggested that thrombolites may represent burrowed stromatolites. Nevertheless, domical thrombolites from the Lower Cambrian of Morocco have well-developed clotted fabrics produced by masses of the calcified botryoidal genera *Renalcis* and *Tarthinia* (Latham and Riding 1988), which may be cyanobacteria.

2.2.5 Tufa

Cyanobacteria are common in calcareous streams and lakes where rapid carbonate deposition on plant surfaces creates porous tufa. Some cyanobacteria, such as *Calothrix* and *Phormidium*, possess some degree of specificity for carbonate precipitation which results in impregnation of their mucilaginous sheaths by $CaCO_3$ crystals (see Pentecost and Riding 1986). In other cases, cyanobacteria, together with green algae and mosses, merely provide a substrate upon which essentially inorganic surficial mineral encrustation occurs. Where the substrate is occupied by a variety of these organisms the interaction of different sizes, shapes, and growth rates results in a thick and generally unstructured deposit. However, where cyanobacteria are the dominant organisms their small size and coordinated seasonal growth result in a layered deposit which may be termed tufa stromatolites and tufa oncoids (Riding, this Vol.). These occur commonly as oncoids in temperate freshwater calcareous streams and rivers, but they may also create domical biohermal structures in lacustrine environments (Eggleston and Dean 1976; Braithwaite, Casanova, Frevert and Whitton 1989; Winsborough et al. in prep.).

3 Classification

Classifications of calcified cyanobacteria have been included in Pia (1927), Maslov (1956), Johnson (1961), Vologdin (1962), Korde et al. (1963), Roux (1985), and Chuvashov et al. (1987). In addition, broad groupings of Palaeozoic calcareous algae and cyanobacteria have been outlined by Chuvashov and Riding (1984). However, no stable classification has yet emerged. Pia's (1927) concept of the Porostromata has been useful in generally defining small tubiform fossils, some of which form microfabrics within stromatolites, but this has included genera such as *Rothpletzella* Wood, which probably are not cyanobacteria. On the other hand the Porostromata, despite Pias' (1927) inclusion of *Epiphyton* in it, does not readily encompass solid (as opposed to tubiform) fossils such as *Angulocellularia* which are likely to be cyanobacterial in origin. Consequently, the Porostromata is a heterogeneous group which, while encompassing some non-cyanobacterial microproblematica, nevertheless, cannot incorporate botryoidal (such as *Renalcis*) and solid calcified cyanobacteria.

At generic level the principal problems for classification concern the status of taxa such as: (1) *Garwoodia* Wood / *Mitcheldeania* Nicholson, which have also been regarded as chlorophytes; (2) *Rothpletzella* and *Wetheredella* Wood, which have at times been assigned to the cyanobacteria; (3) *Renalcis* and *Epiphyton*, which are abundant in the Cambrian and have relatives in the Devonian, but for which no good Recent analogues have yet been clearly demonstrated; and (4) *Tubiphytes*, an important component of some Permian and Triassic reefal facies which has at times been referred to the cyanobacteria. With these reservations, which are examined further below, the following principal groups are recognized here: *Angulocellularia* and *Epiphyton* groups, which are respectively bushy and dendritic in general form; *Girvanella* group, filamentous and unbranched; the *Hedstroemia* group, filamentous and branched; and the *Renalcis* group, which is botryoidal to bushy in external morphology. In addition, there is the *Garwoodia*/*Mitcheldeania* group whose members are similar to the *Hedstroemia* group in being filamentous and branched, but generally larger, and an assortment of microproblematica which have often been attributed to the cyanobacteria. It should be noted that the *Epiphyton*, *Girvanella* and *Hedstroemia* groups are quite diverse and could readily be further subdivided.

3.1 Angulocellularia Group

Features: Micritic, solid, bush-like form, erect or pendant growth.

Example: *Angulocellularia* Vologdin.

Affinities: Calcification of the sheath of *Schizothrix calcicola* (Oscillatoriaceae) produces the micritic, bush-like dendritic skeleton known as *Angulocellularia* (Riding and Voronova 1982a; Fig. 6). The divergent trichomes of the cyano-

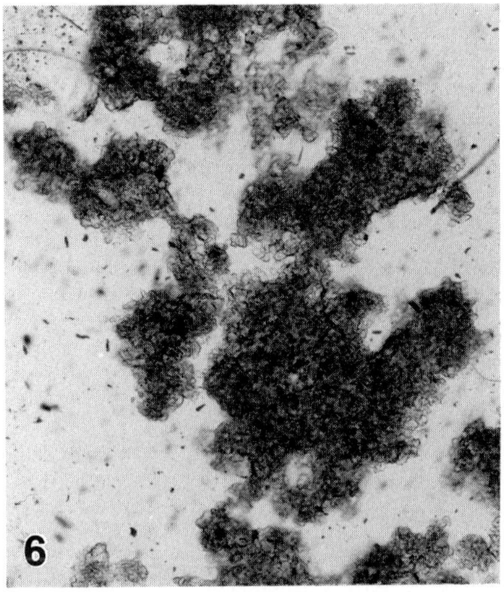

Fig. 6. *Angulocellularia* Vologdin. Recent. Squaw Island, Conandaigua Lake, New York State. x115

bacterium within a common sheath produce the bush-like form, and impregnation of the sheath by precipitation of micritic crystallites, apparently during life, results in its calcification. The narrow tubes originally occupied by the trichomes become obscured, possibly due to infilling by micritic cement.

Occurrence: As individuals and also as massive dendrolites in association with archaeocyaths and sponges in Cambrian and Lower Ordovician reefs (see references in Riding and Voronova 1982a). *Angulocellularia*-like dendritic clusters are also present in cryptic environments in some Cretaceous shallow marine carbonate sediments (e.g. see crusts inside dasycladalean algae figured by Barattolo 1983, Plate 5), and it is likely that more will be reported that have hitherto been overlooked as being simply irregular micritic crusts. *Frutexites* Maslov is an iron-impregnated fossil similar to *Angulocellularia* (see Riding and Voronova 1982a, p. 108). *Frutexites* was originally described from the Ordovician (Maslov 1960, pp. 60–61) and is also known from the Devonian (Playford et al. 1976) and Jurassic (Szulczewski 1963). A Proterozoic example (Walter and Awramik 1979) is silicified. The only Recent report of *Angulocellularia* is the freshwater lacustrine oncoid occurrence described by Riding and Voronova (1982a).

3.2 Epiphyton Group

Features: Micritic, narrow dendritic filaments which may be rod-like, apparently chambered, or tubular; erect or pendant growth.

Examples: *Epiphyton* Bornemann, *Gordonophyton* Korde, *Tubomorphophyton* Korde.

Affinities: Uncertain, due to the lack of recognized Recent analogues; but the size and morphology are consistent with a cyanobacterial affinity, as Pia (1927, p. 39) first proposed. A more specific, stigonematalean, affinity has been proposed for *Epiphyton* (Luchinina 1975, p. 21) and *Tubomorphophyton* (Riding and Voronova 1982b). The view of Korde (1959) that *Epiphyton* is a red alga has not received support in recent years. For a discussion of these problems see Riding and Voronova (1982b). Pratt (1984) suggested that *Epiphyton* is a diagenetically calcified coccoid cyanobacterium.

Occurrence: *Epiphyton* group genera first appear close to the Precambrian-Cambrian boundary and are common and widespread in Cambrian shallow marine reef limestones. They are closely associated with archaeocyaths in Lower Cambrian reefs, but they also occur as massive dendrolites in their own right. The group reappears prominently, but briefly, in the Upper Devonian in stromatoporoid reefs (Fig. 7), although the genera are usually smaller, and probably different, than in the Cambrian. *Epiphyton* has been reported in the Lower Cretaceous (Masse 1979, Plate 1, Fig. 5) but the bush-like dendritic fossil in this latter occurrence more closely resembles *Angulocellularia*.

3.3 Girvanella Group

Features: Thin-walled tubes, tangled, coiled or aligned; prostrate growth.

Examples: *Girvanella* Nicholson and Etheridge, *Obruchevella* Reitlinger, *Subtifloria* Maslov.

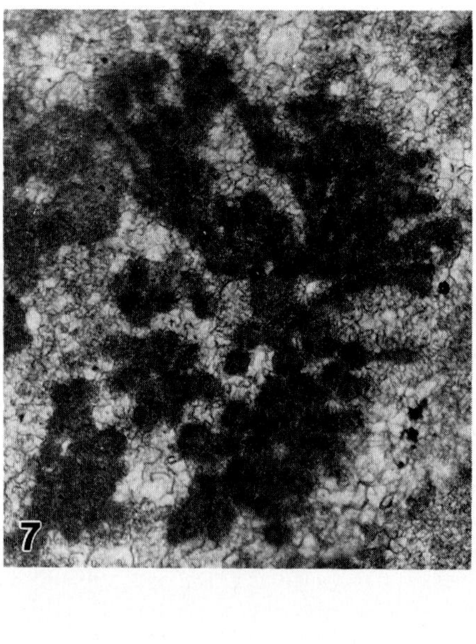

Fig. 7. *Epiphyton* Bornemann. Upper Devonian. Mount Haultain, Alberta. x125. (Specimen courtesy of E.W. Mountjoy)

Affinities: Recent calcified *Plectonema* (Oscillatoriaceae) closely resembles *Girvanella* (Riding 1977b), and it is likely that other *Girvanella*-group fossils may also be calcified oscillatoriacean sheaths.

Occurrence: *Girvanella*-group fossils typically occur as individuals or in small clusters (Fig. 8). *Girvanella*, which is one of the most widely known calcified cyanobacteria in the Palaeozoic (see Danielli 1981), may also occur in oncoids and in crusts on reef-surfaces. *Subtifloria* and *Obruchevella* are only known from the Cambrian, but *Girvanella* is present in marine limestones from Cambrian to Cretaceous, as well as being present in some Recent freshwater environments.

3.4 Hedstroemia Group

Features: Branching clusters of tubes, clusters with fan-like longitudinal section; radial erect growth. Adjacent tubes may share common walls or may be separate.

Examples: *Hedstroemia* Rothpletz, *Botomaella* Korde, *Cayeuxia* Frollo, *Ortonella* Garwood.

Affinities: *Hedstroemia*, *Botomaella* and *Cayeuxia* closely resemble Recent calcified rivulariacean cyanobacteria. *Ortonella* resembles some species of *Scytonema*. However, comparisons have also been made with codiacean (udoteacean) green algae (see Sect. 3.6 below).

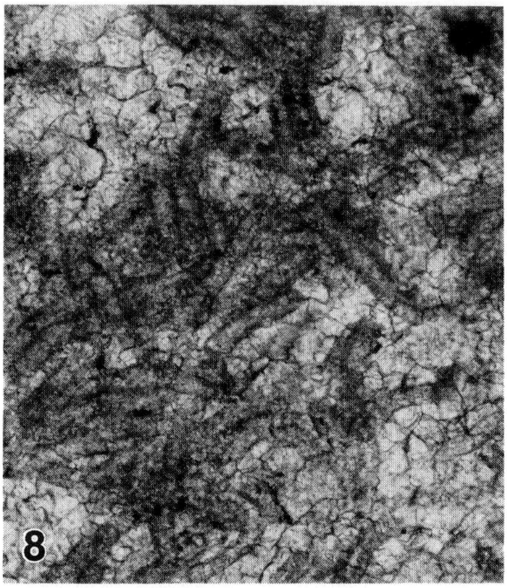

Fig. 8. *Girvanella* Nicholson and Etheridge. Caradoc (Ordovician). Aldon's Quarry, near Girvan, Scotland, x115

Occurrence: *Hedstroemia* group fossils are common in marine environments from the Cambrian (*Botomaella*, *Bija* Vologdin), through the Silurian (*Hedstroemia*) (Fig. 9) and Carboniferous (*Ortonella*), to the Jurassic and Lower Cretaceous (*Cayeuxia*), as well as being represented by Recent analogues. In addition, it is now evident that a number of fossils hitherto included in the Solenoporaceae should also be placed in the *Hedstroemia* group.

3.5 Renalcis Group

Features: Clusters of hollow reniform bodies, usually with thick walls which may be micritic, fibrous or clotted; irregular or erect growth.

Examples: *Renalcis* Vologdin, *Gemma* Luchinina, *Izhella* Antropov (Fig. 10), *Tarthinia* Drosdova.

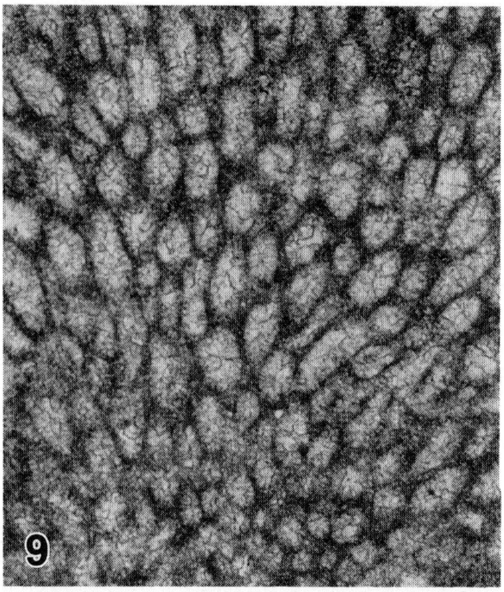

Fig. 9. *Hedstroemia* Rothpletz. Wenlock (Silurian), Högklint Formation, Gotland, Sweden. x46

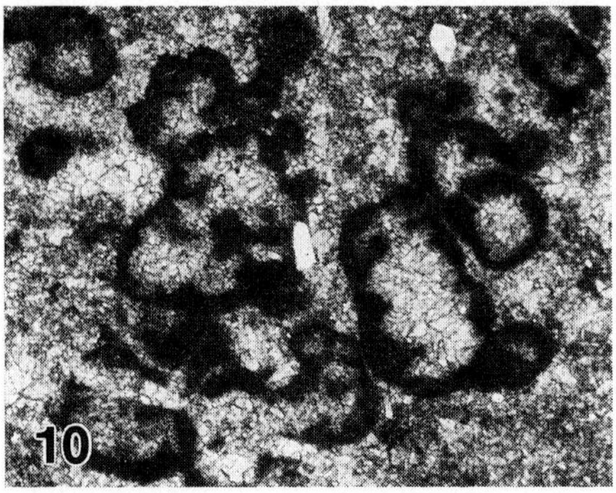

Fig. 10. *Izhella* Antropov. Upper Devonian. Mount Haultain, Alberta. x50. (Specimen courtesy of E.W. Mountjoy)

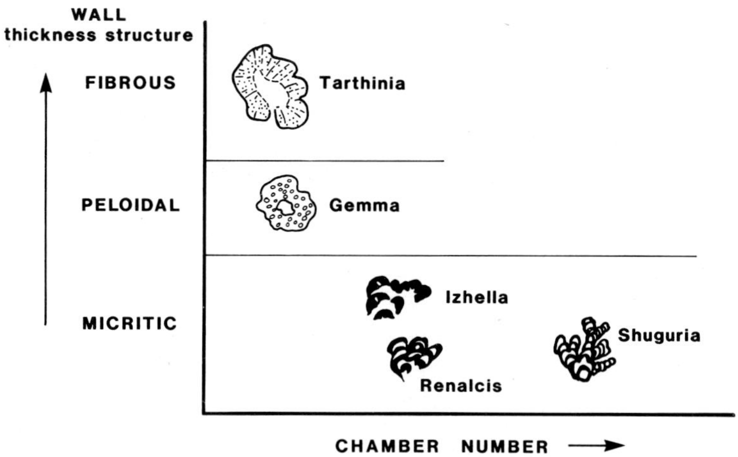

Fig. 11. Morphological characters of representative genera in the *Renalcis* Group, emphasizing differences in wall structure and variations in wall thickness and chamber number. Also see Riding and Voronova (1985, Fig. 4 and Table 1)

Affinities: *Renalcis*-group fossils (Fig. 11) have been included in the cyanobacteria by many authors, on the basis of their general size and morphology, but in the absence of Recent analogues it is difficult to systematically place them with confidence. Hofmann (1975) suggested that some *Renalcis*-group fossils represent calcification of gelatinous colonies of chroococcalean cyanobacteria. He interpreted the thick, dark walls in *Renalais* as the outer, deeply pigmented portions of the colonies, and the chambers as the lighter inner portions (Hofmann 1975, p. 1131; also see Poncet 1976 and Pratt 1984). *Izhella*, common in the Devonian (where it has often been confused with *Renalcis*, e.g. *R. turbitus* Wray belongs with *Izhella*) is characterized by V-shaped clefts in the inner side of the thick micritic wall, which resemble desiccation cracks in calcified sheath. *Gemma* (see Fig. 4) also fits this type of interpretation of the presence of a thick sheath. Its wall contains peloidal clots of dense micrite, but the clots appear too large to represent individual cells. Luchinina (in Zhuravleva et al. 1982) compared *Gemma* in general size and morphology with *Chlorogloea* Wille. *Tarthinia* appears to have a fibrous wall and is thus quite different in structure to both *Renalcis* and *Gemma*, although it also possesses the external botryoidal form. The apparently fibrous wall of *Tarthinia* could suggest a filamentous organization, and the external similarities of these genera could be a misleading guide to their precise affinities within the cyanobacteria. *Shuguria* Antropov, which together with *Izhella* is common in the Devonian, has also been confused with *Renalcis* (e.g. *R. devonicus* Wray belongs with *Shuguria*). Riding and Brasier (1975) compared *Shuguria* (misidentified as *Renalcis*) with foraminifers (see also Loeblich and Tappan 1964).

Occurrence: Renalcis-group fossils are equally as common as those of the *Epiphyton*-group and share a similar stratigraphic distribution: they first appear close to the Precambrian-Cambrian boundary, are common through to the Lower Ordovician, and reappear in the Devonian. In addition, *Renalcis*-group genera occur rarely in the Silurian (Riding and Watts 1983) and also are present in the Lower Carboniferous (Vachard 1974; Bless et al. 1976, p. 138). *Renalcis* is most common in the Lower Cambrian, and *Gemma* and *Tarthinia* are only known from the Lower Cambrian. *Izhella* and *Shuguria* are widespread in the Upper Devonian. The overall similarity of geological distribution of the *Renalcis* and *Epiphyton* groups emphasizes the way in which these fossils fluctuate in abundance through time; a feature which suggests that temporal variations in sea-water chemistry may have acted to periodically enhance marine calcification in cyanobacteria (Riding 1982).

Tarthinia and *Renalcis* are known to be present in Lower Cambrian thrombolites (Latham and Riding 1988) but *Renalcis*-group fossils generally are most widely seen in reef-associations with archaeocyaths and stromatoporoids, as well as with *Epiphyton*-group fossils in dendrolites.

3.5.1 Microproblematica

The following microfossils are commonly regarded as possessing a cyanobacterial affinity, but this attribution is not regarded here as being so strong as it is for the preceding groups. Indeed, some of the following are not, in my opinion, cyanobacterial, but they have so often been suggested to be so that they are included here for completeness. Features and examples given for some groups follow Chuvashov and Riding (1984, Table 1).

3.6 Garwoodia/Mitcheldeania Group

Features: Coarse, thin-walled, radiating to tangles clusters of tubes which are commonly juxtaposed.

Examples: *Garwoodia* Wood/*Mitcheldeania* Nicholson

Affinities: These fossils resemble some of those of the *Hedstroemia* group, but the tubes are larger and their size makes it difficult to consider them as cyanobacteria. It has been suggested that they are codiacean (i.e. udoteacean), an attribution also made for several members of the *Hedstroemia* group (Johnson 1961, pp. 97–98). This view has been developed, for example, by Mamet and Roux (1975), Dragastan (1985, 1988), and Mamet and Préat (1986). Possibly it originates in Rothpletz's (1891) suggestion that *Girvanella* may be a codiacean (see Riding 1975, p. 174). Calcified halimedaceans, such as *Halimeda*, which Johnson (1961) placed together with *Garwoodia* and *Mitcheldeania*, are much larger and morphologically quite different from the latter. Thus, the main support for the view that some coarsely tubed porostromates may be udoteaceans appears to come only from the superficially similar organization of extant uncalcified green algal genera such as *Codium* Stackhouse. However, it must be admitted that modern analogues have also not yet been found for many of these fossils in calcified cyanobacteria either.

At present it seems best to place *Garwoodia/Mitcheldeania* and allies in a separate group to await clarification of their systematic position. The question of whether *Garwoodia* and *Mitcheldeania* are synonymous or are separate genera is a minor but intricate problem stemming from the history of the erection of these taxa which remains to be clarified.

3.7 Rothpletzella Group

Features: Flat, curved or encrusting sheets of juxtaposed tubes which branch dichothomously in one plane.

Examples: *Rothpletzella* Wood (Fig. 12), *Flabellia* Shuysky.

Affinities: Wood (1948, pp. 17–18) regarded *Sphaerocodium gotlandicum*, which he named *Rothpletzella gotlandica*, as a probable alga and commented: "If this form is an alga the habit, and especially the absence of septation of the filaments, would seem to indicate that it belongs to the Myxophyceae (Cyanophyceae) or possibly to the group Siphonales of the Chlorophyceae." However, his reasons for considering an algal affinity were not compelling: "Their walls are composed of "algal dust", their threads branch dichotomously, and their diameter is below that which would accommodate any multicellular animal. Algae of *Girvanella* type are commonly found in association with them, sometimes building composite nodules, so that the conditions were such that algae could flourish" (Wood 1948, pp. 17–18). Although these arguments are inconclusive, *Rothpletzella* has generally been considered to be a cyanobacterium. However, Wood's effort to clarify the taxonomy of *Sphaerocodium* Rothpletz has been less successful. Many subsequent references to the fossil which he named *Rothpletzella* have called it *Sphaerocodium*, mainly due to

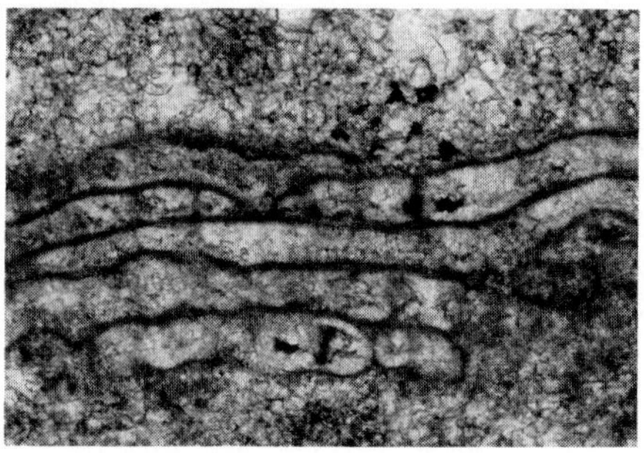

Fig. 12. *Rothpletzella* Wood. Wenlock (Silurian), Wenlock Limestone, May Hill, Gloucestershire, England. x115

Wray's (1967, pp. 36–37) rejection of Wood's decision, and this point still requires clarification.

Rothpletzella is common as reefal crusts, and also forming oncoids and stromatolites, in the middle Palaeozoic (Chuvashov and Riding 1984, Figs. 3, 4). A notable occurrence is in the construction of fore-reef stromatolites in the Upper Devonian Canning Basin of Western Australia (Playford et al. 1976, p. 557, Fig. 11b).

3.8 Wetheredella Group

Features: Short encrusting tubes, possibly branched, with hemispherical or rounded cross-sections; fibrous wall structure sometimes with additional micritic layer.

Examples: *Wetheredella* Wood (Fig. 13), *Aphralysia* Garwood, *Asphaltina* Mamet, *Sphaeroporella* Antropov.

Affinities: In his revision of Silurian "*Sphaerocodium*", Wood (1948) recognized not only *Rothpletzella* (see *Rothpletzella* Group, above) but also *Wetheredella*. He argued that the latter could be a foraminifer, and noted its resemblance to *Nubecularia* Defrance (Wood 1948, pp. 18 and 20–21). Loeblich and Tappan (1964, p. C787) suggested the reverse: "doubtfully a foraminifer, probably algal". Copper's (1976) attribution of a cyanobacterial affinity to *Wetheredella* was not accepted (Riding 1977c), but Mamet and Roux (1975, pp. 156–170) tentatively referred *Aphralysia*, *Sphaeroporella*, *Asphaltina*, *Wetheredella*, and other similar genera to the green algae. Ischenko and Radionova (1981) went further, and drew comparisons between *Wetheredella*

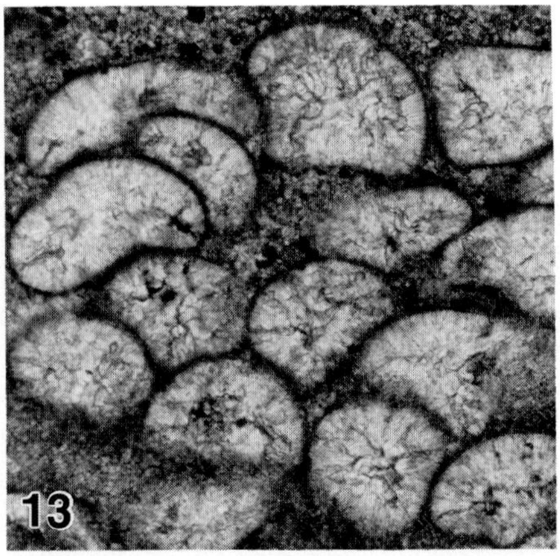

Fig. 13. *Wetheredella* Wood. Wenlock (Silurian), Wenlock Limestone, May Hill, Gloucestershire, England. x115

and *Rhabdoporella* Stolley (a dasycladalean, according to Pia, 1927, p. 62), and suggested a transitional series from these genera to *Rothpletzella*. These interesting observations deepen the systematic problems surrounding these fossils.

3.9 Tubiphytes Group

Features: Erect, irregular, or encrusting skeletons with a smooth exterior and dense, dark, flocculent internal structure showing concentric bands; central tubes often present.

Example: *Tubiphytes* Maslov (Fig. 14).

Affinities: Maslov (1956, p. 82) placed *Tubiphytes* in Algae Incertae Sedis, with the suggestion that it might belong to the Schizophyta (cyanobacteria), but hydrozoan (for details see Babcock 1977, pp. 17–18) and sponge (Ott in Kraus and Ott 1968, p. 271) affinities have also been proposed. *Tubiphytes* ranges from Carboniferous to Cretaceous, and is especially important in reefal environments during the Permian.

3.10 Lithocodium/Bacinella Group

Features: Encrusting masses of irregular open micritic networks usually becoming more orderly externally where a thicker surface layer may be pierced by numerous pores.

Examples: *Lithocodium* Elliott, *Bacinella* Radoičić.

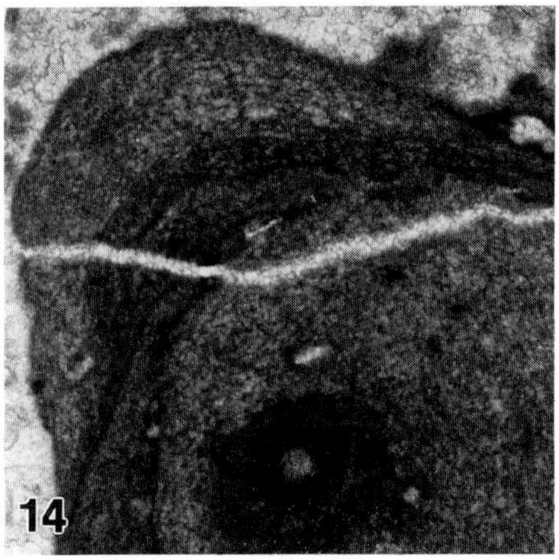

Fig. 14. *Tubiphytes* Maslov. Sakmarian (Early Permian), Trogkofel Formation, Trogkofel, Carnic Alps, Austria. x16. (Specimen courtesy of David Edwards)

Affinities: These genera are often seen juxtaposed and may well represent the same organism. Elliott (1956) regarded *Lithocodium* as a codiacean (i.e. udoteacean), but it has been suggested that these irregular meshwork-like crusts and masses may be microbial in origin (Maurin et al. 1985; Camoin and Maurin 1988). However, similarities also exist with foraminifers, sponges, and hydrozoans. *Lithocodium/Bacinella* ranges from Upper Permian to Cretaceous and is particularly common in shallow marine environments in the upper Jurassic and the Cretaceous.

4 Geological History

Cyanobacteria have the longest well-defined fossil record of all major groups of organisms. Only bacteria may have a longer record and their remains are at present much less well-known. Three principal periods can be recognized in the geological history of cyanobacteria, on the basis of the development of calcification in marine environments, with the middle period defined by the common presence of calcified cyanobacteria in shallow marine carbonate environments from the Early Cambrian (ca. 590 million years) to the late Mesozoic (ca. 65 million years; Fig. 15). Both before this, during most of the Precambrian, and afterwards, during the Cenozoic, there is no clear evidence of common cyanobacterial calcification in the sea.

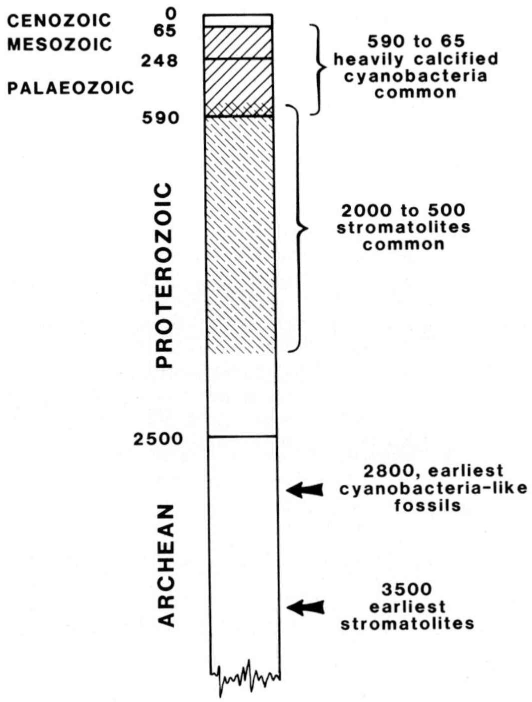

Fig. 15. Distribution of calcified cyanobacteria and stromatolites in marine environments through geological time. Scale in millions of years from Harland et al. (1982)

4.1 Precambrian

The earliest known occurrence of cyanobacteria-like fossils is in the 2800-million-year-old Fortescue Group, Western Australia (Schopf and Walter 1983; Fig. 15). These are present in cherty stromatolitic limestones and also occur in similar younger deposits such as the well-known Gunflint Chert Formation (ca. 2000 million years) of Ontario (Tyler and Barghoorn 1954) and Bitter Springs Formation (ca. 900 million years) of Australia (Schopf 1968). Precambrian stromatolites are associated with extensive, apparently marine, dolomite and limestone deposits in the Proterozoic (ca. 2500–550 million years) and reach a peak of diversity and importance near 800 million years (Awramik 1971) which has never since been equalled. Nevertheless, there is considerable uncertainty concerning their processes of formation. They are obviously calcareous deposits, but it is important to know whether they formed by precipitation of calcium carbonate on or within microorganisms, or by trapping and binding of particulate sediment, or by a combination of these processes.

Gebelein (1976, p. 505–506) has suggested that precipitation is the principal process involved, on the basis of (1) lack of preserved detrital carbonate in most Precambrian stromatolites; and (2) the presence of forms with more than 10 cm of relief which would, it is argued, require cementation for support. However, despite intensive examination of Proterozoic stromatolites by many workers, few tubiform or bush-like microfabrics characteristic of calcified filamentous cyanobacteria have been found, other than poor coarse 'vermiform' microstructures (Bertrand-Sarfati 1976, p. 255), although there is now evidence of better preserved "tussocky" fabrics resembling calcified rivulariaceans from the latest Proterozoic (Bertrand-Sarfati and Pentecost 1989). Consequently, it has been suggested that precipitation of calcium carbonate on or within cyanobacterial sheath material is only a minor aspect of stromatolite formation before the late Precambrian (Riding and Voronova 1982c). This in turn implies that trapping and binding of particulate sediment was the principal process involved.

It is important to check this apparent absence of calcified cyanobacteria in the Precambrian. It may simply be due to lack of detailed searching. However, if it can be established, it will then imply either an inability of sheaths at that time to provide suitable crystal nucleation sites, or lower $CaCO_3$ precipitation rates relative to the Palaeozoic (see Discussion, below), or both.

4.2 Cambrian-Cretaceous

4.2.1 Cambrian-Lower Ordovician

Heavily calcified cyanobacteria appear dramatically near the Precambrian-Cambrian boundary (Riding and Voronova 1984). The presence of facies change from restricted to open marine conditions at this level in key Siberian sections containing carbonate rocks makes it possible that the sudden appearance of calcified cyanobacteria (and other skeletal fossils too, for that matter) is artificially accentuated. Nevertheless, the overall change from the Vendian to the Cambrian is striking. Re-evaluation of the Cambrian marine calcified flora suggests that most Lower Cambrian "algae" are probably cyanobacterial (Riding, this Vol.). Certainly the following important genera are cyanobacteria: *Angulocellularia*, *Botomaella*, *Girvanella*, *Hedstroemia*, and *Subtiflora*. These occur as individuals, dendrolites, reefal crusts and as oncoids. Even more common than these are members of the *Epiphyton* and *Renalcis* groups which are prominent in dendrolites and reefs. They may well be cyanobacteria, although their precise affinities are still unclear.

The abundance and diversity of these fossils continues virtually undiminished into the Lower Ordovician, where sponge-cyanobacterial reef associations (Riding and Toomey 1972) are very similar to those of archaeocyaths and cyanobacteria in the Lower Cambrian. This Early Palaeozoic phase represents not only the beginning, but also the height, of the development of calcified cyanobacteria in the Phanerozoic. At this time cyanobacteria are as conspicuous in reef formation as are sponges and metazoans.

4.2.2 Middle Ordovician to Permian

The Early Palaeozoic importance of calcified cyanobacteria was not sustained through the middle Palaeozoic, but various groups reappear sporadically in the Silurian and Devonian. *Hedstroemia*, in association with probable cyanobacteria previously assigned to *Solenopora*, constructs thin but massive crests on Lower Silurian patch reefs in Gotland (Riding and Watts 1981). *Epiphyton* and *Renalcis*-group genera reappear abundantly in Upper Devonian reefs (Wray 1967), although they differ in detail from those present in the Cambrian. Both groups disappear by the end of the Lower Carboniferous.

Girvanella is common as individual masses, reefal crusts, and in oncoids throughout the middle Palaeozoic, but other prostrate tubiform genera, such as *Subtifloria* are not known from post-Cambian rocks. The type specimen of *Girvanella*, from the Ordovician of south-west Scotland, occurs as very small tangled masses which may have been semi-planktonic rather than benthic. Recent *Plectonema* which is analogous with *Girvanella* also occurs as floating clusters of filaments (Riding 1977b, p. 36).

Microproblematica which have been attributed to the cyanobacteria, such as *Rothpletzella* and *Wetheredella*, are common throughout the middle Palaeozoic as reefal crusts and in oncoids (Chuvashov and Riding 1984, Figs. 3, 5).

During the Carboniferous *Ortonella*, *Bevocastria* and *Garwoodia*/*Mitcheldeania* are locally important in oncoid and stromatolite formation in marginal marine environments (Garwood 1931) but generally the Carboniferous is a period of decline in the importance of calcified marine cyanobacteria.

There is no clear evidence to indicate that this situation alters significantly in the Permian, but assessment of the importance of cyanobacteria is complicated by the inclusion of microproblematica, notably *Tubiphytes* Maslov, in this group. *Tubiphytes* is common in Permian reefs, both as an encrusting organism and as a primary framebuilder. Whatever the nature of *Tubiphytes* (see Microproblematica, above) it does not seem to resemble any extant calcified cyanobacteria, nor does it appear to be algal.

Hedstroemia is present in the upper part of the Capitan reef complex (lower Upper Permian) of the south-western United States, and *Girvanella* is present in the Lower Permian of the Carnic Alps together with occasional *Ortonella* and *Garwoodia* (Table 2 in Flügel 1977), but in general calcified cyanobacteria are not conspicuous in the Permian.

4.2.3 Triassic to Mid-Cretaceous

Calcified cyanobacteria become much more prominent in reef and back-reef/lagoon facies in the Middle and Upper Triassic (Flügel 1975), and this situation continues until the mid-Cretaceous. In the Alps, the Dolomites and in Greece, Middle to Late Triassic sponge and coral reefs have crusts and bushy growths of *Bacinella*, *Bevocastria*, *Cladogirvanella*, *Garwoodia/Mitcheldeania*, *Girvanella*, *Hedstroemia*, and *Ortonella*, while nodules formed by *Apophoretella* Elliott, *Cayeuxia*, *Polytrichella* Schäfer and Senowbari-Daryan, and *Zonotrichites* are common in lagoonal environments (Fois and Gaetani 1980; Schäfer and Senowbari-Daryan 1983; Fois 1981). *Tubiphytes* is also present in Triassic and some Jurassic reefs.

Cayeuxia and its relatives continue to be common through the Jurassic (Flügel 1979) and into the Cretaceous. Dragastan (1985) has correctly compared these with *Rivularia*, although it would probably be best to retain the names *Cayeuxia*, *Zonotrichites*, etc. to distinguish these calcified skeletons from extant cyanobacteria whose diagnoses are based on details of soft organic tissues and not on hard parts.

Micritic crusts, possibly of cyanobacterial origin, are present in Middle Triassic (Fois and Gaetani 1984) and Middle Jurassic (Gwinner 1976) sponge reefs. More dendritic, but also micritic, crusts resembling Cambrian *Angulocellularia* occur in cavities in Lower Cretaceous grainstones (see Masse 1979).

The problematic porous micritic crusts known as *Lithocodium* Elliott and *Bacinella* Radiočić are locally important in reefs from the mid-Triassic to Lower Cretaceous. They have not generally been regarded as being cyanobacterial, and they do not resemble extant calcified cyanobacteria, although a microbial origin as well as various other suggestions have been made concerning their affinity (see Microproblematica, above).

4.3 Late Mesozoic and Cenozoic Decline

The decline in calcified marine cyanobacteria probably began in the Upper Cretaceous. By the early Cenozoic there were few or none present in the shallow carbonate environments in which they had hitherto been conspicuous during most periods since the early Palaeozoic, and this scarcity continues to the present day. The few reports of calcified cyanobacteria, as opposed to those which trap and bind sediment, in Recent shallow marine carbonate environments include (1) calcified sheath material associated with partly cemented oöids from Great Bahama Bank (Winland and Matthews 1974, Fig. 3c); (2) a circular structure, which may be a cross-section of a similar filament from Red Sea reefs (Friedman et al. 1974, Fig. 15); and (3) *Rivularia mesenterica* and R. *polyotis* with rounded carbonate inclusions from the Bahamas, Adriatic Sea, and Shark Bay (Golubic and Campbell 1981). The latter *Rivularia* specimens, however, are from intertidal localities which do not appear to have normal marine salinities. Brackish water *Rivularia* has been reported from the Pleistocene of Greece (Richter et al. 1979). However, calcified cyanobacteria continue to be common in freshwater carbonate environments in the Recent where they are present in lacustrine and fluviatile oncoids, travertines and freshwater marshes.

5 Discussion

5.1 Analogues

Recent analogues are of crucial importance in establishing the identities of fossil calcified cyanobacteria. Despite the fact that calcified cyanobacteria are now virtually absent in marine environments, they are common in freshwater calcareous lakes and streams and, provided that they are not too heavily encrusted (as in tufa deposits), they are convincing analogues for Palaeozoic and Mesozoic marine fossils. Consequently, examination of Recent calcified freshwater cyanobacteria has confirmed that fossils such as *Girvanella* Riding (1977b), *Angulocellularia* (Riding and Voronova 1982a), and *Hedstroemia* are calcified sheaths of filamentous cyanobacteria. Further comparative studies are needed to extend this systematic work.

One limitation of this approach which can be anticipated is that a variety of Recent taxa may produce similar sheaths which, when calcified, are so alike that they come to constitute a single fossil taxon. This is almost certainly true in some cases, but, surprisingly, there are examples which also show the opposite. *Hedstroemia* and *Botomaella*, for example, are distinct fossil genera, but they both possess analogues in extant calcified rivulariacean sheaths (Riding, unpubl.). These sheaths thus exhibit greater morphological variation than do the trichomes which they enclose and by which they are produced. And these differences in sheath character are preserved by calcification. As a result, a Recent taxon may represent several fossil taxa, rather than the reverse.

Variability in sheath character is the main feature reflected in fossil cyanobacteria. Ironically, culturing of these organisms in an attempt by workers on modern cyanobacteria to stabilize and simplify taxonomy (e.g. Rippka et al. 1979) often leads to absence of the sheath altogether. As a result there is a tendency by these workers to undervalue sheath characters in taxonomy which may make it more difficult to relate modern cyanobacteria to fossil calcified sheath material. However, it is relatively easy to study sheath variability in Recent oncoids. The occurrence of the same filamentous cyanobacterium on both the upper and lower surfaces of a stabilized oncoid results in a gradient of sheath morphologies. On the upper surfaces filaments grow erect, elongate, and in open tufts, whereas on the lower surface they are compressed and compact due to more constrained growth. This results in a transitional series of sheath morphotypes ranged between two end-members, one open and elongate, the other dense and compact; but all apparently originating from the same species of cyanobacterium. This information is valuable in interpreting the morphological series exhibited by cyanobacteria, for example some of those noted by Riding and Voronova (1985).

5.2 Timing of Calcification

Study of Recent analogues indicates that cyanobacterial calcification begins during life (Pentecost and Riding 1986). In the case of filamentous forms this does not appear to inhibit growth because the ends of the filaments remain open. There is less information concerning coccoid forms but calcification will not necessarily prevent diffusion of water or light through the enveloping sheath. Post-mortem calcification has been reported in association with the desiccation of *Entophysalis* mats at Laguna Mormona (Horodyski and Von der Haar 1975) and Shark Bay, but the general result is that "mineral incorporation obliterates the biological structure" (Golubic 1983, p. 313). However, these examples are from evaporative intertidal environments. Post-mortem calcification in subtidal environments has been reported in the green alga *Ostreobium* (Lukas 1974; Kobluk and Risk 1977). Calcification takes place within 2–3 months and precipitation of micritic calcium carbonate, on both the interior and exterior of the filaments, preserves the shape of the alga. Similar effects in cyanobacteria have not been reported, but could contribute to their preservation. Pratt and James (1982) and Pratt (1984) have suggested that diagenetic calcification of coccoid cyanobacteria is the main process by which *Renalcis* and *Epiphyton* group microfossils are preserved, and this possibility requires further study.

5.3 Skeletons

The suggestion that stromatolites formed by calcified fossils should be termed *skeletal* stromatolites (Riding 1977a) has from time-to-time aroused criticism. It has been suggested that cyanobacteria do not form skeletons (Monty 1981, p. 4) although Golubic and Campbell (1981, pp. 223–225) described calcification in *Rivularia* and came to the conclusion that: "with respect to biological control over calcification and specific morphology of the resulting concretions, *Rivularia* is comparable to *Penicillus* and *Rhipocephalus*, and shows more organismic control than the intercellular aragonite precipitation of *Halimeda*". But they did go on to say that "this criterion, however, is not fulfilled in the case of the nonspecific micrite that surrounds the "filaments" of *Girvanella* (Golubic and Campbell 1981, p. 225).

Nevertheless, a skeleton or a skeletal element can be regarded as a hard mineralized layer or component formed on or within an organism during its life. Cyanobacterial calcification takes place within or on the mucilaginous sheaths during life. It results in a hard-part which will have offered protection and support to the organism and which reflects some aspects of its morphology. It is possible to argue backwards and forwards about the degree of control involved and whether or not this kind of hard part is comparable with that of algae, invertebrates and vertebrates. In these circumstances it seems necessary to accept that there are different types of skeleton related to different biomineralization processes. Those who oppose the use of the term skeleton for calcified cyanobacteria are likely to encounter problems in attempting to distinguish these hard parts from those of algae such as *Halimeda*, and probably also from those of invertebrates such as some foraminifers. It is unlikely that all organic hard parts are formed in the same way or for the same purpose, but until we have a system of clearly discriminating between them then it seems reasonable to refer to them all as skeletons or skeletal elements.

5.4 Seawater Chemistry

The pattern of cyanobacterial calcification in marine environments through time shows variation on two scales. In the long term, it appears that heavily calcified cyanobacteria are only present during the Palaeozoic and Mesozoic. Within this long period of time they are most common during the Cambrian-Lower Ordovician, Upper Devonian-Lower Carboniferous, and Middle Triassic-mid-Cretaceous.

Monty (1973, p. 615) has suggested that the lack of calcification in Recent subtidal marine stromatolites could be due to some unspecified change in sea-water composition together with a response, on the part of cyanobacteria, to ecological pressure from calcareous red algae which has led cyanobacteria to utilize sediment trapping and binding to occupy niches in which direct competition with rhodophytes is avoided. However, Gebelein (1976, p. 512) pointed out that there is no clear evidence for general competition of this sort between cyanobacteria and rhodophytes, although he agreed that red algae may limit the development of stromatolites in reef communities (Gebelein 1976, p. 514). He went on to develop the idea of a change in sea-water chemistry by suggesting that "rapid rise of the calcareous plankton during the Mesozoic . . . may have adjusted the carbonate equilibria so as to make it virtually impossible for blue-green algae to precipitate carbonate in normal marine waters" (Gebelein 1976, p. 514). With regard to changes near the Precambrian-Cambrian boundary, Monty (1973) and Gebelein (1976) both took the view (questioned here) that Precambrian stromatolite-building biotas lost the ability to precipitate carbonate. Monty suggested that this was due to extinction of carbonate precipitating cyanobacteria, linked to decrease in CO_2 pressure: "such a decrease would have affected the rates of photosynthesis as well as the abilities of blue-green algae to precipitate carbonate" (Monty 1973, p. 605). Gebelin (1976, p. 506) postulated a "lowering of the carbonate super-saturation due to removal of bicarbonate ion by metaphytes and metazoans".

This view of a two-stage reduction in ability to calcify in marine environments, first at the end of the Precambrian and then near the end of the Mesozoic is inconsistent with the appearance of heavily calcified cyanobacteria in the early Cambrian. The overall, long-term decline in stromatolite abundance and diversity from the late Precambrian to the present is plausibly explained by the hypothesis of direct and indirect competition from metazoans (Garrett 1970) and changes in sedimentation (Pratt 1982). But changes in calcification require a different explanation. Monty (1973) and Gebelein (1976) invoked factors such as changes in CO_2 pressure and carbonate equilibria at both the Precambrian-Cambrian and the Mesozoic-Cenozoic boundaries to explain what was perceived at both boundaries to be a loss of calcification in cyanobacteria.

Instead, it is possible to propose changes in the Mg^{2+}:Ca^{2+} ratio of seawater and/or in atmosphere-hydrosphere pCO_2 near the Precambrian-Cambrian boundary stimulating (rather than inhibiting) calcification, to account for the Palaeozoic-Mesozoic occurrence of calcified marine cyanobacteria, followed by a reversed change in Mg^{2+}:Ca^{2+} and/or pCO_2 near the Cretaceous-Tertiary boundary to account for the disappearance of calcification in Cenozoic marine cyanobacteria (Riding 1982). The Cretaceous-Tertiary decline coincides with the change from calcitic to aragonitic oöids reported by Sandberg (1983). Fluctuations in abundance of calcified cyanobacteria through the Palaeozoic-Mesozoic show some broad similarities with the aragonite-facilitating and calcite-facilitating events recognized by Sandberg (with marine cyanobacterial calcification being most conspicuous during calcite-facilitating periods, i.e. Cambrian to Devonian and Triassic to mid-Cretaceous). However, the most generally held view of Proterozoic seawater chemistry is that it would have readily facilitated carbonate precipitation (Holland 1984, p. 422; Grotzinger 1989). Consequently, it is puzzling that calcified cyanobacteria have not been more widely reported from Proterozoic carbonates (see Geological History, Precambrian, above). Some of the assumptions that have been made concerning Precambrian life and environments may need to be reexamined in this light. It is possible that the sudden appearance of diverse calcified cyanobacteria in the earliest Cambrian reflects an evolutionary event for prokaryotes which is comparable with that observed in metazoans at this time. However, the subsequent history of marine cyanobacterial evolution is overprinted by temporal variation in carbonate precipitation rates, as is well seen in the long-term fluctuations in abundance of calcified cyanobacteria during the Phanerozoic.

6 Conclusions

1. With the aid of Recent analogues from freshwater environments it is possible to clarify the affinities of calcareous microfossils which have been suggested to be cyanobacteria. This shows that the principal filamentous groups involved are oscillatoriaceans and rivulariaceans, and that the site of calcification is the protective extracellular mucilaginous sheath surrounding the cells.

2. The main fossil groups cluster around the following typical genera:

Angulocellularia, micritic; bush-like; in the Recent formed by *Shizothrix*.

Epiphyton, micritic, tubular, or chambered; dendritic; bearing a resemblance to stigonemataleans.

Girvanella, tubiform; tangled, coiled, or aligned; in the Recent formed by *Plectonema*.

Hedstroemia, branched, radial tubes; in the Recent formed by rivulariaceans.

Renalcis, botryoidal clusters of thick-walled chambers; may be calcified coccoid colonies.

In addition, the *Garwoodia*/*Mitcheldeania* and *Rothpletzella* groups have also commonly been regarded as cyanobacterial, but this is here considered to be unlikely.

3. In the Recent, calcified cyanobacteria occur in lacustrine and fluvial environments as discrete skeletons and tufa oncoids and as tufa stromatolites. But in ancient marine environments they are most conspicuous as dendrolites, thrombolites and skeletal stromatolites.

Dendrolites and thrombolites are particularly abundant in the Cambrian and Lower Ordovician. In contrast, the Proterozoic shows little evidence of well-calcified cyanobacteria despite the abundance and diversity of stromatolites. Thus, the appearance of calcified rock-forming cyanobacteria near the Precambrian-Cambrian boundary was an event immediately followed by the acme of these organisms in the marine environment. Their subsequent Phanerozoic history is one of overall decline in importance, albeit interspersed by periods of relative abundance in the Upper Devonian to Lower Carboniferous and Middle Triassic to mid-Cretaceous. During the Cenozoic calcified cyanobacteria appear to be virtually absent from the sea. The most likely explanation for this pattern of appearance and disappearance of calcified cyanobacteria in marine environments through time is that prokaryote evolutionary and extinction events are superimposed on fluctuations in environmentally determined calcification (Riding 1989).

Acknowledgements. It is a pleasure to thank Ernst Ott, who encouraged my interest in modern calcified cyanobacteria both before and during tenure of an Alexander von Humboldt-Stiftung research fellowship at the Technical University, Munich. Additional support has been given by the Natural Environment Research Council. I am grateful to Allan Pentecost and Brian Pratt for providing helpful comments on the manuscript.

References

Aitken JD (1967) Classification and environmental significance of cryptalgal limestones and dolomites, with illustrations from the Cambrian and Ordovician of southwestern Alberta. J. Sediment Petrol 37:1163–1178

Awramik SM (1971) Precambrian columnar stromatolite diversity: reflection of metazoan appearance. Science 174:825–827

Babcock JA (1977) Calcareous algae, organic boundstones, and the genesis of the upper Capitan Limestone (Permian, Guadalupian), Guadalupe Mountains, West Texas and New Mexico. In: Hileman ME, Mazzullo SJ (eds) Upper Guadalupian facies, Permian Reef Complex, Guadalupe Mountains, New Mexico and West Texas. SEPM Permian Basin Sec Publ 77–16, pp 3–44

Barattolo F (1983) Osservazioni su *Triploporella steinmanni* n. sp. (alghe verdi, Dasycladali) del Cretacico Messico. Boll Soc Nat Napoli 91:1–35

Bertrand-Sarfati J (1976) An attempt to classify late Precambrian stromatolite microstructures. In: Walter MR (ed) Stromatolites. Developments in sedimentology, vol 20. Elsevier, Amsterdam, pp 251–259

Bertrand-Sarfati J, Pentecost A (1989) "Tussocky" microstructure in stromatolites: a biological event recorded in the Upper Proterozoic sediments of the West African craton. Terra Abstr 1(1):197

Bless MJM, Bouckaert J, Bouzet Ph, Conil R, Cornet P, Fairon-Demaret M, Groessens E, Longerstaey PJ, Meessen JPMTh, Paproth E, Pirlet H, Streel M, van Ameron HWJ, Wolf M (1976) Dinantian rocks in the subsurface north of the Brabant and Ardenno-Rhenish massifs in Belgium, the Netherlands and the Federal Republic of Germany. Meded Rijks Geol Dienst NS 27(3):81–195

Braithwaite CJR, Casanova J, Frevert T, Whitton BA (1989) Recent stromatolites in landlocked pools on Aldabra, western Indian Ocean. Palaeogeogr Palaeoclimatol Palaeoecol 69:145–165

Buchbinder B (1981) Morphology, microfabric and origin of stromatolites of the Pleistocene precursor of the Dead Sea, Israeli. In: Monty CLV (ed) Phanerozoic stromatolites. Springer, Berlin Heidelberg New York, pp 181–196

Camoin G, Maurin A-F (1988)Rôles des micro-organismes bactéries, (cyanobactéries) dans le genèse des "Mud Mounds". Examples du Turonien des Jevels Bireno et Mirhila (Tunisie). CR Acad Sci Paris 307:401–407

Chafetz HS, Folk RL (1984) Travertines: depositional morphology and the bacterially constructed constituents. J Sediment Petrol 54:289–316

Chuvashov BI, Riding R (1984) Principal floras of Palaeozoic marine calcareous algae. Palaeontology 27:487–500

Chuvashov BI, Luchinina VA, Shuysky VP et al. (1987) Fossil calcareous algae. Acad Sci USSR, Siberian Sec, Inst Geol Geophys, Novosibirsk 674:225 pp (in Russian)

Copeland JJ (1936) Yellowstone thermal myxophyceae. Ann NY Acad Sci 36:5–225

Copper P (1976) The cyanophyte *Wetheredella* in Ordovician reefs and off-reef sediments. Lethaia 9:273–281

Danielli HMC (1981) The fossil alga *Girvanella* Nicholson & Etheridge. Bull British Museum (Nat Hist). Geology Ser 35(2):79–107

Dragastan O (1985) Review of Tethyan Mesozoic algae of Romania. In: Toomey DF, Nitecki MH (eds) Paleoalgology: contemporary research and applications. Springer, Berlin Heidelberg New York Tokyo, pp 101–161

Dragastan O (1988) New "Porostromata" algae of the Mesozoic. Revista Española de Micropaleantologia 20(3):355–388

Eggleston JR, Dean WE (1976) Freshwater stromatolitic bioherms in Green Lake, New York. In: Walter MR (ed) Stromatolites. Developments in sedimentology, vol 20. Elsevier, Amsterdam, pp 479–488

Elliott GF (1956) Further records of fossil calcareous algae from the Middle East. Micropaleontology 2(4):327–334

Flügel E (1975) Kalkalgen aus Riffkomplexen der alpin-mediterranen Obertrias. Verh Geol Bei A 2–3:297–346

Flügel E (1977) Environmental models for Upper Paleozoic benthic calcareous algal communities. In: Flügel E (ed) Fossil algae, recent results and developments. Springer, Berlin Heidelberg New York, pp 314–343

Flügel E (1979) Paleoecology and microfacies of Permian, Triassic and Jurassic algal communities of platform and reef carbonates from the Alps. Bull Cent Rech Explor Prod Elf-Aquitaine 3(2):569–587

Fois E (1981) The Sass de Putia carbonate buildup (western Dolomites): biofacies succession and margin development during the Ladinian. Riv It Paleontol 87(4):565–598

Fois E, Gaetani M (1980) The northern margin of the Civetta buildup. Evolution during the Ladinian and the Carnian. Riv It Paleontol 86:469–542

Fois E, Gaetani M (1984) The recovery of reef-building communities and the role of cnidarians in carbonate sequences of the Middle Triassic (Anisian) in the Italian Dolomites. Palaeontogr Am 54:191–200

Friedman GM, Amiel AJ, Schneidermann N (1974) Submarine cementation in reefs: examples from the Red Sea. J Sediment Petrol 44:816–825

Garrett P (1970) Deposit feeders limit development of stromatolites. Am Assoc Petrol Geol Bull 54:848

Garwood EJ (1931) The Tuedian Beds of northern Cumberland and Roxburghshire east of the Liddel Water. Q J Geol Soc London 87:97–159

Gebelein CD (1976) The effects of the physical, chemical and biological evolution of the earth. In: Walter MR (ed) Stromatolites. Developments in sedimentology, vol 20. Elsevier, Amsterdam, pp 499–515

Golubic S (1983) Stromatolites, fossil and Recent: a case history. In: Westbroek P, de Jong EW (eds) Biomineralization and biological metal accumulation. Reidel, Dordrecht, pp 313–326

Golubic S, Campbell SE (1981) Biogenically formed aragonite concretions in marine *Rivularia*. In: Monty CLV (ed) Phanerozoic stromatolites. Springer, Berlin Heidelberg New York, pp 209–229

Grotzinger JP (1989) Facies and evolution of Precambrian carbonate depositional systems: emergence of the modern platform archetype. In: Crevello P, Sarg R, Read JF, Wilson JL (eds) Controls on carbonate platforms and basin development. Soc Econ Paleont Mineral, Spec Publ 44:79–109

Gwinner MP (1976) Origin of the Upper Jurassic limestones of the Swabian Alb (southwest Germany). Contrib Sediment 5:1–75

Harland WB, Cox AV, Llewellyn PG, Pickton CAG, Smith AG, Walters R (1982) A geologic time scale. Univ Press, Cambridge, 131 pp

Hofmann HJ (1975) Stratiform Precambrian stromatolites, Belcher Islands, Canada: relations between silicified microfossils and microstructure. Am J Sci 275:1121–1132

Holland HD (1984) The chemical evolution of the atmosphere and oceans. Univ Press, Princeton, 582 pp

Horodyski RJ, Von der Haar SP (1975) Recent calcareous stromatolites from Laguna Mormona (Baja California) Mexico. J Sediment Petrol 45:894–906

Ischenko AA, Radionova EP (1981) On the morphological features and systematic position of the genus *Wetheredella* Wood, 1948. In: Questions of micropalaeontology. Trans Acad Sci USSR Geol Inst 24:140–151 (in Russian)

Johnson JH (1961) Limestone-building algae and algal limestones. Johnson, Boulder, Col, 297 pp

Kennard JM, James NP (1986) Thrombolites and stromatolites: two distinct types of microbial structures. Palaios 1:492–503

Kobluk DR, Risk MJ (1977) Calcification of exposed filaments of endolithic algae, micrite envelope formation, and sediment production. J Sediment Petrol 47:517–528

Korde KB (1959) Morphology and systematic position of representatives of the genus *Epiphyton*. Acad Sci USSR Rep (Dokl) 126:1087–1089 (in Russian)

Korde KB, Maslov VP, Krilova IN (1963) Division Cyanophyta. In: Orlov Yu A (ed) Fundamentals of palaeontology. Acad Sci USSR, Moscow, pp 29–54 (in Russian)

Kraus O, Ott E (1968) Eine ladinische Riff-Fauna in Dobratsch-Gipfelkalk (Kärnten, Österreich)

und Bemerkungen zum Faziesvergleich von Nordalpen and Drauzug. Mitt Bayer Staatsamml Paläontol Histol Geol 8:263–298

Latham A, Riding R (1988) Thrombolites formed by calcified cyanobacteria in the Lie de Vin Series, Morocco. In: Int Assoc Sedimentol, 9th Eur Reg Meet, Leuven, Abstr, pp 127–128

Loeblich Jr AR, Tappan H (1964) Sarcodina, chiefly "thecamoebians" and Foraminiferida. In: Moore RC (ed) Treatise on invertebrate paleontology, pt C: Protista 2. Geol Soc Am; and Univ Press, Kansas, 900 pp

Luchinina VA (1975) Palaeoalgological characteristics of the Early Cambrian of the Siberian Platform. USSR Acad Sci Siberian Branch, Inst Geol Geophys Issue 216. Nauka, Novosibirsk, 98 pp (in Russian)

Lukas KJ (1974) Two species of the chlorophyte genus *Ostreobium* from skeletons of Atlantic and Caribbean coral reefs. J Phycol 10:331–335

Mamet B, Préat A (1986) Algues givetiennes du bord sud du Bassin de Dinant et des regions limitrophes. Ann Soc Géol Belg 109:431–454

Mamet B, Roux A (1975) Algues dévoniennes et carbonifères de la Téthys occidentale. Rev Micropaleontol 18(3):134–187

Maslov VP (1956) Fossil calcareous algae of the USSR. Acad Sci USSR, Geol Inst, Moscow 160:302 pp (in Russian)

Maslov VP (1960) Stromatolites. Acad Sci USSR, Works Geol Inst 41:188 pp (in Russian)

Masse J-P (1979) Schizophytoides du Crétacé inférieur. Caractéristiques et signification écologique. Bull Centre Rech Explor Prod Elf-Aquitaine 3, 2:685–703

Maurin AF, Bernet-Rollande MC, Monty CLV, Nazhat S (1985) The microbial nature of bacinellid textures – sedimentological bearings. In: Int Assoc Sedimentol, 6th Eur Reg Meet, Lerida, Abstr pp 285–287

Monty CLV (1973) Precambrian background and Phanerozoic history of stromatolitic communities, an overview. Ann Soc Geol Belg 96:585–624

Monty CLV (ed) (1981) Spongiostromate vs. porostromate stromatolites and oncolites. In: Phanerozoic stromatolites. Springer, Berlin Heidelberg New York, pp 1–4

Mountjoy EW, Riding R (1981) Foreslope stromatoporoid-renalcid bioherm with evidence of early cementation, Devonian Ancient Wall reef complex, Rocky Mountains. Sedimentology 28:299–319

Pentecost A (1987) Growth and calcification of the freshwater cyanobacterium *Rivularia haematites*. Proc R Soc Lond B232:125–136

Pentecost A (1988) Growth and calcification of the cyanobacterium *Homeothrix crustacea*. J General Microbiol 134:2665–2671

Pentecost A, Riding R (1986) Calcification in cyanobacteria. In: Leadbeater BSC, Riding R (eds) Biomineralization in lower plants and animals. Syst Assoc Spec Vol 30:73–90

Pia J (1927) Thallophyta. In: Hirmer M (ed) Handbuch der Paläobotanik, vol 1. Oldenbourg, München, pp 31–136

Playford PE, Cockbain AE, Druce EC, Wray JL (1976) Devonian stromatolites from the Canning Basin, Western Australia. In: Walter MR (ed) Stromatolites. Developments in sedimentology, vol 20. Elsevier, Amsterdam, pp 543–563

Poncet J (1976) Hypothèse relative à la morphogenèse du thalle de *Renalcis* (algue calcaire – paléozoïque) et affinité possible avec les Rivulariacées actuelles. Geobios 9 (3):345–351

Pratt BR (1982) Stromatolite decline – a reconsideration. Geology 10:512–515

Pratt BR (1984) *Epiphyton* and *Renalcis* – diagenetic microfossils from calcification of coccoid blue-green algae. J Sediment Petrol 54:948–971

Pratt BR, James NP (1982) *Epiphyton* and *Renalcis* – diagenetic microfossils from calcification of coccoid blue-green algae (abs.). Am Assoc Petrol Geol Bull 66:619

Richter DK, Herforth A, Ott E (1979) Pleistozäne, brackische Blaugrünalgenriffe mit *Rivularia* haematites auf der Perachorahalbinsel bei Korinth (Griechenland). N Jahrb Geol Palaontol Abh 159:14–40

Riding R (1975) Girvanella and other algae as depth indicators. Lethaia 8:173–179

Riding R (1977a) Skeletal stromatolites. In: Flügel E (ed) Fossil algae: recent results and developments. Springer, Berlin Heidelberg New York, pp 57–60

Riding R (1977b) Calcified *Plectonema* (blue-green algae), a Recent example of *Girvanella* from Aldabra Atoll. Palaeontology 20:33–46

Riding R (1977c) Systematics of *Wetheredella*. Lethaia 10:94

Riding R (1982) Cyanophyte calcification and changes in ocean chemistry. Nature (London) 299:814–815

Riding R (1983) Cyanoliths (cyanoids): oncoids formed by calcified cyanophytes. In: Peryt TM (ed) Coated grains. Springer, Berlin Heidelberg New York, pp 276–283

Riding R (1988) Classification of microbial carbonates. In: 6th Int Coral Reef Symp: Benthic microbes and reefs, Discussion Session. Townsville, Abstr, p 5

Riding R (1989) Early history of calcification in prokaryotes and algae. Terra abstracts 1, (1):198–199

Riding R, Brasier M (1975) Earliest calcareous foraminifera. Nature (London) 257:208–210

Riding R, Toomey DF (1972) The sedimentological role of *Epiphyton* and *Renalcis* in Lower Ordovician mounds, southern Oklahoma. J Paleontol 46:509–519

Riding R, Voronova L (1982a) Recent freshwater oscillatoriacean analogue of the Lower Palaeozoic calcareous alga *Angulocellularia*. Lethaia 15:105–114

Riding R, Voronova L (1982b) Affinity of the Cambrian alga *Tubomorphophyton* and its significance for the Epiphytaceae. Palaeontology 25:869–878

Riding R, Voronova L (1982c) Calcified cyanophytes and the Precambrian-Cambrian transition. Naturwissenschaften 69:498

Riding R, Voronova L (1984) Assemblages of calcareous algae near the Precambrian/Cambrian boundary in Siberia and Mongolia. Geol Mag (Tokyo) 121(3):205–210

Riding R, Voronova L (1985) Morphological groups and series in Cambrian calcareous algae. In: Toomey DF, Nitecki MH (eds) Paleoalgology: contemporary research and applications. Springer, Berlin Heidelberg New York Tokyo, pp 56–78

Riding R, Watts N (1981) Silurian algal reef crest in Gotland. Naturwissenschaften 68:91–92

Riding R, Watts N (1983) Silurian *Renalcis* (?cyanophyte) from reef facies in Gotland (Sweden). N Jahrb Geol Paläontol Mh 4:242–248

Rippka R, Dernelles J, Waterbury JB, Herdman M, Stanier RY (1979) Generic assignments, strain histories and properties of pure cultures of cyanobacteria. J Gen Microbiol 111:1–61

Rothpletz A (1891) Fossile Kalkalgen aus den Familien der Codiaceen und der Corallineen. Z Dtsch Geol Ges 43:295–322

Roux A (1985) Introduction à l'étude des Algues fossiles paléozoïgues (de la Bactérie à la tectonique des plaques). Bull Centres Rech Explor Prod Elf-Aquitame 9(2):465–699

Sandberg PA (1983) An oscillating trend in Phanerozoic non-skeletal carbonate mineralogy. Nature (London) 305:19–22

Schäfer P, Senowbari-Daryan B (1983) Die Kalkalgen aus der Obertrias von Hydra, Griechenland. Palaeontographica B, 185:83–142

Schopf JW (1968) Microflora of the Bitter Springs Formation, late Precambrian, central Australia. J Paleontol 42:651–688

Schopf JW, Walter MR (1983) Archean microfossils: new evidence of ancient microbes. In: Schopf JW (ed) Earth's earliest biosphere, its origin and evolution. Univ Press, Princeton, NJ, pp 214–239

Szulczewski M (1963) Stromatolites from the high-tatric Bathonian of the Tatra Mountains. Acta Geol Pol 13:142–145

Tyler SA, Barghoorn ES (1954) Occurrence of structurally preserved plants in pre-Cambrian rocks of the Canadian Shield. Science 119:606–608

Vachard D (1974) Contribution à l'étude stratigraphique et micropaléontologique (algues et foraminifères) du Dévonien-Carbonifère inférieur de la partie orientale du versant meridional de la Montagne Noire (Hérault, France). Thesis, 3ème Cycle, Univ Paris VI

Vologdin AG (1962) The oldest algae of the USSR, pt 1 and 2. Acad Sci USSR, Moscow, 656 pp (in Russian)

Walter MR, Awramik SM (1979) *Frutexites* from stromatolites of the Gunflint Iron Formation of Canada, and its biological affinities. Precambrian Res 9:23–33

Walter MR, Heys GR (1985) Links between the rise of the Metazoa and the decline of stromatolites. Precambrian Res 29:149–174

Winland HD, Matthews RK (1974) Origin and significance of grapestone, Bahama islands. J Sediment Petrol 44:921–927

Winsborough BM, Seeler J-S, Golubic S, Folk RL, Maguire Jr B (in prep) Fresh-water stromatolites, oncoids and wated pebbles from a spring-fed lake in northeastern Mexico

Wood A (1948) *"Sphaerocodium"*, a misinterpreted fossil from the Wenlock Limestone. Geol Assoc Proc 59:9–22, 4 pls

Wray JL (1967) Upper Devonian calcareous algae from the Canning Basin, Western Australia. Prof Contrib Col School Mines 3:76 pp

Zhuravleva IT, Meshkova NP, Luchinina VA, Kashina LN (1982) Late Precambrian and early Cambrian biofacies of the Anabarian Sea. In: Environment and life in the geological past, palaeolandscape and biofacies. USSR Acad Sci, Siberian Sec, Inst Geol Geophys, Novosibirsk, Trans 510:74–103 (in Russian)

Chapter 4
The Solenoporaceae: A General Point of View

A.F. Poignant[1]

Abstract

The classification of the Solenoporaceae is very difficult to establish, as the criteria which may be used are scarce and not very reliable. There is a great deal of confusion surrounding generic and specific allocation and classification systems. Even today, it is still not possible, e.g. between green algae and red algae, for one to determine a definite assignment for this group.

There are only a few generic or specific determination tests, and even then they seem far from convincing. The outer structure seems to alter very slightly in relation to time; the inner structure is more complex: hypothallus (?), perithallus, partitions, walls, pores, and reproductive organs (?) can be examined.

Fifty years ago, numerous phylogenic theories from Solenoporaceae to Corallinaceae were suggested, but of course, all these phylogenic ideas must be restudied if Mamet and Roux's (1977) new classification is to be accepted.

As long as a certain taxonomic order is not established, the stratigraphical distribution is meaningless. Now is the time to reestablish those taxa which make up one distinct family.

1 Introduction

The Solanoporaceae represents an extinct group of marine organisms which has its climax during the Jurassic. The organisms are nodular or encrusting and are formed by closely packed threads of vertically divergent cells. Hypothallus and reproductive organs are still uncertain; the cells are much longer than those of the Corallinaceae. The Solenoporaceae were more or less adapted to their environment and were gradually replaced by the Corallinaceae during the Cretaceous.

At first, these organisms were classed as animals such as Tabulatae, Bryozoa, Hydrozoa, and Stromatoporoidae and subsequently as algae due to resemblance to the Corallinaceae. Pia (1930) was not totally convinced by this and Vialli (1938) even went as far as to state that it is all a question of a stage between Chaetetidae and Corallinaceae, or in other words, a transition between the animal and plant kingdoms. However, Nicholson (1888) stated:

"If evidence can be obtained, proving decisively the existence of a cellular, and not a tubular structure in *Solenopora*, then the reference of the genus to the calcareous algae will follow as a matter of course."

[1]Université Pierre et Marie Curie, Centre d'Etude des Algues Fossiles, Boite 2000–4, Place Jussieu, 75252 Paris Cédex 05, France

Several arguments have been put forward to confirm that these forms belong to algae: organs claimed to be reproductive, cell size, the presence of partitions in tubes, the preserved violet coloring, etc.

There is a great deal of confusion surrounding generic and specific allocation and classification systems. This confusion is due to descriptions, admittedly genuine, which are, however, too brief, too hurried, uncritical and without size criteria. This confusion is also created by a lack of comparisons of species, and a lack of photographs, being only partially compensated for by sketches which serve only as an explanatory illustration of the facts.

2 Historical Background

One must take note of the important work carried out by Deninger (1906), Dybowsky (1879), and Haug (1883), who created their own genera. Not so distant in time, we have Elliott (1963, 1965, 1973), Flügel (1960, 1961, 1972), Garwood (1945), Johnson (1960a,b, 1961, 1963, 1966, 1969), Maslov (1956, 1961), Peterhans (1928, 1929a,b, 1930), Riding (1977), Rothpletz (1908, 1913), Wood (1944), Wray (1967, 1970, 1977a,b), and Yabe (1912), who put forward criticisms and syntheses.

Certain research, such as that by Brown (1894), whose first study suggests that the Solenoporaceae were the ancestral form of the Corallinaceae, by Pia (1930), Wood (1944), and above all by Mamet and Roux (1977), must not be overlooked. Mamet and Roux (1977), after many observations, question the classification generally acknowledged since Pia (1939). The reasoning of the criticism and the simplicity of their observations appear obvious. It seems that the works of the first observers, Deninger (1906), Dybowsky (1879), and Haug (1883), were poorly read and badly interpreted. There are more than 200 species, extending from the Cambrian through the Miocene, corresponding to more than 300 bibliographical references.

3 Determining Criteria

There are only a few generic or specific determination tests, and even these seem far from convincing.

3.1 The Outer Structure

The outer morphology hardly changes from one genus to another, and there is an even smaller change from one species to another. It seems to alter very slightly in relation to time. In general, we are concerned with nodular masses, rounded ovoid, hemisperical or conical, encrusting or strongly branching. It should be noted that the nodular forms seem to be the oldest and have been replaced by ramified forms during the Mesozoic.

3.2 The Inner Structure is More Complex

Several authors have pointed out the presence of two tissues: Hypothallus and perithallus with a rightful doubt concerning the former.

1. *Hypothallus.* The presence of an authentic hypothallus has not been proved for several reasons: (1) its thinness; (2) its shape does not make it possible to distinguish it from the perithallus; and (3) it is not calcified. Very few authors draw attention to its presence and then do not give many details pertaining to it.

2. *Perithallus.* The fairly thick perithallus composes, in fact, the entire tissue. It consists of closely packed tubes, closed packed in the sense that the cell threads are in contact with one another with little or no vacant space between, forming fan-shaped structures.

3. *Partitions.* The tubes are, more often than not, divided by horizontal partitions, almost always concave, delimiting large cells, which are longer than 100 μm.

The considerable height of the cells is a unique feature which means they can be classified at the edge of the plant and animal kingdoms, unless the genus *Marinella* (Barattolo and del Re 1984) is included in this family (cells smaller than 10 μm). Moreover, it must be remembered that *Lithoporella* (Corallinaceae) also has large cells, but these are definitely narrower.

The frequency and positioning of the partitions, which appear to be neither a result of change nor of diagenesis, provide, in my opinion, some good diagnostic criteria.

In Mamet and Roux's (1977) classification, two genera do not have horizontal partitions, and this is not a question of a lack of original calcification nor a consequence of diagenesis. In fact, the horizontal partition could be nothing more than a trail made by longitudinal partitions; thus, two longitudinal partitions facing each other would make it appear as if there are horizontal partitions, equivalent to Peterhans' (1928) meanderings or to the indentations noted by other authors.

4. *Walls.* The tubes always contain vertical walls, sometimes sinuous, of which a thorough investigation has never been made. Every tube possesses walls, but no wall is common to any two tubes. When cross-sections are taken, the tubes can be seen to be circular or polygonal in shape. This appearance is meaningless in systematics and it may even change in the same individual.

5. *Pores.* Perforations have been alluded to by the earliest authors. For some time, these pores were given a fair amount of importance since they were used as determining criteria and as arguments to back up classifications. However, no explanation has been offered, since most authors prefer to deny their presence. They could possibly be a question of "primary synapses" and/or "secondary

synapses" or could even be the result of diagenetic phenomena in the cases where original structures are not being considered.

Other puzzling cell shapes have been pointed out, such as those which are star-like or resemble miliolids. No explanation has been suggested for them.

6. *Indentations*. Here, we are talking about a special and original feature in the internal structure of the tubes. This was acknowledged a long time ago but has since been forgotten. These indentations were noted e.g. by Nicholson (1888), Rothpletz (1908), Peterhans (1929a,b), and Høeg (1961).

With regard to these, we owe certain pertinent remarks, which totally change the determining criteria and systematics, to Mamet and Roux (1977).

The first observations on *Solenopora* made by Dybowsky (1879) show that in cross-sections, the wall of the cell is corrugated and affected by septiform processes (up to six), which are directed inwards as if the wall were invaginating. From this, Johnson (1960a) saw a branching of the tubes, whereas Høeg (1961) for his part, saw longitudinal and incomplete septa located above a bifurcation. Peterhans (1928, 1929a,b, 1930) pointed out, without elaborating further, that certain cells show indentations. This interpretation has been very uncertain for a long time.

Nicholson (1888) recognized implicitly the systematic interest provoked by these indentations:

"no radiating 'septa' are developed, but the type species exhibits more or less inwardly directed septiform processes."

In regards to *Solenopora compacta*, more or less numerous septiform processes, and in *Solenopora*? *filiformis* "the tubes do not exhibit inward septiform processes. I have some doubts about . . . this fossil to the genus." Mamet and Roux (1977) have explained remarkably well what these indentations mean. It is a question of longitudinal, incomplete, and more or less developed partitions, which extend the length of the tubes and which are independent from branching. Thus, the study of cross-sections at random can give the impression of horizontal septa; however, it is rather a question of longitudinal septa on which, depending on the genus, an authentic horizontal septation can superimpose itself.

This apparent or real septation is a determining criterion which will play a role in the classification system.

7. *Reproductive Organs*. This is a somewhat controversial subject. Hasty observations have led to uncertain conclusions; the discovery of actual reproductive organs would greatly simplify the problem of systematic attributions and phylogenetic problems.

I stipulated in 1977 the suggestion that the Solenoporaceae had sufficiently large cells for the reproduction of spores to take place without the need of a favorable, noncalcified external location.

A claim favoring the existence of reproductive organs in Solenoporaceans was made for the first time by Brown (1894) and then by Rothpletz (1913), Garwood (1945), Pia (1930, 1940), Öpik and Thomson (1933), Peterhans (1929b), Maslov (1956), and Elliott (1965, 1973). Elliott's (1965, 1973) examples cannot all be accepted: *Neosolenopora patrinii* was proved to be a bryozoan and *Solenopora sardoa* an anthozoan.

There are certain authors who disagree with this theory and some who have changed their standpoint. On the whole, authors remain doubtful:

Wray (1977a): "unequivocal reproductive organs have not been found in Paleozoic species."

Elliott (1965): "definite reproductive structures are usually missing."

Garwood and Goodyear (1919): "occasionally small elongated fusiform spaces occur in the thallus, filled with crystalline calcite. Similar spaces have been described by Rothpletz (1913) as sporangia. They suggested rather due to the irregular solution of partitions of cell walls."

Garwood (1913): "a case where this is a definite association of these (pseudosporangia) bodies with growth lines,
- their parallelism with the trend of the cell walls;
- their regular spacing. . . I do not wish to assert that the section here shown proves conclusively the presence of true sporangia. . ."

Elliott (1965): "may be sporangial structure . . . may well be sporangia, but this is still uncertain."

Cuif (1969) finally found in the Triassic of Turkey some Solenoporaceae with areas presenting a remarkable resemblance to reproductive organs; the cells were created by the differentiation of cells of the thallus.

In short, there is apparently no certainty as to whether a structure can be seen in cells without any regular shape, similar to that of reproductive organs. These "rosette", "star-like", or "miliolid"-shaped cells are only vegetative cells. There is nothing to show reproductive organs in the cases mentioned. The geometric arrangement in the tissue in no way corresponds to that already known about the Corallinaceae. The relationship between cellular perithallus and reproductive organs is not clear. Undoubtedly we are concerned here with foreign bodies, present in the tissue during its growth, or with traces of decaying organisms, of various scars, and of pre-, pene-, or post-diagenetic dissolution.

4 Systematics and Classification

The classification of the Solenoporaceae is very difficult to establish as the criteria which may be used are scarce and not very reliable. Johnson (1961) stated that there are:

"Difficulties in identifying and interpreting Solenoporaceae, both because of the nature of the calcification of the original plants and because of the effects of solution and recrystallization in the course of fossilization."

Indeed, the distinction between hypothallus and perithallus is not clear because, e.g. the presence of reproductive organs is not certain, the pores and the rosettes cannot be used as criteria which play a part in the classification system, and the polygonal or circular appearance of the tubes in cross-sections cannot be used to back up an argument.

Yet, other characteristics, although not very accurate, can be used; Elliott (1965) stated:

"It should be evident that the Solenoporaceae family does not include any representatives which have very loosely packed filamentous cells such as are seen in *Pycnoporidium*. All the genera of the Solenoporaceae have a compact tissue."

Indeed, the positioning of partitions form the basis of Pia's (1930) classification system:

A) Shapes with pores in rows
I. Cells irregularly dispersed, few horizontal partitions
Solenopora = *Metasolenopora*
II. Cells in rows, horizontal partitions clearly represented
Parachaetetes

B) Shapes with isolated pores, rarely found in rows
I. Cells irregularly dispersed, horizontal partitions clearly represented
Pseudochaetetes = *Solenoporella*
II. Cells in rows, horizontal partitions clearly represented
Petrophyton

Maslov (1961), as a continuation of Peterhans (1929b), suggested:

A) Pores in concentric rows
– cells irregularly arranged
not very many horizontal partitions
Solenopora = *Metasolenopora*
– Cells in concentric rows
clearly defined horizontal partitions, on the one level
Parachaetetes

B) Isolated pores, rarely found in concentric rows
– horizontally arranged cells
clearly defined horizontal partitions
Pseudochaetetes – *Solenoporella*
– cells in concentric rows
horizontal partitions on one level
Petrophyton

Later, various classifications were suggested, varying from author to author and even varying with the same authors. There is no point in referring to all the different theories. The various authors are quick to deny use of pores and cell shapes in cross-section when making a classification, and only make use of the arrangement of the partitions. If these horizontal partitions are on the same level, we are dealing with the genus *Parachaetetes*; otherwise the genus in question is *Solenopora*. Mamet and Roux (1977) have shown the possible existence in cross-section of pseudosepta (identations); Peterhans (1929b) had already drawn attention to this fact. It concerns characteristic structural elements, from which one sees the trail of longitudinal, incomplete partitions, more or less developed inside the tubes. There may also be some real bifurcations.

According to Mamet and Roux (1977):

A) Shapes form with longitudinal partitions
– without horizontal partition
  Solcnopora
– with horizontal partitions
  Parachaetetes

B) Shapes without longitudinal partitions
– without horizontal partition
  Pseudosolenopora
– with rare and irregular horizontal partitions
  Pseudochaetetes
– with frequent and regular horizontal partitions
  Petrophyton

According to Wray (1977): *Rhodophyta*
Rhodophyceae
  Floridae
Cryptonemiales
  Solenoporaceae
  Gymnocodiaceae
  Squamariaceae
  Corallinaceae

4.1 Phylogenetic Theories

For 50 years, numerous phylogenic theories for transition from Solenoporaceae to Corallinaceae were suggested:
Maslov (1956)
Johnson (1960b): The common ancestor splits into two at the Ordovician-Silurian boundary:

S1 near *Archaeolithothamnium*, *Mesophyllum*, and *Lithothamnium*;
S2 near other Corallinaceae.

Endo (1961): The ancestral Cambrian form (not named) divides three times:

Solenopora, then *Petrophyton*, *Pycnoporidium*, and *Nipponophycus*;
Parachaetetes;
Pseudochaetetes.

It is unfortunate that this original idea has not been elaborated on since.

Johnson (1960a): There is a separation at the lower part of the Silurian into *Solenopora* (*Lithoporella*, *Lithothamnium*, *Archaeolithothamnium*) and *Parachaetetes* (the other *Corallinaceae*).

Johnson (1963): This theory is different since, from the separation, we see *Solenopora* without descendants and *Parachaetetes* (*Mesolithon*, *Archaeolthothamnium*, *Mesophyllum*, and *Lithothamnium*). The other Corallinaceae have a phylogenesis which is independent of the Solenoporaceae.

Wray (1970): On the one hand, the separation of the common ancestor ends in *Solenopora* going on to *Archaeolithothamnium*, *Lithothamnium*, and *Mesophyllum* and, on the other hand, it ends in *Parachaetetes* from which more or less directly all the other Corallinaceae originate.

Wray (1977a): The separation starting from the Silurian shows the continuation on the Solenoporaceae and the derivation of the Squamariaceae and the Corallinaceae.

Poignant (1976, modified in 1979): *Solenopora* gave rise to *Paleothamnium* and *Lithothamnium*.

Of course, all these proposed phylogenic ideas must be restudied if Mamet and Roux's new classification is to be adopted.

Indeed, *Solenopora* and *Parachaetetes* seem to be linked by the presence of indentations, whereas other forms do not have any; for example, indentations are unknown in the Corallinaceae. Perhaps, one must look for the ancestor of the Corallinaceae in forms which have no indentation (*Pseudosolenopora*, *Pseudochaetetes*, and *Petrophyton*).

In short, as long as a certain system is not established, the stratigraphical distribution is meaningless.

5 Conclusion

Now is the time to identify those genera which make up one distinct family. The Solenoporaceae have been in existence for 500 million years and, throughout this length of time, they have been important rock formers.

The apparent fall-off of the Solenoporaceae seems to have begun during important ecological changes at the beginning of the Mesozoic. The Corallinaceae benefitted from this decline in Solenoporaceae and filled biotopes which are slightly different. Indeed, the almost certain absence of the hypothallus gives the Solenoporaceae certain growth faculties which differ from those of the

Corallinaceae. One may also add that the type of reproduction, unknown in the case of the Solenoporaceae, was without doubt less efficient than that found in their rivals.

There is now a growing need for the knowledge based on Mamet and Roux's classification. The microscopic structure of these algae must be examined using the electron microscope. One must also strengthen the classification system and put it into some kind of order.

References

Barattolo F, Re del MC (1984) Alghe rosse Solenoporaceae dei dintori de Caserta (Appennino meridionale, Italia). Boll Soc Nat Napoli 93:1–53, 13 pl

Brown A (1894) On the structure and affinities of the genus *Solenopora*, together with descriptions of new species. Geol Mag 4, 1:145–151, 195–203, 13 figs

Cuif JP (1969) Différenciation de tissu reproducteur chez les Solénoporacées triasiques. CR Acad Sci 268:776–778

Deninger K (1906) Einige neue Tabulaten und Hydrozoen aus den mesozoischen Ablagerungen. N Jahrb Mineral, pp 61–70, 2 pl

Dybowsky W (1879) Die Chaetetiden der ost-baltischen Silurformation. Verh Russ Kaisermine Ges St Petersburg 2 Ser 14: 134 pp, 4 tabs

Elliott GF (1963) A Liassic *Pycnoporidium* (calcareous algae). Eclogae Geol Helv 56, 1:179–181, 2 pl

Elliott GF (1965) Tertiary solenoporacean algae and the reproductive structures of the Solenoporaceae. Paleontology 7, 4:695–702, pl 104, 108

Elliott GF (1973) A Miocene solenoporoid algae showing reproductive structures. Paleontology 16, 2:223–230, pl 21–22

Endo R (1961) Phylogenetic relationship among the calcareous algae. Sci Rep Saitama Univ Ser B:52 pp

Flügel E (1960) Solenoporaceen (Algen) aus den Zamblach-Schichten (Rhät) der Fischerwiese bei Alt-Aussel, Steiermark. N Jahresh Geol Paläontol Mb 8:339–354

Flügel E (1961) Algen (Solenoporaceen) aus den Cassianer-Schichten (Ober-Ladin) der Südalpen. N Jahrb Geol Paläontol 7:339–347

Flügel E (1972) Mikroproblematika in Dünnschliffen von Trias-Kalken. Mitt Ges Geol Berg 21:957–988

Garwood EJ (1913) Presidential address to section C (Geology). "Rock Building Algae" British Association, Report for 1913, pp 453–472

Garwood EJ (1945) Note on the organs of reproduction in *Solenopora gracilis*. Proc Geol Assoc London 56:147–148, 1 pl

Garwood EJ, Goodyear E (1919) On the geology of the Old Radnor District, with special reference to an algal development in the Woolhope limestone. Q J Geol Soc London 74:12–29, 1 pl

Haug E (1883) Über sogenannte Chaetetes aus mesozoischen Ablagerungen. N Jahrb Mineral 1:171–179, tab X

Høeg OA (1961) Ordovician algae in Norway. Col Sch Mines Q 56, 2:102–116

Johnson JH (1960a) Paleozoic Solenoporaceae and related red algae. Col Sch Mines Q 55, 3:1–77, 23 pl

Johnson JH (1960b) The Jurassic Algae. Col Sch Mines Q 59, 2:129,45 pl

Johnson JH (1961) Review of Ordovician algae. Col Sch Mines Q 56, 2–1:1–101

Johnson JH (1963) Pennsylvanian and Permian algae. Col Sch Mines Q 58, 3:1–211

Johnson JH (1966) A review of the Cambrian algae. Col Sch Mines Q 61, 1:1–162

Johnson JH (1969) A review of the Lower Cretaceous algae. Prof Contrib Col Sch Mines 6:1–10, 68 pl

Mamet B, Roux A (1977) Algues rouges dévoniennes et carbonifères de la Téthys occidentale, pt 4. Rev Micropaeontol 19, 4:215–266, 9 pl

Maslov VP (1956) Algues calcaires fossiles de l'U.R.S.S. Acad Sci URSS 160; 296 pp (French translation BRGM 3511)
Maslov VP (1961) Algues rouges fossiles d'U.R.S.S. et leur rapport avec les faciēs. 22 pp, 128 fig, 36 pl (French translation BRGM 384)
Nicholson HR (1888) On certain anomalous organisms which are concerned in the formation of some of the Paleozoic limestones. Geol Mag (Tokyo) 3, 5:12-24
Öpik A, Thomson PW (1933) Über Konzeptakeln von *Solenopora*. Publ Geol Inst Univ Tartu 36:1-8, 3 pl
Peterhans E (1928) Etude de l'algue jurassique *Parachaetetes*. Eclogae Geol Helv 22:41-43, 1 pl
Peterhans E (1929a) Les Algues jurassiques *Solenoporella* et *Pseudochaetetes*. Bull Soc Geol Fr 4, 29:3-10, 2 pl
Peterhans E (1929b) Algues de la famille des Solénoporacées dans le Malm du Jura bâlois et soleurois. Mem Soc Paleontol Suisse 1L:1-15, 7 pl
Peterhans E (1930) Une nouvelle Solenoporacée du Tithonique de Sardaigne. Eclogae Geol Helv 23:37-39, 2 pl
Pia H (1930) Neue Arbeiten über fossile Solenoporaceae und Corallinaceae. N Jahrb Mineral Geol Paläontol 3:122-147
Pia J (1939) Sammelbericht über fossile Algen; Solenoporaceae 1930 bis 1938, mit Nachträgen aus früheren Jahren. N Jahrb Mineral Geol Paläontol 3:731-760
Pia J (1940) A new fossil alga (Solenoporaceae) from the Jurassic rocks of Western Australia. J R Soc West Aust 26:29-33, 3 pl
Poignant AF (1976) La phylogénie des Algues rouges. C R Acad Sci Paris 282:1677-1680
Poignant AF (1979) Les Corallinacées mésozoïques et cénozoïques: hypothèses phylogénétiques. Bull Cent Rech Explor Prod Elf-Aquitaine 3,2:753-755
Riding A (1977) Problems of affinity in Paleozoic calcareous algae. Fossil Algae Spec Vol, pp 202-214
Rothpletz A (1908) Über Algen und Hydrozoen im Silur von Gotland und Oesel. Kungl Svensk Vetenskap 43:1-25, 6 pl
Rothpletz A (1913) Über die Kalkalgen, Spongiostromen und einige andere Fossilien aus dem Obersilur Gotlands. Sver Geol Unders 10:1-57, 9 pl
Vialli V (1938) Su Taluni fossili liassici del Monte Peller. Stud Trentini Sci Nat 19:1, 2 pl
Wood V (1944) Organs of reproduction in the Solenoporaceae. Geol Assoc London 55:107-113, pl 5-6
Wray JL (1967) Upper Devonian calcareous algae from the Canning Basin, Western Australia. Prof Contrib Col Sch Mines 3:1-76, 11 pl
Wray JL (1970) Algal in reefs through time. Proc N Am Palaeontol Chicago 1969:1359-1373
Wray JL (1977a) Calcareous algae. Developments in palaeontology and stratigraphy, vol 4. Elsevier, Amsterdam, 185 pp, 170 fig
Wray JL (1977b) Late Paleozoic calcareous red algae. Fossil Algae Spec Vol, pp 167-176
Yabe H (1912) Über einige gesteinbildende Kalkalgen von Japan und China. Sci Rep Tohoku Imp Univ Ser 3 Geol Sendai 1, 1:1-8, 3 pl

Chapter 5

Coralline Algae: Mineralization, Taxonomy, and Palaeoecology

D.W.J. BOSENCE[1]

Abstract

This critical review focusses on three areas of recent research on the crustose coralline algae:

Calcification and Diagenesis

Recent work has clarified the calcification of corallines as being a two-stage process; firstly, tangential and secondly radial calcite is deposited within the matrix of the cell walls. This internal calcification of cell walls is unique to the coralline algae and should be used to interpret early problematic red algae in which other characteristic structures are not present. Recent descriptions of subcrustal aragonite from corallines show that this form of mineralization is not restricted to the squamariaceans, as has been generally considered.

Taxonomic Revisions of Genera

New investigations on Recent coralline algae by workers in Australia and Britain is leading to important revisions of well-known Recent and fossil genera. The time is appropriate to assess the implications of this work for palaeontology. Important questions are the preservation of taxonomic characters and the relationships between fossil taxa and groups of Recent taxa. Specific determination of many fossil corallines suffers from excessive splitting on the basis of too few characters. The importance of detailed measurements and statistics is emphasized.

Ecology and Palaeoecology of Coralline Algae

The longevity of coralline genera, together with their ecological restrictions and plasticity of growth forms in relation to environmental parameters, makes the corallines a good group for palaeoenvironmental analyses. Recent work has shown the strong relationships between growth form and hydraulic energy; and the generic composition of floras and water depths. The latter has exciting potential for erecting quantitative palaeobathymetric zones for the Cenozoic.

1 Introduction

Most of the major advances in corallinology at present are being made by those studying Recent corallines, particularly in the study of calcification, generic classification and morphology. An excellent review of this recent work is to be found in the text by Woelkerling (1988). Palaeontologists, meanwhile, have made some progress in generic concepts and in palaeoecology. In this post-Harlan-Johnson era there still remain major problems in the validity of fossil

[1]Department of Geology, Royal Holloway and Bedford New College, University of London, Egham, Surrey TW20 OEX, UK

coralline species and there is little palaeontological data on which to build a phylogeny of the coralline algae.

In this review the palaeontological implications of the recent advances on present day crustose corallines and new developments in ecology and palaeoecology are discussed under the following headings:

1. Calcification and diagenesis;
2. Taxonomic revisions of coralline taxa;
3. Ecology and palaeoecology.

2 Calcification and Diagenesis

The coralline algae are the most consistently and heavily calcified group of red algae and as such have recently been elevated to ordinal status (Corallinales: Silva and Johansen 1986). The calcification involves high magnesium calcite (and occasionally brucite, Weber and Kaufman 1965) precipitation within most cell walls (Figs. 1A-C, 2A-D). Other calcified red algae are the Solenoporaceae, the Squamariaceae (Denizot 1968) and some members of the Nemalionales (*Galaxaura*, *Liagora* and *Nemalion*: Dixon 1973).

Early electron microscope work showed that the calcified cell walls of coralline algae had a two-layered structure: an inner layer of acicular calcite parallel to the cell wall was succeeded by radial, inward-growing calcite crystals (Figs. 1A, 2C; Bailey and Bisalputra 1970; Alexandersson 1974, 1977; Flajs 1977a,b; Garbary 1978). Flajs (1977a,b) described two cell-wall types: the *Lithothamnium*-type with the two layered structure described above and a *Goniolithon*-type with a central noncalcified zone and outer radial crystals. The *Lithothamnium*-type has subsequently been described from a wide range of corallines (*Mesophyllum*, *Lithothamnion*, *Lithophyllum* (Fig. 2C), *Phymatolithon*, *Porolithon*, *Corallina* and *Jania*) by Cabioch and Giraud (1986) and Massieux et al. (1983) and is also seen in the extinct *Solenopora* (Flajs 1977b). However Flajs' (1977b) *Goniolithon*-type has neither been confirmed by other authors nor has it been found in other genera. More recent work in the 1980s using SEM and TEM techniques on calcified tissue has significantly advanced our knowledge of calcification (Massieux et al. 1983; Walker and Moss 1984; Cabioch and Giraud 1986). However the physiological processes involved in calcification within the cell wall remain unclear. Experimental work by Digby (1979) showed how, in the normally reducing environment of the thallus, increased alkalinity is brought about by oxidation through strong oxidase, catalase and carbonic anhydrase activity and the removal of hydrogen ions. More recently Okazaki et al. (1982) have shown that alginic acid is associated with cell wall calcification and that this acid induces calcite deposition and strongly inhibits aragonite deposition. Cabioch and Giraud (1986) have found no ultrastructural differences between calcified and uncalcified floridean cells. In an ultrastructural study of a range of temperate corallines they have shown that calcification is a two-stage process which relates to the two calcified zones

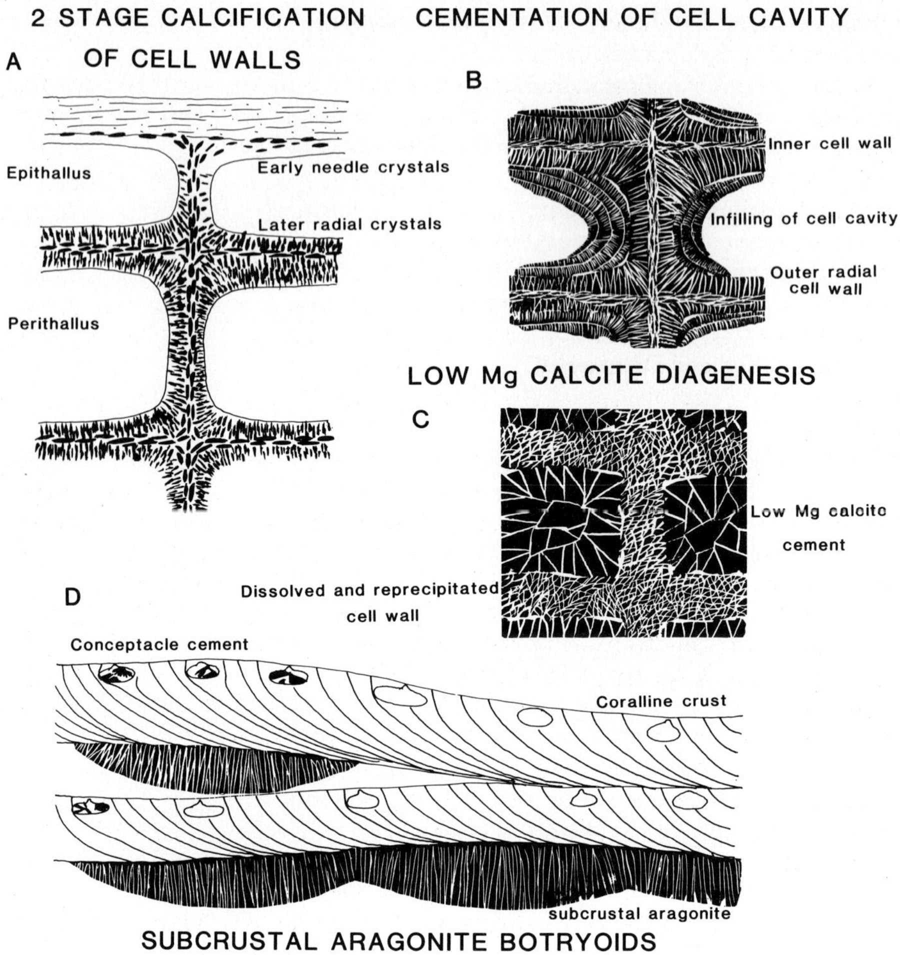

Fig. 1A-D. Calcification and diagenesis. **A** Sketch from electron micrograph (Cabioch and Giraud 1986) illustrating cell wall calcification. Early needle calcite is parallel to cell walls; later radial calcite mineralizes epithallial and perithallial cell walls, x2,000. **B** Sketch indicating two layered coralline cell wall with later infilling of cell cavities by radial calcite after disintegration of cell contents, x2,000. **C** Diagenesis of coralline tissue. Dissolution of high magnesium calcite and reprecipitation of low magnesium calcite in cell wall. Low magnesium calcite cementation of cell cavity, x2,000. **D** Subcrustal aragonite botryoids growing in spaces between crusts and early aragonite cements infilling conceptacles, x200

of the cell wall described above. Whereas Flajs (1977a) considered calcification to be a relatively late-stage process linked to cell degeneration, Cabioch and Giraud (1986) have shown that it occurs around the epithallus which contains the uppermost cells of the thallus (Fig. 1A). The first phase of precipitation involves the growth of needle-shaped crystals within and parallel to polysaccharide fibrils between filaments in a position analogous to the middle lamella of higher plants. With growth, these needle-shaped crystals are replaced or grow into the platey and needle-shaped crystals of the middle cell wall. Later, bladed to fibrous calcite crystals grow radially and parallel to an organic matrix to form the secondary layer at right angles to the primary layer (Figs. 1A, 2C). The mineralogy of these different phases has not been described but earlier work by Flajs (1977b) and Massieux et al. (1983) using microprobe plotting of magnesium and calcium concentrations indicates that the secondary radial layer has a higher magnesium content than the earlier magnesium-poor layer.

When dead tissue is buried within the thallus by outward surface growth, cells are frequently further infilled by layers of micron-sized magnesium calcite crystals (Figs. 1B, 2F) which may eventually fill the cell cavity (Alexandersson 1974; Bosence 1985). This micrite infilling of cells is patchy and is clearly seen in Recent dead, and fossil material (Fig. 2F). Alexandersson (1974) suggests that decaying cell contents may trigger off this later calcification. The supply of calcium carbonate for crystal growth is a problem as there appears to be no local dissolution. Therefore, it would appear to have to come from pore waters passing through the cells and the outer living tissue. An additional or related problem is that whilst most of the cells are infilled with high magnesium calcite others are infilled with aragonite (pers. observations and analyses). This may help to explain earlier reports (Winland 1969; Milliman 1974) of aragonite replacing coralline grains in shallow tropical environments. In extensive surveys of coralline skeletons from the Caribbean I have never seen any replacive textures (e.g. Bosence 1983, 1984, 1985). Oti and Muller (1985) have studied the fine-scale alteration of the high magnesium calcite skeleton to low magnesium calcite both experimentally and in Cenozoic fossil corallines. They show that the process is a small-scale dissolution and reprecipitation replacement of the original cell walls such that the wall microstructure is lost but the walls are recognizable from later low magnesium calcite cements in the cell cavities (Fig. 1C).

Extracellular cements are also common in the corallines (Massieux et al. 1983; Bosence 1985) and the Squamariaceae (Denizot 1968). The commonest are subcrustal botryoids of aragonite (Fig. 1D) which have been described from living, dead and fossil corallines. When first described these were described as early diagenetic cements (Alexandersson 1974) but recently Walker and Moss (1984) have shown that the subcrustal aragonite crystals contain concentrically arranged organic material within the fans which is in intimate association with the basal layers of the alga. Walker and Moss (1984) consider that the fans may have a functional importance to the algae in re-attaching loosened crusts to their substrates. Although similar, but now calcitized, subcrustal fans are known in fossil material it would be difficult to decide whether they were laid down extracellularly by the algae or as chemical precipitates. Other, similar, prismatic aragonite cements are commonly found in growth cavities, vacated conceptacles, etc. (Alexandersson 1974; Bosence 1985). A variety of subcrustal aragonite fans and palisades has been described from squamariacean algae by Denizot (1968). Recently James et al. (1988) have compared these with the common calcite crusts formed beneath Palaeozoic phylloid algae. These calcite crusts

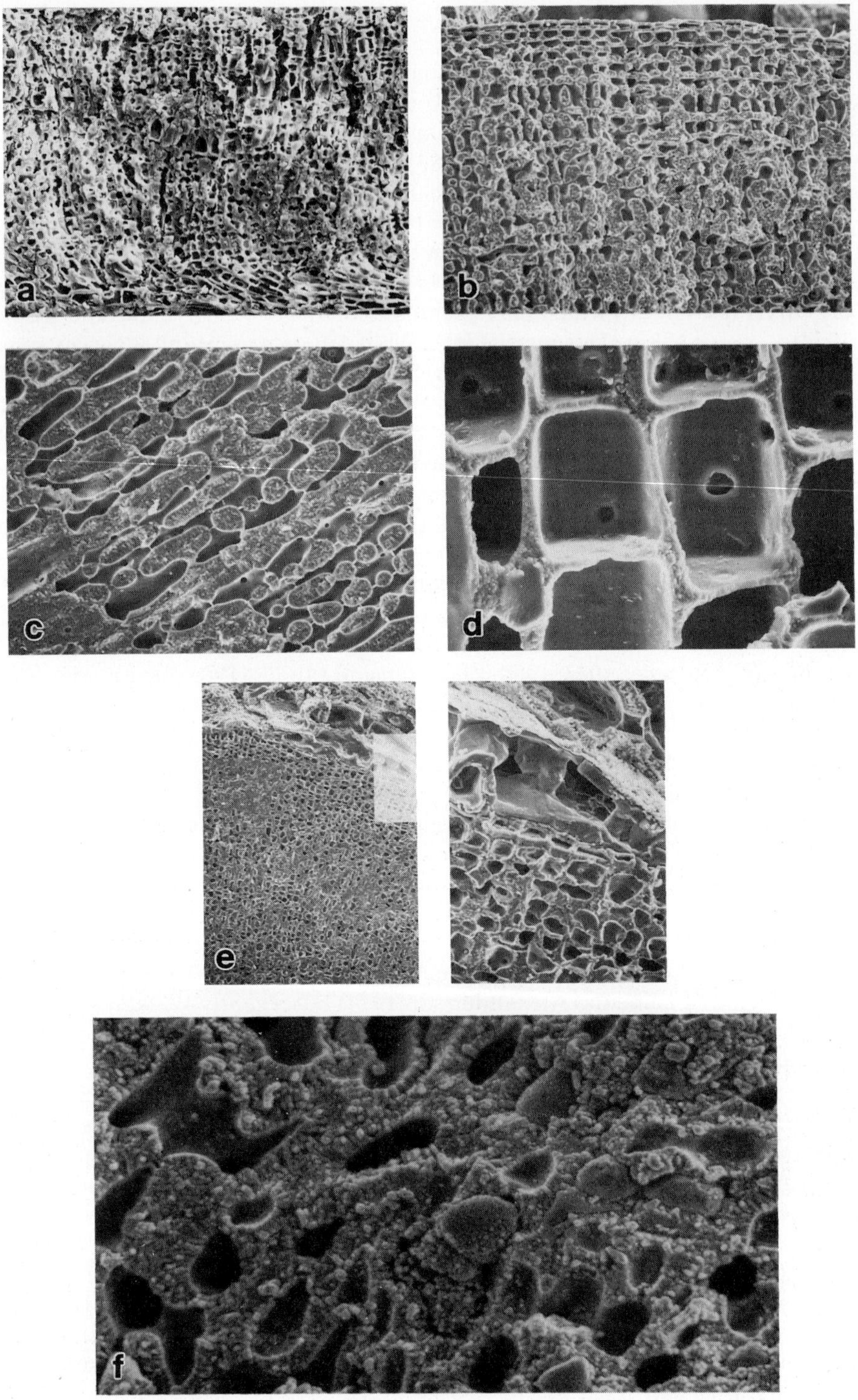
a
b
c
d
e
f

may now also be compared with subcrustal cements found below coralline algae.

Other palaeontological implications of this work on calcification include the recognition of coralline algal cells and filaments in fossil material when differentiated thallus (Fig. 2 a) or conceptacles are absent. The red algae are the only algal group in which the cell and filament walls are calcified (intracellular calcification). In addition, the cell walls have a differentiation into primary and secondary layers (Figs. 1, 2f). It is therefore reasonable, in the absence of other morphological characters, to use this as an identifying character for fossil calcified red algae. Thus, the Solenoporaceae have been shown to have calcified cell and filament walls which are two-layered (Flajs 1977b), and, in the absence of differentiated tissue and reproductive structures, it seems most appropriate to view this group as an extinct family of the red algae. Similarly, in the absence of any other data it is reasonable to assume that *Cuneiphycus* and *Archaeolithophyllum* are fossil red algae as has been suggested by Wray (1977). Such problematic algal groups as the ungdarellids would repay detailed study with polished and etched sections under the SEM, which should reveal the nature of the calcified tissue.

3 Taxonomic Revisions of Coralline Taxa

3.1 Morphological Characters

An important aspect of the systematics of the coralline algae is the relationship between Recent and fossil taxa, the most stable and fundamental taxon in this context being the genus. In defining genera many of the taxonomically important characters can be recognized in both fossil and Recent forms. However in Recent material there are additional morphological features such as developmental and staining characteristics (Table 1). The commonly used characters for describing fossil genera according to Lemoine (1977), and

Fig. 2a-f. Coralline algal microstructure. **a** Vertical section through crust of *Porolithon* illustrating basal hypothallus with horizontal filaments and upper perithallus with vertical filaments and with horizontal row of trichocytes in centre. Recent, St. Croix, US Virgin Islands, x220. **b** Vertical section through crust of *Lithophyllum* illustrating perithallial tissue and single layered outer layer of epithallus. Note increase in cell wall calcification with depth in thallus. Recent, Jeddah, Saudi Arabia, x500. Coll. L. Montaggioni. **c** Detail of **b** illustrating cell wall calcification in *Lithophyllum*. Thin central zone with outer radial cell wall crystals. Note pit connections between contiguous filaments, x880. **d** Detail of calcified cell wall in vertically arranged perithallial filaments with primary pit connection between adjacent filaments and cells. *Lithophyllum*. Recent, Jeddah, Saudi Arabia, x2,500. Coll. L. Montaggioni. **e** Preservation of epithallus beneath overgrowing foraminifera in Recent *Lithophyllum*. Note patches of cells infilled with micrite. *Lithophyllum*. Recent, Jeddah, Saudi Arabia. General view, *left*, x180, zoom, *right* x720. Coll. L. Montaggioni. **f** Fractured surface of *Lithothamnion* with perithallial filaments arranged *lower left* to *top right*. Note secondary cell fusions (*upper left*) and micrite infilled cells (*right*). Recent, Co. Galway, Ireland, x2000

Table 1. List of characters used in the description and assignment of Recent and fossil coralline algae

Generic characters	Occurrence of characters[a]		
	Recent	Fossil	
		Commonly used	Potentially usable
Morphology			
General			
Crustose	●	●	●
Branching	●	●	●
Columnar	●	●	●
Mamillate	●	●	●
Articulated	●	●	●
Articulate branching	●	○	○
Epithallus			
Thickness	●	○	●
Cell Shape	●	○	●
Meristem	●	○	●
Perithallus			
Presence/absencc	●	●	●
Filaments/rows	●	●	●
Secondary cell fusions	●	○	●
Secondary pit connections	●	○	●
Double perithallus	●	●	●
Trichocytes	●	●	●
Medullary perithallus			
Presence	●	●	●
Thickness	●	●	●
Cell height	●	●	●
Cell shape	●	●	●
Genicula	●	○	○
Hypothallus			
Single layered	●	●	●
Palisade	●	●	●
Thickness	●	●	●
Coaxial/non-coaxial	●	●	●
Median hypothallus	●	●	●
Asexual Conceptacles			
Location	●	●	●
Perforations	●	●	●
Shape	●	●	●
Primordia	●	○	●
Tetraspores	●	○	●
Sexual conceptacles			
Location	●	●	●
Shape	●	●	●
Spermatangium	●	○	●
Carpogonium	●	○	●
Carpospore	●	○	●
Biochemical			
Staining of sporangial caps	●	○	○
Developmental			
Spore cell division	●	○	○
Conceptacle development	●	○	●

[a] ● Yes; ○ no.

Poignant (1984, 1985) are also listed (Table 1) and they represent some 64% of characters available to algologists. However, I argue below that it may be possible, with more detailed thin section and SEM observations, to increase the morphological characters available to palaeoalgologists.

Epithallus. It has been generally assumed by palaeoalgologists (Wray 1977; Lemoine 1978a) that the epithallus is uncalcified and, as such, is not a preservable character. The reason for this may be that until recently all present-day material had been illustrated from decalcified material. However, with the publication of SEM micrographs of fractured surfaces of fresh and dried museum specimens (e.g. Woelkerling et al. 1985, 1988), it is clear that the epithallus can be well-calcified and it is difficult to see why it should not be preserved. The most likely circumstance for its preservation is where crusts are overgrown by later crusts and are therefore protected from bioerosion or abrasion (Fig. 2e). Such an occurrence has been recently illustrated by Voigt (1981). Recognition will require careful SEM work, together with a knowledge of epithallial tissue and location of the perithallial meristem. If this tissue can be differentiated in fossils then it will have important implications because epithallial characters are used to differentiate between *Lithothamnion, Phymatolithon* and *Leptophytum* (Adey 1966, 1970). These closely related genera can only be differentiated in the Recent by epithallial cell shape, length of meristematic cells, and the phosphotungstic haematoxylin staining of the sporangial caps. At our present state of knowledge they are all grouped into the genus *Lithothamnion* in fossil material.

Secondary Cell Connections. Connections between contiguous cells are clearly seen in living tissue (Fig. 2) and occur as two types. Secondary pits are small (1–2 μm) circular pits on opposing sides of cell walls separated by a plug (Figs. 2 c-d). They are considered by Adey (1970), Johansen (1981) and Woelkerling (1987b, 1988) to be of fundamental importance in coralline systematics and of ordinal significance in the red algae (Pueschel and Cole 1982; Silva and Johansen 1986; Gabrielson and Garbary 1986). Cell fusions are larger, irregular connections between rows of cells or along filaments of cells (Fig. 2f). The subfamilial classification of Recent Corallinaceae by Johansen (1981) recently modified by Woelkerling (1987a; Table 2), emphasises the taxonomic importance of these characters. As far as I am aware, neither of these structures has been described from fossil corallines although they are clearly seen in fractured Recent specimens (Fig. 2; Woelkerling et al. 1985) and in diagenetically altered Recent rhodoliths from the Skagerak (Alexandersson 1974) and the Red Sea (Flajs 1977b). If diagenesis involved small-scale solution and reprecipitation associated with the change from high to low magnesium calcite, and the cell lumen filled with a coarser calcite cement, then these cell wall characters should be visible in fossils under the SEM as they are in diagenetically altered Recent material.

Table 2. Outline of subfamily classification of Corallinaceae (after Johansen 1981: Woelkerling 1987a)

	Thallus	Cell fusions joining contiguous filaments	Secondary pit connections joining contiguous cells	Conceptacles; tetraspores
Amphiroideae	Geniculate with genicula of one or more tiers of cells	no	yes	Uniporate; tetraspores lacking plugs
Corallinoideae	Geniculate with genicula of single tier of cells	yes	no	Uniporate; tetraspores lacking plugs
Metagoniolithioideae	Geniculate with genicula not aligned in tiers	yes	no	Uniporate; tetraspores lacking plugs
Lithophylloideae	Non-geniculate	no	yes	Uniporate; tetraspores lacking plugs
Mastophoroideae	Non-geniculate	yes	no	Uniporate; tetraspores lacking plugs
Melobesoideae	Non-geniculate	yes	no	Multiporate; tetraspores with plugs
Choreonematoideae	Non-geniculate	no	no	Uniporate; tetraspores with plugs

3.2 Location and Development of Conceptacles and Spores

The details of conceptacle development and of tetraspores and carpospores is only rarely considered in fossil material even though spores often remain in conceptacles and become mineralized (e.g. Bosence 1983). Therefore, it is urged that more detailed descriptions of the position and development of conceptacles and the nature of tetraspores and carpospores be given in future.

Geniculae. The decalcified joints (genicula) of articulated corallines are not normally preserved. However, in unusual circumstances articulated corallines may be preserved in situ by overgrowing crustose corallines (e.g. Bosence 1984) and the genicula and margins of the intergenicula thus preserved. Although this is the only case I am aware of, if articulating corallines encrust other colonial organisms they can potentially be overgrown by their host and preserved intact.

If the above criteria can be considered and fossils carefully examined under the SEM, then the proportion of preservable characters which might be recognized in both fossil and Recent material rises from 64% to a respectable 89%. The number of fairly broadly defined characters used in identifying genera rises to 34, which is a substantial figure for any fossil group and should form the basis for an accurate linkage between fossil and Recent taxa.

3.3 Recent Taxonomic Revisions

At present, corallinologists in Australia, Britain and America are substantially revising the morphology, types and taxonomy of coralline genera. Many of the original type specimens have been located and examined under optical microscopy, and with an SEM on fractured surfaces, and the morphology redescribed in detail. Much of this work is contained in the excellent new text by Woelkerling (1988). Whilst it is not possible here to review the large number of papers which have been published in the last 3 years, the results so far indicate important revisions of coralline genera. I briefly discuss below the major revisions published to date and palaeontological implications of this work.

Dermatolithon. Woelkerling et al. (1985) suggest that this thin, crustose form cannot be differentiated from *Titanoderma* as both have a dorsiventrally arranged prostrate crust with a single layered palisade hypothallus, secondary pit connections and unipored conceptacles.

Leptophytum. Woelkerling and Irvine (1986) consider that this genus requires clarification as a taxon which may or may not be distinct from the very similar *Phymatolithon.*

Litholepis. Woelkerling (1986) shows that this genus is a heterotypic synonym of *Titanoderma* as they both have secondary pit connections, unipored tetrasporangial conceptacles, single layered palisade hypothallus and dorsiventral crust organisation. Secondary cell fusions are absent in both genera.

Lithothamnion. (Heydrich 1897) is now conserved against *Lithothamnium* Philippi (1837) as Woelkerling (1983, 1985a) has shown that all Philippi's type material bears unipored conceptacles and is a mixture of species of *Goniolithon, Lithophyllum* and *Pseudolithophyllum.*

Paragoniolithon. Woelkerling (1987b) has recently reexamined this genus and found it to be a heterotypic synonym of *Neogoniolithon*, as both have trichocytes singly or in vertical stacks in an otherwise similar thallus.

Pseudolithophyllum. This genus has had a varied history but present views are that it should be considered in the sense of Lemoine (1978b) and Mendoza and Cabioch (1984) rather than that of Adey (1970). The thallus has a single-layered hypothallus and the perithallus has rectangular cells arranged in filaments joined by secondary pit connections.

Spongites (Kützing). This genus has been resurrected by Woelkerling (1985b) to include crustose corallines with a multilayered, non-coaxial hypothallus and a perithallus with trichocytes occurring singly or in vertical stacks.

Tenarea. Woelkerling et al. (1985) consider that this genus, with its isobilaterally arranged two rows of hypothallial cells, is clearly distinct from *Titanoderma*. They also suggest a similarity with the isobilaterally arranged fossil genus *Distichoplax*.

These taxonomic revisions are particularly important to palaeoalgologists as the characters which are now being used by algologists are those which are in the main preservable and are visible in fossils. In addition, genera such as *Litholepis*, *Paragoniolithon* and, probably, *Leptophytum* which have not to my knowledge been recognized in fossil material are no longer considered separate genera. Hopefully this will give confidence to palaeoalgologists in that the criteria that have traditionally been used are now being used by algologists. However, when examining fossil material the additional characters should be looked for (Table 1) which may only be recognizable under the SEM.

There are very large problems concerning the identification of fossil species. Most specific identifications require careful measurements of crust and branch thickness, hypothallus thickness, hypothallial and perithallial cell sizes and conceptacle shapes and sizes. Because these characters are naturally variable the mean and standard deviation as well as the ranges are required if specimens are to be compared. Many palaeontologists have followed Johnson (in Johnson and Adey 1965) who only used size ranges and many species described by Johnson are from incomplete material. It is therefore very difficult to compare specimens with published material and much work is required redescribing fossil types.

4 Ecology and Palaeoecology

Work in recent years on the variable morphology of present day coralline algae and the generic floral associations in different climates and water depths has provided a good database for palaeoecological and palaeoenvironmental analysis. This information can be confidently applied to ancient examples because of the generally good preservation of coralline thallus morphology and the relative ease of identification of corallines to generic level. In addition most coralline genera are extant and have an average longevity of 85 Ma (data from Wray 1977). This increases the confidence with which comparisons can be made between present-day coralline ecology and ancient corallines.

4.1 Relationships Between Morphology and Turbulence

The morphology of coralline crusts and branches in present day reefs was considered by Bosence (1983, 1985). I showed then that high energy (i.e. reef front or algal ridge) locations are typified by thick crusts which are closely

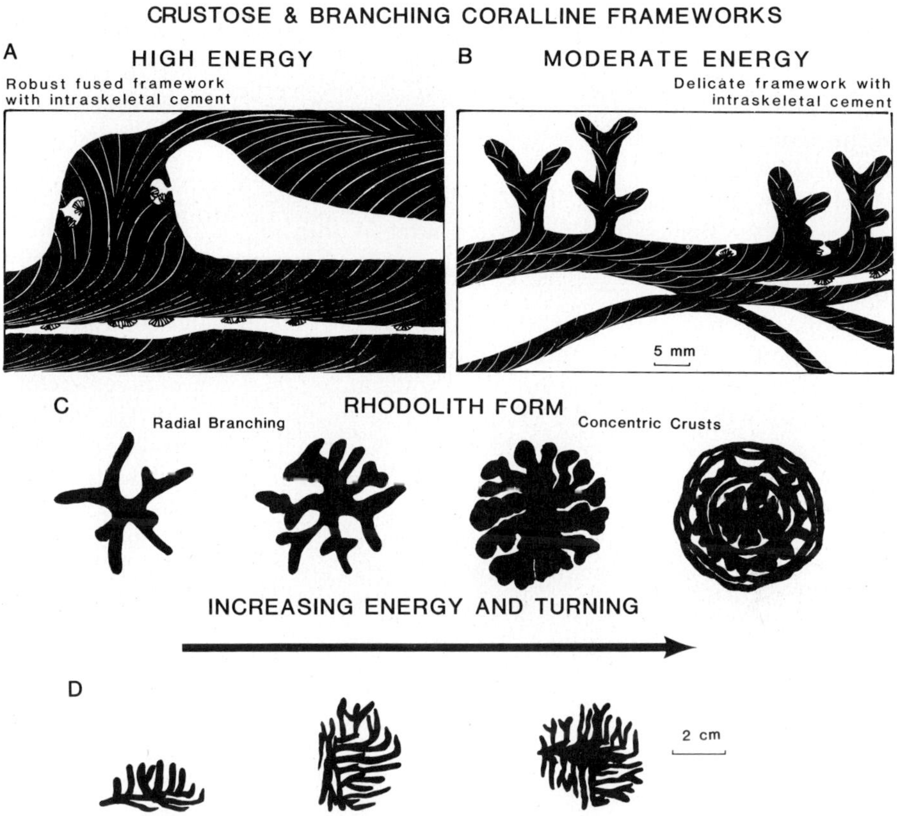

Fig. 3A-D. Relationships between water energy and coralline thallus. **A** Typical growth forms of crusts and branches in high energy coralline reef frameworks. **B** Typical growth forms of crusts and branches in coralline reef frameworks in moderate energy settings (after Bosence 1985). **C** Increase in lateral growth of branches and density of branching of rhodoliths with increasing turbulence and development of dense concentric crusts (after Bosence 1976). **D** Increase in numbers of branching directions with turning of rhodoliths (after Bosence 1983)

superposed (Fig. 3A). Branches and columns are thick and perithallial tissue from neighbouring branches may fuse to further strengthen the framework. The framework is also made rigid by the extensive extracellular cementation in this environment. Coralline genera commonly forming such frameworks are *Lithophyllum* and *Porolithon*. Reef frameworks from medium energy reefs and buildups such as maerl, back-reef mounds, and the deeper water coralligène (Bosence 1983, 1985) typically have thinner branches and crusts, are more delicate in their construction (Fig. 3B) and are formed of leafy crusts which may leave the substrate and overgrow neighbouring crusts. Adjacent branches do not fuse together. Corallines found forming such leafy and branching frameworks are *Mesophyllum*, *Lithophyllum*, *Titanoderma*, *Neogoniolithon*, *Lithothamnion* and *Phymatolithon*.

In a recent discussion of coralline ecology and functional morphology Steneck (1986) describes similar ranges of coralline morphologies as described above. He discusses experimental work which suggests that the thicker crusts are better able to resist physical and biological disturbance. The major biological disturbance is considered to be herbivory by parrot fish, sea-urchins, limpets and chitons (Steneck 1986).

Free-living rhodoliths (rhodolites) can be very sensitive indicators of turbulence in the shallow marine environment. Within any one species the morphology of free living forms can vary from open-branched through to densely branched forms (Cabioch 1966; Bosellini and Ginsburg 1971; Bosence 1976 (Fig. 3C). Open-branched forms are found in the quieter water environments where there is little transport of rhodoliths. If rhodolith transport is very infrequent in comparison to branch growth rates, then zones of redirected, upward growth is seen in the rhodoliths after turning (Fig. 3D). The densely branched forms are typical of rhodoliths growing in higher energy conditions. The increase in branch density is achieved by lateral branch growth and lateral growth following apical abrasion during transport. This may be followed by concentric crustose growth of the original algae or other encrusters. Stabilization results in reversion to branching growth (Bosence and Pedley 1982). In addition to turning by currents, rhodoliths may also be turned by organisms, as was suggested for rhodoliths from a wackestone facies in the Miocene of Malta in which crustacean debris is common (Bosence and Pedley 1982).

4.2 Palaeobathymetry

The establishment of depth/abundance charts for coralline floras from the Caribbean and Pacific sites by Adey (1979) indicates the great potential that the coralline algae have for palaeobathymetry. Since present-day crustose genera have existed from the Miocene, the following bathymetric ranges should be applicable for the Neogene:

- Intertidal-20 m; *Neogoniolithon*, *Porolithon*, *Lithophyllum* and *Hydrolithon*.
- 20–40 m; *Neogoniolithon*, *Lithophyllum*, *Hydrolithon*, *Titanoderma* and *Mesophyllum*.
- 40–60 m; *Mesophyllum* dominant with *Archaeolithothamnium*, *Lithothamnion* and *Lithophyllum*.
- 60–100 (possibly 200 m); *Mesophyllum* and *Lithothamnion* with *Archaeolithothamnium* and *Lithophyllum*.

The only way of confirming these depth sequences back through time is to analyze the distribution and abundance of in situ crusts where the palaeodepth can be measured. Such sites are on very well exposed reef fronts where depositional units can be traced down from the reef flat. Such sites should be investigated to ascertain the depth distributions of Neogene coralline floras.

5 Conclusions

This review concludes with two important points for palaeoalgologists. Firstly, that there has been a history of confusion with respect to the taxonomy of coralline genera which is only recently being corrected. The characters which are being used to redefine genera are, in the main, preservable but palaeoalgologists must use the SEM to obtain additional characters concerning the epithallus and cell wall perforations to increase confidence in their determinations. Thus, at the generic level it is possible for algologists and palaeoalgologists to work closely together to establish evolutionary relationships and higher taxonomic groupings. However, at the species level a large amount of work is required on fossil types to provide good quantitative descriptions and illustrations.

Secondly, with respect to palaeoecology it is now realized that thallus shape together with the generic associations give very good palaeoenvironmental information. Thus, coralline thalli are amongst the most sensitive skeletons to changes in hydraulic energy and should now be used in detailed palaeoenvironmental analysis. Similarly, generic associations of floras which can be accurately assessed in fossil sequences give the opportunity for quantitative palaeobathymetry which has wide-ranging uses in basin analysis.

References

Adey WH (1966) The genera *Lithothamnium*, *Leptophytum* (nov. gen.) and *Phymatolithon* in the Gulf of Maine. Hydrobiologia 28:321–370

Adey WH (1970) A revision of the Foslie crustose coralline herbarium. Kon Norske Vidensk Selsk Skr 1:1–46

Adey WH (1979) Crustose coralline algae as microenvironmental indicators for the Tertiary. In: Gray J, Boucot AJ (eds) Historical biogeography. Oregon State Univ Press, pp 459–464

Alexandersson ET (1974) Carbonate cementation in coralline algal nodules in the Skagerrak, North Sea: biochemical precipitation in undersaturated waters. J Sediment Petrol 44:7–26

Alexandersson ET (1977) Carbonate cementation in Recent coralline algal constructions. In: Flügel E (ed) Fossil algae: recent results and developments. Springer, Berlin Heidelberg New York, pp 261–269

Bailey A, Bisalputra T (1970) A preliminary account of the application of thin-sectioning, freeze-etching and scanning microscopy to the study of coralline algae. Phycologia 9:83–101

Bosellini A, Ginsburg RN (1971) Form and internal structure of recent algal nodules (rhodolites), Bermuda. J Geol 79:669–682

Bosence DWJ (1976) Ecological studies on two unattached coralline algae from western Ireland. Palaeontology 19:365–395

Bosence DWJ (1983) Coralline algal reef frameworks. J Geol Soc London 140:365–376

Bosence DWJ (1984) Construction and preservation of two recent algal reefs, St. Croix, Caribbean. Palaeontology 27:549–574

Bosence DWJ (1985) Preservation of coralline-algal reef frameworks. In: Proc 5th Int Coral Reef Congr, Tahiti 6:623–628

Bosence DWJ, Pedley HM (1982) Sedimentology and palaeoecology of a Miocene coralline algal biostrome from the Maltese Islands. Palaeogeogr Palaeoclimatol Palaeoecol 39:9–43

Cabioch J (1966) Contribution à l'étude morphologique anatomique et systématique de deux Mélobésiées: *Lithothamnium calcareum* (Pallas) Areschoug et *Lithothamnium corallioides* Crouan. Bot Mar 9:33–53

Cabioch J, Giraud G (1986) Structural aspects of biomineralization in the coralline algae (calcified Rhodophyceae). In: Leadbetter BSC, Riding R (eds) Biomineralization in lower plants and animals. Univ Press, Oxford, pp 141–156
Denizot M (1968) Les algues floridèes encroutantes (à l'exclusion des Corallinacées). Thesis, Univ Paris
Digby PSB (1979) Reducing activity and the formation of base in the coralline algae: an electrochemical model. J Mar Biol Assoc UK 59:455–477
Dixon PS (1973) Biology of the Rhodophyta. Oliver & Boyd, Edinburgh
Flajs G (1977a) Skeletal structures of some calcifying algae. In: Flügel E (ed) Fossil algae: recent results and developments. Springer, Berlin Heidelberg New York, pp 225–231
Flajs G (1977b) Die Ultrastrukturen des Kalkalgenskeletts. Palaeontographica 160:69–128
Gabrielson PW, Garbary D (1986) Systematics of red algae (Rhodophyta). CRC Crit Rev Plant Sci 3, 4:325–366
Garbary DJ (1978) An introduction to the scanning electron microscopy of the red algae. Syst Assoc Spec Publ 10:205–222
James NP, Wray JL, Ginsburg RN (1988) Calcification of encrusting aragonitic algae: implications for the origin of late Palaeozoic reefs and cements. J Sediment Petrol 58:291–303
Johansen HW (1981) Coralline algae, a first synthesis. CRC, Boca Raton
Johnson JH, Adey WH (1965) Studies of *Lithophyllum* and related algal genera. Col Sch Mines Q 60(2):1–105
Lemoine P (1978a) Les difficultés de la phylogénie chez les algues corallinacées. Bull Soc Geol Fr 8, 19:1319–1325
Lemoine P (1978b) Typification du genre *Pseudolithophyllum* Lemoine. Rev Algol N S 13, 2:177
Massieux M, Coudray J, Denizot M (1983) Premieres résultats d'une étude de la calcification de *Porolithon onkodes* (Heydrich) Foslie (Rhodophycée, Méolbésiée, actuel) à la microsonde électronique. CR Acad Sci Paris 297, 3:579–582
Mendoza ML, Cabioch J (1984) Redefinition comparée de deux espèces de corallinacées d'Argentine: *Pseudolithophyllum fueginum* (Heydrich) comb. nor. et *Hydrolithon discoideum* (Foslie) nov. comb. Crytogamie Algol 4:141–154
Milliman JD (1974) Marine carbonates. Springer, Berlin Heidelberg New York, 375 pp
Okazaki M, Furunga K, Tsukayama K, Nisizawa K (1982) Isolation and identification of alginic acid from a calcareous red alga *Serraticardia maxima*. Bot Mar 25:123–131
Oti M, Muller G (1985) Textural and mineralogical changes in coralline algae during meteoric diagenesis. An experimental approach. N Jahrb Mineral Abh 151, 2:163–195
Poignant A (1984) La notion de genre chez les algues fossiles A. Les corallinacées. Bull Soc Geol Fr 26, 4:603–604
Poignant A (1985) Présentation simplifiée d'une nouvelle classification générique des rhodophycées mésozoiques et cénozoiques. Bull Soc Geol Fr 8, 1(4):603–605
Pueschel CM, Cole K (1982) Rhodophycean pit plugs: an ultrastructural survey with taxonomic implications. Am J Bot 69:703–720
Silva PC, Johansen HW (1986) A reappraisal of the order Corallinales (Rhodophyceae) Br Phycol J 21:245–254
Steneck RS (1986) The ecology of coralline algal crusts: convergent patterns and adaptive strategies. Annu Rev Ecol Syst 17:273–303
Voigt E (1981) Erste fossile Nachweise des Algen-Genus Fosiella Howe. Facies 5:265–282
Walker R, Moss B (1984) Mode of attachment of six epilithic crustose Corallinaceae (Rhodophyta) Phycologia 23:321–329
Weber JN, Kaufman JW (1965) Brucite in the calcareous algae *Goniolithon*. Science 149:996–997
Winland HD (1969) Stability of calcium carbonate polymorphs in warm shallow sea water. J Sediment Petrol 39:1579–1587
Woelkerling WJ (1983) A taxonomic reassessment of *Lithothamnium* Philippi (Corallinaceae, Rhodophyta) based on studies of RA Philippi's original collections. Br Phycol J 18:165–197
Woelkerling WJ (1985a) Proposal to conserve *Lithothamnion* against *Lithothamnium* (Rhodophyta: Corallinaceae). Taxon 34, 2:302–3
Woelkerling WJ (1985b) A taxonomic reassessment of *Spongites* (Corallinaceae, Rhodophyta) based on studies of Kützing's original collections. Br Phycol J 20:123–153

Woelkerling WJ (1986) The genus *Litholepis* (Corallinaceae, Rhodophyta): taxonomic status and disposition. Phycologia 25:253–261

Woelkerling WJ (1987a) The genus *Choreonema* in southern Australia and its subfamilial classification. Phycologia 26:111–127

Woelkerling WJ (1987b) The status and disposition of *Paragoniolithon* (Corallinaceae, Rhodophyta). Phycologia 26:144–148

Woelkerling WJ (1988) The coralline red algae: an analysis of the genera and subfamilies of non-geniculate Corallinaceae. British Museum (Natural History) Oxford University Press, London, 268 pp

Woelkerling WK, Irvine LM (1986) The typification and status of *Phymatolithon* (Corallinaceae, Rhodophyta). Br Phycol J 21:55–80

Woelkerling WJ, Chamberlain YM, Silva PC (1985) A taxonomic and nomenclatural reassessment of *Tenarea*, *Titanoderma* and *Dermatolithon* from types and other critical specimens. Phycologia 24:317–337

Wray JL (1977) Calcareous algae. Elsevier, Amsterdam, 190 pp

Chapter 6
Cyclocrinitids

S.C. BEADLE[1]

Abstract

The cyclocrinitids are a small group of middle Ordovician to early Silurian macrofossils. They are usually regarded as an extinct tribe of dasycladacean algae, since they are morphologically very similar to the living dasycladacean *Bornetella*. Cyclocrinitids had a central main axis bearing numerous lateral branches, which radiated outwards in all directions. The branches expanded at their tips to form thick lateral heads, which were generally hexagonal in outline and usually 1–3 mm wide. The lateral heads were tightly packed together to form a faceted outer cortex, which was spherical to claviform in shape and typically 1–4 cm in size. The thallus was encrusted by calcium carbonate, particularly around the lateral heads. Cyclocrinitid preservation is quite variable, and this has often confused taxonomists; new descriptions of the European and Asian forms are needed. Cyclocrinitids are important associates of many benthic invertebrate palaeocommunities. They are often abundant locally and may be significant rock-builders; several rock units have been informally named after them. Cyclocrinitids were restricted to shallow, quiet waters at low latitudes. Ecophenotypic thallus-size variation within Silurian species has been used to indicate relative palaeodepths. Ordovician cyclocrinitids have much potential value as palaeoenvironmental indicators.

The Tribe Cyclocriniteae was named by Pia (1920) and emended by Bassoullet et al. (1979). It includes the Ordovician and Silurian genera *Cyclocrinites* Eichwald, *Mastopora* Eichwald, *Coelosphaeridium* Roemer, and *Apidium* Stolley. It should be noted that the genus name *Cyclocrinites* (not *Cyclocrinus*) has priority, so the tribe is the Cyclocriniteae (not Cyclocrineae). Furthermore, *Mastopora* is now commonly regarded as a junior synonym of *Cyclocrinites* (as are *Nidulites* Salter, *Pasceolus* Billings, *Cerionites* Meek and Worthen, and *Lunulites* Owen). Nitecki (1970, pp. 73–75) elaborates on these points.

The history of cyclocrinitid research is discussed by Osgood and Fischer (1960) and Nitecki (1970). Early workers variously regarded cyclocrinitids as foraminifers, sponges, corals, bryozoans, gastropod eggs, cystoids, crinoids, or tunicates. The first comprehensive study was that of Stolley (1896), which reviews all previous work. Stolley showed convincingly that cyclocrinitids resembled dasycladacean green algae, and this idea soon gained wide acceptance in Europe. Pia (1920, 1927) reviewed the cyclocrinitids, and described them as a dasycladacean tribe. Unfortunately, Stolley's work had a very limited distribution in North America, so it was largely overlooked; North American workers commonly regarded cyclocrinitids as aberrant sponges, rather than as algae. However, J.H. Johnson and Konishi (1959) and J.H. Johnson (1961) reviewed the cyclocrinitids and placed them with dasycladaceans. The second

[1] Dept. of Earth & Planetary Sciences, The Johns Hopkins University, Baltimore, MD 21218, USA

comprehensive study was the monograph of Nitecki (1970). This work is the foundation of modern research; it has clarified much of the confusion regarding the morphology, preservation, and taxonomy of cyclocrinitids. The valuable annotated bibliography of Nitecki et al. (1987) lists all works on cyclocrinitids to 1980.

The cyclocrinitids are now widely accepted as an extinct dasycladacean tribe (Beadle 1988), and that interpretation is followed here. However, Nitecki (1986) suggested that they were problematic algae related to receptaculitids.

1 Morphology

Cyclocrinitids had a central main axis, from which lateral branches radiated outward in all directions (Fig. 1b). The lateral branches expanded at their distal tips to form bulbous lateral heads, which were more or less hexagonal in outline. The lateral heads were tightly packed together to form a faceted outer cortex. The living dasycladaceans *Bornetella* Munier-Chalmas (Fig. 1a) and *Neomeris* Lamouroux have very similar morphologies (Stolley 1896; Nitecki 1970; Nitecki and Johnson 1978; Beadle 1988). The resemblances are apparent even at early growth stages (Elliott 1972). It is possible to reconstruct some cyclocrinitids in considerable detail (Pia 1927; Osgood and Fischer 1960; Nitecki and Johnson 1978). The standard anatomical terminology is that of Nitecki (1970).

Cyclocrinitid thalli vary from claviform (Fig. 2a) to spherical (Fig. 3b) in shape. It is often difficult to determine their original shapes, as they were very vulnerable to compaction (Nitecki and Johnson 1978). The unusual 'bilobate' thalli described by Lee and Caldwell (1977) may well have been compacted. Specimens in shales are often completely flattened (Fig. 3c). Cyclocrinitid

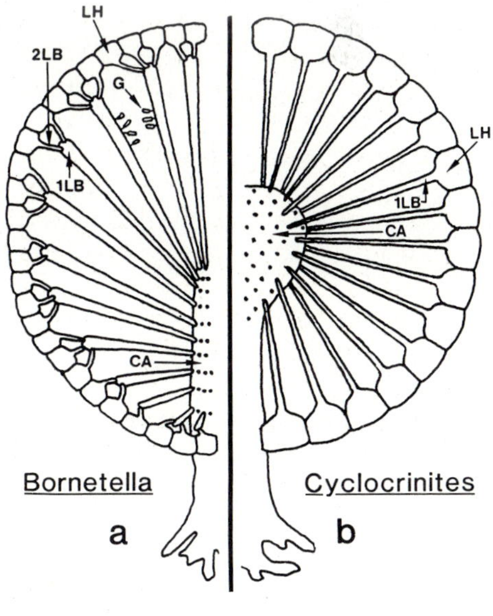

Fig. 1a,b. Schematic longitudinal cross-sections through a Recent dasycladacean and a Silurian cyclocrinitid. *CA* Central axis; *1LB* primary lateral branch; *2LB* secondary lateral branch; *LH* lateral head; *G* gametangia. **a** Recent *Bornetella capitata* (Harvey). Thallus size = 1.0–1.5 cm. After Valet [1968, Pl. 10(7), Fig. 5]; **b** Lower Silurian *Cyclocrinites dactioloides* (Owen). This species had no secodary branches, but other cyclocrinitids did. Position of gametangia unknown. Thallus size = 1–4 cm

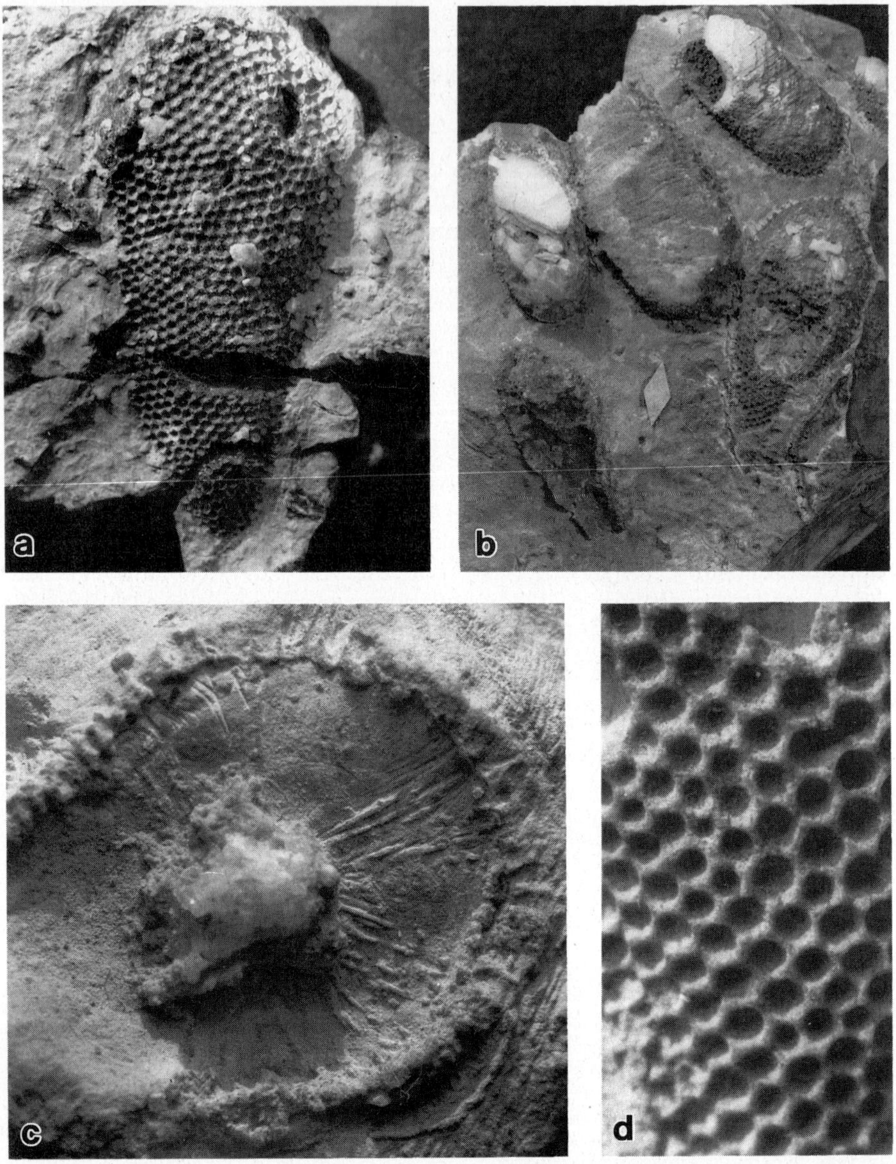

Fig. 2a-d. Middle Ordovician *Cyclocrinites pyriformis* (Bassler). **a** Holotype. Note claviform shape. The preserved calcareous material encrusted the proximal surfaces of the lateral heads. USNM 56625. Chambersburg Limestone; Strasburg, Virginia. x1.75, **b** Specimens from the '*Nidulites* Bed'. Thalli are filled with sparry calcite. USNM 97551. Same locality as **a** x1, **c** Horizontal cross-section, showing central main axis, lateral branches, and lateral heads. USNM 111806. Ward Cove Formation; Staffordsville, Virginia. x3.4, **d** Enlargement of calcareous layer. The lateral heads were generally hexagonal, and tended to show hexagonal closest packing. USNM 111789. Chambersburg Limestone; Hagerstown, Maryland. x6.25

Fig. 3a-d. **a-b** Lower Silurian *Cyclocrinites dactioloides* (Owen). Farmers Creek Member, Hopkinton Dolomite; Delaware Country, Iowa. **a** Longitudinal cross-section, showing proximal surfaces of lateral heads and remnant of central axis. FMNH UC 59064. x1.5, **b** Internal mould. Note spherical shape. FMNH UC 59458. x1.25, **c-d** Lower Silurian *Cyclocrinites favus* (Salter). Photographs courteously provided by Dr. Atle Mørk. **c** Flattened thallus in shale. PMO 105837. Solvik Formation; Bjørkøya, Norway. x1.6. **d** Cyclocrinitid thalli in shale, probably swept together by currents. PMO 105839. Solvik Formation; Avløs, Norway. x0.25

preservation is quite variable, even within a single species (Nitecki 1970). Many specimens have complex diagenetic histories (Osgood and Fischer 1960; Nitecki and Johnson 1978; Mørk and Worsley 1980). Preservational differences have often confused taxonomists.

Most cyclocrinitids are only 1–4 cm in length; the largest approach 6 cm. They probably had a siphonous cellular organization like other dasycladaceans, and this restricted them to small sizes (Beadle 1988). Certain forms, notably *Apidium*, were only 1–10 mm in length (Kummerow 1937; Lee and Caldwell 1977; R.E. Johnson and Sheehan 1985). Some small thalli probably represent immature individuals (Elliott 1972; Beadle and Johnson 1986, p. 597).

The main axis was rooted to a solid substrate; attachment scars and 'anchoring rhizoids' have been reported (Nitecki 1970; Lee and Caldwell 1977). The main axis and lateral branches are only rarely preserved; they were normally very thin, as in other dasycladaceans (Figs. 2c, 3a). In some forms, notably *Coelosphaeridium*, the axes and branches were thicker, but the increased width was generally balanced by decreased length. In some, but not all, species, the lateral branches divided to form secondary branches. However, there was generally less branching than in Recent dasycladaceans, and so cyclocrinitids typically had fewer, larger lateral heads (Beadle 1988). The lateral heads are usually 1–3 mm in width; they increased in size as the thallus grew (Beadle and Johnson 1986).

The arrangement of the lateral branches is not well established. Pia (1920) regarded cyclocrinitids as 'aspondylous', with branches randomly arranged. However, Nitecki (1970) found that *Cyclocrinites welleri* Nitecki had lateral branches in whorls. The lateral heads of many species are arranged in a strikingly regular manner.

Cyclocrinitids were encrusted by surficial deposits of calcium carbonate. The calcification typically encrusted the proximal or distal surfaces of the lateral heads; these are often the only structures preserved (Fig. 2d). The calcification was probably aragonitic (Osgood and Fischer 1960; Beadle 1988). Some species apparently had calcareous external membranes (Nitecki 1970, Fig. 19). The problematic *Amia hexagona* Fry may well represent an uncalcified cyclocrinitid (Fry 1983).

Certain forms had small, radially-arranged ridges or knobs on the proximal or distal surfaces of the lateral heads (Fig. 4). This unusual ornamentation is especially variable and well preserved among the Ordovician species of the Baltic area (Stolley 1896, 1898; Neben and Krueger 1973, 1979). This feature has not been reported among living dasycladaceans, and so it merits further study.

No definitive cyclocrinitid reproductive structures are known. The 'sporangia' of Currie and Edwards (1942) appear to be inorganic crystals (Elliott 1972). The ovoid 'gametangia' of Osgood and Fischer (1960) resemble peloids of faecal pellet origin. Nitecki (1972a) reported 'gametocysts' in *Cyclocrinites dactioloides* (Owen); these structures apparently occur only in dolomitized specimens, and may be diagenetic mineral growths. Cyclocrinitids may have been 'endosporate', with uncalcified reproductive structures that were formed within the main axis (Pia 1920; Elliott 1972, p. 368).

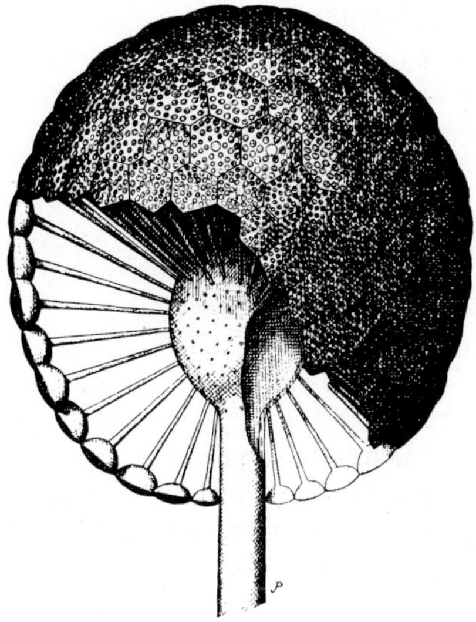

Fig. 4. Reconstruction of *Cyclocrinites porosus* Stolley, from the Ordovician of the Baltic area. The distal surfaces of the lateral heads had complex ornamentation, termed 'Skulptur' by Pia (1927). Thallus size = 8–10 mm. From Pia (1927, Fig. 43b)

2 Geographic Distribution

2.1 North America

Cyclocrinitids occur widely in the USA, Canada, and Greenland. Nitecki (1970) reviews all previous work. Lee and Caldwell (1977) list the 'Arctic Ordovician' localities. Beadle and Johnson (1986) review all Silurian localities. Other recent studies include Nitecki (1972b), Nitecki and Johnson (1978), M.E. Johnson and Campbell (1980), R.E. Johnson and Sheehan (1985), and Frey (1987).

2.2 Europe

Cyclocrinitids are common in southern Norway (Størmer 1953; Høeg 1961; Skjeseth 1963; Mørk and Worsley 1980; Baarli 1987) and in northern Estonia (Raymond 1916; Bekker 1924; Roomusoks 1970). They probably also occur in submarine beds in the northern Baltic Sea (Størmer 1953, p. 136). Cyclocrinitids from these deposits have been widely distributed in glacial erratics (Kiesow 1894, 1899; Stolley 1896, 1898; Hucke and Voigt 1967; Neben and Krueger 1973, 1979). Cyclocrinitids occur at a few localities in Scotland and Wales (Currie and Edwards 1942; Elliott 1972); an additional report, by Spjeldnæs (1955), remains unconfirmed. A problematic Shropshire fossil, *Sphaerospongia hospitalis* Salter, resembles a cyclocrinitid (Salter 1873). The many Ordovician species of Europe need review and revision. Silurian forms are reviewed by Beadle and Johnson (1986). Nitecki et al. (1987) list many additional references.

2.3 Asia

Ordovician cyclocrinitids occur at several localities in central and southern Asia, including eastern Kazakhstan (Gnilovskaya 1972), northern India (Reed 1912), and western China (Mu 1982a,b). The Asian forms most closely resemble those found in Europe.

3 Stratigraphic Distribution

The oldest reported cyclocrinitid is *Cyclocrinites welleri* Nitecki, from the lower middle Ordovician of California (Nitecki 1970). Cyclocrinitids were most abundant and diverse during the Caradoc, when they were widely distributed in North America, northern Europe, and central Asia. They are less common in Ashgill deposits, and they declined throughout the Llandovery (Beadle and Johnson 1986). They were apparently extinct by the end of Llandovery time.

The radiation and decline of the cyclocrinitids closely paralleled that of other dasycladaceans. The decline of these warm-water algae at the end of the Ordovician may reflect the simultaneous cooling and glaciation (Beadle 1988).

4 Palaeoecology

Cyclocrinitid palaeoecology is discussed by Nitecki (1970), Nitecki and Johnson (1978), Mørk and Worsley (1980), Beadle and Johnson (1986), and Beadle (1988). Cyclocrinitids were ecologically very similar to other dasycladaceans.

Cyclocrinitids typically occur with normal marine invertebrates. However, they have also been found in high-salinity restricted-water environments (M.E. Johnson and Campbell 1980). Palaeogeographic reconstructions show that most, and perhaps all, cyclocrinitids lived within 30° of the palaeo-equator (Beadle and Johnson 1986; Beadle 1988). They are usually found in carbonate rocks, but they are also common in certain siltstones and shales. Cyclocrinitids probably colonized soft substrates by attaching themselves to small solid objects; living dasycladaceans are often found on soft bottoms attached to pebbles, shells, or coral fragments.

Cyclocrinitid thalli were relatively fragile; they were largely restricted to quiet-water environments, either below wave-base or in protected lagoons (Nitecki 1970; Beadle and Johnson 1986). Cyclocrinitids were probably very vulnerable to strong currents. Comparison with the anatomically-similar Recent dasycladacean *Bornetella sphaerica* (Zanardini) suggests that their thalli may have been easily torn away from their holdfasts (Nitecki and Johnson 1978). Furthermore, they probably grew on small objects which were readily transported; living dasycladaceans in Texan lagoons are often moved about in this manner (Conover 1964, p. 14). Thus, large accumulations of cyclocrinitids may have been formed by rare storm-generated currents, which penetrated into normally quiet waters and swept the thalli together (Nitecki and Johnson 1978).

Cyclocrinitid thalli often show evidence of local transport (Osgood and Fischer 1960; Mørk and Worsley 1980).

Cyclocrinitids were most common at relatively shallow depths, below wave-base but within the photic zone. Comparison with living dasycladaceans suggests that they lived at depths of less than 100 m (Beadle and Johnson 1986). Silurian benthic communities are commonly regarded as depth-zoned; cyclocrinitids are most common in *Cryptothyrella*, *Pentamerus*, and *Stricklandia* communities, or their equivalents (Beadle and Johnson 1986). They have also been found in very shallow protected-water environments (M.E. Johnson and Campbell 1980). Cyclocrinitids rarely ranged below the *Stricklandia* community, but there are a few reports of them in *Clorinda* (or equivalent) communities (Cocks and Toghill 1973; Baarli 1987).

Some Silurian cyclocrinitids show apparent ecophenotypic variation: certain populations have significantly smaller thalli than stratigraphically adjacent populations. This variation most likely represents differences in ambient light intensity, and it has been used to indicate relative palaeodepths (Beadle and Johnson 1986). This approach is of interest to palaeoceanographers, since it can be used to determine the bathymetric relationships of the associated invertebrate communities (M.E. Johnson 1987, Fig. 2).

5 Stratigraphic Significance

The distribution of cyclocrinitids is sporadic. However, they are often extremely abundant locally, and they may be significant rock-builders. Some units could well be termed 'cyclocrinitid limestones'; for instance, their thalli comprise 25–40% of the rock in parts of the Thumb Mountain Formation of Cornwallis Island, Canada (Lee and Caldwell 1977). They are also very common in certain shaly units, such as the Solvik Formation of Norway (Fig. 3d) and the Gasworks Mudstone of Wales (Beadle and Johnson 1986).

Cyclocrinitids are often used as local guide fossils because of their abundance and distinctive appearance. Several rock units have been informally named for them, including the 'Cyclocrinites Beds' (or 'Cerionites Beds') of Iowa (M.E. Johnson 1983), the 'Nidulites Beds' of Virginia (Fig. 2b) (Cooper and Cooper 1946), the 'Coelosphaeridium Beds' and 'Cyclocrinus Beds' of Norway (Skjeseth 1963), and the 'Coelosphaeridium Zone' of Estonia (Bekker 1924). Cyclocrinitid-bearing glacial erratics in northern Europe have been described as 'Coelosphaeridiengesteine' and 'Cyclocrinus-Kalke' (Kiesow 1894; Hucke and Voigt 1967, Table 2).

Some species may be useful for interregional correlation. *Cyclocrinites pyriformis* (Bassler), from the Champlainian of the Appalachians, may be synonymous with *Mastopora parva* (Nicholson and Etheridge) from the Caradoc of Scotland (Nitecki 1970). Spjeldnæs (1955) claimed that *Coelosphaeridium cyclocrinophilum* Roemer occurred in middle Caradoc strata in Wales, Norway, and Estonia. In both Norway and Estonia, these *Coelosphaeridium*-bearing beds are directly overlain by *Cyclocrinites*-bearing beds (Raymond

1916, p. 245; Størmer 1953, p. 136). *Cyclocrinites favus* (Salter) characterizes lower Llandovery strata in Norway, Scotland, and Wales (Beadle and Johnson 1986); it may also be synonymous with the contemporary North American species *C. gregarius* (Billings).

6 Further Research

Ideally, all cyclocrinitids should be well-defined taxonomically, precisely dated with respect to a global standard, and placed into a palaeoecological context, along with the associated invertebrates. Silurian cyclocrinitids approach this ideal. However, considerable work is needed on the Ordovician forms, which are much more abundant, widespread, and diverse.

Nitecki (1970) has recently revised all North American species; a similar review of the European and Asian forms is needed. Many of these 'species' may represent only different states of preservation. Furthermore, many are based upon specimens from glacial erratics, so their stratigraphic ranges are somewhat uncertain. New descriptions would be welcome, especially for the Estonian and Norwegian forms, which are quite important stratigraphically.

Cyclocrinitids may prove useful in reconstructing Ordovician palaeogeography and palaeoecology. Cyclocrinitids appear to be reliable indicators of low palaeolatitudes. They can also be used as relative depth indicators, and this could help to establish the bathymetric relationships of Ordovician palaeocommunities.

References

Baarli BG (1987) Benthic faunal associations in the Lower Silurian Solvik Formation of the Oslo-Asker Districts, Norway. Lethaia 20:75–90

Bassoullet J-P, Bernier P, Deloffre R, Génot P, Jaffrezo M, Vachard D (1979) Essai de classification des Dasycladales en tribus. Bull Cent Rech Explor Prod Elf-Aquitaine 3:429–442

Beadle SC (1988) Dasyclads, cyclocrinitids and receptaculitids: comparative morphology and palaeoecology. Lethaia 21:1–12

Beadle SC, Johnson ME (1986) Palaeoecology of Silurian cyclocrinitid algae. Palaeontology 29:585–601

Bekker H (1924) Stratigraphical and palaeontological supplements on the Kukruse Stage of the Ordovician rocks of Eesti (Estonia). Acta Comment Univ Tartu Ser A 6, 1:13–20

Cocks LRM, Toghill P (1973) The biostratigraphy of the Silurian rocks of the Girvan District, Scotland. J Geol Soc London 129:209–243

Conover JT (1964) The ecology, seasonal periodicity, and distribution of benthic plants in some Texas lagoons Bot Mar 7:4–41

Cooper BN, Cooper GA (1946) Lower Middle Ordovician stratigraphy of the Shenandoah Valley, Virginia. Bull Geol Soc Am 57:35–114

Currie ED, Edwards WN (1942) Dasycladaceous algae from the Girvan area. Q. J Geol Soc London 98:235–240

Elliott GF (1972) Lower Palaeozoic green algae from southern Scotland, and their evolutionary significance. Bull Br Mus Nat Hist Geol 22:355–376

Frey RC (1987) The palaeoecology of a Late Ordovician shale unit from southwest Ohio and southeastern Indiana. J Paleontol 61:242–267

Fry WL (1983) An algal flora from the Upper Ordovician of the Lake Winnipeg region, Manitoba, Canada. Rev Palaeobot Palyno 39:313–341
Gnilovskaya MB (1972) Middle and Late Ordovician Calcareous Algae of Eastern Kazakhstan. Akad Nauk SSSR Inst Geol Geokhronol Dekembr, Leningrad, 196 pp [in Russian]
Høeg OA (1961) Ordovician algae in Norway. Col Sch Mines Q 56, 2:103–116
Hucke K, Voigt E (1967) Einführung in die Geschiebeforschung (Sedimentärgeschiebe). Ned Geol Ver, Oldenzaal, 132 pp
Johnson JH (1961) Review of Ordovician algae. Col Sch Mines Q 56, 2:1–101
Johnson JH, Konishi K (1959) A review of Silurian (Gotlandian) algae. Col Sch Mines Q 54, 1:1–114
Johnson ME (1983) New member names for the Lower Silurian Hopkinton Dolomite of Eastern Iowa. Proc Iowa Acad Sci 90:13–18
Johnson ME (1987) Extent and bathymetry of North American platform seas in the Early Silurian. Paleoceanography 2:185–211
Johnson ME, Campbell GT (1980) Recurrent carbonate environments in the Lower Silurian of Michigan and their inter-regional correlation. J Paleontol 54:1041–1057
Johnson RE, Sheehan PM (1985) Late Ordovician dasyclad algae of the Eastern Great Basin. In: Toomey DF, Nitecki MH (eds) Paleoalgology: contemporary research and applications. Springer, Berlin Heidelberg New York Tokyo, pp 79–84
Kiesow J (1894) Die Coelosphaeridiengesteine und Backsteinkalke des westpreussischen Diluviums, ihre Versteinerungen und ihr geologisches Alter. Schr Naturforsch Ges Danzig NF 8, 3-4:67–96
Kiesow J (1899) Bermerkungen zu den Gattungen *Cyclocrinus*, *Coelosphaeridium* und *Apidium*. Schr Naturforsch Ges Danzig NF 10, 1:77–93
Kummerow E (1937) Die Bruteinrichtungen paläozoischer Ostracoden, sowie über *Receptaculites* und einige ordovizische Kalkalgen der Gattung *Apidium*. Jahrb Preuss Geol Landesanst. Berg Akad 57:464–474
Lee DG, Caldwell WGE (1977) A new dasycladacean alga associated with the 'Arctic Ordovician' fauna on Corwallis Island. Can J Bot 55:52–60
Mørk A, Worsley D (1980) The environmental significance of algae in the middle Llandovery succession of the central Oslo Region. Lethaia 13:339–346
Mu X-N (1982a) Some calcareous algae from Xizang. In: Paleontology of Xizang, vol 5. Science Press, Beijing, pp 205–240 (in Chinese with English summary)
Mu X-N (1982b) Discovery of Ordovician calcareous algae in Batang, Sichuan. In: Stratigraphy and palaeontology in West Sichuan and East Xizang, China, vol 2. Sichuan People's Press, Chengdu, pp 1–6 (in Chinese with English summary)
Neben W, Krueger HH (1973) Fossilien ordovicischer und silurischer Geschiebe. Staringia 2: 69 pp
Neben W, Krueger HH (1979) Fossilien kambrischer, ordovizischer und silurischer Geschiebe. Staringia 5: 63 pp
Nitecki MH (1970) North American cyclocrinitid algae. Fieldiana Geol 21: 182 pp
Nitecki MH (1972a) Gametangia of Silurian *Ischadites hemisphericus* (Receptaculitaceae, Dasycladales). Phycologia 11:1–4
Nitecki MH (1972b) North American Silurian receptaculitid algae. Fieldiana Geol 28: 108 pp
Nitecki MH (1986) Receptaculitids and their relationship to other problematic fossils. In: Hoffman A, Nitecki MH (eds) Problematic fossil taxa. Oxford Univ Press, New York, pp 27–34
Nitecki MH, Johnson ME (1978) Internal structures of *Cyclocrinites dactioloides*, a receptaculitid alga from the Lower Silurian of Iowa. Fieldiana Geol 39:1–15
Nitecki MH, Bradof KL, Nitecki DV (1987) Literature of the receptaculitid algae:1805–1980. Fieldiana Geol NS 16:215 pp
Osgood RG, Fischer AG (1960) Structure and preservation of *Mastopora pyriformis*, an Ordovician dasycladacean alga. J Paleontol 34:896–902
Pia J (1920) Die Siphoneae verticillatae vom Karbon bis zur Kreide. Abh Zool Bot Ges Wien 11, 2:263 pp
Pia J (1927) Thallophyta. In: Hirmer M (ed) Handbuch der Paläobotanik, vol 1. Oldenbourg, München Berlin, pp 31–136
Raymond PE (1916) Expedition to the Baltic Provinces of Russia and Scandinavia. Part 1. The correlation of the Ordovician strata of the Baltic basin with those of eastern North America. Bull Mus Comp Zool Harv 56:177–286

Reed FRC (1912) Ordovician and Silurian fossils from the central Himalayas. Mem Geol Surv India Palaeontol Indica Ser 15, 7, 2:168 pp
Roomusoks A (1970) Stratigraphy of the Ordovician Viru and Khar'yu Series of Northern Estonia, 1. Tartu Gos Univ, 346 pp Tallin (in Russian)
Salter JW (1873) A catalogue of the collection of Cambrian and Silurian Fossils contained in the Geological Museum of the University of Cambridge. Cambridge Univ, 204 pp
Skjeseth S (1963) Contributions to the geology of the Mjøsa districts and the classical Sparagmite area in southern Norway. Norg Geol Unders 220:126 pp
Spjeldnæs N (1955) *Coelosphaeridium* (Chlorophyta Dasycladaceae) from the Caradocian beds of N. Wales. Norsk Geol Tidsskr 35:151–153
Stolley E (1896) Untersuchungen über *Coelosphaeridium*, *Cyclocrinus*, *Mastopora* und verwandte Genera des Silur. Arch Anthropol Geol Schleswig-Holstein 1, 2:177–282
Stolley E (1898) Neue Siphonen aus baltischem Silur. Arch Anthropol Geol Schleswig-Holstein 3, 1:40–65
Størmer L (1953) The Middle Ordovician of the Oslo region, Norway 1. Introduction to stratigraphy. Norsk Geol Tidssk 31:37–141
Valet G (1968) Contribution à l'étude des Dasycladales. 1-Morphogenese. N Hedwigia 16:21–82

Chapter 7

Dasycladalean Algae of the Palaeozoic and Mesozoic

G.F. Elliott[1]

Abstract

General compilative publications with good bibliographies for the Dasyclades of the Palaeozoic and Mesozoic are mentioned, followed by brief discussion of published literature on classification and changing trends in the characters to be considered for this, and their evaluation. The absence of available dasycladalean evidence between the Upper Permian and Middle Trias is discussed, and suggestions made for possible future research to remedy this. The influence of good and bad preservation on interpretation is illustrated. Examples are given of the usefulness of dasycladaleans in palaeoecology and palaeogeography. The classification of Deloffre is discussed.

1 Introduction

Dasyclads (the general noun for members of the chlorophyte algal order Dasycladales) are well represented in the Palaeozoic and Mesozoic. They appear to have been largely confined to shallow marine warm clearwater facies, as interpreted by comparison with the habitats of living survivors. Nevertheless, they are common enough at many levels to be useful microfossils for stratigraphic purposes, and they have been the subject of very numerous studies. Inevitably the dasyclads of the different parts of this long time span have received somewhat unequal treatment. Best served are the Jurassic and Cretaceous, where the monograph of Bassoullet et al. (1978) offers a ready entry and detailed guide to dasyclads of these periods. The Trias, in great part the subject of Julius Pia's earlier pioneer studies (1920, 1927) is not now served by a comparable contemporary work, though some regional studies such as that of Bystricky (1964) on the dasyclad limestones of eastern Czechoslovakia, go some way towards this, and Flügel's summarized study on the diversity and environments of Permian and Triassic Dasycladacean Algae, though not a taxonomic-descriptive study, contains a wealth of useful information and interpretation (Flügel 1985). For the Palaeozoic, the summary work of Roux (1985) deals with all algae of this age, thus including dasyclads. A mine of information, it can be supplemented by other studies; one thinks especially of the long series of regional papers by Mamet and his collaborators, largely but not exclusively on the Upper Palaeozoic. Finally, the published results of the *Groupe français d'étude des algues fossiles* (Bassoullet et al. 1975), on the different values of various dasyclad structures in classification, are leading to a revised classifi-

[1]Department of Palaeontology, British Museum (Natural History), Cromwell Road, London SW7 5BD, UK

cation long overdue. It is noteworthy that dasyclads, from the limited nature of the fossilized remains, do not usually lend themselves to an easy evaluation of phylogeny. Valet's study of living dasyclads (1968, 1969) shows how much is not normally preserved in the fossil state. The *Groupe Français* attempted to group all dasyclads in tribes, on the basis of their morphological similarities (Bassoullet et al. 1979). It is interesting that when the time-ranges are compared, a progression in time, admittedly with much overlap, is readily observed (Bassoullet et al. 1979, p. 431); i.e. the successive range-terminations progress from older to younger.

With all this in mind, we know that dasyclads appeared in the Lower Palaeozoic, proliferated in the Permian, and again in the Middle-Upper Triassic, Upper Jurassic, Lower Cretaceous, surviving in facies-abundance to the Palaeogene, after which they declined to their present-day "relict-flora" position. Throughout, they appear to have favoured the same general environment, so far as we can tell: shallow warm clear-water marine conditions, usually coastal, often lagoonal. Beginning as aspondyl (the branches set non-verticillate on the central axial body), they proceed to the verticillate branch arrangement, sometimes leading to adult terminal reproductive discs, or other structures. The branches themselves show an astonishing number of patterns, which are invaluable taxonomically. Güvenç (1979) has carefully analyzed those with the metaspondyl type of branch, to show the affinities between the Palaeozoic and Triassic genera to indicate possible relationships. The basic trends towards this kind of evolution may have occurred separately in different stocks and at different times, as recognized by Kochansky-Devidé and Gušić (1971) for dasyclad structures generally. A good example is the Ordovician *Archaeobatophora* (Nitecki 1976), with branches comparable to those of the living *Batophora*, although contemporaneous with much more primitive genera. Our colleagues the French Group, in a major review of dasyclad classification characters, inclined towards a morphological-structure taxonomy (Bassoullet et al. 1975), replacing the older ideas of Julius Pia, who placed a principal emphasis on the positions of the reproductive structures, whether within the stem-cell, in swollen primary branches, or in special structures borne on these branches. Pia wrote before much later work, and the more recent changes in classification were long overdue. Indeed, there have been indications recently of different interpretation of reproduction structures, as in Skompski's (1984) account of the growth and function of *Kulikia* in the Carboniferous. Obviously, there is a great deal still to be discovered. What can be done, however, is to suggest what is desirable in future research, and this fits with the title wording of this chapter: Palaeozoic and Mesozoic dasyclads. How do the two fit together?

2 The Palaeozoic/Mesozoic Junction

It is generally accepted that the level at which top Permian underlies bottom Trias is the record of a time of profound changes for life on Earth. Many and various explanations, of varying degrees of credibility, have been advanced to account for this. So far as dasyclads are concerned, a rich flora with *Mizzia* as its key plant ceases abruptly as the Permian stops. The detailed summarized study of Flügel (1985), whilst not primarily on the component taxa of the Upper Permian/Triassic dasyclad floras, contains much useful information on them. Above the Permian there comes in most places a lower Triassic of largely arenaceous rocks with molluscs, but apparently without dasyclads. Then, in the right facies of the Middle Triassic limestones, dasyclads appear again, as diplopores in incredible abundance (they were the *Halimeda* of their time, ecologically speaking). Whilst it is true that some diplopores have been described from the Upper Palaeozoic, and their possible ancestry and relationships evaluated, the contrast between the two floras is striking. One asks, what were the dasyclads doing, and where?

This is, I think, a very real gap in our knowledge. To fill it, one needs to find a Lower Trias in algal-limestone facies. Two possibilities can be suggested. One is in the Trucial Oman area of Arabia. About 25 years ago, the now defunct Iraq Petroleum Group, by whom I was then employed, prospected there and carried out a basic geological survey of the local succession. Both palaeontologically determined Permian and Triassic occur, in a long succession of limestones. The key formation for our problem is the Ghail Limestone, ca. 600 m thick (Hudson 1960). Although I described an alga (not a dasyclad) from the Upper Ghail (Elliott 1964), the lower part was not intensively sampled, although noted as algal. Moreover, such samples as were taken never reached me, and it was too early in my algal career to realize fully the possible importance of these. The other area, which I have never worked on, is the Western Himalayan/Central Asian region, where the impact of the drifting Indian plate on the Asian mass has possibly preserved a similar rock.

If one takes the conventional, unexciting but likely view that late-Permian shallow-marine life suffered from a geologically sudden dearth of suitable shelf-sea facies, and that this continued in the Lower Trias, then occasional dasyclad survivors had a wonderful opportunity to explode evolutionary-wise when good times returned again, and this they seem to have done. If a dasyclad-yielding Lower Trias exists, then it may hold the key to the dasyclad transition from Upper Permian to Middle Trias, and it may profoundly affect thinking regarding the title of this chapter. One would then see which genera survived into Triassic time, which of them did not flourish, and the nature of these early Triassic dasyclads in relation to the later floras known to us.

By comparison, the change from Mesozoic to Cenozoic was apparently much less abrupt for algae. Such dasyclad genera as *Cymopolia* and *Neomeris*, both surviving today, are well represented in both Upper Cretaceous and Palaeocene.

3 Some Problems

Apart from this major question, the study of new dasyclads (and of better preserved material of known genera), builds up our knowledge of these fascinating plants. That it sets us new problems as well is part of the attraction of these fossils. Skompski (1987), working on exceptionally well-preserved materials, has re-examined and discussed the problematic organisms known as Palaeoberessellae, which have been referred by different workers to the dasyclad algae, to the Foraminiferidae and to the extinct Ischyrospongia (summarized in Skompski 1987). He concludes that they are dasyclads, and discusses their position within the order Dasycladales. Preservation after diagenesis plays a great part in the appearance of certain structures, and the necessity for allowing for this has not always been obvious to investigators. We look forward to extending our knowledge from well-preserved known taxa, as well as from new taxa.

It will be seen from the above that our knowledge of the Dasyclads of the Palaeozoic and Mesozoic is in an exceptional state of flux. Much is unknown, with insufficient certainty to avoid controversy in an overall view. But the progress of new researchers should, we hope, remedy this.

4 Palaeogeography, Palaeoecology

As mentioned above, dasyclads are useful fossils for palaeoecology (and palaeogeography). Their occurrence globally in Mesozoic Tethyan sediments has been used by Elliott, not merely to show the probable origins of their current distribution (Elliott 1981), but also their occurrence and distribution in seas marginal to the Tethys in what is now northwestern Europe (Elliott 1977, 1984, 1986). In the Palaeozoic, Poncet (1987) has studied the palaeogeography of *Vermiporella* in the Middle and Upper Ordovician.

5 Deloffre's Classification, 1988

The dasyclad classification of Deloffre (1988) is an important landmark in dasyclad studies. He divided the Dasycladales into five families: Beresellaceae, Seletonellaceae, Diploporaceae, Dasycladaceae and Acetabulariaceae, covering about 180 genera arranged in 30 tribes. The two new families are welcome, as recognizing the special characters of their component genera. Beresellaceae form a group somewhat set apart from other dasyclads, and the character given as distinctive for the family is "Tubular algae with or without septation". It will be interesting to see if this definition is widely accepted or not. In contrast, the criteria for separation of Diploporaceae from Dasycladaceae – metaspondyle thallus, and euspondyle thallus with verticils similar, respectively – is in accord with recent acceptance of primary branch form as significant in classification. It should be borne in mind that whilst logically one uses the same uniform criteria for classification in different related taxa, the extent of evolution varies in these characters in different groups, necessitating some modification if classification is to reflect the evolution that has taken place. However, the separation of Diploporaceae from Dasycladaceae is a useful step towards the understanding of dasyclad evolution of the Upper Palaeozoic/Lower Mesozoic. The characters used for classification are selected-morphological, and follow naturally from the work of the Groupe Français.

Many students of the dasyclads will find genera with whose tribal allocation by Deloffre they will not agree; they will from their personal knowledge place them elsewhere, or at any rate query them. This does not detract from the value of his attempt, in view of our incomplete knowledge of so many genera. He recognizes the possibility of an eventual phylogenetic classification: the most likely family to show this is the Acetabulariaceae, whose component genera may well be of different ancestries, so making the family polyphyletic, but we do not know this for a fact.

It is usual for *Dasycladus*, which gives its name to the Order, Family and subtribe, to give its name to the appropriate tribe also, rather than *Triploporella*, even if this is a well-calcified fossil genus.

Deloffre's work is a useful reference paper for all workers on dasyclads, and should be a useful tool for future efforts as knowledge grows.

References

Bassoullet JP, Bernier P, Deloffre R, Génot P, Jaffrezo M, Poignant AF, Segonzac G (1975) Réflexions sur la systématique des Dasycladales fossiles. Etude critique de la terminologie et importance relative des critères de classification. Geobios 8, 4:259–290

Bassoullet JP, Bernier P, Conrad MA, Deloffre R, Jaffrezo M (1978) Les algues Dasycladales du Jurassique et du Crétacé. Geobios Mem Spec 2:330 pp

Bassoullet JP, Bernier P, Deloffre R, Génot P, Jaffrezo M, Vachard D (1979) Essai de classification des Dasycladales en tribus. Bull Cent Rech Explor Prod Elf-Aquitaine 3, 2:429–442

Bystricky J (1964) Slovensky kras Stratigrafia a Dasycladaceae mesozoika Slovenského krasu. Ustredny Ustav Geol, Bratislava, 204 pp

Deloffre R (1988) Nouvelle taxonomie des algues dasycladales. Bull Cent Rech Explor Prod Elf-Aquitaine 12:165–217

Elliott GF (1964) *Zonotrichites* (calcareous algae) from the Arabian Triassic. Eclogia Geol Helv 57 2:567–570

Elliott GF (1977) Inferred isocrymal distribution of Jurassic dasycladacean Algae in Europe, north Africa and southwestern Asia. J Geol Soc London 133:363–373

Elliott GF (1981) The Tethyan dispersal of some chlorophyte algae subsequent to the Palaeozoic. Palaeogeogr Palaeoclimatol Palaeoecol 32:341–358

Elliott GF (1984) Climatic tolerance in some aragonitic green algae of the post-Palaeozoic. Palaeogeogr Palaeoclimatol Palaeoecol 48:163–169

Elliott GF (1986) English Jurassic dasyclads: seaweed indicators of marine climate. Geol Today 2, 1:20–23

Flügel E (1985) Diversity and environments of Permian and Triassic dasycladacean Algae. In: Toomey DF, Nitecki MH (eds) Paleoalgology: contemporary research and applications. Springer, Berlin Heidelberg New York Tokyo, pp 344–351

Güvenç T (1979) Dasycladacées metaspondyles du Paléozoique supérieur et du Trias. Bull Cent Rech Explor Prod Elf-Aquitaine 3, 2:625–637

Hudson RGS (1960) The Permian and Trias of the Oman Peninsula, Arabia. Geol Mag 97:299–308

Kochansky-Devidé V, Gušić I (1971) Evolutionstendenzen der Dasycladaceen mit besonderer Berücksichtigung neuer Fünde in Jugoslavien. Paläontol Z 45, 1–2:82–91

Nitecki MH (1976) Ordovician Batophoreae (Dasycladales) from Michigan. Fieldiana Geol 35, 4:29–40

Pia J (1920) Die Siphoneae Verticillatae vom Karbon bis zur Kreide. Abh Zool Bot Ges Wien 11, 2:263 pp

Pia J (1927) Thallophyta I. Abteilung. In: Hirmer M (ed) Handbuch der Paläobotanik. Oldenbourg Munich Berlin

Poncet J (1987) Paléobiogéographie du genre *Vermiporella* (algue verte calcaire) à l'Ordovicien moyen et supérieur. Ann Soc Geol Nord 106 3:279–283

Roux A (1985) Introduction à l'étude des algues fossiles paléozoiques (de la bactérie à la tectonique des plaques). Bull Cent Rech Explor Prod Elf-Aquitaine 9, 2:465–699

Skompski S (1984) The functional morphology of the Carboniferous dasycladacean genus *Kulikia*. N Jahrb Geol Paläontol Mh 7:427–436

Skompski S (1987) The dasycladacean nature of late Paleozoic palaeoberesellid algae. Act Geol Pol 37, 1–2:21–31

Valet G (1968) Contribution à l'étude des Dasycladales. 1. Morphogenèse. N Hedwigia 16:21–82

Valet G (1969) Contribution à l'étude des Dasycladales. 2. Cytologie et reproduction. 3. Révision systématigue. N Hedwigia 17:551–644

Chapter 8

Cenozoic and Recent Dasycladales

P. GÉNOT[1]

Abstract

Cenozoic Dasycladales include about 43 genera and 200 species. The number of living genera is reduced to 8, with about 40 species.

Calcification encloses sterile and fertile organs to a varying degree, according to genera, species, and individuals. Consequently, preservation of fossils mainly depends on the extent of the initial calcification around the thallus.

Main characters commonly used to define taxa include the type and the position of the reproductive organs, absence or presence of articulation, and division of branches.

A large number of species, belonging to the genera *Neomeris*, *Cymopolia*, and *Acicularia*, is known in Palaeogene sediments. The richest cenozoic assemblages have been found in Sardinia (Palaeocene) and in the Paris basin (Palaeocene and Eocene).

Living representatives have a discontinuous geographical distribution. They are confined to shallow warm marine or brackish waters.

1 Morphology

The thallus of Cenozoic and living Dasycladales (Table 1) is usually cylindrical or club-shaped, undivided or dichotomously branched. Other species are spherical or appear like a disc at the top of an erect axis. In some genera, the thallus is composed of alternate calcified segments and uncalcified joints. The whole plant is attached to the substrate by means of uncalcified rhizoids that are never preserved in fossils.

Primary branches are arranged in regular whorls along the tubular axis (Figs. 3,10–11). Most Cenozoic and Recent species bear two orders of branches: each primary branch produces one (*Montiella*) to eight (*Bornetella*, *Cymopolia*) second order sterile branches. Dilated tips of primaries (*Zittelina*) or secondaries (*Cymopolia*, *Neomeris*, *Bornetella*) may form a faceted outer surface (Fig. 5). Some genera bear more than two orders of branches, exceptionally seven (Table 2).

In Dasycladaceae, fertile organs are spherical or ovoid vesicles usually located at the top or on the sides of primary sterile branches. They are considered as modified branches, assigned to the reproductive function. Each ampulla contains one to several cysts or produces gametes directly (Table 3).

In Acetabulariaceae, fertile ampullae are elongated and radially arranged in specialized verticils. Many cysts occur inside reproductive chambers (Tables 4–6) (Valet 1969; Tappan 1980).

[1] Laboratoire de Biogéologie et Biostratigraphie, Université de Nantes, 44072 Nantes Cedex, France

Table 1. Shape of thallus in Cenozoic and Recent Dasycladaceae

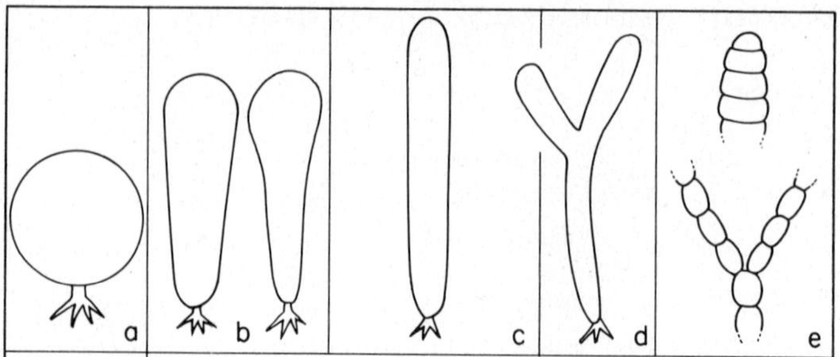

		spherical	club - shaped to cylindrical			
			non - divided		divided	articulated
Photographs		Fig.9	Figs.7-8	Figs.1-2		Figs.5-6
GENERA (examples)	RECENT	Bornetella		Dasycladus Neomeris	Dasycladus	Cymopolia
	CENOZOIC		Dactylopora Trinocladus Zittelina	Dissocladella Jodotella Neomeris Trinocladus Zittelina		Belzungia Broeckella Cymopolia Uteria

Fig. 1. *Neomeris annulata* Dickie, Recent, Rose Island, Bahamas. x7,5. External view of the thallus. **Fig. 2.** *Neomeris fragilis* (Defrance), Lutetian, Grignon, Paris Basin. x35. Transversal section showing axial calcification **Fig. 3.** *Neomeris annulata* Dickie, Recent, Rose Island, Bahamas. x35. Central axis (*right*), primary branches and fertile ampullae (*left*) **Fig. 4.** *Neomeris* sp., Lutetian, Campbon, Brittany. x95. Calcification of fertile ampullae **Fig. 5.** *Cymopolia barbata* (Linné), Recent, Andros Island, Bahamas. x25. Segments of the thallus **Fig. 6.** *Cymopolia* sp., Stampian, Ormoy-la-Rivière, Paris Basin. x40. Isolated segment **Fig. 7.** *Bornetella oligospora* Solms-Laubach, Recent, Cauda, Vietnam. x6. External view of the club-shaped thallus **Fig. 8.** *Dactylopora cylindracea* Lamarck, Auversian, Le Fayel, Paris Basin. x11. External view of the club-shaped thallus **Fig. 9.** *Bornetella sphaerica* (Zanardini), Recent, Oahu, Hawaii. x7. Central axis, branches and fertile ampullae (dry specimen) **Figs. 10–11.** *Neomeris auversiensis* (Morellet), Auversian, Auvers- sur-Oise, Paris Basin. Calcification of the central axis at the top (Fig. 10, x110) and at the base (Fig. 11 x190) of the thallus

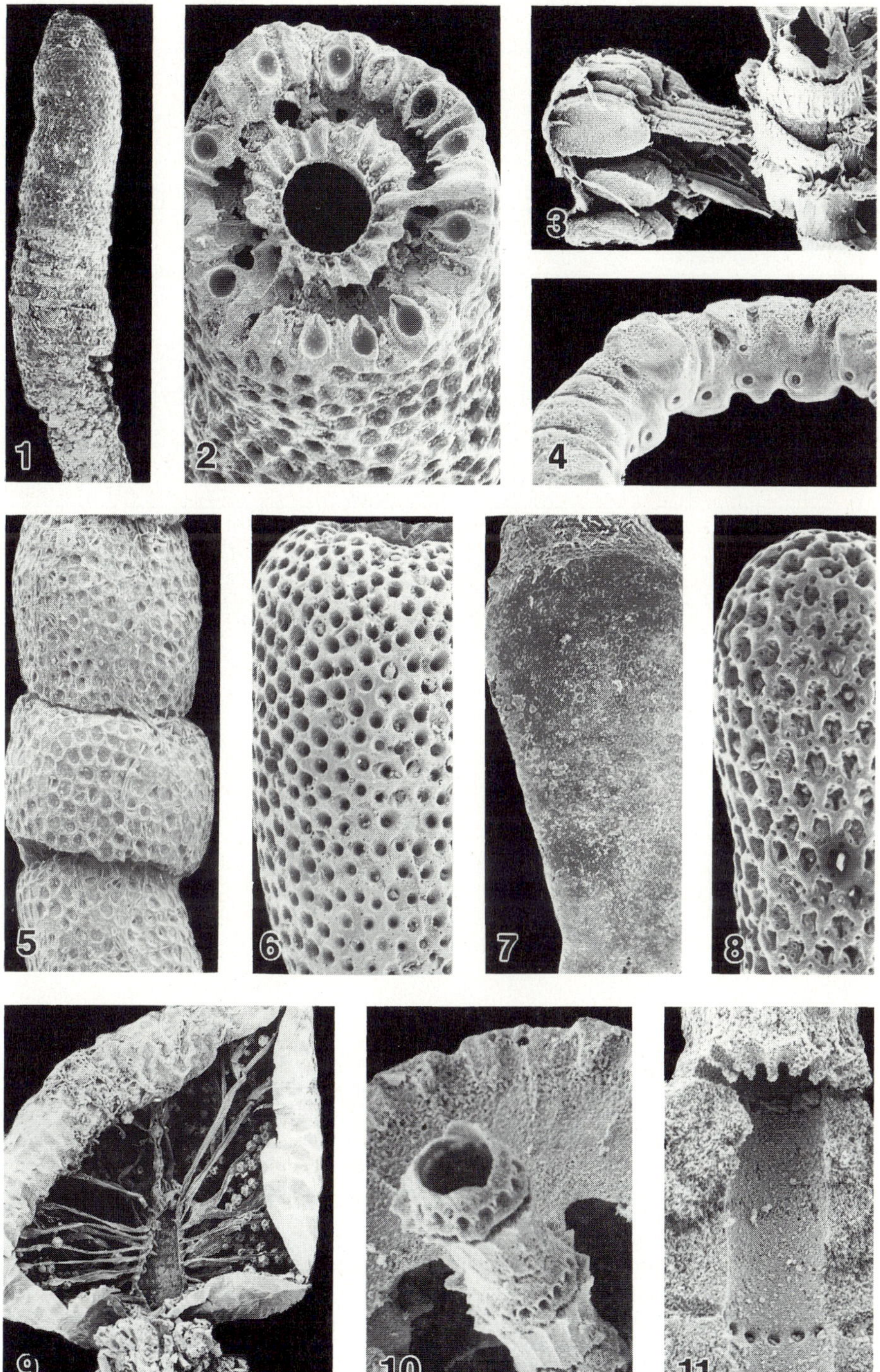
1
2
3
4
5
6
7
8
9
10
11

Table 2. Division of branches in Cenozoic and Recent Dasycladaceae

	a	b	c	d
	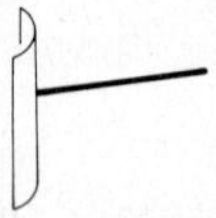 no division	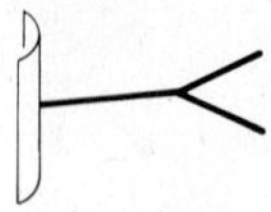 1 division	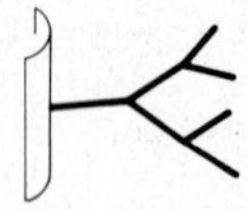 2 divisions	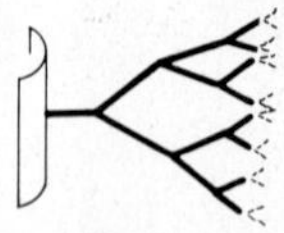 more(up to 6)
Photographs	Fig.12	Figs.13,16,17		Fig.14
GENERA (examples) RECENT		Bornetella Cymopolia Neomeris	Dasycladus	Batophora
GENERA (examples) CENOZOIC	Dactylopora Zittelina	Broeckella Cymopolia Dissocladella Jodotella Neomeris	Trinocladus	Belzungia Thyrsoporella

Fig. 12. *Zittelina dactyloporoides* (Morellet), Auversian, Baron, Paris Basin. x130. Primary branch **Fig. 13.** *Cymopolia elongata* (Defrance), Lutetian, Coislin, Brittany. x250. Primary and secondary branches **Fig. 14.** *Belzungia terquemi* Morellet, Lutetian, Campbon, Brittany. x220. Numerous divisions of the branches **Fig. 15.** *Bornetella nitida* (Harvey), Recent, Palawan, Philippines. x170. External view of the upper faces of two secondary branches **Fig. 16.** *Cymopolia barbata* (Linné), Recent, Andros Island, Bahamas. x190. Primary branch, secondary branches and fertile ampulla **Fig. 17.** *Cymopolia zitteli* Morellet, Auversian, Le Fayel, Paris Basin. x125. Primary branches, secondary branches and fertile ampullae **Fig. 18.** *Neomeris larvarioides* (Morellet), Lutetian, Le Bois-Gouët, Brittany. x110. Primary branch, secondary branches (partially preserved) and fertile ampulla **Fig. 19.** *Neomeris herouvalensis* Steinmann, Cuisian, Hérouval, Paris Basin. x90. Secondary branches and fertile ampullae (primary branches are not calcified) **Fig. 20.** *Batophora oerstedii* Agardh, Recent, Key West, Florida. x45. Fertile ampulla **Fig. 21.** *Parkerella* sp., Montian, Mons, Belgium. x190. Fertile ampullae arranged in pairs **Fig. 22.** *Carpenterella jonesi* Morellet, Lutetian, Campbon, Brittany. x250. Tuft of fertile ampullae

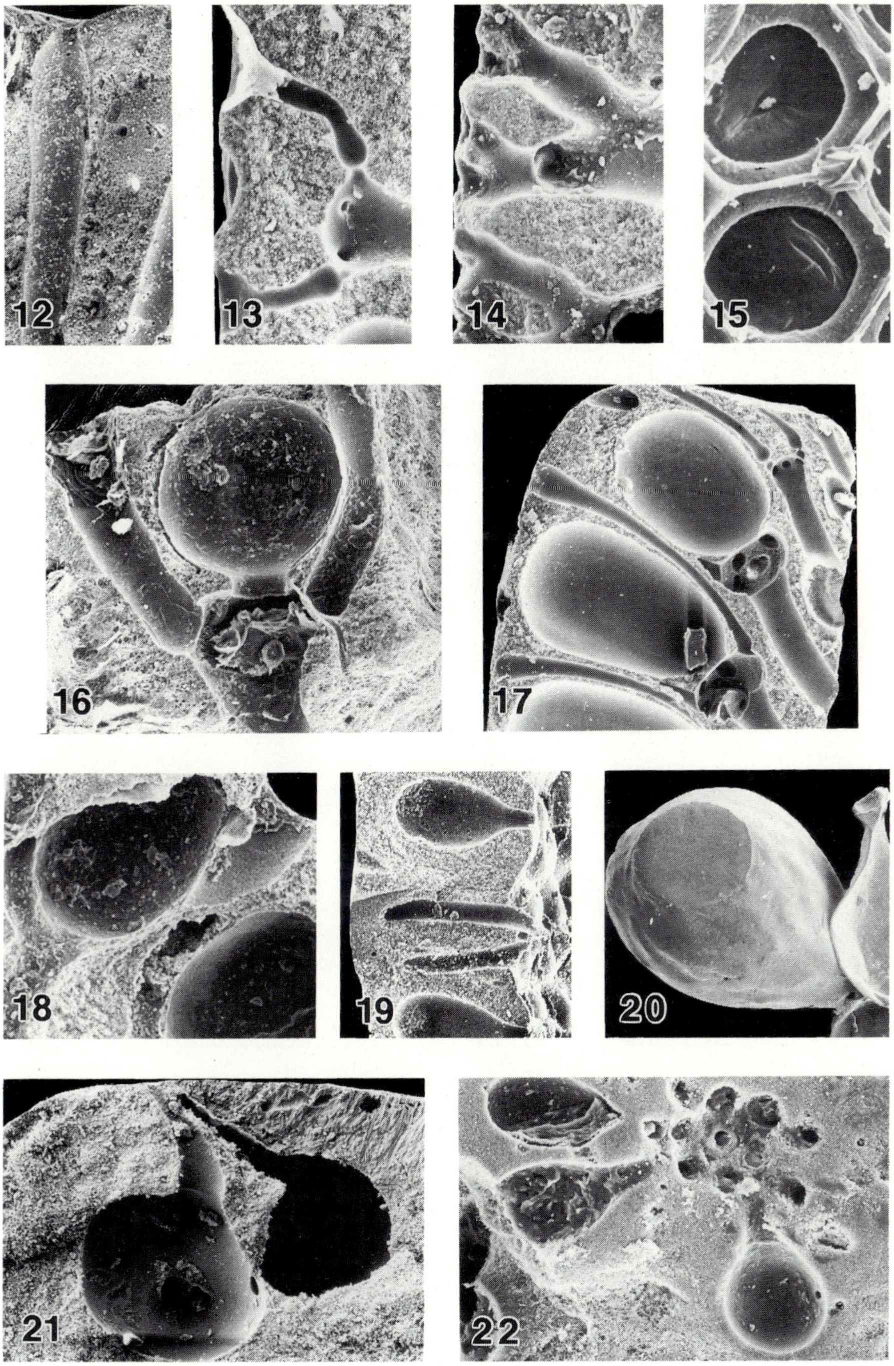
12
13
14
15
16
17
18
19
20
21
22

Table 3. Fertile organs in Cenozoic and Recent Dasycladaceae

A - Fertile organs at the top of branches	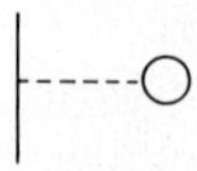 a	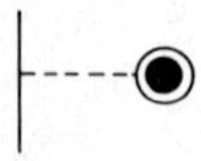 b	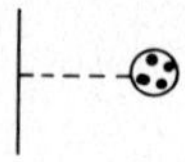 c	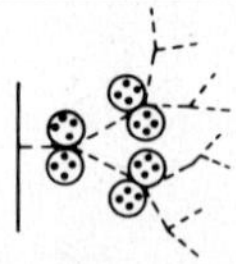 d
	one ampulla on each primary branch: no cyst	one cyst	cysts	numerous ampullae and cysts
Photographs	Figs.16-17	Figs.3,18,19		Fig.20
GENERA RECENT	Cymopolia Dasycladus	Neomeris (cyst unknown in fossils)	Chlorocladus	Batophora
GENERA CENOZOIC	Cymopolia Montiella			

B - Fertile organs around the sides of branches	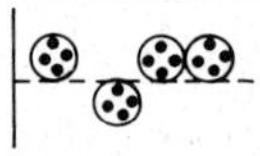 e	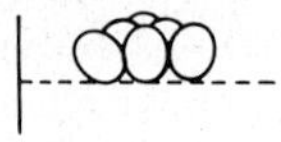 f
	Numerous cysts	Tufts of organs without cyst
Photographs	Figs.9,23	Figs.21-22
GENERA RECENT	Bornetella	
GENERA CENOZOIC	Zittelina (including Maupasia, Digitella)	Carpenterella, Jodotella Parkerella

C - Fertile organs inside branches	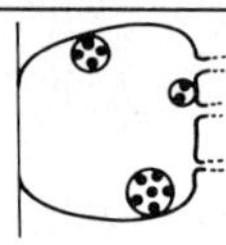 g	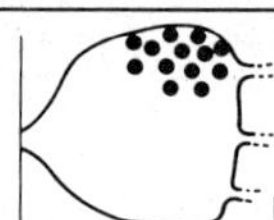 h	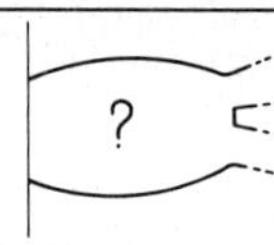 i
	Ampullae and cysts	Cysts	Non-calcified organs ?
Photographs	Fig.26	Fig.27	Fig.28
CENOZOIC GEN.	Uteria	Broeckella	Thyrsoporella

D - Particular cases	Grouped or scattered cysts around branches 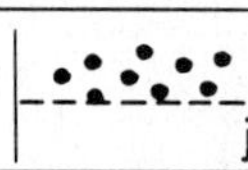 j	Isolated ampullae  k
Photographs	Fig.24	Fig.25
CENOZOIC GEN.	Dactylopora	Terquemella Frederica

Fig. 23–32. (see pp 138 for caption)

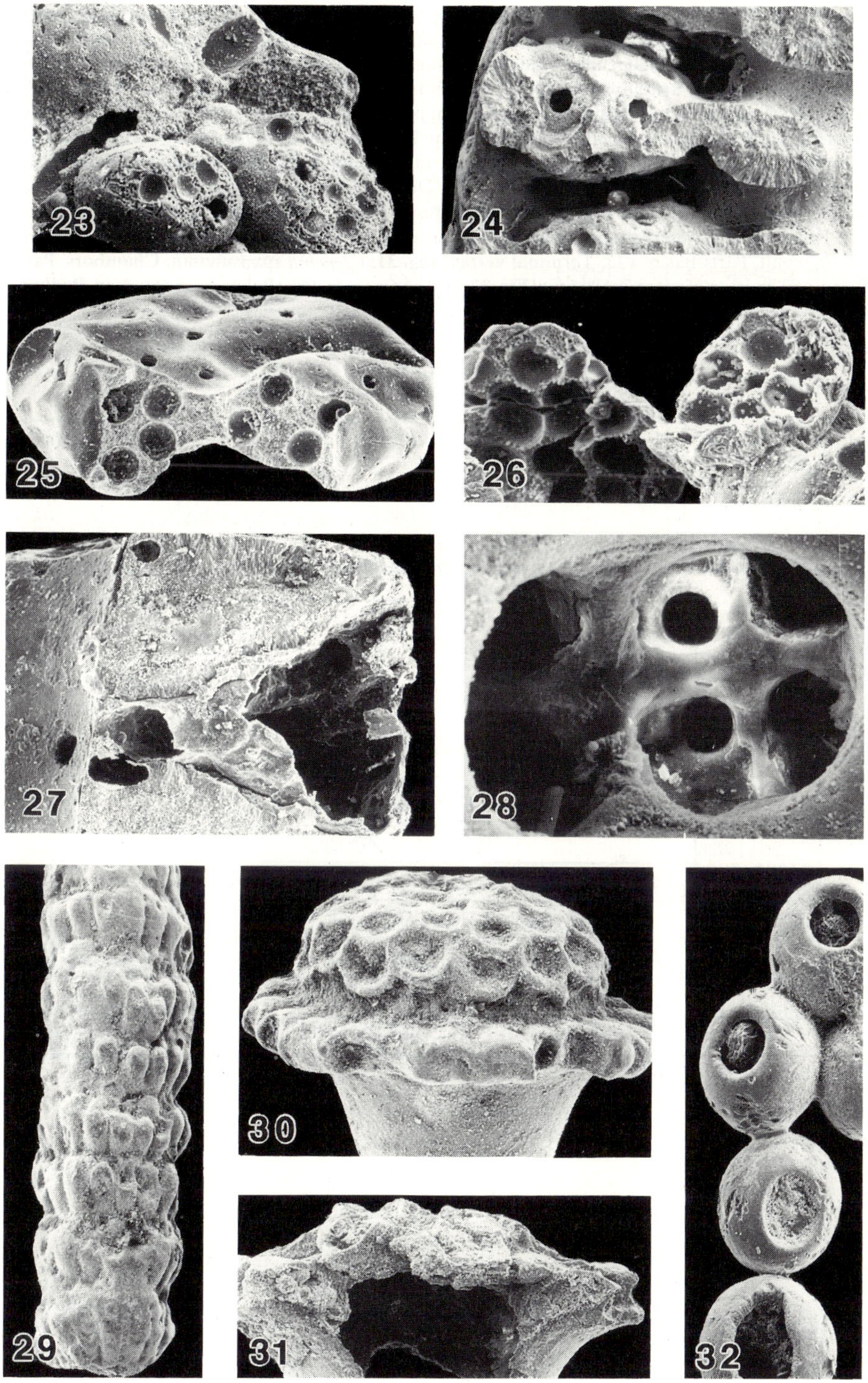
23
24
25
26
27
28
29
30
31
32

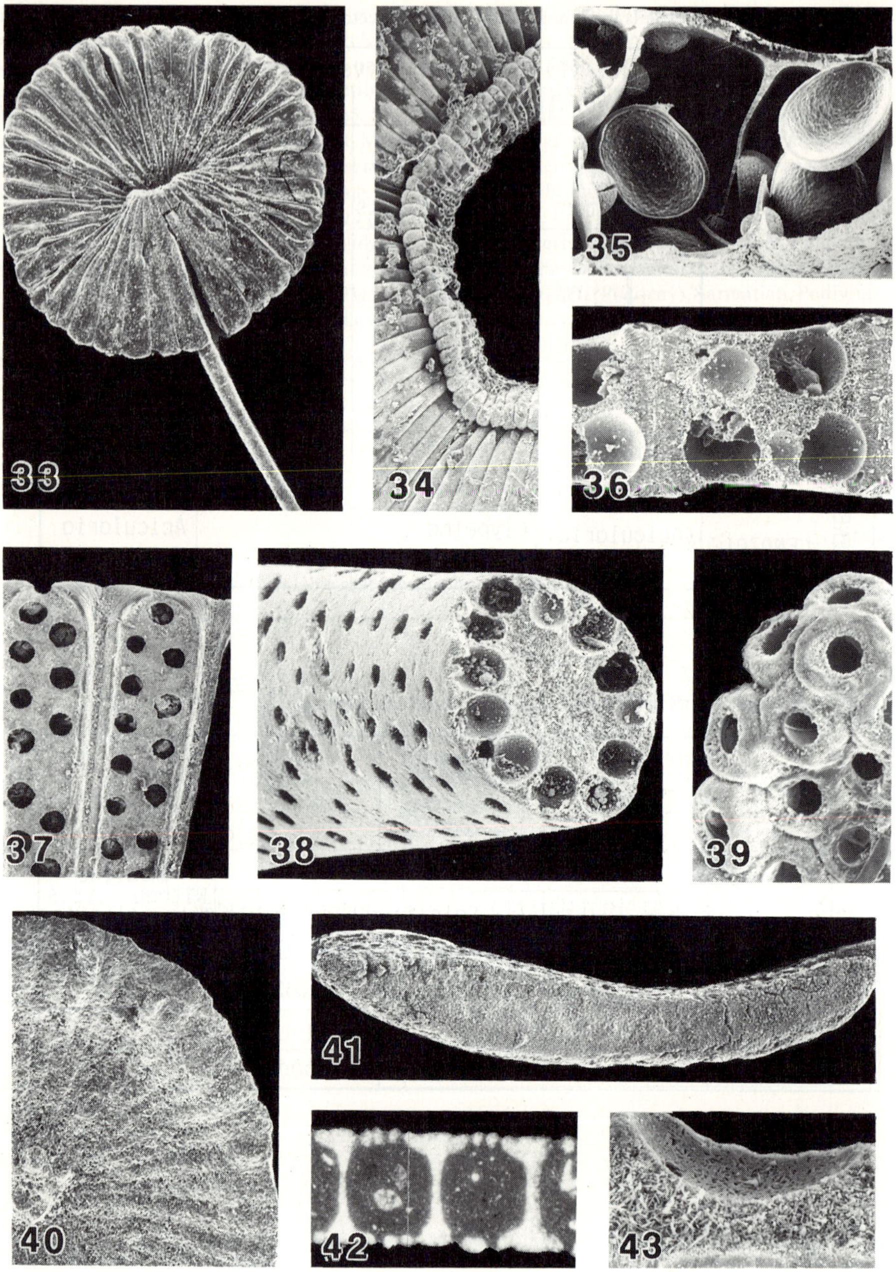
33
34
35
36
37
38
39
40
41
42
43

When fertile organs are unknown in fossils, thick primary branches are usually presumed to be reproductive structures.

In some particular cases the calcified wall may contain groups of cysts or scattered cysts along the branches (Table 3D).

2 Calcification

Aragonitic calcification covers the outer layer of the organs. In fossil species, aragonite may be diagenetically replaced by calcite.

Calcification encloses organs to a varying degree; for example, in *Neomeris*, calcium carbonate impregnates different parts of the plant (Figs. 2–4, 10–11): almost the entire thallus in *N. larvarioides*, only reproductive organs in *N. pseudo-eruca* (Génot 1987).

Some living Dasycladales are never calcified (*Batophora*).

3 Definition of Taxa

The main characters commonly used to define Cenozoic and Recent subfamilies, tribes, and genera are: type of reproductive organs (single or clustered ampullae without cysts, or containing one or several cysts); position of reproductive organs; type of thallus (absence or presence of articulation); division of branches. Taxonomic importance of the shape of the thallus and aspect of the outer surface is variable according to authors (specific or generic level).

4 Stratigraphical and Geographical Distributions

A few genera that arose during the Mesozoic (*Acicularia*, *Acroporella*, *Clypeina*, *Cymopolia*, *Dissocladella*, *Neomeris*, *Praturlonella*, *Trinocladus* and *Triploporella*) have Cenozoic representatives. Two of them are known in present seas:

Fig. 33. *Acetabularia crenulata* Lamouroux, Recent, Barranquilla, Colombia. x6. Axis and reproductive cap **Figs. 34–35.** *Acetabularia acetabulum* (Lamouroux), Mediterranean coast, France. Upper corona in the central part of the cap (Fig. 34, x40) and internal view of two rays showing the cysts (Fig. 35, x140) **Fig. 36.** *Acicularia munieri* Morellet, Lutetian, Grignon, Paris Basin. Internal view of a ray showing the location of the cysts **Fig. 37.** *Acicularia* cf. *munieri* Morellet, Lutetian, Villiers-Saint-Frédéric, Paris Basin. x50. External view of two rays **Fig. 38.** *Acicularia pavantina* d'Archiac, Auversian, Le Guépelle, Paris Basin. x90. Transverse section of a ray showing the location of cysts **Fig. 39.** *Acetabularia schenckii* Moebius, Recent, Key Largo, Florida. x130. Group of calcified cysts **Figs. 40–42.** *Orioporella villattae* Segonzac, Thanetian, Carla-de-Roquefort, Pyrenees. External view of a reproductive cap (Fig. 40, x13,5). Longitudinal section (Fig. 41, x22,5). Transverse sections of two rays (Fig. 42, x32,5). **Fig. 43.** *Orioporella bonieri* Morellet, Montian, Mons, Belgium. x580. Detail of the internal view of a ray showing the location of a cyst

Cymopolia and *Neomeris*. About 25 genera are restricted to the Cenozoic, each one including very few species (Deloffre and Génot 1982).

Acicularia, *Cymopolia* and *Neomeris* have the most numerous species, particularly during Palaeocene and Middle Eocene (Tables 7–8).

The number of species greatly decreases after the Eocene. Some genera are short-lived, such as *Broeckella*, which is widespread in the Tethyan Palaeocene.

Richest Cenozoic assemblages (Table 9) have been discovered in the Palaeocene of Belgium (Mons basin), France (Aquitaine – Pyrenees area and Paris basin), Italy (Sardinia), USSR (Ukraine), Yugoslavia (Slovenia), Czechoslovakia (Carpathian Mountains), Middle East (Iraq, Iran) and China (Tibet), in the Eocene of France (Paris basin, Brittany, Cotentin), southern England (Sussex), Hungary and China (Xizang), and in the Miocene of Romania and Poland.

Only eight genera are living in modern oceans and seas, including about 40 species, mainly of *Neomeris* and *Acetabularia* (Table 7). They have a discontinuous geographical distribution in the Indo-Pacific, Atlantic (tropical America and West Indies) and Mediterranean areas.

5 Environmental Distribution

Most living Dasycladales are growing in tropical and subtropical marine waters. Few species occur in warm-temperate seas (Valet 1979). They are confined to shallow environments, usually less than 5 metres depth. Maximum abundances occur in quiet areas sheltered from wave action. Individuals are attached to rocks, small stones, shell or coral fragments, mangrove roots, etc. Some species, such as *Halicoryne spicata* and *Acetabularia mediterranea*, tolerate substantial salinity fluctuations. *Batophora oerstedii* is found in marine and brackish environments, even sometimes in freshwater.

Ecological requirements of Cenozoic Dasycladales seem to be the same as those of living representatives (Elliott 1984). Richest assemblages have been found in sediments and with other organisms that suggest shallow warm marine environments: lagoonal waters behind reefs, coastal bays, etc. [Palaeocene-Lower Eocene of the Middle-East; Lutetian of the Paris basin (Génot 1987); Palaeocene of Sardinia (Dieni et al. 1985)]. No fossil species is known in freshwater deposits.

Table 7. Proportional representation of species number in Cenozoic and Recent Dasycladales genera

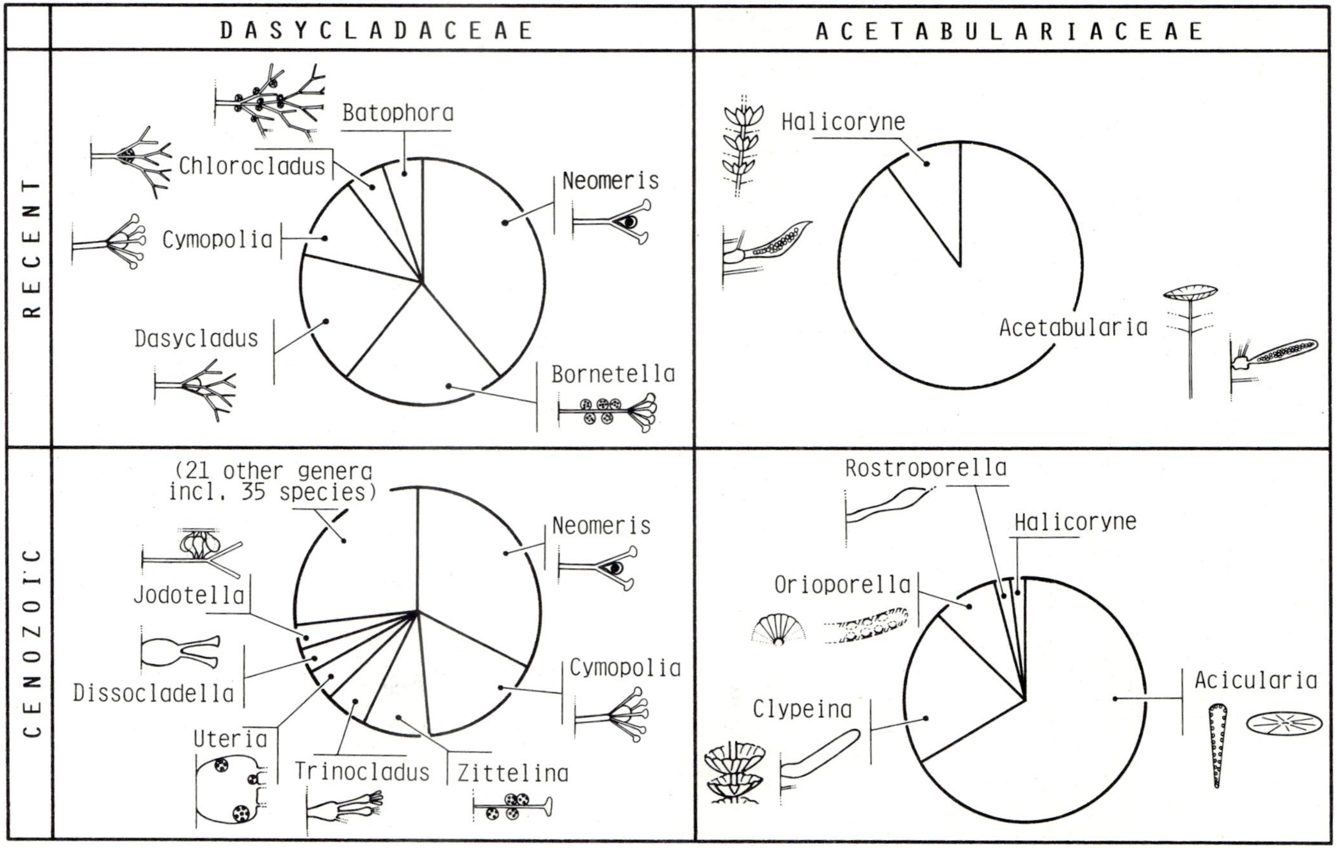

Table 8. Main Dasycladales genera: evolution of the number of species during the Cenozoïc

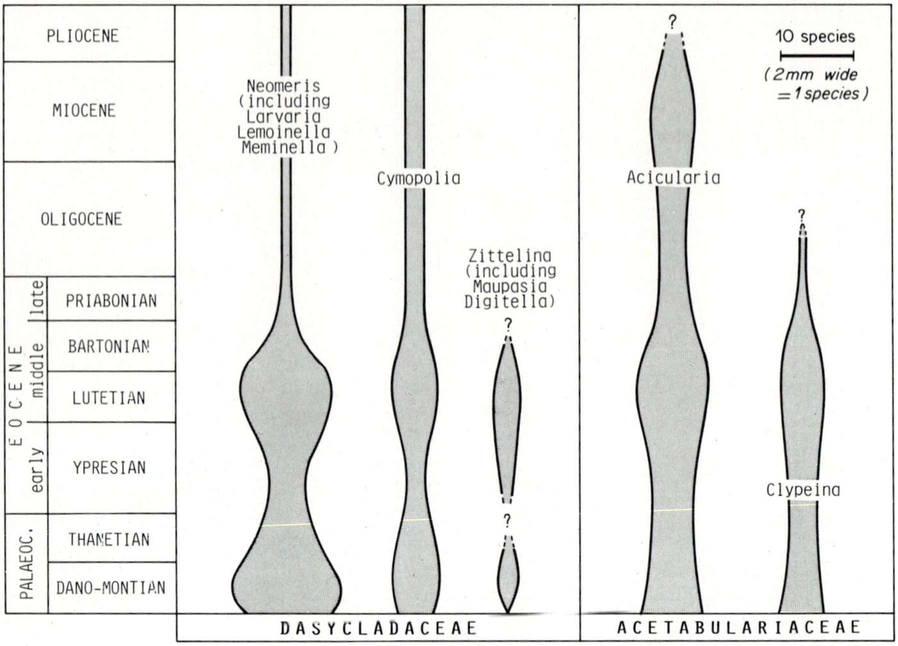

Acknowledgements. I am very grateful to M. Hauray and A. Barreau for scanning electron micrographs and A. Cossard for drawings (University of Nantes).

References

Deloffre R, Génot P (1982) Les Algues Dasycladales du Cénozoique. Mem Cent Rech Explor Prod Elf-Aquitaine 4:247 pp, 20 pl

Dieni I, Massari F, Radoičić R (1985) Palaeocene dasycladalean algae from Orosei (eastern Sardinia). Mem Sc Geol Inst Geol Miner Univ Padova, XXXVIII, 22 pl

Elliott GF (1984) Climatic tolerance in some aragonitic green algae of the Post-Palaeozoic. Palaeogeogr Palaeoclimatol Palaeoecol 48:163–169

Génot P (1987) Les Chlorophycées calcaires du Paléogène d'Europe Nord-Occidentale (Bassin de Paris, Bretagne, Cotentin, Bassin de Mons). Thèse Doctorat d'Etat, Nantes, 518 pp, 48 pl

Tappan H (1980) Order Dasycladales. In: Freeman W (ed) The paleobiology of plant protists. Freeman San Francisco, pp 860–912

Valet G (1969) Contribution à l'étude des Dasycladales, pt 2,3. N Hedwigia 17:551–644, 133–162 tab

Valet G (1979) Approche paléoécologique du monde des Dasycladales à partir de l'écologie des formes actuelles. Bull Cent Rech Explor Prod Elf-Aquitaine 3, 2:859–866

Table 9. Dasycladales: main discoveries of Palaeogene species in the world

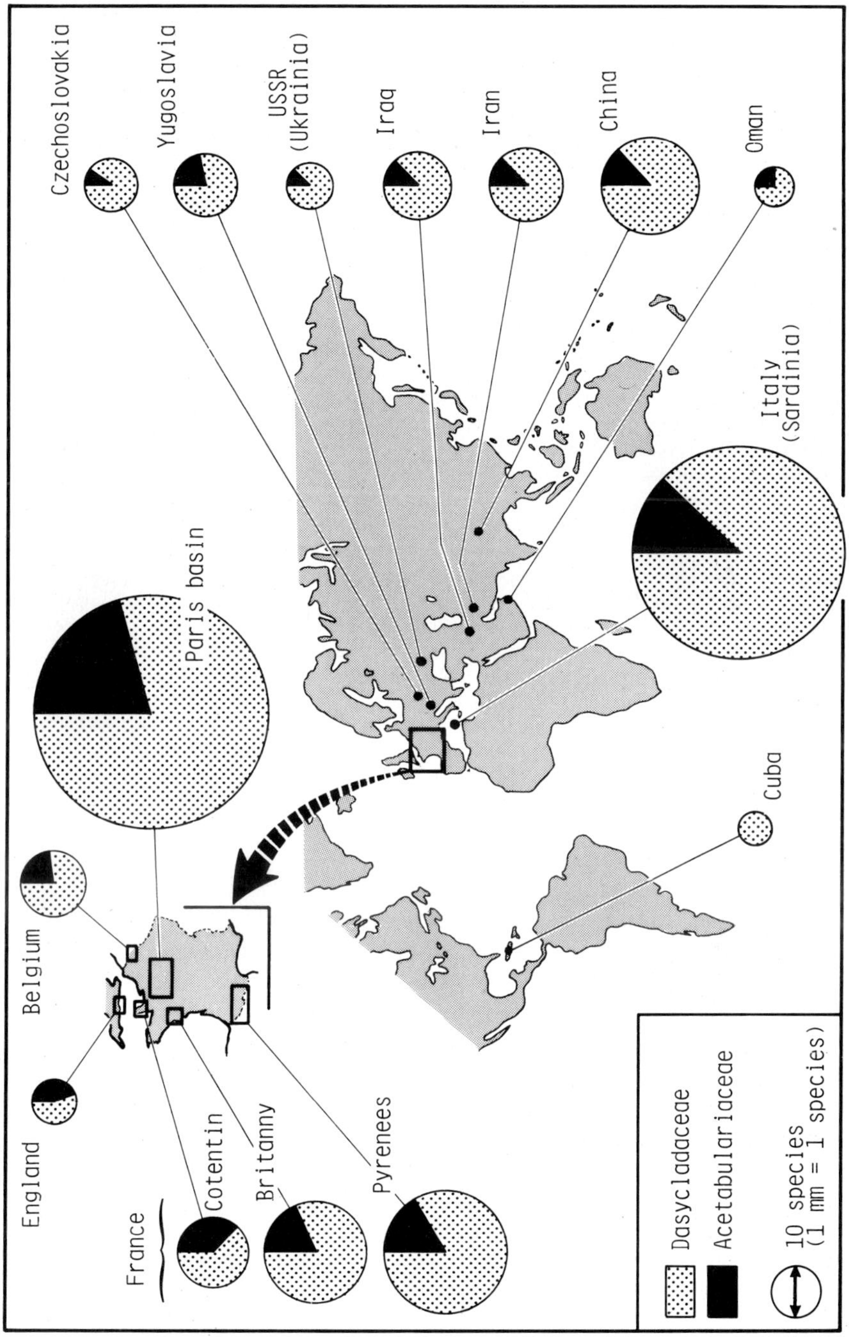

Chapter 9

Fossil Udoteaceae and Gymnocodiaceae

X. Mu[1]

Abstract

The Gymnocodiaceae and fossil Udoteaceae (erect) are similar to each other both in growth form and in vegetative structure. They differ in that the Gymnocodiaceae contain internal reproductive organs which are usually absent in the Udoteaceae. It is generally accepted that the Gymnocodiaceae belong to the red algae and the Udoteaceae to the green algae. Well-preserved silicified Gymnocodiaceae can reveal some details of cell filaments, reproductive organs and parasitism. However, the systematic position of this family still remains to be resolved. The published records of fossil Udoteaceae are more abundant, but their biological characters are less well known and much discrepancy exists in their nomenclature, definition and content. In view of the fact that fossil material can only provide limited data on the anatomy of the algae and no information is available at present about pigments, storage products, chemical characters of cell wall or reproduction which are critical to the classification of living algae, and that parallelism is frequently found in different groups of living algae, the Gymnocodiaceae and fossil Udoteaceae are considered heterogeneous and in practice it is often difficult to distinguish between them. In order to advance our understanding of these important groups, search for better preserved permineralized fossil material and close cooperation between students of fossil and living algae are badly needed.

1 Introduction

Udoteaceae and Gymnocodiaceae are important groups of calcareous algae. The former are green algae, represented by both fossil and Recent forms widely distributed in tropical (or subtropical) shallow seas, and their history can be traced back to the Ordovician; the latter is an extinct family, which is generally considered to belong to the red algae, ranging from the late Paleozoic to the Early Cenozoic(?) with an ecological distribution roughly similar to that of the Udoteaceae.

Gymnocodiaceae are characterized by an erect, branched or unbranched, segmented or nonsegmented, thallus composed of medullary and cortical zones with reproductive organs located in the peripheral part. The erect Udoteaceae, which predominate in the family, have much in common with Gymnocodiaceae except that the reproductive organs are usually not preserved in the former and in practice it is often difficult to distinguish between them. For this reason they are treated here together.

[1]Nanjing Institute of Geology and Palaeontology, Academia Sinica, Chi-Ming-Ssu, Nanjing, People's Republic of China

2 Fossil Udoteaceae

Although the history of studies on fossil Udoteaceae (long called Codiaceae) can be traced back more than a hundred years, it was not until 1926–1927, when Pia incorporated *Boueina* and *Palaeoporella* together with *Gymnocodium* into the Codiaceae and pointed out their importance as rock builders, that the concept of a group of fossil Udoteaceae became rooted in the literature. Since then the fossil records of the family have steadily increased with new genera and species being added, until now more than 30 genera have been described as Udoteaceae; some of which, however, may not belong to the family. Konishi (1961) first summarized Palaeozoic Udoteaceae. Recently, Roux (1985) has made a synthesis for Palaeozoic algae in which fossil Udoteaceae are included. A comprehensive summary of fossil Udoteaceae (erect) ranging from Palaeozoic to Cenozoic has been compiled by Bassoullet et al. (1983). The latter is an important source, although it is not exhaustive and some genera included may not belong to the Udoteaceae.

Morphologically, the fossil Udoteaceae are naturally represented by three categories: filamentous (nodular), phylloid, and erect. The following will centre on the last group, with only a brief account of filamentous forms when necessary.

3 Morphology and Anatomy

The morphology and anatomy of Recent Udoteaceae vary greatly, ranging from simple filamentous to erect segmented forms with tissue differentiated into medulla and cortex. The erect segmented forms are exemplified by *Halimeda* although some forms of *Halimeda* have a prostrate habit which is represented by both Recent and fossil taxa, and its morphology and anatomy have been used to make comparisons by analogy with other fossil forms. Hillis (1959) detailed the morphology and anatomy of living *Halimeda*, and Bassoullet et al. (1983) made a generalized schematic diagram of fossil erect Udoteaceae.

4 Growth Form

The growth forms of fossil erect Udoteaceae which also resemble those of the Gymnocodiaceae, may be branched (dichotomously or trichotomously) or unbranched, and segmented or non-segmented. Segments are variously shaped and may be cylindrical, oval, spherical, etc. The node (Hillis 1959), an important feature in living *Halimeda*, has not yet been recorded in fossil Udoteaceae.

5 Anatomy

The thallus is generally composed of a medulla and cortex (Fig. 1a). The medulla may be monoaxial or multiaxial. The medullary filaments are arranged

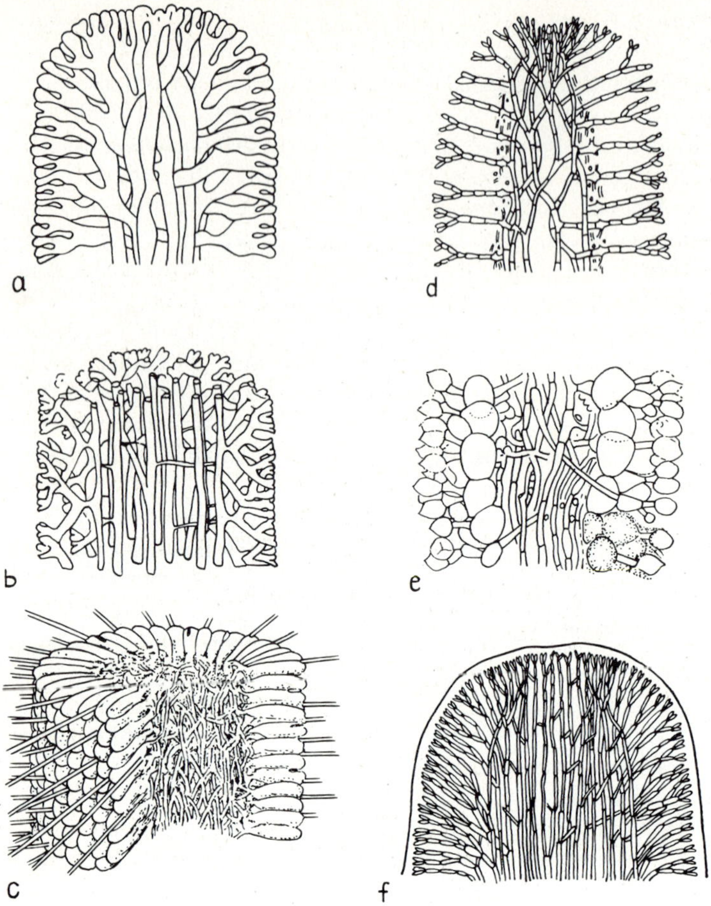

Fig. 1a-f. Convergence of internal organization in some living algae. **a** *Halimeda* (Udoteaceae, Chlorophyta); **b** *Penicillus* (Udoteaceae, Chlorophyta); **c** *Codium* (Codiaceae, Chlorophyta); **d** *Cumagloia* (Helminthocladiaceae, Nemalionales, Rhodophyta); **e** *Galaxaura* (Chaetangiaceae, Nemalionales, Rhodophyta); **f** *Furcellaria* (Furcellariaceae, Gigartinnales, Rhodophyta). **a** after Hillis-Colinvaux (1980); **b** after van den Hoek (1981); **c** after Tseng et al. (1962); **d**, **f** after Dixon, 1973; **e** after Svedelius (1953)

parallel to the axis of the thallus or interwoven or tangled, and bifurcate or trifurcate to give rise to lateral cortical filaments. The greatest variation in fossil Udoteaceae is represented by the cortical filaments, which are straight, swollen, or constricted, or show a combination of these (Fig. 2). They may branch several times before reaching the surface. The termination of lateral filaments may be represented by small utricles (swellings).

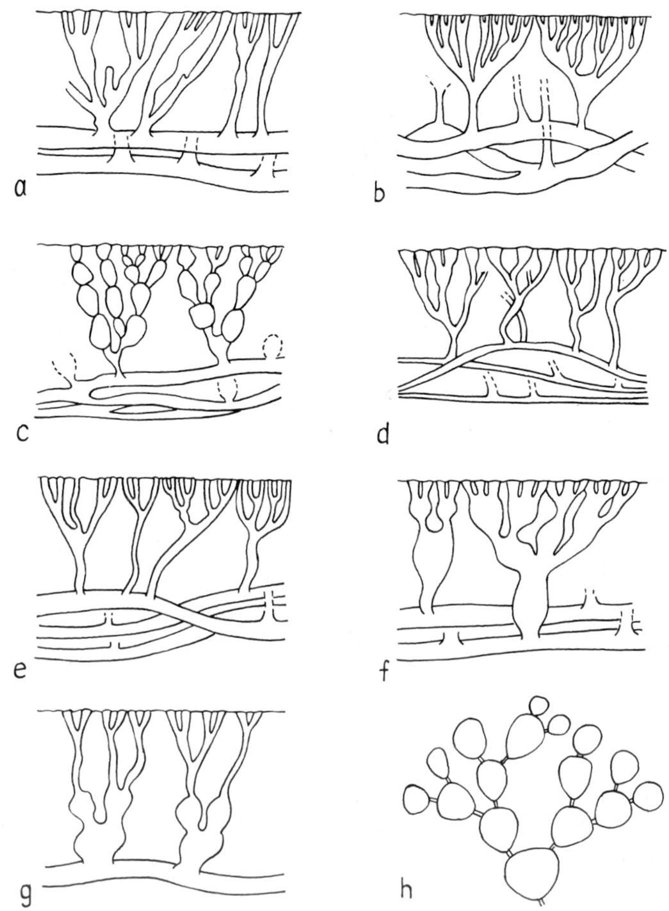

Fig. 2a-h. Representations of filament and utricle structure in various udoteacean genera, mostly based on materials in the collections of the British Museum (Natural History): **a** *Dimorphosiphon* (Ordovician); **b** *Aphroditicodium* (Permian; BM(NH) Dept. Palaeont. reg. no. V.59461); **c** *Tauridium* (Permian); **d** *Arabicodium* (Jurassic-Cretaceous); **e** *Boueina* (Triassic-Cretaceous); **f** *Halimeda incrassata* (Ellis) Lamx (Recent) (Hillis 1959); **g** *Leckhamptonella llewellyae* Elliott (Jurassic); **h** diagrammatic growth plan of serial-segmented udoteacean. See also comparisons for some other Palaeozoic genera in Obrhel (1968, Fig. 1) and Guilbault and Mamet (1976, Fig. 2) (after Elliott 1982)

6 Reproductive Structures

The reproductive structures in Recent Udoteaceae are both internal and external (Hillis-Colinvaux 1984), and the latter have no potential for fossilization.

In most fossil Udoteaceae no reproductive structure can be seen. However, rounded cavities located in the calcareous wall of the thallus, which are interpreted as reproductive organs, have been reported from a few Palaeozoic genera such as *Orthriosiphon* (Johnson and Konishi 1956), *Orthriosiphonoides* (Petryk and Mamet 1972) *Litanaia* (Maslov 1956), *Uva*? (Maslov 1956). They also exist in some Mesozoic forms, e.g. *Arabicodium elongatus* (Dragastan 1971 Pl.7, *3,4,7*), *A. texana* (Johnson 1969, Pl.2, *5,7*), and possibly *Halimeda elliotti* (Conrad and Rioult 1977, Pl.1, *7,9,10*, etc.). Since all the so-called "reproductive structures" in fossil material show no remaining cytological features and it is impossible to determine their biological nature, further comparison of them with Recent algae cannot be made at present.

7 Criteria for Classification

Wray (1977) proposed three criteria for generic classification of the erect as well as phylloid Udoteaceae, which I have modified as follows:

1. Growth form: branched, unbranched, segmented, non-segmented, and bladed.
2. Vegetative structure: size, shape, number, arrangement and branching mode of medullary and cortical filaments.
3. Reproductive structure: shape, size and location.

8 Calcification

The calcification of living Udoteaceae, e.g. *Halimeda*, has been well documented. It is characterized by precipitation of aragonite crystals, predominately needle-shaped, on the surface of filament walls within the thallus (Flajs 1977a,b; Hillis-Colinvaux 1980). In most of the fossil Udoteaceae their calcareous skeletons are composed of calcite which are the replacement products of the original aragonite. Two kinds of calcite may be recognized in fossil Udoteaceae: the more common is sparite, the other is dark micrite, which is only seen in a few genera, e.g. *Lowvillia* and *Dimorphosiphonoides* (Guilbault and Mamet 1976).

In fossil Udoteaceae, calcification mainly occurs in the cortex and the medulla is usually poorly calcified. As a result, the fossil is mostly represented by a calcareous tube perforated by many canals (pores). In this case, they are easily confused with Gymnocodiaceae and sometimes it is also difficult to distinguish them from fossil Dasycladaceae. Conrad and Rioult (1977) discussed the micritization, sparitization and silicification of *Halimeda elliotti* which, however, may not belong to the Udoteaceae (see below).

9 Systematic Position

Although the published records of fossil Udoteaceae are much more abundant than those of the Gymnocodiaceae, the biology of the material is less well known and their taxonomic assignments, at least for some of genera, have not been resolved with certainty. The reproductive structures deciphered in certain forms, as mentioned above, throws doubt upon their udoteacean affinities. They are more similar in shape, size and preservation to those commonly seen in the Gymnocodiaceae, rather than to Recent Udoteaceae. They resemble to some extent the gametangia of *Codium* (Codiaceae) and *Pseudocodium* (Pseudocodiaceae, Hillis-Colinvaux 1984) in being internal, but do not open to the surface of the thallus. This kind of reproductive organ is common in a number of Recent red algae, so the possibility that these fossil forms may not belong to the Udoteaceae cannot be ruled out.

The major arguments for udoteacean affinities of fossil forms lies in their gross morphology and coenocytic nature. Since the information obtained from fossil material is limited and parallelism is a common phenomenon among different algal groups (Fig. 1), it is likely that the assignment of a fossil form to the Udoteaceae may some times exceed the evidence available in the fossil state.

10 On the Classification of Fossil Udoteaceae

The classification of fossil Udoteaceae is a matter of controversy (see Table 1). In the 1920s, Pia referred erect forms, such as *Boueina* and *Palaeoporella*, together with *Gymnocodium*, to the Codiaceae; and filamentous (or nodular) forms such as *Ortonella*, *Mitcheldeania*, and *Hedstoemia*, were assigned to his artificial group Porostromata in the Cyanophyta (Pia 1927). However, some authors have suggested codiacean affinities for the latter and a new subfamily Mitcheldeanieae was created by Pia for them (Pia 1937). After Pia's work, some authors accepted the codiacean affinity (Elliott 1956; Johnson 1961; Konishi 1961; Bourque et al. 1981, etc.), but a new subfamily name Garwoodioideae was proposed to replace Mitcheldeanieae (Elliott 1956; Konishi 1961). Others still preferred a cyanophyte affinity (Maslov 1956, 1963; Shuysky 1973; Emberger 1976, 1978, 1979; Wray 1977, etc.); or considered that some belonged to the Cyanophyta and others to Chlorophyta. Riding (1977) suggested that the informal taxonomic name Porostromata be retained to cover these filamentous forms together with *Girvanella*. Subsequently Dragastan (1985) distributed between the Cyanophyta and Codiaceae. It should be noted that the family name "Codiaceae" used by the above authors is synonymous with the current usage of Udoteaceae. It is evident that the filamentous (or nodular) and erect groups are not homogeneous, and hence should belong to different algal divisions. For example, there are good reasons to compare *Ortonella* with the blue-green alga *Scytonema* (Monty 1967), whereas *Paragarwoodia* (Poncet 1974) may be related to the Udoteaceae. Considering that parallelism also occurs in filamentous forms among different divisions of Recent algae, caution must be made in determining their systematic position.

Konishi (1961) tried to establish a system of classification to incorporate both Recent and fossil Udoteaceae. Because the classification of Recent examples relies heavily on cytology and reproduction, details of which are rarely available under fossil conditions, this approach is limited. The same problems apply to the approach of Roux (1985), in which filamentous forms are assigned

Table 1. Classification schemes of filamentous or nodular "Udoteaceae" (boxed)

Pia 1927	Maslov 1963	Wray 1977	Pia 1937	Johnson 1961; Bourque, Mamet & Roux 1981
Schizophyceae	Cyanophyta	Cyanophyta	Codiaceae	Codiaceae
Porostromata			Mitcheldeanieae	
Hedstroemia	*Hedstroemia*	*Hedstraemia*	*Hedstroemia*	*Hedstroemia*
Mitcheldeania	*Garwoodia*	*Garwoodia*	*Mitcheldeania*	*Garwoodia*
Ortonella	*Ortonella*	*Ortonella*	*Ortonella*	*Ortonella*
Epiphyton	*Cayeuxia*	*Cayeuxia*	*Bevocastria*	*Bevocastria*
Zonotrichites	*Rothpletzella*	*Bevocastria*		*Dimorphosiphon*
------------------	------------------		------------------	*Palaeoporella*
Chlorophyceae				------------------
Codiaceae	Codiaceae	Codiaceae		
Halimeda	*Boueina*	*Palaeoporella*		
Boueina	*Calcifolium*	*Dimorphosiphon*		
Gymnocodium	*Abacella*	*Eugonophyllum*		
Palaeoporella	*Litanaia*	*Anchicodium*		
Ovulites	*Lancicula*	*Calcifolium*		
	Uva	*Ivanovia*		
------------	------------	------------		

Dragastan 1985	Roux 1985	Riding 1977
Cyanophyta	Codiaceae	Porostromata
Rivulariaceae	filamentous Codiaceae	(blue green + green algae)
Ortonella (Proparte)	*Hedstroemia*	*Girvanella*
Cayauxia	*Garwoodia*	*Ortonella*
Rivularia	*Mitcheldeania*	*Hedstroemia*
------------	*Ortonella*	*Garwoodia*
Scytonemataceae	*Gaspesiella*	*Cayeuxia*
Ortonella		*Bevocastria*
Paraortonella		
Mitcheldeaniaceae	Udoteaceae	
Mitcheldeania	erect Udoteaceae	
Chlorophyta	*Palaeoporella*	
Codiaceae	*Dimorphosiphon*	
Garwoodia	*Litanaia*	
Hedstroemia	*Lowvillia*	
Bevocastria	*Succodium*	
Arabicodium	*Maslovina*	
----------------	----------------	

to the Codiaceae, and erect ones to the Udoteaceae. Without information on cytology and reproduction it is very difficult to distinguish between fossil Codiaceae and Udoteaceae.

Studies of fossil Udoteaceae owe a great deal to the advances in our knowledge of Recent Siphonales. They suffer from these as well. Several decades ago when *Codium* and *Halimeda* were grouped into the same family Codiaceae (e.g. Fritsch 1935) it was convenient for palaeophycologists to follow this classification. Recently it has been realized that essential differences exist in the biological nature of *Halimeda* and *Codium*; they have been assigned to different families in different suborders (Hillis-Colinvaux 1984), and fossil Codiaceae, following *Halimeda*, have been transfered to the Udoteaceae or Halimedaceae. Incidentally, it has been shown recently (Hillis-Colinvaux 1984) that the family name Halimedaceae has priority over Caulerpaceae and

Udoteaceae (Silva 1980). For reasons mentioned above these changes complicate the subdivision of fossil Udoteaceae. The most probable difference between the two families which could be preserved in the fossil state is calcification. The known fossil Udoteaceae are calcified, whereas the fossil Codiaceae are not. However, calcification seems less important than has been suggested by some authors (Poncet 1982; Hillis-Colinvaux 1984) because calcareous and non-calcareous forms can be found in the same family, e.g. in the Dasycladaceae. It has also been suggested that a Recent non-calcareous group may have calcareous ancestors and vice-versa (Elliott 1978). Therefore, it is very difficult, if not impossible, to distinguish between fossil Codiaceae and Udoteaceae.

11 Geological Importance and Evolution

Recent members of the Udoteaceae (e.g. *Halimeda*, *Penicillus*) are among the most important carbonate producers and bioherm builders in warm seawater (Milliman 1974; Neumann and Land 1975; Hillis-Colinvaux 1980; Drew 1983; Hudson 1985; Drew and Abel 1988; Hine et al. 1988; Marshall and Davies 1988; Multer 1988; Orme and Salama 1988; Payri 1988; Phipps and Roberts 1988; Roberts et al. 1988; Hillis-Colinvaux, this Vol.). Their fossil representatives could be important limestone or bioherm builders too (e.g. Høeg 1927; Johnson 1961; Jux 1966; Segonzac 1986; Flügel 1988; Mankiewicz 1988).

Recent Udoteaceaens are a major source of $CaCO_3$-mud from post-mortem disintegration of their skeletons. It has been speculated that many Palaeozoic carbonate rocks rich in mud may be formed by fossil Udoteaceae (Wray 1977). In the study of Ordovician algae from Canada, Mamet et al. (1984) pointed out the role of the Udoteaceae in sedimentation of carbonate: *Dimorphosiphon* contributes to carbonate production and *Lowvillia* to bufflestones. Recent Udoteaceae are distributed in tropical (or subtropical) seas with low to moderate energy. Although most of them are shallow-water inhabitants, they may extend to 140 m with relatively high density and diversity (Hillis-Colinvaux 1986). They are important constituents of the chlorozoan assemblage (Lees and Buller 1972), commonly in association with Dasycladaceae and hermatypic corals. However, the lower limit of their depth distribution is greater than the latter two.

It is believed the same is true for their ancestors. The palaeoecology of fossil Udoteaceae is discussed in a general way by Elliott (1978, 1979, 1984) and he (1981, 1984) demonstrated their Tethyan distribution during Mesozoic and Cenozoic time. Poncet (1982) used Devonian occurrences of fossil Udoteaceae to locate palaeolatitudes of continents. Information about the characteristic representatives and their palaeoenvironmental distributions in Palaeozoic time can be found in the summary by Roux (1985).

Little is known of the evolution of Udoteaceae. A level of organization similar to that of Recent *Halimeda* was achieved in *Dimorphosiphon* in Ordovician times. It is suggested that *Halimida* arose by the hybridization of *Boueina* and *Arabicodium* in the Early Mesozoic (Elliott 1965). In a review of the evolution of the "serial segmented" Udoteaceae from Palaeozoic to Recent, Elliott (1982) considered that swollen cortical branches is a more advanced character than that of straight ones.

12 Gymnocodiaceae

Gymnocodium, the type genus of the Gymnocodiaceae Elliott, was established by Pia on the basis of *Gyroporella bellerophontis* Rothpletz 1894 (Pia 1920), which was previously referred by him to the genus *Macroporella*, a dasyclad acean (Pia 1912). Afterwards, he compared it with *Halimeda* and *Boueina*, and transferred it to the Codiaceae (Pia 1920, 1927). Having discovered reproductive structures in some specimens, he compared it with *Galaxaura*, a Recent red alga of the family Chaetangiaceae, and transferred it once more to the Chaetangiaceae (Pia 1937). In 1955, Elliott created the Family Gymnocodiaceae, and in doing so he made a further comparison with *Galaxaura* which supported that of Pia, and established a new genus *Permocalculus*. Since Elliott's work the red algal affinities of the Gymnocodiaceae have been generally accepted although some authors still consider them as Codiaceae (Bilgütay 1959; Konishi 1961; Maslov 1963, etc.).

During the last three decades, studies of the Gymnocodiaceae have expanded both geographically and geologically: from Europe to Asia, America and Africa; from the Permian to Jurassic and Cretaceous and Early Tertiary? The following works of regional floras are some of the publications concerned, to differing degrees, with the family: Accordi (1956), Bilgütay (1959), Elliott (1955, 1958), Güvenç (1966), Herak and Kochansky-Devidé (1960, 1963), Johnson (1951, 1968), Kochansky-Devidé (1954), Konishi (1954a), Korde (1965), Mu (1981, 1982), Ogilvie-Gordon (1927), Rao and Varma (1953), Rezak (1959), Srivastava (1982), Termier et al. (1977), Vachard (1980). Emberger (1976) compiled bibliographies of the Carboniferous and Permian algae, in which both Gymnocodiaceae and Udoteaceae are included.

In the study of Permian Gymnocodiaceae from SW China, I have found beautifully preserved silicified material. Using orientated thin-sections of isolated specimens, which reveal some details of the anatomy of the algae I established a new subgenus *Pyrulites* in *Permocalculus* (Mu 1981), and for the first time discovered fossil fungi parasitic in Gymnocodiaceae (Mu 1977). Mu and Riding (1983) used the SEM to study acid-extracted silicified material from the Permian of China, and established a new genus *Nanjinophycus*, which was compared with *Galaxaura* and tentatively placed in the Gymnocodiaceae.

13 Morphology and Anatomy

1. Growth Form

The growth form of the Gymnocodiaceae is erect, branched (dichotomously or trichotomously) or unbranached, segmented or non-segmented. The shape of the segments is cylindrical, oval, spherical of pyriform. No uncalcified node has been found between segments.

2. Anatomy
The thallus is composed of filaments arranged in central (medullary) and peripheral (cortical) zones. The medullary filaments can rarely be seen because of poor calcification in this part of the thallus (Pl. 1, *6*). They may be straight, or swollen and constricted. They run longitudinally along the axis, at times bifurcating and giving rise to lateral filaments at right or acute angles. Lateral filaments are generally similar to medullary ones in shape, and branch several times to reach the surface of the thallus where they form polygonal structures. The diameters of the lateral filaments may be thinner than those of the medullary ones, but in some forms they terminate in the form of a funnel or bowl with diameters which may exceed those of the medulla.

The filaments are usually preserved in limestone in the form of canals or pores perforating calcareous walls, which are generally filled with dark matrix (micrite). None of the soft parts (e.g. cell wall) can been seen and hence no cross-partitions have been found. The filaments seems to be coenocytic. This will be discussed in some detail below.

14 Reproductive Structures

In the cortex, or the area between cortex and medulla, oval or spherical cavities are frequently found which are filled with matrix (micrite) or cement (spar). The pattern of their preservation falls into two types:

1. Rounded cavities scattered in the calcareous wall filled with dark matrix (Pl. 1, *6*);

2. Rounded or oval bodies each with a thin dark membrane (Pl. 1, *1,3*). In the best specimens, it can seen that they are each located on a short lateral filament with the terminal part tapering into a thin tube which is open on the surface of the thallus. Sometimes they are arranged in layer along the peripheral part of the thallus (Pl. 1, *1,3,5,6*; Mu 1981, pl. I, *11*; pl. II, *2,3,6*).

The reproductive organs of the Gymnocodiaceae have been compared with conceptacles of the Chaetangiaceae (Elliott 1956; Mu and Riding 1983) or tetrasporangia of Corallinaceae (Mu 1981). In the study of *Permocalculus ampulancea*, Elliott (1961) recognized asexual and sexual (male and female) reproductive organs, which were suggested to represent asexual and sexual generations comparable with those in living *Galaxaura*. However, it was found out later that the "male" plant was another alga (Elliott, personal communication). Korde (1965) elaborated a number of reconstructions of reproductive organs of Gymnocodiaceae, which are based only on a few thin-sections. It seems to me that the preservation of her material does not allow such imaginative reconstructions.

The reproductive structures seen in the Gymnocodiaceae are the strongest argument for red algal affinities, which led to the transfer of *Gymnocodium* from

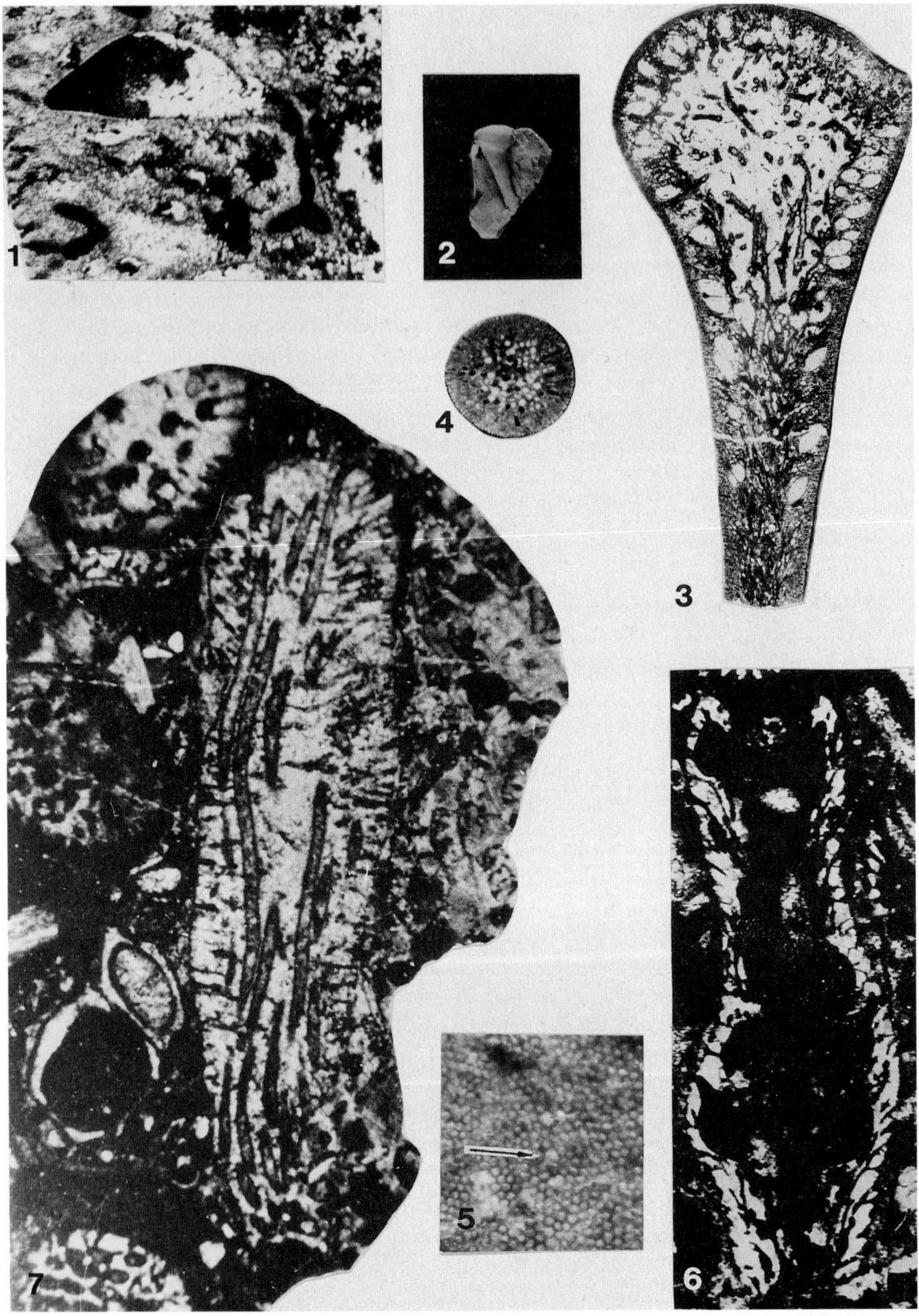

Plate I

the green to the red algae (Pia 1937). Subsequent work seems to support this interpretation. However, the systematic position of Gymnocodiaceae is not finally resolved. This will be discussed below.

15 Criteria for Classification

The following criteria can be useful for generic classification in the family:

1. Growth form: branched, unbranched, segmented, non-segmented, pyriform.

2. Vegetative structure: size, shape, number and arrangement of medullary and cortical filaments; and the ratio of diameter of medulla to outer diameter of the thallus.

3. Reproductive structure: shape, size and location.

16 Calcification and Silicification

The most commonly found skeletons of the Gymnocodiaceae are calcite. Since the family is compared with Recent *Galaxaura* or *Liagora*, the calcareous skeletons of which are aragonitic (Flajs 1977a,b), it is suggested by analogy that the same is true for the Gymnocodiaceae (Mu and Riding 1983). Calcification usually takes place in the cortical zone; the medullary zone is generally not calcified. During diagenesis their aragonitic skeletons are transformed into calcitic ones.

Silicification is common in the Gymnocodiaceae. In the early diagenetic stage, the cell wall of the algae may be preserved (see Mu 1977, 1981). In late silicification, which is common in carbonate rocks and frequently characterized by the selective silicification of fossil skeletons, no soft parts of the algae are preserved; Mu and Riding (1988) studied the pattern and process of silicification of *Nanjinophycus*, a possible Gymnocodiacean, and proposed a model for the silicification process. They suggested that water insoluble organic material in the matrix controls the selective silicification, by preventing the calcite in the matrix from silicifying.

Plate I. 1–5. *Permocalculus* (*Pyrulites*) *sinicus* Mu. Changhsingian, Upper Permian; W. Guizhou, China (Mu 1981). **1** A sporangium terminal on a short lateral filament (x50); **2** lateral view of a solid silicified specimen (x1); **3** longitudinal section of the same specimen (**2**) (x8); **4** transverse section of the initial part of the same specimen (**2**); (x15); **5** surface view of epidermal cells of the cortex in an isolated silicified specimen. Note the larger openings (*arrows*) representing the outlets of discharge tubes of sporangia (x45); **6** *Gymnocodium bellerophontis* (Rhothpletz). Longitudinal section; Bellerophon Formation, Upper Permian, Alleghe area (Belluno), W. Dolomites, Italy. Note oval sporangia with terminal discharge tubes (x20; Fois 1977) **7** *Dimorphosiphon rectangulare* Høeg. A longitudinal section and some transverse sections; Middle Ordovician, Lake Mjøsa, Norway (x20; Elliott 1972)

17 Systematic Position

The growth form, internal structure with differentiation into medulla and cortex, together with the shape, size and location of reproductive organs, all suggest that the Gymnocodiaceae are more similar to red algae than to any others. However, their red algal nature has not conclusively been confirmed. An important difficulty which remains is that until now no cross partitions have been found within the filaments of *Gymnocodidium* and *Permocalculus*, which are typical of the family, even in the best materials where the cell wall of the algae probably is preserved (see Mu 1977). If the cell filaments are really non-septate, the algae are more likely to be green algae. It may be, however, as I have suggested (Mu 1981) that the cross septa were entirely represented by pit-plugs (septa plugs) which are characteristic of red algae. In Recent red algae the septal plug is chiefly composed of protein (Pueschel 1981) which is less resistant and cannot survive fossilization. Considering that no direct evidence is available at present, assignment of the Gymnocodiaceae to the red algae is not conclusive.

Certain genera relevant to the Gymnocodiceae need discussion of their taxonomic assignment. *Nipponophycus* Yabe et Toyama was originally referred to the Furcellariaceae (red algae) (Yabe and Toyama 1928). Since then it has been assigned to the Solenoporaceae (Endo 1961) and Udoteaceae (= Codiaceae) (Johnson 1964, 1969; Bassoullet et al. 1983). Tappan (1980) retains it in the Furcellariaceae. I showed the presence of reproductive structures like those of the Gymnocodiceae and tentatively transferred it to the family (Mu 1982). Although its precise systematic position remains to be resolved, it seems more reasonable to ascribe it to the red algae than to any other division. *Succodium* Konishi is generally considered as belonging to the Udoteaceae. However, the presence of structures interpreted as reproductive organs led Korde (1965) to transfer it to the Gymnocodiaceae. This was followed by Vachard (1980) and Vachard and Montenat (1981), but rejected by some others (Bassoullet et al. 1983). Mu and Riding (1983) suggested that if its internal reproductive organs can be substantiated, *Succodium* Konishi should be assigned to the Gymnocodiaceae like *Nanjinophycus* Mu and Riding. The systematic position of *Succodium* is not conclusive at present. *Diversocallis* Dragastan was erected as a gymnocodiacean alga (Dragastan 1967). However, its irregular form and compact internal structure suggest that it is very likely to be a synonym of *Pseudolithothamnium* Pfender (Squamariaceae). *Oligoplagia* Herak was initially referred to calcisponges (Herak 1944) and subsequently to microproblematica (Ott 1967) and then Gymnocodiaceae (Flügel 1971). Owing to poor preservation of its internal structure, and presence of cross partitions it cannot be assigned to the Gymnocodiaceae and its systematic position remains to be clarified. *Abatea* Senobari-Daryan and Schäfer was classed as a genus of the Gymnocodiaceae (Senowbari-Daryan and Schäfer 1980). But the existence of cross septa and "sporangia" in the medullary zone, which are quite different from the characters of the Gymnocodiaceae, make this assignment doubtful.

18 Geological Importance of Gymnocodiaceae

The records of *Gymnocodiaceae* range from the Permian to Tertiary? with the most abundant occurrence in Permian rocks. They are also abundant at some levels in the Cretaceous. The Triassic genera *Abatea* and *Oligoplagia*, as mentioned above, might not belong to the family. Gymnocodiaceans are widespread within the Tethys and indicative of warm, shallow sea waters with low-moderate energy. They are important carbonate producers, and gymnocodiacean limestones have been widely found in Tethyan Permian and some Cretaceous rocks. During Permian time, their importance geologically and biologically, includes: baffle sedimentation, carbonate production, and habitats for invertebrates such as foraminifers, ostrocodes, bivalves and gastropods. They also have stratigraphical value. The *Permocalculus* (*Pyrulites*) zone, which is roughly the equivalent of *Palaeofusulina* zone, indicative of the Changhsing stage of the Upper Permian, can be recognized in a vast area of S. China, (Mu, 1981).

19 General Remarks

The biological features of soft tissue of algae are rarely preserved in the fossil state. For most Gymnocodiaceae and Udoteaceae, even the cell wall, which is the most resistant part, usually could not survive fossilization. What we are left with are only their altered skeletons, which are in the form of moulds. The original filaments and reproductive organs of the algae are replaced by cavities filled with matrix or cement (micrite or sparite). In such material it is impossible to identify the nature of reproductive organs and also impossible to prove whether the alga is coenocytic or not (Fig. 3). The affinities of fossil algae are only

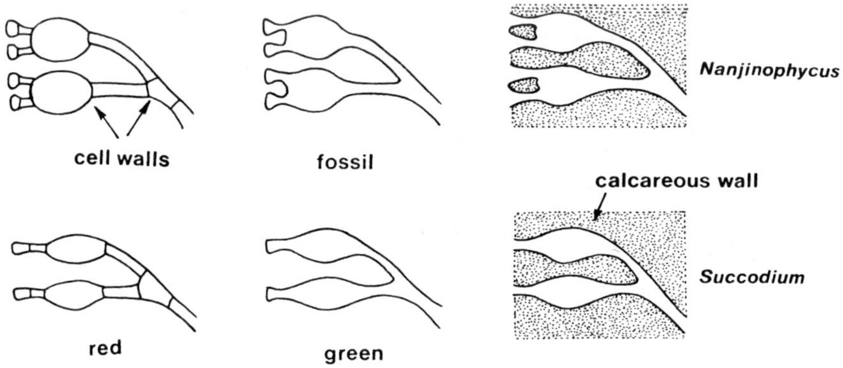

Fig. 3. Comparison between possible original organization of cortical filaments in *Nanjinophycus* and *Succodium* and the resulting fossil morphology. If cross-partitions occur but are not calcified the final appearance will be similar to specimens lacking original cross-partitions; yet this feature, if preserved, could help to distinguish between a red and green algal affinity. (after Mu and Riding 1983)

based on the superficial similarity between them and living ones. It is well-known that parallelism or convergence is common among different groups of living algae, i.e. representatives of different groups may have similar habit and internal organization. For example, the external and internal vegetative structures of *Halimeda*, (Udoteaceae) are in many ways similar to those of *Codium* (Codiaceae), they can also be comparable with those of some Chaetangiaceae (red algae) and other algae (Fig. 1). Thus, the assignment of many fossil genera to the Udoteaceae in the literature is uncertain. The only generally available criterion to distinguish between the Gymnocodiaceae (red algae) and the Udoteaceae (green algae) is that of the reproductive structures: ovoid or spherical in form, and located in the peripheral part of the thallus in the Gymnocodiaceae, but usually absent in the Udoteaceae. In Recent Udoteaceae the reproductive organs are external in *Halimeda* and possibly internal in *Penicillus* (Hillis-Colinvaux 1984). Their shape, size and location are not comparable with those found in some genera of fossil Udoteaceae, which suggests that these latter genera may not belong to Udoteaceae. They possibly belong to red algae like the Gymnocodiaceae. When reproductive organs are not seen in fossil algae, it is very difficult, if not impossible, to distinguish between the Gymnocodiaceae and Udoteaceae. However, it is also likely that some ancestors of Udoteaceae may have reproductive organs dissimilar to those of Recent ones. Thus, the possibility of their green algae affinities cannot be ruled out at present.

20 Suggestions for Future Work

From the foregoing it is evident that problems in determining the systematic position and classification of Gymnocodiaceae and fossil Udoteaceae are the result of the inadequate information available regarding fossil material. In studying fossil calcareous algae we are, of course, not dealing with complete organisms, but only with their calcareous skeletons which are generally in the form of moulds. The most important information on the biological nature of the algae, such as pigments, storage products, chemical characters of the cell wall, and reproductive organs, which are critical to the classification of living algae, have been lost during fossilization. Diagenesis of fossils makes the case worse (Elliott, 1984). As a result it is very difficult, if not impossible, to determine the systematic position of fossil algae based only upon their skeletons which are also often altered, just as we cannot recognize the health of a person by his clothes.

To advance our understanding of Gymnocodiaceae and fossil Udoteaceae, and other groups of fossil algae as well, some suggestions are proposed as follows:

1. Search for better preserved permineralized material. The fossil material of calcareous algae preserved in limestones is not good enough to reveal the details of their biological nature, due to the fact that algal-bearing limestones are usually laid down under oxidizing conditions which permit the soft parts of the

algae to decay rapidly after death. The ideal materials are permineralized, e.g. silicified. Silicified specimens formed in an early diagenetic stage may preserve some details of algal structures, even of their soft parts, (e.g. cell walls, etc.). Beautifully preserved material may be found in chert beds or concretions in shallow marine carbonate sequences. These deposits should be attentively examined and sampled in the field.

2. A thorough comprehension of the structure of the fossil is the first step in taxonomic work. Although this sounds commonplace, in practice it too often ignored. Many algal taxa in the literature have been created based on only one section, or a few random sections from different individuals which are often poorly preserved. This may lead to mistakes in reconstruction of the algae and to confusion of one taxon with another. To avoid this, the adoption of orientated thin-sectioning is recommended (see Mu 1981). The acid dissolving approach is useful for obtaining complete individuals which can reveal the gross morphology and surface nature of the algae. This method is most suitable for selectively silicified material in limestones (Mu and Riding 1983). Acid maceration methods should be used together with sectioning.

3. Close co-operation between experts on fossil and living algae is urgently needed. The present is the key to the past. Palaeobiological and palaeoecological studies on fossil algae heavily rely on information about Recent algae. There is a wide scope for joint work in comparative anatomy, ecology, ultrastructure, and chemotaxonomy, for which multidisciplinary approaches are needed. The results of such cooperation will be rewarding and can lead to a better understanding of the *Gymnocodiaceae* and fossil *Udoteaceae*.

21 Conclusions

Fossil Udoteaceae and Gymnocodiaceae are important groups of fossil calcareous algae. The former ranges from Palaeozoic to Cenozoic with abundant living descendants; the latter is an extinct group ranging from Palaeozoic to early Tertiary? It is generally accepted that the Gymnocodiaceae belong to the red algae. Well-preserved silicified Gymnocodiaceae can reveal some details of cell filaments, reproductive organs and parasitism. However, the systematic position of this group still remains unresolved. The records of fossil Udoteaceae are more abundant, but their biological features are less well known and much discrepancy exists in their nomenclature, definition and classification.

The erect group of fossil Udoteaceae resembles the Gymnocodiaceae in many ways: growth form, vegetative structure, palaeoecological and geographical distribution, etc. The major difference between them which can be deciphered is that the gymnocodiaceans contain internal reproductive organs which are usually absent in the Udoteaceans. In view of the fact that fossil material can only provide limited data on the anatomy of the algae, and no information is available at present about pigments, storage products, chemical characters of the cell wall, or reproduction, which are critical to the classification of living algae, and that parallelism is common among different groups of living algae at different taxonomic levels, the Gymnocodiaceae and fossil Udoteaceae are considered heterogeneous. In addition, their biological assignments are not conclusive, and in practice it is often difficult to distinguish between them. In order to advance our understanding of these important groups search for better preserved permineralized fossil material, and close cooperation between students of fossil and living algae by using multidisciplinary approaches are badly needed.

Acknowledgements. I am grateful to Dr. G.F. Elliott for critically reading the manuscript and giving helpful advice.

References

Accordi D (1956) Calcareous algae from the Upper Permian of the Dolomite (Italy) with stratigraphy of the "Bellerophon-zone". Palaeontol Soc India J 1, 1:75–84

Bassoullet JP, Bernier P, Deloffre R, Genot P, Poncet J, Roux A (1983) Les Algues Udoteacées du Paleozoique au Cenozoique. Bull Cent Rech Explor Prod Elf-Aquitaine 7, 2:449–621

Bilgütay U (1959) The Permian calcareous algae from southeastern Anatolia. Bull Mineral Res Explor Turkey 52:48–58

Bourque A, Mamet B, Roux A (1981) Algues siluriennes du synclinorium de la Baie des Chaleurs, Québec, Canada. Rev Micropaleontol 24, 2:83–126

Conrad M, Rioult M (1977) *Halimeda elliotti* nov. sp. Algue calcaire (Chlorophyceae) du Turonien des Alpes-Maritimes (S.E. France). Geol Mediter 4, 2:83–98

Dixon PS (1973) Biology of the Rodophyta. Hufner, New York

Dragastan O (1967) Algae calcarease in Jurasicul superior si Cretacicul inferior din Muntii Apuseni, St. Si. Cerc Geol Geofiz Geogr Ser Geol Bucuresti 12, 2:441–454

Dragastan O (1971) New algae in the Upper Jurassic and Lower Cretaceous in the Bicaz valley, East Carpathian (Romania). Rev Esp Micropaleontol 3, 2:155–192

Dragastan O (1985) Review of Tethyan Mesozoic algae of Romania. In: Toomey DE, Nitecki MH (eds) Paleoalgology. Springer, Berlin Heidelberg New York Tokyo, pp 101–161

Drew EA (1983) *Halimeda* biomass growth rate and sediment generation on reefs in the Central Great Barrier Reef Province. Coral Reefs 2:101–110

Drew EA, Abel KM (1988) Studies on *Halimeda*. I. The distribution and species composition of *Halimeda* meadows throughout the Great Barrier Reef Province. Coral Reefs 6:195–205

Elliott GF (1955) Fossil calcareous algae from the Middle East. Micropaleontology 1, 2:125–131

Elliott GF (1956) Algues calcaires codiacées fossiles d'Iraq nouvelles et peu connues. Bull Soc Geol Fr 6, 6:789–795

Elliott GF (1957) New calcareous algae from the Arabian peninsula. Micropaleontology 3, 3:227–230

Elliott GF (1958) Algal debris facies in the Cretaceous of Middle East. Palaeontology 1, 3:254–259

Elliott GF (1961) The sexual organization of Cretaceous *Permocalculus* (calcareous algae). Palaeontology 4, 1:82–84

Elliott GF (1965) The interrelationships of some Cretaceous Codiaceae (calcareous algae). Palaeontology 8, 2:199–203

Elliott GF (1978) Ecological significance of post-Paleozoic green calcareous algae. Geol Mag 115, 6:437–442

Elliott GF (1979) Influence of organic reefs on the evolution of post-Paleozoic algae. Geol Mag 116, 5:375–383

Elliott GF (1981) The Tethyan dispersal of some Chlorophyte algae subsequent to the Paleozoic. Palaeogeogr Palaeoclimatol Palaeoecol 32:341–358
Elliott GF (1982) A new calcareous green alga from the Middle Jurassic of England: its relationships and evolutionary position. Palaeontology 25, 2:431–437
Elliott GF (1984) Modern developments in the classification of some fossil green algae. In: Irvine DEG, John DM (eds) Systematics of the Green Algae. Academic Press, New York London, pp 297–302
Emberger J (1976) Les Algues (Chlorophyceae, Prasinophyceae, Rhodophyceae) de Carbonifère et du Permien. Essai D'un inventaire bibliographique, géographique, stratigraphique. Bull Inst Géol Bassin Aquitaine Num Spéc 168 pp
Emberger J (1978) Les Algues (Chlorophyceae, Prasinophyceae, Rhodophyceae) du Dévonien. Essai D'un inventaire bibliographique, géographique, stratigraphique. Bull Inst Géol Bassin Aquitaine Num Spéc 94 pp
Emberger J (1979) Les Algues (Chlorophyceae, Prasinophyceae, Rhodophyceae) du Trias. Essai d'un inventaire bibliographique, geógraphique, stratigraphique. Bull Inst Géol Bassin Aquitaine Num Spéc 157 pp
Flajs G (1977a) Skeletal structure of some calcifying algae. In: Flügel E (ed) Fossil algae. Springer, Berlin Heidelberg New York, pp 225–231
Flajs G (1977b) Die Ultrastukturen des Kalkalgenskeletts. Palaeontolographica B 160, 4–6:62–128
Flügel HW (1971) *Oligoplagia* Herak 1944, eine Gymnocodiaceae. N Jahrb Geol Paläontol Mh 9:532–536
Flügel E (1988) *Halimeda*: paleontological record and palaeoenvironmental significance. Coral Reefs 6:123–130
Fois E (1977) Due nuovi affioramenti della Formazione a Bellerophon nel Cordevole (Belluno). Riv Ital Paleontol 83, 4:675–686
Fritsch FE (1935) The structure and reproduction of the algae. I. University Press, Cambridge 191 pp
Guilbault JP, Mamet B (1976) Codiacées (Algues) ordoviciennes des Basses-Terres du Saint-Laurent. Can J Earth Sci 13, 5:636–660
Güvenç T (1966) Description de quelques espèces d'algues calcaires (Gymnocodiacées et Dasycladacées) du Carbonifère et du Permian des Taurus occidentaux (Turquie). Rev Micropaleontol 9, 2:94–103
Herak M (1944) Zur Kenntnis triadischer Kalkschwamme (Sycones). N Jahrb Mineral Geol Palaontol B 88:107–135
Herak M, Kochansky-Devidé V (1960) Gymnocodiacean calcareous algae in the Permian of Yugoslavia. Geol Vjesn 13:185–195
Herak M, Kochansky-Devidé (1963) Jungpaläozoische Kalkalgen aus dem Bükk-Gebirge (Nordungarn). Geol Hung Ser Palaeontol 28:45–78
Hillis LW (1959) A revision of the genus *Halimeda* (Order Siphonales). Inst Mar Sci Univ Texas 6:321–403
Hillis-Colinvaux L (1980) Ecology and taxomy of *Halimeda*: Primary producer of coral reefs. In: Blaxter JHS, Russel FS, Young M (eds) Advances in marine biology, vol 17. Academic Press, New York London, pp 1–327
Hillis-Colinvaux L (1984) Systematics of the Siphonales. In: Irvine DEG, John DM (eds) Systematics of the Green Algae. Academic Press, New York London, pp 271–296
Hillis-Colinvaux L (1986) *Halimeda* growth and diversity on the deep fore-reef of Enewetak Atoll. Coral Reefs 5, 1:19–21
Hine AC, Hallock P, Harris MW, Mullins HT, Belknap DF, Jaap WC (1988) *Halimeda* bioherms along an open seaway: Miskito Channel, Nicaraguan Rise, SW Caribbean Sea. Coral Reefs 6:173–178
Høeg OA (1927) Preliminary note on a new Codiaceae from the Ordovician of Norway. Avh Nors Vidensk Akad Oslo, Mat Naturviss Kl 4:1–15
Hudson JH (1985) Growth rate and carbonate production in *Halimeda opuntia*: Marquesas Keys, Florida. In: Toomey DF, Nitecki MH (eds) Paleoalgology. Springer, Berlin Heidelberg New York Tokyo, pp 257–263

Johnson JH (1951) Permian calcareous algae from the Apache Mountains. J Paleontol 25, 1:21–30
Johnson JH (1961) Limestone-building algae and algal limestones. Col Sch Mines Spec Publ 297 pp
Johnson JH (1964) The Jurassic algae. Col Sch Mines Q 59, 2:129 pp
Johnson JH (1968) Lower Cretaceous algae from Taxas. Prof Contrib Col Sch Mines 4
Johnson JH (1969) A review of the Lower Cretaceous Algae. Prof Contrib Col Sch Mines 6: 180 pp
Johnson JH, Konishi K (1956) Mississippian algae from the Western Canada basin and Martana. Studies of Mississippian algae, pt 2. Col Sch Mines Q 51, 4:85–107
Jux U (1966) Palaeoporella in Boda-Kalk von Dalarne. Palaeontographica. B 118, 4–6:153–165
Kochansky-Devidé V (1954) Permische Foraminiferen und Kalkalgen aus der Umgebung von Bar in Montenegro. Geol Vjesn 5:295–298
Kochansky-Devidé V, Slišković T (1969) Oberpermische Mikrofossilien von Han Orahavica und Suljci in Bosnien mit Besonderer Berucksichtigun der Alga *Permocalculus fragilis*. Geol Vjesn 22:105–110
Konishi K (1954a) A new species of *Gymnocodium* and its algal associates in the Permian Kosaki Formation of southern Kuyshu, Japan. J Geol Geogr Trans 25, 1/2:1–19
Konishi K (1954b) *Succodium*, a new codiacean genus, and its algal associates in the Late Permian Kuma Formation of southern Kyushu, Japan. In: Studies on the Paleozoic marine algae of Japan, 2. J Fac Sci Univ Tokyo (2)9, 11:225–240
Konishi K (1961) Studies of Paleozoic Codiaceae and allied algae, Part I, Codiaceae (excluding systematic descriptions). Sci Rep Kanazawa Univ 7, 2:159–261
Korde KB (1965) Razvitie i smena morskikh organismov na rubezhe Paleozoia i Mesozoia. Algae. Trud Paleontol Inst Akad Nauk SSSR 108:87–92 (in Russian)
Lees A, Buller AT (1972) Modern temperate-water and warm-water shelf carbonate sediments contrasted. Mar Geol 19:159–198
Mamet B, Roux A, Shalaby H (1984) Rôle des Algues calcaires dans la sédimentation Ordovicienne de la Plateforme du Saint-Laurent. Geobios Mem Spec Vol 8:261–269
Mankiewicz C (1988) Occurrence and paleoecologic significance of *Halimeda* in late Miocene reefs, southeastern Spain. Coral Reefs 6:271–279
Marshall JF, Davies PJ (1988) *Halimeda* bioherms of the northern Great Barrier Reef. Coral Reefs 6:139–148
Maslov VP (1956) Fossil calcareous algae of the USSR. Acad Sci USSR, Inst Geol Sci Trud 160, 301 pp (in Russian)
Maslov VP (ed) (1963) Vodorosli. Osnov Paleontol 14:17–266 (in Russian)
Milliman JD (1974) Marine carbonates, pt 1. Springer, Berlin Heidelberg New York
Monty CLV (1967) Distribution and structure of Recent stromatolitic algal mats, Eastern Andros Island, Bahamas. Ann Soc Geol Belg 90:55–100
Mu X-N (1977) Upper Permian fossil fungi from Anshun of Guizhou. Acta Palaeontol Sin 16, 2:151–158 (in Chinese with English abstract)
Mu X-N (1981) Upper Permian calcareous algae from Western Guizhou. Acta Palaeontol Sin 20, 1:33–48 (in Chinese with English summary)
Mu X-N (1982) Some calcareous algae from Xizang. In: The series of the scientific expedition to the Qinghai-Xizang Plateau: palaeontology of Xizang, vol 5. Science Press, Beijing, pp 205–240 (in Chinese with English abstract)
Mu X-N, Riding R (1983) Silicified gymnocodiacean algae from the Permian of Nanjing, China. Palaeontology 26, 2:261–276
Mu X-N, Riding R (1988) Silicification of Permian calcareous algae from Nanjing, China. Geol Mag 125:123–139
Multer HG (1988) Growth rate, ultrastructure and sediment contribution of *Halimeda incrassata* and *Halimeda monile*, Nonsuch and Falmouth Bays, Antiqua, WI. Coral Reefs 6:179–186
Neumann AC, Land LS (1975) Lime mud deposition and calcareous algae in the Bight of Abaco, Bahamas: a budget. J Sed Petrology 45, 4:763–786
Obrhel J (1968) *Maslovina meyenii* n.g. et n. sp. neue Codiacea aus dem Silur Bohmens. Vest Ustred Ustavu Geol 43:367–370
Ogilvie-Gordon MM (1927) Das Grödner-, Fassa-, und Enneberggebiet in den südtiroler Dolomiten, 3: Paläeontologie. Abh Geol Bundesanst 24, 2:1–89

Orme GR, Salama MS (1988) Form and seismic stratigraphy of *Halimeda* banks in part of the northern Great Barrier Reef Province. Coral Reefs 6:131–137

Ott E (1967) Segmentierte Kalkschwamme (Sphinctozoa) aus der alpinen Mitteltrias und ihre Bedeutung als Riffbildner in Wettersteinkalk. Abh Bayr Akad Wiss Math Naturwiss Kl NF 131:205–226

Payri CE (1988) *Halimeda* contribution to organic and inorganic production in a Tahitian reef system. Coral Reefs 6:251–262

Petryk AA, Mamet BL (1972) Lower Carboniferous algal microflora, southwestern Alberta. Can J Earth Sci 9, 7:767–802

Pfender J (1936) Sur un organisme constructeur des calcaires crétacés et nummulitiques: *Pseudolithothamnium album* nov. gen. et sp. Bull Soc Geol Fr (5)6, 4–5:303–308

Phipps CVG, Roberts HH (1988) Seismic characteristics and accretion history of *Halimeda* bioherms on Kalukalukuang Bank, eastern Java Sea (Indonesia). Coral Reefs 6:149–159

Pia J von (1912) Neue Studien über die triadischen *Siphoneae Verticillatae*. Beitr Palaeontol Geol Österr Ung Or 25, 1:25–81

Pia J von (1920) Die Siphoneae verticillatae vom Karbon bis zur Kreide. Abh Zool Bot Ges Wien 11, 2:1–263

Pia J von (1926) Pflanzen als Gesteinbildner. Bornträger, Berlin

Pia J von (1927) Thallophyta. In: Hirmer M (ed) Handbuch der Paläobotanik, vol 1. Oldenbourg, Mu Berlin, pp 31–136

Pia J von (1937) Die wichtigsten Kalkalgen des Jungpaläozoikums und ihre geologische Bedeutung, Congr Av Etude Strat Car 2nd, Heerlen, 1935, CR 2:765–856

Poncet J (1974) Description de quelques Algues calcaires éodévoniennes du Nord-Est du Massif Armoricain. Bull Soc Geol Fr (7) 16, 2:105–229

Poncet J (1982) L'apport des Udoteaceae (Algues vertes calcaires) dans la paléogéographie mondiale éodévonienne. Bull Soc Geol Fr (7) 24, 5–6:1087–1091

Pueschel CM (1981) A reappraisal of the cytochemical properties of Rhodophycean pit plugs. Phycologia 19:210–217

Rao SRN, Varma CP (1953) Permian algae from the middle Productus beds. Palaeobotanist 2:19–23

Rezak R (1959) Permian algae from Saudi Arabia. J Paleontol 33, 4:531–539

Riding R (1977) Problem of affinity in Paleozoic Calcareous Algae. In: Flügel E (ed) Fossil algae. Springer, Berlin, Heidelberg, New York, pp 202–211

Roberts HH, Aharon P, Phipps CV (1988) Morphology and sedimentology of *Halimeda* bioherms from the eastern Java Sea (Indonesia). Coral Reefs 6:161–172

Roux A (1985) Introduction à l'étude des Algues fossiles paléozoiques (de la bactérie à la tectonique des placque). Bull Cent Rech Explor Prod Elf-Aquitaine 9, 2:465–699

Segonzac G (1986) Aspects concernant les Algues de type *Halimeda* (Chlorophycées), actuelles et fossiles. Bull Soc Hist Nat Toulouse 122:125–129

Senowbari-Daryan B, Schäfer P (1980) *Abatea culleiformis* n.g., n. sp., eine neue Rotalge (Gymnocodiaceae) aus den "oberrhatischen" Riffkalken südlich von Salzburg (Nördliche Kalkalpen, Österreich). Verh Geol BA 3:393–399

Shuysky VP (1973) Calcareous reef-forming algae from the Lower Devonian of Ural. Acad Sci USSR Sci Cent Ural, Inst Geol Geochem AN Zavaritsky. Nauka, Moscow (in Russian)

Silva PC (1980) Names of classes and families of living algae. Regnum Veg 103:1–156

Srivastava NK (1982) Calcareous algae from Jandaira Formation (Potiquar Bassin), Brazil, pt 1. An Acad Brasil Ciec 54, 1:219–231

Svedelius N (1953) Critical studies on some species of *Galaxaura* from Hawaii. N Acta Soc Sci Upsalla (4) 15, 9:1–92

Tappan H (1980) The paleobiology of plant protists. Freeman, San Francisco 1028 pp

Termier H, Termier G, Vachard D (1977) Monographie paléontologique des affleurements Permiens du Djebel Tebaga (Sud-Tunisien). Palaeontolographica A 156, 1–3:1–109

Tseng CK, Zhang DR, Zhang JF, Xia EZ, Xia BM, Dong ML, Yang DZ (1962) Manual of Chinese economic seaweeds. Science Press, Beijing (in Chinese)

Vachard D (1980) Tethys et Gondwana au Paleozoicque superieur, les donnees afganes. Doc Trav IGAL 2:463 pp

Vachard D, Montenat C (1981) Biostratigraphie, Micropaléontologie et Paléogeographie du Permien de la région de Tezak (Montagnes Centrales d'Afghanistan) Palaeontographica B 178 1-3:1-88

van den Hoek C (1981) Chlorophyta: Morphology and Classification. In: Lobban CS, Wynne MJ (eds) The biology of seaweeds. Blackwell, Oxford, pp 86-132

Wray L (1977) Calcareous algae. Elsevier, Amsterdam

Yabe H, Toyama S (1928) On some rock-forming algae from the younger Mesozoic of Japan. Sci Rep Tohoku Univ (2) 12, 1:141-152

Chapter 10

Recent Calcified Halimedaceae

L. HILLIS[1]

Abstract

Five Recent genera of Halimedaceae, namely *Halimeda*, *Penicillus*, *Rhipocephalus*, *Tydemania*, and *Udotea* calcify. Only *Halimeda* and *Udotea*, the genera with more than 10 species, are widely distributed in the tropical Atlantic, Indian and Pacific oceans, but *Halimeda* alone produces extensive populations **worldwide** making it globally significant. Because of the essentially tropical distribution of all 5 taxa they can be considered strong indicators of tropical environments.

Halimeda is also widely distributed across the reef, with some of its 33 species indicating different reefal conditions. Approximately three-quarters of its species are associated with consolidated or gravelly habitats as opposed to sand/mud substrata. Large populations may occur on the deep fore-reef where *Halimeda* grows at depths of over 140 m, in the extreme shallows of back reef moderate energy environments, and in lagoonal regions. Some lagoonal populations overlay *Halimeda* bioherms which date from at least the early Pleistocene. Extensive present day populations together with *Halimeda* bioherms provide dramatic evidence of the importance of calcareous Halimedaceae in reef construction, and offer support for calling reefs **algal** rather than **coral**. Although *Halimeda* is a principal component of many tropical reef systems, non-*Halimeda* reefs are also part of the global reef picture.

1 Definition of the Group

The largest living green marine algae belong to the order Bryopsidales, formerly Siphonales, and of the modern algae they are, paradoxically, among the least known. Six genera out of a total of twenty-four calcify. Five of the taxa, namely *Halimeda* Lamouroux, *Penicillus* Lamarck, *Rhipocephalus* Kützing, *Tydemania* Weber-van Bosse, and *Udotea* Lamouroux (Figs. 1–5) have been assigned to the family Halimedaceae: *Pedobesia* MacRaild and Womersley has been placed in the family Bryopsidaceae (Hillis-Colinvaux 1984). Assignment to these families is based principally on differences in morphology, reproduction and distribution, although the data available for some genera are still very limited. Selection of the family name Halimedaceae (Hillis-Colinvaux 1984) is based on the epithet having priority over both Caulerpaceae and Udoteaceae (Silva 1980).

The most taxonomically diverse genus of the Halimedaceae is *Halimeda* with 33 living species. Five sections of this genus, Rhipsalis, Opuntia, Halimeda, Micronesicae, and Crypticae, were delimited in 1980 by Hillis-Colinvaux using behaviour of filaments at the nodes, or intersegmental regions, as the basis of separation. Of the other genera, *Udotea* has approximately 26 species, *Penicillus* five species, *Rhipocephalus* two species, and *Tydemania* two species. The em-

[1]Zoology Department, The Ohio State University, 1735 Neil Ave., Columbus, Ohio 43210, USA

Fig. 1. Grove of the sand-growing species *H. incrassata* at a depth of approximately 1 m on the north shore of Jamaica. (magnification ca. 0.2x; photograph by P.A. Colinvaux)

Fig. 3. *Rhipocephalus phoenix* (Photograph courtesy of W. Fenical; magnification ca. nat. size)

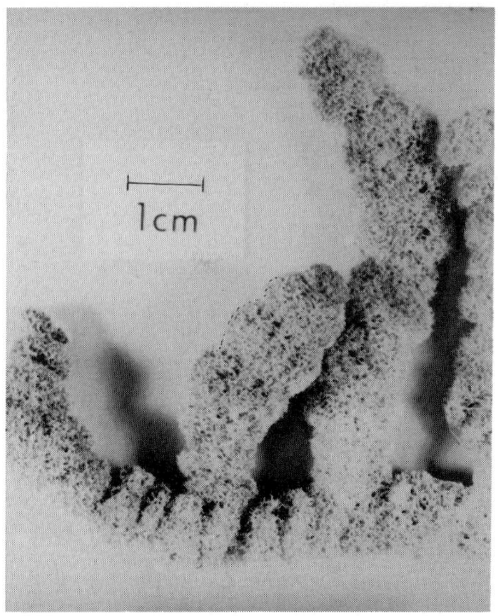

Fig. 4. *Tydemania expeditionis.* A small portion of a sprawling thallus collected in the lagoon of Enewetak Atoll

Fig. 2. *Penicillus* meadow, north shore of Jamaica. Thallus stages range from pre-capitulum to senescent capitulum. Predominant species is *capitatus* (magnification approx. 0.2x)

Fig. 5. *Udotea* sp. from the Great Barrier Reef. (Photograph courtesy of D. Meyer; magnification approx. 0.66x)

phasis of this paper is on the living representatives, and on features other than purely morphological ones that are of present or potential interest in interpreting the geological record.

1.1 Architecture and Calcification of the Halimedaceae

The coenocytic filament is the basic structural unit of the order. Different branching patterns and organization of the filaments produce distinctive capitate (*Penicillus*), globuliferous (*Tydemania*), phylloid (*Rhipocephalus* and *Udotea*) and catenulate (*Halimeda*) forms, some of which have associated stipes, rhizomes and much enlarged holdfasts. Gepp and Gepp (1911), Hillis (1959), and Hillis-Colinvaux (1980, 1984) provide details of organization and taxonomic criteria as well as additional references.

All five genera deposit calcium carbonate in the conspicuous phase of their life history, in the form of aragonite (McConnell and Colinvaux 1967). Calcification is **external** to the filament, and for *Halimeda* begins in the newly formed (apical) segment, in association with the filament wall, when the segment is about 36 hours old (Wilbur et al. 1969; Hillis-Colinvaux 1980). The crystals are first deposited among the fibrils of the pilose outer layers of the walls, thereby forming a lining of the interfilamental spaces of the segment (Wilbur et al. 1969). In species of *Rhipocephalus*, *Penicillus* and *Udotea* aragonite deposits are formed within a sheath. A recent useful review of calcification in algae is provided by Borowitzka (1982).

Extent of calcification varies with age, taxonomy and environment. In general *Penicillus*, *Rhipocephalus* and *Udotea* are more lightly calcified, with mature thalli containing ca. 29–59 percent $CaCO_3$ (Hillis-Colinvaux 1980, Tables pp. 268, 269, 271) whereas the range for the mature *Halimeda* specimens sampled is 47–90% if the lightly calcified, subtropical *H. cuneata* Hering is excluded. Highest percentages were obtained for recently shed segments and the oldest segments on the plants. When analyzing such data sets it is important to realize that although apical segments of *Halimeda* are the youngest (Colinvaux et al. 1965), they may, nonetheless, contain carbonate deposits.

There have been few studies of calcium carbonate content with depth. Böhm (1973) examined seven *Halimeda* species growing on the north shore of Jamaica over depths from < 1 to ca. 57 m. Percent calcium carbonate decreased in *simulans* Howe and *goreauii* W. R. Taylor at depth, remained essentially the same for *tuna* (Ellis and Solander) Lamouroux, and increased for *discoidea* Decaise, *gracilis* Harvey *ex* J. Agardh, *opuntia* (Linnaeus) Lamouroux and *copiosa* Goreau and Goreau.

1.2 Geographic Distribution: Genera as Indicators of Tropical Environments

The calcareous Halimedaceae are almost entirely tropical in distribution, as defined by water temperatures of 25 °C and higher rather than by degrees of latitude (Hillis 1959). This definition produces asymmetric bands across the Pacific and Atlantic oceans (Fig. 6) that are distinctly broader in the west than in the east, and makes the presence of some of these genera in Bermuda and Japan, as well as the absence of all of them from Peruvian waters readily understandable. Exceptions to the tropical pattern include two species of *Halimeda*. *H. cuneata* appears to be restricted to the subtropics of both hemispheres of the Indian and Pacific oceans (Fig. 6), whereas *H. tuna* grows in the subtropical Mediterranean but also has an extensive tropical distribution, being reported for the Atlantic, Indian and Pacific oceans. Presence in three oceans has been attributed to only six other species of the genus, namely *incrassata* (Ellis) Lamouroux, *simulans*, *opuntia*, *copiosa*, *discoidea*, and *gracilis*. Some species of *Penicillus* and *Udotea* also grow in the Mediterranean as well as in the tropics. The Mediterranean presence of these three taxa may have resulted from a recolonization event following massive extinctions during the Messinian crisis (Hillis-Colinvaux 1980).

The genus *Udotea* may be less stenothermic than the others, with, for example, three species growing in subtropical waters off North Carolina in the northwestern Atlantic (Schneider 1976). No representatives of *Halimeda* or

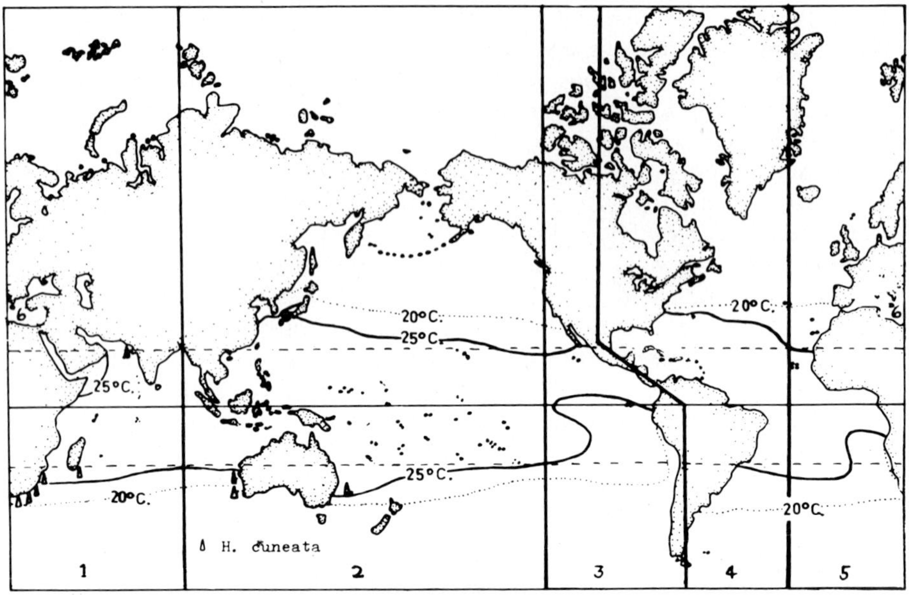

Fig. 6. Generalized global distribution of the Halimedaceae. Genera are principally tropical, as delimited by 25°C isotheres, one of the exceptions being the subtropical species *H. cuneata*, which has an authenticated distribution of S. Africa, southern Australia, and Gulf of Kutch, India. The six biogeographic regions shown are: western and central Indian Ocean, eastern Indian Ocean and western Pacific, eastern Pacific, western Atlantic, eastern Atlantic, and Mediterranean (Hillis-Colinvaux 1980)

Penicillus have been found in these waters (Searles, Schneider, personal communication).

Only two genera, *Halimeda* and *Udotea*, occur in the Atlantic, Indian and Pacific oceans. The known distribution of *Penicillus* outside the Mediterranean is essentially Atlantic and restricted Indo-Pacific; *Rhipocephalus* is limited to the Caribbean, and *Tydemania* to the Indo-Pacific. Because all five genera have a fairly strict tropical range they can be considered strong indicators of tropical environments. This characteristic has been used by Elliott (e.g. 1981) in analyses of the distribution of fossil green algal floras.

1.3 Within-Reef Distribution: Genera and Species as Indicators of Specific Reef Environments

There are very few data on the distribution of the different genera or of their biomass within specific reef systems, even for the most studied genus *Halimeda*. In this section I present an overview of reefal distribution for the five genera. More detailed information, especially for *Halimeda*, is provided in the subsequent discussion of sediment production, as well as by Hillis-Colinvaux (1977, 1980, 1985, 1986a,b,c).

Rhipocephalus, *Penicillus* and *Udotea* generally grow as discrete plants in regions of unconsolidated substrata, 'rooted' with a distinctive, often very large holdfast of free uncalcified filaments and adhering sand particles. All three genera can be locally abundant (Stockman et al. 1967). *Tydemania* also develops dense local populations (Hillis-Colinvaux 1977, 1980), although both its habit and habitat, succinctly described as sprawling over sand and rock surfaces, are different.

In comparison, when conditions are suitable *Halimeda*, a genus with a variety of types of holdfast (Fig. 7) such as the large bulbous style similar to *Penicillus*, *Rhipocephalus* and *Udotea* (found in the members of section Rhipsalis), but also relatively small matted ones with few adhering extraneous particles, or else discrete patches of loose filaments, can develop very dense populations at many sites across the reef on both sand and rock substrata. Consequently, *Halimeda* is one of the most versatile of algal colonizers. For those accustomed to thinking of the genus as predominantly a plant with a conspicuous holdfast for growing in lagoonal sands and muds, it is noteworthy that approximately three-quarters of its species grow principally on substrata other than sand (Hillis-Colinvaux 1985, 1986a). Sprawler species (mostly Section Opuntia) are associated with rock or sand-gravel substrata and have a habit rather similar to that of *Tydemania*; other *Halimeda* species grow readily

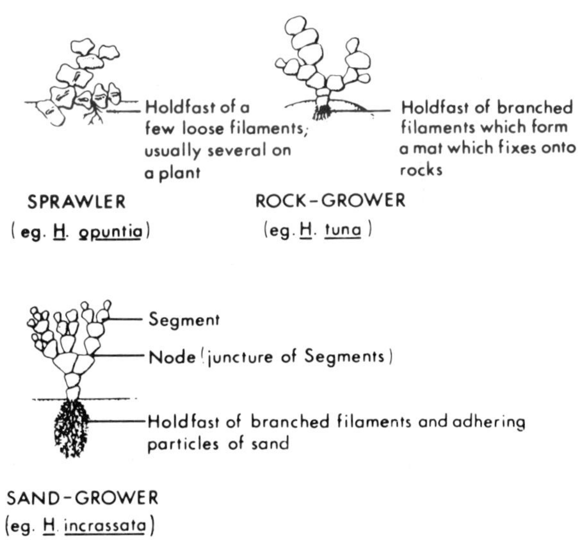

Fig. 7. Basic holdfast systems of *Halimeda* are associated with the major reef habitats of this genus. Species which sprawl over rock or sand usually have several inconspicuous points of attachment; strictly rock or sand-growing thalli generally possess a single holdfast (Hillis-Colinvaux 1980)

on rock surfaces. Where hard substrata predominate and the surface is essentially vertical as on many fore-reefs, draperies of *Halimeda* may develop over much of the reef wall.

None of the five genera achieves sustained space in high energy environments such as the algal ridge (Fig. 8), but *Halimeda* does develop close to 100% cover in regions of moderate energy such as the *H. opuntia* zone of inter-island channels at Enewetak Atoll (Hillis-Colinvaux 1980, 1985). Its populations are also uncommon to rare on most spur and grooves (Fig. 8); for this region Hillis-Colinvaux (1977, 1980) reported about 15% cover by *Halimeda* off a leeward islet of Enewetak Atoll in 1975, but essentially none for sites examined on the north shore of Jamaica in 1967–1969.

All five genera have broad bathymetric ranges (Taylor 1960), with *Halimeda* extending from the intertidal to at least –150 m (Hillis-Colinvaux 1982, 1985, 1986a; Blair and Norris 1985, 1988). Depth tolerances among the species, where known, are generally broad (Hillis-Colinvaux 1986a), although one Caribbean species, *H. cryptica* L.H. Colinvaux and Graham appears to be restricted to deep water, that is, to 25 m and deeper (Colinvaux and Graham 1964).

Present data indicate, therefore, that neither the genera of Halimedaceae nor species of *Halimeda* with the exception of *H. cryptica* and possibly one or two others, are precise depth indicators within the euphotic zone (Hillis-Colinvaux 1980, Fig. 90; Johns and Moore 1988). However, within an entire reef system genera and species can function as indicators, as the following examples show:

1. An abundance of rhizophyte ('rooted') taxa usually indicates back-reef environments, since species with well-developed holdfasts (*Penicillus*, *Rhipocephalus*, the larger *Udotea*, and the taxonomic section Rhipsalis of *Halimeda*) generally are uncommon to rare on the deep fore-reef;

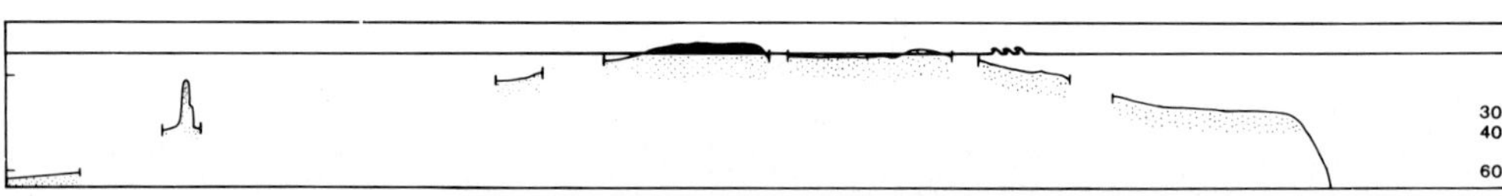

LAGOON REEF				ISLET or INTER-ISLAND CHANNEL	BEACH	SEAWARD REEF				
LAGOON BASIN	LAGOON SLOPE / PINNACLE (2000+)		LAGOON TERRACE			BACK REEF	REEF-CREST		FORE-REEF	
							ALGAL RIDGE	SPUR & GROOVE	TERRACE	SLOPE (WALL)
POPULATIONS MAY BE VERY DENSE CLOSE TO SHORE AND ON TOP OF PINNACLES				APPROX. 100% COVER, SOME SITES	0	PROBABLY LOW, IN CREVICES	0	15% COVER ON 1 SITE LEEWARD – TRANSITIONAL REEF	AMID CORAL HEADS (?+IN SAND)	HIGH SP. DIVERSITY AND COVER GROWTH TO >140m.
ROCK & SAND STRATEGIES	ROCK-GROWERS SOME SPRAWLERS	ROCK & SAND STRATEGIES	ROCK & SAND STRATEGIES	*Halimeda opuntia* ZONE	0	ROCK-GROWERS	0	ROCK-GROWERS AND SPRAWLERS	ROCK & ?SAND STRATEGIES	ROCK-GROWERS SOME SPRAWLERS

Fig. 8. Distribution of Halimedaceae within a reef. Transverse section of Enewetak Atoll shows distribution of the different growth strategies of Halimedaceae across the reef (*bottom line*), and general *Halimeda* biomass (*second line from botton*) (Modified from Hillis-Colinvaux 1986a)

and Pacific Oceans. For understanding entire reef systems additional analyses of reef regions away from the ridge are needed; they are especially desirable for the Indian Ocean. As studies of the unconsolidated regions of reefs increase we may expect many additional sites to be found with dense present-day populations of Halimedaceae.

3.1 Three-Dimensional Within-Reef Space Occupied by the Halimedaceae

The quantity of live *Halimeda* in reefs is generally deceptive, made so by its cryptic. When photographed against rock surfaces resolution may be so poor that identification even as a *Halimeda* is precluded. Reefs therefore need to be examined with an emphasis on this genus if we are to obtain good data on the space occupied. At present there are analyses of only three reef systems (Enewetak Atoll, north shore of Jamaica, and the Great Barrier Reef at the latitude of Townsville), which include this genus across most of a reef system, that is from back-reef or lagoon to deep fore-reef, and which therefore can provide a statement of distribution within entire reef systems (Gilmartin 1960, 1966; Colinvaux and Graham 1964; T.F. Goreau and Graham 1967; T.R. Goreau and N.I. Goreau 1973; Moore et al. 1976; Hillis-Colinvaux 1977, 1980, 1985, 1986a,c,d, 1988; Drew 1983; Drew and Abel 1985, 1988a; Colin 1986; Colin et al. 1986). For the other Halimedaceae there seem to be no similar analyses of distribution, although extensive information on the distribution of *Tydemania* at Enewetak is provided by Gilmartin (1960, 1966) and Hillis-Colinvaux (1977). Scuba diving has been an important tool in these studies, but it is only with the availability of submersibles and remote photographic systems that there has been the opportunity to examine considerably wider expanses of entire systems, and thus obtain for broad areas of lagoon or fore-reef a baseline quantitative analysis of distribution.

The densest covers of living *Halimeda* found so far occur in three principal places of a reef: 1) relatively shallow to very shallow (ca. 0.3 m) rock platforms, 2) lagoonal or back reef areas with an unconsolidated substratum of *Halimeda* gravels, sand (sometimes mixed with fine rubble to which some of the plants may be attached), and muds, and 3) vertical walls as represented by the fore-reef slope, or sides of bommies (pinnacles). The first two are the locations of meadows, the third of draperies. Percent cover at any of these sites can be very high.

Halimeda Meadows

The biomass of *Halimeda* over many horizontal reef surfaces is so dense as to resemble terrestrial meadows. The *H. opuntia* zone found in moderate energy interislet or back reef environments only a few centimetres deep (Hillis-Colinvaux 1980, 1985, 1986a) is an example in the first of the three preceding categories of habitat. In such an environment *H. opuntia* produces tightly branched clumps, and cover is 80–100%. Other species, with habit similarly modified to small-segmented compact bushes, provide dense covers in similar habitats (Hillis-Colinvaux 1980). *Halimeda* communities on the flattish tops of lagoon pinnacles provide a second example of a dense cover over lithified substrata; where observed at Enewetak these communities were of mixed species, although *H. macrophysa* Askenasy provided most of the cover (Hillis-Colinvaux 1977, 1980).

Dense populations of *Halimeda* growing on **unconsolidated habitats** include *Halimeda* meadows of the northern part of the Great Barrier Reef (Drew and Abel 1985, 1988a), meadows in the lagoons of several atolls of the Marshall Islands (Colin 1986 and personal communication; Hillis-Colinvaux 1988) and an atomic bomb crater at Enewetak (Hillis-Colinvaux 1980). Meadows of the northern Great Barrier Reef, occurring principally at depths of 20–40 m but with smaller ones at 100 m or more, cover 2000 km^2 or more of sea bed, at least some of which is a post Pleistocene *Halimeda* bioherm (Davies and Marshall 1985; Orme 1985; Phipps et al. 1985). Although the communities include *Penicillus* and *Udotea*, the predominant genus for the sites examined is *Halimeda* (Drew and Abel 1985, 1988a). The three commonest species belong to section Opuntia, with *copiosa* the most prominent. *Halimeda* biomass of up to 4 kg dry weight m^{-2} is reported.

In the Marshall Islands, *Halimeda* meadows on unconsolidated substrata have been discovered in the lagoons of Rongelap (ca. 36 m), Kwajalein (ca. 32 m and deeper) and Enewetak (ca. –21 and –25 m) Atolls (Fig. 12). All these

Fig. 12. *Halimeda* meadow of mixed species, in lagoon of Enewetak Atoll at approximately 22 m depth (Photo by P. Colin; magnification approx. 0.1x)

meadows, like the Australian ones, contain a mixture of *Halimeda* species. The most prominent taxa in the Enewetak meadows are *incrassata*, *cylindracea* Decaisne, near *gigas* W.R. Taylor/*xishaensis* Dong and Tseng, *discoidea*, *opuntia* f. *opuntia*, and a small segmented, fine form of *opuntia*. Where the substratum was sand and small fragments of coral rubble a lax *incrassata* often predominated. Where it was sand or gravel mixed with small rocks the taxon presently delimited as *gigas*/*xishaensis* was the most abundant macroalga. Other Halimedaceae were not obvious. Percent cover was high (Hillis-Colinvaux 1980, 1986c, colour Fig. p. 47), and the contribution by the holdfast alone to total carbon investment was also often high: over 40% where the rhizophyte *H. cylindracea* predominated (Hillis-Colinvaux 1988). Also at Enewetak, but at an interisland-back reef site on the north end of Runit Islet, a *single species* meadow on a mud-silt substratum was located at the bottom (ca. –11 m) of the atomic bomb crater 'Cactus'. There the rhizophyte species *H. incrassata* provided close to 100% cover (Hillis-Colinvaux 1980, Fig. p. 260).

Extensive *Halimeda* banks recently discovered elsewhere, such as east of the central Sunda shelf margin in the Java Sea (Roberts et al. 1987a,b) and the south west Caribbean (Hine et al. 1988) suggest that dense groves of living *Halimeda* will be found for many other parts of the tropical world ocean.

Penicillus Meadows

I have observed dense growth of *Tydemania* in the western Indian Ocean and at Enewetak, but the areas covered have not been sufficiently dense or extensive to be called meadows. Hence, the only other genus of the Halimedaceae known to produce extensive dense stands is *Penicillus*. Sizeable groves were found at some of my field sites in Jamaica (Fig. 2), and Stockman et al. (1967) record densities of 76–108 thalli m^{-2} and higher in southern Florida. It seems likely that similar stands occur elsewhere in the Caribbean.

Halimeda Draperies

Vertical, pendant colonies and communities of *Halimeda* may, in some respects, be considered vertical meadows. They represent a very prominent kind of *Halimeda* development on reef slopes and walls, regions that traditionally, and especially in the deep fore-reef, have not been considered habitats for *Halimeda*. The term 'drapery' is used because the extensive pendant colony forms of members of section Opuntia, especially the species *copiosa*, which may well prove to be the predominant wall species globally, suggest traditional wall hangings. Their long and trailing fronds drape down and across rock surfaces, caverns and crevices, to develop festoons that can be the major benthic carbonate source of a slope.

Other high biomass taxa on slopes and walls include the *Halimeda* species *distorta* (Yamada) L.H. Colinvaux and *cryptica*. As in the meadows a single *Halimeda* species may predominate, or there may be two to several prominent *Halimeda* taxa. On the fore-reef of Jamaica, at –25 to –100 m *H. cryptica*, an unusual uniaxial species predominates, but is often associated with *copiosa* (T.F.

Goreau and N.I. Goreau 1973). At San Salvador, Bahamas, *H. copiosa* provides up to 39% cover from −117 to −130 m but is associated with three other *Halimeda* species (Littler et al. 1986). *Halimeda copiosa* also develops extensive hangings on the sloping walls of bommies (pinnacles) associated with Davies Reef in the Great Barrier Reef (personal observation), and was one of the prominent species collected from the deep fore-reef of Enewetak Atoll (Hillis-Colinvaux 1985, 1986a,d).

Other distinctive habits are exhibited by the wall-growing halimedas (Hillis-Colinvaux 1985). Especially noteworthy in terms of biomass represented were the large single species clumps or tussocks produced by *H. distorta* at about −100 m (Hillis-Colinvaux 1986c, colour photograph p. 47). A single tussock collected by submersible had a dry weight of 682 g.

Overall vertical cover of the fore-reef can be high. At Enewetak *Halimeda* populations were found to cover 10–50% or more of the atoll slope down to 110 m, with at least a third of the species growing to more than 90 m (Hillis-Colinvaux 1986a,d). At this depth cover by hermatypic corals did not exceed 1% (Colin et al. 1986; Hillis-Colinvaux 1986a). The genus continued down to at least −140 m. The full extent of its cover was, however, difficult to determine during submersible observations and from photographs because sediments blanketed extensive areas of the fore-reef slope. Sediment cover was conspicuously heavier on the leeward side of the atoll.

There are essentially no data on distribution or biomass of the other Halimedaceae on vertical surfaces. At Enewetak only *Udotea* and *Tydemania* are known (Taylor 1950). Two species of *Udotea* grew to at least −138 m, and the diminutive *U.* near *javensis* (Montagne) A. and E.S. Gepp was relatively abundant on likely substrata. *Udotea* biomass, however, was insignificant compared to the *Halimeda* contribution. No *Tydemania* thalli were located on the fore-reef wall.

3.2 Yield of Carbonate Sediments

Most of the material shed as debris by these genera has a high as well as relatively uniform carbonate content (Hillis-Colinvaux 1980, Tables pp. 268–269, 271) but the rate at which a thallus yields sediments varies. Consequently, the major question of concern to workers deriving their data from **living** material is the number of generations, or of 'turnovers', produced annually. Turnover time herein is considered to be the length of time taken to replace an individual. In the context of carbonate production this definition, based on Lindeman's analysis of energy flow in a Minnesotan lake (1942), does not have the problem inherant in Lindeman's application of it to primary production since respiratory losses are not coupled to carbonate production. Losses of parts of a plant from grazing or from other causes must, however, be considered.

Data on generation times for the Halimedaceae are very limited. In laboratory aquarium systems I found that *Penicillus capitatus* Lamarck replaced itself about every 1.5–2 months (Colinvaux et al. 1965), a generation time in accord with the field results of Stockman et al. (1967). *Udotea flabellum* (Ellis and Solander) Lamouroux in the same laboratory system replaced itself in 6–9 months. *Halimeda* field census data from Jamaica and laboratory observations, mostly on sand-growing species, indicate an annual turnover time of 1/2–3 generations, although some laboratory *H. incrassata* thalli replaced themselves in 2 months. Field data based on number of segments gained rather than entire thalli provide turnover times of 39 days of populations of *H. incrassata* and *H. monile* (Ellis and Solander) Lamouroux in Nonesuch Bay, Antigua W.I. (Multer 1986, 1988); about 1 year for *H. opuntia* in the Marquesas Keys, Florida, depth < 2 m, (Hudson 1985); and 15 days for *H. opuntia* and *H. copiosa* combined at Davies Reef, Great Barrier Reef, depth 5–8 m (Drew 1983). The different approaches used to obtain generation-turnover times for *Halimeda* complicate the evaluation of the resulting data; some of the techniques are mandated, however, by the different habits of the plants. Species that are predominantly sprawlers such as *H. opuntia* and *H. copiosa* do not have the discrete life form usually exhibited by rhipsalian (rhizophyte) halimedas or *Penicillus* which makes distinguishing different generations more straightforward. For turnover figures to be used with confidence it is obvious that many more data are needed, and that more species and a much wider range of habitats should be examined.

3.3 How Much Carbonate Does *Halimeda* Produce?

Data on carbonate *Halimeda* has become available in recent years from two major sources: from biologists and geologists studying the growth of living thalli and calculating total productivity based on biomass and turnover or generational data (Merten 1971; Hillis-Colinvaux 1974, 1980; Drew 1983; Abel and Drew 1985; Drew and Abel 1985; Hudson 1985; Multer 1988), and from geologists analyzing sedimentary deposits with seismic profiling and side-scan sonar techniques, together with radiocarbon dating of reef cores to determine vertical accretion (Davies and Marshall 1985; Orme 1985; Phipps et al. 1985; Roberts et al. 1987a,b). Translated into rates of vertical accretion most of the figures are in the range of 0.18–5.9 m 1000 years^{-1}.

Different methodologies, environments, population sizes, and possibly species differences account for this broad diversity, with the lowest of the figures given (I have excluded figures one to two orders of magnitude below the above range) resulting in part from a relatively low percent cover (= ca. 21%, Hillis-Colinvaux 1974, 1980). Of the data available only those based on 'annual contribution per single thallus' (Hillis-Colinvaux 1980; Multer 1988) provide a ready basis for modelling and prediction. Thus, if a single rhipsalian thallus produces 0.8 g carbonate year^{-1} (Hillis-Colinvaux 1980, considered *conservative*), a 60% cover (= 600 *H. incrassata* thalli, cf. Hillis-Colinvaux 1980, p. 274) by the same species under similar conditions, a density which is still below that of *Halimeda* meadows, would produce 0.4 m of carbonate vertical accretion in 1000 years. The porosity used for the calculations is 60%, following Stockman et al. (1967). In the same way, a cover of 100% *Halimeda*, which approximates cover in the *H. opuntia* zone of Enewetak and the *Halimeda* meadows at –21m in the lagoon west of Enewetak Islet, would yield a vertical accretion of 0.7 m in 1000 years. (The reader should be aware of two textual errors in earlier publications which may be confusing. Hillis-Colinvaux 1980, p. 274, text line 12: 4066 mm $year^{-1}$ should read 0.066 mm $year^{-1}$. Hillis-Colinvaux (1986c, p. 47, first column last line) should read 7 cm 1000 $years^{-1}$; and second column lines 1–2 should be 'years for a low cover, by sand growing species, to 3 meters'.)

The highest figures in the range provided in paragraph one are based on cores taken from *Halimeda* banks east of the central Sunda Shelf margin in the eastern Java Sea (Roberts et al. 1987a,b), and the northern Great Barrier Reef (Davies and Marshall 1985). Both groups of workers obtained data indicating rates of vertical accretion that increased throughout the Holocene, with average rates for the time spans represented in the cores analyzed being 2.8 and 1.75 m 1000 $years^{-1}$ respectively. Both groups also suggest that the implied rapid growth rates of *Halimeda* are related to nutrient enrichment from upwellings.

Although export or import of sediments may be involved in the development of some *Halimeda* bioherms, those of the northern Great Barrier Reef are considered to represent in situ accumulations of *Halimeda*, some of which are presently topped by meadows of living *Halimeda* (Drew and Abel 1985). They provide strong evidence for the reef-building activity and ecological success of at least one genus of the Halimedaceae throughout the Holocene.

3.4 Major Accretion Events

Biotic and abiotic factors producing widespread or catastrophic death of the plants lead to major episodes of accretion. Likely factors include temperature and salinity stresses, disturbances resulting from tropical storms, sexual reproduction, and possibly epidemic infestation by cyanobacteria. Only a brief elaboration of two biotic factors can be included in this paper, but it will serve to underscore the potential of certain events. The first example, derived from field studies in Bermuda, Jamaica and Enewetak, is the observation of extensive populations of *Halimeda* shrouded by cyanobacteria (Hillis-Colinvaux 1980, Fig. 72). Since similar events in culture have led to the death of the plants or else a shedding of branches with subsequent regrowth, both entailing sediment production, these same patterns are predicted for field populations.

The second example is based on field and laboratory observations of sexual reproduction, an event which now has been reported for all five genera and many *Halimeda* species (Hillis-Colinvaux 1959, 1973, 1980, 1984; Meinesz 1980 and personal communication; Drew and Abel 1988b), although its very existence for some of the genera was questioned until relatively recently (Dawson 1966). Death follows immediately after the formation of gametangia (Figs. 9, 10), the outpouring of the energy reserves of the entire plant into the gametes, and their subsequent release (holocarpy). The process from the first readily observable stages of gametangial formation to release of gametes occurs in ~36h. A widespread occurrence of sexual reproduction would therefore be a major sediment-producing event. And, since the cytoplasmic contents of the algae are channelled into gamete production, only the framework materials of the filaments remain. Shed segments from sexually reproducing plants, therefore, are essentially skeletal.

3.5 Impact of Halimedaceae on the Reef Mass of Modern and Holocene Reefs

As one moves across a classical tropical reef composed of lagoon or back reef, algal ridge-spur and groove, and fore-reef, the principal halimedacean (and other) components change (Fig. 8). As noted earlier, percent cover by Halimedaceae is essentially zero on the ridge and relatively low on most spur and grooves. Cover by the Halimedaceae and principally by *Halimeda* can, however, be very high in the other two major regions.

Reef mass ultimately depends on the rates of accretion in these three major regions which, over geological time, are affected by sea level. During lowered sea levels, for example, lagoons may become extinct, with a corresponding decreased contribution by the halimedas living on the floor and on pinnacle walls. Delimitation of different physiographic regions in a carbonate model of a reef allows differences in rates of accretion to be considered; it also recognizes the importance of the extent of their areas. Thus, in reefs with lagoons of considerable size such as Enewetak and Bikini Atolls, lagoonal populations are likely to contribute more than half the reef mass, even with accretion rates significantly lower, as obtained by Smith and Kinsey (1976; 0.8 kg m^{-2} $year^{-1}$ for lagoon, 4 kg m^{-2} $year^{-1}$ for reef ridge).

In reef systems where accretions of lagoon and fore-reef are most likely to be from *Halimeda* it is reasonable to suggest that the origin of reef mass depends on the relative rates of accretion on lagoon floor, reef ridges (by their populations of corals and coralline algae), and fore-reef as expressed by:

$$R_m = (H_sL + R_vR_h + H_sC_dF)t,$$

where R_m = reef mass, H_s = deposition of *Halimeda* segments per unit area, L = area of lagoon, R_v and R_h = vertical and horizontal components of reef ridge growth, C_d = deposition of coral debris per unit area, F = area of fore-reef wall, and t = time. For *Halimeda* bioherms, since the ridge and fore-reef elements are absent, the expression can appropriately be shortened to:

$$R_m = (H_sL)t.$$

4 A Classification of Tropical Reefs by Principal Components

The predominant components of tropical reef systems are photoautotrophs, so it seems especially quixotic to continue to use the term 'coral reefs' in scientific publications. Recently I argued that 'algal reefs', a name which includes not only calcareous algae but also the symbiotic components and fleshy algae, more accurately reflects both the ecological and physical basis of reef systems (Hillis-Colinvaux 1986c). The evidence presented in this chapter, drawn from both present and historical records, provides additional support for the use of **algal reef**, or alternatively of **coralgal reef** in naming this tropical carbonate system.

Halimeda, although one of the principal components of many Recent reefs, is not, however, equally abundant in all tropical algal reefs. For reasons not understood but which may be related to problems of successful dispersal, both species diversity and biomass of *Halimeda* (and the other calcareous Halimedaceae) are greatly reduced on certain reefs. There are, in other words, ***Halimeda*** **and non-*Halimeda* reefs**. The absence of *Halimeda* from some reef systems has important implications in the ecological and accretionary processes of these reefs.

Acknowledgements. A significant portion of this paper was written while I was a Bunting Institute Science Scholar of Radcliffe College at Harvard University. The scholarly and convivial support of the Harvard and Bunting communities, and the financial support of the Office of Naval Research for the Science Scholar Program are gratefully acknowledged. I also thank Robert Riding for thoughtful discussion of the early history of the calcareous Bryopsidales, and the stimulus thereby provided me for developing an analytical interface between living and fossil Halimedaceae.

References

Abel KM, Drew EA (1985) Response of *Halimeda* metabolism to various environmental parameters. In: Proc 5th Int Coral Reef Congr 5:21–26

Blair SM, Norris JN (1985) Studies of deep-water Caribbean species of *Halimeda* (Udoteaceae): Observations on depth distribution and morphology. J Phycol 21 (Suppl):5

Blair SM, Norris JN (1988) The deep-water species of *Halimeda* Lamouroux (Halimedaceae, Chlorophyta) from San Salvador Island, Bahamas: species composition, distribution and depth records. Coral Reefs, 6, 3/4:227–236

Böhm EL (1973) Studies on the mineral content of calcareous algae. Bull Mar Sci 23:177–190

Borowitzka MA (1982) Morphological and cytological aspects of algal calcification. Int Rev Cytol 74:127–162

Colin PL (1986) Benthic community distribution in the Enewetak Atoll lagoon, Marshall Islands. Bull Mar Sci 38:129–143

Colin PL, Devaney DM, Hillis-Colinvaux L, Suchanek TH, Harrison JT III (1986) Geology and biological zonation of the reef slope, 50–360 m depth, at Enewetak Atoll, Marshall Islands. Bull Mar Sci 38:111–128

Colinvaux LH, Graham EA (1964) A new species of *Halimeda*. N Hedwigia 7:5–10 + pl

Colinvaux LH, Wilbur KM, Watabe N (1965) Tropical marine algae: growth in laboratory culture. J Phycol 1:69–78

Davies PJ (1986) *Halimeda* bioherms and *Halimeda* biostromes: low energy accumulations in the Great Barrier Reef. Int Soc Coral Reef Stud Marburg, FRG, Abstr, p 13

Davies PJ, Marshall JF (1985) *Halimeda* bioherms. Low energy reefs, Northern Great Barrier Reef. In: Proc 5th Int Coral Reef Congr 5:1–8

Dawson EY (1966) Marine botany. Holt, Rinehart & Winston, New York, 371 pp

Drew EA (1983) *Halimeda* biomass, growth rates and sediment generation on reefs in the central Great Barrier Reef Province. Coral Reefs 2:101–110

Drew EA, Abel KM (1985) Biology, sedimentology and geography of the vast inter-reefal *Halimeda* meadows within the Great Barrier Reef Province. In: Proc 5th Int Coral Reef Congr 5:15–20

Drew EA, Abel KM (1988a) Studies on *Halimeda*. I. The distribution and species composition of *Halimeda* meadows throughout the Great Barrier Reef Province. Coral Reefs 6:195–205

Drew EA, Abel KM (1988b) Studies on *Halimeda*. II. Reproduction, particularly the seasonality of gametangia formation, in a number of species from the Great Barrier Reef Province. Coral Reefs 6:207–218

Elliott GF (1981) The Tethyan dispersal of some chlorophyte algae subsequent to the Palaeozoic. Palaeogeogr Palaeoclimatol Palaeoecol 32:341–358

Flügel E (1988) *Halimeda*: paleontological record and palaeoenvironmental significance. Coral Reefs 6:123–130

Folk RL, Robles R (1964) Carbonate sands of Isla Perez, Alacran Reef complex, Yucatan. J Geol 72:255–292

Gepp A, Gepp ES (1911) The Codiaceae of the Siboga expedition including a monograph of Flabellarieae and Udoteae. Brill, Leiden, 150 pp

Gilmartin M (1960) The ecological distribution of the deep water algae of Eniwetok Atoll. Ecology 41:210–221

Gilmartin M (1966) Ecology and morphology of *Tydemania expeditionis*, a tropical deep-water siphonous green alga. J Phycol 2:100–105

Goreau TF, Graham EA (1967) A new species of *Halimeda* from Jamaica. Bull Mar Sci 17:432–441

Goreau TF, Goreau NI (1973) The ecology of Jamaican coral reefs. II. Geomorphology, zonation and sedimentary phases. Bull Mar Sci 23:399–464

Heckel PH (1974) Carbonate buildups in the geologic record. In: Laporte LF (ed) Reefs in time and space: selected examples from the recent and ancient. Soc Econ Paleontol Mineral Tulsa, Okl Spec Publ 18

Hillis LW (1959) A revision of the genus *Halimeda* (order Siphonales). Inst Mar Sci Univ Texas 6:321–403

Hillis-Colinvaux L (1973) Reproduction in the calcareous green algae of coral reefs. J Mar Biol Assoc India 14:328–334

Hillis-Colinvaux L (1974) Productivity of the coral reef alga *Halimeda* (Siphonales). In: Proc 2nd Int Coral Reef Symp, vol 1, pp 35–42

Hillis-Colinvaux L (1977) *Halimeda* and *Tydemania*: distribution, diversity and productivity at Enewetak. In: Proc 3rd Int Coral Reef Symp, vol 1, pp 365–370

Hillis-Colinvaux L (1980) Ecology and taxonomy of *Halimeda*: Primary producer of coral reefs. Adv Mar Biol 17:1–327

Hillis-Colinvaux L (1982) Submersible study of the deep macrophyte populations of Enewetak Atoll, Marshall Islands, with emphasis on the calcareous genus *Halimeda*. Joint Oceanogr Assemb, Halifax, Poster Abstr, vol 2, p 20

Hillis-Colinvaux L (1984) Systematics of the Siphonales. In: Irvine DEG, John DM (eds) Systematics of the green algae. Academic Press, New York London, pp 271–286

Hillis-Colinvaux L (1985) *Halimeda* and other deep fore-reef algae at Enewetak Atoll. In: Proc 5th Int Coral Reef Congr, vol 5, pp 9–14

Hillis-Colinvaux L (1986a) Deep water populations of *Halimeda* in the economy of an atoll. Bull Mar Sci 38:155–169

Hillis-Colinvaux L (1986b) Distribution patterns of some Bryopsidales in the geologic past: their bearing on present distributions. Bot Mar 29:217–277

Hillis-Colinvaux L (1986c) Historical perspectives of algae and reefs: have reefs been misnamed? Oceanus 29:43–48

Hillis-Colinvaux L (1986d) *Halimeda* growth and diversity on the deep fore-reef of Enewetak Atoll. Coral Reefs 5:19–21

Hillis-Colinvaux L (1988) Characteristics of *Halimeda* meadows, with emphasis on a meadow near Enewetak Islet, Enewetak Atoll (Marshall Islands). In: Proc 6th Int Coral Reef Symp 3:119–125

Hillis-Colinvaux L, Orme GR (1988) Trends in research on the biology and geology of *Halimeda*. In: Proc 6th Int Coral Reef Symp. Status review. 1:193–198

Hine AC, Hallock P, Harris MS, Mullins HT, Belknap DF, Jaap WC (1988) *Halimeda* bioherms along an open seaway: Miskito Channel, Nicaraguan Rise, SW Caribbean Sea. Coral Reefs 6:173–178

Hudson JH (1985) Growth rate and carbonate production in *Halimeda opuntia*: Marquesas Keys,

Florida. In: Toomey DF, Nitecki MH (eds) Paleoalgology: contemporary research and applications. Springer, Berlin Heidelberg New York Tokyo, pp 257–263
Johns HD, Moore CH (1988) Reef to basin sediment transport using *Halimeda* as a sediment tracer, Grand Cayman Island, West Indies. Coral Reefs 6:187–193
Lindeman RL (1942) The tropic dynamic aspects of ecology. Ecology 23:399–418
Littler MM, Littler DS, Blair SM, Norris JN (1986) Deep-water plant communities from an uncharted seamount off San Salvador Island, Bahamas: distribution, abundance, and primary productivity. Deep Sea Res 33:881–892
Mankiewicz C (1988) Occurrence and paleoecologic significance of *Halimeda* in late Miocene reefs, southeastern Spain. Coral Reefs 6:271–279
Marshall JR, Davies PJ (1988) *Halimeda* bioherms of the northern Great Barrier Reef. Coral Reefs 6:139–148
McConnell D, Hillis-Colinvaux L (1967) Aragonite in *Halimeda* and *Tydemania* (order Siphonales). J Phycol 3:198–200
Meinesz A (1980) Connaissances actuelles et contribution à l'étude de la reproduction et du cycle des Udotéacées (Caulerpales, Chlorophytes). Phycologia 19:110–138
Merten MJ (1971) Ecological observations of *Halimeda macroloba* Decaisne (Chlorophyta) on Guam. Micronesica 7:27–44
Moore CH Jr, Graham EA, Land LS (1976) Sediment transport and dispersal across the deep fore-reef and island slope (-55m to -305m), Discovery Bay, Jamaica. J Sediment Petrol 46:174–187
Multer HG (1986) Growth rate and sediment contribution of *Halimeda incrassata* and *Halimeda monile* from different environments, Antigua WI. In: Int Soc Coral Reef Stud, Marburg, FRG, Abstr, p 34
Multer HG (1988) Growth rate, ultrastructure and sediment contribution of *Halimeda incrassata* and *Halimeda monile*, Nonsuch and Falmouth Bays, Antigua, WI Coral Reefs 6:179–186
Orme GR (1985) The sedimentological importance of *Halimeda* in the development of back reef lithofacies, Northern Great Barrier Reef (Australia). In: Proc 5th Int Coral Reef Congr, 5:31–38
Orme GR, Salama MS (1988) Form and seismic stratigraphy of *Halimeda* banks in part of the northern Great Barrier Reef Province. Coral Reefs 6:131–137
Phipps CVG, Roberts HH (1988) Seismic characteristics and accretion history of *Halimeda* bioherms on Kalukalukuang Bank, eastern Java Sea (Indonesia). Coral Reefs 6:149–159
Phipps CVG, Davies PJ, Hopley D (1985) The morphology of *Halimeda* banks behind the Great Barrier Reef east of Cooktown, QLD. In: Proc 5th Int Coral Reef Congr 5:27–30
Roberts HH, Phipps CV, Effendi L (1987a) *Halimeda* bioherms of the eastern Java Sea, Indonesia. Geology 15:371–374
Roberts HH, Phipps CV, Effendi L (1987b) Morphology of large *Halimeda* bioherms, eastern Java Sea (Indonesia): a side-scan sonar study. Geo Mar Lett 7:7–14
Schneider CW (1976) Spatial and temporal distributions of benthic marine algae on the continental shelf of the Carolinas. Bull Mar Sci 26:133–151
Silva PC (1980) Names of classes and families of living algae. Regnum Veg 103:1–156
Smith SV, Kinsey DW (1976) Calcium carbonate production, coral reef growth and sea level change. Science 194:937–939
Stockman KW, Ginsburg RN, Shinn EA (1967) The production of lime mud by algae in South Florida. J Sediment Petrol 37:633–648
Suchanek TH, Colin PL, McMurtry GM, Suchanek DS (1986) Bioturbation and redistribution of sediment radionuclides in Enewetak Atoll lagoon by callianassid shrimp: biological aspects. Bull Mar Sci 38:144–154
Taylor WR (1950) Plants of Bikini and other northern Marshall Islands. Univ Mich Press, Ann Arbor, 227 pp
Taylor WR (1960) Marine algae of the eastern tropical and subtropical coasts of the Americas. Univ Mich Press, Ann Arbor
Wilbur KM, Hillis-Colinvaux L, Watabe N (1969) Electron microscope study of calcification in the alga *Halimeda* (order Siphonales). Phycologia 8:27–35

Chapter 11

The Genus Concept in Charophyta: Evidence from Palaeozoic to Recent

M. Feist[1] and N. Grambast-Fessard[1]

Abstract

In fossil Charophyta, the generic criteria are based on characters of the calcified fructifications (gyrogonite or utricle). However, the classification of living forms is based mainly on vegetative parts. An alternative system employing gyrogonite characters might permit the classification of both living and fossil members of the Characeae. Thus, the systematics of fossil Charophyta may be considered as founded on natural taxa and not on organ-genera. Examples of Palaeozoic to Recent representatives demonstrate the various criteria used to distinguish genera in the different families. Phylogenetic lineages of species allow the recognition of phyletic genera, especially in the Clavatoraceae. Relationships between genera outline the main phylogenetical trends in Charophyta. Two new suborders are established as well as two new combinations concerning species figured in the present paper: Palaeocharineae *nov. subordo*, Charineae *nov. subordo*; *Nitellopsis* (*Tectochara*) *morulosa n. comb.*. *Porochara sardiniae n. comb.* In addition, the revision of all charophyte genera has led us to put nine genera into synonymy: *Minhechara* Wei = *Latochara* Mädler; *Paracuneatochara* Wang = *Leonardosia* Sommer; *Caucasuella* Kyansep-Romashkina = *Aclistochara* Mädler; *Gobichara* Karczewska & Ziembinska-Twordzydlo = *Microchara* Grambast; *Piriformachara* Liu & Wu = *Mesochara* Grambast; *Qinghaichara* Yang = *Nitellopsis* Hy; *Retusochara* Grambast = *Peckisphaera* Grambast; *Sinochara* Lin & Wang = *Peckichara* Grambast; *Songliaochara* Wang, Lu & Zhao = *Nodosochara* Mädler; *Turbochara* Wang = *Nodosochara* Mädler; *Tolypella* (Al. Braun) Leonhardi, pro parte = *Sphaerochara* Mädler emend. Horn af Rantzien & Grambast.

1 Introduction

In Charophyta, the systematics of fossil forms are based on the morphology of the calcified female gametangium (oosporangium), called the gyrogonite, whereas the classification of extant species is based on the vegetative parts. Gyrogonites range from 200 to 2000 μm and can be investigated only by means of optical and scanning electron microscopy. On the other hand, the size of the thallus, from 1.5 to 200 cm (Wood and Imahori 1965), permits direct observations. The vegetative parts, always present in the living plant, even in species that do not calcify or reproduce vegetatively, are commonly missing in the fossil which predominantly comprises isolated gyrogonites. According to Horn af Rantzien (1959), the gyrogonites do not represent the entire plant and should therefore be classified as organ genera and organ species.

However, the fundamental structure of the oosporangium is similar in extant and fossil representatives (Fig. 1). Oosporangia are spherical bodies with

[1] Laboratoire de Paléobotanique, URA n° 327 du CNRS, Université des Sciences et Techniques du Languedoc, 34095 Montpellier, France

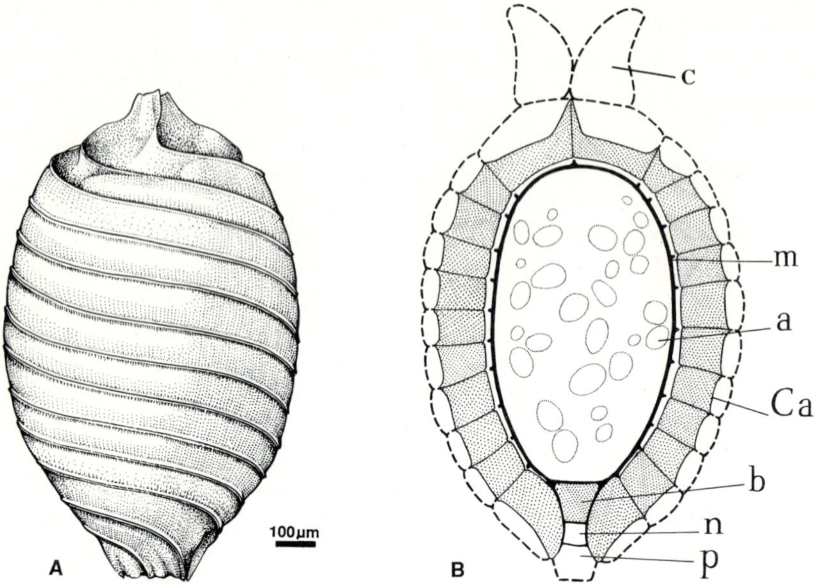

Fig. 1A,B. *Chara hispida* L. (Characeae). From Grambast 1958 (unpublished thesis). **A** Gyrogonite (calcified oogonia) of a Recent specimen (published in Soulié-Märsche 1973); **B** section of the same: *m* oospore membrane; *a* starch grains; *Ca* calcified part of the spiral cells; *b* basal plate (calcified sister-cell of the oosphere); *n* nodal cell; *p* basal pore. In *dotted lines*, uncalcified part of the spiral cells and of the coronular cells (*c*), not preserved around the gyrogonite (unpublished)

a multicellular wall enclosing the oospore. The basal pore is obstructed by one to three sister cells of the oosphere which constitute the basal plate. In species which calcify, the multicellular wall, as well as the sister cells of the oosphere become particularly resistant. In addition, the thickened membrane surrounding the oospore may be preserved in the fossil state both in calcified species, such as *Rhabdochara praelangeri* Castel (Fig. 4c), and in uncalcified ones, such as *Nitellites sahnii* Horn af Rantzien from the Jurassic of India (Horn af Rantzien 1957). The vegetative parts are seldom preserved as fossils, but when they are found, they do not differ significantly from those of extant species and occur as long internodal cells alternating with short nodal cells. A new approach to the genus concept is given by the use of electrophoresis in the living forms. Applied by P. Gemayel (1988) to the genera *Chara*, *Nitella* and *Tolypella*, this method totally confirms the validity of the generic criteria based on the gyrogonite morphology.

In this study, the first part is a survey of the successive morphological types, from the Upper Silurian to Recent, which will give an outline of the main distinctive criteria on which orders and families are based. Secondly, the various generic criteria are defined by taking examples from different families and periods, including examples of living Characean genera also represented in the fossil state. This will help us to decide if the fossil charophyte genera should be considered either as parataxonomic or as natural taxa.

2 Bases of Classification

The classification of charophytes is based on hierarchical characters which reflect the phylogeny of the group. Evolution started with the originally vertical gyrogonite cells spiralling either dextrally or sinistrally and reducing in number. Subsequent modifications led to improvement in the enclosing and protection of the egg. This took place in one of two different ways, either by the acquisition of a supplementary cover around the gyrogonite or by the closing of the apical pore of the gyrogonite.

In the current classification (see Appendix), the orders are based on the orientation of the gyrogonite cells, i.e. vertical, dextral spiral or sinistral spiral. The distinction of families is based on the number of the cells and, among the Charales, either on the presence of an utricle (Clavatoraceae), an apical pore always open (Porocharaceae), an operculum (Raskyellaceae) or a closed apex (Characeae). The generic criteria are based on particular characters of the gyrogonite apex, base, basal plate and the general outline of the gyrogonite. Species are based on special characters of gyrogonite shape, ornamentation and dimensions. In the Clavatoraceae, characters of the utricle are also taken into account.

3 Successive Morphological Types

Praesycidium siluricum (Sycidiaceae), described by T. Ishchenko and A. Ish chenko (1982) from the Upper Silurian (Pridoli) of Podolia (Ukraine, USSR) is the oldest charophyte species presently known. The first and most important stage of diversification occurred during the Devonian, when the three orders of Charophyta were already represented (Fig. 2). During the Permian, Triassic and Jurassic, gyrogonites were discrete forms, generally of small dimensions and devoid of any ornamentation, but with high evolutionary potential. Thus, the gyrogonite of the Triassic Porocharaceae Stellatocharoideae type, with its apical pore at the end of a neck, or collar, is found again inside the utricle of the Jurassic-Cretaceous Clavatoraceae. On the other hand, Porocharaceae gave rise to the Characeae and Raskyellaceae by two different methods of closing the apical pore (Feist and Grambast-Fessard 1984), namely by joined spiral cells or by intercalation of supplementary opercular cells.

A second stage of diversification took place in the Lower Cretaceous, with the rapid radiation of the Clavatoraceae which are characterized by the evolution of the utricle. Grambast (1974) put forward three evolutionary lineages and presented a phylogeny of the family. These lineages are amongst the most striking of the plant kingdom.

A third diversification occurred in the Upper Cretaceous (Campanian-Maastrichtian) with the expansion of the Characeae. Following the extinctions at the Cretaceous-Tertiary boundary (Feist 1979) and a last diversification during the Paleogene, the group regressed until the present state, with only six extant genera, grouped in one family, the Characeae. In contrast to the fossil

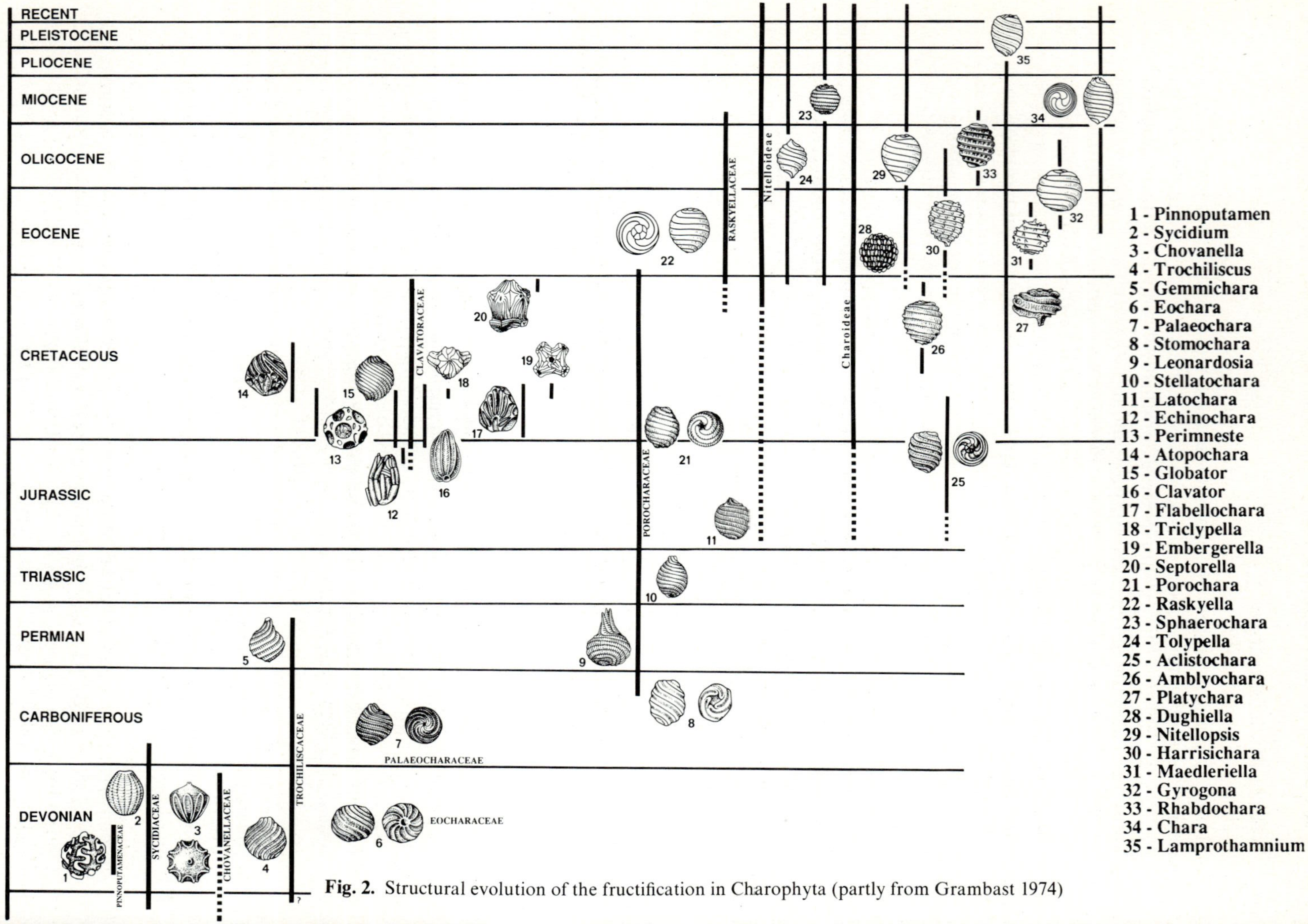

Fig. 2. Structural evolution of the fructification in Charophyta (partly from Grambast 1974)

record, the extant charophytes show structural and taxonomic impoverishment (although the number of living species reaches 80 according to Wood and Imahori 1965) and loss of calcified structures, such as ornamentation and calcified multipartite basal plate.

4 The Generic Concept from the Fossil Record

The generic criteria vary in importance and significance in the different periods and families as is shown by the following examples taken from the Palaeozoic families as well as from the Clavatoraceae and Characeae.

4.1 Palaeozoic Families

In two monotypic Palaeozoic families, the genera *Eochara* Choquette (Fig. 2, No. 6) and *Palaeochara* Bell (Fig. 2, No. 7) differ in the number of gyrogonite spiral cells (being more than six and just six respectively). In all post-Palaeozoic charophytes, which are here placed in the new Suborder Charineae (see Appendix), the number of spiral cells is fixed at five.

4.2 Clavatoraceae

The seventeen genera of the Clavatoraceae are well-defined by the characters of the utricle. Grambast (1966, 1967, 1970, 1974) reconstructed inter- and intrageneric lineages linking successive stages of evolution. These lineages may be illustrated by two examples:

1. Perimneste-Atopochara lineage (from Grambast 1967, 1974; see Fig. 3a,b,c). The two genera *Perimneste* Harris and *Atopochara* Peck seem at first sight quite different, both in shape and size as well as in structure. In the Berriasian *P. horrida* Harris (Fig. 3a), the antheridial casts are recognizable in hollows at the utricle surface. Dissolving the calcite with weak acetic acid shows the position of the utricle cells, which are grouped in three distinct clusters. From the Berriasian to the Cenomanian, these clusters progressively condense and the number of antheridial casts reduces. This proceeds until the typical structure of *Atopochara* is realized in *A. trivolvis* Peck of the Barremian, where the central basal cell of each cluster has disappeared and the antheridia sites are reduced to one sterile cell (Fig. 3b). From the Barremian to the Cenomanian, the cells of antheridial origin disappear altogether and the upper cells become strongly spiralized. The end member of this progressive evolution is seen in *A. multivolvis* Peck (Fig. 3c).

This progressive series of structural modifications shows how these forms may be linked; the separation between the two genera being placed, perhaps arbitrarily, at the appearance of *A. trivolvis*, when the cells are joined together and the antheridia really disappear.

Fig. 3a-f. Clavatoraceae. **a** *Perimneste horrida* Harris; Hils, NW Germany, Berriasian; basal view, x33. **b** *Atopochara trivolvis* Peck ssp. *triquetra* Grambast; Warlingham Borehole, S England, Lower Barremian; lateral view, x40 (BGS. MPK 5875). **c** *Atopochara multivolvis* Peck; Le Revest, SE France, Cenomanian, lateral view, x32 (Feist 1981). **d** *Globator trochiliscoides* Grambast: El Mangraner, E Spain. Barremian: lateral view, X32. **e** *Globator nurrensis* (Pecorini) Grambast: Cala d'Inferno, NW Sardinia. Berriasian: lateral view, X30. **f** *Globator maillardi* (Saporta) Grambast: Broadoak Borehole, S England. Berriasian: lateral view, X50 (BGS. MPK5876)

2. Globator lineage (from Grambast 1966, 1974) (Fig. 3d,e,f). This series shows the evolution from the Berriasian *Globator maillardi* (Saporta) Grambast to the Barremian *G. trochiliscoides* Grambast, by the reduction in the number of the utricle cells and by the progressive acquisition of spiralling.

In this case, all the species are attributed to the same genus, because the successive evolutionary stages represent modifications of the same basic structure. However, if the end-members of the lineage had been found independently, they would probably have been attributed to two different genera as was the case for *Perimneste* and *Atopochara*. *Globator* may be considered as a phylogenetic genus, in the sense that it is composed of a continuous lineage of successive species.

5 Characeae

In the previous examples, generic distinction did not present particular difficulties. This is not the case with the very homogeneous family Characeae which includes about fifty genera, comprising nearly half of the presently recognized genera of charophytes. The family is first undoubtedly known from the Middle Jurassic (Feist and Grambast-Fessard 1982), and persists as the only extant family.

In this family, the genera are distinguished not according to evolutionary stages as in the Clavatoraceae, but on the basis of a series of more or less important morphological characters. The first most important character relates to details of the structure of the apical part of the gyrogonite, such as the presence or absence of a periapical groove, the presence or absence of apical nodules, and variations in the width of the cells around the apex. Another important character is the structure of the basal plate, which represents the calcified sister-cells of the oosphere. Observing these characters is not easy because of their minute size and the problem is often made more difficult by the degree of calcification or preservation of the gyrogonite. However, some of the genera of the Characeae are well characterized and appear to constitute natural taxa, such as in the following examples:

5.1 Genus *Harrisichara* Grambast

The gyrogonites in this genus are characterized by an undifferentiated apex, a distinct basal column and, in most species, ornamented spiral cells. *Harrisichara*, which includes more than twenty species, is reported from the Palaeocene to the Lower Oligocene. In the Palaeogene sequences from the Isle of Wight (southern England), an evolutionary sequence from *H. vasiformis* to *H. tuberculata* can be clearly recognized (Feist-Castel 1977, Pl. 21, Figs. 1–5).

In this case, only fossil forms are concerned. Connections have, however, also been established between some fossil gyrogonites and those of extant genera, as is shown by the following examples.

5.2 *Nitellopsis* and *Tectochara* (Fig. 4a-b, d-f)

The former genus *Tectochara* Grambast, established for fossil species and known from Palaeocene to Quaternary, is characterized by a decrease in both the thickness and width of the spiral cells in the periapical zone and by a very thin basal plate.

The extant, but rather rare, *Nitellopsis* Hy was poorly known for a long time. The plant lives in relatively deep water (4–12 m), is dioecious and reproduces mostly vegetatively by means of bulbils, so the female reproductive organs are rarely encountered. Krassavina (1971), studying fragments of thallus and calcified oogonia coming from the gut contents of ducks killed along the Azov

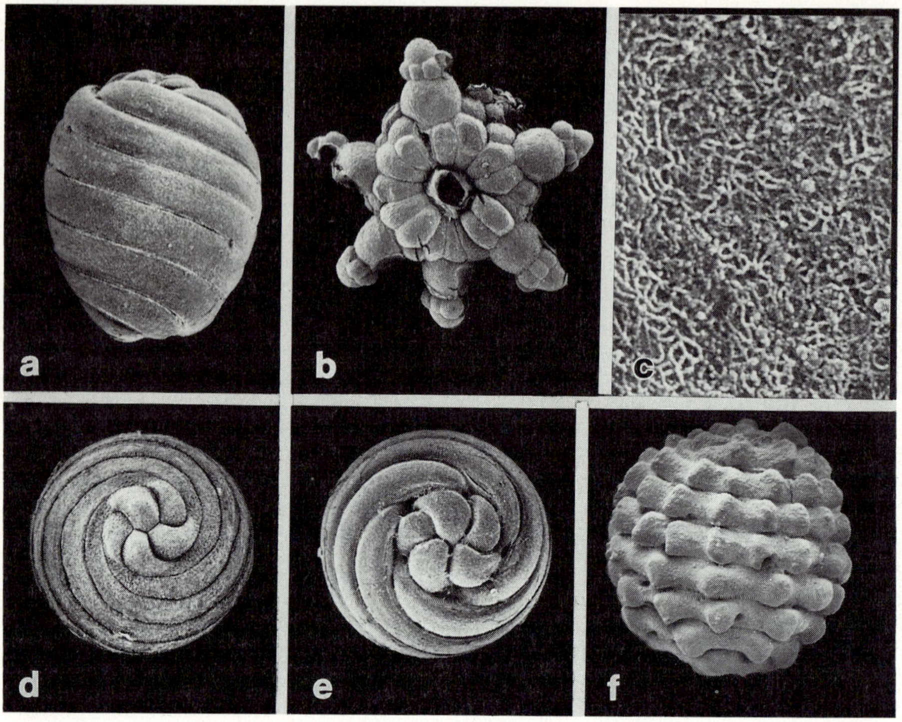

Fig. 4a-f. Characeae. **a** *Nitellopsis obtusa* (Desvaux) Groves; Sarthe, NW France, Recent; lateral view, x47. **b** *Nitellopsis obtusa* (Desvaux) Groves; Bordeaux, SW France, Recent; bulbil, x92. **c** *Rhabdochara praelangeri* Castel; Marseille, SE France, Chattian; outer view of the oospore membrane (ectosporostine), x600. **d** *Nitellopsis (Campaniella) helicteres* (Brongniart) Grambast & Soulié-Märsche; Mont-Bernon, NE France, Sparnacian; apical view, x45. **e** *Nitellopsis obtusa* (Desvaux) Groves; same locality as **a**, Recent; apical view, x47. **f** *Nitellopsis (Tectochara) morulosa nov. comb.*; Saint-Gély-du-Fesc, S France, Cuisian-Lutetian; lateral view, x40; fossil representative of the genus with spiral cells bearing tubercles (Feist-Castel 1972, Pl. 1, Fig. 5)

Sea (Ukraine, USSR), recognized the identity of these gyrogonites, belonging to *Nitellopsis obtusa* (Desvaux) Groves, and of *Tectochara diluviana* (Mädler) Grambast.

In this example, it has been possible to connect the living representative with its fossil ancestors that are the Tertiary species formerly assigned to *Tectochara*; this taxon is now considered as a subgenus of *Nitellopsis*.

5.3 *Tolypella* and *Sphaerochara* (Fig. 5d-f)

Another example of the relationships between fossil and living charophytes can be seen by examining the characters of the basal plate of the gyrogonite. Grambast (1956) showed that the morphological aspects of this plate could be

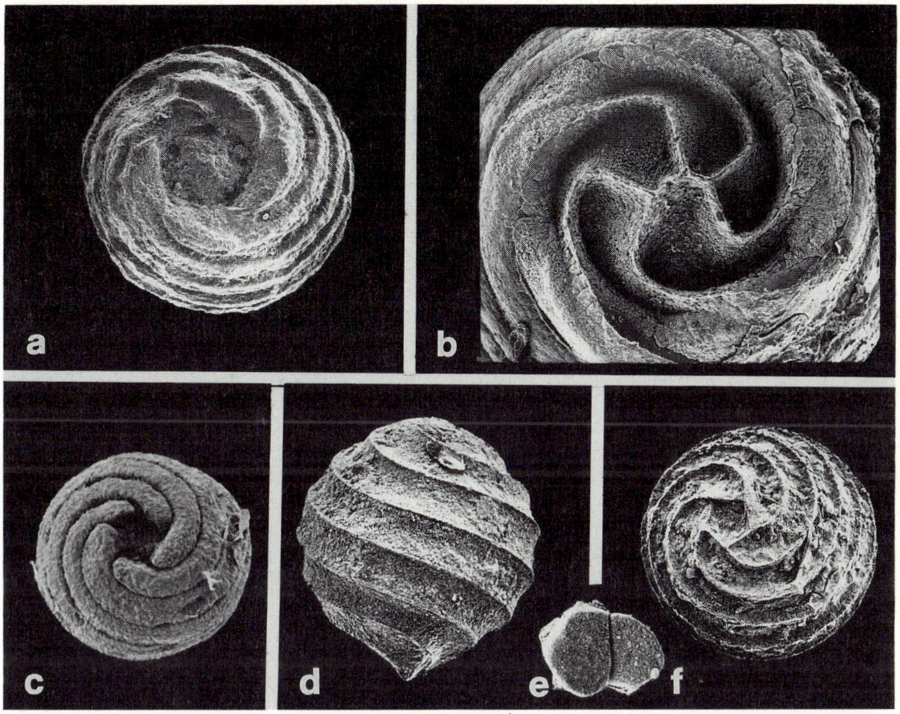

Fig. 5a-f. Porocharaceae and Characeae. **a** *Aclistochara bransoni* Peck; Elk Creek, USA, Upper Jurassic; apical view, x66. **b** *Lamprothamnium papulosum* Groves; Marseillan, S France, Recent; apical view, x150. **c** *Porochara sardiniae nov comb.*; Cala d'Inferno, NW Sardinia, Berriasian; apical view, x50 (Colin et al. 1984). **d** and **e** *Tolypella pumila* Grambast; Hoogbutsel, Belgium, Lower Oligocene; lateral view, x110 and isolated bipartite basal plate, x324. **f** *Sphaerochara intricata* (Trent. ex Roth) Leonhardi n. comb.; Juignée, W France, basal view showing the undivided basal plate, x90

used to recognize the genera within the Characeae. Subsequently Daily (1969) and Sawa and Frame (1974) showed that the basal plate could be either simple or multipartite in the living species of *Tolypella*, each of these types of basal plate being characteristic of a section of the genus. Thus, this character is clearly of taxonomic value; in the present classification of the living forms, however, it can define only subgeneric divisions. Applying this criterion, the fossil genus *Sphaerochara* Mädler emend. Horn af Rantzien and Grambast must be included in the Nitelloideae (Feist and Grambast-Fessard 1982) since its gyrogonites are identical to those of *Tolypella* (Al. Braun), corresponding to the section *Rothia* R.D. Wood; both have a simple calcified basal plate, and the two taxa should thus be considered as synonymous, with *Tolypella* section *Rothia* falling into the synonymy of *Sphaerochara* (see Appendix).

5.4 *Lamprothamnium* and *Aclistochara* (Fig. 5a-c)

The gyrogonites of both the Cretaceous-Recent *Lamprothamnium* Groves and the Jurassic-Cretaceous *Aclistochara* Peck possess a deep peripheral groove at the apex and are cylindrical and unornamented. Because of these similarities, Soulié-Märsche (1989) considered the two genera to be identical. However, the narrower apical zone of *Aclistochara*, resembling the small size of the apical opening of the Porocharaceae (Fig. 5c), favours the maintenance of two distinct genera (Feist and Grambast-Fessard 1982).

The examples taken from the Characeae (see also Fig. 6) demonstrate that the study of the gyrogonites of fossil forms may be useful for the recognition and definition of present genera. Care must be taken, however, that rationalization of the nomenclature of fossil and extant genera through the recognition of apparent synonymies in temporally widely separated gyrogonite taxa does not mask distinct stages in evolution.

6 Conclusions

As in the higher plants, the classification of the charophytes can be based on characters of the female reproductive organs, which represent the main part of their fossil remains. The classification used here reflects the phylogeny of the group and it expresses the main evolutionary tendencies as well as relationships between genera and species. The gyrogonites may thus be considered as characteristic of natural taxa rather than merely representing organ-genera. For this reason, it is possible to try to harmonize the systematics of fossil and living forms and to look for a natural classification for the whole phylum.

Appendix: Systematic Section

1. Taxonomy

Since the last publication of a complete classification (Tappan 1980, p. 955), two new families have been formed (Pinnoputamenaceae and Karpinskyaceae). In the present paper, a new name is proposed for the Karpinskyaceae and two new suborders are established: Moellerinaceae nom. nov. = Karpinskyaceae Wang and Lu (1980, p. 196). Moellerinaceae is proposed to serve as a substitute for Karpinskyaceae established by Wang and Lu (1980) as it is based on the earliest described genus *Moellerina* Ulrich 1886 which also possesses the most typical characters of the family; the presence of coronular cells in *Karpinskya* is a particular character. Two new suborders, based on the number of cells composing the gyrogonite, are introduced in order to present a more logical classification: these are Palaeocharinae and Charinae (see Charophyta, below).

There are two new combinations relating to specimens figured in this present paper:

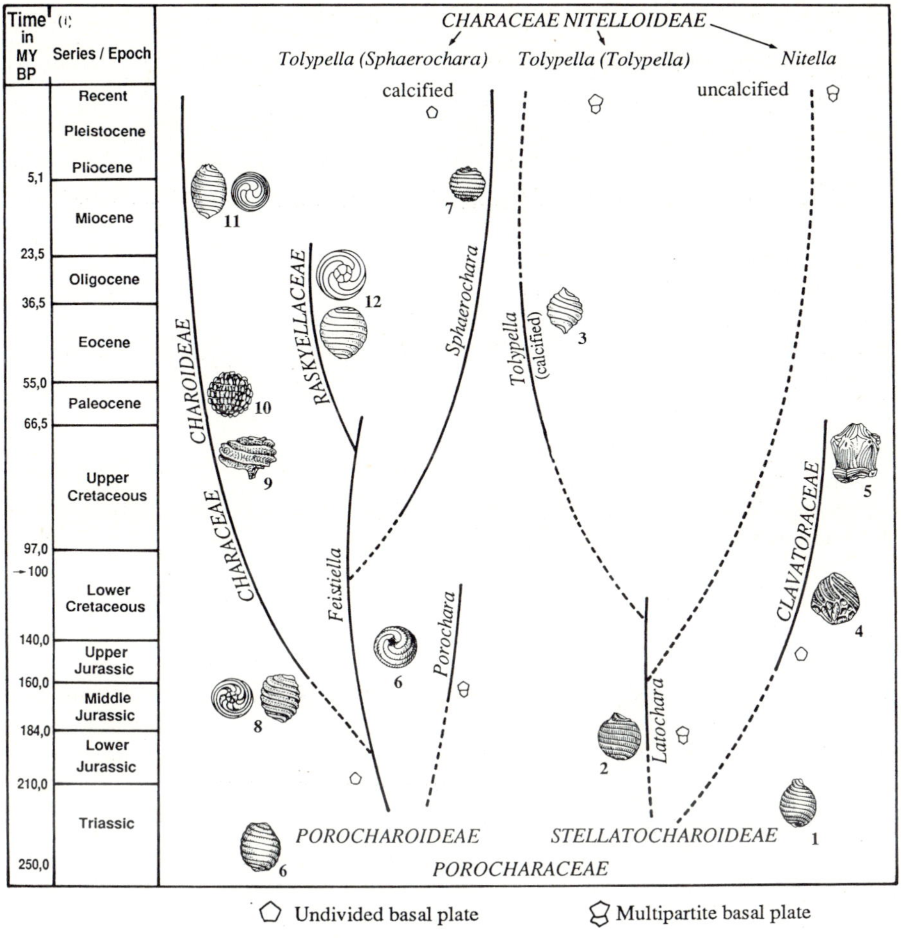

Fig. 6. Phylogeny of the post-palaeozoic charophytes. Apical structure and basal plate are considered the main criteria; thus *Tolypella* is interpreted as polyphyletic. *1* Geological scale from Haq and Van Eysinga (1987) *1–12* Lateral and apical views of selected genera. *1 Stellatochara*; *2 Latochara*; *3 Tolypella*; *4 Atopochara*; *5 Septorella*; *6 Feistiella*; *7 Sphaerochara*; *8 Aclistochara*; *9 Platychara*; *10 Dughiella*; *11 Chara*; *12 Raskyella*

– *Nitellopsis (Tectochara) morulosa nov. comb.* = *Tectochara morulosa* Feist-Castel 1972, p. 2, Pl. 1, Figs. 1–7.
– *Porochara sardiniae nov. comb.* = *Musacchiella sardiniae* Feist and Grambast-Fessard, in Colin et al. 1984, p 348, Pl. 1, Figs 6–13.

On the other hand, the reexamination of all charophyte genera for the "Treatise on Invertebrate Paleontology" has led us to put several genera into synonymy; they are indicated as follows, with new combinations based on the type-species of each genus:

a. Porocharaceae

– *Minhechara* Wei, based only on the wider apical pyramidal projection, is not distinguishable from *Latochara* Mädler 1955. *Latochara columelaria nov. comb.* = *Minhechara columelaria* Wei in Hao et al. 1983, J. Wuhan Coll. Geol., p. 174, Pl. 43, Figs. 11–17.
– Species from the Permian of China assigned to *Paracuneatochara* by Z. Wang (1984) consist of isolated gyrogonites with the same characters as the Brazilian *Leonardosia* Sommer 1954, known from casts only. Both have the same apical morphology, with a long apical neck ending with the apical pore. In addition they both are of Permian age.

  Leonardosia jinxiensis nov. comb. = *Paracuneatochara jinxiensis* Z. Wang 1984, Acta Micropalaeontol Sin., p. 56, Pl. 2, Figs. 1–15.

b. Characeae

– Charoideae
– *Caucasuella* Kyansep-Romashkina 1980 is not distinguishable from *Aclistochara* Mädler 1955 because both have the same apex, which is sunken and convex in its centre. In addition, the genus *Jurella*, which is considered by Kyansep-Romashkina as closely related to *Caucasuella*, has been attributed by Liu (1982) to *Aclistochara*.

  Aclistochara gulistanica nov. comb. = *Caucasuella gulistanica* Kyansep-Romashkina 1980, Limnobios, Akad Nauk USSR, p. 81, Pl. 1, Figs. 4–7.
– *Gobichara* Karczewska and Ziembinska-Twordzydlo 1972 must be considered as a junior synonym of *Microchara* Grambast 1959, because they are based on similar diagnostic characters: charoid apex, small size, common development of ornament.

  Microchara deserta nov. comb. = *Gobichara deserta* Karczewska and Ziembinska-Twordzydlo 1972, Palaeontol. Pol., p. 73, Pl. 15–16.
– *Piriformachara* Liu and Wu 1985 differs from *Mesochara* Grambast 1962 only by spiral convexity, a character which may depend on the degree of calcification and which cannot be considered alone as a generic criterion. *Mesochara gumudiensis nov. comb.* = *Piriformachara gumudiensis* Liu and Wu 1985, Bull. Inst. Geol. Chin. Acad. Sci., p. 148, Pl. 1, Figs. 4–7.
– *Qinghaichara* Yang in Hao et al. (1983) presents the same diagnostic characters as *Nitellopsis*: spiral cells "becoming narrower and remarkably thinner at apical periphery, apical rosette produced by inflated ends of spiral cells", basal plate thin ("plate-like"), "funnel-shaped outer depression around the basal orifice".

  Nitellopsis ovalis nov. comb. = *Qinghaichara ovalis* Yang in Hao et al. 1983, J. Wuhan Coll. Geol., p. 156, Pl. 36, Figs. 1–4.
– *Retusochara* Grambast 1971 and *Peckisphaera* Grambast 1962 have similar apical structures, with spiral endings which are slightly enlarged and a thick basal plate. The larger size of *Retusochara* cannot be considered significant. *Peckisphaera macrocarpa nov. comb.* = *Retusochara macrocarpa* Gramb. 1971, Paleobiol. Continent, p. 31, Pl. 14–15.

- All the salient features of *Sinochara* Lin and Z. Wang are found in *Peckichara* Grambast 1957, namely an apex with nodules, surrounded by a weak periapical furrow.
  Peckichara rudongensis nov. comb. = *Sinochara rudongensis* Lin and Z. Wang 1982, in: S. Wang, Huang Z. Wang et al. Geol. Publ. House, Peking, p. 32, Pl. 15, Figs. 5–9.
- *Songliaochara* Z. Wang, Lu and Zhao 1983 and *Turbochara* Z. Wang 1978, which differ only in the smaller size of the latter, are to be considered as junior synonyms of *Nodosochara* Mädler 1955, being based on the same criteria: presence of an apical rosette in the center of the apex with periapical narrowing of the cells, tapered base, thick basal plate.
  Nodosochara heilongjiangensis nov. comb. = *Songliaochara heilongjiangensis* Z. Wang, Lu and Zhao 1985, Ed. Sci. Technol. Heilong-Jiang, p. 65, Pl. 28, Figs. 2–8.
  Nodosochara specialis nov. comb. = *Turbochara specialis* Z. Wang 1978, Mem. Nanjing Inst. Geol. Paleontol., p. 78, Pl. 6, Figs. 1–9.

– Nitelloideae

- *Tolypella* (A. Braun 1857) Leonhardi, pro parte = *Sphaerochara* Mädler 1952 emend. Horn af Rantzien and Grambast 1962.
  Daily (1969) and Soulié-Märsche (1979), taking into account the characters of the gyrogonites, revealed the identity of the living *Tolypella* section *Rothia* Wood and the fossil *Sphaerochara*. This section includes a single living species, *T. intricata*.
  Sphaerochara intricata n. comb. = *Tolypella intricata* Trentepohl *ex* Roth 1797, Cat. Bot., p. 125; Leonhardi, emend. Wood 1965, p. 735.

2. *Classification of the Charophyta*

Order Sycidiales Mädler 1952: gyrogonite composed of vertical cells.
  Chovanellaceae Grambast 1962: vertical cells not subdivided.
  Sycidiaceae Karpinsky 1906: numerous vertical cells, subdivided.
  Pinnoputamenaceae Z. Wang and Lu 1980: cells forming two-branched opposite systems.

Order Trochiliscales Mädler 1952: gyrogonite composed of dextral spiral cells.
  Trochiliscaceae Karpinsky 1906 emend. Z. Wang and Lu 1980: numerous (often 18) dextral spiral cells.
  Moellerinaceae nom. nov. not very numerous (5–12) dextral spiral cells.

Order Charales Lindley 1836: gyrogonite composed of sinistral spiral cells.
  Palaeocharineae *nov. subordo* more than five sinistral spiral cells.
    Eocharaceae Grambast 1959: numerous (more than six) spiral cells.
    Palaeocharaceae Pia 1927: six spiral cells.
  Charineae *nov. subordo* five sinistral spiral cells.
    Porocharaceae Grambast 1962: apical pore always open.
    Clavatoraceae Pia 1927: gyrogonite surrounded by a utricle.
    Raskyellaceae Grambast 1957: apical pore closed by an operculum of five

cells leaving a rosette opening after being shed.
Characeae Agardh 1824: oogonium and gyrogonite with spiral cells joined at the summit.

Acknowledgements. We are grateful to the Manager, Biostratigraphy Research Group, British Geological Survey, Nottingham, for giving permission to figure specimens from Broadoak and Warlingham Boreholes, and to C.J. Wood for improving the English text.

References

Colin JP, Feist M, Grambast-Fessard N, Cherchi A, Schroeder R (1984) Charophytes and ostracods from the Berriasian (Purbeckian facies) of Cala d'Inferno (Nurra region, NW Sardinia). Boll Soc Paleontol Ital 93, 2:345–354, 2 pl

Daily FK (1969) Some Late Glacial Charophytes compared to modern species. Proc Indiana Acad Sci 1968, 78:406–412

Feist M (1979) Charophytes at the Cretaceous-Tertiary boundary. New data and present state of knowledge. In: Christensen W, Birkelund T (eds) Proc Symp Cretaceous-tertiary boundary events, vol 2, Copenhagen, pp 88–94

Feist M (1981) Charophytes du Crétacé moyen et données nouvelles sur les Clavatoracées. Cretaceous Res 2:319–330

Feist M, Grambast-Fessard N (1982) Clé de détermination pour les genres de Charophytes. Paleobiol Continent 13, 2:1–28

Feist M, Grambast-Fessard N (1984) New Porocharaceae from the Bathonian of Europe: phylogeny and palaeoecology. Palaeontology 27, 2:295–305

Feist-Castel M (1972) Charophytes éocènes de la région montpelliéraine. Paleobiol Continent 3, 1:1–22, 6 pl

Feist-Castel M (1977) Evolution of the Charophyte floras in the Upper Eocene and Lower Oligocene of the Isle of Wight. Palaeontology 20 1:143–157

Gemayel P (1988) Application de l'électrophorèse aux Charophytes: mise au point, premiers résultats. Actes 9ème Colloque International de Cytologie de la Reproduction sexuée des Plantes Supérieures, Reims, 1986. PUR, pp 1–86

Grambast L (1956) La plaque basale des Characées. CR Acad Sci Paris Ser D 242:2585–2587

Grambast L (1962) Classification de l'embranchement des Charophytes. Nat Monspel 14:63–86

Grambast L (1966) Un nouveau type structural chez les Clavatoracées; son intérêt phylogénétique et stratigraphique. CR Acad Sci Paris Ser D 262:1929–1932

Grambast L (1967) La série évolutive *Perimneste-Atopochara* (Charophytes). CR Acad Sci Paris Ser D 264:581–584, 4 pl

Grambast L (1970) Origine et évolution des *Clypeator* (Charophytes). CR Acad Sci Paris Ser D 271:1264–1267, 4 pl

Grambast L (1974) Phylogeny of the Charophyta. Taxon 23, 4:463–481

Grambast L, Soulié-Märsche I (1972) Sur l'ancienneté et la diversification des *Nitellopsis* (Charophytes). Paleobiol Continent 3, 3:1–14

Haq BU, Van Eysinga FWB (1987) Geological time table. Elsevier, Amsterdam

Horn af Rantzien H (1957) Nitellaceous gyrogonites in the Rajmahal Series (Upper Gondwana) of India, with notes on the flora and stratigraphy. Stockholm Contrib Geol 1, 1:1–29, 3 pl

Horn af Rantzien H (1959) Morphological types and organ-genera of Tertiary Charophyte fructifications. Stockholm Contrib Geol 4:45–197, 21 pl

Ishchenko TA, Ishchenko AA (1982) Novaia nakhodka kharophitov v verkhniem silure Podolii. In: Sistématika i êvolutsia drevnik rastenii Ukraini. Nauk Dumka, Kiev, pp 21–32, pl 5–6

Krassavina LK (1971) Sravnitelnoie izutchenié sovremenikh i iskopaemikh kharophitov: plodonoshiénia *Nitellopsis obtusa* i guirogoniti vidov Tectochara. Bot J 56, 1:106–117, 2 pl

Liu J (1982) Jurassic Charophytes from Sichuan province. Bull Inst Geol Chinese Acad Geol Sci (5):97–110 (in Chinese)

Sawa T, Frame PW (1974) Comparative anatomy of Charophyta. I. Oogonia and oospores of *Tolypella*, with special reference to the sterile oogonial cell. Bull Torrey Bot Club 101, 3:136–144

Soulié-Märsche I (1973) Données générales sur les Charophytes du Plio-Pléistocène. Bull Assoc Franç. Etude Quaternaire (1973) 2:69–77

Soulié-Märsche I (1989) Etude comparée de gyrogonites de charophytes actuelles et fossiles et phylogénie des genres actuels. Edit Révisée Thèse Univ Montpellier 1979. Imprimerie des Tilleuls, Millau, Fr, 237 p, 45 pl

Tappan H (1980) The paleobiology of plant protists, vol 1. Freeman, San Francisco, 1028 pp

Wang Z, Lu HN (1980) New discovery of Devonian Charophytes from South China with special reference to classification and gyrogonite orientation of Trochiliscales and Sycidiales. Acta Paleontol Sin 19, 3:190–200, 2 pl (in Chinese)

Wood RD, Imahori K (1965) A revision of the Characeae. Edit. Cramer, Weinheim, vol 1:904 pp, vol 2:395 pl

Chapter 12

Calcification of the Charophyte Oosporangium

A.R. LEITCH[1]

Abstract

The oosporangium is a unique and complex structure consisting of a central reproductive cell (the oospore) surrounded by vegetative cells. In the subfamily Chareae, six vegetative ensheathing cells (i.e. five spiral cells and one basal cell), completely surround the oospore and are in intimate contact with it. After fertilization, a thick multilayered wall, the compound oosporangial wall (COW) forms around the oospore. This wall is derived by the simultaneous deposition of three layers by the ensheathing cells and three layers by the oospore. A calcified layer is deposited onto the COW by the ensheathing cells. Calcification occurs outside the plasmalemma of each ensheathing cell but within the confines of the cell wall; this is 'extracytoplasmic' calcification. The ensheathing cells secrete an organic matrix that nucleates calcite development and is presumably involved in crystal shaping. The calcite crystals of *Chara* are tabular and arranged in stacks, forming columns (like gastropod shell). Columns of calcite are found in *Lamprothamnium* calcine, although no substructure to these columns has been resolved. The differences seen in the calcified layer of *Chara* and *Lamprothamnium* using light microscopy are largely a matter of perspective as to which is being seen, the columns of calcite or the organic matrix.

1 Introduction

The family Characeae is a group of macrophytic green algae, anchored to organic-rich sediments by colourless rhizoids. They occur in lakes and the backwaters of streams and rivers, often forming high-density 'charophyte meadows'. They favour oligotrophic calcareous water to a depth of 15 m (Stross 1979; Spence 1982) disappearing from lakes when they become eutrophic. A few species are found in brackish water. One genus, *Lamprothamnium* Groves, is exceptional in that it can survive highly saline conditions, although it requires brackish water to complete its life cycle (DeDeckker and Geddles 1980).

The taxonomic affinities of the family Characeae remain controversial. Their supposed intermediate position between green algae and bryophytes has led workers to propose that they represent a subdivision of higher plants (Bremer and Wantorp 1981). However, this is not widely accepted. Mattox and Stewart (1984) considered that the Characeae belonged to a class of green algae (Division Chlorophyta) while Round (1984) elevated the group to divisional status (the Charophyta). Within the single extant family Characeae there are six extant genera belonging to two subfamilies, the Nitelleae comprising *Nitella* Ag., and *Tolypella* Braun, and the Chareae comprising *Chara* L., *Nitellopsis* Hy, *Lychnothamnus* (Rupr.)Leonh., and *Lamprothamnium*. The genus *Tolypella*

[1]J.I. Centre for Plant Science Research, Colney Lane, Norwich, NR4 7UJ, England

has two sections, acutifolia and obtusifolia, with such different oosporangial morphologies (see Sawa and Frame 1974; and Daily 1975) that the validity of the genus should be questioned.

The family Characeae is unique amongst the algae in that the reproductive cell (termed the oosphere before fertilization and oospore after fertilization) is completely ensheathed by vegetative cells. This forms an integral structure called the oosporangium (or oogonium). The complex development and morphology of the oosporangium is well documented (Sundaralingam 1954, 1962). Briefly, the oosphere/oospore is ensheathed by five elongate, sinistrally spiralling cells, called spiral cells, and one (in Chareae and Tolypella section acutifolia) or three (in *Nitella* and *Tolypella* section obtusifolia) small cells, called sterile cells. In this work these cells are collectively called the ensheathing cells. Subtending each spiral cell is either one cell (Chareae) or two cells (Nitelleae) which together form a crown of five or ten cells, called coronula cells, at the apex of the oosporangium. In addition, the oosporangium has two small cells at its base that anchor the structure to the algal thallus. The ensheathing cells and the oospore are of interest here because these cells are involved in post-fertilization development.

Before fertilization the wall of the oosphere is in intimate contact with the walls of the ensheathing cells. Both walls are microfibrillar and are considered to be primary cell walls (Leitch 1986, 1989). After fertilization, a thick multilayered wall, called the compound oosporangial wall (formerly known as the oospore membrane), is deposited around the oospore. The compound oosporangial wall (COW) is derived partly from the oospore and partly from the ensheathing cells by the deposition of secondary wall layers. The oospore deposits three secondary wall layers against the inside of its primary wall. Simultaneously the ensheathing cells deposit three secondary wall layers against the inside of their primary walls. The resulting COW is an integral unit that surrounds the oospore and is built up of eight layers (Leitch 1986, 1989).

In the subfamily Chareae and *Tolypella* (section acutifolia), after the deposition of the COW, a calcified 'shell-like' layer called calcine is deposited (Horn af Rantzien 1956; Daily 1975). The calcine has a mineral component of low magnesium calcite in freshwater species or high magnesium calcite in brackish water species (Daily 1975; DeDeckker and Geddles 1980; Leitch 1986). The calcine is secreted by the ensheathing cells directly onto the COW. Horn af Rantzien (1956) reported that the first calcine deposited (i.e. closest to the COW) was distinctive from later formed calcine, calling the two types endo- and ecto-calcine respectively. The calcine of *Chara* and *Lamprothamnium* are of particular interest because Soulié-Märsche (1979) and Feist and Grambast-Fessard (1984) reported that these genera typify two distinct forms of calcification. In this paper techniques in light microscopy (LM), scanning electron microscopy (SEM), and transmission electron microscopy (TEM) are used to examine the differences between these calcine forms. Calcine from *Lamprothamnium papulosum* Groves, *Chara hispida* L. and *Chara delicatula* Agardh. is examined.

2 Materials and Methods

Plant material of *Chara delicatula* and *Chara hispida* was collected from drainage ditches in the Gordano Valley, Avon (ST 444 735) where they were growing in clear, calcareous water to a depth of 1 m. *Lamprothamnium papulosum* was collected from a small lake near a salt marsh at Lymington, Hampshire (SZ 328 939) where the plants were growing in turbid, brackish water to a depth of 1.5 m.

The oosporangia were fixed for TEM in 2.5% glutaraldehyde, 0.1 M formaldehyde in 0.2 M sodium cacodylate buffer (pH 7.6) for 3 h at room temperature. They were then post-fixed in 2% osmium tetroxide (in sodium cacodylate buffer) for 3 h at room temperature and block stained overnight in 0.25% uranyl acetate. The oosporangia were embedded in LR white (medium) resin and polymerized at 70°C overnight. Sections were cut with a diamond knife, stained with lead citrate and uranyl acetate and examined at 80 kV in a Philips 300 transmission electron microscope.

The oosporangia used for SEM were freeze-dried prior to examination. The calcine, was isolated from the oosporangium using fine entomological needles and air dried. All specimens were coated with gold using a Polaron Sputter coater and examined at 15 kV in a Philips 501B scanning electron microscope.

The oosporangia for light microscopy were fixed and embedded as for TEM and then ground to thin sections using carborundum powders. They were then examined using an Olympus BHB compound microscope.

3 Development of the COW and the Calcine in *Chara* and *Lamprothamnium*

Following fertilization the COW develops around the oospore. The oospore itself secretes three secondary wall layers in the following sequence: an amorphous layer; a helicoidal layer (i.e. a characteristically ordered microfibrillar wall; see Neville (1986); and a random microfibrillar layer (Plate I.1). Simul-

Plate I.1 A section through a COW. *1* the random microfibrillar layer; *2* the oospore helicoidal layer; *3* the amorphous layer; *4* the oospore primary wall; *5* the spiral cell primary wall; *6* the crystalline layer; *7* the pigmented layer. The ornamentation layer produces an irregular surface to the COW (*arrow*). *Chara hispida* TEM x6700

Plate I.2. Section through the lateral wall of a spiral cell of *Chara hispida* showing the calcine (*c*) deposited on each side of the wall (*w*). Electron-dense fragments in the calcine are calcite crystals. TEM x1100

Plate I.3. A weakly calcified oosporangium of *Lamprothamnium papulosum* showing concave spiral profiles (*arrows*). Note that the apex is the most weakly calcified. SEM x50

Plate I.4. A strongly calcified oosporangium of *Lamprothamnium papulosum* showing flat spiral profiles (*arrows*). Note that the apex is the most weakly calcified. SEM x50

Plate I.5. Section through a calcifying spiral cell of *Chara delicatula*. An organic matrix (*o*) is seen outside the plasmalemma (*p*). Note the electron-dense crystal nuclei (*n*). The holes (*h*) are sites where crystals have been pulled out of the section. TEM x 7000

Plate I.6. Section through the calcine of *Chara hispida* showing a complex organic matrix (*o*). Note the organic layers in the endocalcine (*arrowheads*). TEM x3300

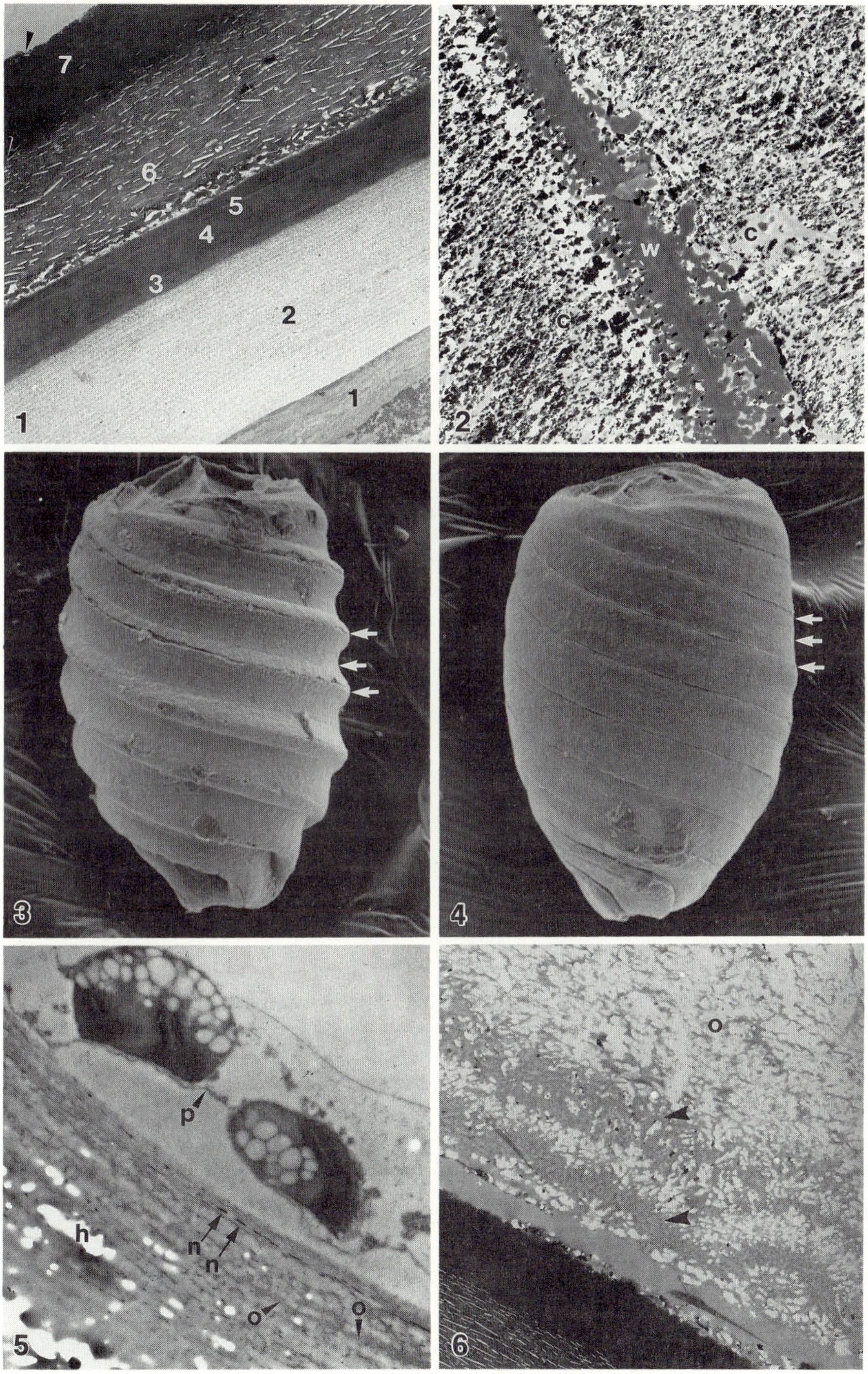

Plate I.1–I.6

taneously, the ensheathing cells deposit three secondary wall layers in the following sequence: a crystalline layer, possibly of calcium sulphate (Leitch 1986), embedded in an electron-dense matrix; a highly pigmented helicoidal layer; and an amorphous layer (Plate I.1; Leitch 1986).

After the development of the COW the ensheathing cells deposit the calcine. Calcine is initially deposited against the COW and the inner half to two-thirds of the lateral walls of the ensheathing cells (Fig. 1). Further calcification increases the thickness of the calcine (Fig. 1). Because calcine is deposited within the confines of the walls of the ensheathing cells, the form of the calcine reflects the cell shape and so occurs as five spiralling units and a small rounded/pentagonal unit of calcine. Each unit of calcine lies against the COW and is separated by a cell wall (Plate I.2). The spiralling units are called spirals and the basal unit is called the basal plate. With the death of the ensheathing cells the calcification stops. The outer wall of the ensheathing cells then decays

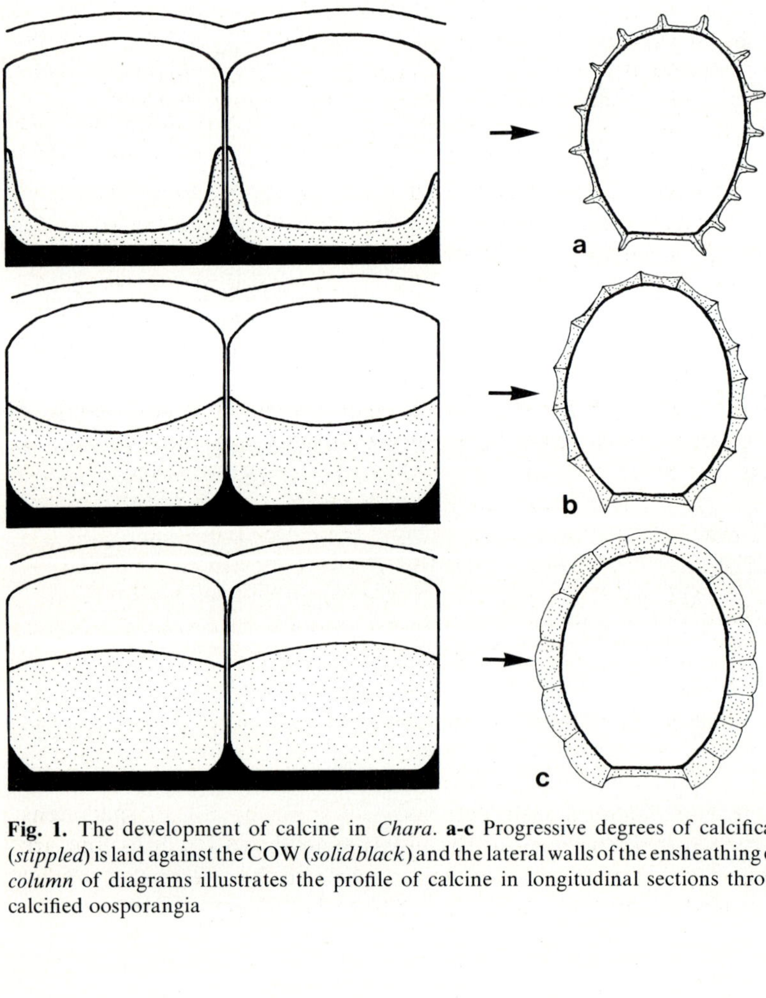

Fig. 1. The development of calcine in *Chara*. **a-c** Progressive degrees of calcification. Calcine (*stippled*) is laid against the COW (*solid black*) and the lateral walls of the ensheathing cells. The *right column* of diagrams illustrates the profile of calcine in longitudinal sections through varyingly calcified oosporangia

away to reveal the underlying calcine. When this has happened a weakly calcified oosporangium shows spirals with a concave profile (Plate I.3) and a strongly calcified oosporangium shows spirals with a flat or convex profile (Plate I.4). The degree of calcification depends on the species, the environmental conditions, and the physiological condition of the alga. The spiral apices are often more weakly calcified than the rest of the spiral. This is particularly noticeable in *Lamprothamnium* (Plates I.3–I.4). On oospore germination the weakly calcified apex cracks and falls away by the growth of the emerging germling.

Fixing, embedding, and sectioning calcifying oosporangia for TEM is difficult because the thick walls and the calcine reduce infiltration of the fixative and resin. However, it is clear that calcification occurs outside the ensheathing cells' plasmalemma (Plate I.5) and within the confines of the cell wall. The process reduces the volume of the cytoplasm. This site of calcification cannot be considered as intracellular calcification, as reported by Daily (1975), because calcification is outside the plasmalemma, it is 'extracytoplasmic' but within the cell wall.

Calcification occurs by the secretion of an organic matrix, probably in the form of precursors, that polymerizes into sheets, outside the plasmalemma of the ensheathing cells (Plate I.5). Under TEM the organic matrix that is closest to the plasmalemma shows electron-dense patches (Plate I.5). These patches are interpreted as calcite nucleation sites. The organic matrix that is furthest from the plasmalemma (hence older) shows electron-opaque cavities (Plate I.5). These cavities are sites once occupied by calcite crystals, the futher from the plasmalemma the larger the cavities and hence the larger the crystals. Fully developed calcine has calcite crystals ramified by an organic matrix (Plate I.6). Unfortunately, the staining requirement (uranyl acetate and lead citrate) to show the organic matrix decalcifies the calcine. However, omitting uranyl acetate reveals electron-dense calcite crystal fragments (Plate I.7).

In a strongly calcified oosporangium of *Chara* the calcine that is closest to the COW (hence older) is of a different form from younger, later formed, calcine (Plate I.6). First-formed calcine has a much higher organic matrix to calcite ratio than the later-formed calcine and the matrix appears as thick, tightly packed bands. First-formed calcine represents Horn af Rantzien's (1956) endocalcine and accounts for less than 10% of the calcine in strongly calcified oosporangia. However, very weakly calcified oosporangia show only calcine that is typical of endocalcine (Plate I.8). Bands of organic matrix are found in younger calcine (ectocalcine) but they are more diffuse and are seen in transverse section as concave bands arising from the lateral wall (Plate I.9). Bands of organic matrix are found in *Lamprothamnium* but they are much thinner and a distinct endocalcine cannot be resolved.

A transverse fracture of a spiral from *Chara* shows in SEM a demarcation line between the endocalcine and the ectocalcine in some, but not all, specimens (Plate I.10). A transverse fracture of the ectocalcine shows columns that are long, thin, closely packed and radiate diagonally from the lateral walls of the spirals and vertically from their inner wall (Plate I.11). These columns may be

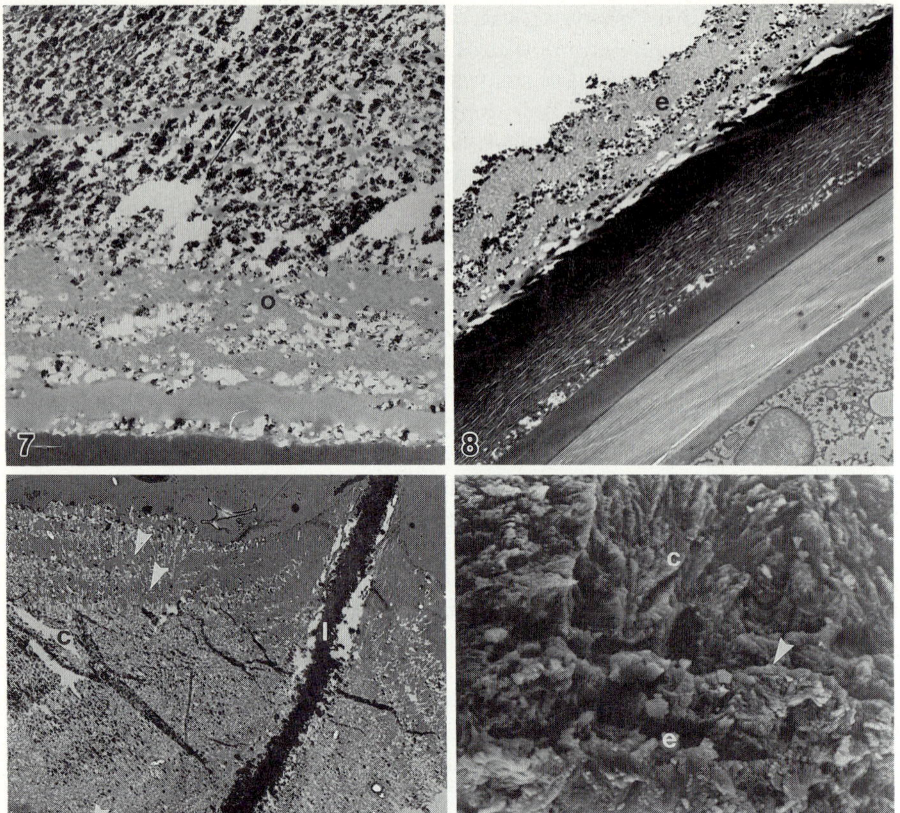

Plate I.7–I.12

simple or may form fans. Vertically aligned columns appear to blend into diagonal columns. *Lamprothamnium* calcine (Plate I.12) differs from *Chara* calcine (Plate I.11) in that the columns are more defined, they have a more uniform direction, and vertical columns from the base form a distinct suture with diagonal columns from the lateral wall (Plate I.13).

In *Chara* at high resolution the columns are seen to be composed of stacked layers of tabular crystals (Plate I.14). This is similar to the crystal layers seen in gastropod shell (Wilbur and Saleuddin 1983). The youngest calcine, which is closest to the plasmalemma in the calcifying oosporangium, shows stacks of crystals above each column (Plate I.15). Greater resolution of each stack shows layers of crystals that diminish in size towards the top (Plate I.16). At the apex of each stack is the smallest crystal and it represents a crystal growing after its nucleation. Each crystal in the stack grows laterally forming tabular crystals. The growth of the crystal is probably influenced by the organic matrix, as occurs in molluscs (Degens 1976). With the lateral growth of the cyrstals the stacks fuse together to form the columns. The layers of tabular crystals comprising the columns were not resolvable in *Lamprothamnium*.

Soulié-Märsche (1979) and Feist and Grambast-Fessard (1984) noted that the calcine from *Chara* and *Lamprothamnium*, when ground into thin-section using carborundum powders, and examined under light microscopy appeared different. They called the *Lamprothamnium* calcine the 'Y' form of calcification because of an inverted 'Y' visible in transverse section (Plate I.17). Likewise, a concave substructure to *Chara* calcine gave rise to the term 'U' form of calcification (Plate I.18). An examination of the inverted 'Y' of *Lamprothamnium* calcine (Plate I.17) shows a strong resemblance to the fractured face of calcine examined under SEM (Plate I.12). It is concluded therefore, that the 'Y' form is generated by the suture between the change in column direction. Further, the columns can be seen under light microscopy as radiating lines. An examination of the calcine of *Chara* under light microscopy (Plate I.18), and comparing this with calcine examined under TEM (Plate I.9), shows clearly that the concave bands are bands of organic matrix found within the calcine. A close

---

Plate I.7. Section through the calcine of *Chara hispida* showing the organic layers in the endocalcine (*o*). Note the array of crystal fragments in the direction of the *arrow*. TEM x2700

Plate I.8. Section through a weakly calcified oosporangium of *Chara hispida*. Note the endocalcine (*e*) on the COW. TEM x2000

Plate I.9. Section through two spirals of *Chara hispida* showing concave bands of organic matrix (*arrowheads*) in the calcine. The central area of calcine (*c*) sections least well. Note the lateral wall (*l*). TEM x830

Plate I.10. Fractured spiral of *Chara hispida* showing the endocalcine (*e*) and the calcine (*c*). Note that in this specimen there is a distinct demarcation between calcine types (*arrowhead*). SEM x830

Plate I.11. Fractured spiral of *Chara hispida* showing the columns (*c*) and the endocalcine (*e*) at the fracture surface. SEM x450

Plate I.12. Fractured spiral of *Lamprothamnium papulosum* showing columns (*c*) at the fracture surface. SEM x620

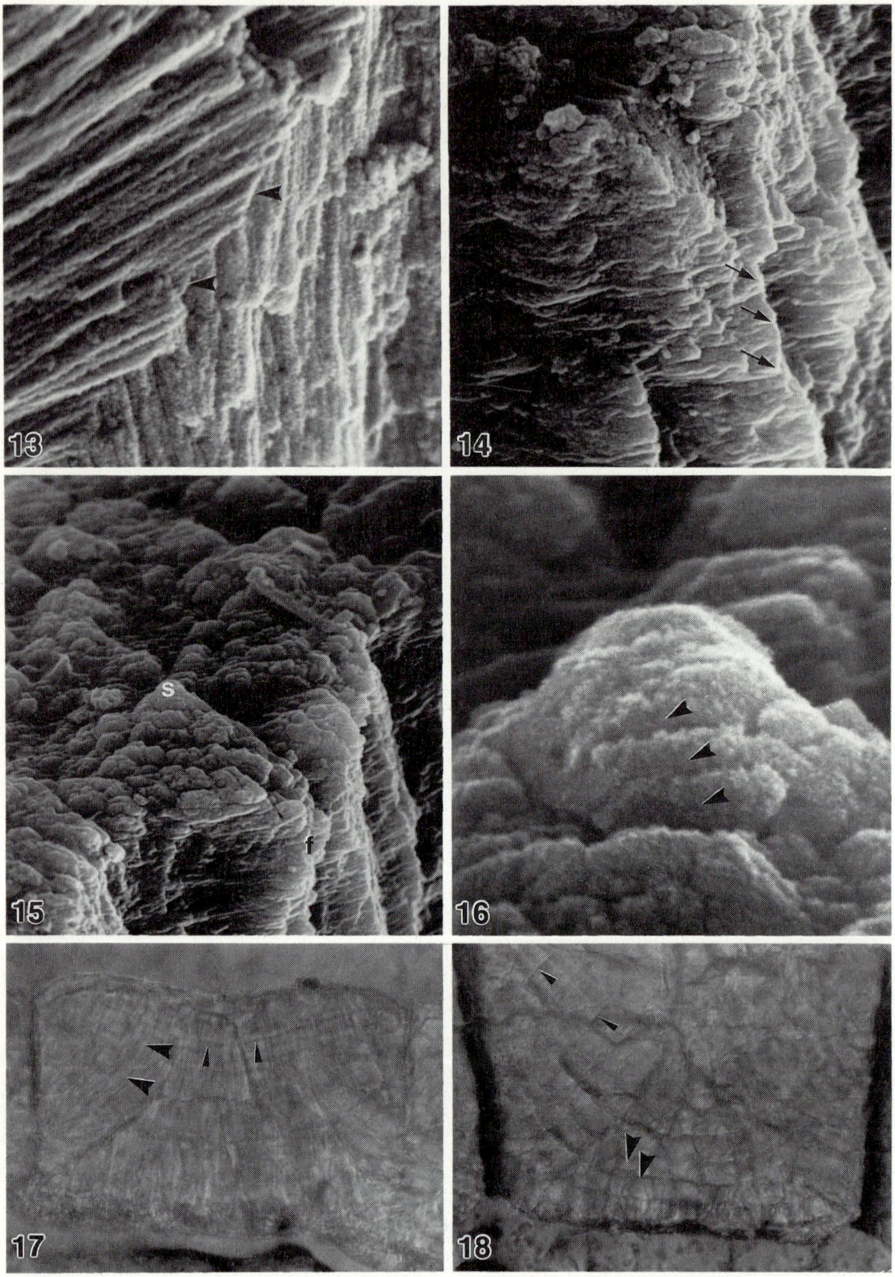

Plate I.13–I.18

Chapter 13

Calcareous Nannofossils

S.D. Houghton[1]

Abstract

Calcareous nannofossils (mainly coccoliths) have been important constituents of pelagic carbonates since the Early Jurassic, accounting for over 50% of some Tertiary sediments and around 30% of Recent oozes. Coccolith biosynthesis by calcareous nannoplankton forms an important role in marine biogeochemical cycles, contributing to the fluxes of carbon and calcium from the hydrosphere to the lithosphere. Most coccoliths are thought to descend through the water column 'sealed' within the faecal pellets of zooplankton grazers. This mode of sedimentation ensures that a constant supply of nannofossil carbonate, largely free from dissolution effects, reaches the ocean floor. Although post-depositional dissolution and diagenesis may modify coccolith biocoenoses, patterns of accumulation of coccoliths in sediments do reflect the oceanographic regimes of the overlying surface waters. Calcareous nannofossils are useful as tools for studying palaeoclimatic change and also form very good biostratigraphic markers.

1 Introduction

Coccoliths are the micron-sized (ca. 1–25 μm) calcitic plates secreted by unicellular haptophyte marine algae, the coccolithophorids. In the living cell, coccoliths are extruded to form a cover termed collectively the coccosphere, which lies outside the cell membrane (Fig. 1). Between 10 and 100 coccoliths comprise the coccosphere, which may range in size from 5 to 50 μm. The term nannoplankton, as defined by Lohmann (1909) for the very small forms, 60 μm or less, that pass through phytoplankton nets, therefore includes both the coccolithophorids and their skeletal elements. Individual dispersed coccoliths are usually all that is observed in fossil assemblages because of post-mortem disintegration of the coccosphere. However, intact or partially intact coccospheres of the more robust forms may occasionally be found (Figs. 2, 3a). Fossil coccoliths (Fig. 3), together with other calcite bodies of unknown origin which occur in the same size fraction, constitute the heterogeneous group, known as calcareous nannofossils. These other forms, termed nannoliths by some workers, display a wide range of elaborate structures and often form a substantial component of the nannofossil assemblages. The most notable of these forms are *Discoaster* species.

[1]Department of Earth Sciences, The Open University, Milton Keynes, Buckinghamshire, MK76AA, UK

Present address: Adams Environmental The Business Centre 103–109, Lavender Hill, London SW11 5QF, UK

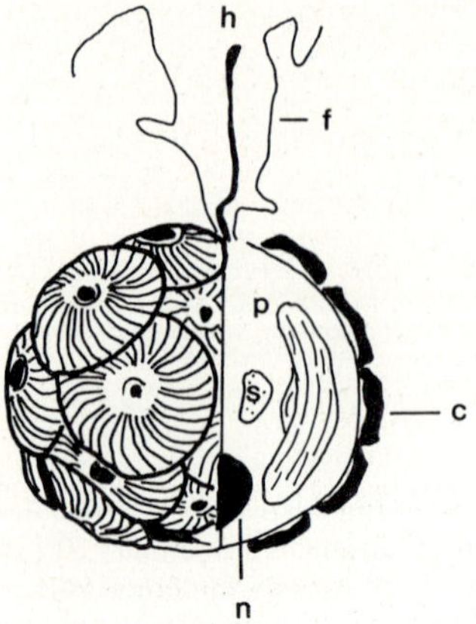

Fig. 1. Generalized diagram of a motile coccolithophorid cell showing exterior at *left*, and section at *right* (after Tappan 1980). *c* Coccolith; *f* two flagella; *h* haptonema; *p* plastids (chloroplasts); *n* nucleus; *s* storage body of leucosin

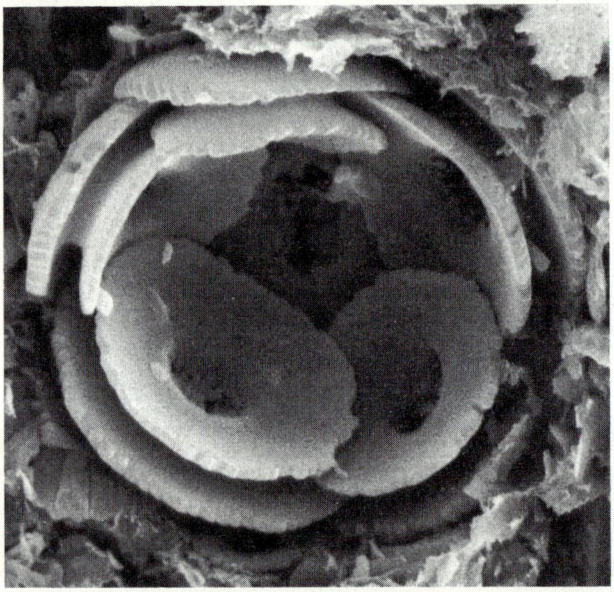

Fig. 2. Cross-sectional view of a coccosphere of *Reticulofenestra pseudoumbilica*, showing interlocking coccolith construction (x10 700); sample from Site 223 DSDP, Arabian Sea (Middle Miocene age)

Fig. 3a-f. SEM micrographs of some common Late Cenozoic calcareous nannofossils. **a** Coccosphere of *Coccolithus pelagicus* (x2600). Small coccolith at *centre* of coccosphere *Gephyrocapsa ericsonii*, other coccolith at *right side* of coccosphere *Calcidiscus leptoporus*. **b** *Gephyrocapsa caribbeanica*, distal view (x15 000); **c** *Braarudosphaera bigelowii*, proximal view (x10 500); **d** *Calcidiscus leptoporus*, distal view showing slight etching of individual crystallines (x9100); **e** *Rhabdosphaera clavigera*, side view (x9300); **f** *Syracosphaera pulchra*, proximal view (x14 800)

Species of calcareous nannofossils are defined on morphological characteristics of fragments or components (coccoliths) of the original organism. Some species are therefore probably no more than artificial categories. Studies on Recent coccolithophorids have shown that some individuals bear two morphologically distinct coccoliths on the same coccosphere (dimorphism). Additionally, some coccolithophorids display dithecatism: in which the coccosphere consists of a double layer of coccoliths, an inner layer (endotheca) composed of caneoliths and an outer layer (exotheca) composed of cyrtoliths. Further taxonomic problems are caused by coccolithophorid individuals secreting morphologically distinct coccoliths during successive phases of their life cycle. The dual plant and animal nutritional behaviour of some coccolithophorids (see ecology, below) has led to the group being claimed by both botanists and zoologists. However, most micropalaeontogists have used the Code of Biological Nomenclature for the systematic description of the taxa. There is no general agreement on the classification of Mesozoic and Cenozoic calcareous nannofossils. Many authors prefer to arrange nannofossil genera in alphabetical order for their systematic descriptions. Tappan (1980) included coccolithophorids in Class Coccolithophyceae and Division Haptophyta of the Plant Kingdom, however, more recent studies (e.g. Westbroek et al. 1986) include coccolithophorids in the phylum Prymnesiophyta (class Prymnesiophyceae).

This chapter reviews:

1. The main biological and ecological aspects of Recent coccolithophorids;
2. The main calcareous nannofossil evolutionary trends;
3. Calcareous nannofossil biostratigraphy;
4. Dissolution and diagenetic effects on nannofossil assemblages;
5. Recent advances in the use of nannofossils as tools in the elucidation of palaeoclimatic; and palaeo-oceanographic change.

2 Biology of Recent Coccolithophorids

Most haptophytes (including coccolithophorids) are characterized by a haptonema which lies between two smooth, whip-like flagella (Parke and Adams 1960). Internally, the coccolithophorid cell contains a single sub-spherical nucleus and two golden-brown chloroplasts. The photosynthetic pigments in the cell are chlorophyll-a and carotenoids (Sherwood 1974). The cell is enclosed by a membrane, which in turn has an outer mucilaginous covering. External to the cell membrane are the coccoliths. The life cycles of coccolithophorids are complex, and for many species they are still poorly understood. Detailed laboratory studies have been limited to a few species, mainly because of the practical difficulties in obtaining and maintaining monospecific cultures. Biomineralization aspects of coccolith formation have been most extensively studied on the species *Emiliania huxleyi* (Lohmann) Hay and Mohler and *Pleurochrysis carteri* (Westbroek et al. 1986; de-Vrind et al. 1986). These two species may be cultured easily on synthetic seawater or semi-synthetic media in large batches, and coccolith formation may to some extent be regulated by chemical and other environmental parameters (Dorrigan and Wilbur 1973). Coccoliths of *E. huxleyi* are formed in a specialized organelle, located near the nucleus and consisting of a coccolith vesicle (which surrounds the growing coccolith) and is joined to a system of anastomosing tubes, the reticular body. Prior to calcification, the specialized organelle develops a high concentration of Ca^{2+}-binding, water-soluble polysaccharides relative to the other cell constituents (van der Wal et al. 1983). Initial calcification occurs on the rim of a base plate within the coccolith vesicle, and as calcification proceeds, complete coccoliths are extruded to the periphery of the cell where they interlock to form the coccosphere. Pure coccoliths can be isolated by ultrasonic disintegration of the cell and centrifuging the resultant mixture in a silica or sucrose gradient (de Jong 1975). Laboratory studies have shown that three different cell types of *E. huxleyi* are commonly recorded:

1. Coccolith-bearing cells (C-cells);
2. Naked cells which lack conspicuous extracellular scales or coccoliths (N-cells);
3. Flagellated cells that have organic scales on their cell surface but no calcified coccoliths (S-cells).

However, it is not known what processes trigger the three cell types. Coccolithophorid division occurs inside the cell and the shell is either abandoned or shared equally by the two daughter cells, with subsequent reconstruction of a complete cell. When environmental conditions are favourable, cell division is rapid, and in the logarithmic growth phase the cells may divide every 24 h.

Many haptophytes show an alternation of two distinct stages in their life history (a dimorphic life cycle; Tappan 1980). The motile cell alternates during the life history with a non-motile stage which may be a pelagic cyst. Most coccolithophores are thought to be free-swimmers during one stage of their life cycle, using the pair of flagella for propulsion. Coccolithophorids, however, are probably passive plankton, as laboratory studies (Leadbeater 1971) have shown that they move on average 5 to 18 mm/min (at 18 °C) which is certainly less than the rate of most current movements. Studies by Black (1972) have shown that coccolithophorids have a flexible skeleton during the free-swimming stage and a rigid skeleton resulting from membrane calcification during the encysted phase of the life cycle. Planktonic and benthic stages, both with and without coccoliths, have been observed in several species, and although their significance has yet to be fully understood, the benthic stages are thought to occur in hostile environments. The ability of the coccolithophorids to secrete coccoliths on their organic scales differentiates them from all other algae.

3 Coccolith Morphology and Crystal Structure

X-ray diffraction patterns and crystal forms have indicated that in nature coccoliths are composed of low-magnesium calcite. Most coccoliths are constructed of rhombohedral calcite crystals in a radial arrangement, a few have hexagonal prisms, and some have a combination of these two forms in a single crystal (Tappan 1980). On the basis of crystallization two main types of coccolith are distinguished in coccolithophorids: (1) holococcoliths, (2) heterococcoliths.

The basis of this subdivision is the varying degree of crystal growth. Holococcoliths have unmodified crystal forms with retention of their geometric faces and angles and are precipitated outside the cell membrane (Parke and Adams 1960). Crystallolith holococcoliths consist of simple rhombohedral crystals scattered over the surface of the organic scale, although they do not completely cover the scale (e.g. motile phase of *Coccolithus pelagicus*). Calyptroliths include holococcoliths constructed of hexagonal crystals in the form of proximally open caps, e.g. *Sphaerocalyptra papillifera* (Halldal and Markali) Halldal. Zygoliths holococcoliths include forms which have an elliptical ring with a slightly or strongly arched crossbar (e.g. *Zygosphaera divergens* Halldal and Markali). Holococcoliths, because of their small crystal size and their loosely packed construction, are susceptible to rapid post-mortem disintegration and are rarely found in the fossil record.

Heterococcoliths are internally formed and elaborately constructed. The crystal faces and angles are modified, and the form of the coccolith is completely controlled by the cell. Heterococcoliths are made up of numerous crystals, commonly with radial construction, although rarely achieving exact radial symmetry. These are termed heliolithae. The most commonly occurring heliolith coccoliths are placoliths, which comprise discs built of two elliptical or circular shields constructed of radially arranged crystallites. The central area of placoliths may comprise an open pore [e.g. *Umbilicosphaera sibogae* (Weber-van Bosse) Gaarder], which may be crossed by a bridge (e.g. *Gephyrocapsa caribbeanica* Boudreaux and Hay, Fig. 3b), or may be filled with a lattice [e.g. *Reticulofenestra pseudoumbilica* (Gartner) Gartner, Fig. 2]. Caneoliths are basket-like coccoliths with a central area or bottom, of lamellae and petaloid upper and lower rims (e.g. *Syracosphaera pulchra* Lohmann, Fig. 3f). Cyrtoliths are basket coccoliths with a central process, which may be very elongated and club-shaped, e.g. *Rhabdosphaera clavigera* Murray and Blackman (Fig. 3e), or may be distally flared and trumpet-shaped as in *Discosphaera tubifera* (Murray and Blackman) Ostenfeld. Helicoliths include coccoliths which have a marginal flange, which increases in height around the periphery of the oval disc, overlapping the early part of the flange in a spiral construction, e.g. *Helicosphaera carteri* (Wallich) Kamptner. Unusual, elongated, diamond-shaped coccoliths with a central area of parallel laths and a raised rim are termed scapholiths and are represented by living *Anoplosolenia* and *Calciosolenia*.

Some calcareous nannofossils, constructed of a single or very few crystals, may have radial symmetry. These forms are termed the Ortholithae and include *Discoaster* species, ceratoliths and pentaliths. *Discoaster* species are discussed under 'other nannofossil groups' (see below). Ceratoliths are large horseshoe-shaped crystals. Each ceratolith acts as a single crystal of calcite, e.g. *Ceratolithus cristatus* Kamptner (Fig. 4). Pentaliths are nannofossils formed of five pieces, each consisting of a single crystal orientated 72° apart, e.g. *Braa-*

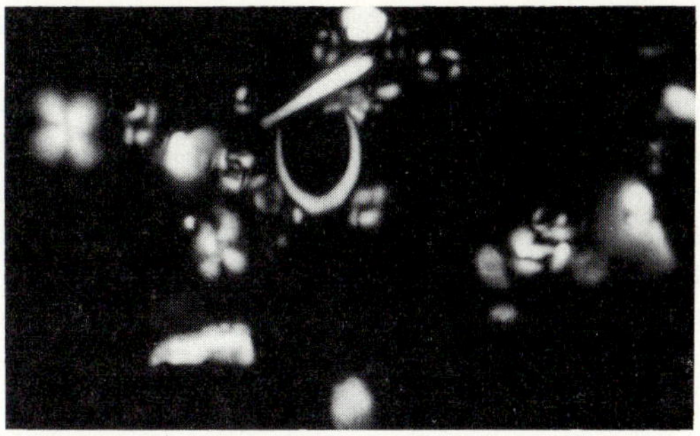

Fig. 4. Optical photomicrograph (using crossed polars) of Recent calcareous nannofossils from the Celtic Sea (x1600). Notice horseshoe-shaped *Ceratolithus cristatus* (*centre*) which acts as a single crystal as opposed to the extinction patterns produced by the surrounding coccoliths

rudosphaera bigelowii (Gran and Braarud) Deflandre (Fig. 3c). For a comprehensive review of coccolith morphology, see Tappan (1980).

Calcite crystals in coccoliths are arranged in such a way that a negative, uniaxal, pseudofigure is seen under crossed polars. This results in each individual coccolith showing a well-defined extinction cross (Fig. 4). Kamptner (1954) observed that each type of coccolith has a diagnostic interference pattern, and that the extinction position of the calcite crystals is important for the recognition of the species. The lines of the interference figure may curve either to the left (laevogyre) or to the right (dextrogyre), as they are traced out from the centre of the coccolith in distal view.

4 Other Calcareous Nannofossil Groups

Although all fossil coccoliths are generally thought to have been secreted by coccolithophorids, other nannofossil groups have different and more obscure affinities. These groups will now be discussed briefly.

Discoaster species (Fig. 4a) comprise nannofossils of star- or rosette-shaped construction. Most forms are flat-lying and remain dark when viewed with crossed nicols (using optical microscopy) because the c-axis of the calcite is vertical. The group is thought to contain numerous genera, most of which are usually included in *Discoaster*. Tan Sin Hok (1927) attempted to separate the *Discoaster* group into the genera *Hemidiscoaster*, forms with fusion of alternate rays, and *Eudiscoaster*, forms in which all the rays occur as independent crystals. This subdivision is considered inapplicable because some otherwise homogeneous species have variants belonging to both the suggested genera (Bramlette and Reidel 1954). The secretion of fused crystal units may also have palaeoecological significance (see Sect. 6). Theodoris (1983) used *Eu-discoaster* and *Helio-discoaster* as a replacement nomenclature and suggested that *Discoaster* is an illegitimate name. However, most workers still retain the name *Discoaster* for the whole group and this terminology is followed here.

The microrhabulids include rod-like cylindrical to pyriform calcareous nannofossils of unknown origin which first appeared in the Late Jurassic although they were most abundant in the Late Cretaceous. The nannoconids include forms with a thick wall of wedge-like calcite plates orientated perpendicular to and spirally surrounding an axial cavity or canal. *Nannoconus* (Fig. 5c) appeared in the latest Jurassic and is particularly common in the Lower Cretaceous of the Tethyan realm, where they occur in fine-grained limestones together with coccoliths, tintinnids, ammonites and planktonic foraminifera. *Nannoconus* had a cosmopolitan distribution as it has been recorded from Spitzbergen and also from the Falkland Plateau.

The schizosphaerellids include forms with subspherical tests (5–30 μm in diameter) constructed of two overlapping or interlocking dissimilar valves (Fig. 8b). The valves are constructed of small tabular crystals regularly arranged in a cubic lattice. The bipartite construction is unlike anything recorded in the coccolithophorids, but many represent an encysted or non-motile phase, similar to the braarudospheres (Tappan 1980). *Schizosphaerella* occur abundantly in Jurassic marine sediments.

Fig. 5. a *Discoaster nobilis* (range Palaeocene) (x6000); **b** *Thoracosphaera granifera* (range Pliocene-Recent) (x6500); **c** *Nannoconus* sp. (x8600); **d** Coccosphere of *Ellipsagelospheara* cf. *keftalrempti* (x6400)

The *Thoracosphaera* (Fig. 5b) are spherical shells composed of a mosiac of interlocking crystal units. The crystals are constructed of polygonal prisms that may be solid or axially perforate. The genus *Thoracosphaera*, now classified as a dinophyte and not a coccolithophorid (Tangen et al. 1982), has been traditionally included in nannofossil studies. Living *Thoracosphaera* are found in the open ocean under normal environmental conditions, however, the *Thoracosphaera* blooms which followed the Cretaceous/Tertiary boundary may reflect an adaption to special environmental conditions because of the absence or scarcity of competing calcareous plankton.

5 Ecology of Recent Coccolithophorids

Recent coccolithophorids, being predominantly autotrophic plankton, are largely restricted to the photic zone of the water column (0–200 m depth) where sunlight can easily penetrate. Their maximum concentrations occur above a depth of 100 m in the tropics and above 50 m in temperate regions; they therefore live above the thermocline in the world's oceans. The vertical distribution of some of the more common coccolithophorid species in the oceans is summarized in Table 1. Upper photic zone species (0–50 m) are typically of low density and include cyrtoliths which are characterized by elongated or distally flared appendices (e.g. *Discosphaera tubifera* and *Rhabdosphaera clavigera*). A smooth, thin-walled form of *Umbellosphaera* (*U. irregularis* Paasche) is also dominant in tropical surface waters. Mid-photic zone species are typically placoliths with an open central pore (e.g. *Umbilicosphaera* spp.,

Table 1. Vertical zonation of Recent coccolithophorid species in the photic zone of the Pacific and Atlantic oceans. Compiled mainly from Okada and Honjo (1973) and Okada and McIntyre (1977)

Species	Vertical distribution in the photic zone
Emiliania huxleyi *Gephyrocapsa oceanica*	Throughout (0–200 m)
Discosphaera tubifera *Rhabdosphaera clavigera* *Umbellosphaera irregularis*	Upper Photic Zone (0–50 m)
Calcidiscus leptoporus *Umbilicosphaera hulburtiana* *Umbilicosphaera sibogae* *Oolithotus fragilis cavum* *Umbellosphaera tenuis*	Middle Photic Zone (ca. 50–150 m)
Anthosphaera quadricornu *Florisphaera profunda* *Thorosphaera flabellata* *Hayaster perplexus* *Oolithotus fragilis*	Lower Photic Zone (150–200 m)

Oolithotus fragilis cavum Okada and McIntyre) and *U. tenuis* (Kamptner) Paasche, a more heavily constructed form of *Umbellosphaera* with ribs on the distal shield. Lower photic zone forms include large coccolithophorid cells, characterized by non-perforate coccoliths (e.g. *Oolithotus fragilis* (Lohmann) Okada and McIntyre and *Hayaster perplexus* (Bramlette and Riedel) Bukry, and forms with multilayered, unusual ellipsoidal to rectangular plate-shaped coccoliths (*Thorosphaera flabella* Halldal and Markali, and *Florisphaera profunda* Okada and Honjo). Two species, *Emiliania huxleyi* and *Gephyrocapsa oceanica*, are commonly reported at all depths in the photic zone, possibly indicating that these forms can live at various depths and temperatures. However, this may not represent a true living distribution, but could represent the disintegration of macroaggregates of the species at depth. Additionally, the small cell size of *E. huxleyi* and *G. oceanica* ($< 10\,\mu m$) may hinder their retention by some filter-feeding zooplankton grazers (see Sect. 8). Some coccolithophorid species have also been reported from depths as great as 4000 m in the Atlantic, such forms are probably heterotrophic or saprophytic.

Coccolithophorids, together with dinoflagellates, predominate in the plankton communities of tropical seas and they occur in lesser numbers in cooler oceanic waters where diatoms are more abundant. Coccolithophorids are particularly dominant in waters of low nutrient content, including regions of 'oceanic convergence', subtropical gyre centres (e.g. the Sargasso Sea) and in marginal seas with anti-estuarine circulation such as the Mediterranean Sea (Berger 1976). Over 300 living coccolithophorid species have now been described, with the most diverse floras being those in the Atlantic. *Emiliania huxleyi* is the most abundant living coccolithophorid and is "probably the most productive lime-secreting organism on Earth" (Westbroek et al. 1986). The species is a cosmopolitan form and has been reported from tropical to subarctic waters.

Cold-water adapted coccolithophorids are also a consistent component of subpolar and polar seas. Thirteen coccolithophorid species are known to survive in the Weddell Sea, Antarctica (Thomsen et al. 1988) and 16 species occur in the seas around West Greenland (Thomsen 1981). Few if any of these Greenland species have a restricted distribution, and most seem indigenous throughout boreal, subarctic and arctic areas. At least two of these species, *Wigwamma annulifer* Manton, Sutherland and Oates and *Papponamonas flabellifera* Manton and Oates occur in both hemispheres, which suggests that relatively recently, in geological times, an exchange of biological material has occurred between the polar latitudes (Thomsen et al. 1988). Species which occur at both poles are perhaps able to withstand transport across the equator by surviving in cold deeper waters, far below the photic zone (Thomsen 1981). Further studies involving sampling of deep-water plankton below the photic zone in low latitudes is needed to test this transport mechanism. *W. annulifera* also occurs in the northern Pacific Ocean, off South Alaska (59° N, 151° W). This geographical distribution suggests that the Northwest Passage could be a significant migration route for cold-water plankton from the Atlantic to the Pacific Ocean (Manton et al. 1976).

Turrisphaera borealis Manton, Sutherland and Oates, *Turrisphaera arctica* Manton Sutherland and Oates and *Coccolithus pelagicus* (Wallich) Schiller are species known to tolerate subzero temperatures. *Turrisphaera* is characterized by unusual tower-shaped coccoliths composed of hexagonal crystallites recticulately arranged, each crystallite having a central perforation. *Turrisphaera* occurs at Resolute Bay, Cornwallis Island in the Northwest Passage at a temperature of ca. –1 °C (Manton et al. 1976). *C. pelagicus* has been reported living under pack ice in the Eastern Greenland Polar Current (between Iceland and Greenland) at temperatures as low as –1.7 °C (Braarud 1979). The motile phase of *C. pelagicus* is reported to ingest bacteria, graphite and nannoplankton cells (< 5 μm in diameter). The dual nutritional behaviour of the species may explain its northern distribution in the Atlantic and may allow the species to survive the low light intensities characteristic of the long dark season of the Arctic.

The distribution of *C. pelagicus* in the plankton of the North Atlantic is limited by the maximum southerly position of the 14 °C isotherm, which lies about 40° N in the winter. In the South Atlantic, the species has been found in postglacial cores south of latitude 25° S, but was absent from all water samples. This absence of *C. pelagicus* in the Recent plankton has led to the suggestion that until latest Pleistocene times, *C. pelagicus* was still living in the South Atlantic and that its disappearance was probably associated with postglacial warming in the Southern Hemisphere (McIntyre and Bé 1967). Although *C. pelagicus* is generally thought to be absent from the Recent plankton of the Southern Hemisphere oceans, Hallegraeff (1984) has recorded coccospheres of the species in waters off western Tasmania (latitude 40° S).

Coccolithophorids, although characteristically oceanic, also bloom seasonally in coastal waters. Plankton studies in shelf waters from northwest Europe have indicated that coccolithophorid blooms only occur in the spring/summer period when the water column is thermally stratified with a well-developed thermocline (Holligan et al. 1983; Murray et al. 1983). Surface blooms of cells of *E. huxleyi* may occur in numbers as high as 8.5 million cells/l in the Celtic Sea (Holligan et al. 1983). There is, however, no record of significant coccolithophorid growth in the tidally mixed shelf waters of northwest Europe, e.g. in the southern North Sea and in the central and eastern English Channel. The abundance and species diversity of coccolith assemblages in the shelf sediments of northwest Europe (Houghton 1988, 1989) mirror the distribution of coccolithophorids in the overlying water. Figure 6 shows the species diversity of coccolith assemblages in Recent sediments from the Celtic Sea and English Channel. Highest species diversity assemblages (8 or more species, a maximum of 18) occur in sediments from the Celtic Sea and western English Channel, which are deposited below a seasonal thermocline (Houghton 1988). Low diversity coccolith assemblages (six or fewer species) are deposited from the tidally mixed waters in the central and eastern English Channel.

The restriction of coccolithophorid blooms to stratified shelf waters may be explained by the nutrient and light preferences of the group. Coccolithophorid growth above a thermocline is enhanced by the good light regime and is also

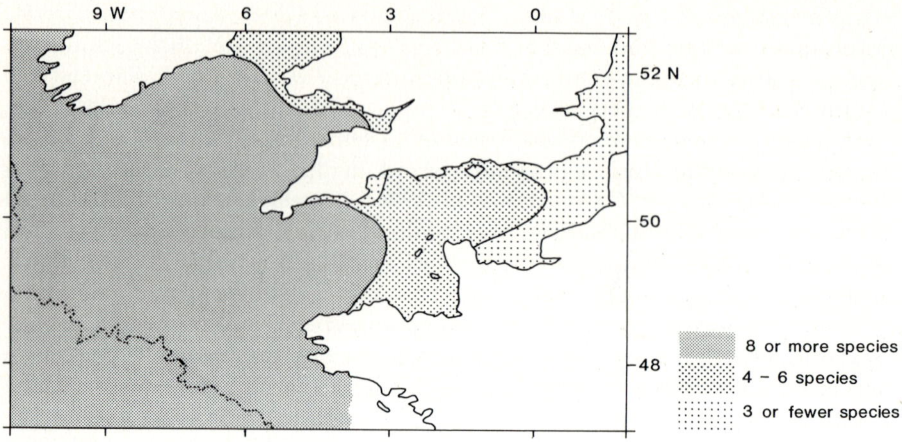

Fig. 6. Species diversity of Recent coccolith assemblages in the surface sediments of the Celtic Sea and the English Channel (after Houghton 1986)

favoured by the low nutrient levels. A diagnostic feature of many nannoplankton species is that the optimum rate of division is generally higher than that of larger-net phytoplankton, following the broad generalization that rate of uptake, growth and division is higher with reduced cell size (Raymont 1963). Oceanic nannoplankton are thought to have a significant advantage over larger-celled species in competition for nutrients, because they generally have lower Ks values (Friebele et al. 1978). The Ks value of a species defines the minimum concentration at which it can divide. Nannoplankton species such as *Emiliania huxleyi*, which has a Ks value of ca. 0.1 μM nitrate/l (Eppley et al. 1969), are able to bloom in the nutrient deficient waters overlying a thermocline, whereas large dinoflagellate cells which typically have large Ks values (e.g. ca. 5 μM nitrate/l) are generally excluded from the surface water plankton. However, in tidally mixed waters, although there are sufficient nutrients for algal growth, coccolithophorids are removed from the surface waters before any significant growth can occur. Other low diversity continental shelf coccolith floras associated with strong tidal mixing include examples from the Bay of Fundy (Gran and Braarud 1935) and in the Torres Strait between Australia and Papua New Guinea (Conley 1979).

Some coccolithophorid species prefer brackish waters and a few even prefer freshwater occurring in rivers and lakes. The most commonly reported freshwater species is *Hymenomonas rosa* Stein (Tappan 1980). The euryhaline characteristics of coccolithophorids as a group are evidenced by the occurrence of 52 species in the Gulf of Elat (Aqaba), where the mean salinity is 41 parts per thousand (ppt) (Winter et al. 1979). *E. huxleyi* has a wide salinity tolerance. It has been found in the Gulf of Elat, thrives in the Black Sea (17–18 ppt), and also occurs in the Sea of Azov (11 ppt) (Pitsky 1963). *Coccolithus pelagicus* probably has the most wide ranging salinity range, as it occurs in the Dead Sea at salinities as high as 250 ppt. Extremely low salinities are tolerated by *Cricosphaera* which occurs in supratidal rock pools influenced by wave splash (0.4->25 ppt). *Cricosphaera* also occurs as a psammophyte, living in the interstitial waters of intertidal sands at Roscoff (Brittany).

Although conventional plankton sampling studies have indicated that coccolithophorids are important members of the plankton from many areas, it is only since the development of satellite imagery that the accurate mapping of the extent and quantitative significance of coccolithophorid blooms has been achieved. Coccolithophorids, because of their external calcitic plates, have high scattering signatures and may be recognized as high-reflectance patches at all wavelengths in the visible band using satellite imagery (Aiken and Bellan 1988). The recognition of coccolithophorid blooms by satellite colour scanners (Holligan et al. 1983) indicates the potential of remote sensing techniques for the identification and tracking of plankton blooms, and these images are essential for the sampling requirements of international oceanographic surveys such as the Global Ocean Flux Study (GOFS).

Coccolithophorids play an important role in marine biogeochemical cycles and contribute to the ocean's effects on the world's climate. The secretion of calcitic plates removes CO_2 (dissolved in the oceans as bicarbonate) from the surface waters and transfers it to bottom sediments through the depositon of the coccoliths. The occurrence of large blooms of coccolithophorids also helps reduce the heat budget of the oceans by reflecting up to 30% of the incident sunlight. Coccolithophorids are known to release volatile dimethyl sulphide (DMS) which adds to atmospheric sulphur and probably to the acid rain of European maritime countries. DMS also acts as a precursor for cloud condensation nuclei (CNN), which consequently causes a decrease in the Earth's albedo (Charlson et al. 1987). The secretion of DMS may represent a Gaian response to climatic warming (i.e. "the greenhouse effect") with the biota blanketing the Earth with cloud and reducing the effect of solar heating (Lovelock 1988; Aiken and Bellan 1988). Present studies (GOFS) on the distribution of coccolithophorid blooms in the eastern North Atlantic will investigate the oceanic uptake of CO_2 in the Northern Hemisphere. The North Atlantic is well suited for such studies, since active processes in the area have significant implications for the global carbon cycle. The Northern Hemisphere is the major site of anthropogenic carbon input. Additionally, the North Atlantic is thought to be a major sink for atmospheric carbon, both by the processes of primary production and by the downwelling of CO_2-rich water along its northern margin. Such studies should help to determine whether coccolithophorids and other marine algae are helping to modify the climatic warming caused by the "greenhouse effect".

6 Evolution of Calcareous Nannofossils

Although there are a few reports of isolated coccolith-like bodies from Palaeozoic sediments, calcareous nannofossils are generally thought to have

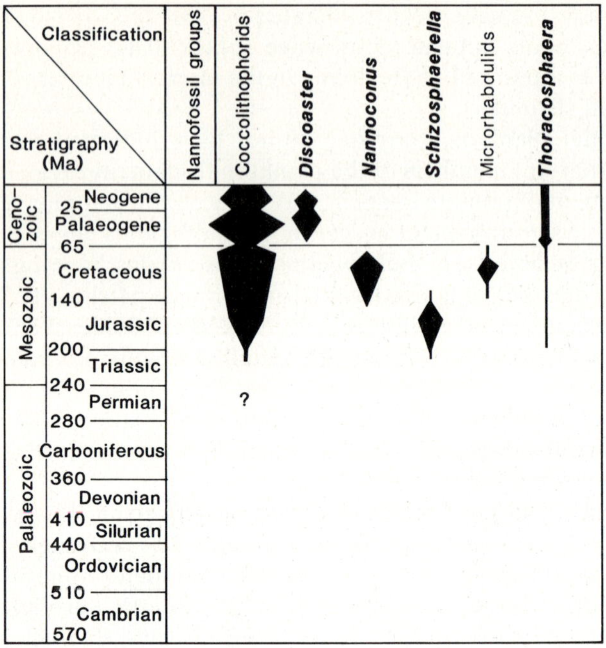

Fig. 7. Generalized stratigraphic range of some groups of calcareous nannofossils (after Bignot 1985)

originated in epicontinental seas during the Late Triassic (Fig. 7). The earliest calcareous nannofossil to appear in the Triassic was *Prinsiosphaera triassica* Jafar (Jafar 1983). This genus comprises "spherical to hemispherical solid nannofossils often containing a depression at one end and consisting of a parallely stacked group of calcitic plates orientated in a random fashion" (Fig. 8a). Jafar recorded these forms from the Carnian to the Rhaetian of the Alps. The assumed organic origin of *Prinsiosphaera* is based on its abundance, stratigraphical occurrence, structural regularity and distribution in the Alpine area. The biological affinity of Prinsiosphaera is unclear, as it has a completely different morphology to coccoliths. However, it is probably the product of some planktonic alga and may be related to a primitive calcareous dinoflagellate (Bown 1987). *Shizosphaerella puncticulata* Deflandre and Dangeard (Fig. 8b) appeared slightly later than *Prinsiosphaera* and is found in Upper Triassic/ Lower Jurassic boundary sediments. *Schizosphaerella* replaced *Prinsiosphaera* as the dominant spherical nannofossil group in the earliest Jurassic and they probably occupied the same niche in the epicontinental seas. Next to evolve in the Late Triassic was *Eoconusphaera zlambachensis* (Moshkovitvitz) Bown and Cooper, a nannofossil with a truncated cone morphology which is subcircular in plan view. The first *Thoracosphaera* (*T. geometrica* Jafar) also appear at this time, although its relationship to younger members of this genus is unclear.

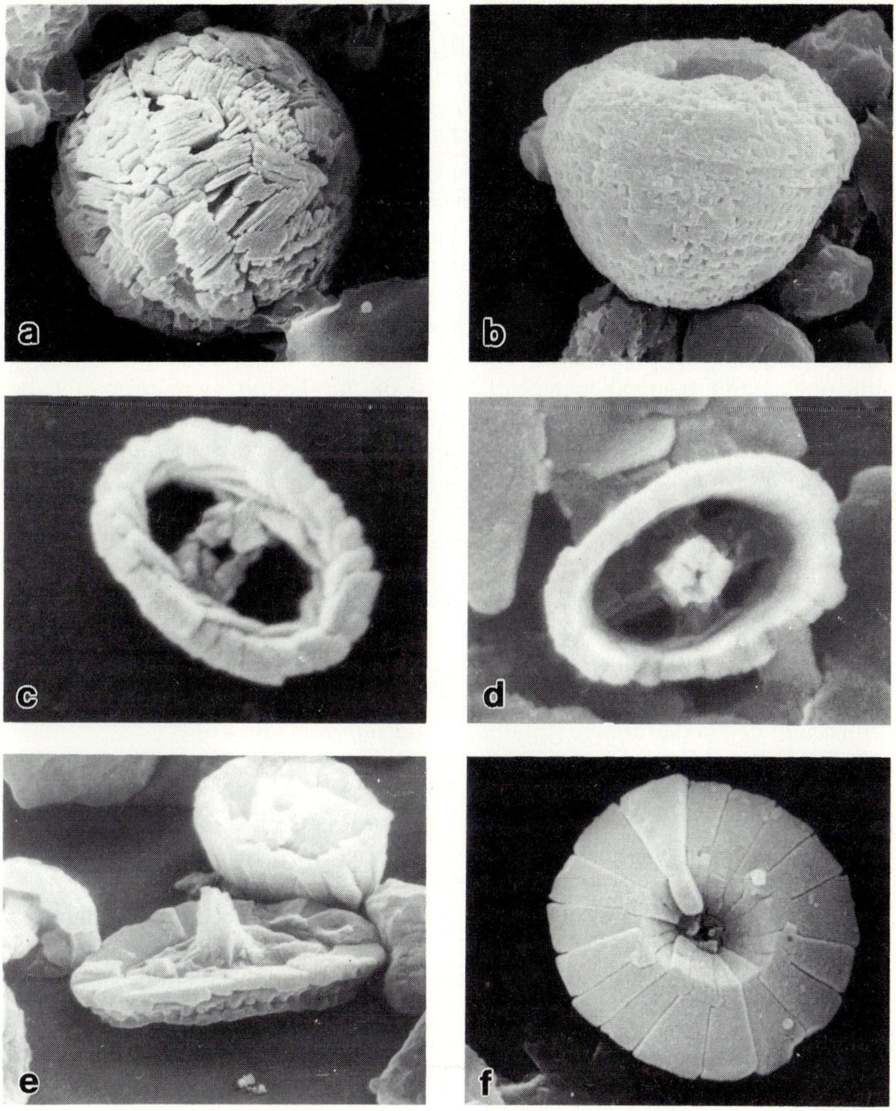

Fig. 8a-f. SEM micrographs of some Late Triassic-Early Jurassic calcareous nannofossils (all figures from Bown 1987). **a** *Prinsiosphaera triassica* (x5700); **b** *Schizosphaerella puncticulata*, side view (x3300); **c** *Archaeozygodiscus koessenensis*, distal view (x19 800); **d** *Crucirhabdulus minutus*, distal view (x17 400); **e** *Mazaganella pulla*, distal view (x7300); **f** *Biscutum novum*, distal view (x10 500)

The first 'true' coccoliths, *Archaeozygodiscus koessensis* Bown (Fig. 8c) and *Crucirhabdus minutus* Prins, ex. Rood, Hay and Barnard (Fig. 8d), are recorded in the Norian *Rhabdoceras suessi* ammonite Zone (Jafar 1983; Bown 1987). The comparatively sudden appearance of coccoliths and other calcareous nannofossils in the Late Triassic may have been because a previously 'naked' group of algae developed the capability to secrete micron-sized calcitic shields and plates on their organic cell wall, with a consequent development of their fossil record. The appearance of calcareous nannofossils followed a period of heavy salt precipitation throughout the Permian and Triassic periods and perhaps the salinity of the Late Triassic epicontinental seas was suitable for the secretion of calcitic plates (Perch-Nielsen 1986). The initial location of coccolith-bearing algae appears from previous records to have been in the Tethys Ocean. The subtropical and tropical waters of much of the Tethys were probably the optimum environment for coccolithophorid growth. It is also possible the appearance of coccoliths in the Late Triassic represented the evolution of a new algal group which was filling a planktonic niche left vacant by the mass mortalities of the Permian.

The next coccolith lineage to appear in the Early Jurassic was the tiered placolith group represented by *Mazaganella* (Fig. 8e), although it has only been reported from the Sinemurian and Pliensbachian of the southern Tethys area (Bown 1987). Placoliths with radiating crystal sutures appeared in the Late Sinemurian and are represented by *Biscutum novum* (Goy) Bown (Fig. 8f). The final lineage to appear in the Early Jurassic (Pliensbachian) were placoliths with imbricated crystal sutures (e.g. *Lotharingius imprimus* Bown).

Throughout the Early Jurassic new coccoliths suddenly appear with little evidence as to their ancestry (Tappan 1980; Bown 1987). This is probably the result of poor preservation, as occasionally sediments are found with a high-diversity assemblage which contains some species which have no known ancestors. Goy et al. (1979) described a remarkably abundant and high-diversity nannofossil assemblage from the Lower Toarcian 'Schistes Carton' of the Paris Basin. This assemblage is the result of an exceptional preservation, a feature rarely noted in Lower Jurassic sediments because of their diagenetic alteration. *Mitrolithus jansae* (Wiegand) Brown and Young, a truncated conical nannofossil, appeared in the early Sinemurian and was a dominant component of Mediterranean-Tethys assemblages until the Early Toarcian (Bown and Cooper 1989).

Species diversity and production initially remained low in the earliest Liassic but then, in spite of minor fluctuation, diversity slowly increased throughout the Early and Middle Jurassic. The first significant diversification occurred during the early part of the Late Jurassic (Oxfordian). Oceanic nannofossils show a peak in diversity in the Tithonian, when assemblages were enriched with the first nannoconids and other nannofossils that flourished in the Early Cretaceous (Roth 1983). Although nannofossil diversity steadily increased throughout the Middle and Late Jurassic, nannofossil productivity remained low, as evidenced by the rarity of pure pelagic carbonate sequences of this age. Coccolithophores were abundant, however, in some Jurassic shelf seas as evidenced by the Kimmeridgian coccolith limestones of the Dorset coast and the German Solnhofen Limestone. In the Kimmeridgian and Tithonian, high latitude nannofloras were dominated by *Ellipsagelosphaera*, whereas in low latitudes *Watznaueria* was the dominant form (Cooper 1989). The migration of calcareous nannoplankton from shelf and epicontinental seas into the oceanic regime in the latest Kimmeridgian to Tithonian is reflected in an abrupt change from impure limestone, marl and chert sequences to pure nannofossil carbonates. Nannofossil productivity was sufficiently high in the pelagic realm during the latest Jurassic to depress the carbonate compensation depth (CCD) in the Tethyan Ocean (Roth 1986). The end of the Jurassic is marked by a slight decrease in nannofossil diversity, although only *Schizosphaerella* became extinct near the Jurassic/Cretacaeous boundary.

The general elevation in sea level accompanied by the development of carbonate platforms throughout the Tethys during the Early Cretaceous promoted the appearance of many new species and genera. Lower Cretaceous sediments of northeast Europe yield high diversity (>40 species/sample) assemblages comprising a mixture of cosmopolitan, endemic, Tethyan and Boreal-Arctic species (Crux 1989). A second marked increase in diversity occurred in the Aptian and Albian, new forms include *Broinsonia*, *Gatnerago* and *Prediscosphaera*. Most Early Cretaceous assemblages contain smaller individual coccoliths than those in the Late Cretaceous and are frequently dominated by *Ellipsagelosphaera*. The widespread pelagic limestones of the Early Cretaceous of southern Europe, and the chalks and marls of the Gulf Coast and Mexico consist of abundant coccoliths together with nannoconids and calpionellid tintinnids.

Maximum diversity, as well as abundance of calcareous nannofossils was achieved during the Late Cretaceous when the worldwide chalks, which were comprised mainly of coccoliths and their skeletal elements, were deposited. Many of the Late Cretaceous nannofossil species were long-ranging forms lasting for ca. 5–10 m.y. Diversity remained high during the Late Cretaceous (> 130 species) until a decrease in the Middle Maastrichtian. At least 155 new species originated during the 80 m.y. of the Cretaceous; a rate of ca. 2 new species/m.y. over the whole of the period (Perch-Nielsen 1986). During the same 80 m.y. around 126 species became extinct before the end of the Maastrichtian, which indicates a rate of ca. 1.5 extinctions/m.y.

At the end of the Maastrichtian came the well-documented extinctions of marine fauna and flora and the calcareous nannoplankton suffered heavy losses with only some 15 to 18 genera (about one-third of the Maastrichtian flora) surviving into the Early Danian (Perch-Nielsen 1982). These species, presumably, must have comprised the total generic reservoir for the Cenozoic evolution. Most of the surviving genera evolved during the Jurassic and have relatively simple morphology, such as *Biscutum*, *Braarudosphaeara* and *Markalius*. Forms with a complex construction such as *Prediscosphaera* disappeared. Perhaps coccolithophorids which secreted simple coccoliths were able to survive the stresses of the Cretaceous/Tertiary (K/T) boundary event, whereas overspecialized species could not adapt to the environment and became extinct (Perch-Nielsen 1986). At low latitudes, *Thoracosphaera* blooms characterized the plankton assemblages shortly after the boundary, although the genus had only a sporadic distribution during the Cretaceous. None of the nannofossil genera which survived the K/T boundary became successful during the Cenozoic (Aubry 1989). Even the long-ranging genera *Braarudosphaera* and *Scapholithus*, which continue through to the Recent, show no evolutionary diversification during the Cenozoic. The other surviving genera (e.g. *Biscutum* and *Chiastozygus*), with the exception of *Markalius* which continued to the Oligocene, underwent progressive extinctions in the Palaeocene and Early Eocene.

The mass extinctions of calcareous nannoplankton and planktonic foraminifera at the boundary drastically reduced the flux of calcium carbonate reaching the ocean floor. The boundary clay is virtually barren of pelagic carbonate. Deposition of carbonate was reduced for at least 0.35 m.y., although in some cases for as long as 1 m.y.. This decrease in carbonate is attributed to a reduction in secretion of carbonate skeletons in the surface waters, rather than a prolonged phase of dissolution on the seafloor. K/T boundary sequences also contain a large negative anomaly in $\delta^{13}C$ (up to 3‰) measured in the foraminiferal and coccolith carbonate. This negative anomaly is considered to be a worldwide event and probably indicates that the surface waters only supported a very impoverished plankton. Minimum values of $\delta^{13}C$ occurred ca. 40–50 kyr after the boundary and did not return to pre-boundary values for at least ca. 300–400 Kyr. Although the magnitude of the $\delta^{13}C$ varies at different sites, it has been detected only in the calcareous plankton. Primary productivity is thought to have been suppressed for ca. 1 m.y. after the K/T boundary.

Various processes and events have been proposed to account for the mass extinctions at the K/T boundary. These include a metorite impact event, a severe regression at the end of the Maastrichtian, development of widespread oceanic anoxia, temperature and salinity variations, changes in solar radiation, nutrient limitation, and excessive volcanism. Perch-Nielsen (1986) has indicated that although an iridium anomaly, microtectite concentrations and/or shocked quartz (all features thought indicative of an extraterrestial impact) occur at both the K/T and Eocene/Oligocene (E/O) boundaries, nannofossils reacted quite differently to these events. Nannofossil abundance dramatically decreased at the K/T boundary; however, only negligible changes occur at the E/O boundary. This difference in the reaction of the nannofossil assemblages may indicate a variation in the magnitude of the impact event, or may indicate that most of the low-stress ('intolerant') forms were eliminated from the plankton by the previous impact event. However, the nannofossil species which became extinct at the K/T boundary represent those forms that were most abundant and widespread, whereas those that survived were extremely rare and had a restricted distribution. These extinction and survival patterns are not thought to be indicative of an extraterrestial impact; such extinctions are expected to occur at random, with the probability of survival being related to the number of individuals in the population (Stanley 1979; Donovan 1987).

The mass extinction of calcareous nannoplankton at the K/T boundary has recently been suggested to have caused a rapid global warming (Rampino and Volk 1988). The elimination of the marine calcareous phytoplankton may have caused a severe decrease in DMS production leading to a dramatic reduction of CCN and a concomittant decrease in marine cloud albedo. The decrease in marine cloud albedo may have caused a rapid global warming calculated to be $> 6°C$, with the highest temperature differences being recorded at higher latitudes. Oxygen isotope analyses of marine sediments have indicated a general warming at the K/T boundary and the resultant higher surface water temperatures may have been a significant factor in the suppression of the algal productivity following the K/T boundary.

Following the major extinction of Cretaceous nannofossils at the K/T boundary, new forms evolved in the Cenozoic, many being significantly larger than the previous forms. Additional new and morphologically unusual forms appeared, including *Sphenolithus*, *Triquetrorhabdulus*, *Ceratolithus* and *Discoaster*. During the early Cenozoic (Fig. 9), new taxa evolved slowly at first, rising to a total of ca. 20 nannofossil species by the end of the Danian, and up to 60 species in the Late Palaeocene. Rapid evolution occurred in the Early Eocene (120 species). This increase was partly due to the appearance of *Discoaster* species in the Late Palaeocene (first occurrence datum at ca. 60 Ma). The evolution of *Discoaster* species occurred during the heaviest values of $\delta^{13}C$ in $CaCO_3$ of any epoch in the Cenozoic. Peak values of $\delta^{13}C$ (3.8 per mil in bulk carbonate) occurred at ca. 60 Ma (Shackleton and Hall 1984) and indicates that *Discoaster* species evolved during a period of very high surface productivity. During the Late Palaeocene diversification of *Discoaster* species was rapid

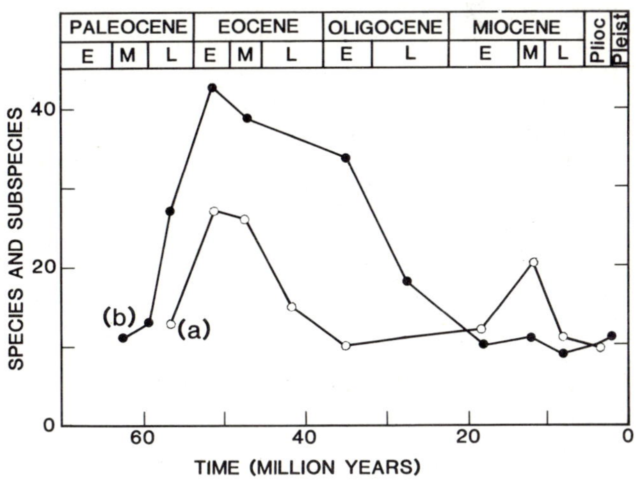

Fig. 9. Diversity of *Discoaster* (*a*) and coccolith species (*b*) in the Cenozoic, shown by subdivision of the epochs (after Haq 1971)

averaging 7.5 FOs/m.y. Overall nannofossil diversity then steadily declined until the end of the Lower Miocene, when ca. 30 recorded species occurred. A second, less dramatic, radiation of nannofossil taxa occurred during the Middle Miocene, resulting in a total of over 50 species. Subsequently, the trend of decreasing diversity continued until the Pleistocene. The Recent coccolithophorid flora is still less diverse than that of the Cretaceous.

Within the Cenozoic, alternating diversification and decline of nannofossil assemblages seems to be temperature related. Greatest diversity occurs in warmer intervals, whereas less diverse assemblages consisting of cosmopolitan taxa occur in cooler periods. Haq (1973) suggested that species with a limited temperature tolerance may spread widely in equable seas, but when climates became more extreme, these species had extremely reduced biogeographic ranges which resulted in their extinction. However, the majority of the Recent coccolithophorid species recorded in the Atlantic and Pacific oceans exhibit eurythermal characteristics (Okada and McIntyre 1979). The wide temperature tolerance of coccolithophorids as a group suggests that effects other than simply water temperature were controlling nannofossil diversity. Changes in nannofossil diversity throughout the Cenozoic may also be explained by variations in the depth of the photic zone in the world's oceans. In Recent oceans, highest diversity coccolithophorid assemblages are found in the subtropics and tropics, where there is a deep (up to 200 m) photic zone, and based on coccolithophorid assemblages, this zone has a tripartite subdivision into upper, middle and lower photic zones (Okada and Honjo 1973). North of ca. 40°N in the North Pacific, the photic zone decreases to around 100 m and has a bipartite subdivision, with loss of the lower photic niche and a contraction of the upper photic niche. The consequent reduction in the availability of ecological niches lowers the diversity of coccolithophore assemblages in these higher latitudes. During the warm intervals of the Cenozoic, it is likely that the depth of the photic zone increased and that the subtropical tripartite subdivision of the photic zone expanded into more northern and southern latitudes, allowing the migration of higher diversity coccoliths assemblages into these region. The patterns of evolution throughout the Cenozoic are reflected in biostratigraphic zonation; nannofossil zones during the cooler Oligocene have a duration of ca. 3 m.y., whereas warm epoch zones average 1.7 m.y.

A major evolutionary trend shown by coccoliths throughout the Cenozoic is a gradual reduction in the volume of calcite secreted per specimen. Palaeocene and Eocene coccolith assemblages contain some very large (10–26 μm diameter) forms, e.g. *Chiasmolithus*. Most Mio/Pliocene coccoliths range in size between 6 and 12 μm, although small recticulofenestrids < 5 μm may dominate Pliocene high latitude floras. *Pseudoemiliania lacunosa* (Kamptner) Gartner, the first coccolith species with slits in one shield, evolved in the latest Early Pliocene (FOD = ca. 3.6 Ma). At about the same time small *Gephyrocapsa* species, forms with a diagonal bar crossing the central pore, evolved. Small coccolith species (2 to 5 μm) occur throughout the Pleistocene, when the genus *Gephyrocapsa* frequently dominated. Most Recent nannofloras are dominated by *Emiliania huxleyi* (Fig. 13), another species with slits in its shields; the warm-water ecophenotype of the species has slits in both shields, and the cold-water ecophenotype has slits in one shield only.

The decrease in volume of calcite secreted per nannofossil specimen has been suggested to be related to climatic cooling trends as carbonate is more available in warmer waters. However, the assumption that large nannofossils with a high density of calcite per specimen are associated with warmer waters is not substantiated by observations of Recent plankton communities. Recent warm-water morphovariants of coccoliths, e.g. *E. huxleyi* contain considerably less calcite per specimen than the cold-water form. Other tropical forms are fragile and of low density (e.g. *U. irregularis* Paasche). Cold-water coccolith species are typically strong, imbricated, high density structures (e.g. *Coccolithus pelagicus*). The same principle of increased calcite encrustment in cold-water forms when compared with warm-water forms, is also strongly apparent in planktonic foraminifera. Tropical forms have characteristically fragile, perforate tests.

The gradual reduction in volume of calcite secreted per coccolith specimen is also reflected in *Discoaster* evolution (Bukry 1971). Most Palaeogene *Discoaster* assemblages are characterized by compact, multirayed, heavily constructed forms with large central knobs. In plan view these early *Discoaster* have a short 'free length' of rays, or the rays are very broad and the area encompassed by calcite is very high (> 80% cover in plan view) (Fig. 10m-p). However, a few more delicately constructed forms with a longer free length of ray also occur and include *D. okadai Bukry* (ca. 28% calcite cover) and *D. lodoensis* Bramlette and Riedel (ca. 40–50%). Mid-Cenozoic *Discoaster* species have a reduced number of rays, although in plan view there is still a substantial area of calcite (Fig. 10i-k). *Discoaster* assemblages in the Late Pliocene are restricted to very narrow-rayed forms (Fig. 10a-e). The skeletons of these represent a mere framework (< 25% cover) of the massive *Discoaster* 'shields' typical of the Palaeogene. Only two *Discoaster* species occur just before their extinction in the latest Pliocene at ca. 1.9 Ma: *D. brouweri*, Tan emend Bramlette and Riedel, a six-rayed form and, *D. triradiatus* Tan, a three-rayed form.

The occurrence of a wide spectrum of *Discoaster* morphologies, even within the same assemblage, was probably ecologically controlled and suggests that the forms were depth-stratified. Test thickening and calcite crust formation is well documented in deep-water planktonic foraminifera (e.g. Orr 1967) and high density 'solid' coccoliths are typical of the lower photic zone in the Recent Oceans (see Sect. 5). A similar depth-control on calcite secretion may have occurred in the *Discoaster* group. It is likely that changes in the amount of calcification of the *Discoaster* skeletons affected the passive bouyancy of the *Discoaster*-bearing algae. Bouyancy in the water column may be accomplished by numerous ways including a reduction in the specific gravity of the cell, by increasing the surface-to-area ratio, or by added resistance. Coccolith calcite has a density of between 2.7 and 2.95, whereas seawater varies between 1.021 and 1.028 (Tappan and Loeblich 1973). A reduction in specific gravity may be achieved by economic use of skeletal secretion. (Parsons and Brasier 1987). Bé (1968) has shown that in Recent planktonic foraminifera populations, surface

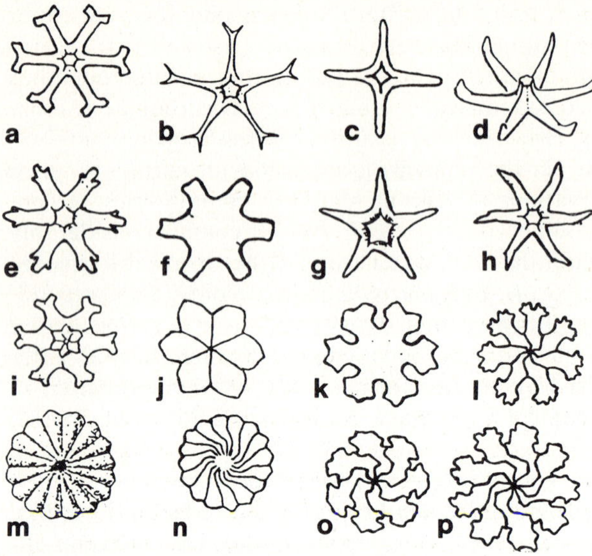

Fig. 10a-p. *Discoaster* shape variation (mainly after drawings by Prins 1971). **a** *Discoaster challengeri* Bramlette and Riedel (M. Miocene- L. Pliocene); **b** *D. pentaradiatus* Tan emend. Bramlette and Riedel (M. Miocene- L. Pliocene); **c** *D. tamalis* Bramlette and Riedel (L. Pliocene); **d** *D. brouweri* Tan emend. Bramlette and Riedel (M. Miocene- L. Pliocene); **e** *D. surculus* Martini and Bramlette (L. Miocene- L. Pliocene); **f** *D. kugleri* Martini and Bramlette (M. Miocene); **g** *D. quinqueramus* Gartner (L. Miocene); **h** *D. neohamateus* Bukry and Bramlette (L. Miocene); **i** *D. bollii* Martini and Bramlette (M.-L. Miocene); **j** *D. adamantus* Bramlette and Wilcoxon (Oligocene- L. Miocene); **k** *D. deflandrei* Bramlette and Riedel (Eocene- M. Miocene); **l** *D. broennimannii* Stradner (L. Eocene); **m** *D. multiradiatus* Bramlette and Riedel (U. Palaeocene- L. Eocene); **n** *D. gemmeus* Stradner (L.-M. Eocene); **o** *D. munitus* Stradner (L. Eocene); **p** *D. mirus* Deflandre (L. Eocene)

dwellers have tests with larger mural pores and pore densities than deeper cold-water forms.

The heavily calcified Late Palaeocene *Discoaster* forms such as *D. bramlettei* (Bukry and Percival) Romein and *D. multiradiatus* Bramlette and Riedel (Fig. 10m), which had greater than 90% calcite cover in plan view, probably evolved and lived deep in the photic zone of the water column. More delicately constructed forms such as *D. okadai* (28% cover) and *D. lodoensis* (40–50%) lived near the surface. The trend towards an increased dominance of narrow-rayed, weakily encrusted forms throughout the Cenozoic may reflect an increased preference of the algal group for a shallow-depth habitat. As the group migrated upwards through the water column, forms preferring deeper water became extinct but were not replaced by new forms. During the cool Oligocene Epoch, *Discoaster* diversity was severely restricted, dropping to less than five species per nannofossil zone. Oligocene *Discoaster* species are dominated by the *D. tanii* group, simple six- and five-rayed forms with untapered arms (ca. 40% calcite cover). Two other common species *D. adamanteus* and *D. deflandre*, are heavily calcified (ca. 70–75% cover) forms in which three of the six-rays are fused to form a single crystal and are separated from the other fused crystal unit by well-defined interradial sutures (Black 1972). The Oligocene represents the only Epoch when weakily encrusted ($< 33\%$ cover) forms do not occur in *Discoaster* assemblages. This may be explained by an absence of *Discoaster* species in an upper photic habit at this time, or perhaps because of the cool temperature of the Oligocene oceans. the *Discoaster*-bearing algae favoured the secretion of fused and heavily calcified forms, even in a shallow-water niche. Alternatively, their absence may be due to a period of enhanced calcite dissolution during the Oligocene.

Most of the broad-ray, heavily encrusted, *Discoaster* species (e.g. *D. deflandrei* Bramlette and Riedel and *D. adamanteus* Bramlette and Wilicoxon) became extinct in the early Miocene, and this suggests that a niche became available for forms that were adapted for a lower photic zone habit. The narrow-rayed *Discoaster* forms dominant in the Late Pliocene were probably restricted to surface-dwelling phytoplankton populations. Many of the *Discoaster* species which appeared in the Miocene and Pliocene deviate from a discoid form and develop arms which are slightly bent (*D. brouweri*, *D. pentaradiatus*, *D. berggrenii* and *D. triradiatus*) and which tend to raise the central structure of the *Discoaster* above the tips of the arms in an umbrella-like construction. Assuming that *Discoaster* specimens represented a skeletal covering for some near spherical-shaped algal cell (Deflandre 1952), the increased curvature displayed by these younger species suggests that the group evolved towards forms with a smaller cell volume.

Detailed studies on the evolutionary processes of calcareous nannofossils are still relatively few. Deep ocean drilling undertaken by the Deep Sea Drilling Project and subsequently by the Ocean Drilling Program should provide sufficient opportunities to evaluate the process of nannofossil evolution and help test the theories of gradualism versus punctuated equilibrium. Oceanic sediments facilitate such studies as sample retrieval is usually good (100/s of specimens of the same species occur in one assemblage), morphological characteristics are well-preserved, and a good latitudinal coverage helps ensure that evolutionary patterns are separated from ecologically controlled migration. Ideal oceanic sequences for evolutionary studies should be characterized by high sediment accumulation rates, little or no dissolution, low amounts of bioturbation and should be integrated with established magnetostratigraphies (Backman 1986).

7 Calcareous Nannofossil Biostratigraphy

The object of biostratigraphy is to group strata into units on the basis of the variety and abundance of their contained fossils. Biostratigraphers use various biozones (usually referred to simply as 'zones') as a means for the calibration of stratigraphic sequences. Four types of biozone are employed in stratigraphical classification:

1. Assemblage Zone: strata characterized by a distinctive fossil assemblage;
2. Range Zone: strata representing the stratigraphic range of a particular fossil;

3. Acme Zone: strata characterized by the acme or maximum abundance of a certain form;
4. Interval Zone: the stratigraphical interval between two biohorizons.

Most nannofossil zonations utilize 'interval zones' whose boundaries are considered synchronous biohorizons, represented by unique events such as the evolutionary first appearance and extinction of particular species.

At present there is no calcareous nannofossil zonation for the Late Triassic (Carnian-Rhaetian). Biostratigraphical zonation for the Jurassic is based mainly on the first occurrence of one nannofossil species relative to the first occurrence of another. Bernard and Hay (1974) produced the first tentative zonation of the Jurassic of southern England and France. Perch-Nielsen (1985a) reviewed the zonation schemes for the Jurassic and concluded that the zonations are far from satisfactory. However, Bown (1987) has recently updated the nannofossil zonation for the Lower Jurassic of northwest Europe. Additional work still needs to be done to improve the nannofossil biostratigraphy of the Middle and Upper Jurassic. Important biostratigraphical markers for the Jurassic include species of *Parhabdolithus*, *Biscutum*, *Discorhabdulus*, *Stephanolithion* and *Stradnerlithus*. The first occurrence of *Nannoconus* is almost synchronous with the Jurassic/Cretaceous boundary, occurring in the latest Tithonian. Species which have their first occurrence just above the boundary, in the earliest Berriasian, include *Rotelapillus laffittei* (Noel) Noel and *Lithraphidites carniolensis* Deflandre.

The nannofossil zonation of the Cretaceous (particularly the Late Cretaceous) is more advanced than that of the Jurassic and has been reviewed by Perch-Nielsen (1985a). Crux (1989) has recently refined the biostratigraphy of the Lower Cretaceous (Ryazanian- Barremian) of northwest Europe into 16 nannofossil zones, and has integrated these zones with the ammonite and belemite zonantions. Important biostratigraphical markers for the Cretaceous include forms of *Reindarditites*, *Eiffellithus*, *Quadrum*, *Micula*, *Tranolithus* and *Prediscosphaera*. Future biostratigraphical studies of the Upper Cretaceous should concentrate on the integration of the calcareous nannofossil and macrofossil (belemnite and ammonite) zones and their correlation with the magnetostratigraphic and chronostratigraphic scales.

The Cretaceous/Tertiary boundary may be placed at the first occurrence of *Bianolithus sparsus* Bramlette and Martini and/or *Cyclagelosphaera alta* Perch-Nielsen (Romein 1979; Varol 1989). The standard zonation schemes of Martini (1971), Bukry (1973), and Okada and Bukry (1980) are usually used for the subdivision of the Cenozoic. Finer subdivisions have been developed for local environments; however, these schemes still use the Martini or Bukry zonations as a framework. Martini (1971) used the letters NP and NN and numbers to designate the Nannoplankton Palaeogene zones and the Nannoplankton Neogene Zones respectively. Okada and Bukry (1980) proposed CP and CN (Coccoliths Palaeogene and Coccolith Neogene) to codify the original zones of Bukry (1973).

The standard Palaeogene nannofossil zonation of Martini (1971) is shown in Fig. 11b. Early Palaeocene subdivision is accomplished by the first occurrence

		MY		Martini, 1971
Pleistocene		0.27	NN21	*Emiliania huxleyi*
Pleistocene		0.47	NN20	*Gephyrocapsa oceanica*
Pleistocene / Pliocene	Late	1.66, 1.89	NN19	*Pseudoemiliania lacunosa*
Pliocene	Late	2.33	NN18	*Discoaster brouweri*
Pliocene	Late	2.43	NN17	*D. pentaradiatus*
Pliocene	Late	3.56	NN16	*D. surculus*
Pliocene	Early	3.7	NN15	*Reticulofenestra pseudoumbilica*
Pliocene	Early	4.1	NN14	*Discoaster asymmetricus*
Pliocene	Early	4.5	NN13	*Ceratolithus rugosus*
Miocene	Late	5.6	NN12	*Amaurolithus tricorniculatus*
Miocene	Late	8.2	NN11	*Discoaster quinqueramus*
Miocene	Late	8.85	NN10	*Discoaster calcaris*
Miocene	Late / Middle	10.0	NN9	*Discoaster hamatus*
Miocene	Middle	10.8	NN8	*Catinaster coalitus*
Miocene	Middle	13.1	NN7	*Discoaster kugleri*
Miocene	Middle	14.4	NN6	*Discoaster exilis*
Miocene	Middle / Early	16.2	NN5	*Sphenolithus heteromorphus*
Miocene	Early	17.4	NN4	*Helicosphaera ampliaperta*
Miocene	Early	21.5	NN3	*Sphenolithus belemnos*
Miocene	Early	23.2	NN2	*Discoaster druggii*
Miocene	Early	23.7, 24.0	NN1	*T. carinatus*

a

		MY		Martini, 1971
Oligocene	Late	25.0		
Oligocene	Late	28.2	NP25	*Sphenolithus ciperoensis*
Oligocene	Middle	30.2	NP24	*Sphenolithus distentus*
Oligocene	Middle	34.6	NP23	*Sphenolithus predistentus*
Oligocene	Early	35.1	NP22	*Helicosphaera reticulata*
Oligocene	Early	36.7	NP21	*Ericsonia subdisticha*
Eocene	Late		NP20	*Sphenolithus pseudoradians*
Eocene	Late	37.8	NP19	*Isthmolithus recurvus*
Eocene	Late	39.8	NP18	*Chiasmolithus oamaruensis*
Eocene	Middle	42.3	NP17	*Discoaster saipanensis*
Eocene	Middle	46.0	NP16	*Discoaster tani nodifer*
Eocene	Middle	49.8	NP15	*Nannotetrina fulgens*
Eocene	Middle	52.6	NP14	*Discoaster sublodoensis*
Eocene	Early	53.7	NP13	*Discoaster lodoensis*
Eocene	Early	55.3	NP12	*Tribrachiatus orthostylus*
Eocene	Early	56.3	NP11	*Discoaster binodosus*
Eocene	Early	57.8	NP10	*Tribrachiatus contortus*
Palaeocene	Late	59.2	NP9	*Discoaster multiradiatus*
Palaeocene	Late	60.0	NP8	*Heliolithus riedelii*
Palaeocene	Late	60.4	NP7	*Discoaster mohleri*
Palaeocene	Late	61.6	NP6	*Heliolithus kleinpellii*
Palaeocene	Late	62.0	NP5	*Fasciculithus tympaniformis*
Palaeocene	Early	63.8	NP4	*Ellipsolithus macellus*
Palaeocene	Early	64.8	NP3	*Chiasmolithus danicus*
Palaeocene	Early	65.9	NP2	*Cruciplacolithus tenuis*
Palaeocene	Early	66.4	NP1	*Markalius inversus*

b

Fig. 11. a Neogene nannofossil zonation of Martini (1971) integrated into the chronostratigraphic framework of Berggren et al. (1985b); **b** Palaeogene nannofossil zonation of Martini (1971) integrated into the chronostratigraphic framework of Berggren et al. (1985a) of coccoliths such as *Cruciplacolithus*, *Chiasmolithus* and *Ellipsolithus*. Within the *Heliolithus kleinpelli* Zone (NP6) the first occurrence of the *Discoaster* group (*D. bramletii*) occurred. Subsequent subdivision of the Late Palaeogene is accomplished by the first occurrence of *Discoaster* species (*D. mohleri* Bukry and Percival, *D. nobilis* Martini and *D. multiradiatus* Bramlette and Reidel). An improved biostratigraphical zonation scheme for the Palaeocene has recently been proposed by Varol (1989). The zonation, applicable in low to mid-latitudes, comprises 20 zones and 24 subzones (prefixed NTp, Nannofossil Tertiary-Palaeocene) and is defined upon last occurrences of species and their acmes. The Palaeocene/Eocene boundary is usually placed at the top of NP9/CP8, which approximates to the last occurrence of *Fasciculithus tympaniformis* Hay and Mohler (Varol 1989) and the first occurrence of the planktonic foraminifera *Pseudohastigerina* Perch-Nielson 1985b). The Eocene (zones NP 10–20) is subdivided by forms of *Tribranchiatus*, *Discoaster*, *Chiasmolithus*, *Isthmolithus* and *Sphenolithus*. Designation of zones NP18-NP20 may be difficult in low latitudes because of the scarcity of the zonal fossils in the assemblages.

The Eocene/Oligocene boundary, defined in terms of nannofossils, is usually placed at the top of NP20 which corresponds to the last occurrence of disc-shaped *Discoaster* species (*D. barbadiensis* Tan and *D. saipanesis* Bramlette and Riedel). This event occurs slightly younger than the disappearance of the planktonic foraminifer *Hantkenina* (Perch-Nielsen 1985b). Throughout the Oligocene species, diversity is low and zonation is accomplished by the first occurrence and last occurrence of forms of *Ericsonia*, *Helicosphaera* and *Sphenolithus*. The Oligocene/Miocene boundary is usually placed at the top of NP 25 or within NN1.

The standard Neogene nannofossil zonation of Martini (1971) is shown in Fig. 11a. The subdivision of the Miocene is based predominantly on species of *Discoaster* and *Sphenolithus*. Biostratigraphic zonation is hindered in high latitudes and in marginal sea assemblages, where *Discoaster* species may be scarce or absent. New, medium to large-sized, placolith species which appeared in the Miocene include *Calcidiscus macintyrei* (Bukry and Bramlette) Loeblich and Tappan (NN1) and *Reticulofenestra pseudoumbilica* (NN2) (Varol 1985). Within the NN9 *Discoaster hamatus* Zone is the first occurrence of new five-rayed *Discoaster species* which include *D. hamatus* Martini and Bramlette, *D. bellus* Bukry and Percival and *D. prepentaradiatus* Bukry and Percival. The first occurrence of *Amaurolithus primus* (Bukry and Percival) Gartner and Bukry, the oldest horseshoe-shaped calcareous nannofossil in the Neogene, occurs in *D. quinqueramus* Zone (NN11). Other species of *Amaurolithus* appear soon after *A. primus*. The upper part of Zone NN12 is characterized by the first occurrence of birefringent ceratoliths (*Ceratolithus*). In some high latitude Pliocene North Atlantic sediments only the last occurrence of *R. pseudoumbilica* (3.56 Ma) and/or the first occurrence of *P. lacunosa* (ca. 3.6 Ma) may be used for subdivision. Zonation of the Late Pliocene is based on a series of extinction datums of *Discoaster* species; in chronological order these include *D. tamilis* Kamptner, *D. asymmetricus* Gartner, *D. surculus* Martini and Bramlette, *D.*

pentaradiatus Tan emend. Bramlette and Riedel and *D. brouweri*. The last 0.18 m.y. of the range of the *Discoaster* group (2.07–1.89 Ma) is characterized by a higher (> 20%) proportion of *D. triradiatus* and this forms a useful indication for the extinction of the group and an important biostratigraphical event, even in the presence of considerable reworking (Backman and Pestiaux 1986).

The Pliocene/Pleistocene boundary, now placed at 1.66 Ma, occurs between the extinction datums of *D. brouweri* at 1.89 Ma and *C. macintyrei* at 1.7–1.45 Ma. The boundary, in terms of nannofossils, is therefore placed towards the base of the *P. lacunosa* (NN19) Zone. The Pleistocene is primarily subdivided by small placolith species and also by *C. macintyeri* and *H. selli*. The NN19 Zone of Martini (1971) can be subdivided by nannofossil events which occurred in the following order; first occurrence of *G. oceanica* s.l., last occurrence of *C. macintyrei* and last occurrence of *H. sellii* Bukry and Bramlette (Fig. 12). The last occurrence of *H. sellii*, however, is diachronous across latitudes, ranging between, 1.04 and 1.42 Ma (Backman and Shackleton 1983; Pujos 1988). Thierstein et al. (1977) studied the global synchroneity of Middle and Late Pleistocene nannofossil datums. The most important nannofossil event in the

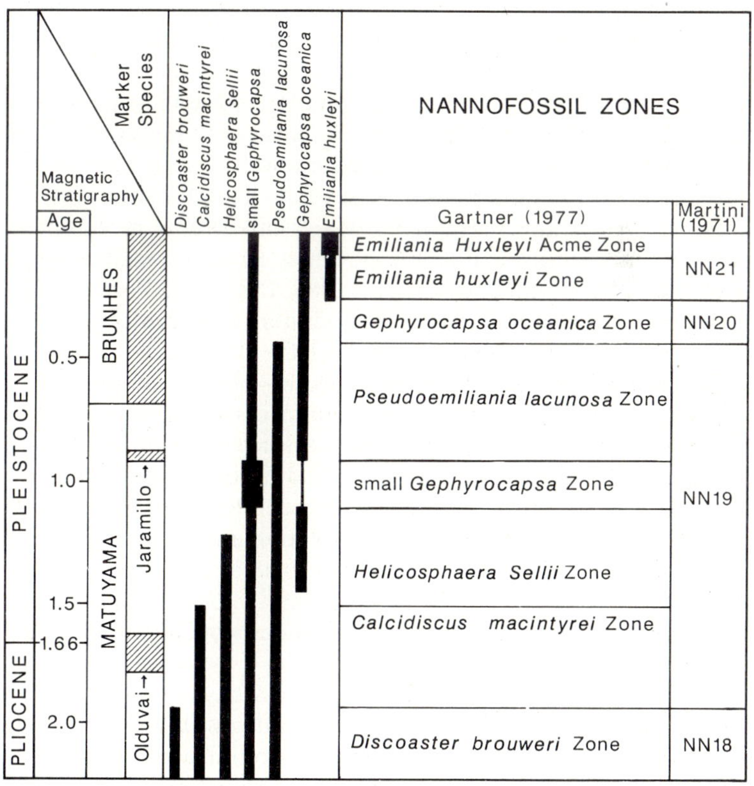

Fig. 12. Calcareous nannofossil zonation schemes for the Pleistocene

Middle Pleistocene is the last occurrence of *P. lacunosa* which marks the top of the NN19 Zone. This occurs as a globally synchronous datum within the middle of oxygen isotope stage 12, dated at 0.458 Ma. The first occurrence of *E. huxleyi*, which designates the NN20/NN21 boundary, occurs consistently late in oxygen isotope stage 8, dated at 0.268 Ma.

The dominance reversal of *Gephyrocapsa spp./E. huxleyi* has been used to subdivide NN21 Zone. During the first two-thirds of its range *E. huxleyi* occurs in relatively low abundances and is subordinate to *Gephyrocapsa*; however, it subsequently becomes the most dominant form. The *Gephyrocapsa/E. huxleyi* reversal in dominance is time-transgressive. In tropical and subtropical sediments it occurs between oxygen isotope stages 5a and 5b (0.085 Ma), whereas in transitional sediments it correlates with oxygen isotope stage 4c (ca. 0.073 Ma). The dominance of *G. oceanica* and the rarity, or even absence of *E. huxleyi* in sediments underlying monsoon-influenced regions of the Indo-Pacific (see Sect. 9.2) may not allow the NN20 and NN21 Zones of Martini (1971) and the *E. huxleyi* Acme Zone of Gartner (1977) to be separated. In such assemblages where *E. huxleyi* is rare because of an ecological control, an overestimation of the age of the sediment may occur. Heckel (1973) also indicated that the absence of *E. huxleyi* from nearshore sediments from the Capricorn Basin was probably a facies-control, and should not necessarily be used as evidence for exculsion of a sediment from the *E. huxleyi* Zone. A potentially useful biostratigraphical species which may be used as an alternative marker for the *E. huxleyi* Zone in the Indo-Pacific region is *Geminilithella subtilis* (Muller) Varol. This species was used by Hughes et al. (1986) to date uplifted, latest Pleistocene sediments in the Solomon Islands.

Calcareous nannofossils have several characteristics which facilitate their use as biostratigraphical markers. Many nannofossil species have been shown to have synchronous extinctions and first appearances (Backman and Shackleton 1983; Backman and Pestiaux 1986; Pujos 1988); therefore, their datums may be used with some degree of confidence, regardless of water mass characteristics. As a group, nannofossils are far more eurythermal than planktonic foraminifera. This enhances their use in oceanic correlation because zonal species frequently have a cosmopolitan distribution. Nannofossils may be recovered from a surprisingly wide spectrum of sediment types, ranging from fine-grained silts and clays to sands and muddy gravels. Their occurrence in coarse-grained sediments may be attributed to their mode of settling, which is primarily within large faecal pellets of zooplankton grazers (see Sect. 8).

8 Sedimentation, Dissolution and Diagenesis of Nannofossil Assemblages

The extent to which a calcareous nannofossil assemblage preserved in bottom sediments reflects the living floral assemblages in the surface water depends upon the processes of sedimentation, dissolution and diagenesis. Dissolution probably is the most important of these, and begins soon after the organism dies and sinks into waters undersaturated with respect to calcium carbonate. Dissolution is inhibited by the fact that most nannofossils are thought to descend to the seafloor within the faecal pellets of zooplankton grazers. Copepods are major grazers of coccolithophorids (Honjo 1976), and daily may consume up to 1200 individuals, ejecting the calcite plates in some 200 faecal pellets. A single pellet may contain up to 100 000 coccoliths, which represents ca. 1 μg of $CaCO_3$ (Tappan 1980). Although the gut of copepods is thought to be acidic, laboratory studies (Honjo and Roman 1978) have shown that coccoliths contained in copepod faecal pellets (produced by *Calanus* and *Acartia* species) showed no signs of mechanical destruction or chemical dissolution. Even fragile coccolith specimens of the macrococcoliths of *Umbellosphaera irregularis* and many species of holococcoliths remained intact. Large coccolithophore cells such as *Syracosphaera* (ca. 30 μm) and *Pontosphaera* (ca. 50 μm) were found as isolated coccoliths within faecal pellets, but *Emiliania* and *Gephyrocapsa* (< 10 μm) were often found to occur as intact coccospheres. This suggests that small cells of ca. 10 μm are inefficiently ingested and may be the lower size limit of copepod retention for filter feeding.

An organic coating envelops faecal pellets and seals the coccoliths from corrosive waters throughout the major part of their descent (Honjo 1976). The descent of the coccoliths within the "faecal pellet express" accelerates deposition dramatically, estimated to be ca. 100 years in the oceans for an isolated coccolith to a few weeks for a mean-sized faecal pellet. This mode of deposition ensures that a constant flux of coccolith carbonate, relatively free from dissolution, reaches the bottom sediment, and also promotes a better correspondence between plankton assemblages and fossil assemblages. In the equatorial Pacific, 92% of the coccolith carbonate produced in the surface waters may reach the bottom within faecal pellets (Honjo 1976).

Faecal pellets, although protecting coccoliths during descent through the water column, rapidly disintegrate on the seafloor because of bacterial action. Once coccoliths spill out of the pellets they become susceptible to dissolution. In deeper waters a coccolith lysocline (the depth below which all coccolith skeletons show signs of dissolution) has been identified between 3- and 4-km depth. Selective dissolution of less resistant species occurs between these depths; dissolution increases dramatically below 4 km resulting in preservation of only solution-resistant nannofossil assemblages dominated by placolith-type coccoliths. Below the carbonate compensation depth (CCD) which separates predominantly calcareous sediments from carbonate poor pelagic sediments, few nannofossils survive. In the North Pacific, the coccolith CCD varies considerably (Roth and Coulbourn 1982), with deepest values (5500 m) being found in the Central Equatorial Pacific, and shallower values occurring towards the margins of the basin, particularly in the Transition Region of the California Current, where it may shoal to a depth of 3300 m.

Sedimentation of coccoliths in macroaggregates other than faecal pellets has been described from several localities. Macroaggregates (ca. 300 μm) consisting of ca. 15 000 intact cells of *E. huxleyi* (Fig. 13) embedded in mucoid material were recorded from a sediment trap in the Fladen Ground, North Sea (Cadee 1985). Other macroaggregates have been described of *Oolithotus fragilis*

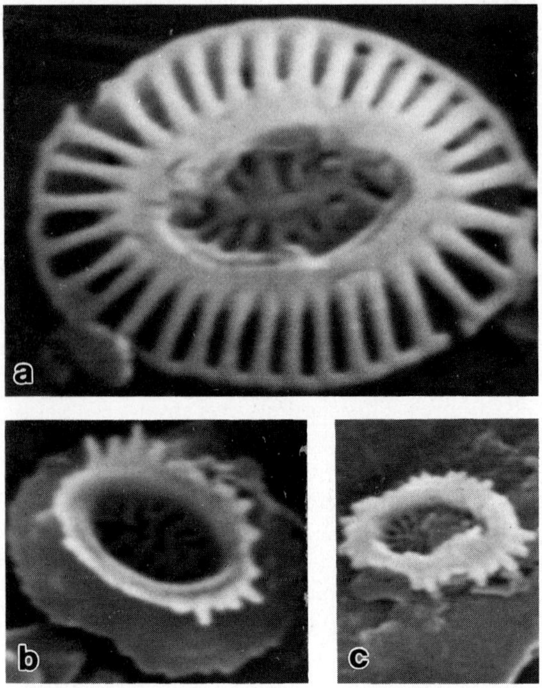

Fig. 13a-c. SEM micrographs of *Emiliania huxleyi*. **a** Distal view of the warm-water ecophenotype showing well-preserved T-shaped elements (x33 000). **b** Distal view, showing solid proximal shield and central ring structure but with distal T-shaped elements absent (x17 000); **c** Both proximal and distal shields absent, remnant central ring preserved alone (x13 000)

('palmellid' stage) from the Mediterranean (Bernard 1963) of *Umbilicosphaera sibogae* (Weber-van Bosse) Gaarder in traps from the Panama Basin after a bloom of the species in the surface water (Honjo 1982), and clusters of *Gephyrocapsa oceanica* cells from Australian coastal waters (Hallegraeff 1984). Such aggregates of algal cells are thought to facilitate rapid removal from inhospitable environments that may occur in the surface waters (Smetacek 1985). Sedimentation of coccoliths in macroaggregates is probably an important mode of transport of coccoliths to the seafloor and may be widespread among coccolithophorids, particularly in shelf seas where environmental conditions are liable to fluctuate rapidly.

Berger (1970) has estimated that about four-fifths of the calcite supplied to the ocean bottom is being redissolved; however, nannofossil skeletons are more resistant to this effect than other microfossil tests. Part of this increased resistance to dissolution is attributed to the incorporation of acid-resistant, cellulose-like polysaccharide compounds within the skeletal calcite (Franke and Brown 1971). Organic coatings on the surface of coccoliths are also thought to inhibit dissolution of coccolith carbonate. The dissolution characteristics of nannofossils are utilized by species of tintinnid and planktonic foraminifera which incorporate nannofossils into their wall structure (Fig. 14a).

Initial dissolution of coccoliths is reflected in etching of individual crystallites resulting in rounded corners and serrate margins of shields (Fig. 3d).

Fig. 14. a Detailed view of planktonic foraminifera test of *Globigerina (Zeaglobigerina) aperta* Cushman showing incorporation of placolith coccoliths into the wall structure (x800). **b** Well-preserved distal view of *Coccolithus pelagicus* showing no overgrowth (x4200). **c** Distal view of *Coccolithus pelagicus* showing overgrowth on radial elements, although most of the crystallites can still be distinguished (x5200). **d** Distal view of severely overgrown coccolith of *Coccolithus pelagicus*, some of the radial crystallites have overgrown at the expense of others (x6000)

Usually coccoliths constructed of smaller crystallites are more susceptible to dissolution. More advanced stage dissolution isolates the individual elements of coccoliths and removes loosely attached plates. Finally, disintegration of the coccolith may result in the production of a mass of micron-sized coccolith debris. Figure 13a-c shows coccoliths of *Emiliania huxleyi* in various stages of dissolution. Initial dissolution of coccoliths of *E. huxleyi* occurs at the rim of the structure within the T-shaped elements. High resolution electron microscopy of the T-shaped elements has shown that they consist of a mosaic of microdomains of 300–500 A in diameter, whereas the crystal lattice of the more resistant radially arranged elements is continuous (Parker et al. 1983). The central ring of these coccolith is the most resistant structure in this species (Fig. 13c).

Post-depositional diagenetic overgrowth on nannofossil specimens commonly occurs in Cenozoic and Cretaceous sediments (Bukry et al. 1971; Roth 1973). Such processes, although they may be used as a tool for identifying early diagenetic processes within sediments, often cause taxonomic problems, and may alter the apparent diversity of assemblages. In the deep ocean, overgrowth on nannofossil specimens seems to be related to the early lithification of calcareous sediments. Diagenesis of nannofossil oozes, which consist almost exclusively of low Mg calcite, occurs by deposition of syntaxial overgrowth cement on *Discoaster* species, coccoliths, foraminifera and other fragments (Matter 1974). Precipitation of overgrowth onto *Discoaster* rays occurs soon after burial, usually preceding calcite overgrowth on coccoliths which is thought to be initiated at ca. 100 m depth below the seafloor. Wise and Kelts (1972) suggested that these differences in the timing of overgrowth are caused by differences in the construction of nannofossils: *Discoaster* rays are usually single crystals and act as large seed crystals similar to echinoderm fragments, whereas coccoliths consist of many sub-micron sized elements which have varying optical orientation. Laboratory evidence (Adelseck et al. 1973) has indicated that *Discoaster* rays and large placoliths grow at the expense of small coccoliths under simulated diagenetic conditions. In the south and central Pacific, secondary calcite is reported to occur regularly only on the large coccolith species *Calcidiscus leptoporus* and *Umbilicosphaera sibogae*, and is usually restricted to the distal shield (Roth and Berger 1975). Overgrowth on placoliths is usually apparent by the abnormal thickening of the distal shield crystallites (cf. Fig. 14b-d).

Results from the Deep Sea Drilling Project taken from many ocean depths have shown that the most solution-resistant Cenozoic genera, *Coccolithus*, *Calcidiscus*, *Reticulofenestra* and *Dityococcites*, occur in most deep-water and shallow-water calcareous sediments. Solution prone genera absent from deeper water sediments include *Micrantholithus*, *Tranversopontis*, and *Scyphosphaera*.

Overgrowth on nannofossils can also occur in surface sediment from near-shore environments. Greater than 1% of nannofossil assemblages in late Quaternary sediments from the Gulf of Aqaba (Elat) and the Red Sea were reported to have overgrowth (Winter 1982a). Although overgrowth on nannofossils in the deep oceans is usually associated with differential solution, similar processes are not thought to occur in the 1800-m-deep gulf, where the entire water column is supersaturated with $CaCO_3$. Calcite overgrowths on the nannofossil assemblages from the Red Sea were therefore concluded to occur within the sediment because of the supersaturation of the water column and interstitial waters.

9 Calcareous Nannofossils as Palaeoclimatic and Palaeo-Oceanographic Indicators

9.1 Calcareous Nannofossil Assemblages as Indicators of Palaeodepth

Coccolithophorids, because of their predominantly planktonic habit, are not thought to be particularly useful as indicators of the depth of the sediments in which they are found. However, holococcoliths are usually restricted to assemblages deposited at shallow depths up to a few 100 m. Some caution must be applied to this general conclusion, as holococcoliths may survive transport into greater depths by turbidity currents. Okada (1984) has used the abundance of *Florisphaera profunda* Okada and Honjo in marginal seas along the western Pacific Ocean as a tool to determine palaeodepth. *F. profunda* was not recorded in shallow seas but was found to increase to over 50% of the assemblages at ca. 2000 m depth. It is generally accepted that Recent *Braarudosphaera bigelowii* has highest abundances in near-shore areas and despite a few exceptions is very scarce or absent from pelagic sediments. In bottom sediments from Sendai Bay, Japan, the *B. bigelowii* component of the nannofossil assemblages increased from < 10% in deeper water sediments (> 100 m depth) to nearly 50% in near-shore sediments shallower than 24 m (Takayama 1972).

Quantitatively, coccoliths are generally less abundant in shallow, near-shore sediments than in open-oceanic sediments. Heckel (1973) reported that in fine fractions of sediments from the Capricorn Basin deposited above 40-m water depth coccoliths were rare (< 2%); the coccolith component increased to over 10% in water depths greater than 100 m. In the seas around Britain, Recent coccolith abundances decrease from 10^9 coccolith per gram (> 10% of the fine fractions) in deep water sediments (> 100 m depth) from the Celtic Sea to < 10^7 coccolith per gram (< 0.1%) in sediments shallower than 30 m in the English Channel (Houghton 1986, 1988).

9.2 Calcareous Nannofossil Assemblages as Indicators of Marginal Sea/Open Oceanic Sediments

Thierstein (1976) recorded the differences between Cretaceous nannofossil assemblages in the Tethys and those found in Pacific oceanic sites. Species present in the epicontinental and marginal seas of the Tethys but absent from the oceanic Pacific sites include *Nannoconus* spp., *Conusphaera mexicana* Trejo, *Micrantholithus obtusus* Stradner and *Lithastrinus floralis* Stradner. Atlantic Cretaceous marginal sea and shelf assemblages contain abundant *Broinsina* and *Nannoconus*, whereas oceanic Atlantic assemblages are dominated by *Watznaueria barnesae* (Black) Perch-Nielsen (Roth and Bowdler 1981). In the Cenozoic, genera which seem to be more common in shelf areas rather than in oceanic conditions, include *Pontosphaera*, *Scyphosphaera*, *Braarudosphaera*, *Micrantholithus* and *Pemma*. *Discoaster* is a typical component of oceanic Cenozoic assemblages particularly at low mid latitudes. The genus has not been recorded in the Upper Pliocene shelf assemblages from the Capricorn Basin (Hecked 1973) and is also absent in the near-shore Upper Pliocene St. Erth Beds and Coralline Crag of England (Jenkins and Houghton 1987; Houghton and Jenkins 1988).

Okada and Honjo (1975) investigated the Recent coccolithophorid distribution in plankton samples from the Pacific marginal seas of Asia. They concluded that the dominance of *Gephyrocapsa oceanica* and the very sparse occurrence of *Emiliania huxleyi* was diagnostic of coccolith assemblages in these marginal seas. With minor exceptions, *G. oceanica* was noted to predominate in the plankton of the East China Sea, South China Sea, Java Sea, Timor Sea and

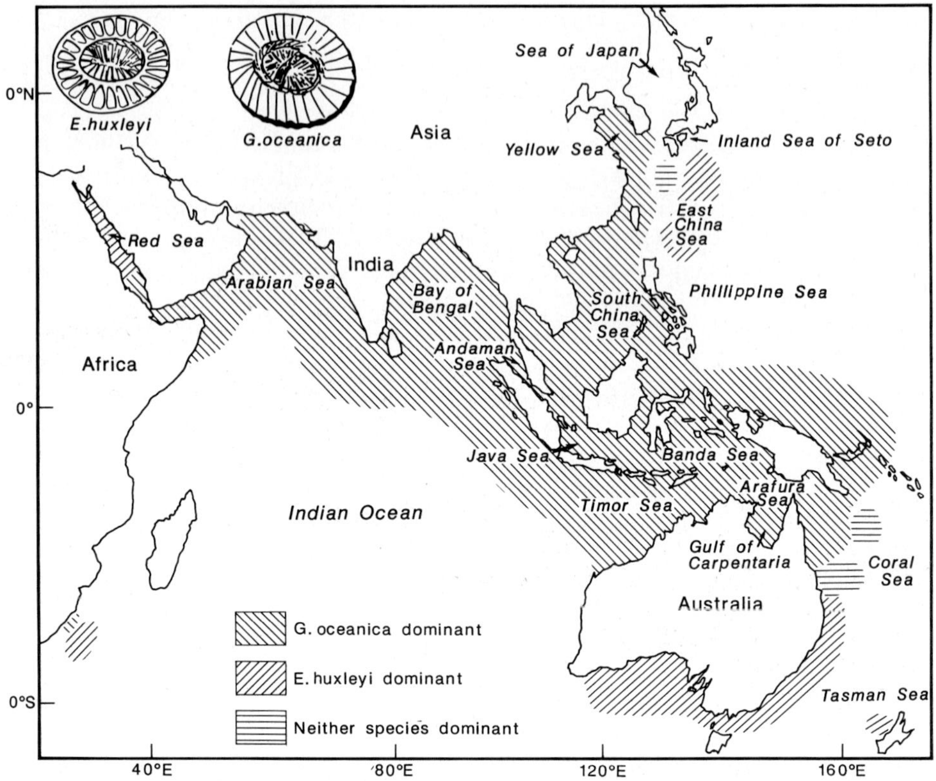

Fig. 15. Map showing areas of dominance of *G. oceanica* and *E. huxleyi* in Recent coccolith assemblages from the marginal seas of the Indian Ocean and western Pacific Ocean, including both sediment and plankton samples, compiled from many sources. Marginal seas with *no shading* represent areas with no suitable coccolithophorid data

off the Northern coast of Australia (Fig. 15). The species was most dominant in coccolithophorid samples from the South China Sea (95–100% of nannoflora). A trend of increasing dominance of the species in the plankton towards land was also recorded. Similar reversals in dominance of *E. huxleyi* over *G. oceanica* towards the coast have also been noted in sediments from the Great Barrier Reef (Heckel 1973; Conley 1979), the west Tasman Sea (Burns 1975) and in the East China Sea (Wang and Samtleban 1983). *G. oceanica* is also the most abundant species in assemblages from the Bay of Bengal, the Arabian Sea and in the Adaman Sea in the northern Indian Ocean (Guptha 1976, 1981). Coccolithophorid assemblages in Australian coastal waters have a latitudinal dominance shift from *G. oceanica*-dominated in the north (North West Shelf, Coral Sea and northern Great Barrier Reef), to *E. huxleyi*-dominated in the south (Tasman Sea, Great Australian Bight; Hallegraeff 1984).

The abundance of *G. oceanica* in the marginal seas of the Indian and Pacific oceans is certainly not characteristic of shelf waters from the North Atlantic, where from the tropics to the subarctic, *E. huxleyi* is by far the most abundant species (sometimes forming a monospecific flora) and *G. oceanica* is usually rare. *G. oceanica*, a warm-water species with a maximum temperature range of between 12 and 30°C (Okada and McIntyre 1979), is not expected to be a common component of high latitude shelf assemblages. The species is replaced in temperate waters by *G. caribbeanica*, a cold-water preferring *Gephyrocapsa* species. However, in shelf regions of the North Atlantic, *G. caribbeanica* tends to be rare and is certainly not so adaptive as *G. oceanica* to the near-shore environment. The rarity and absence of *E. huxleyi* in the Asian marginal seas occurs in waters which are well within the temperature and salinity tolerance of the species, and perhaps the distribution pattern of the species is better explained by reference to changes in water mass fertility. Winter (1982b) correlated the percentage abundance of *E. huxleyi* and *G. oceanica* in bottom sediments along a traverse from the Gulf of Eden-Red Sea to the Gulf of Aqaba, with changes in the nutrient characteristics in the overlying watermass. *G. oceanica* was suggested to favour high fertility waters and *E. huxleyi* low fertility waters. Studies on cores from the Gulf of Aqaba indicated that during interglacial periods, particularly near major Wadi systems, when the Gulf had a higher fertility caused by nutrient supply from the land, *G. oceanica* was found in highest percentages. A preference for higher nutrient levels may also explain the *E. huxleyi*/*G. oceanica* dominance reversal recorded towards land from several marginal seas.

The dominance of *E. huxleyi* and *G. oceanica* in Atlantic and Indo-Pacific marginal seas generally reflects their abundance patterns found in oceanic sediments. *E. huxleyi* is more abundant in the Atlantic than in the Pacific, and this difference is especially discerniable when comparing equatorial assemblages. Schneiderman (1977) noted *E. huxleyi* to be ubiquitous in the Atlantic Ocean, the species forming over 50% of the flora in tropical regions and increasing to near 100% in some subpolar sediments. Okada and McIntyre (1979) recorded the average annual species composition of coccolithophorid populations from five weather stations in the North Atlantic, ranging from tropical to subarctic environments. *E. huxleyi* had a mean abundance (data compiled from combined surface water and 100-m levels) which ranged from 69.7 to 79.5%, whereas *G. oceanica* was rarely recorded in the Atlantic survey, and had a maximum value of 6.1% at the tropical locality. In the Pacific Ocean, low-latitude, high abundances of *E. huxleyi* ($>$40%) are generally restricted to assemblages underlying waters of low fertility and include the East Pacific Rise, near the Galapagos and below the Central Pacific gyre at 20° N to 30° N latitude (Roth and Coulbourn 1982). *G. oceanica* is most abundant in assemblages deposited below the high-fertility equatorial water mass.

The variations in abundance of *G. oceanica* and *E. huxleyi* in the Atlantic and Indo-Pacific regions may be related to major sedimentary facies patterns and oceanic circulation fractionation. There is a distinct tendency for carbonates to accumulate in the Atlantic and for siliceous oozes to accumulate in the Indian and Pacific oceans (Sverdrup et al. 1942). These broad differences in biogenous output are related to the fractionation of both silica and carbonate between the ocean basins as controlled by deep-circulation patterns. The North Atlantic is dominated by an anti-estuarine, nutrient-poor circulation, the North Pacific by a nutrient-rich, estuarine circulation (Redfield et al. 1963). These difference in fertility are related to the nutrient elements being concentrated in the direction of major circulation of deep-water from the North Atlantic through the South Atlantic, Indian and South Pacific to the North Pacific. The dominance of *E. huxleyi* at all latitudes in the North Atlantic suggests that the nutrient-deficient anti-estuarine circulation is the optimum oceanic environment for the species.

The dominance of *G. oceanica* in the low-latitude marginal seas of the Indo-Pacific occurs in a region of strong seasonal variation in phytoplankton crop, related to monsoonal influences. These changes are most pronounced in the northwest Indian Ocean during the (winter) southwest monsoon (roughly May-August) when there is strong upwelling and nutrient-enrichment in the surface waters (McGill 1973). The richness in nutrient elements during the monsoon upwelling regions, contrasts markedly with the general paucity of surface nutrients over wide areas of the open tropical and subtropical oceans. Monsoonal-influenced upwelling occurs off Somalia, in the northern Arabian Sea, Bay of Bengal, off the Adaman Islands, off northern Australia and in the marginal seas of southeast Asia, all regions where the nannofossil assemblages contain dominant *G. oceanica*. The high nutrient-preference of *G. oceanica* explains its dominance pattern in monsoonal-influenced regions and also explains the change to *E. huxleyi*-dominated assemblages in the subtropical regions of the southern Great Barrier Reef, Tasman Sea and southern coast of Australia where there is no monsoonal influence. Changes in the ratio of *E. huxleyi*/*G. oceanica* in low to mid-latitude assemblages may therefore be used as an index of water-mass fertility in both marginal and oceanic environments. These ecological controls on the abundance of *E. huxleyi* and *G. oceanica*, however, may prove to be a hinderance in the resolution of Late Pleistocene nannofossil biozonation (see Sect. 7).

Okada and Honjo (1975) also reported the common occurrence of malformed coccoliths in the plankton of the Asian marginal seas. Such coccoliths are susceptible to dissolution and have rarely been reported in bottom sediments from the same area; malformed coccoliths are therefore unlikely to be of palaeoecological significance. Additional studies on other shelf and marginal sea areas, particularly from the western Indian Ocean and South Atlantic, may further help to identify those features which are characteristic of such environments.

Coccoliths are also found in Recent estuarine and tidal flat sediments. Coccolith assemblages are transported into such environments as suspended silt on flood-tides and are comparable in species-composition with those assemblages found in offshore sediments (Houghton 1986, 1988). Such processes were probably common place during interglacials when sea-level standings were high and will allow interglacial sediments to be integrated into the biostratigraphic framework established for oceanic sediments. The Pleistocene marker species *E. huxleyi*, *Gephyrocapsa oceanica*, and *Pseudoemiliania lacunosa* have all been recorded in various interglacial deposits of northwest Europe (Houghton, unpubl. work). Coccoliths therefore have good potential to form a biostratigraphic link between the oceanic regime and the continental record.

9.3 Calcareous Nannofossils as Palaeotemperature Indicators

McIntyre and Bé (1967) mapped the occurrence of the more common coccolith species in both plankton and bottom sediment samples from the Atlantic Ocean. From this study the coccolith species were divided into five latitudinal climatic assemblages: tropical, subtropical, transitional, subarctic and subantarctic (Fig. 16). The diagnostic species of the climatic assemblages are listed in Table 2, although all samples containing coccoliths were dominated by the species *Emiliania huxleyi* and *Gephyrocapsa* species. Tropical and subtropical assemblages usually contain three times more species than subpolar assemblages, confirming the strong latitude control on species diversity.

Biogeographic ranges of coccoliths determined from bottom sediment and plankton samples show that living species have a slightly wider distribution than those preserved in the sediment. This is attributed to the rapid warming of the Atlantic during post glacial times. However, post mortem dispersion of coccolith species beyond their living biogeographic range by ocean currents is negligible. The present-day latitudinal control on coccolith distribution indicates that fossil

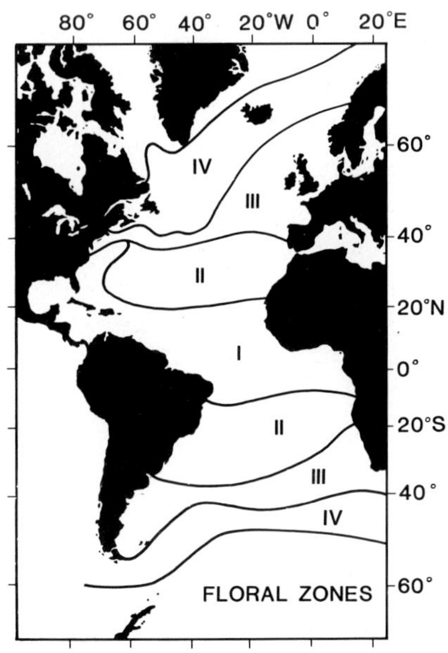

Fig. 16. The coccolithophorid climatic zones of Atlantic Ocean sediments. *I* Tropical; *II* subtropical; *III* transitional; *IV* subarctic, subantarctic (after McIntyre and Bé 1967)

Table 2. Diagnostic Recent coccolith species in Atlantic climatic sediment assemblages

Tropical	Subtropical
Umbellosphaera irregularis	*Umbilicosphaera sibogae*
Neosphaera coccolithomorpha	*Rhabdosphaera clavigera*
Oolithotus fragilis	*Calsiosolenia murrayi*
Umbellosphaera tenuis	*Syracosphaera pulchra*
Discosphaera tubifera	*Umbellosphaera irregularis*
Rhabdosphaera stylifera	*Umbellosphaera tenuis*
Transitional	Subpolar
Calcidiscus leptoporus, type C	*Coccolithus pelagicus*[a]
Polar	
Barren of coccoliths	

[a] Species found today in North Atlantic but absent from South Atlantic.
After McIntyre and Bé (1967) and Ruddiman and McIntyre (1976).

coccoliths should be excellent tools for the elucidation of past climatic and water-mass fluctuations.

Laboratory studies on cultures of *E. huxleyi* have shown that the species synthesizes long (C_{37}-C_{39}) straight-chain, unsaturated ketones (Volkman et al. 1980). The degree of unsaturation of the ketones is controlled by variations in the environmental temperature of the algal culture (Dias and Bradshaw 1988). Identical compounds have also been identified from marine sediments which range from Recent to Miocene in age. The distinctive chemical structure of the ketones and their resistance to biological degradation suggests that the occurrence of these compounds in marine sediments may be used as a tool to reconstruct palaeoenvironments. Additional studies are needed to determine whether such ketones are synthesized by other algal groups and to identify the extent of the ketones' stratigraphical distribution. If these ketones are found to be uniquely synthesized by coccolithophorids, then it may also be possible to recognize coccolithophorid input into sediments even in areas of strong dissolution, such as below the carbonate compensation depth (Volkman et al. 1980).

McIntyre et al. (1972) studied Pleistocene nannofossil and planktonic foraminiferal assemblages in sediments from the North Atlantic to delineate the intensity and position of polar-front movements. Pleistocene polar, subpolar, transitional and subtropical water masses were identified and latitudinal shifts of these assemblages southwards were correlated with polar front movements. During the last 225 000 years six southward polar front migrations were delineated with up to 10° latitude climatic shift being recorded from the open ocean. Latitudinally related coccolith assemblages have also been described from the Cretaceous (Thierstein 1976), and Palaeogene (Haq and Lohmann 1976).

Discoaster species are thought to have great potential as palaeotemperature indicators. From their first appearance in the Late Palaeocene until their extinction in the Late Pliocene, *Discoaster* species have showed an ecological preference for tropical and subtropical environments, i.e. for warm water masses. Bukry (1973) has used the *Discoaster/Chiasamolithus* ratio as an indicator of palaeotemperature. Early Eocene sediments at latitude 14° S yielded 76% warm-water *Discoaster* and 24% *Chiasmolithus*, whereas higher latitude assemblages (37° S) contain 46% *Discoaster* and 54% *Chiasmolithus*.

Oxygen isotope variations, microfossil data, and cyclic variations in the carbonate content of sediments have been used to demonstrate that the Cenozoic palaeoclimate varied at periodicities correlatable to the orbital parameters (e.g. Hays et al. 1976; Pisias and Leinen 1984). These data suggest that the Earth's surface temperature varies in response to periodic variations in the Earth's orbit and axis. Over ca. 100 Kyr the shape of the Earth's orbit is known to oscillate from a circular path to an elliptical one (eccentricity period). A second variable results from tilting of the Earth's axis from ca. 21.5° to 24.5° and back over a period of ca. 42 Kyr (obliquity period). A third variable is the precession of the equinoxes which results from the 'wobble' of the Earth's axis, and has a period of ca. 21 Kyr. All these variables affect the amount of solar radiation reaching the Earth's surface, and patterns of change can be calculated from astronomical data.

Backman and his co-workers have initiated detailed quantitative studies to determine the usefulness of the genus *Discoaster* as an indicator of sea-surface palaeotemperature change. These studies are based on a quantitative analysis of *Discoaster* abundance from closely spaced samples in deep sea cores. The abundance values obtained have been shown to be proportional to the accumulation rate of the species in question (Backman and Shackleton 1983). Backman and Pestiaux (1986) have determined the variations in the total abundance of *Discoaster* species for the Late Pliocene time interval between 1.9 and 3.6 Ma (Ocean Drilling Programme Site 606, latitude 37° N, longitude 36° W).

Two patterns of accumulation are revealed in the time-dependent variation in *Discoaster* accumulation (Fig. 17):

1. A long-term trend of decreasing total *Discoaster* abundance with decreasing age;
2. An overprint of short-term oscillations.

Spectral analysis of the *Discoaster* abundance plots was carried out to test if palaeoclimatic forcing of all the main orbital elements (eccentricity, obliquity, precession) is imprinted on the Discoaster record. This study showed that the 413-Kyr eccentricity period, the 41-Kyr obliquity period and the precessional periods could be identified, although the eccentricity period was the dominant cycle. The long-term trend of decreasing accumulation of *Discoaster* specimens was interpreted by Backman and Pestiaux (1986) to indicate a progressive decrease in sea-surface temperature in the Late Pliocene. A similar investigation was carried out by Backman et al. (1986) from Deep Sea Drilling Programme Hole 522A (56° N, 23° W) cored on the Rockall Plateau. From this higher latitude site a similar long-term trend of decreasing *Discoaster* abundance with decreasing age overprinted by short-term abundance oscillations was identified.

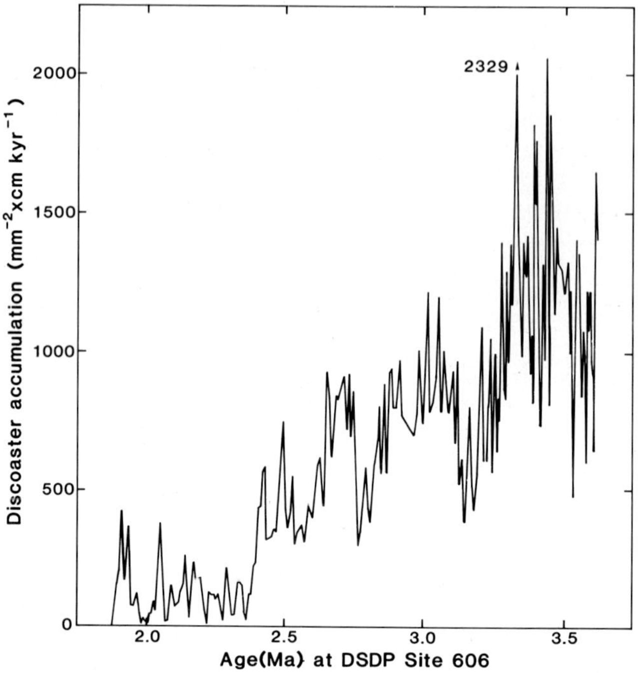

Fig. 17. Plots of total *Discoaster* accumulation versus time in Upper Pliocene sediments from Deep Sea Drilling Project Site 606. Notice the trend of decreasing accumulation with decreasing age and the overprint of short-term oscillations (after Backman and Pestiaux 1986)

Spectral analysis of the abundance oscillations for the preglacial period revealed a dominant quasiperiodicity associated with obliquity-induced temperature variations in the surface waters. Spectral peaks corresponding to the 100 Kyr eccentricity and 21 Kyr precession periodicites were also detected.

The interpretation that the long-term trend of declining *Discoaster* abundance throughout the Late Pliocene is a reflection of a prolonged cooling of northern subtropical waters in the North Atlantic may be treated with some caution. Since the Middle Miocene, the *Discoaster* group had been undergoing a long-term decline in importance which continued until they eventually became extinct in the latest Pliocene (Fig. 18). The Late Pliocene decline of *Discoaster* species recorded from sites 522A and 606 could merely be the 'tail' in the final decline of the group. With more *Discoaster* species becoming extinct in the Pliocene (up to 5.5 extinctions/m.y.) than new species evolving (max. 2.5 first occurrences/m.y.), the extinction of the group seemed inevitable, regardless of temperature change. The decline in *Discoaster* importance to almost negligible values in the plankton communities of the North Atlantic just prior to their extinction is reflected in *Discoaster*/coccolith ratios of the order of 1/10 000 to 1/5000 (Backman 1984).

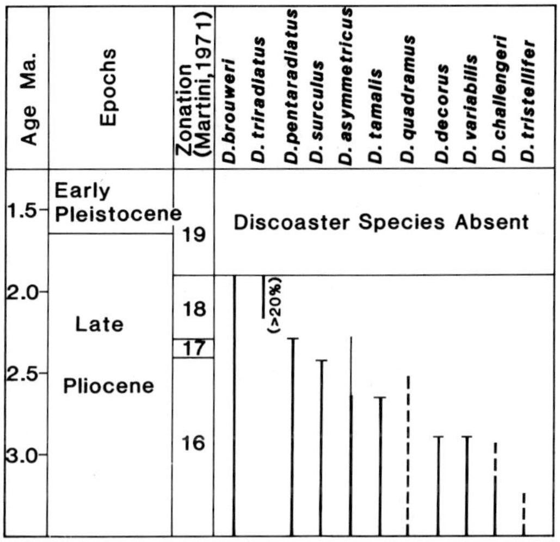

Fig. 18. Distribution of *Discoaster* species in the Late Pliocene (compiled from Backman and Shackleton 1983; Perch-Nielsen 1985b; Backman and Pestiaux 1986)

Backman and Pestiaux (1986) have argued that Early Pliocene surface water temperatures were considerably higher than than in Late Pliocene. The basis of this hypothesis was the greater (x4) abundance of *Discoaster* species and total specimens in the Early Pliocene than in the Late Pliocene interval. However, this difference may merely reflect the general decline in dominance of *Discoaster* species in the plankton of tropical and subtropical waters. Confirmation of higher surface water temperatures in the Early Pliocene may be provided by a similar decline in importance of other warm-water-preferring nannofossil groups (e.g. ceratoliths) throughout the Pliocene. Alternatively, cold-water forms should show an increase in dominance over the time interval. Shoaling of the Central American seaway between 4.5 and 3 Ma is thought to have intensified the surface water circulation of the North Atlantic subtropical gyre system (Backman and Pestiaux 1986). However, the introduction of warm water into the North Atlantic did not appreciably alter the overall decline in *Discoaster* abundance.

If temperature variation is assumed to be the main control on *Discoaster* decline and extinctions, it might be expected that *Discoaster* extinction datums would not be synchronous, but would vary with latitude with species dying out in cooler northern waters as their niche contracted, the remaining forms surviving in the south. Most of the available evidence in the North Atlantic suggests this is not the case. The final extinction datum of the *Discoaster* group has been dated at 1.89 Ma for both high- and low-latitude sites. Bukry (1978) reported that within the *Discoaster* group, different *Discoaster* species had different temperature optima, with *D. surculus* Martini and Bramlette dominating in cooler areas. If a decline in temperature was the main cause of *Discoaster* extinctions and decline, then *D. surculus* (Fig. 7e), best adapted to cooler waters, might have been expected to be the last survivor of the group as palaeotemperature deteriorated in the Late Pliocene. However, *D. surculus* became extinct at ca. 2.42–2.46 Ma (Backman and Pestiaux 1986) and therefore appears not to have survived the first major phase of Northern Hemisphere Glaciation (based on isotopic values and on the first evidence of ice-rafted debris) dated at 2.55–2.44 Ma (Shackleton et al. 1984). This extinction datum suggests that effects other than temperature were responsible for the removal of the species from the plankton. Ecological factors, such as competition for nutrients from other warm-water plankton groups or perhaps increased predation of the *Discoaster*-bearing algae, probably influenced the decline of the group.

Backman's pioneer orbital forcing studies on *Discoaster* abundance have so far been restricted to the North Atlantic. Similar studies should also be carried out in the Pacific and Indian Oceans to see how orbital forcing varies with different palaeoceanographic regimes and latitudes. Orbital forcing studies from Pleistocene examples will have to be determined from other temperature-sensitive nannofossil groups.

9.4 Oxygen Isotopes in Nannofossils

The estimation of palaeotemperature by use of oxygen-isotope analysis of calcitic microfossil skeletons has provided a detailed insight into how the Earth's climate has varied. Such studies document ratios of oxygen isotopes as a relative deviation ($\delta^{18}O$ per mil) from the mean ratio of a standard:

$$\delta\,^{18}O = 1000\left[\frac{(18O/16O)\ \text{sample}}{(^{18}O/^{16}O)\ \text{standard}}\right]-1\ .$$

The usual reference standard is PDB-1, which represents the rostrum of a Cretaceous belemnite from the Pee Dee Formation, South Carolina. The variation in isotopic composition of the ocean waters over time is documented by the changing ratio of $^{18}O/^{16}O$ in fossil carbonates found in oceanic sediments. Foraminifera, both planktonic and benthic forms, have been commonly used for such studies and have yielded information on the oceanographic regimes on the seafloor and in the upper water column. Isotopic analysis of coccolith carbonate is also a potential tool in the elucidation of palaeo-oceanographic change in the world's oceans. Coccoliths are used for isotopic analysis because they have several advantages over foraminifera. Recent coccolithophorids grow and secrete calcite only in the photic zone and therefore their coccoliths should incorporate an isotopic signal characteristic of the upper part of the water column (ca. 0–200 m depth). Foraminifera, however, not only secrete calcite from waters of specific density and at a depth corresponding to the chlorophyll maximum in the water column but also may continue secreting calcite during gametogenesis far below the euphotic zone. Secondly, coccolith species have a wide biogeographic distribution, being found preserved in all but polar sediments. Thirdly, coccoliths have a lower magnesium content and larger individual crystallites in their skeletons than foraminifera and are consequently less susceptible to dissolution and recrystallization. Coccoliths are also more abundant in many Tertiary and Cretaceous pelagic carbonates. Despite these advantageous properties, coccoliths, until relatively recently, have not been widely used in isotope studies. Unfortunately, because of the small size of nannofossils it is thought to be impossible to isolate individual species of coccolith from sediments. Isotope analysis of coccolith carbonate is therefore usually carried out on fine fractions of sediments (ca. $< 25\ \mu M$) which contain polyspecific assemblages.

The oxygen isotope composition of coccolith carbonate from a Quaternary core in the Caribbean (Anderson and Steinmetz 1981) follows the same pattern of glacial/interglacial oscillations characteristic of those recorded in planktonic foraminifera (Fig. 19), including the asymmetrical sawtoothed pattern indicative of rapid deglaciation events. However, although the isotopic variations in coccolith and foraminifera carbonate are generally synchronous the nannofossil $\delta^{18}O$ values are typically offset from planktonic foraminifera values by up to +2 per mil during glacial cycles and ca. +1 per mil during interglacials. Other studies have produced contrasting results, with the oxygen isotope signal of coccoliths isotopically lighter relative to equilibrium values (Anderson and Cole 1975; Dudley 1976).

Dudley et al. (1986) investigated oxygen isotope fractionation in Recent coccolithophorids grown in culture. This study showed two distinct groups of non-equilibrium but temperature-controlled fractionation in coccolith carbonate precipitation: ^{18}O-rich coccoliths including the species *Emiliania huxleyi* and *Gephyrocapsa oceanica* (+1 per mil relative to equilibrium calcite) and ^{18}O-depleted coccoliths (–2 to –3 per mil relative to equilibrium calcite) in-

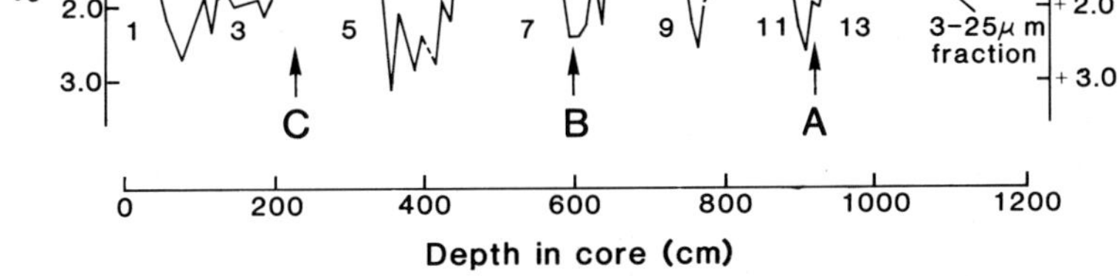

Fig. 19. Oxygen isotopic results on the 3–25 μm fraction of core P6304–4. Emiliani's oxygen isotope data on the planktonic foraminifera species *G. sacculifera* is also shown. The *numbers* beneath the 3–25 μm ^{18}O curve identify the interglacial isotopic stages. *A*, *B*, *C* indicate three globally synchronous Late Quaternary coccolith datum levels: *A* Extinction of *Pseudoemiliania lacunosa* (0.458 Ma); *B* first appearance of *Emiliania huxleyi* (0.268 Ma); *C* reversal in dominance of *G. oceanica* and *E. huxleyi* (0.085 Ma) from Anderson and Steinmetz (1981)

cluding the species *Calcidiscus leptoporus*, *Umbilicosphaera sibogae*, *U. hulburtiana* Gaarder and *Syracosphaera pulchra*. These data imply that the direction and magnitude of deviation of the isotopic record of coccoliths relative to foraminifera may be due to variation in species composition, rather than preservational characteristics of the coccolith assemblages. Biological fractionation ("vital effect") is therefore the most likely cause of resulting non-equilibrium values.

Although there are limitations to the isotope analysis of coccoliths, it remains an excellent tool for the study of Late Cenozoic climatic change. As well as providing an indirect indicator of palaeotemperature oscillations, the technique also provides a record of the glacial/interglacial cycles, of changing global ice volumes, and of glacio-eustatic variations of sea level. The identification of the isotopic stages in deep-sea cores serves as an additional stratigraphic framework which can be integrated with the palaeomagnetic and biostratigraphic records. Much work still needs to be done in this comparatively new field; the mechanisms that control isotopic fractionation in coccolithophorids are still poorly understood. Future studies should also concentrate on the extent to which the $\delta^{13}C$ in nannofossils reflect the $\delta^{13}C$ in the surface ocean waters. Such studies may throw light on enhanced phases of local upwelling of ^{12}C and nutrient-enriched waters. Fractionation of polyspecific assemblages may also improve the quality of isotopic nannofossil results. Separation of individual species or genera could be facilitated if detailed studies were carried out on the settling properties of nannofossils specimens. *Discoaster* species, because of their stellate shape and generally large size (ca. $> 10\ \mu m$) should have different settling properties from coccoliths thus allowing fractionation as a monogeneric assemblage.

10 Conclusions

Calcareous nannofossils, since their emergence in the latest Triassic, have been important members of the phytoplankton and significant contributors to the carbonate budget of pelagic sediments. Their rapid mode of sedimentation in faecal pellets and other macroaggregates and their solution resistant skeletons, ensure that there is usually a good correspondence between fossil and living assemblages. Calcareous nannofossils are good biostratigraphical markers and excellent tools in the elucidation of palaeoclimatic and palaeoceanographic change.

Acknowledgements. K. Perch-Nielsen, D.G. Jenkins, J. Jenkins, L. Harper, and N. Swinburne are acknowledged for their critical comments on an early draft of the manuscript. O. Varol is thanked for valuable discussions on Cenozoic nannofossil biostratigraphy. The figures were drafted by A. Bach-Lloyd and J. Taylor. P. Bown is thanked for supplying the SEM micrographs for Fig. 8.

References

Adelseck CG Jr, Geeham GW, Roth PH (1973) Experimental evidence for the selective dissolution and overgrowth of calcareous nannofossils during diagenesis. Bull Geol Soc Am 84:2755–2762

Aiken J, Bellan I (1988) Interpreting ocean colour – towed instrument backs up satellite images. NERC News 7:28–30

Anderson TF, Cole S (1975) The stable isotope geochemistry of marine coccoliths: a preliminary comparison with planktonic foraminifera. J Foramin Res 5:188–192

Anderson TF, Steinmetz JC (1981) Isotopic and biostratigraphical record of calcareous nannofossils in a Pleistocene core. Nature (London) 294:741–744

Aubry M-P (1989) Phylogenetically based calcareous nannofossil taxonomy: implications for the interpretation of geological events. In: Crux JA, van Heck SA (eds) Nannofossils and their applications. Horwood, Chichester, pp 21–40

Backman J (1984) Cenozoic calcareous nannofossil biostratigraphy from the northeast Atlantic. In: Roberts D, Schnitker D et al. (eds) Init Rep DSDP, vol 4. US Gov Print Off, Washington, DC, pp 903–909

Backman J (1986) Accumulation patterns of Tertiary calcareous nannofossils around extinctions. Geol Rundsch 75:185–196

Backman J, Pestiaux P (1986) *Discoaster* abundance variations from DSDP Site 606: biochronology and paleoenvironmental implications. In: Ruddiman WF, Kidd RB (eds) Init Rep DSDP, vol 94. US Gov Print Off, Washington, DC, pp 903–910

Backman J, Shackleton NJ (1983) Quantitative biochronology of Pliocene and Early Pleistocene calcareous nannoplankton from the Atlantic, Indian and Pacific oceans. Mar Micropaleontol 8:141–170

Backman J, Pestiaux P, Zimmerman H, Hermelin O (1986) Palaeoclimatic and palaeoceanographic development in the Pliocene North Atlantic: *Discoaster* accumulation and coarse fraction data. In: Summerhays CP, Shackleton NJ (eds) North Atlantic palaeoceanography. Geol Soc Spec Publ 21, pp 231–242

Bé AWH (1968) Shell porosity of Recent Planktonic Foraminifera as a climatic index. Science 161:881–884

Berger WH (1970) Biogenous deep sea sediments: Fractionation by deep-sea circulation. Geol Soc Am Bull 81:1385–1402

Berger WH (1976) Biogenous deep sea sediments: production, preservation and interpretation. In: Riley JP, Chester R (eds) Chemical oceanography, vol 5, 2nd edn. Academic Press, New York, London, pp 266–388

Berggren WA, Kent DV, Flynn J (1985a) Jurassic to paleogene, pt 2. Paleogene geochronology and chronostratigraphy. In: Snelling NJ (ed) The chronology of the geological record. Mem Geol Soc London 10:141–195

Berggren WA, Kent DV, van Couvering J (1985b) Neogene geochronology and chronostratigraphy. In: Snelling NJ (ed) The chronology of the geological record. Mem Geol Surv London 10:211–260

Bernard F (1963) Vitesse de chute en mer des amas palmelloïdes de *Cyclococcolithus*. Ses conséquences pour le cycle vital des mers chaudes. Pelagos 1:5–34

Bernard T, Hay WW (1974) On Jurassic coccoliths: a tentative zonation of the Jurassic of southern England and north France. Eclog Geol Helv 67:3 563–85

Bignot G (1985) Elements of micropalaeontology. Graham & Trotman, London, 217 pp

Black M (1972) Crystal development in the Discoasteraceae and Braarudosphaeraceae (planktonic algae). Palaeontology 15:476–489

Bown PR (1987) Taxonomy, evolution, and biostratigraphy of Late Triassic-Early Jurassic calcareous nannofossils. Spec Pap Palaeontology 38:118

Bown PR, Cooper KE (1989) Conical calcareous nannofossils in the Mesozoic. In: Crux JA, van Heck SA (eds) Nannofossils and their applications. Horwood, Chichester, pp 98–106

Braarud T (1979) The temperature range of the motile stage of *Coccolithus pelagicus* in the North Atlantic region. Br Phycol J 14:349–352

Bramlette MN, and Reidel WR (1954) Stratigraphical value of discoasters and some other microfossils related to recent coccolithophorids. J Paleontol 10:291–322
Bukry D (1971) *Discoaster* evolutionary trends. Micropaleontology 17:43–52
Bukry D (1973) Coccolith and silicoflagellate stratigraphy, Tasman Sea and southwest Pacific Ocean, Deep Sea Drilling Project Leg 21. In: Burns RE, Andrews JE et al. (eds) Init Rep DSDP, vol 21. US Gov Print Off, Washington, DC, pp 885–893
Bukry D (1978) Biostratigraphy of Cenozoic marine sediments by calcareous nannofossils. Micropaleontology 24:44–60
Bukry D, Douglas RG, Kling SA, Krasheninnikov V (1971) Planktonic microfossil biostratigraphy of the northwest Pacific Ocean. In: Fisher AG et al. (ed) Init Rep DSDP, vol 6. US Gov Print Off, Washington, DC, pp 1253–1300
Burns DA (1975) The abundance and species composition of nannofossil assemblages in sediments from the continental shelf to offshore basin, western Tasman Sea. Deep Sea Res 22:425–431
Cadee GC (1985) Macroaggregates of *Emiliania huxleyi* in sediment traps. Mar Ecol Prog Ser 24:193–196
Charlson RJ, Lovelock JE, Andreae MO, Warren SG (1987) Oceanic phytoplankton, atmospheric sulphur, cloud albedo and climate. Nature (London) 326:655–661
Conley SM (1979) Recent coccolithophorids from the Great Barrier Reef-Coral Sea region. Micropaleontology 25:20–43
Cooper KE (1989) Nannofossil provincialism in the Late Jurassic-Early Cretaceous (Kimmeridgian to Valanginian) Period. In: Crux JA, van Heck SA (eds) Nannofossils and their applications. Horwood, Chichester, pp 223–246
Crux JA (1989) Biostratigraphy and palaeogeographical applications of Lower Cretaceous nannofossils from north-western Europe. In: Crux JA, van Heck JA, van Heck SA (eds) Nannofossils and their applications. Horwood, Chichester, pp 143–211
Deflandre G (1952) Classe des cocolithophorides. In: Grassé P-P (ed) Traité de zoologie, vol 1. Masson, Paris, pp 439–470
de Jong EW (1975) Isolation and characterization of polysaccharides associated with coccoliths. Ph D Thesis, Univ Leiden
de-Vrind JPM, Borman AH, Thierry R, Westbroek P, Gruter M, Kamerling JP (1986) Calcification in the coccolithiophorids *Emiliania huxleyi* and *Pleurochrysis carterae* II. Biochemical aspects. In: Leadbeater BSC, Riding R (eds) Biomineralization in lower plants and animals. Syst Assoc Spec Publ 30. Clarendon, Oxford, pp 205–217
Dias I, Bradshaw S (1988) 5000 mile plankton sample – a transect of the Atlantic Ocean. NERC News 7:26–27
Donovan SK (1987) How sudden is sudden? Nature (London) 238:109
Dorrigan JL, Wilbur KM (1973) Calcification and its inhibition in coccolithophorids. J Phycol 9:450–456
Dudley WC (1976) Paleoceanographic applications of oxygen isotope analyses of coccoliths grown in culture. Thesis, Univ Hawaii, 168 pp
Dudley WC, Blackwelder P, Brand L, Duplessy J-C (1986) Stable isotope composition of coccoliths. Mar Micropaleontol 10:1–8
Eppley RW, Rodgers JN, McCarthy JJ (1969) Half-saturation constants for the uptake of nitrate and ammonia by marine phytoplankton. Limnol Oceanogr 14:912–920
Franke WW, Brown RM Jr (1971) Scale formation in Chrysophycean algae, III. Negatively stained scales of the coccolithophorid *Hymenomonas* Arch Mikrobiol 77:12–19
Friebele ES, Correl DL, Faust MA (1978) Relationship between cell size and rate of orthophosphate uptake: In situ observations of an estuarine population. Mar Biol 45:39–52
Gartner S (1977) Calcareous nannofossil biostratigraphy and revised zonation of the Pleistocene. Mar Micropaleontol 2:1–25
Goy C, Noel D, Busson G (1979) Les conditions de sédimentation des schistes-carton (Toarcien inf.) du bassin de Paris déduites de l'étude des nannofossiles calcaire et des diagraphies. Doc Lab Geol Fac Sci Lyon 75:33–57
Gran HH, Braarud T (1935) A qualitative study of phytoplankton of the Bay of Fundy and the Gulf of Maine. Can Biol Board J 1:279–467

Guptha MVS (1976) Calcareous nannofossils from sediments of the continental slope off Bombay (India). Riv Ital Paleontol 82:417–430

Guptha MVS (1981) Nannoplankton from recent sediments off the Adaman Islands. Indian J Mar Sci 10:293–295

Hallegraeff GM (1984) Coccolithophorids (calcareous nannoplankton) from Australian waters. Bot Mar 17:229–247

Haq BU (1971) Paleogene calcareous nannoflora Part IV. Paleogene nannoplankton biostratigraphy and evolutionary rates in Cenozoic calcareous nannoplankton. Stockholm Contrib Geol 25:29–158

Haq BU (1973) Transgressions, climatic change and the diversity of calcareous nannoplankton. Mar Geol 15:M25–M30

Haq BU, Lohmann GP (1976) Early Cenozoic calcareous nannoplankton biogeography of the Atlantic Ocean. Mar Micropaleontol 1:119–194

Hays JD, Imbrie J, Shackleton NJ (1976) Variations in the Earth's orbit; pacemaker of the ice ages. Science 194:1121–1132

Heckel H (1973) Late oligocene to recent nannoplankton from the Capricorn Basin, Great Barrier Reef area. Geol Surv Queensl Publ 359, Palaeontol Pap 33:1–23

Holligan PM, Vollier M, Harbour DS, Camus P, Champagne-Phillipe M (1983) Satellite and ship studies of coccolithophore production along a continental shelf edge. Nature (London) 304:339–342

Honjo A (1976) Coccoliths: production, transportation and sedimentation. Mar Micropaleontol 1:65–79

Honjo A (1982) Seasonality and interaction of biogenic and lithogenic particulate flux at the Panama Basin. Science 218:883–884

Honjo A, Roman MR (1978) Marine copepod fecal pellets, production, preservation and sedimentation. J Mar Res 36:45–57

Houghton SD (1986) Coccolith assemblages in Recent marine and estuarine sediments from the continental shelf of northwest Europe. Ph D Thesis, Univ Southampton, 465 pp

Houghton SD (1988) Thermocline control on coccolith diversity and abundance in Recent sediments from the Celtic Sea and English Channel. Mar Geol 83:313–319

Houghton SD (1989) Coccolith sedimentation and transport in the Irish Sea. Mar Geol 86:67–74

Houghton SD, Jenkins DG (1988) Subtropical microfossil indicators from the Late Pliocene Celtic Sea. Mar Geol 79:119–126

Hughes GW, Varol O, Dunkley PN (1986) Pleistocene uplift of Tetrepare Island, Solomon Group as implied by foraminifera and calcareous nannofossil evidence. R Soc NZ Bull 24:397–408

Jafar SA (1983) Significance of the Late Triassic calcareous nannoplankton from Austria and southern Germany, N Jahrb Geol Paleontol Abh 166:218–259

Jenkins DG, Houghton SD (1987) Age, correlation and paleoecology of the St. Erth Beds and the Coralline Crag of England. Meded Werkgr Tert Kwart Geol 24:147–156

Kamptner E (1954) Untersuchungen über den Feinbau der Coccolithen. Arch Protistenkd 100:1–90

Leadbeater BSC (1971) Observation by means of cine' photography on the behaviour of the haptonema in the plankton flagellates of the class Haptophyceae. J Mar Biol Assoc UK 51:207–217

Lohmann H (1909) Die Gehäuse und ihre Bedeutung für die Erforschung des Lebens im Meer. Verh Dtsch Zool Ges 19:200–239

Lovelock J (1988) The ages of Gaia: a biography of our living Earth. Oxford Univ Press, London, 252 pp

Manton I, Sutherland J, Oates K (1976) Arctic coccolithophorids: two species of *Turrisphaera* gen. nov. from West Greenland, Alaska, and the Northwest Passage. Proc R Soc London Ser B 194:179–194

Martini E (1971) Standard Tertiary and Quaternary calcareous nannoplankton zonation. Proc 2nd Plankton Conf Roma 1970, 2:739–785

Matter A (1974) Burial diagenesis of pelitic and carbonate deep-sea sediments from the Arabian Sea. In: Whitmarsh RB, Weser OE, Ross DA et al. (eds) Init Rep DSDP, vol 23. US Gov Print Off, Washington, DC, pp 421–470

McGill DA (1973) In: Zeitschel B (ed) The biology of the Indian Ocean. Chapman & Hall, London; and Springer, Berlin Heidelberg New York, pp 53–102
McIntyre A, Bé AWH (1967) Modern Coccolithophoridae of the Atlantic Ocean. I. Placoliths and crytoliths. Deep Sea Res 14:561–597
McIntyre A, Ruddiman WF, Jantzen R (1972) Southward penetrations of the North Atlantic Polar Front: faunal and floral evidence of large-scale surface water mass movements over the last 225,000 years. Deep Sea Res 20:355–374
Murray JW, Weston J, Sturrock S (1983) Sedimentary indicators of water movement in the western Approaches to the English Channel. Continent Shelf Res 1:330–352
Okada H (1984) Modern nannofossil assemblages in sediments of the coastal and marginal seas along the Western Pacific Ocean. Utrecht Micropaleontol Bull 30:171–187
Okada H, Bukry D (1980) Supplementary modification and introduction of code numbers to the low-latitude coccolith biostratigraphic zonation (Bukry 1973, 1975) Mar Micropaleontol 5:321–325
Okada H, Honjo S (1973) The distribution of oceanic coccolithophorids in the Pacific. Deep Sea Res 20:355–374
Okada H, Honjo S (1975) Distribution of coccolithophorids in marginal seas along the western Pacific and in the Red Sea. Mar Biol 31:271–285
Okada H, McIntyre A (1977) Modern coccolithophores of the Pacific and North Atlantic Ocean. Micropalaeontology 23:1–55
Okada H, McIntyre A (1979) Seasonal distribution of modern coccolithophorids in the western North Atlantic. Mar Biol 54:319–328
Orr WH (1967) Secondary calcification in the foraminiferal genus *Globoratalia*. Science 157:1554–1555
Parke E, Adams I (1960) The motile (*Crystallolithus hyalinus* Gaarder and Markali) and non-motile phases in the life history of *Coccolithus pelagicus* (Wallich) Schiller. J Mar Biol Assoc UK 39:263–274
Parker SB, Skarnulis AJ, Westbroek P, Williams RJP (1983) The ultrastructure of coccoliths from the marine alga *Emiliania huxleyi* (Lohman) Hay and Mohler: an ultra-high resolution electron microscope study. Proc R Soc London Ser B 219:111–117
Parsons DG, Brasier MD (1987) Changes in planktonic and benthonic foraminifera through the Campanian-Maastrichtian phosphogenic cycles, southwest Atlas. In: Hart MB (ed) Micropalaeontology of carbonate environments. Horwood, Chichester, pp 111–120
Perch-Nielsen K (1982) Maastrichtian coccoliths in the Danian: survivors or reworked "dead bodies"? INA Newslett 6:42–46
Perch-Nielsen K (1985a) Mesozoic calcareous nannofossils. In: Bolli HM, Saunders JB, Perch-Nielsen KP (eds) Plankton stratigraphy. Univ Press, Cambridge, pp 329–426
Perch-Nielsen K (1985b) Cenozoic calcareous nannofossils. In: Bolli M, Saunders JB, Perch-Nielsen KP (eds) Plankton stratigraphy. Univ Press, Cambridge, pp 427–554
Perch-Nielsen K (1986) Geological events and the distribution of calcareous nannofossils – some speculations. Bull Centres Rech Explor Prod Elf-Aquiaine 10:421–432
Pisias NG, Leinen M (1984) Milankovitch forcing of the oceanic system. Evidence from the northwest pacific. In: Berger A, Imbrie J, Hays J, Kukla G, Satzman B (eds) Milankovitch and climate I. NATO ASI Ser C, vol 126. Reidel, Dordrecht, pp 397–330
Pitsky GK (1963) O kachestvennom sostave fitoplanktona Azovskogo morya (On the qualitative composition of the phytoplankton of the Azov Sea). Akad Nauk Ukr SSR, Sevastopol, Biol Stants Tr 16:71–89
Prins B (1971) Speculations on relations, evolution and stratigraphic distribution of discoasters. In: Farinacci A (ed) Proc 2nd Planktonic Conf, Rom 1970, vol 2, pp 1017–1037
Pujos A (1988) Spatio-temporal distribution of some Quaternary coccoliths. Oceanol Acta 11:65–78
Rampino MR, Volk T (1988) Mass extinctions, atmospheric sulphur and climatic warming at the K/T boundary. Nature (London) 332:63–65
Raymont JEG (1963) Plankton productivity in the oceans. Pergammon, Oxford, 660 pp
Redfield AC, Ketchum BH, Richards FA (1963) In: Hill MN (ed) The sea, vol 2. Wiley Interscience, New York, p 26

Rio D, Backman J, Raffi I (1987) Calcareous nannofossil biochronology and the Pliocene/Pleistocene Boundary. In: The neogene/quaternary boundary. Fin Rep IGCP Proj 41: 31 pp
Romein AJT (1979) Lineages in Paleogene calcareous nanoplankton. Utrecht Micropaleontol Bull 22:545–605
Roth PH (1973) Calcareous nannofossils – Leg 17. Deep Sea Drilling Project. In: Winterer EL, Ewing J et al. (eds) Init Rep DSDP, vol 17. US Gov Print Off, Washington, DC, pp 695–795
Roth PH (1983) Jurassic and lower Cretaceous nannofossils in the western Atlantic (Site 534): biostratigraphy, preservation and observations on biogeography and palaeoceanography. In: Sherida RE, Gradstein FM et al. (eds) DSDP Init Rep, vol 76. US Gov Print Off, Washington, DC, pp 587–621
Roth PH (1986) Mesozoic palaeoceanography of the North Atlantic and Tethys Oceans. In: Summerhays CP, Shackleton NJ (eds) Geol Soc London Spec Publ 21:299–320
Roth PH, Berger WH (1975) Distribution and dissolution of coccoliths in the South and Central Pacific. In: Slitter WV, Bé AWH, Berger WH (eds) Dissolution of deep-sea carbonates. Cushman Found Foram Res Spec Publ 13:87–113
Roth PH, Bowdler JL (1981) Middle Cretaceous calcareous nannoplankton biogeography and oceanography of the Atlantic Ocean. Spec Publ Soc Econ Paleontol Mineral 32:517–546
Roth PH, Coulbourne RT (1982) Floral and solution patterns of coccoliths in surface sediments of the North Pacific. Mar Micropaleontol 7:1–52
Ruddiman WF, McIntyre A (1976) Northeast Atlantic paleoclimatic changes over the past 600,000 years. Mem Geol Soc Am 145:111–146
Schneiderman N (1977) Selective dissolution of Recent coccolithophorids in the Atlantic Ocean. In: Ramsay ATS (ed) Oceanic micropalaeontology. Academic Press, New York London, pp.1009–1053
Shackleton NJ, Hall MA (1984) Carbon isotopic data from Leg 74 sediments. In: Moore TC, Rabinovitz PD et al. (eds) Init Rep DSDP, vol 74. US Govt Print Off, Washington, DC, pp 613–620
Shackleton NJ, Backman J, Zimmerman HG, Kent DV, Hall MA et al. (1984) Isotope calibration of the onset of ice-rafting and history of glaciation in the North Atlantic. Nature (London) 307:320–323
Sherwood RW (1974) Calcareous nannofossil systematics, paleoecology and biostratigraphy of the Middle Eocene Weches Formation of Texas. Tulane Stud Geol Paleontol 11:1–79
Smetacek VS (1985) Role of sinking in diatom life history cycles: ecological, evolutionary and geological significance. Mar Biol 84:239–251
Stanley SM (1979) Macroevolution. Freeman, San Francisco
Sverdrup HU, Johnson MW, Flemming RH (1942) The oceans – their physics, chemistry and general biology. Prentice-Hall, Englewood Cliffs, NJ, 1087 pp
Takayama T (1972) A note on the distribution of *Braarudosphaera bigelowii* (Gran and Braarud) Deflandre in the bottom sediments of Sendai Bay, Japan. Trans Proc Palaeontol Soc Jpn NS 87:429–435
Tangen K, Brand L, Blackwater L, Guillard R (1982) *Thoracosphaera heimii* (Lohmann) Kamptner is a donophyte: observations on its morphology and life cycle. Mar Micropaleontol 7:193–212
Tan Sin Hok (1927) Over de samenstelling en het onstaan van krijt-en mergel-gesteenten van de Molukken: Jaarrb Mijnw Ned Oost-Indie, 1926:165 pp
Tappan H (1980) The paleobiology of plant protists. Freeman, San Francisco, 1028 pp
Tappan H, Loeblich AA Jr (1973) Evolution of the oceanic plankton. Earth Sci Rev 9:207–240
Thierstein HR (1976) Mesozoic calcareous nannoplankton biostratigraphy of marine sediments. Mar Micropaleontol 1:325–362
Thierstein HR, Geitzenauer KR, Molifino B, Shackleton NJ (1977) Global synchroneity of coccolith datum levels: validation by oxygen isotopes. Geology 5:400–404
Theodoris S (1983) Calcareous nannofossil biozonation of the Miocene and revision of the helicoliths and Discoasters. Utrecht Micropaleontol Bull 32:1–271
Thomsen HA (1981) Identification by electron microscopy of nannoplanktonic coccolithophorids (Prymnesiophycea) from West Greenland, including the description of *Papposphaera sarion* sp. nov. Br Phycol J 16:77–94

1. Outside the outer calcareous wall layer by crests, tubercles, and sutures: e.g. *Calciodinellum* (Fig. 2), *Bicarinellum* Deflandre 1948 *emend.* Keupp 1984, *Orthopithonella "paratabulata"* Keupp 1980c (Fig. 7);
2. Outside the inner calcareous wall layer only by crests: e.g. *Tetramerosphaera* Willems 1985;
3. Outside both the inner and outer calcareous wall layer: e.g. *Alasphaera* (Fig. 6);
4. Inside the organic membranes only (Compare Fig. 1):
   periphragm: e.g. *Obliquipithonella patriciagreeleyae* (Bolli 1974); Keupp 1980a; (Fig. 8);
   endophragm: *Obliquipithonella loeblichi* (Bolli 1974); Keupp 1981, Pl. 48/7;
   autophragm: *Obliquipithonella organica* (Hultberg 1975)

Except for a few specimens representing single species of the genera *Orthopithonella* (*O. "paratabulata"*), *Obliquipithonella* (*O. patriciagreeleyae*, cf. *thayeri*, *organica*), and *Echinodinella* Keupp 1980b (*E. erinacea*, Keupp 1980b), the most spherical to ovoid cysts, the prominent specimens of the so-called "Mesozoic and Cenozoic calcispheres", do not show any paratabulations. But their wall structures and apical archeopyles (= openings for excystment) prove close affinity to true Calciodinellaceae (Figs. 9–13). Corresponding to this, calcareous cysts of the Recent motile dinoflagellate *Ensiculifera* sp., as sampled by Tangen and Dale from plankton of the Oslofjorden in August of 1982 (Figs. 14–17), are congruent morphologically with common Mesozoic cysts (Jurassic, Cretaceous) of the *Obliquipithonella quiltyi* (Bolli 1974)-type. In contrast, there is no evidence for a dinoflagellate nature for the "Palaeozoic calcispheres",

Fig. 8. Broken specimen of *Obliquipithonella patriciagreeleyae* (Bolli 1974) from early Barremian of Speeton/England showing the ventral area of a complete ortho-hexa-paratabulation inside the organic periphragm only. Size of fragment (cyst 3878) 25 μm

Figs. 9–13. The three calciodinellacean subfamilies are characterized by the structure of their (at least outer) calcareous walls

Fig. 9. Fragment of the single-layered calcareous wall (autophragm inside) of *Echinodinella erinacea* Keupp 1980b showing the orthopithonelloid structure with perpendicularly orientated calcite crystals. Thickness of wall: 6 μm. Early Albian from Altwarmbüchen near Hannover/NW Germany

Fig. 10. Wall of *Obliquipithonella multistrata* (Pflaumann and Krasheninnikov 1978). The calcified pericoel (= inner wall layer) show an irregular microgranular structure, while the calcified ectocoel (= outer wall layer) has been constructed by irregularly oblique crystal-orientation, the so-called crossing-bars-structure (Keupp 1979b). Hauterivian/Barremian-boundary from Sarstedt (Gott)/NW Germany. ×2000

Fig. 11. Double-layered calcareous wall of obliquipithonelloid *Bicarinellum calvum* Keupp 1979a. The inner wall layer shows a loosely arranged "crossing-bars structure" due to poor calcification. Early Barremian from Sarstedt (Gott)/NW Germany. ×2500

Figs. 12–13. The *Wallia* (Keupp 1990) from the late Cretaceous, South Dakota/USA, belongs to the Pithonelloideae: its wall-crystals show a uniquely oblique orientation. Diameter of cyst: 90 μm, thickness of wall: 7 μm

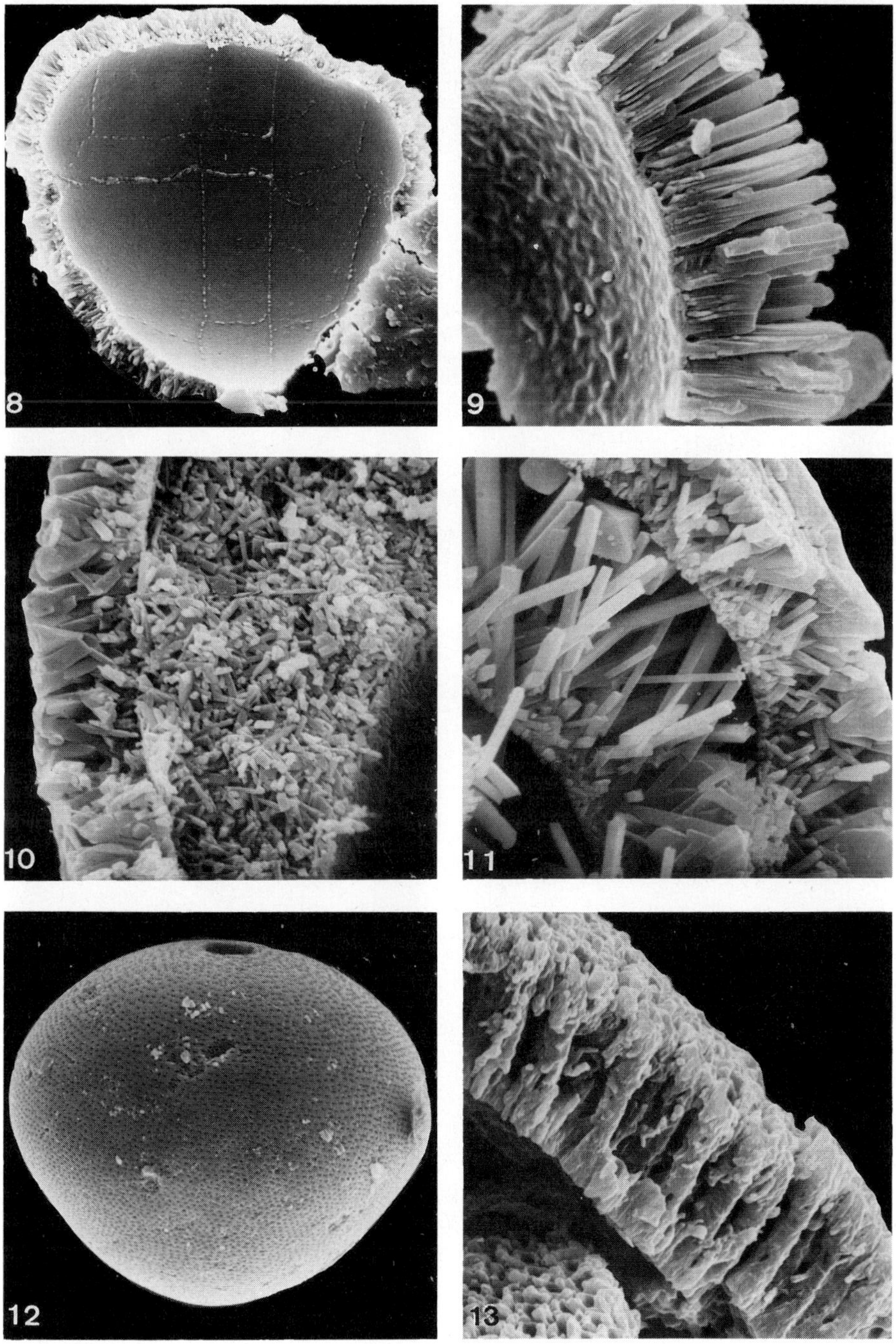
8
9
10
11
12
13

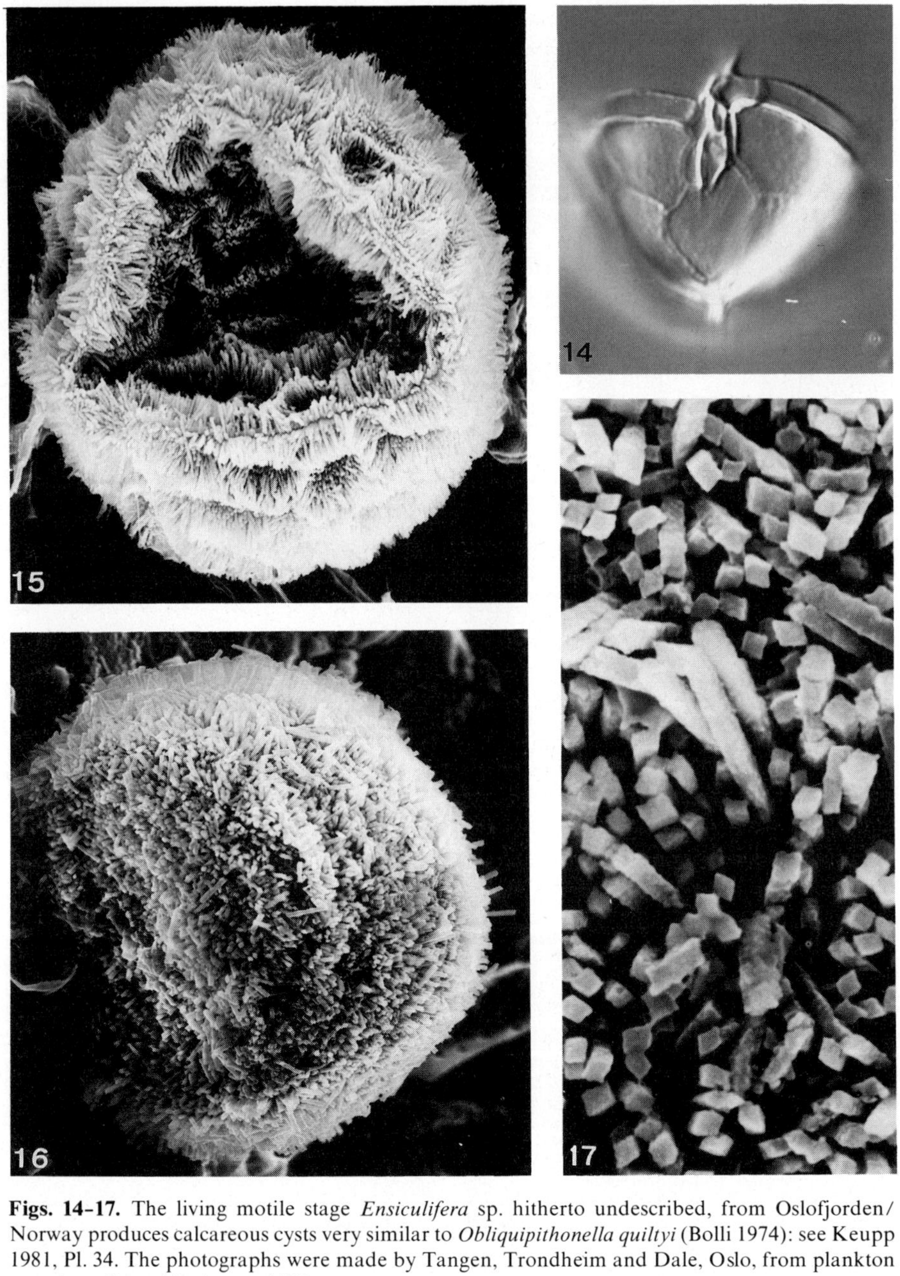

Figs. 14–17. The living motile stage *Ensiculifera* sp. hitherto undescribed, from Oslofjorden/Norway produces calcareous cysts very similar to *Obliquipithonella quiltyi* (Bolli 1974): see Keupp 1981, Pl. 34. The photographs were made by Tangen, Trondheim and Dale, Oslo, from plankton samples collected in August 1982

Fig. 14. Empty hypotheca of the motile stage including sulcus and girdle which show the genus character, namely a long spine protruding from the girdle (= ensicule)

Fig. 15. Partly collapsed calcareous cyst with rigid surface which is probably caused by immaturity, since these morphotypes have been found together with motile thecae in the plankton. Diameter of cysts: 48 μm

Fig. 16–17. Apparently smooth cyst, 43 μm diameter, representing the characteristic *Obliquipithonella quiltyi*-type, and detail of its surface (× 10 000)

which represent a heterogeneous group of algal spores and microproblematica from a more or less restricted environment.

The archeopyle in general lies apically and includes plate equivalences to apical or intercalary series or both. Sometimes the number of plates included is quite variable within one species [see for example *Orthopithonella "paratabulata"* (Keupp 1980c, pl. 31)]. In Mesozoic cysts, the archeopyles are mostly small using the apical plate equivalences, often 3′ only. In some orthopithonelloideans, a large archeopyle including both apical and intercalary plates seems to become dominant in the Tertiary (e.g. *Calciodinellum*, *Orthopithonella albatrosiana*, Kamptner 1963; *Calcipterellum*, Deflandre 1948; *Calcigonellum infula*, Deflandre 1948; see Keupp 1984; Keupp and Versteegh 1989), perhaps in connection with the change of some cysts to oceanic environments (see below).

3 The Wall of Calcareous Cysts

The various reaons why dinoflagellates make cysts seasonally are still under discussion (Sarjeant et al. 1987).

According to Evitt 1985, the formation of fossilizable cysts occur in modern marine peridinians during sexual reproduction (hypnozygotes) rather than in reaction to unfavourable environmental conditions (resting cysts). Observations of a few non-calcareous species have shown (Pfiester 1975, 1984; Spector et al. 1981; Dale 1983) that the process of encystment varies specifically. In *Peridinium cinctum* (Müller) Ehrenberg 1932, for example, three separate organic membranes are constructed step by step, beginning with a progressively enlarged planozygote covered with a theca of single cellulose plates (this is very similar to the vegetative motile stage), and ending within the nonmotile hypnozygote. In contrast to most other species, which shed the theca after forming the cyst, the theca of the planozygote in *Peridinium cinctum* has been made resistant by impregnation with chitin. In other species four more or less resistant wall membranes arise during cyst formation (von Stosch 1973).

In calciodinelloids, the nature of the cysts (hypnozygotes or resting cysts or both) is still unknown. Cenozoic pelagic species with thin calcareous walls probably represent hypnozygotes (Dale 1986), while thick walled, more or less neritic cysts may be chiefly resting cysts. It appears that before the formation of calcareous cysts two to four organic membranes are in existence (Fig. 1). While the outermost "ectophragm" is normally not fossilizable, the inner membranes ("autophragm" or instead "periphragm" + "endophragm" – sometimes an additional "mesophragm" too) are often preserved consisting of acid insoluble material, probably sporopollenin.

Taxonomic confusion originates from isolated phragms of calciodinelloids in palynological preparation procedures (Hultberg 1985). For example, the new organic-walled dinoflagellate cyst *Neuffenia*, created by Bremer and Dürr (1986) from Oxfordian limestones rich in calcispheres, seems to be such an isolated phragm. By analogy with organic-walled cysts, it can be assumed that the interspaces between the poreless membranes are filled with a polysaccharide mucus. Due to a strong buffer effect from the polysaccharides, a considerable supersaturation of dissolved calcium carbonate is possible. Similar buffer solutions prevent the calcification of the mucus covering photosynthetically active cyanobacterians. But the mineralization of this mucus takes place very rapidly when the maximum effect of buffering is weakened (S. Golubic, pers. comm.). In a similar way, calcite crystals may grow to form the wall in many calciodinelloids. Independent of the primary number of organic membranes, one to three separate calcareous layers can be found. Different crystalline structures in each layer suggest their independent mineralization. Despite crystal growth inside the organic membranes, the mode of mineralization appears to be extracellular. Therefore, golgiderived vesicles characteristic of genetically governed mineralizations could not be observed during wall formations of thoracospheres (Tangen et al. 1982). The common formation of dendritic, sometimes hollow, crystals in calciodinelloids support the interpretation of rapid mineralization from supersaturated solutions. The primary orientation of crystals seems to be determined by an epitactic overgrowth of distinct membrane structures or by an organic matrix (specific structure of the polysaccharides) (Bandel and Keupp 1985). Further crystal growth appears as an inorganic pattern (Keupp 1981).

The phenotypic development of wall-forming crystals seems to be controlled by environmental factors, such as temperature and salinity. For example, the analysis of sedimentary rocks and fossils from the Late Aptian/Early Albian of the Lower Saxonian Basin (northwestern Germany) allows reconstruction of a coastal upwelling system (Kemper 1982). Comparison of synchroneous calcareous dinoflagellate associations from the central basin and the marginal upwelling area show clear trends of morphological change within cosmopolitan cyst species (Keupp 1981, 1982). Particularly the following features changed with reduced temperature conditions:

1. Progressive reduction of inner calcareous wall layers;
2. Increase in the size of crystals, producing a rough cyst surface;
3. Disappearance of ovoid to elongate cysts.

On the other hand, similar trends have been shown in the vertical calcisphere succession of the D-Beds (Berriasian to Early Hauterivian) of Speeton, England, which are characterized by contemporaneous freshwater influence during the Early Valanginian (Keupp and Mutterlose 1984).

Taxonomic problems at genus and species level arise more in the Obliquipithonelloideae than in both of the other subfamilies Orthopithonelloideae and Pithonelloideae. They contain a wide range of primary individual variability, and, additionally, morphological control by environmental factors. Often all intermediate morphotypes can be shown in a single sample of cysts. Environmental influences on cyst morphology are recognizable through reversible morphological trends in vertical successions of cyst associations (Keupp 1980b, 1981). The following variable features seem to be influenced by temperature and salinity:

A Obliquipithonelloideae (for explanations see e.g. Keupp 1981, Pl 19•20, 24–25, 37–39, 46–50):

Number of mineralized wall layers: normally two separate calcareous walls exist, but in well-preserved material from the Lower Cretaceous, cysts with only poorly or noncalcified pericoels occur (Fig. 1; Keupp 1981, Pl. 20/10–14;

Structure of the inner wall layer, independent of the primary rate of crystal nucleation: microgranular to "crossing bars structure" (Figs. 10–11; Keupp 1979b);

Size and shape (more or less dendritic) of calcite crystals modifying the cyst surface: for example, the size of simple rhombohedric crystals of the *Obliquipithonella loeblichi* group, is not correlated with the diameter of cysts (Keupp 1981, Fig. 19; Pl. 46, 50). Their sizes range from 3 to 14 μm (diameter of cysts about 45 μm in the middle) in the Lower Cretaceous, with one sample containing nearly the same sizes;

Intensity of surface sculptures in connection with paratabulations, which are independent of the content of calcium carbonate, e.g. the vertical succession of *Bicarinellum calvum* Keupp 1979a in the Hauterivian/Barremian of Speeton (Keupp 1981, pp. 47–50) shifts from elongated strong nodular morphotypes to smooth, nearly spherical cysts. This may be due to decreasing Tethyan influence of the Boreal Realm;

Cyst shape of some species (elongated versus spherical: see example above: *Bicarinellum calvum*).

B Orthopithonelloideae:

Size and shape of wall crystals in some species (e.g. groups of *Orthopithonella gustafsonii* and *O. porata* (Keupp 1982): Keupp 1980a, 1981);

Intensity of surface sculptures in connection with paratabulations (e.g.: morphotype succession of *Praecalcigonellum polymorphum* from ssp. *dentatum*, with high sutural crests, to ssp. tenue, virtually without sculpture, in the Aptian/Albian (Keupp 1980b). This corresponds to a contemporaneous restriction of benthonic organisms, dominated by textulariid foraminifera, suggested to be caused by cold water conditions (Michael 1979, pp. 314–316);

Cyst shape of some species analogous to the Obliquipithonelloideae (e.g. the elongate derivation from *Praecalcigonellum mutterlosei* (Keupp 1979a), the morphospecies *P. sulcatum* (Keupp 1979b), marks, during the Hauterivian/Barremian, the warmer water influences of the Tethyan: Keupp 1987; Fig. 3).

Number of calcareous walls and structure of inner calcareous wall (only in two observed cases):

1. *Orthopithonella* cf. *tuberosa* (Kamptner 1963) characterized mostly by a single layered wall, occurs sometimes with an additional small inner calcareous layer (Keupp 1981, pl. 12);

2. In Oligocene *Orthopithonella* sp. (Weiler 1982) two double-walled morphotypes occur: (a) with the inner layer constructed by radial fibrous, partly hollow crystals (obviously dominant in the Rupelton facies) and (b) with the inner layer of irregular crystal orientation (dominant in the carbonate bearing sandy facies).

C Pithonelloideae

Up until now, there are few observations on which to base definitive statements.

4 Ecological Aspects

In the Recent, calcareous dinoflagellate cysts are known from a wide climatic range of neritic and pelagic marine environments [Wall and Dale (1968a): latitude 9°-42° N; in the Arctic and Antarctic oceans also: Dale (1983); Akselman and Keupp (in press)]. The relatively rare neritic cysts prefer tropical rather than temperate regions (Dale 1975, p 49), while oceanic forms (in the Recent, they dominate the pelagic dinoflagellate assemblages) including the thoracospheres, reach into the subarctic regions (Dale 1986). Calcareous cysts are probably known since the Late Triassic (*Orthopithonella* and *Obliquipithonella*: Janofske 1987). They existed during the Mesozoic in both Mediterranean and Boreal Realms. Mesozoic calcispheres have been found predominantly in sediments of open shelf to slope environments, sometimes of the inner shelf also (Villain 1981). This palaeogeographic distribution corresponds with the increasing importance of cyst formation in neritic dinoflagellates in general (Reid 1978).

After the extinction of the Pithonelloideae at the Cretaceous/Tertiary boundary, a migration of calcareous dinoflagellates into oceanic environments seems to have occurred. This change presupposes new survival strategies for pelagic species. Thoracospheres created a coccoid non-motile (planktonic) vegetative life phase, instead of the flagellate thecate stage present in all other dinoflagellates (Dale 1986). Some Recent pelagic calciodinellaceans (e.g. *Rhabdothorax*, *Calciodinellum*) favour controlled sinking strategies in their calcareous cysts: motile stages leave the cyst during its suspension in the water column before it sinks to a depth from which the emerging cell can not re-enter the photic zone (Tangen et al. 1982; Dale 1986).

The morphogenetic reactions to some environmental conditions by the calcareous dinoflagellates listed above, in particular those of the Obliquipithonelloideae, make this fossil group of considerable use for palaeoecological analysis. The calcisphere assemblages of the Lower Cretaceous moderate Boreal Realm (NW Germany, North Sea Basin) allow palaeooceanographic and palaeogeographic interpretations, as well as ecostratigraphic correlations, of regional value. This should explained by a few paradigms already published:

1. Periodic connections between the Tethyan and Boreal Realms via the "Polish Street" to the east during the early Hauterivian and the "Paris

Rio D, Backman J, Raffi I (1987) Calcareous nannofossil biochronology and the Pliocene/Pleistocene Boundary. In: The neogene/quaternary boundary. Fin Rep IGCP Proj 41: 31 pp

Romein AJT (1979) Lineages in Paleogene calcareous nanoplankton. Utrecht Micropaleontol Bull 22:545–605

Roth PH (1973) Calcareous nannofossils – Leg 17. Deep Sea Drilling Project. In: Winterer EL, Ewing J et al. (eds) Init Rep DSDP, vol 17. US Gov Print Off, Washington, DC, pp 695–795

Roth PH (1983) Jurassic and lower Cretaceous nannofossils in the western Atlantic (Site 534): biostratigraphy, preservation and observations on biogeography and palaeoceanography. In: Sherida RE, Gradstein FM et al. (eds) DSDP Init Rep, vol 76. US Gov Print Off, Washington, DC, pp 587–621

Roth PH (1986) Mesozoic palaeoceanography of the North Atlantic and Tethys Oceans. In: Summerhays CP, Shackleton NJ (eds) Geol Soc London Spec Publ 21:299–320

Roth PH, Berger WH (1975) Distribution and dissolution of coccoliths in the South and Central Pacific. In: Slitter WV, Bé AWH, Berger WH (eds) Dissolution of deep-sea carbonates. Cushman Found Foram Res Spec Publ 13:87–113

Roth PH, Bowdler JL (1981) Middle Cretaceous calcareous nannoplankton biogeography and oceanography of the Atlantic Ocean. Spec Publ Soc Econ Paleontol Mineral 32:517–546

Roth PH, Coulbourne RT (1982) Floral and solution patterns of coccoliths in surface sediments of the North Pacific. Mar Micropaleontol 7:1–52

Ruddiman WF, McIntyre A (1976) Northeast Atlantic paleoclimatic changes over the past 600,000 years. Mem Geol Soc Am 145:111–146

Schneiderman N (1977) Selective dissolution of Recent coccolithophorids in the Atlantic Ocean. In: Ramsay ATS (ed) Oceanic micropalaeontology. Academic Press, New York London, pp.1009–1053

Shackleton NJ, Hall MA (1984) Carbon isotopic data from Leg 74 sediments. In: Moore TC, Rabinovitz PD et al. (eds) Init Rep DSDP, vol 74. US Govt Print Off, Washington, DC, pp 613–620

Shackleton NJ, Backman J, Zimmerman HG, Kent DV, Hall MA et al. (1984) Isotope calibration of the onset of ice-rafting and history of glaciation in the North Atlantic. Nature (London) 307:320–323

Sherwood RW (1974) Calcareous nannofossil systematics, paleoecology and biostratigraphy of the Middle Eocene Weches Formation of Texas. Tulane Stud Geol Paleontol 11:1–79

Smetacek VS (1985) Role of sinking in diatom life history cycles: ecological, evolutionary and geological significance. Mar Biol 84:239–251

Stanley SM (1979) Macroevolution. Freeman, San Francisco

Sverdrup HU, Johnson MW, Flemming RH (1942) The oceans – their physics, chemistry and general biology. Prentice-Hall, Englewood Cliffs, NJ, 1087 pp

Takayama T (1972) A note on the distribution of *Braarudosphaera bigelowii* (Gran and Braarud) Deflandre in the bottom sediments of Sendai Bay, Japan. Trans Proc Palaeontol Soc Jpn NS 87:429–435

Tangen K, Brand L, Blackwater L, Guillard R (1982) *Thoracosphaera heimii* (Lohmann) Kamptner is a donophyte: observations on its morphology and life cycle. Mar Micropaleontol 7:193–212

Tan Sin Hok (1927) Over de samenstelling en het onstaan van krijt-en mergel-gesteenten van de Molukken: Jaarrb Mijnw Ned Oost-Indie, 1926:165 pp

Tappan H (1980) The paleobiology of plant protists. Freeman, San Francisco, 1028 pp

Tappan H, Loeblich AA Jr (1973) Evolution of the oceanic plankton. Earth Sci Rev 9:207–240

Thierstein HR (1976) Mesozoic calcareous nannoplankton biostratigraphy of marine sediments. Mar Micropaleontol 1:325–362

Thierstein HR, Geitzenauer KR, Molifino B, Shackleton NJ (1977) Global synchroneity of coccolith datum levels: validation by oxygen isotopes. Geology 5:400–404

Theodoris S (1983) Calcareous nannofossil biozonation of the Miocene and revision of the helicoliths and Discoasters. Utrecht Micropaleontol Bull 32:1–271

Thomsen HA (1981) Identification by electron microscopy of nannoplanktonic coccolithophorids (Prymnesiophycea) from West Greenland, including the description of *Papposphaera sarion* sp. nov. Br Phycol J 16:77–94

Thomsen HA, Buck KR, Coale SL, Garrison DL, Gowing MM (1988) Nannoplanktonic coccolithophorids (Prymnesiophyceae), Haptophycea from the Weddel Sea, Antarctica. N J Bot 8:419–436

van der Wal P, de Jong EW, Westbroek P, de Bruyn WC (1983) Ultrastructure polysaccharide localization in calcifying and naked cells of the coccolithophorid *Emiliania huxleyi*. Protoplasma 118:157–168

Varol O (1985) Miocene calcareous nannofossils from the Mut Basin, southern Turkey. J Micropalaeontol 4:127–139

Varol O (1989) Palaeocene calcareous nannofossil biostratigraphy. In: Crux JA, van Heck SA (eds) Nannofossil and their applications. Horwood, Chichester, pp 267–326

Volkman JK, Eglington G, Corner EDS, Sergent JR (1980) Novel unsaturated straight-chain and ethyl ketones in marine sediments and a coccolithophorid *Emiliania huxleyi*. In: Douglas AG, Maxwell JR (eds) Advances in organic geochemistry 1979: physics and chemistry of the earth, vol 12. Pergammon, Oxford, pp 219–227

Wang F, Samtleban C (1983) Calcareous nannoplankton in surface sediments of the East China Sea. Mar Micropaleontol 8:249–259

Westbroek P, van der Wal P, van Emburg PR, de-Vrind JPM, de Bruijn WC (1986) Calcification in the coccolithophorids *Emiliania huxleyi* and *Pleurochrysis carterae*. I. Ultrastructure aspects. In: Leadbeater BSC, Riding R (eds) Biomineralization in lower plants and animals. Syst Assoc Spec Publ 30. Clarendon, Oxford, pp 189–203

Winter A (1982a) Post-depositional shape modification in Red Sea coccoliths. Micropaleontology 28:319–323

Winter A (1982b) Paleoenvironmental interpretation of Quaternary coccolith assemblages from the Gulf of Aqaba (Elat) Red Sea. Rev Esp Micropaleontol 14:291–314

Winter A, Reiss Z, Lutz B (1979) Distribution of living coccolithophorid assemblages in the Gulf of Elat (Aqaba). Mar Micropaleontol 4:197–223

Wise SW, Kelts KR (1972) Inferred diagenetic history of a weakly silified deep sea chalk. Trans Gulf Coast Assoc Geol Soc 22:177–204

Chapter 14

Fossil Calcareous Dinoflagellate Cysts

H. KEUPP[1]

Abstract

Present knowledge of fossil calcareous dinoflagellate cysts is summarized briefly. All genera of the cyst family Calciodinellaceae Deflandre 1947, including most of the Mesozoic "Calcisphaerulidae" Bonet 1956, known from the fossil record since the late Triassic, are listed here and subdivided into three subfamilies.

Several specimens show a more or less complete paratabulation which is expressed by different modes. A monophyletic character of the traditional peridinian family can be proposed owing to a corresponding tabulation pattern (ortho-hexa-tabulation of the general formula 4′,3a,7″, 4–6c,5‴,2⁗). A more or less strong control over cyst morphology by environmental factors is postulated due to the quasi-extracellular mineralization. This is obviously problematic for systematics. But on the other hand, these mineralization features can be used for environmental and stratigraphic interpretations.

Two new names (*Pentadinellum*, *Wallidinellum*) are introduced here for the calcareous cysts of Recent *Ensiculifera* cf. *mexicana* and *Scrippsiella sweenyae*.

1 Calcareous Cysts vs Thecae

The study of fossil dinoflagellates is the domain of palynology due to their acid-insoluble, sporopollenin cysts. In general, the motile stages of the Peridiniales with cellulose thecae are not preserved. Since the work of Deflandre (1947, 1948), calcareous cysts (which he had described as thecae) have been known from Mesozoic and Cenozoic sedimentary rocks. Their sizes range from 10 to 100 μm (mostly about 40 μm). Despite their global abundance, especially of the *Pithonella*-like calcispheres during the Cretaceous and Jurassic (Beschoren 1926; Wanner 1940; Locker 1967; Keupp 1981, a.o.), our knowledge of this fossil group is still very poor. Two main reasons are responsible for this:

1. Most micropaleontologists isolate their material using sieves (63 μm), washing out most of the calcareous cysts. And, in addition, the investigators of calcareous nannoplankton deal with the fraction smaller than 20 μm; therefore, the middle size range is missed.
2. The sample preparation techniques of palynologists involve the use of acids which destroy calcareous dinoflagellates.

Two groups of Recent dinophyceans containing calcareous walls can be distinguished:

[1] Institute of Palaeontology, Freie Universität, Schwendener Str. 8, D-1000 Berlin 33, Germany

1. Order Peridiniales Haeckel 1894
   Family Calciodinellaceae Deflandre 1947 *emend.* Bujak and Davies 1983: *calcareous cysts only*
2. Order Thoracosphaerales Tangen 1982: *coccoid vegetative stages*

The first group, with many genera and species, is reported from sedimentary rocks from Late Triassic onwards, especially from the Cretaceous. Only one species representing the order Thoracosphaerales, *Thoracosphaera heimii* (Lohmann 1920), is known from the fossil record since the Palaeocene. The calcareous shell of *Thoracosphaera* does not show the tabulation patterns characteristic for most motile stages of dinophyceans. Therefore, the separation of thoracospheres (including some fossil and Recent species of, as yet, unconfirmed affinity) from resting cysts of the Calciodinellaceae, which often do not have signs of the thecate tabulation patterns (= paratabulations), is still a problem in fossil and Recent material. Both calcareous shells sometimes show more or less complete morphological conformity (*Th. heimii* with some Obliquipithonelloideans). According to Tangen et al. 1982, the coccoid vegetative stages of *Thoracosphaera* have evolved from resting cysts.

This may have happened in connection with the migration of calcareous dinoflagellates from more neritic to pelagic paleoenvironments (see below). Before the culturing of isolated clones of *Thoracosphaera heimii* was achieved, the differentiation of both *Thoracosphaera* and *Pithonella*-like calcispheres was determined according to size: about 40–100 μm for the pithonelloids and 7–24 μm for the thoracospheres (Bolli 1974; Keupp 1979a). However, later findings involving small calcareous cysts show that this criterion is useless in the differentiation of fossil calcareous bodies (Keupp 1981). Many calcareous cysts are probably included among the numerous Recent and fossil species described as Thoracosphaera (e.g. Stradner 1961; Kamptner 1963, 1967; Bramlette and Martini 1964; Fütterer 1976, 1977; Keupp 1978, 1979a; Jafar 1979; Gilbert and Clark 1983). All species of these "thoracospheres", which represent calcareous cysts, should be placed in the cyst genera *Obliquipithonella* Keupp 1984 and *Orthopithonella* Keupp 1984.

2 Taxonomic Aspects

The taxonomy of dinoflagellates used currently is ambivalent. The classificatory system of Recent dinoflagellates is based on motile stage morphology, while that of fossil specimens is based on cyst morphology. In peridinians the morphological development of both thecae and cysts (with organic as well as calcareous walls) seems in many cases to be independent (Sarjeant 1974, pp. 86–87; Tappan 1980). For example, the modern motile stages of *Scrippsiella sweenyae* (Balech 1959) and *Scrippsiella trochoidea* (Stein 1983) show a very close affinity owing to the corresponding tabulation pattern of their thecae. But, using cyst taxonomy, their calcareous cysts (*Wallidinellum n.g.* resp. *Rhabdothorax* Kamptner 1958) belong to different subfamilies. Due to practical reasons, palynologists use separate classifications for thecae (often with a conservative morphology) and cysts (using more advanced development as a criterion).

All calcareous genera of certain and probable dinoflagellate cysts from Late Triassic to Recent are listed in Table 1 showing their stratigraphic ranges. Three subfamilies of the Calciodinellaceae can be distinguished owing to their different wall structures (Keupp 1987). Their definition is based on the crystallographic orientation of calcite crystals forming the calcareous wall (or at least its outer part; Figs. 9–12):

1. Perpendicular to cyst surface:	Orthopithonelloideae
2. Irregularly oblique to tangential:	Obliquipithonelloideae
3. Uniformly oblique:	Pithonelloideae (only fossil)

Some cyst genera show more or less clear paratabulations which reflect, from the motile theca, a complete or a reduced "orthohexa-tabulation" of the general formula 4′, 3a, 7″, 4–6c, 5‴, 2⁗. In this way the theca is built by different cycles of single plates: three of the four apical (′) and three intercalary plates (3a) form the apical area round the centrally positioned plate 3′. The ventral sulcus (position of the longitudinal flagellum) interrupts the circular cycles as a small area between plate 3′ and the antapex. It includes the hexagonal plate 1′ which looks symmetrical due to the onlap of six other plates (= "ortho-hexa"). After the cycle of seven precingular plates (″), the girdle (circular furrow of the transversal flagellum), built by 4 to 6 small plates, follows. The hypocyst below the girdle is built by a cycle of 5 postcingular plates (‴). The antapex shows two large pentagonal plates intersected ventrally by the end of the sulcus. Mechanisms which are responsible for the replication of this thecate plate pattern on the cysts are still unknown. This described plate pattern represents a quite conservative tabulation of the Peridiniineae (Bujak and Davies 1983). Often a reduced number of paraplates can be observed which results from fusion of single plate equivalents. This decrease of plate number characterizes only the cysts and does not occur in the motile stages. The realization of paratabulations varies (for terminology see Sarjeant 1982):

1. Holotabulate: sutures, crests (e.g. *Calciodinellum* Deflandre 1947; Fig. 2), or edges of the cyst (e.g. *Heptasphaera* Keupp 1979a, Fig. 3) correspond to the thecate tabulation;
2. Intratabulate: processes arise from centre of thecate plates (e.g. *Alasphaera* Keupp 1979c; Fig. 6);
3. Cingotabulate: expressed only by indications of the cingulum and ventral sulcus (e.g. *Carinellum* Keupp 1981; Figs. 4–5);
4. Cryptotabulate: only by shape and position of the archeopyle (e.g. partim *Orthopithonella*; Fig. 7).

A high variability in cyst shape results from both specific paraplate reductions and from the different modes of paratabulations. In addition, the localization of plate-equivalent structures also shows a wide range in the cysts:

Table 1. Genera of Calciodinellaceae and their stratigraphic ranges[a]

Cyst genus	Stratigraphic distribution	Theda
Subfamily Orthopithonelloideae Keupp 1987:		
Calciodinellum Deflandre 1947	Miocene-Recent	&
Calcigonellum Deflandre 1948	Miocene-Quaternary	
Praecalcigonellum Keupp & Versteegh 1989	Hauterivian-Quaternary	
Calcipterellum Deflandre 1948	Miocene	
Echinodinella Keupp 1980b	Hauterivian-Albian	
Keuppisphaera (Keupp 1987) Lentin & Williams '89	Late Albian	
Orthocarinellum Keupp 1987	Late Albian	
Orthopithonella Keupp 1984	? Late Triassic, Jurassic-Recent	?
Sphaerodinella Keupp & Versteegh 1989	Miocene-Recent	?
Rhabdothorax Kamptner 1958	Albian-Recent	Scrippsiella trochoidea (Stein)
"tetrahedral cyst type" Wall & Dale '68	Recent	?
Subfamily incertas sedis:		
Dimorphosphaera Keupp 1979a	Barremian-Albian	
Subfamily Obliquipithonelloideae Keupp 1987:		
Obliquipithonella Keupp 1984	? Late Triassic, Jurassic-Recent	Scrippsiella patagonica Akselman & Keupp
incl. "spherical cyst" Wall & Dale 1968		Ensiculifera "rex" Dale & Tangen (partim)
Alasphaera Keupp 1979c	Valanginian/Hauterivian	
Bicarinellum (Defl. 1948) Keupp 1984	Oxfordian-Eocene	
? incl. "bicarinate cyst" Wall & Dale	Recent	?
Calcicarpinum Deflandre 1948	Eocene	
Carinellum Keupp 1981	Late Hauterivian-Eocene	
Cubodinellum Keupp 1987	Albian-early Cenomanian	
Gonellum Keupp 1987	Albian	
Heptasphaera Keupp 1977a	Late Jurassic-early Barremian	
Nephrodinella Keupp 1981	Late Hauterivian-Barremian	
Pentadinellum n.g.	Recent	Ensiculifera cf. mexicana Balech
Sliteria Krasheninnikov & Basov 1983	Turonian-early Campanian	
Tetramerosphaera Willems 1985	Maastrichtian	

Wallidinellum n.g.	Pleistocene-Recent	Scrippsiella sweenyae Balech
"clavate cyst type" Wall & Dale '68	Recent	?
"keeled cyst type" Wall & Dale '68	Quaternary (fossil)	
"pyriform cyst type" Wall et al. '73	Holocene	?
"wedged cyst type" Wall & Dale '68	Quaternary (fossil)	
Subfamily Pithonelloideae Keupp 1987		
Pithonella (Lorenz 1901), Villain '77	Late Barremian-Maastrichtian	
incl. "Palinosphaera Borza 1972"		
? Bonetocardiella Dufour 1968	Albian-Danian	
Wallia Keupp (1990)	Campanian	
? Centosphaera Wise & Wind 1976	Maastrichtian	

[a] The taxa of motile stages of Recent calcareous cysts are given where possible.

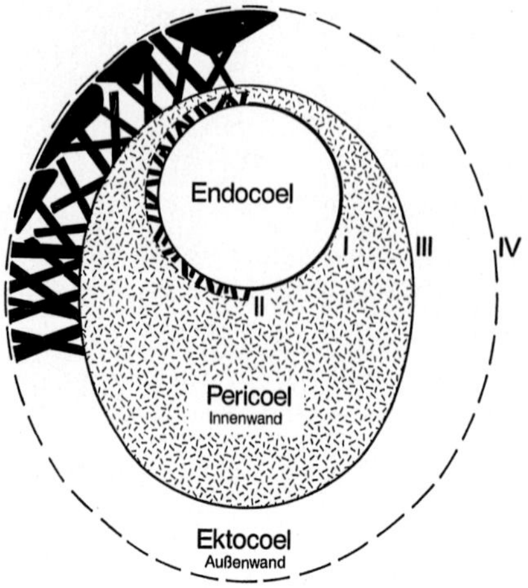

Fig. 1. *Obliquipithonella* showing the three to four possible organic phragms (*I* endophragm; *II* mesophragm; *III* periphragm; *IV* ectophragm). Instead of the phragms *I* (incl. *II*) and *III*, an autophragm can exist. The mineralization of one, two, or in some cases three calcareous wall layers can occur independently in each interspace of two phragms. (Keupp 1981)

Figs. 2–7. Some different modes of paratabulations developed in calcareous dinoflagellate cysts (terminology according to Sarjeant 1982)

Fig. 2. Holotabulate paratabulation expressed by surface-crests in *Calciodinellum operosum* Deflandre 1947 from late Miocene, El Medhi, Algeria (original sample material collected by G. Deflandre). Diameter of cyst: 30 μm

Fig. 3. Holotabulate reflection of an ortho-hexa-tabulation expressed by edges of the cyst in *Heptasphaera michaeli* Keupp 1979a. Early Barremian of Sarstedt (Moorberg), NW-Germany. Diameter of cyst: 50 μm

Figs. 4–5. Stereopair of *Carinellum parasolis* Keupp 1984 (view on the epicyst) showing a cingotabulate paratabulation. Eocene from Caussy/France (sample collected by G. Deflandre). Diameter of cysts: 45 μm

Fig. 6. The processes built by both inner and outer calcareous wall-layers of *Alasphaera caudata* Keupp 1979c reflect intratabularly the orthoperidinian tabulation. Lower Hauterivian from Sarstedt (Moorberg)/NW-Germany. Diameter of cyst: 65 μm

Fig. 7. *Orthopithonella gustafsonii* (Bolli 1974) with cryptotabulate paratabulation expressed by polygonale archeopyle (= *Orthopithonella "paratabulata"* Keupp 1980b). Late Aptian from Mittellandkanal near Schwiecheldt/NW-Germany. Diameter of cyst: 42 μm

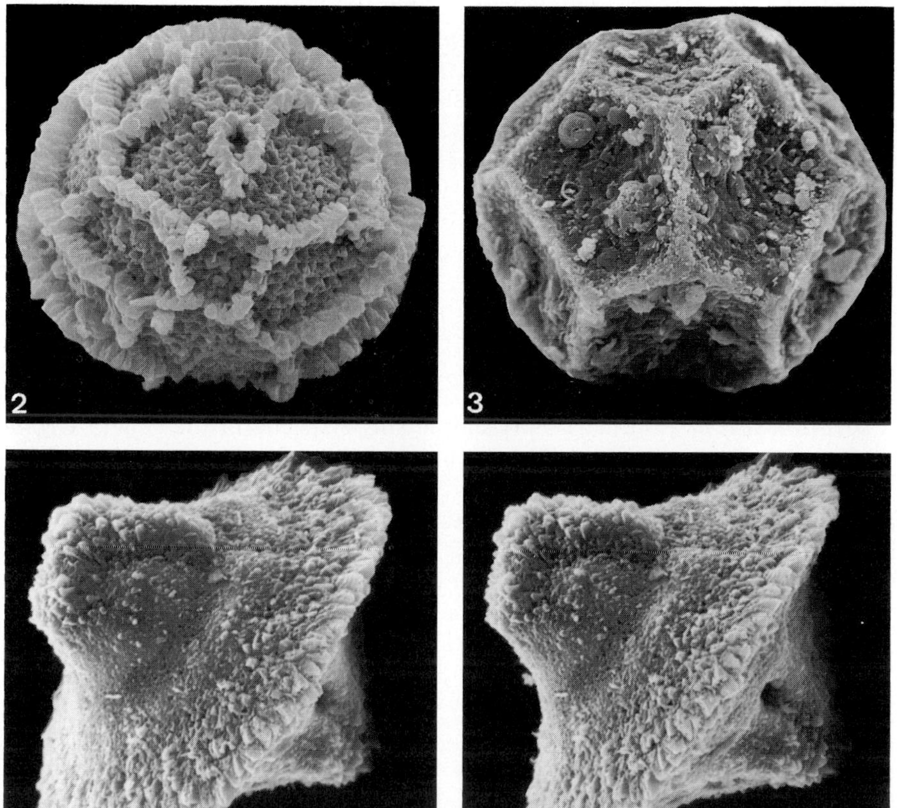
2
3
4
5
6
7

1. Outside the outer calcareous wall layer by crests, tubercles, and sutures: e.g. *Calciodinellum* (Fig. 2), *Bicarinellum* Deflandre 1948 *emend.* Keupp 1984, *Orthopithonella "paratabulata"* Keupp 1980c (Fig. 7);
2. Outside the inner calcareous wall layer only by crests: e.g. *Tetramerosphaera* Willems 1985;
3. Outside both the inner and outer calcareous wall layer: e.g. *Alasphaera* (Fig. 6);
4. Inside the organic membranes only (Compare Fig. 1):
   periphragm: e.g. *Obliquipithonella patriciagreeleyae* (Bolli 1974); Keupp 1980a; (Fig. 8);
   endophragm: *Obliquipithonella loeblichi* (Bolli 1974); Keupp 1981, Pl. 48/7;
   autophragm: *Obliquipithonella organica* (Hultberg 1975)

Except for a few specimens representing single species of the genera *Orthopithonella* (*O. "paratabulata"*), *Obliquipithonella* (*O. patriciagreeleyae*, cf. *thayeri*, *organica*), and *Echinodinella* Keupp 1980b (*E. erinacea*, Keupp 1980b), the most spherical to ovoid cysts, the prominent specimens of the so-called "Mesozoic and Cenozoic calcispheres", do not show any paratabulations. But their wall structures and apical archeopyles (= openings for excystment) prove close affinity to true Calciodinellaceae (Figs. 9–13). Corresponding to this, calcareous cysts of the Recent motile dinoflagellate *Ensiculifera* sp., as sampled by Tangen and Dale from plankton of the Oslofjorden in August of 1982 (Figs. 14–17), are congruent morphologically with common Mesozoic cysts (Jurassic, Cretaceous) of the *Obliquipithonella quiltyi* (Bolli 1974)-type. In contrast, there is no evidence for a dinoflagellate nature for the "Palaeozoic calcispheres",

Fig. 8. Broken specimen of *Obliquipithonella patriciagreeleyae* (Bolli 1974) from early Barremian of Speeton/England showing the ventral area of a complete ortho-hexa-paratabulation inside the organic periphragm only. Size of fragment (cyst 3878) 25 μm

Figs. 9–13. The three calciodinellacean subfamilies are characterized by the structure of their (at least outer) calcareous walls

Fig. 9. Fragment of the single-layered calcareous wall (autophragm inside) of *Echinodinella erinacea* Keupp 1980b showing the orthopithonelloid structure with perpendicularly orientated calcite crystals. Thickness of wall: 6 μm. Early Albian from Altwarmbüchen near Hannover/NW Germany

Fig. 10. Wall of *Obliquipithonella multistrata* (Pflaumann and Krasheninnikov 1978). The calcified pericoel (= inner wall layer) show an irregular microgranular structure, while the calcified ectocoel (= outer wall layer) has been constructed by irregularly oblique crystal-orientation, the so-called crossing-bars-structure (Keupp 1979b). Hauterivian/Barremian-boundary from Sarstedt (Gott)/NW Germany. ×2000

Fig. 11. Double-layered calcareous wall of obliquipithonelloid *Bicarinellum calvum* Keupp 1979a. The inner wall layer shows a loosely arranged "crossing-bars structure" due to poor calcification. Early Barremian from Sarstedt (Gott)/NW Germany. ×2500

Figs. 12–13. The *Wallia* (Keupp 1990) from the late Cretaceous, South Dakota/USA, belongs to the Pithonelloideae: its wall-crystals show a uniquely oblique orientation. Diameter of cyst: 90 μm, thickness of wall: 7 μm

Street" to the west during Late Barremian to Aptian enabled the warmer water faunas and floras to influence the Boreal basins (North Sea Basin, Lower Saxonian Basin, etc.), which are partly separated by swells and islands (Michael 1979). Through these migration two modifications occurred within the calcisphere assemblages: (1) new species immigrated; (2) the more-or-less cosmopolitan cyst species reacted phenotypically. Considering only the second point, nearly congruent contemporaneous vertical sequences of temperature governed morphotypes of obliqui- and orthopithonelloideans described above (two- or single-layered calcareous walls, small or large crystals, ovoid or spherical cyst shapes) within different sections of the Lower Saxonian Basin as well as in the North Sea Basin (Keupp 1981, Figs. 22–24; 1987, Fig. 3). Despite the rare occurrence of locally restricted morphotypes (see following paradigm), the calcispheres can be used here for stratigraphic correlations. Large-scale ecological alterations appear to have been responsible for the synchronous modifications of cyst morphologies.

2. On the other hand, the local distribution patterns of *Obliquipithonella* morphotypes in late Aptian sediments confirm the hypothesis of a coastal upwelling within the Lower Saxonian Basin at this time (see above, Keupp 1982).
3. In the above-mentioned example of vertical calcisphere succession of the D-Beds from Speeton Clay during a regressive-transgressive cycle (Keupp and Mutterlose 1984), the morphotypes obviously correlate to the different salinities. The increasing freshwater influence seems to have modified the cyst morphology in the same way as decreasing temperatures.

5 Phylogenetic Aspects

The phylogeny of the Peridiniineae sensu Bujak and Davies (1983), including the cyst family Calciodinellaceae, is still obscure. The earliest certain peridiniinean tabulations are reported inside the organic phragms of the Lower Cretaceous calcispheres *Obliquipithonella patriciagreeleyae* and *Obliquipithonella* cf. *thayeri* as described by Keupp (1979c, 1981, Pl. 18/7). On the other hand, the number and arrangement of tubercles modifying the cyst surface of the Oxfordian *Bicarinellum jurassicum* (Deflandre 1948) also reflect a peridiniinean pattern (Keupp 1984 p. 20). Therefore, the Peridiniineae can be traced back at least to the Late Jurassic. With the proviso of a peridiniinean affinity to non-paratabulated cysts of Late Triassic age, which correspond to the wall-structure of calcareous dinoflagellate cysts from the Jurassic and Cretaceous (Janofske 1987), the roots of the Peridiniineae suborder may reach into the Triassic. The conservative Calciodinellaceae seem to be a monophyletic family according to their unique paratabulation and their corresponding archeopyle position. Probably they represent an ancestral group of the suborder Peridiniineae.

6 Prospects

Despite some findings concerning morphology, wall structures palaeobiogeography, and palaeoecology of calcareous dinoflagellate cysts over the last 20 years, our knowledge of this important group remains poor. The outline of possibilities and interpretations in calcisphere research given here describe qualitative trends and feelings rather than objective statements. The paradigms used here have still to be proved through further investigations in other localities and stratigraphic positions. Furthermore, we need to understand the biology of the Recent calciodinelloideans, including their different life cycles, conditions for the formation of their calcareous shells, and the environmental conditions controlling their morphology, before objective systematics of fossil forms and definitive palaeoecological analysis will be possible.

Acknowledgements. I am grateful to Karl Tangen, who sent me photographs and information on the new species of Recent *Ensiculifera* (Figs. 14–17). I am indebted to Barrie Dale for some stimulating discussions concerning Recent calcareous dinoflagellates and to both Elisabeth Gierlowski-Kordesch and Robert Riding for reading the English manuscript.

Appendix

Wall and Dale (1968a) described for the first time some Quaternary calcareous cysts. These authors named the Recent cysts of *Scrippsiella trochoidea* and of *Scrippsiella sweenyae* according to their thecate stage. All other calcareous cysts of which the motile stages are not known (except of *Calciodinellum operosum*), have been documented in an open descriptive nomenclature. The so called "oblate cyst type" could later be identified belonging to *Ensiculifera* cf. *mexicana* Balech 1967 (Wall et al. 1970). Using a consistent mode of classification independently for cysts and thecae, which is essential to correlate fossil and modern dinoflagellates, the cysts have to be named separately.

The more or less spinose calcareous cysts of *Scrippsiella trochoidea* (see Wall et al. 1970) correspond with the genus *Rhabdothorax* (Gaarder and Heimdal 1973; Fütterer 1977, p. 718; Keupp 1987, p. 44). Kamptner (1937, 1958) supposed the Recent species *R. erinacea* and *R. nigra* to be coccolithophorids with affinities to *Thoracosphaera* sensu Kamptner 1928. *Rhabdothorax* now has been classified within the subfamily Orthopithonelloideae due to the perpendicular orientation of its calcitic wall crystals. The affinities of subjectively distinguishable *Rhabdothorax* morphotypes to one or several motile species are still uncertain (see Dale 1976, Pl. 1/19; 1977).

For the calcareous cysts of *Scrippsiella sweenyae* described by Wall and Dale (1968a, pp. 1401, 1403) I here introduce the new obliquipithonelloid genus name *Wallidinellum* n.g.:

Name: In honour of David Wall
Diagnosis of the genus: Microgranular calcitic cyst showing an incomplete

ortho-hexa-tabulation expressed by external rounded ridges, dominated by the precingular and antapical homologues. Typespecies: *Wallidinellum dalei* n.sp.

Holotype: Single grain preparation C216 (Woods Hole Oceanographic Institution) from surface sediments of the Western Arabian Area, figured by Wall and Dale 1968a, Pl. 172, Figs. 5–6.

Name: In honour of Barrie Dale
Diagnosis of the species: see genus -diagnosis

Description: For morphologic details see the original description given by Wall and Dale (1968a). In contrast to the opinion of these authors, the seven nodular plate areas near the equatorial zone seem to reflect the precingular plates rather than the cingular plates, analogous to other calcareous cysts (e.g. *Bicarinellum*: Keupp 1981, Fig. 11; 1987, Fig. 7). The zigzag margin of the operculum result from a large apical archeopyle including both apical and intercalary plates.

Differential diagnosis: *Wallidinellum* shows similarities with *Heptasphaera*, having a single layered microgranular wall and a more or less tabulate paratabulation expressed by edges (especially of the hypocyst) without a clear paracingulum. But in contrast to the apical flattened *Heptasphaera* with small archeopyle (equivalence of 3′ only) the apical area of *Wallidinellum* is convex and possesses a large attached operculum. *Sliteria* (Krasheninnikov and Basov 1983 is similar to *Heptasphaera* in cysts shape, paratabulation mode, and the small archeopyle. It differs from both *Heptasphaera* and *Wallidinellum* by its coarse wall structure.

The "oblate cyst type" of Wall and Dale (1968a) recognized as a calcareous cyst of *Ensiculifera* cf. *mexicana* by Wall et al. (1970) is here given the new obliquipithonelloid genus name *Pentadinellum* n.g.:

Name: According to the pentagonal outline in polar view
Diagnosis of genus: A strongly oblate autophragm is surrounded by a bicarinate calcareous wall of microgranular structure.
Typespecies: *Pentadinellum oblatum* n.sp.

Holotype: Single grain preparation C721 (Woods Hole Oceanographic Institution) from Recent sediments of Bermuda, figured by Wall and Dale (1968a, Pl. 172, Fig. 20).

Name: (lat.) due to the oblate morphology.
Diagnosis of the species: see genus-diagnosis

Description: For details see Wall and Dale (1968a, p. 1406). Both circular keels around the (?) ventrally tapered cyst are created by pre- and postcingular plate homologues. The outline of the cyst in polar view is pentagonal, due to the morphological dominance of the lower (= postcingular) keel. Similar reductions to a pentagonal cyst-groundplan (in contrast to seven precingulars), conditional on morphological assimilation of the precingular equivalences on the postcingulars, has been reported from other cysts (e.g. *Praecalcigonellum polymorphum*: Keupp 1980b; *Carinellum* ssp.: Keupp 1981, 1987; *Gonellum kurti* Keupp 1987). In congruence with the genus *Bicarinellum*, the wide transversal band seems to reflect the fused homologues of precingular and cingular plates (Keupp 1987). The number of recognizable septa dividing the transversal band into 7 poorly defined areas represent the sutures of precingulars rather than of cingular plates (compare: *Calciodinellum operosum*: Keupp 1984, Fig. 8/6; *Praecalcigonellum polymorphum*: Keupp 1980b, Pl. 15/7, 11). The apical archeopyle (operculum) shows a subcircular to polygonal outline. Weak ridges of the more or less flattened hypocyst probably mark the ventral sulcus and the two antapical plates.

Differential diagnosis: Among the Obliquipithonelloideae, *Pentadinellum* show morphologically close similarities to the genus *Bicarinellum*, except for the oblate autophragm. The pentagonal groundplan, the prominent carinae, and the more or less flattened hypotract especially correspond with *Bicarinellum cristatum* Keupp (1982) from sediments of the lowermost Barremian to early Albian. But its spherical endocoel and lack of an operculum both distinguish it.

References

Akselman R, Keupp H (in press) Recent obliquipithonelloid calcareous cysts of *Scrippsiella patagonica* sp. nov. (Peridiniaceae, Dinophyceae) from Plankton of the Golfo San Jorge (Patagonia, Argentina). Mar Micropaleontol

Bandel K, Keupp H (1985) Analoge Mineralisationen bei Mollusken und kalkigen Dinoflagellaten Zysten. N Jahrb Geol Paläontol Mh 2:65–86

Beschoren B (1926) Cenoman und Turon der Kreidemulde von Sack bei Alfeld. N Jahrb Min Geol Paläontol Beil B 55:87–132

Bolli HM (1974) Jurassic and Cretaceous Calcisphaerulidae from DSDP leg 27, Eastern Indian Ocean. Init Rep DSDP 27:843–907

Bonet F (1956) Zonificatión microfaunistica de las Calizas Cretácias del Este de México. In: 20th Congr Geology Int Mex, pp 3–102

Borza K (1972) Neue Arten der Gattung Cadosina Wanner, Pithonella Lorenz und Palinosphaera Reinsch aus der oberen Kreide. Geol Sb Geol Carp Slov Acad Vied 23:139–150

Bramlette MN, Martini E (1964) The great change in calcareous nannoplankton fossils between the Maastrichtian and Danian. Micropaleontology 10:291–322

Bremer W, Dürr G (1986) *Neuffenia willei* n.g.n.sp., eine neue Peridiniaceae aus dem oberen Oxfordien Südwestdeutschlands. N Jahrb Geol Paläontol Mh 1:11–15

Bujak JP, Davies EH (1983) Modern and fossil Peridiniineae. AASP Contrib Ser 13:203 pp

Bukry D, Bramlette MN (1969) Some new and stratigraphically useful calcareous nannofossils of the Cainozoic. Tulane Stud Geol Paleontol 7:131–142

Dale B (1976) Cyst formation, sedimentation, and preservation: Factors affecting dinoflagellate assemblages in recent sediments from Trondheimsfjord, Norway. Rev Palaeobot Palynol 22:39–60

Dale B (1977) New observations on *Peridinium faeroense* Paulsen (1905), and classification of small orthoperidinioid dinoflagellates. Br Phycol J 12:241–253

Dale B (1983) Dinoflagellate resting cysts: "benthic plankton". In: Fryxell GA (ed) Survival strategies of the Algae. Univ Press, Cambridge, pp 69–139

Dale B (1986) Life Cycle strategies of Oceanic dinoflagellates. Unesco Tech Pap Mar Sci 49:65–72

Deflandre G (1947) Calciodinellum nov.gen., premier représentant d'une famille nouvelle de Dinoflagellés fossiles à thèque calcaire. C R Acad Sci 224:1781–1782

Deflandre G (1948) Les Calciodinellidés Dinoflagellés fossiles à thèque calcaire. Botaniste 34:191–219

Evitt WR (1985) Sporopollenin dinoflagellate cysts. Their morphology and interpretation. AASP Found ICCC B4-7245:333 pp

Fütterer D (1976) Kalkige Dinoflagellaten ("Calciodinelloideae") und die systematische Stellung der Thoracosphaeroideae. N Jahrb Geol Paläontol Abh 151, 2:119-141

Fütterer D (1977) Distribution of calcareous dinoflagellates in Cenozoic sediments of Site 366, Eastern North Atlantic. Init Rep DSDP 61:709-737

Gaarder KR, Heimdal BR (1973) Light and scanning electron microscope observations on *Rhabdothorax regale* (Gaarder) Gaarder nov.comb. Norw J Bot 20:89-97

Gilbert MW, Clark DL (1983) Central Arctic Ocean paleoceanographic interpretations based on late Cenozoic calcareous dinoflagellates. Mar Micropaleontol 7:385-401

Hultberg SU (1985) *Pithonella organica* – a new calcareous dinoflagellate with an inner organic wall. Grana 24:115-120

Jafar SA (1979) Taxonomy, stratigraphy and affinities of calcareous nannoplankton genus *Thoracosphaera* Kamptner. In: Proc 4th Int Palynology Conf, vol 2, Lucknow, pp 1-21

Janofske D (1987) Kalkige Nannofossilien aus der Obertrias (Rhät) der nördlichen Kalkalpen. Berliner Geowiss Abh 86:45-67

Kamptner E (1927) Beitrag zur Kenntnis adriatischer Coccolithophoriden. Arch Protistenkd 58:173-184

Kamptner E (1928) Über das System und die Phylogenie der Kalkflagellaten. Arch Protistenkd 64:19-43

Kamptner E (1937) Neue und bemerkenswerte Coccolithineen aus dem Mittelmeer. Arch Protistenkd 89:279-316

Kamptner E (1955) Fossile Coccolithineen-Skelettreste aus Insulinde. Verh Kon Ned Akad Wetensch Naturkd L 2:5-105

Kamptner E (1958) Betrachtungen zur Systematik der Kalkdinoflagellaten. Arch Protistenkd 103:54-116

Kamptner E (1963) Coccolithineen-Skelettreste aus Tiefseeablagerungen des Pazifischen Ozeans. Ann Naturhist Mus Wien 66:139-204

Kamptner E (1967) Kalkflagellaten-Skelettreste aus Tiefseeschlamm des südatlantischen Ozeans. Ann Naturhist Mus Wien 66:139-204

Kemper E (ed) (1982) Das späte Apt und frühe Alb Nordwestdeutschlands, Versuch einer umfassenden Analyse einer Schichtfolge. Geol Jahrb A65:703 pp

Keupp H (1978) Calcisphaeren des Untertithon der Südlichen Frankenalb und die systematische Stellung von *Pithonella* Lorenz 1901. N Jahrb Geol Paläontol Mh 2:87-98

Keupp H (1979a) Calciodinelloidea aus der Blätterton-Fazies des nordwestdeutschen Unter-Barremium. Ber Naturhist Ges Hannover 122:7-69

Keupp H (1979b) Lower Cretaceous Calcisphaerulidae and their relationship to calcareous dinoflagellate cysts. Bull Centre Rech Explor Prod Elf-Aquitaine 3, 2:651-663

Keupp H (1979c) Alasphaera caudata n.g.n.sp., eine neue Calciodinelloidee aus der Unterkreide (Hauterivium) Nordwestdeutschlands. N Jahrb Geol Paläontol Mh 5:284-290

Keupp H (1980a) *Pithonella patriciagreeleyae* Bolli 1974, eine kalkige Dinoflagellaten-Zyste mit interner Paratabulation (Unter-Kreide, Speeton/SE-England). N Jahrb Geol Paläontol Mh 9:513-524

Keupp H (1980b) *Calcigonellum* Deflandre 1948 und *Echinodinella* n.gen. (kalkige Dinoflagellaten-Zysten) aus der nordwestdeutschen Unter-Kreide. Facies 2:123-148

Keupp H (1980c) *Pithonella paratabulata* n.sp., eine unterkretazische Calcisphaere mit äußerer Paratabulation. Facies 3:239-249

Keupp H (1981) Die kalkigen Dinoflagellaten-Zysten der borealen Unter-Kreide (Unter-Hauterivium bis Unter-Albium). Facies 5:1-190

Keupp H (1982) Die kalkigen Dinoflagellaten-Zysten des späten Apt und frühen Alb in Nordwestdeutschland. Geol Jahrb A65:307-363

Keupp H (1984) Revision der kalkigen Dinoflagellaten-Zysten G. Deflandres, 1948. Paläontol Z 58, 1/2:9-31

Keupp H (1987) Die kalkigen Dinoflagellaten-Zysten des Mittelalb bis Untercenoman von Escalles/Boulonnais (N-Frankreich). Facies 16:37-88

Keupp H (1990) Eine nene pithonelloide Dinoflagellaten. Kalkzyste aus der Oberkreide von South Dakota/USA, Facies 22:47-58

Keupp H, Mutterlose J (1984) Organismenverteilung in den D-Beds von Speeton (Unterkreide, England) unter besonderer Berücksichtigung der kalkigen Dinoflagellaten-Zysten. Facies 10:153–178

Keupp H, Versteegh G (1989) Ein neues systematisches Konzept für kalkige Dinoflagellaten-Zysten der Subfamilie Orthopithonelloideae Keupp 1987. Berliner Geowiss Abh 106:207–219

Krasheninnikov VA, Basov IA (1983) Cretaceous Calcisphaerulids of the Falkland Plateau, Leg. 71. Init Rep DSDP 71:977–997

Locker S (1967) Die Sphären der Oberkreide und die sogenannte Orbulinaritfazies. Geologie 16, 7:850–859

Lohmann H (1920) Die Bevölkerung des Ozeans mit Plankton. Nach den Ergebnissen der Zentrifugenfänge während der Ausreise der "Deutschland" 1911. Arch Biontol 4 (1916–1919):1–617

Michael E (1979) Mediterrane Fauneneinflüsse in den borealen Unterkreide-Becken Europas, besonders Nordwestdeutschlands. Aspekte der Kreide Europas. IUGS Ser A 6:305–321

Pfiester LA (1975) Sexual reproduction of *Peridinium cinctum* f. *ovoplanum* (Dinophyceae). J Phycol 11:259–265

Pfiester LA (1984) Sexual reproduction. In: Spector DL (ed) Dinoflagellates. Academic Press, New York London, pp 181–199

Reid PC (1978) Dinoflagellate cysts in the plankton. New Phytol 80:219–229

Sarjeant WAS (1974) Fossil and living dinoflagellates. Academic Press, New York London, 182 pp

Sarjeant WAS (1982) Dinoflagellate cyst terminology: a discussion and proposals. Can J Bot 60:922–945

Sarjeant WAS, Lacalli T, Gaines G (1987) The cysts and skeletal elements of dinoflagellates: speculations on the ecological causes for their morphology and development. Micropaleontology 33, 1:1–36

Spector DL, Pfiester LA, Triemer RE (1981) Ultrastructure of the dinoflagellate *Peridinium cinctum* f. *ovoplanum*. II. Light and electron microscope observations on fertilization. Am J Bot 68, 1:34–43

Stosch von HA (1973) Observations on vegetative reproduction and sexual life cycles of two freshwater dinoflagellates, *Gymnodinium pseudopalustre* Schiller and *Wolszynskia apiculata* sp. nov. Br Phycol J 8:105–134

Stradner H (1961) Vorkommen von Nannofossilien im Mesozoikum und Alttertiär. Erdöl Z 3:77–88

Tangen K, Brand LE, Blackwelder PL, Guillard RRL (1982) *Thoracosphaera heimii* (Lohmann) Kamptner is a dinophyte: observations on its morphology and life cycle. Mar Micropaleontol 7:193–212

Tappan H (1980) The paleobiology of plant-protists. Freeman, San Francisco, 1028 pp

Villain J-M (1981) Les calcisphaerulidae: intérêt stratigraphique et paléoécologique. Cretaceous Res 2:435–438

Wall D, Dale B (1968a) Quaternary calcareous dinoflagellates (Calciodinellideae) and their natural affinities. J Paleontol 42:1395–1408

Wall D, Dale B (1968b) Modern dinoflagellate cysts and evolution of the Peridiniales. Micropaleontology 14:265–304

Wall D, Guillard RRL, Dale B, Swift E (1970) Calcitic resting cysts in *Peridinium trochoideum* (Stein) Lemmermann, an autotrophic marine dinoflagellate. Phycologia 9, 2:151–156

Wall D, Dale B, Harrada K (1973) Description of new fossil dinoflagellates from the late Quaternary of the Black Sea. *Micropaleontology* 19:18–31

Wanner J (1940) Gesteinsbildende Foraminiferen aus Malm und Unterkreide des östlichen Ostindischen Archipels nebst Bemerkungen über *Orbulinaria* Rhumbler und andere verwandte Foraminiferen. Paläontol Z 22, 2:75–99

Weiler H (1982) Ein Fund von *Pithonella* sp., einer kalkigen Dinoflagellaten-Zyste im Fischschiefer von Bremmelbach, Elsass. Sci Geol Bull Strasbourg 35, 3:129–136

Willems H (1985) *Tetramerosphaera lacrimula*, eine intern gefächerte Calcisphaere aus der Ober-Kreide. Senckenbergiana Laethaea 66:177–201

III Algae and Stromatolites Through Time

cyanobacteria are nevertheless an important component of the microbial community. Although the primacy of cyanobacteria in stromatolite construction from the Archaean to the Recent is a matter of some debate (see Walter 1983 for a discussion on Archaean stromatolites), I view that cyanobacteria are and were the most significant biological agents for stromatolite construction.

2 Definition of a Stromatolite

Although the pre-Phanerozoic was the time of stromatolites' greatest development, the term 'stromatolite' originated from the study of a Mesozoic nonmarine occurrence. In 1908 Ernst Kalkowsky coined the German term "stromatolith" (from the Greek *stroma* meaning bed, mattress, or layer and *lithos* meaning stone) for masses of laminated nonmarine limestone in the Lower Triassic Buntsandstein of the Harz Mountains, Germany. He also introduced the term "stromatoid" for individual growth structures with more or less planar layers but with a tendency toward upward convex doming that occurred within stromatoliths. A microbial origin was attributed to these structures by Kalkowsky. Over the years, the anglicized term stromatolite gained favor; however, the term was applied to individual, convex upward, laminated structures.

A number of researchers prefer a purely descriptive, nongenetic definition of stromatolite and restrict usage to "... attached, laminated, lithified, sedimentary growth structures, accretionary away from a point or limited surface of initiation" (Semikhatov et al. 1979, p. 993 but note difference of opinion in this paper). It is rarely possible to demonstrate a biological origin for ancient stromatolites, hence preference by some researchers for such a nongenetic definition. Adherents to the descriptive definition, nevertheless, apparently recognize that most of the structures they would call stromatolites probably formed as a result of microbial activity. The descriptive definition also includes numerous abiogenic structures such as some calcretes, stalagmites, boiler crusts, and certain geyserites.

On the other hand, a widely used definition of stromatolite (and the one preferred in this chapter) is genetic and defines a stromatolite as an organosedimentary structure produced by the sediment trapping and binding activity of microorganisms, primarily cyanobacteria (from Awramik and Margulis in Walter 1976, p. 1). This definition does not specify lithification, the presence of laminations, or convexity away from the region of attachment. It is a process-oriented definition that reflects the corpus of knowledge that has accumulated over the years, indicating that most structures commonly referred to as stromatolites were built by photosynthetic microorganisms. The definition eliminates the need for separate appellations, some multiple and/or cumbersome, such as algal stromatolite, cyptalgalaminite, potential stromatolite, and thrombolite for microbially produced structures that are here grouped under the umbrella term stromatolite. A new term, microbialite, has been introduced by Burne and Moore (1987, p. 241) for an organosedimentary deposit formed from the interaction between benthic microbial communities and detrital or chemical sediments. The term is virtually identical to the genetic definition of stromatolite.

3 The First Stromatolites

Two Early Archaean greenstone belts, one in the Pilbara region of Western Australia and the other in the Barberton Mountain Land of South Africa, contain stromatolites in rocks radiometrically dated as old as 3500 Ma. Although not morphologically complex, the stromatolites are nevertheless well laminated and range in shape from undulatory, stratiform structures to domical structures, some with a pseudocolumnar geometry (Lowe 1980b; Walter et al. 1980; Buick et al. 1981; Byerly et al. 1986; Fig. 3). Oncolites are also found (Lowe 1983). These Early Archaean stromatolites are found preserved as chert, and the Pilbara stromatolites might have originally been calcium carbonate (Buick et al. 1981). The overall shape and microstructure of the stromatolites are similar to many Proterozoic and younger examples. Apparently, no true columnar stromatolites have yet been found in Early Archaean terranes. Conically-shaped, laterally-linked laminae occur in some Pilbara stromatolites (Lowe 1983). Near both the Western Australian and South African stromatolite localities, filamentous and coccoidal microbial fossils have been discovered in chert (Awramik et al. 1983, 1988; Walsh and Lowe 1985; Schopf and Packer

Fig. 3. Early Archaean stromatolites from the 3500-Ma-old Warrawoona Group, Western Australia

1987). Most of the Western Australian filaments occur within thinner, dark laminae of wavy laminated light-dark couplets and the filaments are commonly oriented parallel to lamination; an organization suggestive of stromatolitic laminae (Awramik et al. 1983, 1988).

The occurrence of stromatolites in 3500-Ma-old, Early Archaean strata has great palaeobiological significance. On one hand, most geologists and palaeontologists would regard stromatolites as evidence for the presence of primitive organisms. The construction of a stromatolite, however, requires microbes that are metabolically, behaviorally, and morphologically quite sophisticated (Awramik 1986). Thus, the presence of stromatolites on the Earth 3500 Ma ago indicates that microbial evolution had already evolved to a rather advanced grade of prokaryote. However, do these structures therefore indicate the presence of cyanobacteria? At a minimum, the Early Archaean stromatolites indicate the presence of cyanobacteria-like microorganisms in terms of their phenotype (in the full biological sense of the term). The filamentous microfossils from chert of the same age present permissive evidence for cyanobacteria (Awramik et al. 1983) while the encapsulated pluricellular coccoids described by Schopf and Packer (1987) argue well for cyanobacteria. The carbon isotopic data support the interpretation that autotrophs were present in the ecosystem (Walter 1983). The cumulative weight of these observations and data on microfossils, stromatolites, and isotopic geochemistry provides presumptive evidence that cyanobacteria had evolved by Early Archaean time. (Whether or not these 'cyanobacteria' where oxygen-releasing photoautotrophs is a separate question because many modern mat-building cyanobacteria can use photosystem I). Phylogenetic reconstructions of prokaryotes based on molecular biological techniques indicate that cyanobacteria are one of the latest prokaryotic phyla to evolve (Woese 1987) and this supports the widely held conclusion based on more traditional lines of research that among prokaryotes, cyanobacteria are phylogenetically one of the last phyla to appear (e.g. see Margulis and Schwartz 1982, p. 24). Therefore, the presence of cyanobacteria in 3500-Ma-old rocks is not a trivial issue for it would imply that prokaryotes had undergone a great deal of evolution prior to 3500 Ma ago and that most major clades of prokaryotes had already appeared. Evolution of prokaryotes during the remainder of the pre-Phanerozoic was at lower taxonomic levels in already established major clades; some of these clades may have gone extinct.

4 Other Archaean Stromatolites

Although the Archaean represents 1440 Ma of geological time and stromatolites appeared early in the Archaean, they are actually rare in the Archaean (Fig. 4). Less than 20 occurrences are currently known; although this number has slowly increased over the past five years, the overall number of localities will probably remain relatively low when compared with Proterozoic examples. This paucity may be due to geological factors rather than biological factors. In the Archaean, sedimentary basins developed within greenstone belts and sedimentation was

Fig. 4. Late Archaean domical stromatolites from the approximately 2700-Ma-old Steep Rock Group, western Ontario, Canada. Large dome in *center* of picture is about 3 m across

characterized by rapid deposition of compositionally immature sediments in subsiding basins or in unstable alluvial settings (Lowe 1980a). Stable, shallow-water environments not deluged with sediments were uncommon and/or short-lived in greenstone belt environments. Carbonates apparently were scarce (Cameron and Baumann 1972). Hence, even phototropic microbes that could grow fast, reproduce quickly, and which possessed rapid motility would have had difficulty maintaining themselves at or near the sediment-fluid interface, influence sedimentation, and produce a stromatolite. The development of sizable, stable, shallow water environments (i.e. continental shelves and platforms) concomitant with Proterozoic-style tectonic regimes toward the end of the Archaean and in the early Proterozoic resulted in a proliferation of stromatolites (Cloud 1976b). The passive margins of continents, when inundated, provided relatively large, stable, shallow shelf environments conducive to stromatolite formation. In addition, depending on climatic and local tectonic conditions, such environments could also be major sites for the production and accumulation of carbonate sediments. The great majority of fossil and recent stromatolites are found as limestone or dolostone constructions. This relationship is not fortuitous. Cementation and lithification of microbiologically accreted sediment are crucial for the preservation of stromatolites and the carbonate geochemical system provides an excellent means of fossilization. Also the uptake of CO_2 during photosynthesis would be expected to induce calcium carbonate precipitation.

Most of the major morphological varieties of stromatolites had appeared by the end of Archaean time. These include wavy laminated (stratiform), nodular, domical, pseudocolumnar, columnar, conical, and oncolitic types. However, in the Archaean, unlike the Proterozoic, columnar stromatolites are rare, few branching columnar stromatolites are found and of these no complexly branched varieties have been discovered. The known Archaean stromatolites have shapes that are common throughout geological time; no distinctive morphology unique to the Archaean has yet been recognized. The microstructure of Archaean stromatolites is similar that found in many Proterozoic stromatolites. Based on studies of Proterozoic and Recent stromatolites, our current understanding on the development of microstructure suggests that many of these Archaean stromatolites were built by filamentous cyanobacteria (Walter 1983).

5 Proterozoic Stromatolites

Palaeontologically, the Proterozoic is known for its stromatolites. It was during this long interval of geological time (some 1930 million years, from 2500 to 570 Ma ago) that stromatolites reached the acme of their diversity and abundance and then underwent drastic reductions in diversity and abundance at the close of the Proterozoic. Basic stromatolite architectures (cones, columns, domes, etc.) were already established by Late Archaean time, and late in the Early Proterozoic, conical, columnar, and columnar-branching stromatolites become conspicuous components in stromatolitic deposits. Many Proterozoic stromatolites are well preserved on both the macroscopic and the microscopic level, and in a few cases, microfossils are found in silicified laminae. A useful way of studying stromatolites is to consider them from all observational levels, from the large metre-sized bioherm or biostrome groupings of individual structures down to the microscopic level of microfossils and microstructure (Fig. 5). A conspicuous component of Proterozoic stromatolites that separates this time interval from the Archaean and Phanerozoic is the presence of numerous columnar-branching and conical stromatolites (Figs. 6, 7). Following the example of James Hall (1883) in giving stromatolites binomial names and taxonomic descriptions, geologists and palaeontologists who describe stromatolites and use them in biostratigraphy have devised several techniques for their study. The serial sectioning – graphical reconstruction method of Krylov (1963) is the most widely used technique for studying the morphology of individual stromatolites and in the naming of stromatolites. Studies on the biostratigraphic distribution of stromatolites for the Proterozoic has produced workable schemes for the subdivision of Proterozoic time (see reviews by Semikhatov 1976; Bertrand-Sarfati and Walter 1981).

A brief analysis of data and the patterns of diversification (using primarily the stromatolite taxonomic rank of form; Fig. 1) presented in Walter and Heys (1985) and an analysis of stromatolite information presented elsewhere for Proterozoic stromatolites results in the following observations [Note: (a) Va-

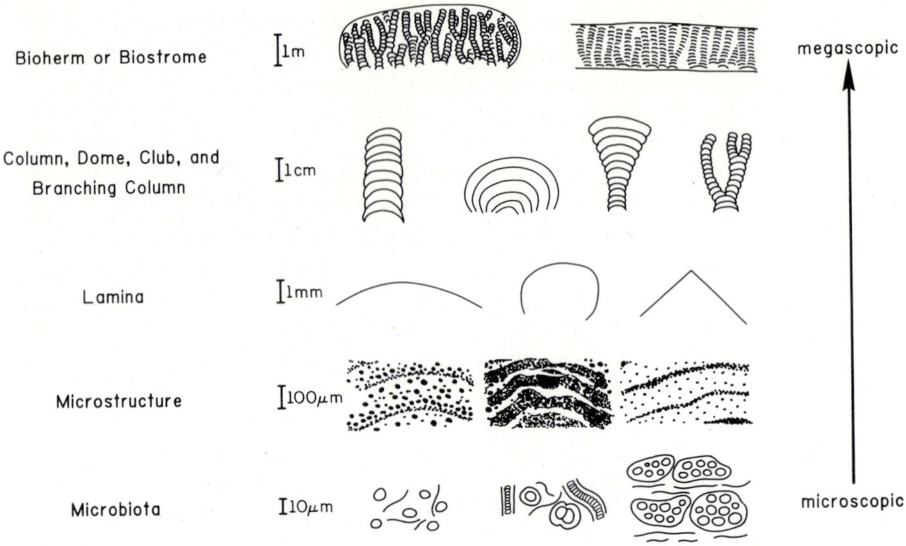

Fig. 5. Levels of observations and organization in stromatolites

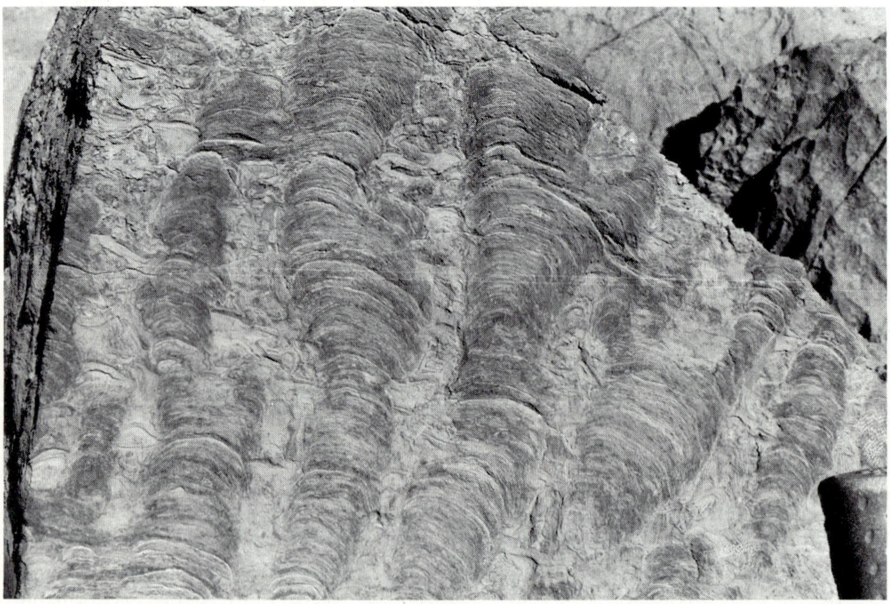

Fig. 6. *Baicalia mauritanica*, a columnar branching stromatolite from the Late Proterozoic Atar Group, Mauritania

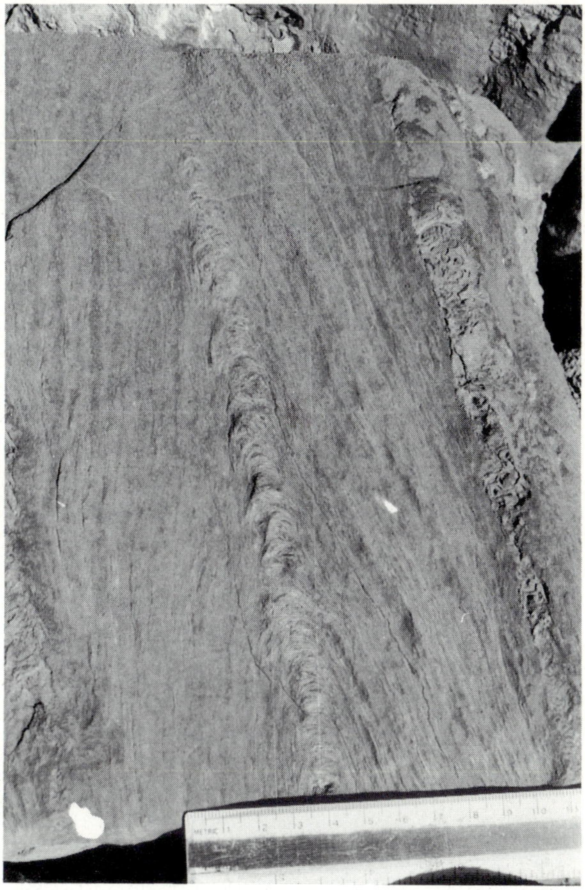

Fig. 7. Weathered, longitudinal section of *Conophyton jacqueti* from the Late Proterozoic Atar Group, Mauritania

riation between the points on the graph of Fig. 2 is impossible to determine because of the imprecise age determinations and difficulties with the taxonomy of stromatolites. Only the broadest or sketchiest of patterns are possible to elucidate; (b) Only data used in Walter and Heys (1985) are used. Taxa described from India are not well represented in these data and several new monographs have been published (most notably Zhu et al. 1987, Early Proterozoic stromatolites) since 1985. These new additions are currently being compiled]: (1) Only six taxa of stromatolites have been described from the first 300 Ma of the Early Proterozoic (2500 to 2200 Ma ago). (2) Two diversifications occurred, one during the late Early Proterozoic, 2200 to 1650 Ma ago, and the other during the Middle Riphean, 1350 to 1050 Ma ago. (3) Two diversity plateaus occurred, one during the late Early Proterozoic and into the early Early Riphean, around 1350 Ma ago, and the other from Middle to Late Riphean, around 1050 Ma ago. And (4), a decrease in stromatolite diversity occurred beginning in the Vendian, about 676 Ma ago, and continuing into the Cambrian.

1. The small number (six) of stromatolite taxa from 2200- to 2500-Ma-old Early Proterozoic strata reflects lack of detailed taxonomic treatment and insufficient study before 1985. This relatively low diversity (and low abundance) of stromatolites (Walter and Heys 1985) may be real and reflect low diversity of cyanobacteria, oxygen deficient conditions, and the small number of sedimentary basins as noted in Walter and Heys (1985). The small number of basins may reflect what Goodwin (1981) recognizes as a transitional phase between the Archaean greenstone belt-type tectonics and the macro-plate tectonics characteristic of Middle Proterozoic through Phanerozoic time when ensialic basins developed in an intra-plate tectonic setting.

Oxygen-deficient conditions could have influenced cyanobacteria during the earliest Proterozoic and Archaean. Many modern cyanobacteria, when operating with photosystem II (oxygen-releasing photosynthesis) are microaerophilic; that is, they are adapted to low oxygen levels. Extending this observation into the past, it can be speculated that the oxygenation of the atmosphere beginning about 2300 Ma ago (see review by Cloud 1983), could have been pivotal in the evolution and spread of cyanobacteria. Prior to this time, many cyanobacteria might have been restricted to oxygenated environments largely of their own creation; localized regions where cyanobacteria were growing and producing O_2. Microaerophily could make cyanobacteria 'pre-adapted' to the oxygenated conditions that were to begin about 2300 Ma ago, and they could easily disperse and readily colonize newly oxygenated shallow water habitats.

The diversity of earliest Proterozoic microfossils found in stromatolites is low when compared to Middle-Late Proterozoic examples. My reading of the Early Proterozoic microfossil record indicates that all the major orders of cyanobacteria were already present; however, diversity within these orders appears to be low. The low diversity could have been the result of competition with other prokaryotes. This low diversity of such microbes might have been a contributing factor to the low diversity of stromatolites (this assumes that there is a strong relationship between microbiota and shape in stromatolites; a relationship that remains to be demonstrated).

2. The first diversification occurred during the late Early Proterozoic (2200 to 1650 Ma ago) when the number of stromatolite taxa increased to 97 (Walter and Heys 1985). This large diversification represents the first significant adaptive radiation of stromatolites and probably the microbes that built them. Macro-plate tectonics began to dominate producing more 'basins' (26 noted in Walter and Heys 1985). Continued oxygenation of the atmosphere and shallow hydrosphere occurred and allowed for greater cyanobacterial dispersal. Cyanobacteria probably underwent a major adaptive radiation. The appearance of new cyanobacteria (species, genera, families within existing orders) in the mat-building community with improved abilities to accrete sediment and form cohesive fabrics, could have resulted in new geometric arrangements in the stacking of laminae. For example, change in the consistency of extracellular gel (sheath and envelope), greater motility, and faster growth could alter the

sediment-microbe plexus in such a way that the interaction of the mat with factors of the physical environment (water depth, currents, wave action, sediment supply, temperature) would result in a new style of branching or some other change in the shape of the stromatolite. The microbial fossil record of cyanobacteria during this interval of time is not good. What few data are available tend to support this speculation.

The second diversification noted above which occurred during the Middle Riphean, approximately 1500 to 1200 Ma ago when the diversity increased from 98 taxa to 182 taxa, is much more problematical. It follows a diversity plateau (see Fig. 2) and there is no suggestion in the available data of a decrease in taxonomic diversity preceding this diversification. In view of the lack of any evidence to suggest otherwise, atmospheric level of O_2 was probably increasing monotonically. Phanerozoic-style plate tectonics were well established, but no clear suturing or breakup of major land masses has yet been determined for this interval of the Proterozoic. In addition to or aside from another phase of cyanobacterial diversification [the microfossil record in stromatolites, although not great, does show an increase in diversity (Schopf 1977), but not the pronounced increase one might associate with the profound increase in stromatolite diversity], the only possible major change in the Earth-biosphere system that might have had some effect on increased stromatolite diversity was the appearance of eukaryotes. Although the subject of considerable debate, the oldest eukaryotes are probably photoautotrophic plankters that appeared about 1500 Ma ago (Vidal 1984). Possible benthic algal eukaryotes appear in the record about 1300 Ma ago (for instance *Beltina* from the Greyson Shale; see review by Hofmann 1985). The competition between benthic eukaryotic algae and cyanobacteria might have resulted in a further cyanobacterial diversification giving rise to an increase in stromatolite taxa. The role that eukaryotes play in stromatolite construction today is not well understood; however, it is possible that benthic, microbial eukaryotes took part in stromatolite construction at this time, contributing to the production of new groups and forms of stromatolites. A problem with this hypothesis is that there is no clear indication in the microfossil record as preserved in stromatolites that eukaryotes participated in stromatolite construction. Fossil cyanobacteria are the major fossil microbes found. The causes for the doubling of diversity from Early to Middle Riphean times remain obscure.

3. The diversity plateau during the late Early Proterozoic through Early Riphean and during the Middle to Late Riphean is another curious feature of the stromatolite diversity curve. The increase in diversification followed by a tapering off (the plateau) resembles the sigmoidal growth phase curve that increasing populations of various organisms follow and it resembles the curve of the rapid diversification of animals at the beginning of the Phanerozoic (see Sepkoski 1979). A plateau or tapering off in diversification curves is usually explained as a result of filling of available ecological zones by the clades (Bambach 1985; Valentine and Walker 1986). Employing Bambach's (1985, p. 247) explanation, this would mean that the adaptive range of the stroma-

tolite-building microbes, primarily cyanobacteria, with its existing phenotypic attributes, reached a diversity balance related to the ecological space the organisms are able to utilize.

4. The sharp decrease in stromatolite diversity during the Vendian and continuing into the Cambrian has been noted and discussed elsewhere (Awramik 1971; Monty 1973; Pratt 1982; Walter and Heys 1985; Cao 1988). The circumstantial evidence still implicates grazing and burrowing by macrofauna and meiofauna that probably appeared in the Vendian. Also, the advent of biomineralization in the Tommotian, which produced coarse-grained carbonate sediment, changed the nature of the sediment available for stromatolite accretion (Awramik and Riding 1988). Metazoans effectively outcompeted the microbes that formed microbial mats/stromatolites in shallow marine environments and brought about the stromatolite crash and prevented re-establishment of major stromatolite-building activities.

6 Summary and Conclusions

Although stromatolites received their name from a Phanerozoic example, it was during the pre-Phanerozoic that stromatolites achieved their greatest development. Until recently, the understanding of Proterozoic and Archaean stromatolites has proceeded along lines comparing these ancient structures to modern stromatolites and microbial mats forming in some of today's shallow marine environments. Hoffman (1973), however, pointed out that the present may not be an accurate analogue for the pre-Phanerozoic Earth. Many stromatolites known from Proterozoic strata have no known morphological counterparts in the Recent or even the Phanerozoic. In addition, most of Earth's history was animal-free; oxygen levels were much lower in the Late Proterozoic and earlier; shallow water, carbonate accumulating environments in the pre-Cambrian (environments where stromatolites have thrived) lacked the abundant bioclastic material found in later times; and the microbial ancestors of today's stromatolite builders were possibly phenotypically as well as genotypically different.

Acknowledgements. Numerous cups of coffee at Nicoletti's with J.W. Valentine helped to sharpen some of the ideas expressed in this paper. Thanks to David Pierce for comments on the manuscript and to Karen Griffin and David Crouch for help with the figures. NATO Research Grant RG84/0176 to R. Riding and S.M. Awramik, and NSF Grant EAR 87-21192 provided funds to help support the research on stromatolites. This is a contribution of the Precambrian-Cambrian Transition Consortium and IGCP Project 261 "Stromatolites".

References

Awramik SM (1971) Precambrian columnar stromatolite diversity: reflection of metazoan appearance. Science 174:825–827

Awramik SM (1986) New fossil finds in old rocks. Nature (London) 319:446–447

Awramik SM, Riding R (1988) Role of algal eukaryotes in subtidal columnar stromatolite formation. Proc Natl Acad Sci USA 85:1327–1329

Awramik SM, Schopf JW, Walter MR (1983) Filamentous fossil bacteria from the Archean of Western Australia. Precambrian Res 20:357–374

Awramik SM, Schopf JW, Walter MR (1988) Carbonaceous filaments from North Pole, Western Australia: are they fossil bacteria in Archean stromatolites? A discussion. Precambrian Res 39:303–309

Bambach RK (1985) Classes and adaptive variety: the ecology of diversification in marine faunas through the Phanerozoic. In: Valentine JW (ed) Phanerozoic diversity patterns. Univ Press, Princeton, pp 191–253

Barghoorn ES, Tyler SA (1965) Microorganisms from the Gunflint chert. Science 147:563–577

Bertrand-Sarfati J, Walter MR (1981) Stromatolite biostratigraphy. Precambrian Res 15:353–371

Black M (1933) The algal sediments of Andros Island, Bahamas. Philos Trans R Soc London Biol 222:165–192

Buick R, Dunlop JSR, Groves DI (1981) Stromatolite recognition in ancient rocks: an appraisal of irregularly laminated structures in an Early Archaean chert-barite unit from North Pole, Western Australia. Alcheringa 5:161–181

Burne RV, Moore LS (1987) Microbialites: organosedimentary deposits of benthic microbial communities. Palaios 2:241–254

Byerly GR, Lowe DR, Walsh MM (1986) Stromatolites from the 3,300–3,500-Myr Swaziland Supergroup, Barberton Mountain Land, South Africa. Nature (London) 319:489–491

Cameron EM, Baumann A (1972) Carbonate sedimentation during the Archean. Chem Geol 10:17–30

Cao Rui-ji (1988) Study on stromatolite decline event in terminal Precambrian. Acta Palaeontol Sin 27:737–750 (in Chinese with English summary)

Cloud P (1976a) Beginnings of biospheric evolution and their biogeochemical consequences. Paleobiology 2:351–387

Cloud P (1976b) Major features of crustal evolution. Geol Soc S Afr Annexure 79:1–33

Cloud P (1983) Banded iron-formations – a gradualist's dilemma. In: Trendall AF, Morris RC (eds) Iron-formation: facts and problems. Elsevier, Amsterdam, pp 401–416

Golubic S (1973) The relationship between blue-green algae and carbonate deposits. In: Carr NG, Whitton BA (eds) The biology of blue-green algae. Blackwell, Oxford, pp 434–472

Golubic S (1976) Organisms that build stromatolites. In: Walter MR (ed) Stromatolites. Elsevier, Amsterdam, pp 113–126

Goodwin AM (1981) Precambrian perspectives. Science 213:55–61

Hall J (1883) *Cryptozoön* n.g.; *Cryptozoön proliferum* n.sp. NY State Museum Ann Rep 36, Plate 6

Hoffman PF (1973) Recent and ancient algal stromatolites: seventy years of pedagogic cross-pollination. In: Ginsburg RG (ed) Evolving concepts in sedimentology. Johns Hopkins Univ Press, Baltimore, pp 178–191

Hofmann HJ (1985) Precambrian carbonaceous megafossils. In: Toomey DF, Nitecki MH (eds) Paleoalgology: contemporary research and applications. Springer, Berlin Heidelberg New York Tokyo, pp 20–33

Kalkowsky E (1908) Oolith und Stromatolith im norddeutschen Buntsandstein. Dtsch Geol Ges 60:68–125

Knoll AH (1985) The distribution and evolution of microbial life in the Late Proterozoic Era. Annu Rev Microbiol 39:391–417

Krumbein WE (1983) Stromatolites – the challenge of a term in space and time. Precambrian Res 20:493–531

Krylov IN (1963) Stolbchaty vetvyashchiesya stromatolity rifeiskikh otlozhenii yuzhnogo urala. Tr Geol Inst Akad Nauk SSSR 69:1–133

Logan BW (1961) *Cryptozoon* and associated stromatolites from the Recent of Shark Bay, Western Australia. J Geol 69:517–533
Lowe DR (1980a) Archean Sedimentation. Annu Rev Earth Planet Sci 8:140–194
Lowe DR (1980b) Stromatolites 3,400-Myr old from the Archean of Western Australia. Nature (London) 284:441–443
Lowe DR (1983) Restricted shallow-water sedimentation of Early Archean stromatolitic and evaporitic strata of the Strelley Pool Chert, Pilbara Block, Western Australia. Precambrian Res 19:239–283
Margulis L, Schwartz KV (1982) Five Kingdoms. Freeman, San Francisco, 338 pp
Monty CLV (1967) Distribution and structure of Recent stromatolitic algal mats, eastern Andros Island, Bahamas. Ann Soc Geol Belg 90:55–100
Monty CLV (1973) Precambrian background and Phanerozoic history of stromatolite communities, an overview. Ann Soc Geol Belg 96:585–624
Pratt BR (1982) Stromatolite decline – a reconsideration. Geology 10:512–515
Schopf JW (1968) Microflora of the Bitter Springs Formation, late Precambrian, central Australia. J Paleontol 42:651–688
Schopf JW (1977) Biostratigraphic usefulness of stromatolitic Precambrian microbiotas. Precambrian Res 5:143–173
Schopf JW, Packer BM (1987) Early Archean (3.3 to 3.5 Ga-old) fossil microorganisms from the Warrawoona Group, Western Australia. Science 237:70–73
Semikhatov MA (1976) Experience of stromatolite studies in the USSR. In: Walter MR (ed) Stromatolites. Elsevier, Amsterdam, pp 337–357
Semikhatov MA, Gebelein CD, Cloud P, Awramik SM, Benmore WC (1979) Stromatolite morphogenesis – progress and problems. Can J Earth Sci 16:992–1015
Sepkoski JJ Jr (1979) A kinetic model of Phanerozoic taxonomic diversity. II. Early Phanerozoic families and multiple equilibria. Paleobiology 5:222–251
Valentine JW, Walker TD (1986) Diversity trends within a model taxonomic hierarchy. Physica 22D:31–42
Vidal G (1984) The oldest eukaryotic cells. Sci Am 250:48–57
Walcott CD (1914) Cambrian geology and paleontology III: Precambrian Algonkian algal flora. Smithsonian Misc Coll 67:77–156
Walsh MM, Lowe DM (1985) Filamentous microfossils from the 3,500-Myr-old Onverwacht Group, Barberton Mountain Land, South Africa. Nature (London) 314:530–532
Walter MR (1976) Introduction. In: Walter MR (ed) Stromatolites. Elsevier, Amsterdam, pp 1–3
Walter MR (1983) Archean stromatolites: Evidence of the Earth's earliest benthos. In: Schopf JW (ed) Earth's earliest biosphere. Univ Press, Princeton, pp 187–213
Walter MR, Heys GR (1985) Links between the rise of the Metazoa and the decline of stromatolites. Precambrian Res 29:149–174
Walter MR, Buick R, Dunlop JSR (1980) Stromatolites 3,400–3,500 Myr old from the North Pole area, Western Australia. Nature (London) 284:443–445
Winsborough BM, Golubic S (1987) The role of diatoms in stromatolite growth: two examples from modern freshwater settings. J Phycol 23:195–201
Woese CR (1987) Bacterial evolution. Microbiol Rev 51:221–271
Zhu Shixing, Xu Chaolei, Gao Jianping (1987) Early Proterozoic stromatolites from Wutai Mt. and its adjacent regions. Bull Tianjin Inst Geol Min Res 17:1–221 (in Chinese with English summary)

Chapter 16

Cambrian Calcareous Cyanobacteria and Algae

R. RIDING[1]

Abstract

The Cambrian calcareous flora is dominated by cyanobacteria which include the *Angulocellularia*, *Botomaella*, *Girvanella* and *Obruchevella* groups. The important generic groups based on *Renalcis* and *Epiphyton* are also probably cyanobacterial. *Proaulopora* is possibly a cyanobacterium. No calcified red algae are known during the period. The earliest calcified dasycladaleans may be represented by the very rare fossils *Yakutina* (Middle Cambrian) and *Seletonella* (Upper Cambrian). The sudden appearance of heavily calcified cyanobacteria near the base of the Cambrian may reflect a change in environmental conditions which enhanced $CaCO_3$ precipitation rates in the sea. Groups such as *Angulocellularia*, *Epiphyton* and *Renalcis* are important in creating reef limestones and they are often so abundant that they impart a dendritic fabric to the rock (dendrolite).

The more common of these fossils appear to have a worldwide distribution but work has been heavily concentrated in the USSR, on Siberian material and on the Lower Cambrian. Only recently have other workers started to make use of the Soviet results, and many genera have not been reported outside the USSR and Mongolia, although Cambrian floras are known in Europe, North America, Australia and Antarctica. The outstanding problem which remains is of confidently interpreting the affinity of many groups. More also needs to be known concerning the ranges of all the genera, particularly in the Middle and Upper Cambrian, and of their spatial and ecological distributions.

1 Introduction

The Cambrian is significant not only for the earliest abundant remains of animal skeletons and shells, but also for the first occurrence of well-calcified algae and cyanobacteria. Biomineralization was therefore acquired by a very wide spectrum of the organisms on Earth in a short space of time. In the case of cyanobacteria, in particular, this was a striking development because this group has a long Precambrian history. It is possible to explain the relatively sudden appearance of skeletons in metazoans as being a result of rapid evolutionary diversification from the late Proterozoic to the Cambrian. Indeed, this could also apply to algae and cyanobacteria. However, the latter group is widely regarded as being conservative, and it is known that a range of cyanobacteria comparable with extant forms existed in the Proterozoic. Cyanobacteria do not closely control their calcification but are dependent upon environmental conditions favouring precipitation of $CaCo_3$. It is for these reasons that the abrupt appearance of well-calcified cyanobacteria near the Precambrian-Cambrian boundary (Riding and Voronova 1982a) has been interpreted to possibly reflect a change in an environmental factor such as seawater chemistry (Riding 1982).

[1]Department of Geology, University of Wales College of Cardiff, Cardiff CF1 3YE, UK

The Cambrian algal and cyanobacterial calcified flora is quite well known in terms of taxonomic diversity, but there are numerous problems of affinity remaining. These make it difficult to confidently generalize concerning the sequence and timing of appearance of major groups such as chlorophytes, rhodophytes, and so on. Earlier workers, for example Korde (1973), were of the opinion that red algae were particularly important in the Lower Cambrian. However, these fossils, such as *Epiphyton*, *Botominella*, *Bija*, etc. could well be cyanobacteria. In fact, at present it seems quite likely that: (1) calcified cyanobacteria overwhelmingly dominate the Lower Cambrian flora; (2) no bona fide red calcareous algae occur in the Cambrian; (3) dasycladalean green algae are rarely represented in the Cambrian (by the problematic genera *Cambroporella* in the Lower Cambrian, *Amgaella* and *Yakutina* in the Middle Cambrian, and *Seletonella* and *Mejerella* in the Upper Cambrian); and (4) no bona fide udoteacean green algae occur in the Cambrian. However, such generalizations are dangerous given the imperfect state of current knowledge of the systematic positions of many of the fossils involved.

Shallow marine limestones of Cambrian age often contain abundant calcareous cyanobacteria and algae. Commonly they are rock-formers, especially in reefs. Although Lower Cambrian bioherms have often been regarded as dominated by archaeocyaths, in reality genera such as *Angulocellulara*, *Epiphyton* and *Renalcis* are often more important and are actually equally conspicuous because they impart a macroscopic dendritic fabric to the rock (dendrolite, Riding 1988). However, the small size of most of these fossils makes them susceptible to diagenetic alteration. Consequently, they are best studied in cratonic areas where the sequences are relatively undisturbed.

The details of the earliest appearance of the calcified flora near the base of the Cambrian are only recorded at a few localities around the world and they have been mainly studied in Siberia, which has more than its fair share of Lower Cambrian limestone sequences. Indeed, current knowledge of Cambrian calcareous cyanobacteria and algae is largely the result of painstaking work in the Soviet Union, mainly since 1930. However, only recently have these results started to be applied and evaluated by workers outside the USSR. The strength of the Soviet contribution has been in taxonomic analysis, and this has been combined with generally good stratigraphic control. There has been some tendency to proliferate taxa. However, the major problems, still unresolved in many cases, concern affinity. There has been conspicuous instability in the assignment of generic groups to higher taxonomic categories, particularly at division (red algae, green algae, etc.) level.

The most reliable guide to affinity lies in recognition of modern analogues. Many of the more common Cambrian genera range beyond the Cambrian and high into the Palaeozoic. Some, all of them cyanobacteria, possess Recent analogues. In the absence of analogues it is difficult to make progress in systematic evaluation. However, it is possible to rationalize taxonomy by sidestepping problems of affinity and concentrating on purely morphological groupings (Riding and Voronova 1985). This is clearly only a partial approach to the problem, but it serves to refine the descriptive base for systematic classifications which will follow.

It is possible to hypothesize that, following the crisis for stromatolites resulting from the appearance and rapid diversification of metazoans in the Late Precambrian, calcification provided a new and more secure mode of life for cyanobacteria. Animals disturbed, grazed and competed for space with stromatolites (Garrett 1970). Calcification provided physical strength and protection for cyanobacteria to resist disturbance and grazing, and it also gave them the possibility of a reef-building role. Cambrian reefs can be seen as habitats in which, in many cases, cyanobacterial calcification provided a firm substrate for sessile animals such as archaeocyaths. This assumes that the conspicuous dendrolite-forming genera (*Angulocellularia, Epiphyton, Renalcis*) are all cyanobacteria. *Angulocellularia* has a Recent analogue in the cyanobacterium *Schizothrix calcicola* (Riding and Voronova 1982b); *Epiphyton* and *Renalcis* are now often assumed to be cyanobacterial. The small pore spaces between the individual filaments of these genera may well have localized early cementation under the stimulus of photosynthetic metabolism, and this would have provided a further stimulus to reef formation.

It could also have been difficult for stromatolite-building cyanobacteria to cope with the coarse bioclastic sediment which animal biomineralization was beginning to produce (Pratt 1982). Again reef-building, by raising the substrate, would have provided an elevated habitat which reduced the possibility of cyanobacteria being drowned by particulate sediment. Thus, for a variety of reasons, calcification by cyanobacteria at the beginning of the Cambrian can be seen as a successful response to metazoan and parazoan diversification and biomineralization. However, whether or not this was the prime stimulus for cyanobacterial calcification remains to be seen.

A number of important algal groups has long been thought to be represented by calcified fossils in the Cambrian. Johnson (1966, p. 1) noted "the high degree of evolutionary development exhibited by the Cambrian flora and its great variety and richness of form". Johnson had in mind all algae, not only calcified ones, but he was without doubt influenced by Soviet workers, such as Vologdin (1962a), who had ascribed a variety of genera (including *Razumovskia*, *Renalcis*, *Epiphyton*, *Chabakovia*, *Angulocellularia*, and *Bija*) to the Rhodophyta. However, there is no clear evidence that any of these is likely to be a red alga and the same applies to many other genera which Korde (1973) placed in the Rhodophyta. *Solenopora* is an example. A *Solenopora*-like organism reported by Priestley and David (1910) from Antarctica, proved to belong to *Epiphyton* (see Gordon 1920, p. 684). However, Maslov (1937) and Vologdin (1955) later reported *Solenopora* from the Upper Cambrian of the Angara River and the Cambrian of the Tien Shan in Kirgizia. Nevertheless, *Solenopora* has often been confused with other genera, and at present no definite reports of *Solenopora*, as it was originally described by Dybowski (1878), are known outside the Ordovician. *Bija*, which resembles so-called *Solenopora* (see Johnson 1966, p. 15), is actually quite distinct and more closely resembles *Hedstroemia* which is probably a rivulariacean cyanobacterium.

Apart from calcified cyanobacteria, the only significant Cambrian group which must be considered here is the Seletonellaceae (Korde 1971, quoted in Korde 1973, p. 239) which Bassoullet et al. (1979) regard as the earliest of the three dasycladalean families, and which has representatives mainly in the Middle and Upper Cambrian.

2 History of Research

Cambrian calcareous algae and cyanobacteria have been studied for little more than 100 years. The main advances have been made during the past 50 years as a result of studies in Russia of Siberian material. Of the 74 genera discussed here (Table 1), 69 have been originally described in the Soviet Union and few of these

Table 1. Cambrian genera mentioned in the text

Acanthina Korde 1973	*Kyzassia* Korde 1973
Amgaella Korde 1957	*Lebedevaella* Vologdin 1958
Amgaina Korde 1973	*Lukaschevella* Vologdin 1958
Amganella Reitlinger 1959	*Marenita* Korde 1973
Angulocellularia Vologdin 1962	*Mejerella* Korde 1950
Bajanophyton Drosdova 1980	*Mucilina* Korde 1973
Batenevia Korde 1966	*Nephelostroma* Dangéard & Doré 1957
Bija Vologdin 1932	*Nicholsonia* Korde 1973
Bogutschania Korde 1965	*Nubecularites* Maslov 1937
Bostrychosaria Vologdin 1958	*Obruchevella* Reitlinger 1948
Botomaella Korde 1958	*Palaeomicrocystis* Korde 1955
Botominella Reitlinger 1959	*Palaeonites* Maslov 1956
Cambrina Korde 1973	*Parachabakovia* Korde 1973
Cambroporella Korde 1950	*Potentillina* Korde 1973
Cavifera Reitlinger 1948	*Proaulopora* Vologdin 1937
Chabakovia Vologdin 1939	*Pseudoanthos* Korde 1973
Chomustachia Korde 1973	*Razumovskia* Vologdin 1939
Cornutula Korde 1973	*Renalcis* Vologdin 1932
Dasycirriphycus Vologdin 1962	*Sajania* Vologdin 1962
Epiphyton Bornemann 1886	*Seletonella* Korde 1950
Epiphytonoides Korde 1973	*Serligia* Korde 1973
Erbina Korde 1973	*Siberiella* Korde 1957
Filaria Korde 1973	*Solenopora* Dybowski 1878
Fistulella Korde 1973	*Sporinula* Korde 1973
Flexurella Vologdin 1958	*Subtifloria* Maslov 1956
Foninia Korde 1973	*Taninia* Korde 1973
Gemma Luchinina 1982	*Tarthinia* Drosdova 1975
Girvanella Nicholson & Etheridge 1878	*Tomentula* Korde 1973
Globuloella Korde 1958	*Tubomorphophyton* Korde 1973
Glomovertella Reitlinger 1948	*Tubophyllum* Krasnopeeva 1955
Gordonophyton Korde 1973	*Tuvaellina* Vologdin 1958
Hedstroemia Rothpletz 1913	*Umbellula* Korde 1973
Kadvoya Korde 1973	*Uranovia* Korde 1958
Kenella Korde 1973	*Vologdinella* Korde 1957
Kordephyton Radugin & Stepanova 1964	*Vologdinia* Korde 1973
Korilophyton Voronova 1976	*Yakutina* Korde 1973
Kundatia Korde 1973	*Zaganolomia* Drosdova 1980

have yet been reported from other parts of the world. Nonetheless, there is a number of other regions where these fossils are known to occur, notably in Antarctica, Australia, China, Europe, Morocco, and North America, but these have not received close attention either because their carbonate sequences are not as well preserved as those in Siberia, or simply because geologists have not been available to do the work. At the time of writing (1988) the history of research divides into pre-Soviet (1886–1932) and Soviet (1932–1988) periods (Fig. 1). As far as description of new genera is concerned, only 5 of the 74 genera listed here (Table 1) were originally described outside the USSR, and since 1932 *Nephelostroma* is the only new genus described from the rest of the world. Vologdin (1932, 1939) started the Soviet spate of generic descriptions, and Korde overwhelmingly dominates the period between 1950 and 1973. In fact, 41 of the 74 genera in Table 1 are Korde genera.

However, the Soviet contribution to this research did not begin until the 1930s, some fifty years after Bornemann's (1886, pp. 16–17) discovery of *Epiphyton*, together with a new *Confervites* species, *C. primordialis*, which is *Epiphyton*-like, from the Lower Cambrian of south-western Sardinia. *Epiphyton* is a widespread and abundant fossil and it was subsequently recognized in Cambrian limestones from Antarctica (Chapman et al. 1914; Chapman 1916) where it had already been reported as a "*Solenopora*-like organism" by Priestley and David (1910). Gordon (1920) described a new species, *E. grande*, also from Antarctica which Korde (1973) was later to use to erect the genus *Gordonophyton*.

Girvanella was originally described from the Ordovician (Nicholson and Etheridge 1878) but it is common in the Cambrian. Blackwelder (1907) and Yabe and Osaki (1930) reported it from China, Chapman (1907–1908) and Mawson and Madigan (1930) from Australia, Hazzard and Crickmay (1933) from California, and Mercier (1934) from France.

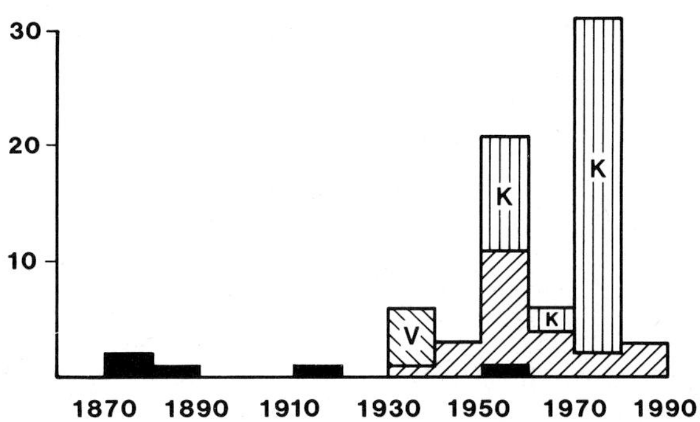

Fig. 1. Publications of genera of calcareous algae and cyanobacteria reported from the Cambrian. Synonyms and incertae sedis included. For list see Table 1. Only 5 (shown in *black*) of the 74 genera shown were originally described outside the Soviet Union: *Solenopora* 1877 (reported, but not actually present, in the Cambrian), *Girvanella* 1878, *Epiphyton* 1886, *Hedstroemia* 1913, *Nephelostroma* 1957. Five of the 6 genera described during the 1930s were by Vologdin (*V*), and 10 of the 21 genera described in the 1950s were by Korde (*K*). All but 2 of the 31 genera described in the 1970's were in Korde's (1973) monograph (*K*)

The third well-known Cambrian genus, which ranks in importance alongside *Epiphyton* and *Girvanella*, is *Renalcis*, which Vologdin described from the Altai Mountains in 1932. Following the First Five Year Plan (1928–1932), development of Siberia moved forward quickly. Cambrian and associated Lower Palaeozoic carbonate sequences are extensively exposed in Siberia, and are also present in the Urals to the west and Altai Mountains to the south (Fig. 2). The largest and best preserved outcrops are on the Siberian Platform (which approximates to the Central Siberian Plateau), particularly in the Anabar region in the north and the middle reaches of the Lena River and its tributaries in the south-east, but also along the Yenisei River on the western side of the Platform and in the Angara River area near Irkutsk. Of all these the prime region is that of the middle Lena above Yakutsk together with its tributaries such as the Amga, Botoma and Sinyaya. This area is more accessible than the remote Anabar region and it has been intensively sampled.

The presence of these extensive and well-preserved Lower Palaeozoic carbonates, and the need to subdivide and correlate them for survey purposes, resulted in the discovery of abundant calcareous algae and cyanobacteria, along with stromatolites and archaeocyaths. All received attention. The two principal early workers were Vologdin and Maslov. Vologdin's chief interest at this time lay in archaeocyaths, but in his papers he also described the algae and cyanobacteria which he found associated with them. These included first descriptions of *Bija* and *Renalcis* in material from the Altai Mountains south-east of Krasnoyarsk (Vologdin 1932) and the first description of *Chabakovia* from Kuvandyk in the southern Urals (Vologdin 1939).

As remote areas became more accessible, palaeontologists often joined survey teams, but before World War II and sometimes afterwards too, it was normal for field geologists making reconnaissance studies, often along rivers, to provide material for study by palaeontologists back in Moscow or Leningrad. Some of these collectors gave their names to the fossils they helped discover; for example N.K. Razumovsky and E.E. Razumovskaya to *Razumovskia* Vologdin (1939). In other cases the river provided the name: the Botoma both to *Botominella* Reitlinger (1959) and to *Botomaella* Korde, the Amga to *Amgaella* Korde (1957b), and so on.

The area of the rivers Aldan, Amga, Botoma and middle Lena upstream from Yakutsk is the largest single outcrop of Lower Cambrian carbonate rocks in Siberia, and probably in the world. It is comparable in size to England. Maslov (1937) studied Lower Palaeozoic strata from this general region of the southern Siberian Platform and described the new genus *Nubecularites* (a junior synonym of *Renalcis*) from the Middle Cambrian and *Epiphyton* and *Girvanella* from the Lower Cambrian. Subsequently this area was to prove a fruitful source of material for Korde and later workers. However, World War II temporarily halted this geological research in the USSR.

Elsewhere, progress in the study of Cambrian algae and cyanobacteria was slow. The main activity in this field during the 1940s was in North America, but it was largely limited to routine reports of *Girvanella* from various localities, including the Grand Canyon (McKee and Resser 1945); Caborca, Mexico

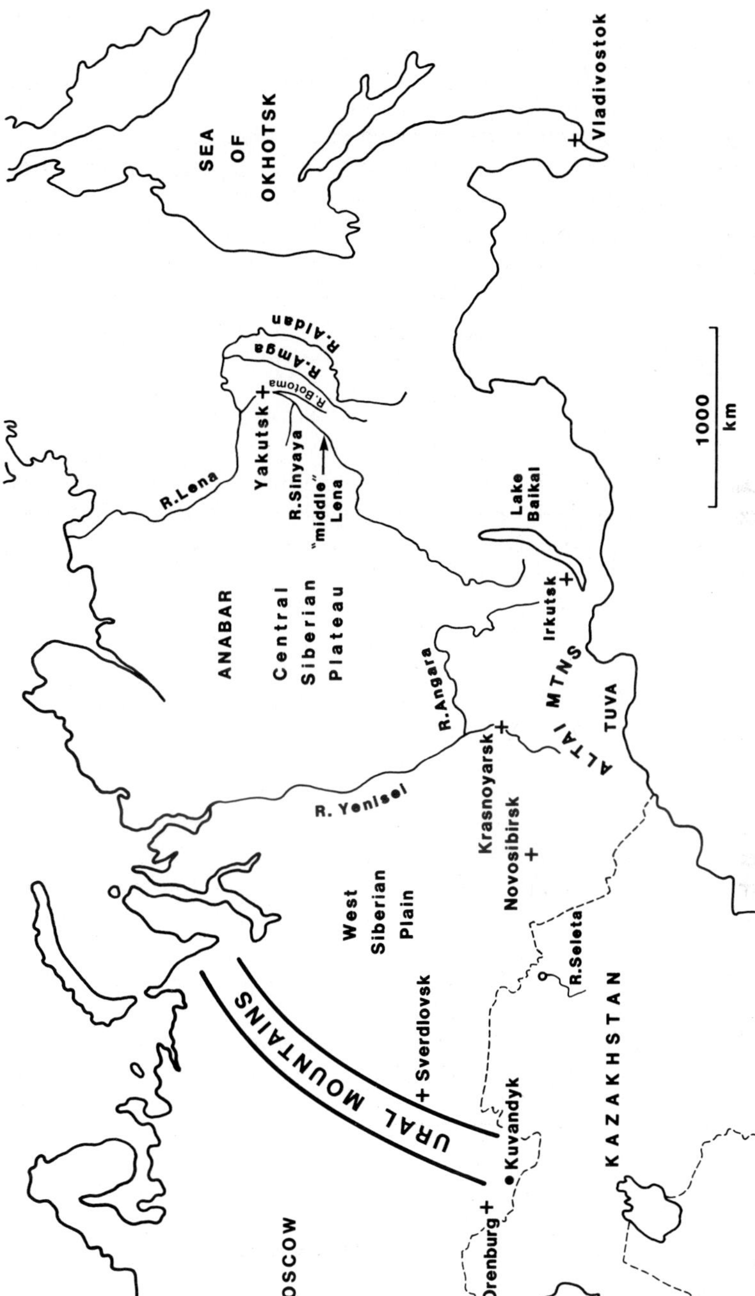

Fig. 2. Localities of Cambrian calcareous cyanobacteria and algae in Siberia and adjacent areas

(Cooper and Arellano 1946; later Johnson 1952 also worked here), and the latest Cambrian-Early Ordovician Ellenburger Group of Texas (Cloud and Barnes 1948; subsequently, Ahr (1971) described the *Girvanella* and *Renalcis* bioherms of the underlying Wilberns Formation in detail).

Following World War II, intensive work on Cambrian calcareous algae and cyanobacteria in the USSR by Korde resulted in a series of publications during the 1950s. Korde (1950a) described the unusual and very rare dasycladalean-like fossils *Seletonella* (named after the River Seleta) and *Mejerella* from the Upper Cambrian on the southern edge of the West Siberian Plain, within northern Kazakhstan; and also the somewhat similar *Cambroporella* from the Lower Cambrian of Tuva, south of the Altai Mountains (Korde 1950b). But subsequent work concentrated on *Botomaella*, *Epiphyton*, *Renalcis* and associated fossils from the Lower Cambrian of the middle Lena region (Korde 1955, 1957a, 1958a,b, 1959), and on more *Seletonella*-like genera (*Amgaella* and *Siberiella*) (later named *Yakutina* by Korde 1973) from the Middle Cambrian of the same area (Korde 1957b). About this time Reitlinger (1959) produced an atlas of Siberian microfossils which included *Obruchevella* from the Upper Cambrian or Ordovician.

This period of Korde's work culminated in an important monograph on "Cambrian algae of the south-eastern Siberian Platform", i.e. the middle Lena region (Korde 1961). In this publication were listed no less than 67 species of *Epiphyton*, 62 of them being new ones. Shortly afterwards, in 1963, Korde and Maslov co-authored sections on rhodophytes, cyanophytes and chlorophytes in the Russian fossil treatise, Osnovy Paleontologii.

Meanwhile, Maslov himself, who had worked widely throughout the Phanerozoic, had produced his impressive book on "Fossil calcareous algae of the USSR (Maslov 1956), before turning his attention to stromatolites (Maslov 1960). Vologdin's work was also moving in that direction and he illustrated much new material from the Sinian (Upper Proterozoic), Cambrian and Ordovician of the Angara (near Irkutsk) and Lena regions, emphasizing stromatolitic microfabrics (Vologdin 1962a). In the same year he published a study of Cambrian archaeocyaths and algae from the Baikal Platform (Vologdin 1962b).

There was nothing happening in the world outside the Soviet Union to compare with this level of innovative research on calcareous algae, cyanobacteria and stromatolites in general, and particularly on those from the late Precambrian and Early Palaeozoic. Dangéard and Doré (1957) in France described *Nephelostroma* (yet another junior synonym of *Renalcis*) together with *Epiphyton* and *Girvanella* from the Cambrian of Normandy (see also Doré 1962). In America, Rezak (1957) discussed the stratigraphic distribution of *Girvanella*. 1961 saw the publication of Johnson's influential book "Limestone-building algae and algal limestones", but this failed to mention, not only *Epiphyton*, but also *Renalcis* and all the other Cambrian genera described from the Soviet Union. However, this serious omission was redressed when Johnson reviewed the Cambrian algae (Johnson 1966), drawing heavily on Soviet work: more than half the illustrations in this publication being taken from either Korde, Maslow or Vologdin. Hill (1964) described *Epiphyton* and *Renalcis* associated with archaeocyaths from the Beardmore Glacier, Antarctica.

Back in the USSR, Korde was preparing her book on Cambrian algae (Korde 1973) which brought together nearly 50 genera in a comprehensive taxonomic treatment which included detailed, cellular-level reconstructions of *Epiphyton* and other fossils (many being interpreted as red algae, for which five orders and seventeen new families were introduced) and a detailed evolutionary scheme for cyanobacteria. However, a new generation of workers was emerging, with interest focusing on the Precambrian-Cambrian boundary on the Siberian Platform (Voronova 1969, 1976, 1979; Stepanova and Luchinina 1982) as well as continuing with work on the Lower Cambrian of the Siberian Platform in general (Luchinina 1975). Drosdova (1980) extended this to a study of algae in Lower Cambrian bioherms in western Mongolia, where algae and cyanobacteria close to the boundary were also described (Voronova in Voronin et al. 1982). Some reassessment of earlier work, such as Korde's proliferation of *Epiphyton* species, took place (Gudymovich 1967) and Luchinina's (1975) classification marked an important new approach by suggesting that all the principal genera belong to the cyanobacteria. But, in general, Soviet work of the 1970s applied and extended preceding lines of research, albeit with stratigraphic refinements and a rather more critical eye for problems of affinity.

Cambrian calcareous algae and cyanobacteria still did not arouse the same level of interest elsewhere, and much of the Soviet work went unutilized; nevertheless, reports of occurrences of these fossils became more common from localities as scattered as western Canada (Handfield 1971: *Epiphyton* and *Renalcis*; McIlraeth 1977: *Epiphyton*), SouthAustralia (Haslett 1976: *Renalcis*; Brasier 1976: *Renalcis*), southern Spain (Zamarreño 1977: *Epiphyton*, *Renalcis*), Labrador (James and Kobluk 1978: *Angulocellularia* identified as *Renalcis* or *Renalcis*-like; Kobluk and James 1979: *Epiphyton*, *Girvanella*, *Renalcis*, *Serligia*); Mali (Bertrand-Sarfati 1979: a *Bija*-like fossil of Cambrian aspect from presumed Vendian); Nevada (Kobluk 1981, Rowland 1981a,b: *Epiphyton*, *Renalcis*); Newfoundland (James 1981: *Epiphyton*, *Girvanella*, *Renalcis*); Virginia (Pfeil and Read 1980: *Epiphyton*, *Girvanella* and *Renalcis*); Markello and Read 1982: *Girvanella* and *Renalcis*); and Antarctica (Rees et al. 1989).

Not until the 1980s was the Soviet work on systematics closely scrutinized. Riding and Voronova (1982b,c, 1984) discussed the affinities of genera such as *Angulocellularia* and *Tubomorphophyton*, and concluded (Riding and Voronova 1982a), in agreement with Luchinina (1975), that most Early Cambrian "algae" were actually likely to be cyanobacteria and reflected enhanced calcification relative to the Late Precambrian. Riding (1982) suggested that this was due to a change in oceanic carbonate chemistry. Pratt and James (1982) and Pratt (1984) proposed that *Epiphyton* and *Renalcis* might be diagenetic products of coccoid cyanobacteria. Riding and Voronova (1985) reviewed the taxonomy of Cambrian calcareous algae and cyanobacteria from the point of view of morphology rather than affinity, and recognized 6 major groups with 19 genus groups, and discussed the morphological series which exist in several of the more common of these.

In the book *Fossil calcareous algae* (Chuvashov et al. 1987) Luchinina (1987, pp. 12–38) provides a classification of calcified cyanobacteria in which notable Cambrian genera are included (pp. 23–24). However, the section by Chuvashov (pp. 109–139) includes *Gordonophyton* and *Tubomorphophyton* and some other *Epiphyton*-like genera in the rhodophytes, which follows Korde's (1973) classification, and *Batenevia* and *Botominella* (which show some resemblances to *Girvanella*) are figured as early representatives of solenoporaceans (Chuvashov et al. 1987, Fig. IV.22). Clearly, discussion of the systematics of these fossils has not finished. Rowland and Gangloff (1988), in a review of Lower Cambrian reefs, suggested that *Renalcis* and *Epiphyton* were facultative heterotrophs which could live either heterotrophically or autotrophically and that this accounted for their occurrence in reef cavities. Luchinina and Tikhomirova (1988) show SEM photographs of *Epiphyton, Renalcis* and *Subtifloria*. It is this kind of new information, combined with Recent analogues, which offers the most promising way to clarify the affinities of Cambrian calcareous algae and cyanobacteria.

3 Systematics

Ideally, morphology should indicate affinity, and both should be reflected in classifications of these fossils. However, the interaction of these two approaches has resulted in muddled schemes. Two reasons stand out: firstly, parallelism of organization in these organisms results in misleading similarities between quite different groups; secondly, workers often appear to have sought to clarify taxonomic difficulties by accepting likely explanations of affinity for a few genera and then attempting to pull the rest into conformity. As a result, these fossils are notorious for changes of attributions of affinity. Just a slight perusal of the literature reveals how *Epiphyton*, *Renalcis* and other genera have been switched about between red algae and cyanobacteria. *Proaulopora* even has the distinction of being moved between greens (Vologdin 1962a, p. 546), reds (Korde 1973, p. 224) and cyanobacteria (Luchinina in Chuvashov et al. 1987, p. 28).

Important classificatory schemes are compared in Table 2, which shows generic groups, but for clarity omits orders, families, etc. into which they have been placed. In some instances in Table 2 the sequence of groupings has been changed to facilitate overall comparison of the schemes, but on the whole the classifications are shown as they are published. Johnson's (1966) classification is largely based upon the work of Korde (1961) and Vologdin (1962a) and, following the Russian authors, places genera such as *Renalcis*, *Angulocellularia*, *Epiphyton* and *Bija* in the red algae, and *Praulopora* and *Seletonella* in the greens. Vologdin (1962a, p. 309) placed *Girvanella* in the cyanobacteria, but Johnson (1966) preferred to also keep a chlorophyte option open for porostromates. Korde's (1973) classification was more comprehensive, notably in including spherical objects such as *Bogutschania*, *Foninia*, and *Acanthina*. Some of these were placed in cyanobacteria, but most of the genera she described were put in the red algae. This included not only *Girvanella*-like taxa such as *Nicholsonia* and *Botominella*, but also *Proaulopora*.

Riding and Voronova (1985) temporarily abandoned affinity in an attempt to reappraise the classification of these fossils on grounds of morphology alone. The result was recognition of six principal groups: (1) spherical (calcispheres, *Uranovia*, etc.); (2) botryoidal (*Renalcis*, *Tarthinia*, etc.); (3) dendritic (*Angulocellularia*, *Chabakovia*, *Epiphyton*, etc.); (4) tubiform, a) fans of tubes (*Botomaella*, *Hedstroemia*, *Solenopora*, etc.), b) tangled, coiled, and straight tubes (*Girvanella*, *Obruchevella*, *Proaulopora*, etc.); (5) tuberous (*Seletonella*, etc.); and (6) cups (radiocyathids). The cup group is omitted from Tables 2 and 3.

Chuvashov et al. (1987), in a comprehensive review of Phanerozoic fossil calcareous algae, attempted again to classify according to affinities, with the work divided between Chuvashov (red algae), Luchinina (cyanobacteria) and Shuysky (green algae). Shuysky safely places the Seletonellaceae in the greens, but there appears to be some conflict in the areas of reds and cyanobacteria. These of course are the two groups that previous workers have chiefly used in attempting to place these fossils systematically. The main difficulties in Chuvashov et al.'s (1987) treatment of Cambrian genera concern the following:

1. *Epiphyton* group; Luchinina (in Chuvashov et al. 1987, p. 23) places *Epiphyton* in the cyanobacteria, but Chuvashov (in Chuvashov et al. 1987, p. 127) puts the very similar genera *Gordonophyton*, *Tubomorphophyton*, *Epiphytonoides*, and *Filaria*, in the red algae.
2. *Girvanella* group; Luchinina places *Girvanella* in the cyanobacteria, but Chuvashov appears to place *Nicholsonia*, a junior synonym of *Girvanella*, along with other kenellaceans in the red algae.
3. *Batenevia* group; Luchinina places *Batenevia* and *Subtifloria* in the cyanobacteria, whereas Chuvashov (in Chuvashov et al. 1987, Fig. 22, p. 139) places *Batenevia* and *Botominella* (junior synonym of *Subtifloria*) in an evolutionary line with solenoporacean supposed red algae.

Despite these continuing problems of affinity (Table 4), which are only likely to be finally resolved by discovery of modern analogues, the comparison of classifications in Table 2 reveals an emerging consensus regarding major groupings. In this paper I follow Riding and Voronova's (1985) scheme based on morphological details, but here I add comments on possible affinities (Table 3). Omitted from detailed consideration here are the spherical and cup-like groups of Riding and Voronova (1985), neither of which are likely to include well-defined algae or cyanobacteria. Of the members of the spherical group, *Uranovia* are small (30 μm) micritic or hollow bodies within stromatolites; *Palaeomicrocystis* is larger (250 μm) and micritic with a clotted microfabric; *Bogutschania* is a small (500 μm) oöid; *Foninia* is large (2.75 mm) and may be an oöid; some specimens attributed to *Acanthina multiformis* (e.g. Korde 1973, Pl. V) appear to be small recrystallized oöids; *Marenita* is micritic 50–300-μm diameter objects, which may be part of oöids or stromatolite fabrics. In contrast, the cup-like group contains radiocyathids which are large, well-defined fossils

Table 2. Comparison of classification schemes, showing generic groupings[a]

VOLOGDIN 1962	JOHNSON 1966	KORDE 1973	RIDING & VORONOVA 1985	CHUVASHOV et al. 1987: LUCHININA pp 23-24	CHUVASHOV et al. 1987: CHUVASHOV pp 126-127	THIS PAPER
Renalcis Razumovskia [1]	Renalcis	Bogutschania Foninia Marenita Renalcis	SPHERICAL Foninia Marenita Globuloella BOTRYOIDAL Renalcis Tarthinia Gemma	Globuloella Palaeomicrocystis Renalcis Chabakovia Gemma Angulocellularia		Renalcis Taninia Chabakovia Tarthinia Gemma
		Pseudoanthos [7] Acanthina [7] Mucilina [7] Kundatia [7]			Pseudoanthos [7] Acanthina [7] Mucilina [7] Kundatia [7]	
Angulocellularia Chabakovia Sajania	Angulocellularia Chabakovia Sajania	Cornutula [8] Razumovskia [1] Kordephyton Kadvoya [9] Serligia [9] Taninia Chabakovia Chomustachia Parachabakovia	DENDRITIC Angulocellularia Kordephyton Taninia Chabakovia Chomustachia Parachabakovia			Angulocellularia
Epiphyton	Epiphyton	Amgaina Epiphyton Cambrina [10] Gordonophyton Tubomorphophyton	Korilophyton Epiphyton Cambrina [10] Gordonophyton Tubomorphophyton	Korilophyton Epiphyton	Gordonophyton Tubomorphophyton Epiphytonoides [16] Filaria [16]	Epiphyton Korilophyton Gordonophyton Tubomorphophyton
Dasycirriphycus [2] Bija 3	Dasycirriphycus [2] Razumovskia [1] Bija [3] Solenopora	Bija [3]	TUBIFORM I Bajanophyton Fistulella [12] (pars) Hedstroemia [3] Solenopora Bija [3]	Bija [3] Zaganolomia [15] Botomaella Hedstroemia [3]	Solenopora	Botomaella Hedstroemia [3] Bajanophyton
Girvanella	Girvanella Nephelostroma [4] Nubecularites [4]	Nicholsonia [11] Kenella Fistulella [12] Botominella [13] Batinevia [14]	TUBIFORM II Girvanella Nicholsonia [11] Batinevia Botominella [13] Subtifloria Razumovskia [1] Botomaella Kenella Obruchevella Cavifera Glomovertella	Girvanella Razumovskia [1] Obruchevella Batenevia Subtifloria	Nicholsonia [11] Kenella Fistulella [12] Botominella [13] Batenevia	Girvanella Batenevia Subtifloria [13] Obruchevella
Proaulopora	Proaulopora	Proaulopora Vologdinella [6]	Proaulopora Amganella 6 Palaeonites Tubophyllum Vologdinella [6]	Proaulopora Tubophyllum	Proaulopora Vologdinella [6]	Proaulopora
	Amgaella Cambroporella Mejerella Seletonella Siberiella [5] Vologdinella [6]	Amgaella [6] Cambroporella Mejerella Seletonella Yakutina [5]	TUBEROUS Amgaella Mejerella Seletonella	SHUYSKY pp 63-64 Amgaella Cambroporella Mejerella Seletonella		Seletonella Yakutina

Table 3. Morphological groups of Cambrian calcareous cyanobacteria and algae (adapted from Riding and Voronova 1985), with possible affinities added

Morphology		Group	Other genera	Affinities
Botryoidal		*Renalcis*	*Chabakovia*, *Chomustachia*, *Gemma*, *Parachabakovia*, *Taninia*, *Tarthinia*	Probably cyanobacteria
Bushlike		*Angulocellularia*		Filamentous cyanobacteria such as oscillatoriaceans
Dendritic		*Epiphyton*	*Gordonophyton*, *Korilophyton*, *Tubomorphophyton*[a]	Probably cyanobacteria, possibly stigonemataleans
Tubiform	branched	*Botomaella*	*Bajanophyton*, *Bija*, *Hedstroemia*, *Kenella*	Filamentous cyanobacteria such as rivularieaceans
	tangled	*Girvanella*	*Batenevia*, *Nicholsonia*, *Subtifloria*	Filamentous cyanobacteria such as oscillatoriaceans
	coiled	*Obruchevella*	*Glomovertella*, *Lukaschevella*	
	straight	*Proaulopora*		Possibly filamentous cyanobacteria
Tuberous		*Seletonella*	*Yakutina*	?Dasycladaleans

[a] And probable synonyms of *Epiphyton* (see No. 10 of Footnote, Table 2).

[a] Higher taxa are omitted for clarity; as are non-calcified and obscure genera and genera based on stromatolitic microfabrics. *1* There is some doubt concerning diagnostic features of *Razumovskia* (see Riding and Voronova 1985, p 65), and consequently it is mobile between classifications. *2 Dasycirriphycus* somewhat resembles *Renalcis*. *3 Bija* may be a junior synonym of *Hedstroemia*. *4* These genera are morphologically misplaced here, *Nephelostroma* is a junior synonym of *Renalcis*. *Nubecularites* may be a catagraph (see Riding and Voronova 1985, p 60) although it has also been tentatively placed in synonymy with *Renalcis* (Reitlinger 1959, p 13). *5 Yakutina* replaces *Siberiella* Korde 1957 (non Radchenko 1955). *6 Vologdinella* and *Amganella* are junior synonyms of *Proaulopora* (see Voronova 1976, p 79). *7 Pseudoanthos* and *Acanthina* are irregular wispy micritic patches; *A. multiformis* appears to be part of spherical objects resembling oöids (see Korde 1973, Pl. 5). *Mucilina* is somewhat similar. *Kundatia* is stick-like with traces of rectangular to ovoid chambers. *8 Cornutula* is oöid-like and about 350 μm in diameter. *9* Kadvoya and *Serligia* are apparently tubular and are not similar to other members of the Cambrinales. *10* Described by Korde (1973), but not listed here, are the following cambrinalean genera, all of which, together with *Cambrina*, are likely to be junior synonyms of *Epiphyton*: *Kyzassia*, *Filaria*, *Umbellula*, *Sajania*, *Vologdinia*, *Potentillina*, *Erbina*, *Tomentula*, *Sporinula* and *Epiphytonoides*. *11 Nicholsonia* is a junior synonym of *Girvanella* (see Danielli 1981, p 96). *12* The type-species of *Fistulella* belongs with *Botomaella* (see Riding and Voronova 1985, p 63). *13 Botominella* is a junior synonym of *Subtifloria* (see Luchinina 1975, p 28). *14* Correct spelling is *Batenevia* (see Voronova 1976, p 68). *15 Zaganolomia* resembles *Praesolenopora* Tsao. *16 Epiphytonoides* and *Filaria* are likely to be junior synonyms of *Epiphyton*. Also see *10*. *xxx* Encloses red algae; *dashes* enclose cyanobacteria; *circles* enclose green algae. Combinations of these symbols, and *question marks*, reflect the views of the original author(s). Note that Riding and Voronova (1985) gave no affinities. See Table 4 for a clearer outline of attributed affinities.

Table 4. Affinities attributed to generic groups shown in the comparison of classifications (Table 2). Riding and Voronova (1985) gave no affinities[a]

VOLOGDIN 1962	JOHNSON 1966	KORDE 1973	RIDING & VORONOVA 1985	CHUVASHOV et al. 1987 LUCHININA pp 23-24	CHUVASHOV et al. 1987 CHUVASHOV pp 126-127	THIS PAPER
		CYAN[a]	SPHERICAL	CYAN		
RED	RED	CYAN[a]	BOTRYOIDAL	CYAN		? CYAN
RED	RED	RED	DENDRITIC	CYAN	RED	CYAN ? CYAN
RED CYAN	RED GREEN OR CYAN	RED	TUBIFORM	CYAN	RED	CYAN ? CYAN
GREEN	GREEN	GREEN	TUBEROUS	SHUYSKY pp 63-64 GREEN	GREEN	? GREEN

which show some similarities with both receptaculitids and archaeocyaths (see Riding and Voronova 1985, p. 75). They do not closely resemble any of the other groups considered here. They are probably not algae.

4 Morphological Series

Riding and Voronova (1985), in describing major morphological groups of Cambrian calcareous algae, recognized the presence of morphological series within a number of them: "In some cases the series are disjunct; in others they are transitional and present serious difficulties in establishing diagnoses and separating taxa. The presence of transitional series shows that confusion in classification of these fossils is not due solely to subjective argument concerning affinity" (Riding and Voronova 1985, p. 57).

Series are most clearly seen in the *Renalcis*, *Angulocellularia*, *Epiphyton*, *Botomaella*, and *Girvanella* groups. The *Girvanella-Obruchevella-Proaulopora* series of unbranched or rarely branched tubiform fossils is discontinuous and does not present serious taxonomic difficulties. The *Botomaella-Hedstroemia-Solenopora* series (Riding and Voronova 1985, p. 71, pp. 74–75) is continuous and therefore more difficult, but elucidation of problems regarding *Solenopora* and its allies will greatly reduce these problems. As has already been noted (Introduction, above), reported Cambrian *Solenopora* are not congeneric with *Solenopora* Dybowski.

The *Renalcis-Epiphyton* ring of Riding and Voronova (1985, p. 71, pp. 73–74) can be simplified for Cambrian specimens (Devonian material may be more difficult) into an Epiphyton series (*Korilophyton-Epiphyton-Tubomorphophyton-Gordonophyton*) which is partially disjunct, and a *Renalcis-Angulocellularia* series (Fig. 3). The latter is the major problem for taxonomy because it shows apparently clear transition along its entire length. The significance of this for interpretations of affinity is still unclear. Saltovskaya (1975) has regarded *Renalcis* and *Epiphyton* as members of a single group, and Pratt (1982) has argued that these genera, and their variants, are products of diagenetic calcification of coccoid cyanobacteria. But *Angulocellularia* compares closely with calcified sheaths of modern *Schizothrix* (Riding and Voronova 1982b) which is a filamentous cyanobacterium. And yet it grades via *Tarthinia* to *Renalcis*, which may well be coccoid in origin (Hofmann 1975). Recognition of morphological series does not solve these difficulties, but it does help us to define the problems more precisely.

◄

[a] *CYAN* Cyanobacteria. Note that the most consistent attribution of affinity has been for genera regarded as green algae. In contrast, many of the Cambrian genera regarded as red algae by Vologdin (1962a) and Korde (1973), are now considered more likely to be cyanobacteria.

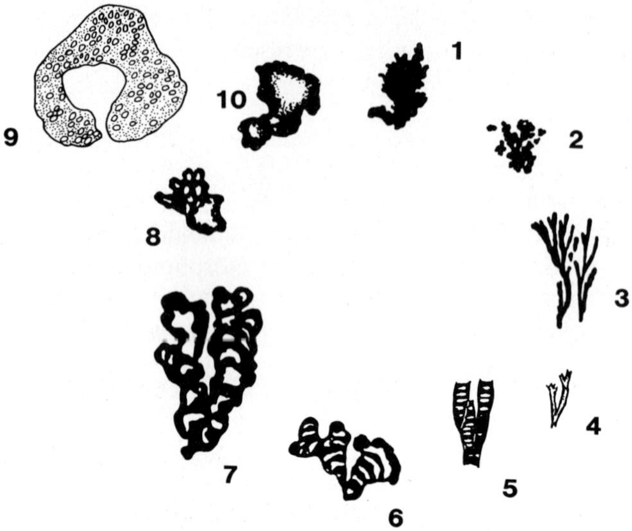

Fig. 3. Morphological series shown by the *Angulocellularia*, *Epiphyton* and *Renalcis* groups (based on Riding and Voronova 1985, Fig. 16). *1 Angulocellularia anisotoma* Vologdin (Riding and Voronova 1982a, Fig. 2c); *2 Korilophyton inopinatum* Voronova (Voronova 1976, Pl. 17, Fig. 5); *3 Epiphyton altum* Korde (Korde 1961, Pl. 17, Fig. 4); *4 Tubomorpophyton botomense* Korde (Riding and Voronova 1982b, Pl. 96, Fig. 5); *5 Gordonophyton distinctum* Korde (Korde 1973, Pl. 40, Fig. 4a); *6 Taninia tomentosa* Korde 1973, Pl. 25, Fig. 3a); *7 Renalcis seriata* Korde (Korde 1961, Pl. 4, Fig. 3); *8 Renalcis erbinatus* Korde (closely resembles *R. granosus* Vologdin) (Korde 1973, Pl. 2, Fig. 4); *9 Gemma inclusa* Luchinina (Riding and Voronova 1985, Fig. 10c); *10 Tarthinia gelatinosus* (Korde) (Korde 1961, Pl. 4, Fig. 5). All specimens are drawn at the same magnification. *Angulocellularia* (*1*) is 750 μm in longest dimension

5 Renalcis Group

Vologdin (1932) described *Renalcis* from limestone believed to be of Middle Cambrian age in the Altai Mountains. The type locality is the Kandalep River, south-east of Krasnoyarsk. The type species, *R. granosus* (Fig. 4f) shows clusters of thick-walled botryoidal chambers with the larger ones enveloping smaller ones. As with so many Soviet Cambrian fossils, the type-specimen appears to be lost and it is unlikely that the type-locality can be accurately pinpointed. Consequently, the original illustration by Vologdin (1932, Pl. 1, Fig. 1) is all that we have, apart from the description, to diagnose the genus. *Renalcis* is one of the best known of Cambrian calcified cyanobacteria and algae. It is very common in Lower Cambrian bioherms, often in association with *Epiphyton* and archaeocyaths. Indeed, Vologdin (1932, p. 79) states "The limestones from the rivers Kandalep, Karagan and Lebed often contain in association with the Archaeocyathinae considerable quantities, nearly equalling those of *Epiphyton*, of small vesicular remains referred by the author to a special genus of algae, *Renalcis* gen. nov.".

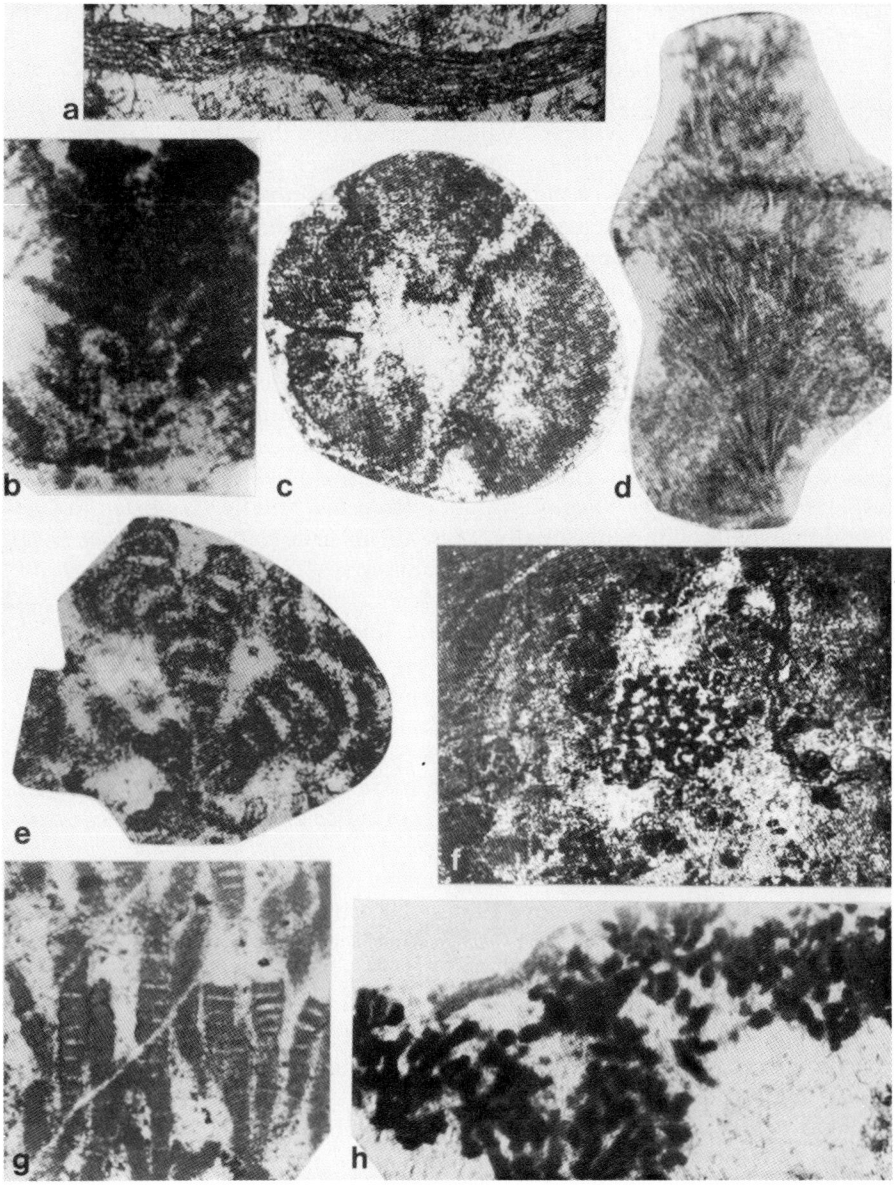

Fig. 4a-h. Original illustrations of some Cambrian calcified cyanobacteria and algae. **a** *Subtifloria delicata* Maslov (1956, Pl. XXVII, Fig. 4). The genotype, Lower Cambrian, Tuva. x27; **b** *Angulocellularia anisotoma* Vologdin (1962a, Pl. 2, Fig. 2). The genotype, Lower Cambrian, River Yangud, Baikal. x15; **c** *Tarthinia rotunda* Drosdova (1980, Pl. III, Fig. 5). Lower Cambrian, western Mongolia. x40; **d** *Botomaella zelenovii* Korde (1961, Pl. I, Fig. 7). Lower Cambrian, River Botoma. x47; **e** *Taninia tomentosa* Korde (1973, Pl. XXV, Fig. 3a). The genotype, Lower Cambrian, Kuznetskii Alatau. x50; **f** *Renalcis granosus* Vologdin (1932, Pl. 1, Fig. 1). The genotype. Middle Cambrian, River Kandalep, Altai Mountains. x7; **g** *Gordonophyton distinctum* Korde (1973, Pl. XL, Fig. 4a). The genotype. Lower Cambrian, Tuva. x36; **h** *Korilophyton inopinatum* Voronova (1976, Pl. XVII, Fig. 6). Nemakit-Daldyn Horizon, River Kotui. x36

This association between *Renalcis*, *Epiphyton* and archaeocyaths has often been called the algal-archaeocyath association, even though it is more likely that both *Renalcis* and *Epiphyton* are cyanobacteria rather than algae. The confusion arises because cyanobacteria have hitherto long been termed blue-green algae. Although the illustration of the type-specimen does not show it, *Renalcis* often has short irregular branches composed of the broad, inflated, thick-walled chambers. A number of somewhat similar genera occur higher in the Palaeozoic, particularly in the Devonian, which have been confused with *Renalcis*. These include *Izhella* Antropov and *Shuguria* Antropov. Riding and Brasier (1975), in suggesting a foraminiferal affinity for *Renalcis*, were actually describing a Devonian specimen of *Shuguria*. Differences in wall thickness and morphology distinguish these genera from *Renalcis*: *Izhella* has very thick walls with internal V-shaped clefts; *Shuguria* has thin walls. In addition, *Shuguria* has more numerous chambers than either *Izhella* or *Renalcis*. Luchinina in Chuvashov et al. (1987, p. 37) shows *Shuguria* ranging from the Middle Cambrian to Carboniferous, although I personally know of no firm records of *Shuguria* (or *Izhella* for that matter) prior to the Devonian.

Nevertheless, other *Renalcis* group genera such as *Taninia*, *Tarthinia* and *Chabakovia* do occur in the Cambrian, and *Taninia* closely resembles *Shuguria* as do two other Cambrian genera: *Chomustachia* and *Parachabakovia*. The internal taxonomy of this group requires refinement, but it forms a distinctive class of chambered fossils with an overall botryoidal to stubbily branched appearance. Two morphological series occur in relation to this group (Fig. 3): (1) botryoidal fossils (e.g. *R. granosus*) grade towards branched fossils (e.g. *Taninia*, Fig. 4e); (2) chambered fossils (the entire group) grade via *Tarthinia* (Fig. 4c) towards solid micritic fossils (e.g. *Angulocellularia*, Fig. 4b). *Tarthinia* includes *T. gelatinosus*, originally described by Korde (1961, p. 57) as a *Renalcis* species (see Riding and Voronova 1984, p. 206), but the genus was described by Drosdova (1975) from the Lower Cambrian of western Mongolia. Latham and Riding (1988) recognized Tarthinia, together with *Renalcis*, forming the macroscopic clots of thrombolites in probable Lower Cambrian of Morocco, and this was the first report of *Tarthinia* outside Mongolia and Siberia.

A distinctive additional genus which can tentatively be assigned to the *Renalcis* group is *Gemma*. In published reports it is only known from the remote Anabar region of north-central Siberia (Luchinina in Zhuravleva et al. 1982, p. 100) and from western Mongolia (Riding and Voronova 1985, Fig. 10c). It has a very thick wall composed of numerous micritic spherules.

The affinities of *Renalcis* and similar genera are a perennial problem. The simplicity, and variability, of form, coupled with relatively small size, suggest cyanobacteria. Although Vologdin (1962a) and Korde (1973) placed *Renalcis* in the Rhodophyta, Korde (1961, p. 55) did refer it to the cyanobacteria and in Maslov and Korde (1963, p. 51) suggested a rivulariacean affinity. Saltovskaya (1975), concentrating on Devonian material, also placed *Renalcis*, along with *Epiphyton*, in the filamentous cyanobacteria (Hormogonophyceae), and Luchinina (1975) working with Cambrian material, also favoured a cyanobacterial affinity for Renalcis. Hofmann (1975) suggested that *Renalcis* represents diagenetic calcification of chroococcalean colonies, an idea developed by Poncet (1976; who actually illustrated Devonian *Izhella*) to include a rivulariacean affinity. Pratt (1984) also supported both a coccoid cyanobacterial and a diagenetic origin for *Renalcis* (and also *Epiphyton*) but concentrated on non-Cambrian material. Luchinina (in Chuvashov et al. 1987, p. 24) also favours a chroococcalean affinity for Renalcis.

Overall, then, opinion supports a cyanobacterial (chroococcalean or rivulariacean) affinity for *Renalcis*, and Hofmann's (1975) suggestion of peripheral calcification of colonies most plausibly accounts for the appearance of this enigmatic fossil. The problem that remains is that *Renalcis* and its allies appear to have a stratigraphic range limited to the Palaeozoic. Is this significant, and do modern analogues exist for *Renalcis*? These questions need to be addressed in future research.

6 Angulocellularia Group

This genus, characterized by being bushy in form and densely micritic in structure, is much more common than published reports indicate. Vologdin (1962a, p. 484; Fig. 4b) described it from the Lower Cambrian of the River Yangud, in the Baikal region. Riding and Voronova (1982b) emended the diagnosis to exclude chambers (which are not visible in Vologdin's photograph of the type-specimen although he gives a drawing showing them (Vologdin 1962a, Fig. 4)), and described a Recent analogue of *Angulocellularia* in calcified sheaths of the oscillatoriacean *Schizothrix calcicola*.

Angulocellularia shows transition towards *Tarthinia* in the *Renalcis* group, and towards *Epiphyton*. For example, *Korilophyton angustum* (Voronova 1976, p. 84), which shows some similarities with *Epiphyton*, has been placed in *Angulocellularia* (Riding and Voronova 1984, p. 206). *Korilophyton* itself, based on *Epiphyton inopinatum*, remains a distinct genus within the *Epiphyton* group. However, it is towards *Korilophyton*, which shows short finger-like branches but is less dendritic than *Epiphyton*, that *Angulocellularia* shows transition (Fig. 3).

7 Epiphyton Group

Epiphyton and its allies constitute perhaps the most important single group of Cambrian calcareous algae or cyanobacteria. *Epiphyton* is so named because Bornemann (1886) found it epiphytically attached to archaeocyaths. *Epiphyton* and its relatives, and also *Renalcis*, can be so abundant that despite their individually small size they form macroscopic dendritic masses (dendrolite, Riding 1988) which may dominate Cambrian reef fabrics.

There are too many species of *Epiphyton*, most of them described by Korde (1961) from the Lower Cambrian of the middle Lena River and the Amga River south-west of Yakutsk, as well as from the Kuznetsky Alatau mountains midway between Novosibirsk and Krasnoyarsk. The species are difficult to distinguish from one another because they are often based on obscure or subtle details of microstructure or branching angle. There has also been a proliferation of genera within the group, again by Korde (1973): *Cambrina*, *Kyzassia*, *Filaria*, *Umbellula*, *Sajania*, *Vologdinia*, *Potentillina*, *Erbina*, *Tomentula*, *Sporinula* and *Epiphytonoides* are all likely to be junior synonyms of *Epiphyton*. These are all based on Lower Cambrian material, and are all from the Altai Mountains (except for *Erbina* from Kuznetsky Alatau); most (the exceptions are *Kyzassia*, *Potentillina* and *Epiphytonoides*) being from the River Bazaikha in the eastern Altai.

Nevertheless, it is not so difficult to recognize several important genera. These are *Epiphyton* (solid micritic, dendritic), *Korilophyton* (solid micritic, but with short branches), *Gordonophyton* (large size, regular discoidal or lenticular transverse chamberlike gaps in the filaments, and often showing gradual distal expansion of filaments), and *Tubomorphophyton* (tubiform or chambered). Of these *Epiphyton* and *Gordonophyton* appear to be most widely distributed. *Gordonophyton* was first described as *Epiphyton* by Gordon (1920) from the Cambrian of Antarctica, and then created by Korde (1973, p. 198) using material (*G. distinctum*) from Tuva (Fig. 4g). It also occurs in the Upper Cambrian of Newfoundland (Pratt 1984, Fig. 5b). *Korilophyton* and *Tubomorphophyton* are not yet known from outside Siberia and Mongolia (see Riding and Voronova 1984). *Korilophyton* (Fig. 4h) is based on material from the Koril Member near the base of the Tommotian in Anabar, which was originally referred to *Epiphyton* (see Voronova 1976, pp. 83–84). *Tubomorphophyton* too, is based on the originally *Epiphyton* species, *E. botomense*, from the middle Lena River (Korde 1973, p. 204).

The idea that *Epiphyton* is a red alga stems from Korde's (1959) report of cells preserved within the filaments, but this observation has never been substantiated. Earlier workers suggested a chlorophyte (Bornemann 1886) or cyanophyte (Pia 1927, p. 39) affinity. Saltovskaya (1975) and Luchinina (1975) both regarded *Epiphyton* as a filamentous cyanobacterium. Riding and Voronova (1982c) compared *Tubomorphophyton* with both red algae and cyanobacteria and concluded that a cyanobacterial affinity is the more likely of the two and that some stigonemataleans, in particular, closely resemble epiphytaceans. Pratt (1984) suggested that *Epiphyton* and *Renalcis* are diagenetic microfossils produced by coccoid cyanobacteria. However, the filamentous form of *Epiphyton*, which is often very well-defined, does not seem to be consistent with such an origin. In particular, it has not been explained how coccoid cells would create such a regularly branched dendritic pattern of filaments, nor how diagenetic calcification (as opposed to calcification during life) would preserve the morphology so well. Luchinina (in Chuvashov et al. 1987, p. 23) places *Epiphyton* and *Korilophyton* in filamentous cyanobacteria, but Chuvashov (in Chuvashov et al. 1987, p. 127) places *Tubomorphophyton* and *Gordonophyton* in the Rhodophyta. Recent analogues are required to resolve these continuing uncertainties concerning the affinities of this important group of Cambrian microfossils.

8 Botomaella Group

Botomaella (Fig. 4d), named by Korde (1958b) after the Botoma River, and *Kenella* have delicate tubular branched filaments characteristically arranged in fan-like radiating clusters. *Hedstroemia* (originally described from the Silurian by Rothpletz 1913) is similar in general form, but has filaments which are both wider and more variable in width. In the Cambrian, *Botomaella* and *Hedstroemia* typically occur as small isolated specimens and do not appear to be associated with reefs. Whereas *Botomaella* has been reported from several areas of central Siberia (see Riding and Voronova 1984), and also from western Mongolia (Drosdova 1980, p. 18), *Hedstroemia* has hitherto mainly been described in the Cambrian as *Bija* or *Rothpletzella* Wood. *Bija* is difficult to clarify because the original material was only illustrated in transverse section (Vologdin 1932, Fig. 11), but it could be a junior synonym of *Hedstroemia*. *Rothpletzella igarcaensis* and *R. flabellata* described by Voronova (1976, p. 82) are not congeneric with *Rothpletzella* Wood, but resemble irregular *Hedstroemia* (see Riding and Voronova 1984, p. 206). *Bajanophyton*, described by Drosdova (1980, p. 43) from the Lower Cambrian of western Mongolia, appears to be intermediate between *Botomaella* and *Hedstroemia*.

Despite these confusions, the affinities of *Botomaella* and *Hedstroemia* can be clarified by comparison with Recent calcified cyanobacteria. These modern analogues (Riding, unpublished) show that the two genera are closely related, both being very similar to species of *Rivularia* (Roth) Agardh. Consequently, and rather surprisingly, we have an example here of species within a single extant genus being more readily separated as fossils into two morphogenera. Normal expectation would be the reverse: that fossil taxa tend to combine several extant ones, as is probably the case with *Girvanella* (Riding 1977, p. 43).

9 Girvanella Group

Girvanella is probably the best known, by name, of all these Cambrian fossils, mainly because of its long post-Cambrian range. But it is also a very widely reported genus within the Cambrian, particularly so in North America (see Sect. 2), although it figures less prominently in Soviet work. Originally described from the Ordovician of Scotland by Nicholson and Etheridge (1878), *Girvanella* characteristically occurs as loosely or densely tangled masses of fairly thin-walled unbranched tubes. It also commonly forms the cortex in oncoids and occurs as crusts within reefs.

Korde (1973, p. 212) distinguished *Nicholsonia* from *Girvanella* and named it after H.A. Nicholson using Lower Cambrian specimens from Tuva. But Danielli (1981, p. 96) placed it in junior synonymy with *Girvanella*. However, it should be noted that one of Korde's (1973) *Nicholsonia* species, *N. grandis*, has tube diameters of 50 μm which is large by *Girvanella* standards.

Two genera which can be included in the *Girvanella* group are *Batenevia* (original spelling *Batinevia*, but see Voronova 1976, p. 68) and *Subtifloria* (Fig.

4a) (senior synonym of *Botominella*, see Luchinina 1975, p. 28). In contrast to *Girvanella* their tubes, although similar in size to *Girvanella* and also unbranched, are aligned in thick cable-like strands, and *Batenevia* has ovoid spaces within the strands. Danielli (1981, p. 96) placed *Batenevia* and *Botominella* in synonymy with *Girvanella*, but their distinctive arrangement of tubes make it useful to continue to recognize them as separate genera.

Another distinctive variant of gross morphology within this group is the development of delicate, wispy prostrate threads which turn upwards at one end and which may be *Razumovskia* (see Riding and Voronova 1985, p. 65).

Cambrian *Girvanella* has been regarded as a cyanobacterium (Vologdin 1962a), a cyanobacterium or chlorophyte (Johnson 1966) and, in the case of *Nicholsonia*, a rhodophyte (Korde 1973) (see Table 4). However, a Recent analogue has been reported in the calcified sheaths of *Plectonema* from freshwater pools in Aldabra Atoll (Riding 1977), and it is likely that members of the *Girvanella* group as a whole have an oscillatoriacean affinity.

10 Obruchevella Group

Another group of tubiform microfossils, whose dimensions suggest that they could be cyanobacterial sheaths, exhibits coiling. Typical of these is *Obruchevella* which Reitlinger (1948) originally described as a foraminifer from the Lower Cambrian of the River Sinyaya, tributary to the middle Lena, although later (Reitlinger 1959) she considered it to be of uncertain systematic position. Elias (1954, p. 52, footnote) regarded it as an alga. Similar genera include *Cavifera* and *Glomovertella*, also described by Reitlinger (1948), from the middle Lena area, and *Bostrychosaria*, *Flexurella*, *Kordeella*, *Lebedevaella*, *Lukaschevella*, and *Tuvaellina* from the Lower Cambrian of Tuva (Vologdin 1958). Further details are given in Loeblich and Tappan (1964, p. C787–C789) who regarded the systematic position of these genera as uncertain. The spiral arrangement of these fossils, particularly well seen in *Obruchevella* and *Lukaschevella*, may be compared with filamentous cyanobacteria, including rivulariaceae such as *Dichothrix* (see Rippka et al. 1979, Fig. 63) and oscillatoriaceans such as *Spirulina* (Luchinina 1975, p. 11; Luchinina in Chuvashov et al. 1987, Fig. II.17, and see Fritsch 1959, p. 832).

Obruchevella is a calcareous fossil (see Reitlinger 1948; Riding and Voronova 1984, p. 206) but phosphatized (e.g. *Spirellus* Jiang, see Peel 1988) and silicified fossils with similar appearance are also known and may occur in the Precambrian, thus complicating evaluation of both the taxonomy and range of these organisms.

11 Proaulopora Group

Proaulopora was originally described by Vologdin (in Krasnopeeva 1937) from the Lower Cambrian of the western Altai Mountains. In contrast to the *Obruchevella* and *Girvanella* groups, the tubes of *Proaulopora* do branch and are straighter and stick-like in appearance. It is also larger, generally having a diameter of 50–100 μm. A distinctive feature is the presence of whorl-like external collars which extend from the outer surface at an acute angle to the stem. Occasionally, *Proaulopora* occurs complete as branched masses (see Korde 1973, Pl. 53, Fig. 2), but more commonly it is broken into short fragments, some of which have received separate names (e.g. *Amganella*). Luchinina (1975, p. 26) regarded *Amganella*, *Palaeonites*, *Tubophyllum*, and *Vologdinella* as junior synonyms of *Proaulopora*. As with *Obruchevella*, *Proaulopora* does not appear to be a very common fossil although it is locally conspicuous as fragments in grainstones.

Proaulopora has the distinction, rare even among Cambrian fossils, of having been referred at various times to the green algae, red algae, and cyanobacteria (Table 4). Luchinina (in Chuvashov et al. 1987, p. 28) compares it with *Calothrix gypsophyla*, a rivulariacean species with similar collar-like projections of the sheath, although she notes that *Proaulopora* is more than twice the diameter of *Calothrix*.

12 Seletonella Group (Fig. 5)

Seletonella is a rare fossil which may be a dasycladalean alga. It has a hollow, tuberous, form with numerous closely packed branches. *Seletonella* is up to 20 mm in size and thus is relatively large when compared with most Cambrian calcareous algae and cyanobacteria. Korde (1950a) described *Seletonella* and *Mejerella* from the Upper Cambrian of northern Kazakhstan, south of Omsk. *Cambroporella* comes from the Lower Cambrian of Tuva (Korde 1950b), and *Amgaella* and *Siberiella* from the Middle Cambrian of the Amga River (Korde 1957b). The name *Siberiella* was preoccupied, and consequently it was renamed *Yakutina* (Korde 1973). In 1971 Korde (quoted in Korde 1973, p. 239) united these genera, which she regarded as dasycladaleans, in the Seletonellaceae. They are all very rare. Korde (1961, p. 63) described a second *Seletonella* species, *S. grandis*, from the Ordovician of northern Kazakhstan.

Elias (1954) reinterpreted *Cambroporella* as a probable bryozoan and compared it with *Coeloclema* Nickles & Bassler. No one seems to have seriously questioned the dasycladalean affinity postulated for the other genera, although the polygonal elements in *Amgaella* (see Korde 1961, Pl. 5, Fig. 2) seem atypical and *Mejerella* (see Korde 1961, Pl. 6, Fig. 6) has unusual long flexuous branches. Bassoullet et al. (1979, p. 432, Footnote) expressed reservations concerning the dasycladalean nature of these Cambrian fossils but nevertheless included them in the Seletonellaceae, which they regarded as one of the three families (the others being Dasycladaceae and Acetabulariaceae) of Dasycladales.

However, *Seletonella* and *Yakutina* may more plausibly be dasycladaleans. These fossils from the Upper and Middle Cambrian respectively must therefore be seriously considered to be not only the oldest known calcified dasyclads, but also the oldest calcified green algae because *Proaulopora* seems unlikely to be a chlorophyte, and at present there are no other serious contenders for this affinity among Cambrian calcareous fossils.

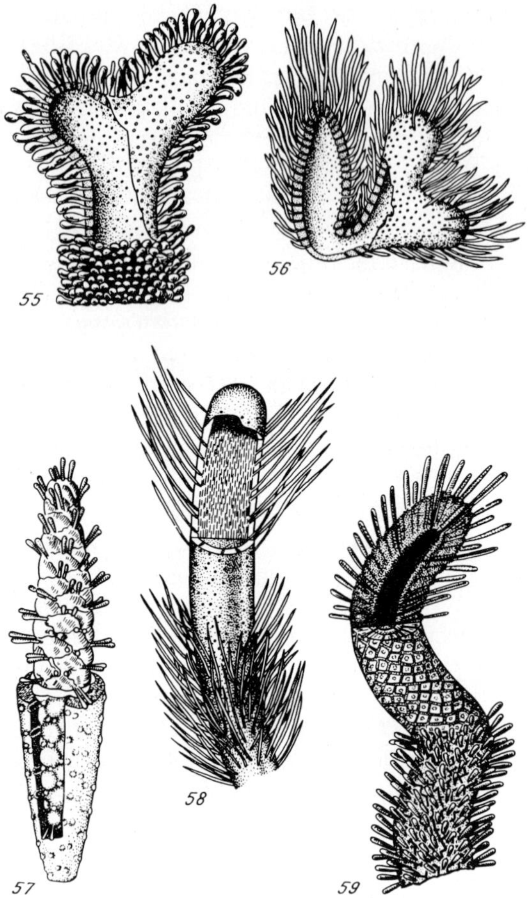

Fig. 5. *Seletonella* group. Reconstructions of *Seletonella mira* Korde (*55*), *Mejerella vamosa* Korde (*56*), *Cambroporella tuvensis* Korde (*57*), *Yakutina aciculata* Korde (*58*), and *Amgaella amganensis* Korde (*59*). Reproduced from Korde (1973, p 73)

13 Conclusions

There are some 20 well-defined, senior-synonym genera of Cambrian calcareous cyanobacteria and algae of a total of more than 70 which have been described. Seventeen of these twenty have originally been recognized in the USSR and most have yet to be reported from elsewhere. Affinities are a major problem, but progress is being made in narrowing down the possibilities. At present it seems likely that all the fossils considered here are either definitely or probably cyanobacteria with the exception of *Proaulopora*, which may only possibly be a cyanobacterium, and *Seletonella* and *Yakutina* which could well be dasycladalean green algae. No calcified red algae are known from the Cambrian. This apparent domination of the flora by cyanobacteria is a striking feature, which is enhanced by the scarcity of algae. It seems reasonable to conclude that calcareous algae, sensu stricto, are not at all common prior to the Ordovician, and that the Cambrian, particularly the Lower Cambrian, represents the major time in the entire stratigraphic record of calcification of marine cyanobacteria. This event may have been switched on by a change in seawater chemistry at the end of the Precambrian which promoted $CaCO_3$ precipitation.

Information concerning the Lower Cambrian flora and the first appearances of genera are quite well documented. The appearance of heavily calcified cyanobacteria at the base of the Cambrian is no less abrupt than the synchronous appearance of diverse invertebrate shelly fossils (Fig. 6). Much of this knowledge comes from work in the Soviet Union on material from Central Siberia and from the Urals and the Altai Mountains. All the major genera, with the exception of *Epiphyton* and *Girvanella*, have been defined in the USSR. The upward limits of the ranges of many genera are poorly known, and there is relatively little detailed information concerning the Middle and Upper Cam-

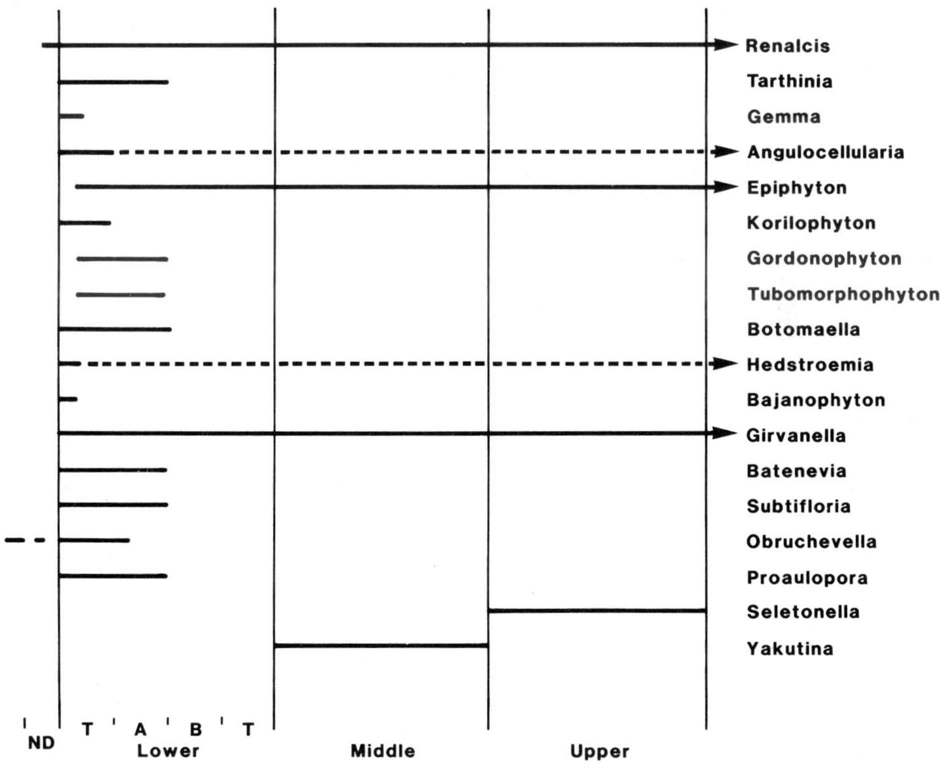

Fig. 6. Ranges of principal genera of Cambrian calcareous cyanobacteria and algae. Based on Riding and Voronova (1984) and other references cited in the text. Note that the precise ranges of *Yakutina* and *Seletonella*, within the Middle and Upper Cambrian respectively, are not known. *ND* Nemakit Daldyn; *T* Tommotian; *A* Atdabanian; *B* Botomian; *T* Toyonian

brian (Fig. 6). Cambrian floras have long been known in other parts of the world, particularly Antarctica, Australia, Europe and North America, but none of these has received the detailed study given to Soviet material. Considerable progress, particularly with regard to palaeobiogeography, can be expected when work in the rest of the world catches up with that in the Soviet Union.

The principal groups can be defined by simple morphological categories: botryoidal (*Renalcis* group), dendritic (*Angulocellularia* and *Epiphyton* groups), tubiform (*Botomaella*, *Girvanella*, *Obruchevella*, and *Proaulopora* groups), tuberous (*Seletonella* group). Discontinuous morphological series occur in several groups and the *Renalcis* and *Angulocellularia* groups show continuous series of morphological variation which make taxonomic definition difficult. Individually, all of these fossils are essentially microscopic, with the exception of the *Seletonella* group. However, members of the *Renalcis*, *Angulocellularia* and *Epiphyton* groups commonly occur in concentrated masses, forming reefs in association with archaeocyaths, which impart a macroscopic dendritic fabric to the rock which can then be termed dendrolite. These reef-forming genera, together with *Girvanella*, are the most widespread and abundant members of the flora. Members of the *Botomaella*, *Obruchevella* and *Proaulopora* groups are less common, and members of the *Seletonella* group are extremely rare, often only being known by single occurrences. The Cambrian represents the first occurrence of a heavily calcified flora, indicating that biomineralization in algae and, particularly, cyanobacteria was broadly synchronous with that in invertebrate groups.

Acknowledgements. This research was strongly supported by The Royal Society and the USSR Academy of Sciences, through their jointly operated exchange programme, and also by the Natural Environment Research Council. I am very grateful to Nina Drosdova, K.B. Korde, Veronica Luchinina and, particularly, Larisa Voronova for advice and discussion. A. Yu. Rozanov and L. Tatarinov helped facilitate my visits to the Palaeontological Institute in Moscow, without which this work would not have been accomplished.

References

Ahr WM (1971) Paleoenvironment, algal structures and fossil algae in the Upper Cambrian of central Texas. J Sediment Petrol 41:205–216

Bassoullet J-P, Bernier P, Deloffre R, Génot P, Jaffrezo M, Vachard D (1979) Essai de classification des Dasycladales en tribus. Bull Centre Rech Explor Prod Elf-Aquitaine 3, 2:429–442

Bertrand-Sarfati J (1979) Une algue inhabituelle verte, rouge ou bleue dans une formation dolomitique présumée d'âge Précambrien supérieur. Bull Centre Rech Explor Prod Elf-Aquitaine 3, 2:453–461

Blackwelder E (1907) Researches in China. Carnegie Inst Washington 1, 2:378–379

Bornemann JG (1886) Die Versteinerungen des Cambrischen Schichten-systems der Insel Sardinien nebst vergleichenden Untersuchungen über analoge Vorkommnisse aus andern Ländern. 1. Nova Acta Kaiserl Leopold-Carol Dtsch Akad Naturforsch 51, 1:147 pp

Brasier MD (1976) Early Cambrian intergrowths of archaeocyathids, *Renalcis*, and pseudostromatolites from south Australia. Palaeontology 19:223–245

Chapman F (1907–1908) On the relationship of the genus *Girvanella* and its occurrence in the Silurian limestones of Victoria. Australas Assoc Adv Sci Rep 11:377–386

Chapman F (1916) Geology. Report of the British Antarctic Expedition, vol 2, pp 81–83

Chapman F, David TWE, Priestley RE (1914) Geology. Report of the British Antarctic Expedition, vol 1, p 241
Chuvashov BI, Luchinina VA, Shuysky VP et al. (1987) Fossil calcareous algae. USSR Acad Sci, Siberian Sec, Trans Inst Geol Geophys 674:225 pp (in Russian)
Cloud PE Jr, Barnes VE (1948) The Ellenberger Group of central Texas. Texas Univ Bur Econ Geol 4621:473 pp
Cooper GA, Arellano ARV (1946) Stratigraphy near Caborca, northwest Sonora, Mexico. Am Assoc Petrol Geol Bull 30:606–611
Dangéard L, Doré F (1957) Observations nouvelles sur les Algues et les Stromatolithes du Cambrien de Carteret (Manche). Soc Geol Fr Bull 7, 8–9:1069–1075
Danielli CM (1981) The fossil alga *Girvanella* Nicholson and Etheridge. Bull Br Mus Nat Hist Geol Ser 35, 2:107 pp
Doré F (1962) Stromatolithes cambriens des synclinaux de May, d'Urville et de la zone bocaine. Soc Geol Fr CR 6:179–181
Drosdova NA (1975) Lower Cambrian algae from Tsakhir Hills (western Mongolia). Fossil fauna and flora of Mongolia. Transactions of the joint Soviet-Mongolian paleontological expedition, vol 2. Nauka, Moscow, pp 303–305 (in Russian)
Drosdova NA (1980) Algae in Lower Cambrian organic mounds of west Mongolia. Transactions of the joint Soviet-Mongolian palaeontological expedition, vol 10. Nauka, Moscow, 140 pp (in Russian)
Dybowski W (1878) Die Chaetetiden der ostbaltischen Silur-Formation. Phys Math Fac Kaiserl Univ Dorpat, 131 pp
Elias MK (1954) *Cambroporella* and *Coeloclema*, Lower Cambrian and Ordovician bryozoans. J Palaeontol 28:52–58
Fritsch FE (1959) The structure and reproduction of the algae, vol 2. Univ Press, Cambridge, 939 pp
Garrett P (1970) Deposit feeders limit development of stromatolites. In: AAPG Annu Convention, Calgary, Abstr. Am Assoc Petrol Geol Bull 54:848
Gordon WT (1920) Scottish National Antarctic Expedition, 1902–1904: Cambrian organic remains from a dredging in the Weddell Sea. Trans R Soc Edinburgh 52:681–714
Gudymovich SS (1967) Calcareous algae from the Anastasyinskaya and Ungutskaya Suites of Upper Precambrian (?)-Lower Cambrian age in the NW of east Sayan. In: Fossil algae of the USSR. USSR Acad Sci Siberian Sec, Inst Geol Geophys. Nauka, Moscow, 148 pp (in Russian)
Handfield RC (1971) Archaeocyatha from the Mackenzie and Cassiar Mountains, Northwest Territories, Yukon Territory and British Columbia. Geol Surv Can Bull 201:119 pp
Haslett PG (1976) Lower Cambrian stromatolites from open and sheltered intertidal environments, Wirrealpa, South Australia. In: Walter MR (ed) Stromatolites. Elsevier, Amsterdam, pp 565–584
Hazzard JC, Crickmay CH (1933) Notes on the Cambrian rocks of the eastern Mohave Desert, California. Univ Cal Publ Bull Dep Geol Sci 23, 2:57–80
Hill D (1964) Archaeocyatha from loose material at Plunkett Point at the head of Beardmore Glacier. Univ Queensl Repr Dep Geol 106:609–622
Hofmann HJ (1975) Stratiform Precambrian stromatolites, Belcher Islands, Canada: relations between silicified microfossils and microstructure. Am J Sci 275:1121–1132
James NP (1981) Megablocks of calcified algae in the Cow Head Breccia, western Newfoundland: vestiges of a Cambro-Ordovician platform margin. Geol Soc Am Bull 92, 1:799–811
James NP, Kobluk DR (1978) Lower Cambrian patch reefs and associated sediments: southern Labrador, Canada. Sedimentology 25:1–35
Johnson JH (1952) *Girvanella* in Cambrian stratigraphy and paleontology near Caborca, northwestern Sonora, Mexico. Smithsonian Misc Coll 119:24–26
Johnson JH (1961) Limestone-building algae and algal limestones. Johnson, Boulder, Col, 297 pp
Johnson JH (1966) A review of the Cambrian algae. Col Sch Mines Q 61, 1:162 pp
Kobluk DR (1981) Earliest cavity-dwelling organisms (coelobionts), Lower Cambrian Poleta Formation, Nevada. Can J Earth Sci 18:669–679
Kobluk DR, James NP (1979) Cavity-dwelling organisms in Lower Cambrian patch reefs from southern Labrador. Lethaia 12:193–218
Korde KB (1950a) Algal remains from the Cambrian of Kazakhstan. USSR Acad Sci Rep N S 73, 4:809–812 (in Russian)

Korde KB (1950b) Dasycladaceae from the Cambrian of Tuva. USSR Acad Sci Rep N S 73, 2:371–374 (in Russian)
Korde KB (1955) Algae from the Cambrian deposits of the rivers Lena, Botoma and Amga. USSR Acad Sci Palaeontol Inst Trans 56:79–91 (in Russian)
Korde KB (1957a) On the recognition of fossil blue-green algae. MOIP Bull Geol 2:164–165 (in Russian)
Korde KB (1957b) New representatives of siphonous algae. Mater Principles Palaeontol 1:67–75 (in Russian)
Korde KB (1958a) On the systematics of fossil Cyanophyceae. Mater Principles of Palaeontol 2:99–111 (in Russian)
Korde KB (1958b) On some aspects of fossil blue-green algae. Mater Principles of Palaeontol 2:112–119 (in Russian)
Korde KB (1959) Morphology and systematic position of representatives of the genus *Epiphyton*. USSR Acad Sci Rep N S 126, 5:1087–1089 (in Russian)
Korde KB (1961) Cambrian algae of the south-eastern Siberian Platform. USSR Acad Sci Trans Paleontol Inst 89:148 pp (in Russian)
Korde KB (1973) Cambrian algae. USSR Acad Sci Trans Paleontol Inst 139:349 pp (in Russian)
Krasnopeeva PS (1937) Algae and archaeocyaths of the oldest layers of the technical survey of Khakassia. Mater Geol Krasnoyarsk Region 3:44 pp (in Russian)
Latham A, Riding R (1988) Thrombolites formed by calcified cyanobacteria in the Lie de Vin Series, Morocco. In: 9th Int Assoc Sedimentologists Regional Meet, Leuven, Abstr, pp 127–128
Loeblich AR Jr, Tappan H (1964) Sarcodina, chiefly "thecamoebians" and Foraminiferida. In: Moore RC (ed) Treatise on invertebrate paleontology, pt C, Protista 2. Geol Soc Am; Univ Kansas Press, 900 pp
Luchinina VA (1975) Palaeoalgological characteristics of the Early Cambrian of the Siberian Platform. USSR Acad Sci Siberian Branch, Inst Geol Geophys Issue 216. Nauka, Novosibirsk, 98 pp (in Russian)
Luchinina VA, Tikhomirova NS (1988) On forms of preservation of Early Cambrian algae and their ultrastructure. In: Calcareous algae and stromatolites. USSR Acad Sci Siberian Sec, Inst Geol Geophys, Novosibirsk, pp 12–14, pls 2–5
Markello JR, Read JF (1982) Carbonate ramp-to-deeper shale shelf transitions of an Upper Cambrian intrashelf basin, Nolichucky Formation, southwest Virginia Appalachians. Sedimentology 28:573–597
Maslov VP (1937) On the distribution of calcareous algae in east Siberia. Probl Palaeontol 2–3: 327–348. (in Russian with English summary)
Maslov VP (1956) Fossil calcareous algae of the USSR. USSR Acad Sci Trans Inst Geol Sci Moscow 160: 302 pp (in Russian)
Maslov VP (1960) Stromatolites. USSR Acad Sci Trans 41:188 pp (in Russian)
Maslov VP, Korde KB (1963) Cyanophyta. In: Orlov Yu A (ed) Principles of palaeontology. USSR Acad Sci, Moscow, pp 29–54 (in Russian)
Mawson D, Madigan CT (1930) Pre-Ordovician rocks of the McDowell Ranges (central Australia). Geol Soc London Q J 86:415–528
McIlraeth IA (1977) Accumulation of a Middle Cambrian, deep-water limestone debris apron adjacent to a vertical, submarine carbonate escarpment, southern Rocky Mountains, Canada. In: Cook HE, Enos P (eds) Deep water carbonate environments. Soc Econ Paleontol Mineral Spec Publ 25:113–124
McKee, ED, Resser CE (1945) Cambrian history of the Grand Canyon region. Carnegie Inst Washington Publ 563:232 pp
Mercier J (1934) Sur la présence d'organismes dans les calcaires inférieurs du cambrien du synclinal de la Brècheau-Diable (Calvados). Soc Geol Fr C R Feb 5, 1934:44–45
Nicholson HA, Etheridge R Jr (1878) A monograph of the Silurian fossils of the Girvan district of Ayrshire with special reference to those contained in the "Gray Collection", 1(1). Blackwood, Edinburgh
Peel JS (1988) *Spirellus* and related helically coiled microfossils (cyanobacteria) from the Lower Cambrian of north Greenland. Rapp Grønlands Geol Unders 137:5–32
Pfeil RW, Read JF (1980) Cambrian carbonate platform margin facies, Shady Dolomite, southwestern Virginia, USA. J Sediment Petrol 50:91–116

Pia J von (1927) Thallophyta. In: Hirmer M (ed) Handbuch der Paläobotanik, vol 1. Oldenbourg, Munich, pp 31–136

Poncet J (1976) Hypothèse relative à la morphogenèse du thalle de *Renalcis* (algue calcaire-paleozoique) et affinité possible avec les Rivulariacées actuelles. Geobios 9, 3:345–351

Pratt BR (1984) *Epiphyton* and *Renalcis* – diagenetic microfossils from calcification of coccoid blue-green algae. J Sediment Petrol 54:948–971

Pratt BR, James NP (1982) *Epiphyton* and *Renalcis* – diagenetic microfossils from calcification of coccoid blue-green algae. Ann Assoc Petrol Geol Bull 66:619

Priestley RE, David TWE (1910) Geological notes of the British Antarctic Expedition 1907–1908. In: C R 11th Congr Geology Int, Stockholm, pp 767–811

Rees MN, Pratt BR, Rowell AJ (1989) Early Cambrian reefs, reef complexes, and associated lithofacies of the Shackleton Limestone, Transantarctic Mountains. Sedimentology 36:341–361

Reitlinger EA (1948) Cambrian foraminifera of Yakutia. Bull Moscow Gen Exam Nat 23, 2:77–81 (in Russian)

Reitlinger EA (1959) Atlas of microscopic organic remains and problematica of Siberia's oldest deposits. USSR Acad Sci Trans Geol Inst Moscow 25: 62 pp (in Russian)

Rezak R (1957) *Girvanella* not a guide fossil to the Cambrian. Geol Soc Am Bull 68:1411–1412

Riding R (1977) Calcified *Plectonema* (blue-green algae), a Recent example of *Girvanella* from Aldabra Atoll. Palaeontology 20:33–46

Riding R (1982) Cyanophyte calcification and changes in ocean chemistry. Nature (London) 299:814–815

Riding R (1988) Classification of microbial carbonates. In: Benthic microbes and reefs. 6th Int Coral Reef Symp, Townsville, Abstr, p 5

Riding R, Brasier M (1975) Earliest calcareous foraminifera. Nature (London) 257:208–210

Riding R, Voronova L (1982a) Calcified cyanophytes and the Precambrian-Cambrian transition. Naturwissenschaften 69:498

Riding R, Voronova L (1982b) Recent freshwater oscillatoriacean analogue of the Lower Palaeozoic calcareous alga *Angulocellularia*. Lethaia 15:105–114

Riding R, Voronova L (1982c) Affinity of the Cambrian alga *Tubomorphophyton* and its significance for the Epiphytaceae. Palaeontology 25:869–878

Riding R, Voronova L (1984) Assemblages of calcareous algae near the Precambrian/Cambrian boundary in Siberia and Mongolia. Geol Mag (Tokyo) 121, 3:205–210

Riding R, Voronova L (1985) Morphological groups and series in Cambrian calcareous algae. In: Toomey DF, Nitecki MH (eds) Paleoalgology: contemporary research and applications. Springer, Berlin Heidelberg New York Tokyo, pp 56–78

Rippka R, Deruelles J, Waterbury JB, Herdman M, Stanier RY (1979) Generic assignments, strain histories and properties of pure cultures of Cyanobacteria. J Gen Microbiol 111:1–61

Rothpletz A (1913) Über die Kalkalgen, Spongiostromen und einige andere Fossilien aus dem Obersilur Gottlands. Sver Geol Unders Ca 10:57 pp, 9 pls

Rowland SM (1981a) Archaeocyathid reefs of the southern Great Basin, western United States. In: Taylor ME (ed) Short papers for the 2nd Int Symp Cambrian system. US Geol Surv Open File Rep 81–743, pp 193–197

Rowland SM (1981b) Archaeocyathid bioherms in the Lower Poleta Formation, Esmeralda County, Nevada. In: Taylor ME, Palmer AR (eds) Cambrian stratigraphy and paleontology of the Great Basin and vicinity, western United States, Guidebook for Field Trip 1. 2nd Int Symp Cambrian system, Denver, Col, pp 44–49

Rowland SM, Gangloff RA (1988) Structure and paleoecology of Lower Cambrian reefs. Palaios 3:111–135

Saltovskaya VD (1975) Genus *Epiphyton* Bornemann (probable synonyms and stratigraphic significance). In: Questions of palaeontology of Tadzhikistan. USSR Acad Sci, Donish, Dushanbe, pp 70–88 (in Russian)

Stepanova MV, Luchinina VA (1982) The lower boundary of the Cambrian based on calcareous algae. In: Boundaries of the principal subdivisions of the Phanerozoic in Siberia. Collected scientific works (in Russian)

Vologdin AG (1932) Archaeocyaths of Siberia, part 2. Cambrian fauna of Altai limestones. State Sci Tech Geol Publ, Leningrad, 106 pp (in Russian)

Vologdin AG (1939) Archaeocyaths and algae of the Middle Cambrian of the southern Urals. Probl Palaeontol 5:210–245 (in Russian)

Vologdin AG (1955) Cambrian *Solenopora* and molluscs of northern Tien-Shan. USSR Acad Sci Rep 105, 2:354–356 (in Russian)

Vologdin AG (1958) Lower Cambrian foraminifera of Tuva. USSR Acad Sci Rep 120, 2:405–408 (in Russian)

Vologdin AG (1962a) The oldest algae of the USSR. USSR Acad Sci Moscow, 656 pp (in Russian)

Vologdin AG (1962b) Archaeocyaths and algae of the Cambrian of the Baikal Platform. USSR Acad Sci Trans 93:3–116 (in Russian)

Voronin Yu I, Voronova LG, Grigoreva NV, Drosdova NA, Shegello EA,, Zhuravlev A Yu, Ragozina AL, Rozanov A Yu, Sayutina TA, Sisoev VA, Fonin VD (1982) The Precambrian-Cambrian boundary in geosynclinal areas (the reference section Salany-Gol, MPR). Transactions of the joint Soviet-Mongolian palaeontological expedition, vol 18. Nauka, Moscow, 150 pp (in Russian)

Voronova LG (1969) Algae. In: Rozanov A Yu, Missarzhevskii VV, Volkova NA, Voronova LG, Krylov IN, Keller BM, Korolyuk IK, Lendzion K, Michniak R, Pykhova NG, Sidorov AD (eds) The Tommotian Stage and the Cambrian lower boundary problem. USSR Acad Sci Geol Inst Trans 206. Nauka, Moscow, pp 189–195 (in Russian) (English edn 1981, Amerind, New Delhi, pp 221–227)

Voronova LG (1976) Calcareous algae of the Precambrian-Cambrian boundary deposits of the Siberian platform. In: Voronova LG, Radionova EP, Palaeozoic algae and microphytolites. USSR Acad Sci Geol Inst Moscow 294:3–85, 156–183, 212–214 (in Russian)

Voronova LG (1979) Calcareous algae of the Precambrian and Early Cambrian. Bull Centre Rech Explor Prod Elf-Aquitaine 3, 2:867–871

Yabe H, Osaki K (1930) *Girvanella* in the Lower Cambrian of south Manchuria. Sci Rep Tohoku Imp Univ Sendai Ser 2 Geol 14:79–83

Zamarreño I (1977) Early Cambrian algal carbonates in southern Spain. In: Flügel E (ed) Fossil algae. Springer, Berlin Heidelberg New York, pp 360–365

Zhuravleva IT, Meshkova NP, Luchinina VA, Kashina LN (1982) Late Precambrian and Early Cambrian biofacies of the Anabarian Sea. In: Environment and Life in the geological past, palaeolandscape and biofacies. USSR Acad Sci, Siberian Sec, Inst Geol Geophys, Novosibirsk, Trans 510:74–103 (in Russian)

Chapter 17

Ordovician Algae and Global Tectonics

A. Roux[1]

Abstract

The evolution of faunal provinces during the Ordovician shows a progressive reduction of the endemisms resulting from the progressive closing of the Proto-Atlantic ocean. The geographic and stratigraphic distributions of some representative algal groups support this interpretation and suggest some modifications for the proposed positions of the landmasses, particularly for North America, the Baltic shield and the Kazakhstan-Tadjikhistan area. The algal migrations seem to be related to the surface currents.

1 Introduction

Most palaeogeographic reconstructions are based on palaeomagnetic data and faunal and floral data are relatively little used, at least not for worldwide reconstructions.

Works using Palaeozoic algae for these research topics are very scarce and limited to some titles concerning mainly the Upper Palaeozoic: Mamet (1972) has shown the striking coincidence of the distribution of the Viséan microflora with the Tethyan belt. Mamet and Roux (1977) and Jansa et al. (1978) have shown that the worldwide distribution of the Devonian-Carboniferous microflora conflicts with the reconstructions which admit a physical contiguity between North-America and Euroafrica, suggesting the presence of a Proto-Atlantic ocean during Carboniferous. Roux (1979, 1986) and Mamet et al. (1987) have shown that the Permo-Carboniferous microflora of the Sverdrup Basin was not equatorial but warm temperate to subtropical, suggesting a migration towards the north and a rotation of the North American block. Mamet and Roux (1983) have confirmed the position of Australia during the Devonian and Carboniferous previously proposed by Keppie (1977). Bildgen and Roux (1982a,b) have shown that the hypothesis of an equatorial palaeoclimatic belt, proposed on the basis of the distribution of karst bauxites, may be confirmed by palaeobotanical data (see a synthesis in Roux 1985). Finally, new positions of the North American block during the Lower Carboniferous and Permian have been proposed (Roux 1986).

This study is a new attempt at reconstruction of the position of the continental blocks (North America and Eurasia) during the Upper Ordovician based on the stratigraphic and geographic distribution of algal microfloras.

---

[1] 51 Avenue de Verdun, 77610 Fontenay-Trésigny, France

2 Ordovician Palaeogeography and Faunal Data

According to Scotese et al. (1979), most of the continents occupied the same latitudinal belt during the Late Cambrian and Middle Ordovician: during the Upper Cambrian, Laurentia, Siberia and the southern portion of Gondwana were in equatorial latitudes; China and Kazakhstania are also supposed to have been within 30° of the equator. According to palaeomagnetic data, Baltica was located at relatively high latitudes. As a consequence, the polar oceans were of wider extent than at any subsequent times. From the Late Cambrian through Middle Ordovician, Laurentia and Gondwana rotated counterclockwise, while Baltica and Siberia moved northward.

The Early Ordovician coincides with a wide transgression over several European and North African regions. The worldwide distribution of trilobites and graptolites has shown a faunal provincialism suggesting that the extreme northwestern part of Europe was part of a Greenlando-Canadian plate, separated from the Gondwano-European area by a Proto-Atlantic ocean called Iapetus. More exactly, the northwestern margin of Iapetus includes Greenland, part of Newfoundland, northwestern Ireland, Scotland, and Norway. The study of trilobite and brachiopod faunas has suggested the existence of two provinces over Europe: a northern province (Baltic area) and a southern province, separated by a wide and deep ocean (see a synthesis in Johnson 1977).

The definition of a Prototethyan realm during this time is problematic; a Prototethys or a Prototethyan realm are intended in this work only to differentiate the North-American realm.

Boucek (1972) has already proposed a probable position of the Lower Ordovician (Tremadoc-Llandeilo) equator, which crossed the North American block from north to south through Hudson Bay and Europe from Finland to the Bay of Bengal. On the basis of that position of the equator, the author distinguished two provinces:

A Pacific tropical northern province including North America, western Ireland, Norway (Trondheim area), the northern part of USSR (Taimyr), China and New Zealand.

A southern province known in Europe, North Africa, the western part of Asia and South America, and divided into:

A tropical southern or Anglo-Scandinavian subprovince including the British Isles (western Ireland excepted), Scandinavia, the Baltic region, part of the USSR (Ural region and Kazakhstan) and the western regions of South America.

A temperate to cold, Mediterranean subprovince gradually linked with the former and approximatively limited by Syria, Florida and Alabama, Wales, Belgium, Bohemia, Poland, Morocco, Sahara and the eastern part of South America.

During the Late Ordovician, a progressive homogeneization of the faunas becomes evident in Europe; during the Lower Ordovician, evolution in each province seems to have been independent; during the Upper Ordovician, the progressive merging of provincialism suggests the disappearance or a reduction of physical barriers, that is, the closing of the Proto-Atlantic ocean.

The progressive closing of Iapetus accounts for more precise faunal exchanges between North America and Baltic Europe. The reduction of endemism seems to be the result of faunal (and probably algal) migrations from North America to Europe.

On the basis of stratigraphic breaks observed in the Ordovician of northern Estonia, Roomusoks (1972) has demonstrated several changes in the faunas: a renewal of bryozoan, brachiopod and trilobite faunas appears at the Volkhov-Kunda Formations boundary (Lower Ordovician); among the immigrants are some representatives of earlier faunal communities of Bohemia and North America.

Near the end of the Middle Ordovician, numerous new genera of brachiopods, bryozoans and trilobites appear in the Keila-Ondu Formation (Viru Group) with a large number of immigrants from North America and Britain among the brachiopods.

During the Upper Ordovician (Ashgillian, Pirgu Formation, Harju Group), a large number of forms appears, including genera characteristic of Middle and Upper Ordovician deposits of North America.

3 Distribution of Some Representative Algal Groups

Vermiporella

Figure 1 shows the worldwide distribution of the species of *Vermiporella*: during the Middle Ordovician, the genus is only known in North America (Canadian Arctic, Hudson Bay, Saint Lawrence Lowlands, New York State) and Scotland, and is unknown in the 'Prototethyan' realm (Scotland belongs during this time to the North American realm). The genus is represented by five species: *V. borealis*, *V. eisenacki*, *V. fragilis*, *V. inconstans* and *V. wesenbergensis*. It is interesting to note that these species were originally described from Upper Ordovician deposits of the 'Prototethyan' realm.

During the Upper Ordovician, the genus is cosmopolitan, distributed from Kazakhstan to the Canadian Arctic and represented by all the previously mentioned species with, however, some species such as *V. acerosa* and *V. diffluens* which are only known in Kazakhstan, and *V. canadensis* which is only known in North-America (Newfoundland, Hudson Bay, Canadian Arctic). Although the genus becomes cosmopolitan during the Upper Ordovician, the former species are 'Prototethyan' and the latter, 'American'.

The presence of *V. borealis*, *V. eisenacki*, *V. fragilis*, *V. inconstans* and *V. wesenbergensis* in North America and Scotland during the Middle Ordovician and the presence of the same species in the 'Prototethyan' realm during the Upper Ordovician, may be explained by migrations from North America and especially the Saint Lawrence Lowlands and the New York State area to the

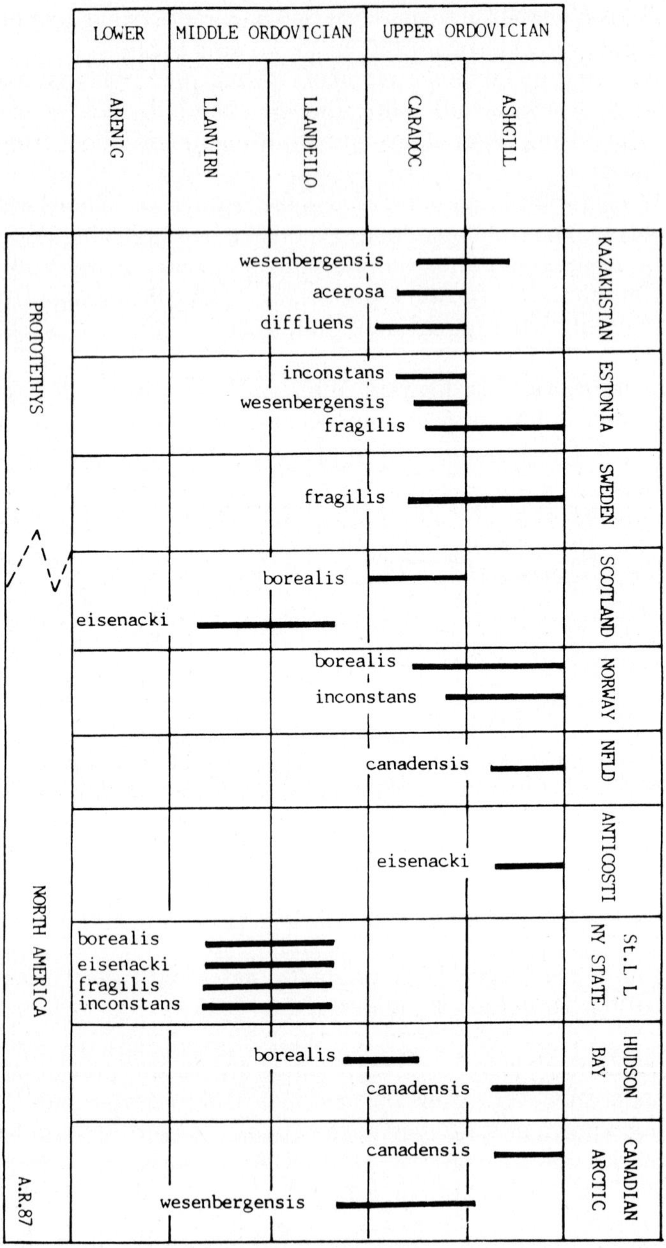

Fig. 1. Geographic and stratigraphic distribution of species of the dasycladacean *Vermiporella*

'Prototethys' through Scotland and Norway, which belong to the North American realm (maps of Johnson 1977; Keppie 1977; Scotese et al. 1979).

These migrations were possible because of the relative proximity of the North American and the Eurasian blocks, due to the more-or-less complete closing of the Proto-Atlantic ocean during the end of the Middle Ordovician and the Upper Ordovician.

Rhabdoporella

Figure 2 shows the worldwide distribution of the species of *Rhabdoporella*: the genus is mainly Silurian, but occurs in Sweden and Anticosti Island during the

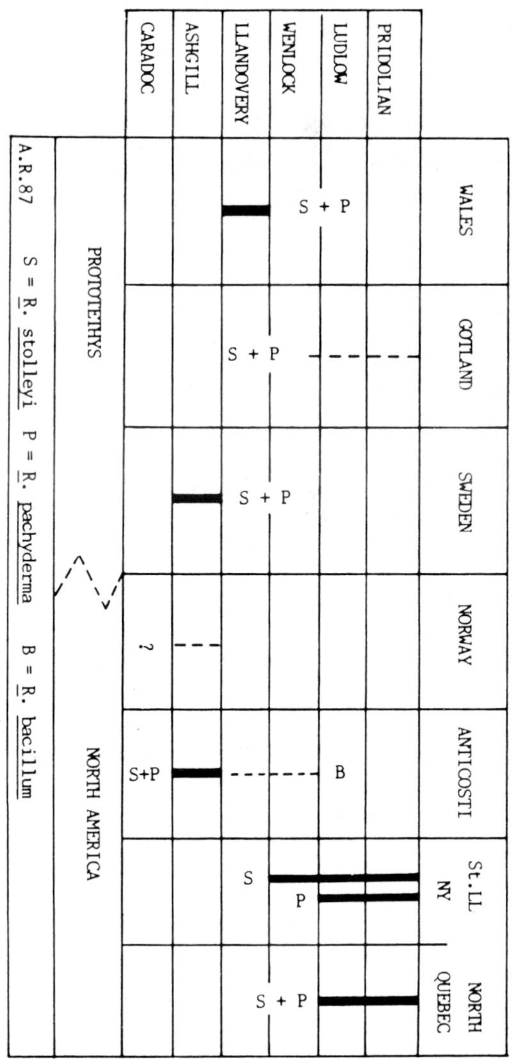

Fig. 2. Geographic and stratigraphic distribution of species of the dasycladacean *Rhabdoporella*

Upper Ordovician (Ashgillian); the Norwegian references are not clear. During the Silurian, the genus is cosmopolitan in North America (Anticosti Island, New York State, Québec, Saint Lawrence Lowlands) and in the 'Prototethys' (Wales, Gotland, Sweden).

It is difficult to establish the pattern of migrations because of the simultaneous occurrence of the species *R. stolleyi* and *R. pachyderma* in Sweden and Anticosti Island during the Upper Ordovician. According to other works (Roomusoks 1972), the migrations may have taken place from Anticosti Island to the 'Prototethys' through Sweden (and possibly Norway) during the end of the Upper Ordovician or the basal part of the Lower Silurian, because of the closing of the Proto-Atlantic ocean which had eliminated or reduced the physical barrier of Iapetus.

Udoteaceae

The stratigraphic and geographic distributions of important genera of erect udoteaceans are shown in Fig. 3: during the Middle Ordovician, the species *Dimorphosiphon rectangulare* is known in the 'Prototethyan' and North American realms (Kazakhstan and Norway and the Saint Lawrence Lowlands respectively). The same species occurs in Scotland and remains present in Kazakhstan during the lower part of the Upper Ordovician (Caradocian). *Dimorphosiphon diadromum* appears in Kazakhstan during the Upper Ordovician and occurs in northern Québec during the Silurian as *D.* cf. *diadromum*. *Dimorphosiphon magnum* seems to be cosmopolitan from Kazakhstan to northern Québec during the Silurian.

Among the genus *Palaeoporella*, the species *P. opuntiformis* appears in Tadjikhistan during the upper part of the Upper Ordovician (Ashgillian) and occurs in Anticosti Island (as *P.* cf. *opuntiformis*) during the Lower Silurian. *P. recta* occurs in Kazakhstan during the lower part of the Upper Ordovician and in northern Québec during the Upper Silurian (as *P.* cf. *P. recta). P. variabilis* occurs in the North American and 'Prototethyan' realms (Norway and Sweden) during the Upper Ordovician.

The simultaneous occurrence of *D. rectangulare* in Kazakhstan and Scotland seems to be indicative of a relative proximity of the North American and north Eurasian blocks during the lower part of the Upper Ordovician (Caradocian). The same scheme is indicated by the simultaneous presence of *P. variabilis* in Sweden and Norway.

The genera *Dimorphosiphonoides* and *Lowvillia* are strictly 'American' and restricted to the Middle Ordovician.

Other Algae

Among the dasycladaceans, it is interesting to note that the genus *Kazakhstanelia*, originally described from the Ordovician of the USSR (central Kazakhstan, Lake Ajakmalajsor and Bajanoul and Pribalkhash regions) is also present in the Middle Ordovician of the Saint-Lawrence Lowlands (Québec and Ontario). Originally described from the Caradocian of Scotland, *Intermurella*

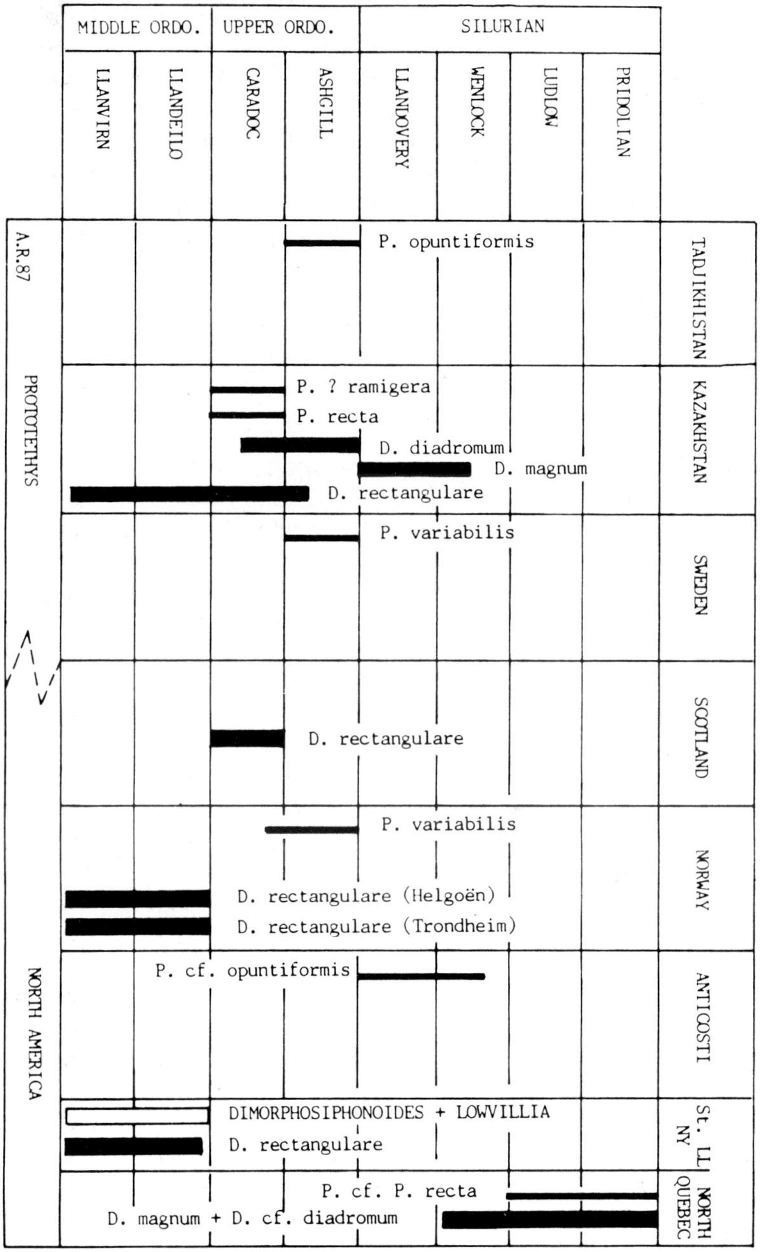

Fig. 3. Geographic and stratigraphic distribution of Udoteaceae

has been recently reported from the Ordovician of the Hudson Bay and Canadian Arctic. Originally described from the Caradocian of Scotland, *Novantiella* has been recently reported from the Silurian of Hudson Bay and Canadian Arctic.

On the other hand, other recently described Ordovician genera such as *Anticostiporella* and *Arcticella* are till now only known from their type-localities; Anticosti Island and Canadian Arctic respectively (see Korde 1957; Elliott 1972; Gauthier 1981; Poncet 1986).

Among the rhodophytes, it is of interest to mention the simultaneous occurrence of *Solenopora spongioides*, originally described from Estonia, in Scotland, Québec and New York State. However, making use of rhodophytes for palaeogeographic reconstructions remains difficult; in fact, the Lower Palaeozoic solenoporaceans need a complete revision.

4 Sedimentary and Palaeoclimatic Data

Middle Ordovician calcareous algae of Québec, Ontario and New York State are abundant but undiversified in shallow-water limestones; the Middle Ordovician of the Saint Lawrence platform is largely dominated by porostromates and codiaceans-udoteaceans; the role of dasycladaceans is accessory.

The microfloras of the Upper Ordovician-Lower Silurian carbonates of Anticosti Island show the same characteristics and suggest neither equatorial nor tropical, but warm-temperate or subtropical waters (Gauthier 1981; Gauthier-Coulloudon and Mamet 1981; Mamet et al. 1984).

The scarcity of dasycladaceans seems to be a characteristic of the Ordovician carbonates of the eastern margin of the North American continent, except in the bioherms of the Chicamanga Group in Alabama (Stock and Benson 1982).

The North American Ordovician is characterized by an important development of bioherms and carbonate mud-mounds. During the Lower Ordovician, the organic buildups occur in the calcareous facies of the external part of the Lower Palaeozoic North American craton: the Great Basin, Texas, Oklahoma and Newfoundland. During the Middle Ordovician (Chazy Group), organic buildups are numerous and diverse in New York, Vermont, Québec (Day Point, Crown Point and Valcour Formations), Tennessee (Carters and Holston Formations), Virginia (Effna Formation) and Newfoundland (Black River Group, Lourdes Formation). According to Wilson (1975), Middle Ordovician buildups developed on a shallow-water shelf westward of the Appalachian Belt in more-or-less agitated waters (see a synthesis in Roux 1985).

The Ordovician microflora of Hudson Bay and Canadian Arctic, which contains five genera of dasycladaceans and three genera of codiaceans-udoteaceans suggests a tropical to subtropical palaeoclimate (Poncet 1986).

In central Sweden, the Upper Ordovician (Ashgill, Boda Limestone) shows an important reef development in which massive beds consist almost entirely of mats of *Palaeoporella variabilis*. Corals and bryozoans occur mainly near the margins of the reef and stromatolites are dominant in the massive limestone of the center. The knoll-forming stromatolites grew on a substrate of corals and were overgrown by *Wetheredella*. The dasycladacean *Rhabdoporella* (with *R. stolleyi* and *R. pachyderma*) formed mats on sandy marls in the off-reef facies (Jux 1966a,b).

In Anticosti Island, *Rhabdoporella* is abundant in the Upper Ordovician (Gamachian, Ellis Bay Formation). Coccoid algae played an important role in the formation of bioherms in the upper part of the Ellis Bay Formation. The substrate of the mud-mounds is constituted by well-sorted packstones with abundant *Rhabdoporella*; the core of the mud-mounds is made up of a *Wetheredella*-coral-bryozoan association bound by coccoid algae. The mud-mounds of the Ellis Bay Formation occur at the maximum of the Upper Ordovician regression (Gauthier 1981).

In Estonia, the Ashgillian is characterized by continuous decrease in water depth. The Early Ashgillian (base of the Pirgu Formation) shows a massive appearance of dasycladaceans ('*Dasyporella*') and the presence of bioherms. The Upper Ashgillian (Porkuni Formation) shows widespread distribution of bioherms on a stable shallow-water shelf (Roomusoks 1972).

Judging from this, the palaeogeographic conditions seem to have been relatively the same in Anticosti Island, central Sweden and Estonia during the Upper Ordovician. The effects of the Late Ordovician glaciation are discussed below.

5 Palaeogeographic Significance

Empirical determination of an equatorial belt

Figure 4 shows the Upper Ordovician distribution of the studied algal groups, plotted on a present-day map of the world.

In North America, the southernmost localities occur in Newfoundland and Anticosti Island, at about 50°N. The northernmost localities occur in Canadian Arctic Archipelago (Ellesmere Island) at about 80°N. If we consider as a first approximation that the dasycladacean-udoteacean association flourished in warm equatorial to tropical or subtropical waters, an assumed position of the equator should have been on the median part of a belt limited by the extreme latitudes of 50°N and 80°N, that is, about 65°N. The palaeoequator ran through the southern part of Baffin Island.

In northwestern Europe, the southernmost localities occur in Estonia, Scotland and southern Sweden, that is, between 55 and 60°N, and the palaeoequator should have passed at about 70–75°N, but more probably 5 to 10° northward off Estonia, Scotland and southern Sweden.

In Kazakhstan and Tadjikhistan, the localities occur between 35 and 45°N, and the palaeoequator should have passed at latitudes 50 to 60°N.

According to this assumed position of the Upper Ordovician equator, Anticosti Island should have been located in a subequatorial position at about 15°S; Scotland, southern Sweden and Estonia should have been located at a

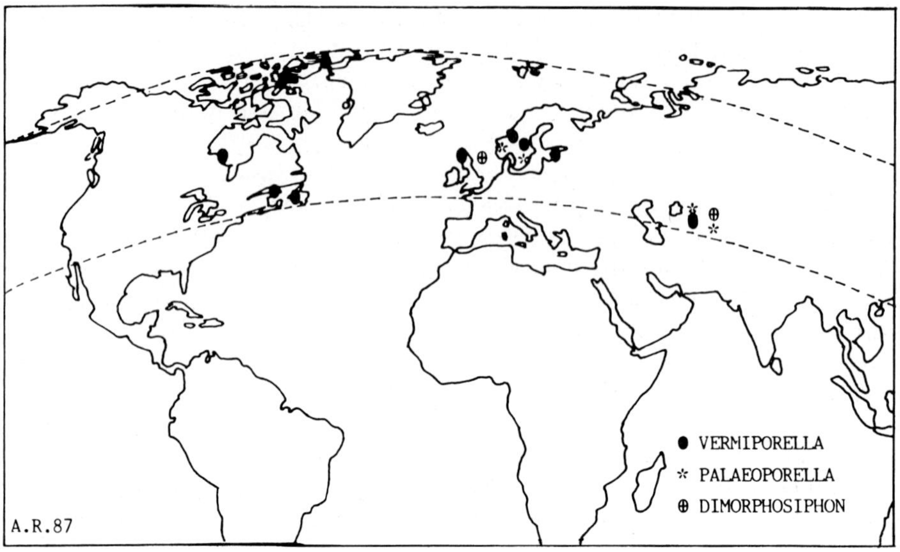

Fig. 4. Upper Ordovician localities plotted on a present-day map of the world. The assumed position of the equatorial belt is indicated by *broken lines*

latitude of 10° S, also in a subequatorial position. Kazakhstan and Tadjikhistan should have been located at latitudes 15 to 20° S in a tropical position.

It is important to note that these conclusions are based only on a comparison of the occurrences of Upper Ordovician localities plotted on a present-day map of the world, but serve as a basis for further investigations concerning particularly the movement of the continents. It is of interest to compare these conclusions with former reconstructions.

6 Comparisons with Former Reconstructions

On the Lower Ordovician map of Smith et al. (1973), the palaeoequator runs through North America and particularly the Canadian Arctic Archipelago which is located between 10° S and 10° N. Estonia and southern Sweden are located at a latitude of about 20° S in tropical position; Scotland is located on the western margin of the Proto-Atlantic ocean at a latitude of about 20° S.

In the previously assumed position of the palaeoequator, the North American continent is located between 20° N and 45° S, that is, within a belt of 65°. On the reconstruction of Smith et al. (1973), the North American block is located between 30° N and 30° S, also within a belt of 60°; the global latitudinal position of North America seems to have been relatively constant.

In northwestern Europe, the palaeoequator of Smith et al. (1973) runs slightly off northern Scandinavia (10°) and Kazakhstan and Tadjikhistan are located at about 30° S in subtropical positions. Kazakhstan, Tadjikhistan and northern Scandinavia are located in a belt of about 20°.

The reconstruction of Scotese et al. (1979, Fig. 13, Mercator projection) shows a map of the world during the Middle Ordovician, from the Llandeilian to the Early Caradocian. Present-day latitude-longitude is shown on each palaeocontinent by a grid of five degrees apart, which can be used to plot data.

The North American block is located within a belt between 30° N and 30° S. The equator runs across Canada from near the USA-Canada border (Vancouver Island, present-day latitude 50° N) to the Hudson Bay (present-day latitude 60° N) and the southern part of Greenland (present-day latitude 60° N). The extreme localities of the distribution of the microflora plotted on a present-day map of the world (Fig. 4, Anticosti and Ellesmere Islands) are located on the reconstruction of Scotese et al. (1979) respectively at about 15° S and 20° N. Scotland is located between 0 and 10° S. Kazakhstan (and probably Tadjikhistan, the position of which is difficult to plot precisely on the reconstructions) is completely separated from the Baltic shield (Baltica) and located between 0 and 10° S. The Baltic shield is located far southward between 60 and 70° S.

7 An Upper Ordovician Map of the World

Figure 5 shows a likely position of the North American continent, the Baltic shield and Kazakhstania-Siberia during the Upper Ordovician, proposed on the basis of the distribution of algal microfloras.

The Proto-Atlantic ocean is nearly completely closed. To be in accordance with the preceding remarks and particularly the sedimentary and palaeoecological data, it seems necessary to rotate the North American block clockwise, in order to draw Anticosti Island into a warm temperate to subtropical position at about 25 to 30° S. Scotland and Norway are then located at nearly the same latitudes, about 20 and 25° S.

The position of Kazakhstan and Tadjikhistan remain the same as on the map of Scotese et al. (1979), that is, about 10° S, but they are united to the Baltic Shield because of the similarity of the microfloras.

The most important modification concerns the Baltic Shield which must be moved northward and located between 10 and 40° S. Estonia and southern Sweden are then located at about 25° S within the same latitudinal belt of Scotland and Anticosti Island.

It is interesting to note that these positions are relatively similar to those of the map of Johnson (1977, Fig. 1A). Ellesmere Island is also located in an equatorial position and the centre of Hudson Bay is at about 15° S. The Canadian Arctic and Hudson Bay are then located in a warmer position than are Anticosti Island and Scotland and Norway.

Cocks and Fortey (1982, Fig. 5) have proposed a configuration of European and neighbouring plates for the Late Ordovician to the Lower Silurian:

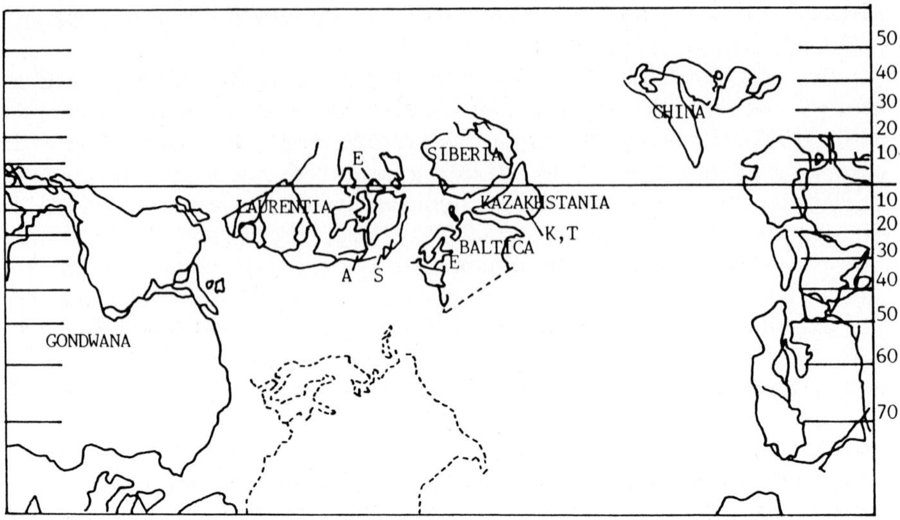

Fig. 5. Upper Ordovician map of the world (Mercator projection) showing a likely position of Laurentia, Baltica and Kazakhstania-Siberia, adapted from Scotese et al. (1979). The Middle Ordovician position of Baltica is shown by *broken lines*

Tornquist's Sea separating the Baltic Continent from the 'Gondwana' Continent has disappeared. Northern Europe, including the southern part of the British Isles and eastern Newfoundland, is separated from the 'Gondwana' Continent by the Rheic Ocean; 'the Iapetus Ocean was evidently no longer the barrier for faunal interchange it had been earlier'; the Bahamitic type of the Baltic limestones 'and the similarity of faunas between Britain and Scandinavia would suggest that both regions were within the tropics'.

On the map of the authors, Anticosti Island is situated at about 18° S. Estonia within 15–20° S and Scotland within 6–11° S. Such a configuration may be explained by the effects of the important Late Ordovician glaciation, causing a general regression and 'marked by a temporary cessation of limestone deposition in all but the most equatorial regions'.

The concept or the effects of glaciations must, however, be used with wariness: for example, limestone deposition in Anticosti Island is continuous from the Upper Ordovician to the Lower Silurian; the Late Ordovician is even marked by the development of bioherms. In Estonia, the Upper Ashgill shows a widespread distribution of bioherms.

Was the Late Ordovician general cooling sufficiently intensive to create subtropical or warm-temperate climatic conditions in the equatorial regions? Was the cooling insufficient to modify significantly the distribution of the climatic belts over the Earth? The debate remains open.

In the first case, the reconstruction of Cocks and Fortey will be correct and then the position of Anticosti, Scotland and Baltica fits relatively well with my Fig. 4. In the second case, the proposed position of the land masses of my Fig.

5 is probably correct and fits relatively well with the Ashgill base map of Scotese (1984, reproduced in Beadle and Johnson 1986).

If one considers that the global direction of oceanic surface currents has not changed during geological times, the microfloral exchanges between North America and the 'Prototethyan' realm may be explained by such surface currents moving counterclockwise.

8 Conclusions

The global palaeogeographic distribution of algal microfloras are closely related to transgressive-regressive cycles and the position of continental blocks; therefore, fossil algae may be useful tools for palaeogeographic reconstructions.

Such studies are, however, still limited for many reasons: an empirical determination of an equatorial belt is restricted by the number of known localities and consequently stratigraphic and geographic distributions require more information; sedimentary and ecological data are often lacking in the literature; finally, the stratigraphic data are not always accurate from one region to another or even from one author to another.

As with any new research, this contribution is certainly incomplete and must be regarded as an introductory study. Such investigations need more careful and extensive taxonomic, sedimentological and ecological studies within accurate stratigraphic frameworks. Nevertheless, the first results are promising.

References

Beadle SC, Johnson ME (1986) Palaeoecology of Silurian Cyclocrinitid algae. Palaeontology 29, pt 3:585–601

Bildgen P, Roux A (1982a) Corrélations entre la répartition des Algues et des bauxites de karst au Paléozoique supérieur. In: 107th Congr Nat Soc Sav, Brest, vol 3, pp 163–171

Bildgen P, Roux A (1982b) Algues, bauxites de karst et tectonique globale au Dévonien-Carbonifère. In: 9th Reun Annu Sci Terre, Paris, p 55

Boucek BV (1972) The paleogeography of Lower Ordovician graptolite faunas: a possible evidence of continental drift. In: 24th Congr Géologie Int, Montréal, Sec 7, pp 256–272

Cocks LRM, Fortey RA (1982) Faunal evidence for oceanic separations in the Palaeozoic of Britain. J Geol Soc London 139:465–478

Gauthier L (1981) Paléoéologie des Algues ordoviciennes et siluriennes de l'Ile d'Anticosti, Québec. M Sc Thesis, Univ Montreal

Gauthier-Coulloudon L, Mamet B (1981) Algues et sédimentation carbonatée dans l'Ordovicien et le Silurien de l'Ile d'Anticosti. In: Lespérance PJ (ed) Subcommiss Silurian stratigraphy, Field Meet Anticosti-Gaspé, Québec, vol 2: Stratigraphy and paleontology, pp 87–90

Elliott GF (1972) Lower Palaeozoic green algae from southern Scotland, and their evolutionary significance. Bull Brit Mus Nat Hist 22, 4:357–376

Jansa LF, Mamet B, Roux A (1978) Visean limestones from the Newfoundland shelf. Can J Earth Sci 15, 9:1422–1436

Johnson GAL (1977) Paleozoic accretion of western Europe. Ann Soc Geol Nord 96(1976)4:347–352

Jux U (1966a) *Palaeoporella* im Boda-Kalk vo Dalarne. Palaeontogr B 118, 4–6:153–165

Jux U (1966b) *Rhabdoporella* im Boda-Kalk sowie in Sandsteinen Dalarnes (Ashgill, Schweden). Paleontogr B 188, 4–6:166–183
Keppie JD (1977) Plate tectonic interpretation of Palaeozoic world maps. N Scot Dep Mines Pap 77–3
Korde KB (ed) (1957) Nouvelles représentantes des Algues siphonées. In: Données pour les 'principes de paléontologie'. Akad Nauk SSSR 1:67–75 (in Russian)
Mamet B (1972) Un essai de reconstitution paléoclimatique basé sur les microflores algaires du Viséen. In: 24th Congr Géologie Int, Montréal, Sec 7, pp 282–291
Mamet B, Roux A (1977) Algues rouges dévoniennes et carbonifères de la Téthys occidentale, pt 4. Rev Micropaléontol 19, 4:216–266
Mamet B, Roux A (1983) Algues dévono-carbonifères de l'Australie. Rev Micropaléontol 26, 2:63–131
Mamet B, Roux A, Shalaby H (1984) Rôle des Algues calcaires dans la sédimentation ordovicienne de la Plate-forme du Saint-Laurent. Geobios Mem Spec 8:261–269
Mamet B, Roux A, Nassichuk W (1987) Algues carbonifères et permiennes de l'Arctique canadien. Geol Surv Can Bull 342
Poncet J (1986) Les Algues calcaires du Paléozoïque inférieur de la Baie d'Hudson et de l'Archipel arctique canadien. Bull Centres Rech Explor Prod Elf-Aquitaine 10, 2:259–282
Roomusoks A (1972) Stratigraphic breaks in the Ordovician of northern Estonia and the corresponding changes in the fauna. In 24th Congr Géologie Int, Montréal, Sect 7, pp 605–611
Roux A (1979) Algues du Paléozoïque supérieur du Bassin de Sverdrup, Archipel arctique canadien. D Sc Thesis, Univ Libre Bruxelles, 354 pp
Roux A (1985) Introduction à l'étude des Algues fossiles paléozoïques (de la bactérie à la tectonique des plaques). Bull Centres Rech Explor Prod Elf-Aquitaine 9, 2:465–699
Roux A (1986) Microflores algaires paléozoiques (tendances évolutives, répartitions) et tectonique globale. Bull Centres Rech Explor Prod Elf-Aquitaine 10, 2:555–563
Scotese CR, Bambach RK, Barton C, van der Voo R, Ziegler AM (1979) Paleozoic base maps. J Geol 87, 3:217–277
Smith AG, Briden JC, Drewry GE (1973) Phanerozoic world maps. In: Hughes NF (ed) Organisms and continents through time. Spec Pap Paleontol, vol 12. Palaeontol Assoc, London, pp 1–42
Stock CW, Benson DJ (1982) Occurrence and distribution of fossils within and adjacent to Middle Ordovician bioherms in the southern Appalachians of Alabama. In: 3rd North American Paleontology Convention, Montreal, Proceedings, vol 2, pp 517–524
Wilson JL (1975) Carbonate facies in geologic history. Springer, Berlin Heidelberg New York

Chapter 18
Ordovician to Devonian Marine Calcareous Algae

A. Roux[1]

Abstract

After the Cambrian, when calcified algae are rare or poorly known, the Ordovician-Devonian microfloras are largely dominated by chlorophytes, and particularly the filamentous (or nodular) codiaceans and the erect udoteaceans.

Long-ranging and non-diagnostic forms are numerous and most taxa have little stratigraphic value. It is possible, however, to determine characteristic associations or assemblages, for each period. In addition, erect udoteaceans and the dasycladaleans show an original pattern of development and are characteristic of systems.

1 Introduction

The first calcareous algae appear by the Lower Cambrian times with *Girvanella* and probable microbial forms such as *Renalcis* and *Epiphyton* which dominate the microflora. According to Wray (1977), the first solenoporaceans appear by the Middle Cambrian and the first udoteaceans (*Palaeoporella*) by the Upper Cambrian. The most important Cambrian event is, however, the appearance of the first dasycladaceans (Seletonellaceae), represented by several morphologically advanced genera such as *Amgaella* and *Siberiella* (Korde 1957) and *Cambroporella*, *Mejerella* and *Seletonella* (Korde 1950).

2 Ordovician

The Ordovician is characterized by the appearance or the diversification of numerous groups, particularly the filamentous codiaceans and the erect udoteaceans. These two groups largely dominate the microflora, but the dasyclads also diversified.

Filamentous Codiaceans

Originally created from the Lower Carboniferous of England, *Bevocastria conglobata* Garwood first occurs in the Middle Ordovician of the Saint-Lawrence Lowlands (New York State, Québec, Ontario) and the Upper Ordovician of Anticosti Island (Guilbault and Mamet 1976).

[1]51 Avenue de Verdun, 77610 Fontenay-Trésigny, France

Ortonella Garwood emend. Mamet and Roux is represented by *O. aequalis* (Høeg) in the Middle Ordovician of the Saint-Lawrence Lowlands, Newfoundland, and Norway. *O. kershopensis* Garwood, which is cosmopolitan during the Dinantian, also already occurs in the Middle Ordovician of New York State.

Garwoodia Wood, which is cosmopolitan from the Ordovician to the Permian, first occurs in the Middle Ordovician of New York State with *G. gregaria* (Nicholson).

Hedstroemia Rothpletz is known by two species: *H. halimedoidea* Rothpletz in the Middle Ordovician of the Saint-Lawrence Lowlands; *H. bifilosa* in the Middle Ordovician of Québec and New York State.

Questionable codiaceans ? or cyanophytes ? are represented by two species of *Sphaerocodium* Rothpletz emend. Wray (= *Rothpletzella* Wood *auct.*, = *Coactilum* Maslov): *S. gotlandicum* Rothpletz emend. Wood, and *S. moniliferum* (Seely). Both appear in the Middle Ordovician.

Erect Udoteaceans

The erect udoteaceans are represented by *Dimorphosiphon* Høeg (tribe Dimorphosiphoneae Shuysky), *Dimorphosiphonoides* and *Lowvillia*, both of Guilbault and Mamet (1976) (tribe Lowvillieae Shuysky), and *Palaeoporella* Stolley. According to Shuysky (1987b), *Palaeoporella* belongs to the Dasycladaceae, tribe Palaeoporelleae Shuysky. The stratigraphic and geographic distributions of these genera are given elsewhere (see Roux, this Vol.).

Dasycladaleans

Although relatively diversified, the Ordovician Dasycladales are represented by simple vermiporellid aspondylous forms, without regular arrangement of the branches, thus belonging to the Seletonellaceae. The most important or common genera are *Dasyporella* Stolley, *Kazakhstanelia* Korde, *Rhabdoporella* Stolley, *Vermiporella* Stolley, and *Novantiella* and *Intermurella*, both of Elliott (1972), etc.

It is important to mention the discovery by Poncet (1986) of the genus *Arcticella* in the Ordovician of the Canadian Arctic Archipelago (Devon Island), which displays euspondylous branches, crowded in successive alternating verticils. *Arcticella* is the first known true dasycladacean and the appearance of the euspondyly must be considered as a very early character in the evolution of the dasyclads (Fig. 2).

Solenoporaceans

Although numerous and diversified, the Ordovician solenoporaceans need careful revision; only important or common taxa are listed below: *Solenopora spongioides* Dybowski is largely distributed in Estonia, Scotland, Saint Lawrence Lowlands, New York State. *S. ouareauensis* Fritz, emend. Sinclair is also abundant in the Middle Ordovician of Québec and New York State and the

Upper Ordovician (Gamachian) of Anticosti Island. *S. trentonensis* Brown is present in Estonia (described as *S. compacta* by Nicholson and Etheridge 1885), the Saint Lawrence Lowlands, New York State and Anticosti Island (Upper Ordovician). *S. canadensis* (Foord), emend. Sinclair is known in the Middle Ordovician of the Saint Lawrence Lowlands. *Parachaetetes embrunensis* (Wilson) has been reported in Estonia, Ontario, Québec and New York State.

Problematic Algae

Although known from the Silurian, *Wetheredella silurica* Wood occurs in the Upper Ordovician (Ashgillian) of Sweden (as *Rhabdoporella glomerata* Jux) and Anticosti Island (as *Wetheredella tumulus* Copper). The genus *Wetheredella* is known as early as Lower Cambrian in Labrador (Kobluk and James 1979).

Halysis Høeg and *Flabellia* Shuysky (Codiaceae ?) play an important sedimentological role. *Halysis* occurs in the Ordovician of Norway, the Saint Lawrence Lowlands, and Anticosti Island. *Flabellia* occurs in the Ordovician of the Saint Lawrence Lowlands and Anticosti Island (Guilbault et al. 1976). These two genera are generally reported to the Hormogonophyceae by Russian authors.

The problematic microorganism *Nuia* Maslov (bacterial or cyanobacterial growth form ?) is globally distributed and ranges up to the Silurian (*Nuia distincta* Ischenko). Gnoli and Serpagli (1980, Fig. 4) have illustrated the Ordovician distribution of *Nuia*; see also Vachard and Tellez-Giron (1986). Numerous species of *Nuia* reported in the Devonian (*N. devonica* and *N.? minuta* Shuysky) do not belong to the genus.

Nuia is an important rock-builder, notably in the Upper Cambrian-Lower Ordovician of Texas and the Middle Ordovician (Chazy Group) of the Lake Champlain area and southern Québec. Although the carbonate mud-mounds are mainly built by coelenterates, sponges, bryozoans or tabulates, *Nuia* has played an active role in the building of the mounds: this form was able to baffle and cause local accumulations of carbonate mud. *Nuia* may, exceptionally, be very abundant and cause the formation of bafflestones. The associated algae mentioned by the authors are only *Girvanella* or *Renalcis*, with solenoporaceans and *Sphaerocodium*. In most cases, *Nuia* is associated with *Girvanella* in the same biotope. In other environments, *Girvanella* and *Nuia* are mutually exclusive. *Renalcis* has been encountered in the mud-mounds of the Arbuckle Group (Oklahoma) without *Nuia*; at least in this case, *Nuia* and *Renalcis* seem to be mutually exclusive.

Other genera of Ordovician *incertae algae* are *Guttoporella*, *Crinitella*, *Palmatoporella*, all of Gnilovskaja (1972), probable *Tharama*, etc. An important event is the first occurrence of probable algal cysts (Radiospheres). Remnants of Cambrian flora are the proauloporales *Proaulopora* Vologdin and *Setula* Luchinina.

Cyclocrineae may be green algae and are very abundant in the Ordovician with *Cyclocrinites* (= *Cyclocrinus*) Eichwald, *Mastopora* Eichwald, *Coelosphaeridium* Roemer, *Striola* Gnilovskaja, and *Apidium* Stolley. *Diversoporella*,

Sinuatoporella, and *Dolioporella*, all of Gnilovskaja (1972) are reported by the author as siphonales, together with *Dimorphosiphon* and *Palaeoporella*. Reported as 'red algae' by Gnilovskaja (1972) but very doubtfully so, are the Moniliporellaceae *Moniliporella*, *Contexta*, *Ansoporella*, *Furcatoporella*, *Plexa*, *Texturata*, and *Villosoporella*.

Characteristic associations

The Middle Ordovician of North America is characterized by the *Dimorphosiphon-Dimorphosiphonoides-Lowvillia-Nuia* assemblage with the species *D. rectangulare*, *D. lesperancei*, *L. multipora*, *L. raripora*, and *L. grandis*. The general lack of *Nuia* in the Black River and Trenton Groups is a local index of the uppermost part of the Chazy Group. The *Solenopora spongioides-S. ouareauensis-S. canadensis-S. trentonensis* assemblage seems to be characteristic of the Ordovician.

3 Silurian

In addition to the very long-ranging and ubiquitous *Girvanella*, the Silurian microflora is still dominated by filamentous codiaceans and erect udoteaceans.

Filamentous codiaceans

Bevocastria conglobata occurs in the Silurian of Estonia and in the Wenlockian to Pridolian of Québec.

Ortonella aequalis is present in the Wenlockian-Ludlovian of Québec (Baie des Chaleurs), with *Garwoodia gregaria* and is also known from the Pridolian of Podolia (Ischenko 1985). Although created from the Viséan, *Garwoodia zonata* Derville is known in the Wenlockian of Québec (Baie des Chaleurs) as *G.* aff. *zonata*. *Ortonella rigida* Lewis occurs in the Llandoverian of Wales and remains present in the Wenlockian-Ludlovian of Québec. *Ortonella humilis* and *O. stricta* have been erected from the Wenlockian and Pridolian of Podolia by Ischenko (1985).

Likewise, *Hedstroemia halimedoidea* and *H. bifilosa* remain present in the Silurian of Gotland, the Pridolian of Podolia (Ischenko 1985) and the Upper Silurian of the Siberian platform (Luchinina 1987). *H. halimedoidea* and *H. bernierensis* Mamet and Roux are present in the Wenlockian-Ludlovian of Québec. *Hedstroemia pulchra* Ischenko is reported from the Pridolian of Podolia.

Among the questionable codiaceans? or cyanophytes?, *Sphaerocodium munthei* Rothpletz, emend. Wood appears in the Lower Silurian and remains present in the Wenlockian to Pridolian (Gotland, Québec, Anticosti Island, Japan). *S. gotlandicum* remains present in the Silurian of Gotland, Sweden, Podolia, Czechoslovakia, Great Britain, Québec, California.

Erect Udoteaceans

The genus *Litanaia* Maslov appears in the Upper Silurian of Québec (Lake Matapedia) as *L. matapediensis* Mamet and Roux. In Podolia, the tribe Litanaiae Shuysky is also represented by *Maslovina* Obrhel (*Maslovina meyenii* Obrhel).

Dimorphosiphon magnum Gnilovskaja remains present in the Lower Silurian of Kazakhstan and the Upper Silurian (Wenlockian to Pridolian) of Québec (Baie des Chaleurs). A new species, *D. pusillus*, was erected by Ischenko (1985).

Likewise, *Palaeoporella* remains present in the Upper Silurian (Ludlovian-Pridolian) of Québec (Baie des Chaleurs) as *P.* sp. cf. *P. recta* Gnilovskaja. The genus disappears in the Upper Silurian.

Dasycladaleans

The Silurian dasycladaleans are represented by abundant *Rhabdoporella* and *Vermiporella* (see Roux, this volume). *Litopora* Johnson occurs in the Ludlovian-Pridolian of Québec (Baie des Chaleurs). *Sokolella* Ischenko is 'hybrid' between *Vermiporella* and *Wetheredella*.

Solenoporaceans

Solenopora sensu stricto (*S. spongioides* Dybowski) is illustrated by Ischenko (1985) from the Late Silurian of Podolia.

Parachaetetes embrunensis remains present in the Lower Silurian of Ontario, the Upper Silurian of Gotland and the Wenlockian-Pridolian of Québec (Baie des Chaleurs). Although mainly a Lower Carboniferous taxon, *Parachaetetes palaeozoicus* (Maslov) is known in the Upper Silurian of Québec (Wenlockian to Pridolian).

Pseudochaetetes filiformis (Nicholson) and *Petrophyton kiaeri* Høeg are present in the Silurian of Japan, the Wenlockian-Ludlovian of Québec (Lake Matapedia), and the Pridolian of Podolia (Ischenko 1985).

Charophytes

First representatives of the Sycidiceae and Trochiliscaeae are encountered in the Ludlovian-Pridolian (*Sycidium* Sandberger, *Praesycidium* Ischenko, *Primochara* Ischenko and Saidakovsky).

Problematic algae

Wetheredella silurica persists in the Silurian of Gotland and the Wenlockian-Pridolian of Québec (Baie des Chaleurs). The other species of the genus, *W. pachytheca* Mamet and Roux is so far only known from the Wenlockian-Ludlovian of Québec (Lake Matapedia).

Halysis and *Flabellia* are reported in the Silurian of Ufa and Podolia (USSR) and the Lower Silurian of Anticosti Island.

The first record of *Nostocites* Maslov is from Ludlovian-Wenlockian of Podolia (as '*Globochaete*').

Characteristic associations

The *S. gotlandicum-S. munthei-Dimorphosiphon magnum-Rhabdoporella* assemblage seems to be characteristic of the Silurian.

4 Devonian

The Devonian microflora is relatively quite prolific. Dasycladaleans are still distinctly subordinate to cyanophytes, rhodophytes, filamentous codiaceans, and erect udoteaceans (Fig. 1).

Filamentous codiaceans

Bevocastria conglobata remains abundant particularly in the Tethyan Realm (Russian platform, Afghanistan, France, Belgium, etc.).

Ortonella balbinia (Poncet) occurs in the Lower Devonian (Siegenian) of France and *O. tenuissima* Garwood, described from the Carboniferous (C_1) of Scotland, is known as *O.* sp. aff. *tenuissima* in the Famennian of New South Wales, Australia.

Hedstroemia is represented by *H. australe* Johnson in the Lower Devonian of Australia.

Sphaerocodium munthei is reported in the Devonian of Kuzbass, the Urals, and France and *S. gotlandicum* persists in the Devonian of Belgium, France, the USSR, Australia and Morocco. The Upper Devonian is characterized by the appearance of *S. exile* Wray and *S. magnum* Wray in the Upper Devonian (Frasnian-Famennian) of Australia (Canning and Bonaparte Gulf basins), where they are often associated with *Bevocastria* and *Ortonella* (Wray 1967).

Erect Udoteaceans

The erect udoteaceans show an acme during the Lower Devonian, a decline in the Middle Devonian, and a strong reduction in the Frasnian.

During the Lower Devonian, five species of *Abacella* Maslov (tribe Abacelleae Shuysky) are present: *A. cornuta* Shuysky (Emsian, Urals); *A. delicatula* Johnson (Lower to Middle Devonian, Australia, Lower Devonian, Urals); *A. dentata* Shuysky (Lower Emsian, Urals); *A. pertusa* Maslov (Lower Devonian, Kuzbass, Lower Emsian, Urals); *A. pustulata* Shuysky (Lower Emsian, Urals).

Litanaia Maslov (tribe Litanaiae Shuysky) is represented by eight species: *L. anirica* Maslov (Lower Devonian, Kuzbass, Lower Emsian, Urals – *L. anirica*

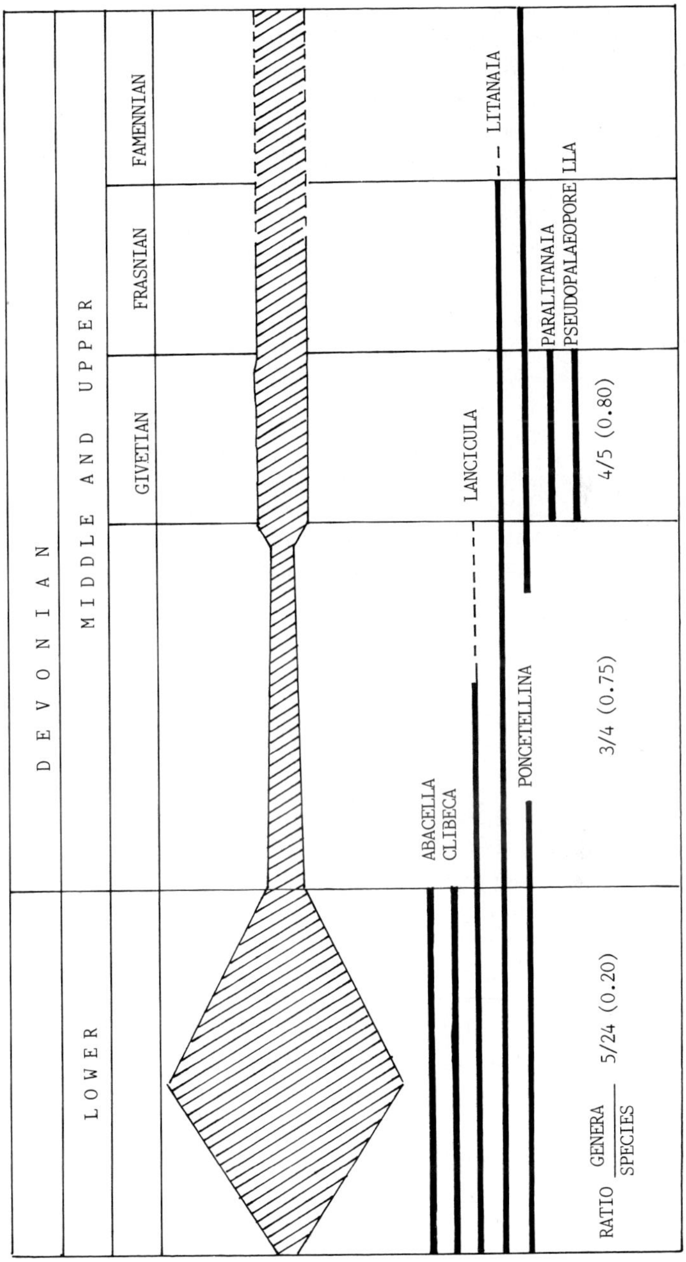

Fig. 1. Stratigraphic distribution and species diversity of Devonian erect udoteaceans. Note: *Lancicula* includes the new genera of Shuysky (1985)

is chosen by Shuysky (1987) as type of the genus *Praelitanaia* Shuysky, tribe Dimorphosiphoneae Shuysky); *L. bella* Pantić (Lower to Middle ? Devonian, Yugoslavia); *L. clivosa* Shuysky (Lower Emsian, Urals); *L. cracens* Johnson (Lower to Middle Devonian, Australia); *L. grandis* Pantić (Lower to Middle ? Devonian, Yugoslavia); *L. ? maslovi* Palla (Lower Devonian, Carnic Alps); *L. mira* Maslov (Lower Devonian, Kuzbass, Urals); *L. robusta* Johnson (Lower Devonian, Australia – *L. robusta* is partly transferred by Shuysky (1987) to a new genus: *Semilancicula* Shuysky).

Lancicula Maslov (tribe Lanciculeae Shuysky) is represented by one species, *L. alta* Maslov (Siegenian, Lower and Upper Emsian of the Kuzbass and the Urals, cited but not figured from the Lower Devonian of Greece, Chio Island). The other species previously described as *Lancicula* are transferred as follows in new genera by Shuysky (1985): *Lanciculella gortanii* (Palla 1966) Shuysky (Lower Emsian, Carnic Alps); *Lepidolancicula lancea* (Shuysky 1973) Shuysky (Siegenian-Lower Eifelian, Urals); *Planolancicula plana* (Shuysky 1973) Shuysky (Siegenian-Emsian, Urals); *Lanciculella sagittata* (Shuysky 1973) Shuysky (Siegenian-Emsian-Eifelian, Urals); *Quasilancicula sergaensis* (Shuysky 1973) Shuysky (Siegenian-Emsian, Urals); *Lepidolancicula valeriae* (Pantić 1973) Shuysky (Lower Devonian, Yugoslavia); *Quasilancicula wolfi* (Johnson 1964) Shuysky (Lower to Middle Devonian, Australia, Lower Emsian, Carnic Alps and Urals); *Lanciculina yugorica* (Shuysky 1973) Shuysky (Lower Emsian, Cap Blanc Nez area, USSR).

The genus *Circella* 'Schirschova, 1984 in press' is related to *Litanaia* but remains a *nomen nudum* at this moment.

Clibeca devoniana Poncet (tribe Clibeciae Shuysky) occurs in the Devonian of France.

The subtribe Voycarellinae Shuysky includes the genera *Voycarella* Shuysky and Schirschova (*V. magna* Shuysky and Schirschova), *Cauculicula* Shuysky and Schirschova (*C. venusta* Shuysky and Schirschova), *Semilancicula* Shuysky with *S. robusta* (Johnson), *pars*.

Also related are the following taxa:

The tribe Ampulliporeae Shuysky with *Ampullipora* Shuysky (*A. subita* Shuysky); *Litanaella* Shuysky and Schirschova (*L. constricta* Shuysky and Schirschova); *Praelitanaia* Shuysky (*P. anulata* Shuysky).

The tribe Botryelleae Shuysky with *Botryella* Shuysky (*B. spinosa* Shuysky and Schirschova).

The genus *Funiculus* Shuysky and Schirschova (*F. venosus* Shuysky and Schirschova) does certainly not belong to the tribe Boueneae Shuysky and is a synonym of *Pseudopalaeoporella* Mamet and Préat.

During the Middle Devonian, all species of *Abacella* and most of those of *Lancicula* and *Litanaia* disappeared, excepting *Lepidolancicula lancea* (Lower Eifelian, Urals), *Quasilancicula wolfi* (Middle? Devonian, New South Wales), *Litanaia cracens* (Middle? Devonian, New South Wales), and *L. grandis* (Middle? Devonian, Yugoslavia). *Litanaia? bijagodensis* Chuvashov (Givetian, Urals) has been withdrawn from the genus (see below). The Eifelian *Litanaia radiosa* Shuysky probably does not belong to the genus.

Palaeoporella was extinct by this time (*P. lummatonensis* Elliott being withdrawn from the genus, see below).

From the Lower Givetian of Belgium, Mamet and Préat (1985a) have described two genera representing a renewal of the Devonian udoteacean-flora:

Paralitanaia (type-species *P. baileuxensis*), including *Litanaia? bijagodensis* Chuvashov (type-species of *Bijagodella* Chuvashov).

Pseudopalaeoporella (type-species *Palaeoporella lummatonensis*) is monospecific and occurs also in the Upper Couvinian (Co_{2d}) of Belgium (the taxon of Elliott has been described from the Givetian of England (= *Funiculus* Shuysky and Schirschova).

Described from the Frasnian of Australia, *Litanaia perisseia* Wray is closely related to *P. baileuxensis* (Mamet and Préat).

The genus *Poncetellina* Mamet and Roux (described as *Poncetella* by Mamet and Roux 1983) occurs from the Lower Devonian (Lower Siegenian of France) with *P. erecta* (Poncet), described as *Solenopora*, to the Upper Devonian (Frasnian-Famennian of Australia, New South Wales) with *P. veeversi* (Mamet and Roux 1983; Fig. 1).

Dasycladaleans

Devonian dasyclads are unevenly represented; for instance, only a questionable vermiporellid and *Litopora spaciosa* are reported from Australia (Upper Devonian of Canning Basin and Lower Devonian of New South Wales respectively).

Forms described as *Rhabdoporella*, such as *R. melekesensis* Kulik and *R. uralica* Shuysky must be withdrawn from the genus.

Other forms, however, certainly represent very important stages of the evolution of the Palaeozoic dasycladaleans:

Unella roquellensis Poncet (Upper Siegenian of France) displays an euspondyl alternating disposition of the branches (R1). This taxon is another example of euspondyly within the Palaeozoic dasyclads.

Diplopora praehexaster Poncet (Middle Siegenian, France) and especially *Diplopora constantini* Poncet (Upper Siegenian, France) are metaspondyl and vestibular. The thallus is unsegmented, the vestibule is simple and the branches not divided (R1). Thus, the first known occurrence of metaspondyl vestibular Dasycladaceae dates as early as the Lower Devonian.

Givetianella described by Mamet and Préat (1982) from the Lower Givetian of Belgium (*G. tsienii*) shows a very large ramose thallus and aspondylous branches (R1) made up of a series of spheroids. From an evolutionary point of view, this dasyclad recalls the Lower Palaeozoic vermiporellids in the shape of the thallus, but the morphology of the branches announces the Carboniferous and Permian *Epimastopora*-like forms with halter-shaped primary branches (Fig. 2).

Other Devonian dasyclads are:

Ulocladia Shuysky and Schirschova with *U. pilosa* Shuysky and Schirschova and *U. calva* Shuysky.

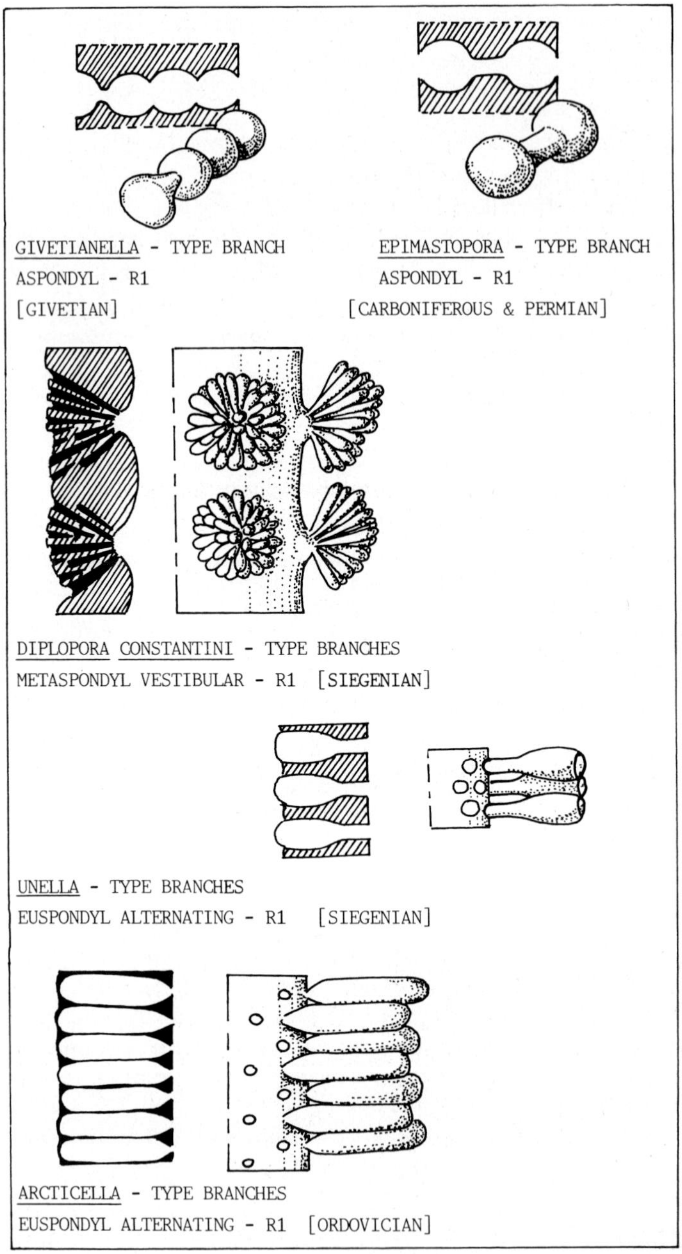

Fig. 2. Shape of thallus, type and arrangement of branches of important Lower Palaeozoic dasycladaleans (partly after Poncet 1965, 1974b, 1986 and Mamet and Préat 1982)

Fig. 3a-f. Some representative genera and species of questionable codiaceans and nodular or filamentous codiaceans. **a** *Sphaerocodium moniliferum* (Seely) 1904. Originally described as *Stromatocerium*. Middle Ordovician, Chazy Group, USA. Vermont. (Seely 1904). x35. **b** *Sphaerocodium gotlandicum* Rothpletz 1908, emend Wood 1948. Lower Devonian, Australia. Note the

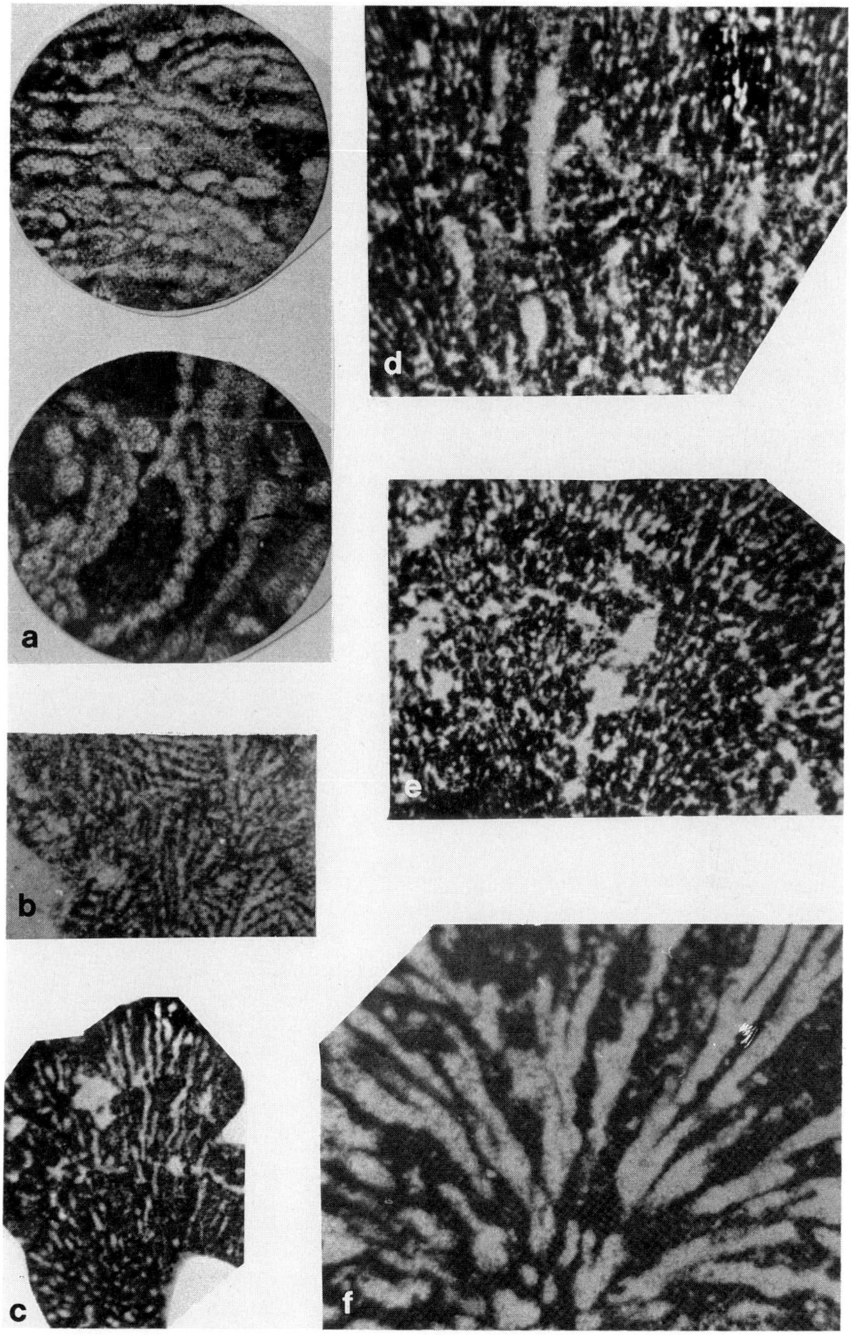

characteristic type of branching. (Johnson 1964). x50. **c** *Garwoodia* aff. *zonata* Derville 1931. Wenlockian, La Vieille Formation, Québec, Baie des Chaleurs. (Bourque et al. 1981). x25. **d,e** *Ortonella rigida* Lewis 1937. Wenlockian, La Vieille Formation, Québec, Baie des Chaleurs. (Bourque et al. 1981). x25. **f** *Hedstroemia halimedoidea* Rothpletz 1913. Silurian, Québec, Lake Matapédia. (Héroux et al. 1977). x30

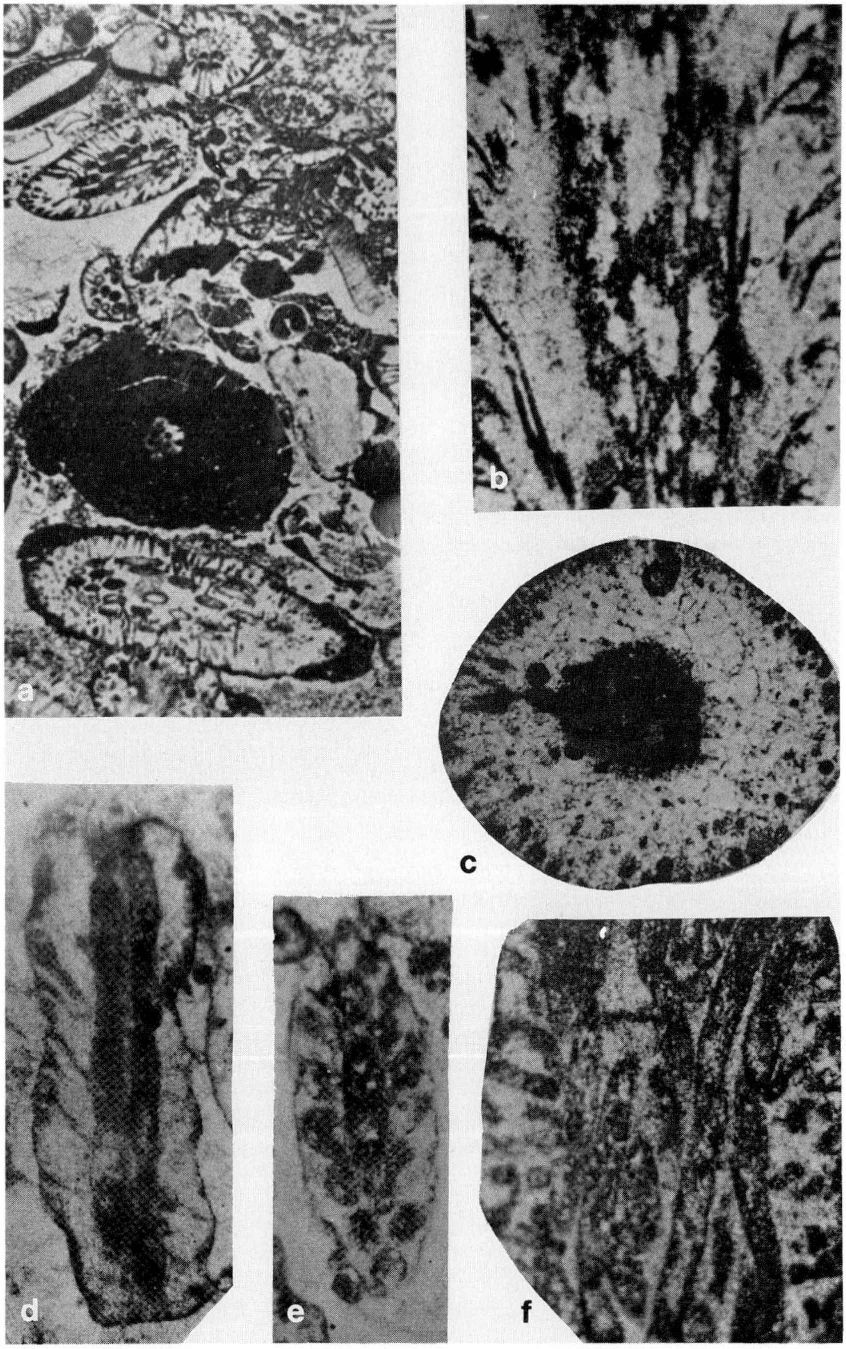
a
b
c
d
e
f

Scasyporella Shuysky with *S. uralica* (Shuysky).

Ivdelipora Shuysky and Schirschova (related to *Albertaporella*) with *I. gloriosa* Shuysky and Schirschova.

Rotella Shuysky and Schirschova with *R. conifera* Shuysky and Schirschova.

Coticula Shuysky and Schirschova with *C. binata* Shuysky and Schirschova.

Thibia Shuysky with *T. proninae* Shuysky.

The genera *Lulipora* Shuysky (type *L. schratovii*) and *Lopsiella* Shuysky (type *L. mansica*) are attributed to the Acetabulariaceae. This attribution is, however, doubtful.

Solenoporaceans

Solenopora spongioides remains present in the Lower Devonian of the Urals with *S.? taravalensis* Shuysky.

Pseudosolenopora Mamet and Roux occurs in the Frasnian-Famennian of Australia (Canning Basin) with *P.? geikiei* (Wray).

Pseudochaetetes Haug, emend. Peterhans is well-represented by *Pseudochaetetes devoniensis* (Delépine) in the Lower Devonian (Siegenian) of France and the Upper Devonian of Turkey (Anatolia).

Parachaetetes johnsoni Maslov is cosmopolitan from the Upper Devonian upwards (Frasnian ?-Famennian of USSR, Belgium, Australia, etc.). Likewise, *P. regularis* Konishi is cosmopolitan from the Frasnian-Famennian upwards (Belgium, Alberta, and Australia where the taxon is very abundant in the Upper Famennian of the Bonaparte Gulf Basin). *Parachaetetes palaeozoicus* remains present in the Upper Famennian of the Urals; *P. concentrica* and *P. russiensis*, both of Maslov (1956) seem to be restricted to the Devonian of the Moscow area (for the taxonomy, see the revision of the Devonian solenoporaceans in Mamet and Roux 1977; Fig. 5).

In addition to common solenoporaceans, there are mainly puzzling but highly characteristic red algae which seem to be restricted to the period:

- The Demidellaceae (with *Demidella* Shuysky).
- The Paralanciculeae (with *Paralancicula* Shuysky).
- The Katavellaceae (with *Katavella* Chuvashov).

Fig. 4a-f. Some representative genera and species of erect udoteaceans. **a** *Dimorphosiphon rectangulare* Høeg, 1927. Ordovician, Norway. Note the medullary tubes and the subperpendicular lateral or cortical filaments. (Høeg 1927). x5. **b** *Palaeoporella variabilis* Stolley 1893. Upper Ordovician (Ashgillian) Sweden. Note the medullary filaments and the dichotomously branching lateral filaments. (Jux 1966). x30. **c** *Poncetellina veeversi* (Mamet and Roux), 1983. Holotype. Frasnian, Cockatoo Formation, Australia, Bonaparte Gulf basin. (Mamet and Roux 1983). x25. **d,e** *Litanaia bella* Pantić, 1973. Lower Devonian, Yugoslavia, Bosnia. (Pantić 1973). x32. **f** *Paralitanaia baileuxensis* Mamet and Préat 1985. Givetian, Belgium. Note the sinuous medullary tubes. (Mamet and Préat 1985[a]). x25

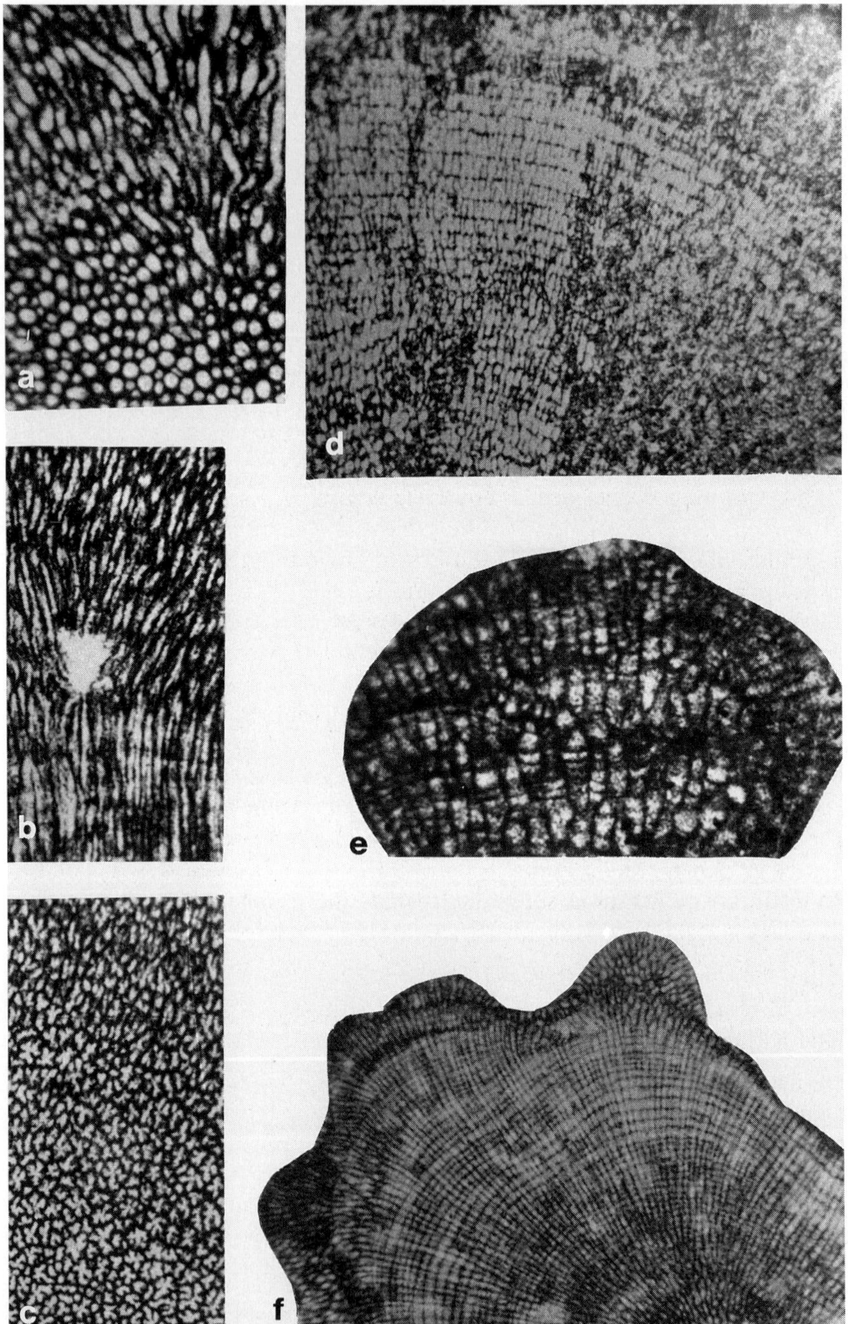
a
b
c
d
e
f

Palaeosiphonocladales (issinellids and palaeoberesellids)

A major event ocurs in the Givetian with the appearance of abundant issinellids and palaeoberesellids: *Issinella devonica* Reitlinger, emend. Mamet and Roux, *Pseudoissinella* Mamet and Rudloff, *Pseudokamaena* Mamet, *Kamaena* Antropov, and *Parakamaena* Mamet and Roux. In Upper Devonian times, *Kamaena* is diversified by the appearance of *K. awirsi* Mamet and Roux and *K. delicata* Antropov. At the same time, the cosmopolitan genus *Palaeoberesella* Mamet and Roux appears, then in Late Devonian *Exvotarisella* Elliott with *P. lahuseni* (von Möller) and *E. index* (Ehrenberg) respectively.

Issinella and *Kamaena* are cosmopolitan in the northern hemisphere from the Middle Devonian onwards. *Exvotarisella index*, originally known from the Viséan to the Lower Namurian of the Tethys, is reported from the Devonian-Carboniferous boundary in eastern Siberia and the Late Famennian of Australia. *Parakamaena* and *Pseudokamaena* are cosmopolitan in the northern hemisphere from the Middle Devonian onwards.

Jansaella ridingi Mamet and Roux, described from the Devonian of Alberta is an incertae algae, probably a chlorophyte.

The palaeoberesellids most probably originate from the issinellids. From *Jansaella* to the palaeoberesellids through the issenellids, one observes a progressive arrangement of the pores ('branches'): in *Jansaella*, they are very thin and numerous, straight and arranged without any order; in *Issinella*, they are apparently less numerous, but perpendicular to the cortex; in the palaeoberesellids, the pores are again less numerous and located only in parts of thallus between two consecutive 'septa'. The pores are simple in *Kamaena*, *Palaeoberesella*, *Pseudokamaena*, *Kamaenella* and *Parakamaena* and dichotomous with three orders of ramification in *Exvotarisella* (Fig. 6).

Kamaena, *Kamaenella*, *Palaeoberesella*, *Parakamaena*, and *Exvotarisella* are included in the Palaeosiphonocladales by Shuysky (1985). The same genera with *Jansaella* and *Pseudokamaena* are attributed to the Dasycladales, Beresellaceae Deloffre, tribe Palaeoberesselleae Mamet and Roux, by Deloffre (1987). *Issinella* and *Pseudoissinella* are attributed to the tribe Issinelleae Deloffre (see Deloffre 1988).

Shuysky (1987b) includes in the Siphonocladales the genera *Calcicaulis* Shuysky and Schirschova (*C. vesiculosum* Shuysky and Schirschova) and

Fig. 5a-f. Some representative rhodophytes. **a** *Parachaetetes embrunensis* (Wilson), 1945. Wenlockian, La Vieille Formation, Québec, Baie des Chaleurs. (Bourque et al. 1981) x25. **b,c** *Solenopora ouareauensis* Fritz 1941, *emend.* Sinclair 1956. Upper Ordovician, Ellis Bay Formation, Anticosti Island. (Gauthier 1981). Note the presence of a possible sporangial structure (**b**) and the "septiform processes" of the longitudinal partition of the tubes (**c**). x25. **d** *Petrophyton kiaeri* Høeg 1932. Ordovician, Norway. (From Høeg 1932). x15. **e** *Parachaetetes johnsoni* Maslov 1962. Upper Famennian, Belgium. Note the irregular diameter of the tubes and the horizontal partitions situated at the same level from one tube to another. (Mamet and Roux 1977). x55. **f** *Parachaetetes regularis* Konishi 1958. Upper Famennian, Australia, Bonaparte Gulf basin. Note the very regular arrangement of the cells. (Mamet and Roux 1983) x62

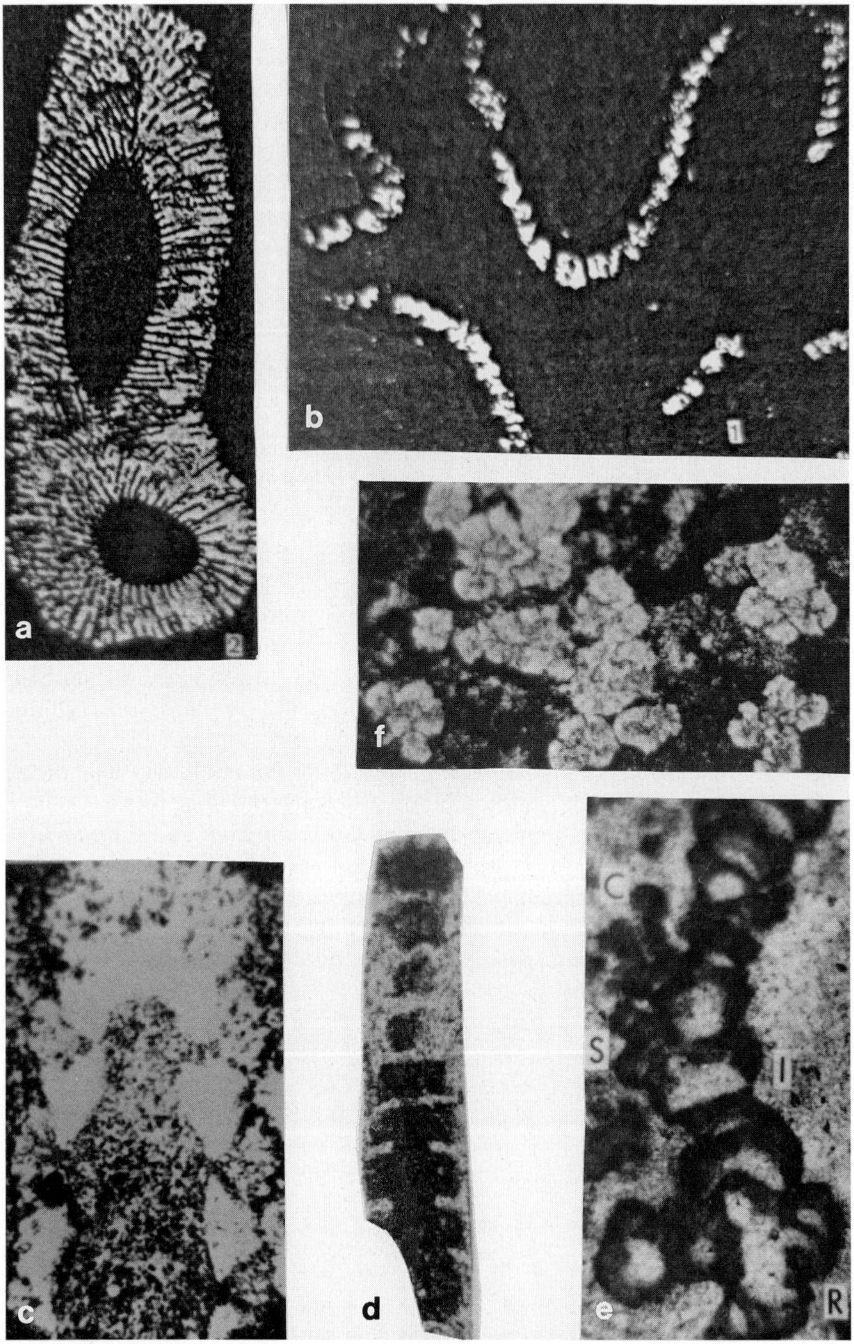
a
2
b
1
f
c
d
C
S
I
e
R

Parmacaulis Shuysky and Schirschova (*P. hulgensis* Shuysky and Schirschova). These genera are, however, radically different from any known Siphonoclades.

Umbellids and charophytes

Charophytes, which find their roots in the Silurian, are very prolific in the Devonian with representatives of Chovanellaceae Grambast, Pinnoputamenaceae Wang and Lu, Sycidiaceae Karpinsky, emend. Peck, Trochiliscaceae Karpinsky, and Karpinskyaceae Wang and Lu.

Many Russian authors include the umbellids among charophytes. They have, however, a particular wall structure (a micritic inner layer and a finely prismatic outer layer) unknown among trochiliscids. They appear in the Lower Devonian, are extremely common in non-marine, brackish and littoral waters, and become extinct in the Early Carboniferous.

Well documented Devonian genera are: *Umbellina* Loeblich and Tappan (= *Umbella* 'Maslov' in Bykova and Polonova 1955); *Elenia* Pojarkov; *Eoumbella* Platonov (= *Costatumbella* Berchenko); *Planoumbella* Platonov; *Spinoumbella* Platonov; *Quasiumbella* Pojarkov; *Biumbella* Mamet (= *Quasiumbelloides* Berchenko); *Kusjaella* Chuvashov; *Lagenumbella* Mamet; *Collumbella* Mamet and Roux.

Problematic algae

Wetheredella silurica remains present in the Upper Devonian (Upper Frasnian) of Belgium and *Halysis* and *Flabellia* are cited from the Lower Devonian of the Urals.

It is important to mention the presence of Labyrinthoconidae in the Givetian of Belgium, Germany and Moravia, whose wall structure and morphology announce the calcified wall of the Carboniferous Stacheiinae, with, however, a different morphology of the cell threads (Mamet et al. 1982).

Proninella Reitlinger in Menner and Reitlinger, and *Asphaltina* Mamet in Petryk and Mamet are present. Calcispheres become very abundant in

Fig. 6a-f. Some representative dasycladaleans, paleoberesellids and problematic algae (renalcids and microcodiacids). **a** *Vermiporella acerosa* Gnilovskaja 1965. Upper Ordovician, USSR, Tarbagatay Range. Note the dichotomy of the thallus and the simple branches, aspondyl R1. (Gnilovskaja 1965). x30. **b** *Vermiporella wesenbergensis* Moskalenko 1956. Upper Ordovician, USSR, Tarbagatay Range. Note the dichotomy of the thallus and the very large central stem. (Gnilovskaja 1965) x30. **c** *Diplopora constantini* Poncet 1965. Holotype. Siegenian, France. Note the continuous thallus and the clusters of branches issuing from simple vestibules. (Poncet 1965). x15. **d** *Kamaena delicata* Antropov 1967. Upper Tournaisian, Great Britain. Note the shape of the medullary cavities. (Mamet and Roux 1974). x135. **e** *Renalcis granosus* Vologdin 1932. Frasnian, Cockatoo Formation, Australia, Bonaparte Gulf Basin. Note the transition *Renalcis* (*R*); *Izhella* (*I*); *Shuguria* (*S*); *Chabakovia* (*C*). (Mamet and Roux 1983). x62. **f** *Palaeomicrocodium devonicum* Mamet and Roux 1983. Holotype. Frasnian, Cockatoo Formation, Australia, Bonaparte Gulf Basin. (Mamet and Roux 1983). x36

Devonian lagoons and restricted environments (*Calcisphaera*, *Parathurammina*, *Palaeocancellus*, *Radiosphaera*, etc.).

Renalcis and related forms

The Lower Palaeozoic microfloras contain long-ranging and non-diagnostic taxa such as *Girvanella*, *Renalcis* or epiphytales of uncertain affinity (microbial? forms).

Renalcis Vologdin, emend. Mamet and Roux is itself a growth-form, including *Izhella* Antropov, *Chabakovia* Vologdin, and *Shuguria* Antropov (see the revision of the genus *Renalcis* in Mamet and Roux 1983).

Epiphyton Bornemann is very common in the Lower and Middle Cambrian and is often a rock-former. Although reported from the Devonian of USSR, no Ordovician or Silurian forms are known.

A related taxon, *Paraepiphyton*, is described by Wray (1967) from the Upper Devonian (Frasnian-Famennian) of Australia. *Paraepiphyton* and *Epiphyton* may be synonymous. *Paraepiphyton* is often associated with *Renalcis*.

Another form, *Tharama*, is described by the same author (Wray 1967) from the Frasnian-?Famennian of Australia, where it is associated with *Renalcis* and *Epiphyton*. *Tharama* is known in the Middle Ordovician of North America (New York State and Ontario).

Forms closely related to *Frutexites* are illustrated by Mamet and Roux (1981, in Bourque et al.) from the Silurian of Québec.

Frutexites was originally described without a type by Maslov (1960) from Ordovician stromatolites of USSR. The 'genus' is known in Devonian stromatolites in Australia and was probably an important rock-former in the Devonian reefs in Alberta. Likewise, forms described as '*Epiphyton*' are important rock-formers in the Devonian reefs of Belgium.

Microcodiaceae

It is important to mention the discovery by Mamet and Roux (1983) of Microcodiaceae Maslov in the Upper Devonian of Australia. *Palaeomicrocodium devonicum* Mamet and Roux is common in marine back-reef environments and often associated with *Renalcis*. Since then, Microcodiaceae have been found in Givetian-Frasnian (Mamet and Préat 1985b).

Characteristic associations

The *Litanaia-Lancicula-Abacella* association is typical of the Lower Devonian. During the Upper Devonian, the *S. exile-P. johnsoni-P. regularis-P. geikiei* assemblage seems to be characteristic.

5 Conclusions

Despite an abundant literature which is often confusing, the Lower Palaeozoic microfloras remain poorly known (Riding 1979). Many problems remain to be solved: the explosion of heavy calcified forms by Lower Ordovician times implies the existence of Cambrian uncalcified stocks; the Silurian dasyclads are poorly known and the affinities of many taxa are uncertain. Deciphering the evolution of the Lower Palaeozoic algae requires further taxonomical and stratigraphical studies.

Acknowledgements. Dr. B.L. Mamet has kindly read the manuscript and made useful comments.

References

Antropov IA (1950) New species of foraminifers from the Upper Devonian of some parts of the Russian Platform. Akad Nauk SSSR, Kazansk Fil Izv Ser Geol 1:21–33 (in Russian)

Antropov IA (1959) Devonian foraminifers from Tataria. Akad Nauk SSSR, Kazansk Fil Izv Ser Geol 7:11–33 (in Russian)

Bassoullet JP, Bernier P, Deloffre R, Génot P, Poncet J, Roux A (1983) Les Algues Udotéacées du Paléozoïque au Cénozoïque. Bull Centres Rech Explor Prod Elf-Aquitaine 7, 2:449–621

Bourque PA, Mamet B, Roux A (1981) Algues siluriennes du Synclinorium de la Baie des Chaleurs, Québec, Canada. Rev Micropaleontol 24, 2:83–126

Chuvashov BI (1965) Foraminifers and algae from the Upper Devonian of the western slope of the middle southern Urals. Akad Nauk SSSR, Uralsk Filial, 74:1–153 (in Russian)

Chuvashov BI (1987a) Place of the calcareous algae in the organic world system. General principles of systematics and methods of study. In: Dubatolov VN (ed) Fossil calcareous algae, morphology, systematics and study methods. Akad Nauk SSSR, Sibirsk Otdel, Tr Inst Geol Geofiz Novosibirsk 674:5–12 (in Russian)

Chuvashov BI (1987b) Red algae (Rhodophyta). In: Dubatolov VN (ed) Fossil calcareous algae, morphology, systematics and study methods. Akad Nauk SSSR, Sibirsk Otdel, Tr Inst Geol Geofiz Novosibirsk 674:5–12 (in Russian)

Copper P (1976) The Cyanophyte *Wetheredella* in Ordovician reefs and off-reef sediments. Lethaia 9:273–281

Deloffre R (1988) Nouvelle Taxonomie des Algues Dasycladales. Bull Centres Rech Explor Prod Elf-Aquitaine 12, 1:165–217

Derville H (1931) Les Marbres du Calcaire carbonifère en Bas-Boulonnais. Boehm, Strasbourg, 322 pp

Elliott GF (1960) A new British Devonian Algae, *Palaeoporella lummatonensis*, and the Brachiopod evidence of the age of the Lummaton Shell-bed. Proc Geol Assoc, London, 72, pt 2:251–260

Elliott GF (1972) Lower Palaeozoic green Algae from Southern Scotland, and their evolutionary significance. Bull Brit Mus Nat Hist Geol 22, 4:358–376

Fritz MA (1941) On *Solenopora compacta* (Billings) and the new variety *Solenopora compacta ouareauensis*. R Can Inst Trans 23, pt 2, 50:157–160

Garwood EJ (1931) The Tuedian beds of northern Cumberland and Roxburghshire east of the Liddel Water. Q J Geol Soc London 87:97–157

Gauthier L (1981) Paléoécologie des Algues ordoviciennes et siluriennes de l'Ile d'Anticosti, Québec. M Sc Thesis, Univ Montréal

Gnilovskaja MB (1972) Middle and Upper Ordovician calcareous algae from eastern Kazakhstan. Akad Nauk SSSR, Inst Geol Geochron Dokembrja. Nauka, Leningrad, 192 pp (in Russian)

Gnilovskaja MV (1965) Ordovician algae of the genus *Vermiporella* from Tarbagatay Range. Int Geol Rev 12:2105–2110 (translated from Russian)

Gnoli M, Serpagli E (1980) The problematical microorganism *Nuia* in the Lower Ordovician of precordilleran Argentina and its paleogeographic significance. J Paleontol 54(6):1245–1251

Guilbault JP, Mamet B (1976) Codiacées (Algues) ordoviciennes des Basses-Terres du Saint-Laurent. Can J Earth Sci 13, 5:636–660

Guilbault JP, Hubert C, Mamet B (1976) *Nuia* et *Halysis*, deux Algues ordoviciennes énigmatiques des Basses-Terres du Saint-Laurent. Nat Can 103:119–132

Héroux Y, Hubert C, Mamet B, Roux A (1977) Algues siluriennes de la Formation de Sayabec (Lac Matapédia, Québec). Can J Earth Sci 14, 12:2865–2908

Høeg OA (1927) *Dimorphosiphon rectangulare*, preliminary note on a new Codiaceae from the Ordovician of Norway. Avh Nors Vid Akad Oslo 1, Mat Naturvid Kl 1:1–15

Høeg OA (1932) Ordovician algae from the Trondheim area. In: The Ovin Group in the Trondheim area II. Paleontological part. Skr Utgitt Nors Vid Akad Oslo 1, Mat Naturvid Kl 4:63–96

Johnson JH (1964) Lower Devonian algae and encrusting foraminifera from New South Wales. J Paleontol 38, 1:98–108

Jux U (1966) Rhabdoporella im Boda-Kalk sowie in Sandsteinen Dalarnes (Ashgill; Schweden). Palaeontographica B 118, 4–8:166–183

Kobluk DR, James NP (1979) Cavity-dwelling organisms in Lower Cambrian patch reefs from southern Labrador. Lethaia 12:193–218

Konishi K (1958) Devonian calcareous algae from Alberta, Canada. Col Sch Mines Q 53 2, pt 2:85–109

Korde KB (1950) On the morphology of the verticillate siphonales from the Carboniferous of the northern Urals. Dokl Akad Nauk SSSR 73, 3:569–571 (in Russian)

Korde KB (ed) (1957) New representatives of the siphoneae algae. In: Korde KB (ed) Data for the principles of palaeontology, vol 1. Akad Nauk SSSR, Leningrad, pp 67–75 (in Russian)

Lewis HP (1937) Calcareous Algae (Ortonella and Rhabdoporella) in the Llandoverian rocks of Wales. Ann Mag Nat Hist 10, 20:120, 617–623

Luchinina VA (1987) Blue green algae (cyanobacteria). In: Dubatolov VN (ed) Fossil calcareous algae, morphology, systematics and study methods. Akad Nauk SSSR, Sibirsk Otdel, Tr Inst Geol Geofiz Novosibirsk 674:12–38 (in Russian)

Mamet B, Préat A (1982) *Givetianella tsienii*, une Dasycladacée nouvelle du Givétien de la Belgique. Bull Soc Belg Geol 91, 4:209–216

Mamet B, Préat A (1985a) Sur quelques Algues vertes nouvelles du Givétien de la Belgique. Rev Micropaléontol 28, 1:67–74

Mamet B, Préat A (1985b) Sur la présence de *Palaeomicrocodium* (Algue ?, *Incertae sedis* ?) dans le Givétien inférieur de Belgique. Geobios 18, 3:389–392

Mamet B, Roux A (1974) Sur quelques Algues tubulaires scalariformes de la Téthys paléozoique. Rev Micropaléontol 17, 3:134–156

Mamet B, Roux A (1975a) Algues dévoniennes et carbonifères de la Téthys occidentale. Rev Micropaléontol 18, 3:134–187

Mamet B, Roux A (1975b) *Jansaella ridingi*, nouveau genre d'Algue? dans le Dévonien de l'Alberta. Can J Earth Sci 12, 8:1480–1484

Mamet B, Roux A (1977) Algues rouges dévoniennes et carbonifères de la Téthys occidentale. Rev Micropaléontol 19, 4:215–266

Mamet B, Roux A (1983) Algues dévono-carbonifères de l'Australie. Rev Micropaléontol 26, 2:63–131

Mamet B, Roux A (1984) *Poncetellina nomen novum* (Algue verte). Rev Micropaléontol 27, 2:138

Mamet B, Préat A, Boulvain F (1982) Sur la structure des Labyrinthoconidae (Algues problématiques, Givétien). Ann Soc Geol Belg 105, 2:283–288

Maslov VP (1929) Microscopic algae from the Carboniferous limestones of the Donetz Basin. Bull Com Geol 48, 10:115–138 (in Russian)

Maslov VP (1956) The fossil calcareous algae of the USSR. Akad Nauk SSSR, Tr Geol Inst, 160:301 p (in Russian)

Maslov VP (1960) The Acicularieae and their significance in the stratigraphy of the USSR. Paleontol Zh 3:115–122 (in Russian)

Maslov VP (1962) Fossil red algae of the USSR. Akad Nauk SSSR, Tr Inst Geol Novosibirsk 53:252 pp (in Russian)

Menner VV, Reitlinger EA (1971) Provincialism of the Middle and Upper Devonian foriminifers in the northern part of the Siberian platform. Akad Nauk SSSR, Vopr Mikropaleontol 14:25–38 (in Russian)

Nicholson HA, Etheridge R (1885) On the synonymy, structure and geological distribution of *Solenopora compacta* Billings sp. Geol Mag Ser 2, 12:529–535

Palla P (1966) *Lancicula gortanii*, nuova specie di Alga Codiacea del devoniano inferiore della Carnia (Alpi orientali) (Chlorophyceae, Codiaceae). Natura 57, 2:89–102

Pantić S (1973) New species of Devonian Algae from Klek, southeastern Bosnia. Ann Geol Penins Balkan 38:259–276

Poncet J (1965) Sur une Dasycladacée nouvelle du Dévonien inférieur: *Diplopora constantini* nov. sp. Bull Soc Géol Fr Ser 7, 7:879–880

Poncet J (1967) Une nouvelle Dasycladaceae dans le Dévonien inférieur du Cotentin. CR Somm Soc Géol Fr 8:374–376

Poncet J (1974a) Description de quelques Algues calcaires éodévoniennes du nord-est du Massif Armoricain. Bull Soc Géol Fr Ser 7, 16:225–229

Poncet J (1974b) *Unella roquellensis* nov. gen., nov. sp., Dasycladacée éodévonienne du Massif Armoricain (France). Observations sur son écologie. Geobios 7, 1:77–80

Poncet J (1986) Les Algues calcaires du Paléozoique inférieur de la Baie d'Hudson et de l'Archipel Arctique Canadien. Bull Centres Rech Explor Prod Elf-Aquitaine 10, 2:259–282

Riding R (1979) Devonian calcareous algae. In: House MR, Scrutton CT, Bassett MG (eds) The Devonian System. Spec Pap Palaeontology, 23:141–144

Riding R, Voronova L (1984) Assemblages of calcareous algae near the Precambrian/Cambrian boundary in Siberia and Mongolia. Geol Mag 121:205–210

Riding R, Voronova L (1985) Morphological groups and series in cambrian algae. In: Toomey DF, Nitecki MH (eds) Palaeoalgology: contemporary research and applications. Springer, Berlin Heidelberg New York Tokyo, pp 56–78

Rothpletz A (1913) Über die Kalkalgen, Spongiostromen und einige andere Fossilien aus dem Obersilur Gotlands. Sver Geol Unders Ser C, Avh Uppsatser 10:54 pp

Roux A (1985)Introduction à l'étude des Algues fossiles paléozoiques (de la bactérie à la tectonique des plaques). Bull Centres Rech Explor Prod Elf-Aquitaine 9, 2:465–699

Roux A (1986) Microflores algaires paléozoïques (tendances évolutives, répartitions) et tectonique globale. Bull Centres Rech Explor Prod Elf-Aquitaine 10, 2:555–563

Seely HM (1904) The *Stromatoceria* of Isle la Motte, Vermont. Vermont State Geol Rep 4:144–165

Shaikin I.M. (1987) Charophyte Algae (Charophyta). In: Dubatolov VN (ed) Fossil calcareous algae, morphology, systematics and study methods. Akad Nauk SSSR, Sibirsk Otdel, Tr Inst Geol Geofiz Novosibirsk 674:140–160 (in Russian)

Shuysky VP (1973) Reef calcareous algae from the Lower Devonian of the Urals. Akad Nauk SSSR, Inst Geol Geochim AN Zavaritsky. Nauka, Moscow, 155 pp (in Russian)

Shuysky VP (1985) On the position of paleoberesellids and other segmented algae in the system of Siphonophyceae. In: New data for geology, biostratigraphy and paleontology of the Urals. Ural Nauchn Tsentr Akad Nauk SSSR, Sverdlovsk, pp 86–95 (in Russian)

Shuysky VP (1986) New representative of dacyclad alga from the Lower Devonian of the Urals. Paleontol J Akad Nauk SSSR 2:118–122 (in Russian)

Shuysky VP (1987a) *Lopsiella mansica*, new representative of the acetabularian alga from the Givetian deposits of the Urals. Paleontol J Akad Nauk SSSR 3:67–75 (in Russian)

Shuysky VP (1987b) Green Algae (Chlorophyta). In: Dubatolov VN (ed) Fossil calcareous algae, morphology, systematics and study methods. Akad Nauk SSSR, Sibirsk Otdel, Tr Inst Geol Geofiz Novosibirsk 674:38–109 (in Russian)

Sinclair GW (1956) *Solenopora canadensis* (Foord) and other algae from the Ordovician of Canada. Trans R Soc Can L, Ser III, 4:65–81

Stolley E (1893) Über silurische Siphoneen. N Jahrb Mineral Geol Paläontol 2:135–146

Vachard D, Tellez-Giron C (1986) El alga *Nuia* en el Ordovicico de Mexico: hipotesis diversas. Rev Inst Mex Petrol 18(2):12–25

Wood A (1948) *Sphaerocodium*, a misinterpreted fossil from the Wenlock Limestone. Proc Geol Assoc 59:9–22

Wood A (1957) The type-species of the genus *Girvanella* (calcareous algae). Palaeontology 1, 1:22–28

Wray JL (1967) Upper Devonian calcareous algae from the Canning Basin, western Australia. Prof Contrib Col Sch Mines 3

Wray JL (1977) Calcareous algae; developments in palaeontology and stratigraphy. Elsevier, Amsterdam, 185 pp

Chapter 19

Carboniferous Calcareous Algae

B. Mamet[1]

Abstract

This compendium records 139 critically assessed valid genera and 22 form-genera of Carboniferous blue-green, green and red algae. It also includes some umbellinids, algal cysts and incertae sedis. For each genus, it lists the valid species and assesses suprageneric attributions. It records stratigraphic range, geographic distribution and abundance. The role in carbonate sedimentation is emphasized.

The vast majority of algae are very long-ranging and cannot be used for precise age determination. Obvious exceptions are *Calcifolium* and *Koninckopora*.

1 Introduction

Carboniferous calcareous algae are diverse, abundant and controversial. The purpose of this compendium is not to add fuel to disputable classifications but to critically record what is securely known, and list valid generic and specific taxa, with their known stratigraphic range, their role in sedimentation and their observed geographic distribution. Suprageneric attributions remain tentative and largely equivocal.

Numerous attempts at classifications have recently been published: Bassoullet et al. (1979) on dasycladales; Bassoullet et al. (1983) on udoteaceans; Chuvashov, Luchinina, Schaikin and Shuysky edited by Dubatolov (1987); and Deloffre (1988). But there is still little concensus on palaeobotanical attributions. Dasyclads are morphologically quite complicated and are therefore only partially understood. On the other hand, simple cyanobacteria remain a maze of puzzling and contradictory taxonomical classes and orders.

These taxonomic limitations admitted, algae are nevertheless of paramount importance for understanding carbonate sedimentation and furnish some useful stratigraphic observations. The purpose of this compilation is therefore to condense taxonomic and stratigraphic information gleaned from numerous articles together with unpublished personal observations.

2 Calcareous Algal Microflora

Nearly two hundred genera have been described in the literature (197 taxa). In this compilation, 35 of these genera are regarded as synonyms. An additional 22

[1]Département de géologie, Université de Montréal, C.P. 6128, succursale A., Montréal, Québec H3C 3J7, Canada

genera are "composite taxa", usually the result of biological activity of two or three organisms (spongiostromids). This leaves us with a grand total of 139 genera and 330 valid species.

For the sake of simplicity, genera are divided into informal groups and plotted on an idealized Carboniferous carbonate platform (Fig. 1).

1. Cyanobacteria (or cyanophytes) are omnipresent, very long ranging through the whole Paleozoic and represented in the Carboniferous by only seven genera of Chabakoviaceae, Epiphytaceae and Girvanellaceae.
2. Twenty-two "genera" of spongiostromids (algal biscuits) are common in Tethysian protected lagoons ("*Chondrostoma*", "*Polymorphocodium*", "*Pycnostroma*") but need further study to establish their real identity.
3. Nodular codiaceans (?) or nodular cyanophytes (?) comprise 8 genera of Garwoodiaceae. *Bevocastria*, *Garwoodia*, *Hedstroemia*, *Mitcheldeania*, *Ortonella*, thrive in Palaeozoic restricted and semi-restricted environments.
4. Carboniferous udoteaceans are poorly understood. Five genera, all strongly endemic could possibly be attributed to the group: *Cummingsella*, *Halenopora*, *Orthriosiphon*, *Orthriosiphonoides* and *Taldykites*.
5. Anchicodiaceae ("phylloid" algae in the restricted sense) comprise only 4 genera, but they can be very abundant and play an important role in building Middle Carboniferous-Early Permian mounds (*Anchicodium-Eugonophyllum-Ivanovia-Neoanchicodium*). Early marine cementation of phylloid blades could indicate a cyanobacterial consortium.
6. The Palaeosiphonocladales are important for carbonate production. They are erect, small, simple cylindrical thalli, with rather simple functional pores or with an outer mucilagenous coating, now preserved as cement. They thrive in dense closely-packed thickets in protected lagoons and mounds. Devonian-Early Carboniferous palaeoberesellids (10 genera including *Exvotarisella*, *Kamaena*, *Palaeoberesella*) and issinellids (3 genera including *Issinella*, *Pseudoissinella*) were succeeded by Late Carboniferous-Early Permian beresellids (6 genera including *Beresella*, *Dvinella*, *Uraloporella*).
7. The most diversified, endemic to a continental block, and often stratigraphically confined chlorophytes are the Dasycladales sensu stricto. Seletonellaceae and Diplororaceae are scarce. The presence of Acetabulariaceae is doubtful. The bulk of the Dasycladaceae is observed in the Tethyan realm. At least 36 genera have been identified (*Albertaporella*, *Anatolipora*, *Anthracoporella*, *Atractyliopsis*, *Borisovella*, *Cabrieropora*, *Clavaporella*, *Coelosporella*, *Columbiapora*, *Connexia*, *Dutroella*, *Eouraloporella*?, *Eovelobitella*, *Epimastopora*, *Epimastoporella*, *Gyroporella*, *Herakella*, *Japhetella*, *Koninckopora*, *Kulikia*, *Macroporella*, *Masloviporella*, *Melporella*, *Nanopora*, *Paraepimastopora*, *Pekiskopora*, *Pseudogyroporella*, *Pseudonanopora*, *Pseudovelebitella*, *Queenslandella*, *Sinarella*, *Sphinctoporella*, *Unjaella*, *Vermiporella*, *Windsoporella*, *Zaporella*).

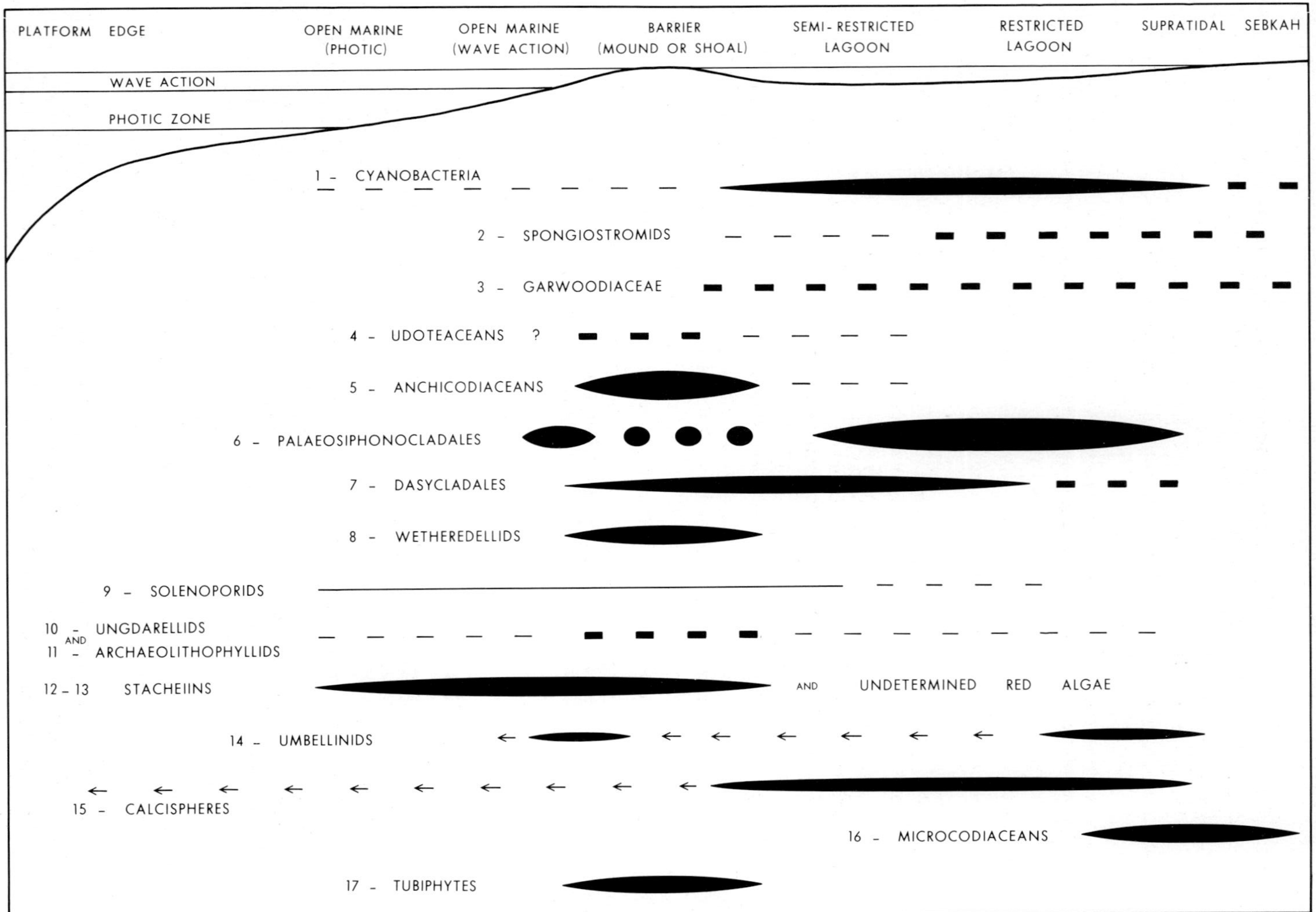

Fig. 1. Idealized reconstruction of a Carboniferous carbonate platform. Distribution and abundance of algal microflora and incertae sedis. Width of *bars* proportional to abundance. *Arrows* indicate reworking

8. Seven genera of incertae chlorophytes could possibly be related to the wetheredellids. *Asphaltina* or *Berestovia* are observed in high energy shoals.

   The next five groups, from 9 to 13, belong to the red algae.
9. The simple forms of rhodophytes are the long ranging solenoporids (4 genera: *Parachaetetes*, *Pseudochaetetes*, *Pseudosolenopora*, *Solenopora*). While there are no records of Carboniferous reef crests built by solenoporids, they are often observed as discrete nodules in open marine lagoons or on open marine ramps.
10. Ungdarellids (5 genera) form important constituents of boundstones and *Komia* and *Ungdarella* are cosmopolitan reef formers.
11. Archaeolithophyllids (2 genera) are crustose or form baffles.
12. Stacheiins are omnipresent in Viséan-Namurian carbonates (10 genera including *Aoujgalia*, *Epistacheoides*, *Mametella*, *Pseudostacheoides*, *Stacheoides*).
13. Six other red algae cannot be attributed to well-defined families. The most widespread is *Cuneiphycus* and related forms.
14. Charales probably include umbellinids, which thrived in the Devonian. They peter out at the Devonian-Carboniferous boundary but 6 genera are still known in the Early Carboniferous (*Biumbella*, *Protoumbella*, *Quasiumbella*).
15. Calcispheres (*Calcisphera*, "*Radiosphaera*" are micritic-walled cysts which proliferated in lagoons. They floated and were easily dispersed into open marine facies and spread as far as the platform margin.
16. About twenty *incertae sedis* (16–17) are tentatively included among the algae? Microcodiaceans (16) (*Microcodium*, *Palaeomicrocodium*) perforated and dissolved marine grainstones. *Tubiphytes* (17) erected Late Carboniferous-Early Permian mounds in association with *Archaeolithoporella*.

3 Algal Microfacies

Algal associations permit recognition of marine microfacies belts in a carbonate ramp or platform.

The deepest algal floras are composed of stacheiins and solenoporids which coincide with the deepest part of the photic zone. Higher up in the open marine facies, rhodophytes are mixed with palaeosiphonocladales. The first dasyclads and wetheredellids thrived in high energy shallow water shoals. Boundstones are only formed by early cemented phylloid anchicodiaceans, *Tubiphytes* or crustose ungdarellids. Slightly deeper water mud-mounds are formed by beresellids.

Semi-restricted lagoons have abundant dasyclads, calcispheres and palaeosiphonocladales, cyanobacteria and nodular codiaceans? Restricted lagoons have cyanobacteria, spongiostromids, few dasycladales, umbellinids and calcispheres.

Microcodiaceans developed in lagoons due to fresh water influx, or in supratidal sabkhas with vadose cementation, and are associated with spongiostromids and cyanobacteria.

4 List of Genera

The following are indicated for each genus:

1. Type of the genus with age and location;
2. Diagnosis;
3. Valid species in the Carboniferous, with stratigraphic range and geographic distribution. Flora is reported from the different basins in the following conventional order. *Western Tethys*: North Africa (Libya, Algeria, Morocco), Western Europe (Spain, France, Belgium, Great Britain, Ireland, Germany, Carnic Alps, Yugoslavia, Poland), Eastern Europe (Byelorussia, Russian Platform, Donbass, Middle and Southern Urals).
   Eastern Tethys: Turkey, Iran, Afghanistan, Central Asia (Kazakhstan, Tadzhikhistan, Tian-Shan), Southern China, Indochina, Australia.
   Siberian block: Kuzbass, Kolyma, Taimyr, Verkoyansk.
   Arctic: Omolon, Arctic Alaska, Arctic Canada.
   America: American Cordillera (British Columbia, Alberta, Idaho, Montana, Utah, Colorado, New Mexico), American Midcontinent (Missouri, Arkansas, Oklahoma), American Appalachians (Tennessee, Alabama, Maritime Provinces of Canada). We have no information on South American microfloras;
4. Suprageneric attributions;
5. Stratigraphic range and distribution for the genus (same convention as above). When the alga plays an important part in sedimentation, this role is emphasized.

The list includes taxa from the Devonian-Carboniferous transition ("Strunian") and from the basal Permian (Asselian).

1. *Albertaporella* (Johnson 1966) Mamet and Roux 1981

Type of genus. Albertaporella involuta Johnson 1966. Latest Tournaisian, Alberta.
Diagnosis. Thallus a succession of spherical verticillated units. Pores phloiophore, simple, in tufts radiating from a vestibule.
Other Carboniferous species.
Albertaporella (?) *gallatinensis* Mamet and Roux 1981. Early Viséan, Montana, Nevada, Utah.
Suprageneric attributions. Albertaporellae (Güvenç 1979) Mamet and Roux 1981 according to Deloffre (1988). Unconvincingly attributed to the Triploporelleae (Pia 1920) Bassoullet et al. 1979 by Shuysky (1987).
Stratigraphic range and distribution. Reported by Deloffre (1988) as Devon-

ian-Carboniferous. This wide range is based on Güvenç's assessment that *Albertaporella* included the Siegenian *Diplopora constantini* Poncet 1965 (= *Ivdelipora* Shuysky and Schirchova 1987) and the Late Permian *Diplopora anatoliana* Güvenç 1970. Correct range is restricted to Late Tournaisian-Early Viséan. Endemic to the American Cordillera (Alberta, Montana, Nevada, Utah).

2. *Amorfia* Rácz 1965

Type of genus. Amorfia jalinki Rácz 1965. Lower Moscovian, Spain.
Diagnosis. Thallus crustose, wavy, branching, flattened cylindrical, with hypo- and perithallic differentiation. Thin polygonal hypothallic tissue and thick rectangular perithallic cells. Cell threads joined in curved bundles. Conceptacles circular.
Suprageneric attribution. Archaeolithophyllaceae Chuvashov 1987.
Stratigraphic range and distribution. Scarce. Same as type of genus.

3. *Anatolipora* Konishi 1956

Type of genus. Anatolipora carbonica Konishi 1956. Late Viséan-Early Namurian, Japan.
Diagnosis. Thallus club-shaped, small with undivided medulla. Branches numerous, phloiophore, with strong oblique insertion and slight curvature.
Suprageneric attribution. Salpingoporellinae Bassoullet et al. 1979 according to Shuysky (1987) and Deloffre (1988).
Stratigraphic range and distribution. Late Tournaisian-Early Namurian, Spain, France, Belgium, Japan, Alaska, Alberta.

4. *Anchicodium* Johnson 1946

Type of genus. Anchicodium funile Johnson 1946. Upper Carboniferous, Kansas.
Diagnosis. Crustose, wavy, phylloid plates. Spongy mass of highly branching threads, terminated in triangular-shaped "utricles".
Other Carboniferous species.
Anchicodium gracile Johnson 1946. Late Carboniferous, Kansas, Urals.
Anchicodium magnum Endo 1952. Namurian-Permian, Algeria, Japan.
Anchicodium nodosum Johnson 1946. Late Carboniferous, Kansas.
Anchicodium plumosum Johnson 1946. Late Carboniferous, Kansas.
Suprageneric attributions. Siphonales, Anchicodiaceae according to Shuysky (1987). Filaments were confused with endolithic algal perforations by Vachard (1981).
Stratigraphic range and distribution. Locally abundant. Forms bafflestones and build-ups (early cementation). Middle Carboniferous-Permian. Spain, Yugoslavia, Russian Platform, Urals, Kazakhstan, Afghanistan, Japan, Alaska, Kansas.

5. *Anthracoporella* Pia 1920

Type of genus. Anthracoporella spectabilis Pia 1920. Upper Carboniferous, Austria.
Diagnosis. Thallus large, branching, vermiporellid-like. Medulla prominent. Branches crowded, numerous, phloiophore, aspondyl, at right angles from the medulla.
Other Carboniferous species.
Anthracoporella girtyi Mamet and Roux 1987. Moscovian-Sakmarian, Canadian Arctic.
Anthracoporella vicina Kochansky-Devidé and Herak 1960. Upper Carboniferous, Yugoslavia.
Suprageneric attributions. Tribe Dasyporelleae (Pia 1920) Bassoullet et al. 1979 according to Deloffre (1988). Tribe Anthracoporellopsiae Shuysky 1985 among the Palaeoberesellaceae.
Stratigraphic range and distribution. Common. Middle Viséan? to Permian. Devonian records are to be disregarded. Forms the "*Anthracoporella* community" of Flügel (1977) in association with phylloid algae. Spain, Yugoslavia, Greece, Turkey, Russian Platform, Urals, Tadzhikhistan, Afghanistan, Kazakhstan, China, Thailand, Sumatra, Japan, Canadian Arctic, Texas.

6. *Anthracoporellopsis* Maslov 1956

Type of genus. Anthracoporellopsis machaevi Maslov 1956. Middle Carboniferous, Donbass (see Plate 3a).
Diagnosis. Thallus cylindrical, rarely ramified. Medulla divided by irregular partitions. Pores aspondyl, straight, perforating both cortex and partitions in all directions.
Suprageneric attributions. Anthracoporellopsiae Shuysky 1985 in Shuysky (1987). Beresellaceae for Deloffre (1988). Palaeoberesselleae? for Mamet and Roux (1974).
Stratigraphic range and distribution. Rather common. Tournaisian to Early Permian. Morocco, Algeria, Spain, Belgium, France, Russian Platform, Urals,

Plate I. a-l Idealized reconstructions of thalli. **a** *Columbiapora*. Reconstruction showing vestibule (in Mamet and Roux 1981[b]). From *Columbiapora johnsoni* Mamet 1974. Approximate diameter of the verticils, up to 1000 mμ. Tournaisian, British Columbia. **b** *Queenslandella* (in Mamet and Roux 1983). From *Queenslandella jenkinsi* Mamet and Roux 1983. Approximate diameter of thallus, up to 700 mμ. Viséan, Australia. **c** *Eovelebitella* (Mamet and Roux 1983). From *Eovelebitella occitanica* Vachard 1974 *emend.* Approximate diameter of verticils, 900–1500 mμ. Late Viséan, France. **d** *Japhetella* (in Mamet and Roux 1983). From *Japhetella boehmi* Mamet and Roux 1983. Approximate diameter of thallus, around 750 mμ. Late Viséan, France. **e** *Microcodium*. Reconstruction of the "corn cob" (in Mamet et al. 1987). Outside diameter of the cob, 400–1000 mμ. Plates dissolve and gouge into surrounding grains. Moscovian, Yukon Territory. **f** *Nansenella*. Reconstruction of the branching habit (in Mamet et al. 1987). Adapted from *Nansenella multifurcata* Mamet and Roux 1987. Diameter of micritized sheaths 25–80 mμ. Bashkirian, Ellesmere. **g** *Issinella* Reconstruction of the branching habit (in Mamet and Roux 1981a). From *Issinella devonica* Reitlinger 1954.

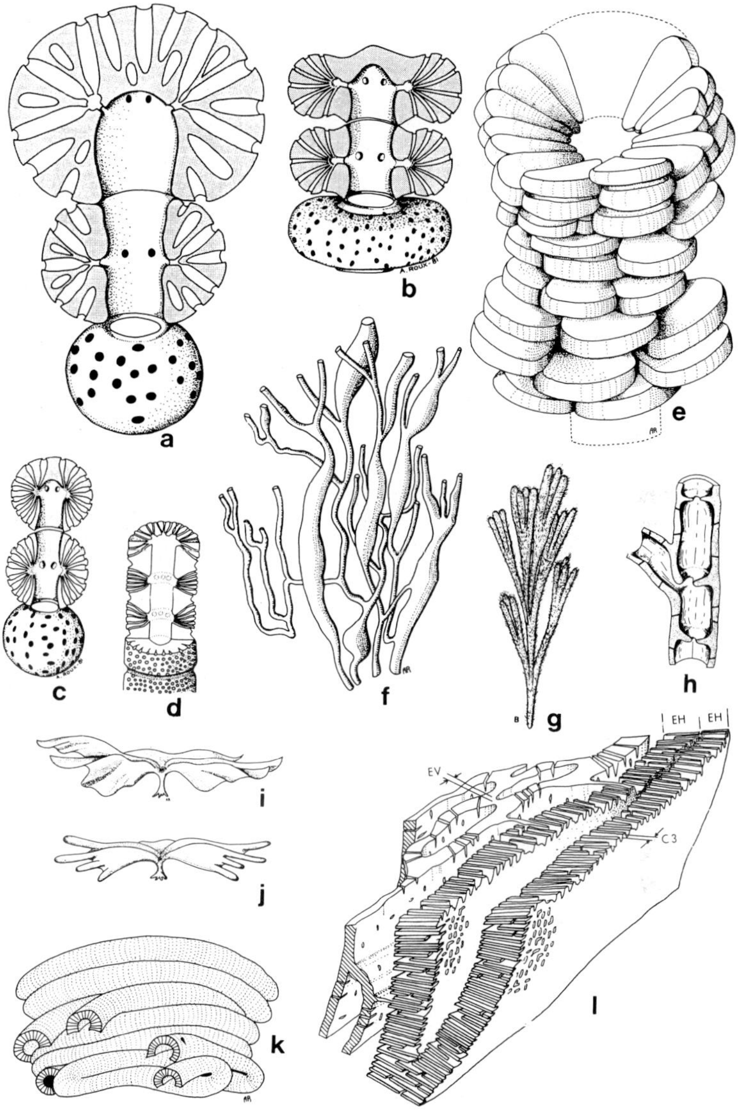

Diameter of thallus 100 to 600 mμ. Late Viséan, Utah. **h** *Kamaenella*. Cross-section showing bifurcation (in Mamet and Roux 1974). From *Kamaenella denbighi* Mamet and Roux 1974. Diameter of thallus 80–140 mμ. Late Viséan, Great Britain. **i–j** Hypothetical views of a phylloid Anchicodiaceae. **i** *Ivanovia* (in Mamet et al. 1987). Based on *Ivanovia tenuissima* Khvorova 1946. Blades reach up to 1 mm in thickness. Thallus pluri-millimetric to centimetric. Moscovian, Canadian Arctic. **j** *Neoanchicodium* (in Mamet et al. 1987). Based on *Neoanchicodium catenoides* Endo in Endo and Kanuma 1954. Same order of magnitude as for **i**. Moscovian, Canadian Arctic. **k** *Richella* (in Mamet et al. 1987). Based on *Richella incrustata* Mamet and Roux 1987). Diameter of tubes around 300–360 mμ. Moscovian, Canadian Arctic. **l** *Epistacheoides* (in Mamet and Roux 1977). Hypothallus (*EV*) with irregular, poorly calcified cells. Perithallus heavily calcified (*EH*), up to 300 mμ with regular cells (*C3*). Based on *Epistacheoides nephroformis* Petryk and Mamet 1972. Viséan, Alberta

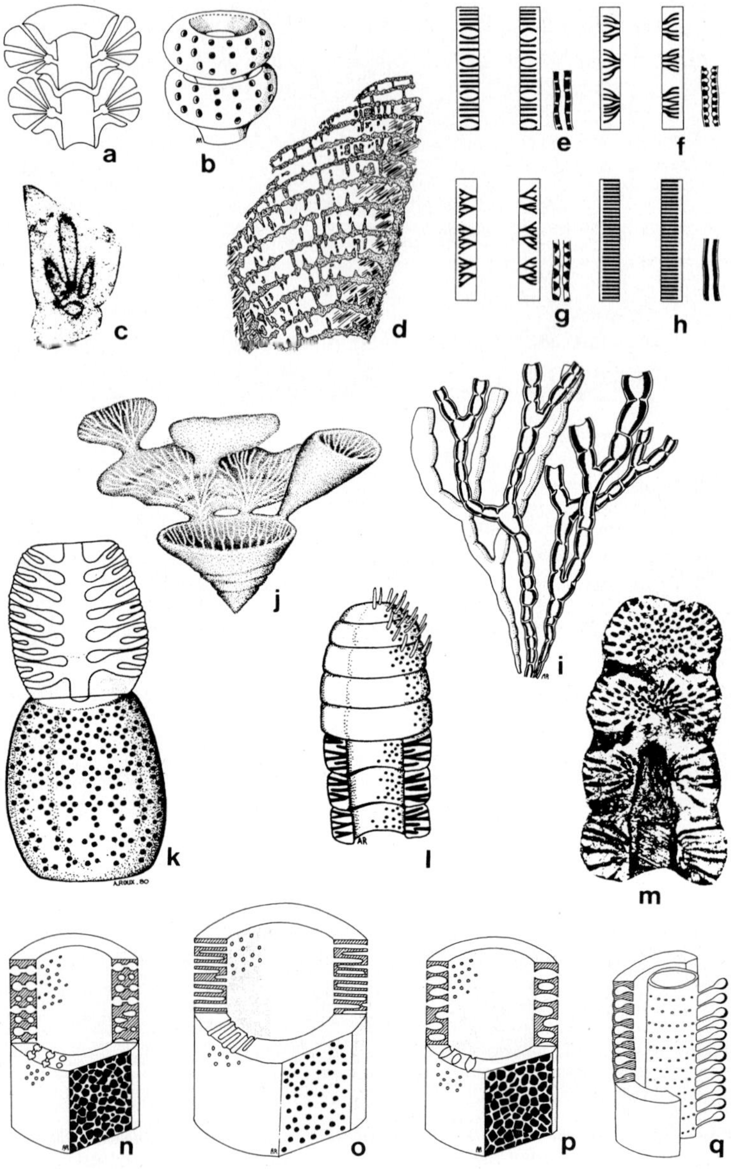

Plate II

Donbass, Kazakhstan, Siberia, (Kuzbass, Omolon), Canadian Arctic, Alaska (Peratrovich Fm).

7. *Aoujgalia* Termier and Termier 1950

Type of genus. Aoujgalia variabilis Termier and Termier 1950. Middle? Viséan, Morocco.
Diagnosis. Thallus encrusting, fusiform. Cells irregular, grossly concentric around support. Wall strongly calcified, hyaline calcite. No thallus differentiation.
Other Carboniferous species.
Aoujgalia elliotti Mamet and Roux 1977. Viséan to Sakmarian. Ireland, Siberia (Kuzbass), Canadian Arctic, Alaska.
Aoujgalia richi Mamet and Roux 1978. Viséan-Baschkirian. Siberia (Kuzbass), Canadian Arctic, Alaska, Tennessee.
Suprageneric attributions. Originally described as a foraminifer, then transferred to the "ordre nouveau d'Ischyrosponges Aoujgaliida" by Termier et al. (1975). Stacheinaceae for Chuvashov (1987) or Stacheiinae for Mamet and Roux (1978).
Stratigraphic range and distribution. Common from Viséan to Namurian. Reaches the Permian in the Canadian Arctic. Morocco, Algeria, Spain, France, England, Belgium, Germany, Donbass, Kazakhstan, Siberia, (Kuzbass, Taimyr, Verkoyansk), Australia, Canadian Arctic, Alaska, American Cordillera, Tennessee.

Plate II. a-q Idealized reconstructions of thalli. **a-c** *Clavaporella.* **a** Cross-section; **b** reconstruction; **c** oblique section through pore and vestibule (in Mamet et al. 1987). Based on *Clavaporella reinae* Rácz 1966. Diameter of verticils 1000 to 1500 mμ. Moscovian, Canadian Arctic. **d** *Masloviporidium.* Foliate thallus with pores in horizontal partitions (in Groves and Mamet 1985 after Maslov 1973). Height of quadratic cells, 20–75 mμ. Namurian, Donbass. **e-h** Shapes and disposition of blind pores among beresellids. All material is Middle Carboniferous. **e** *Beresella* (in Mamet et al. 1987). Based on *Beresella erecta* Maslov and Kulik 1956. **f** *Dvinella* (*Dvinella*) (in Mamet et al. 1987). Based on *Dvinella comata* Khvorova 1949. **g** *Dvinella* (*Trinodella*) (in Mamet et al. 1987). Based on *Dvinella bifurcata* Maslov and Kulik 1956. **h** *Uraloporella* (in Mamet et al. 1987). Based on *Uraloporella variabilis* Korde 1950. **i** *Donezella.* Reconstruction and longitudinal section (in Mamet et al. 1987). Based on *Donezella lutugini* Maslov 1929 *emend.* **j** *Calcifolium.* Reconstruction of thallus by Skompski (1981). Based on *Calcifolium okense* Schvetsov and Birina 1935. Late Viséan-Early Namurian, Poland. **k** *Cummingsella.* Reconstruction and longitudinal section (in Mamet and Roux 1980). Based on *Cummingsella lyoncrossi* Mamet and Roux 1980. Diameter of thallus 400–500 mμ. Namurian, Scotland. **l** *Orthriosiphonoides.* Reconstruction and longitudinal section (in Mamet and Roux 1980). Based on *Orthriosiphonoides salterensis* Petryk (in Petryk and Mamet 1972). Diameter of thallus, 1100–2000 μm. Viséan, Alberta. **m** *Eovelebitella.* Longitudinal section of *Eovelebitella occitanica* Vachard 1974. Late Viséan, Spain. **n** *Epimastopora.* Schematic reconstruction of pores (in Roux 1979). Based on *Epimastopora piai* Korde 1951. Upper Carboniferous, Urals. **o** *Paraepimastopora.* Schematic reconstruction of pores (in Roux 1979). Based on *Epimastopora kansasensis* Johnson 1946. Moscovian, Kansas. **p** *Epimastoporella.* Schematic reconstruction of pores (in Roux 1979). Based on *Epimastopora japonica* Endo 1950. Early Permian, Japan. **q** *Gyroporella.* Reconstruction showing shape of pores (in Mamet et al. 1987). Based on *Gyroporella nipponica* Endo and Hashimoto 1955. Upper Carboniferous-Early Permian, Canadian Arctic

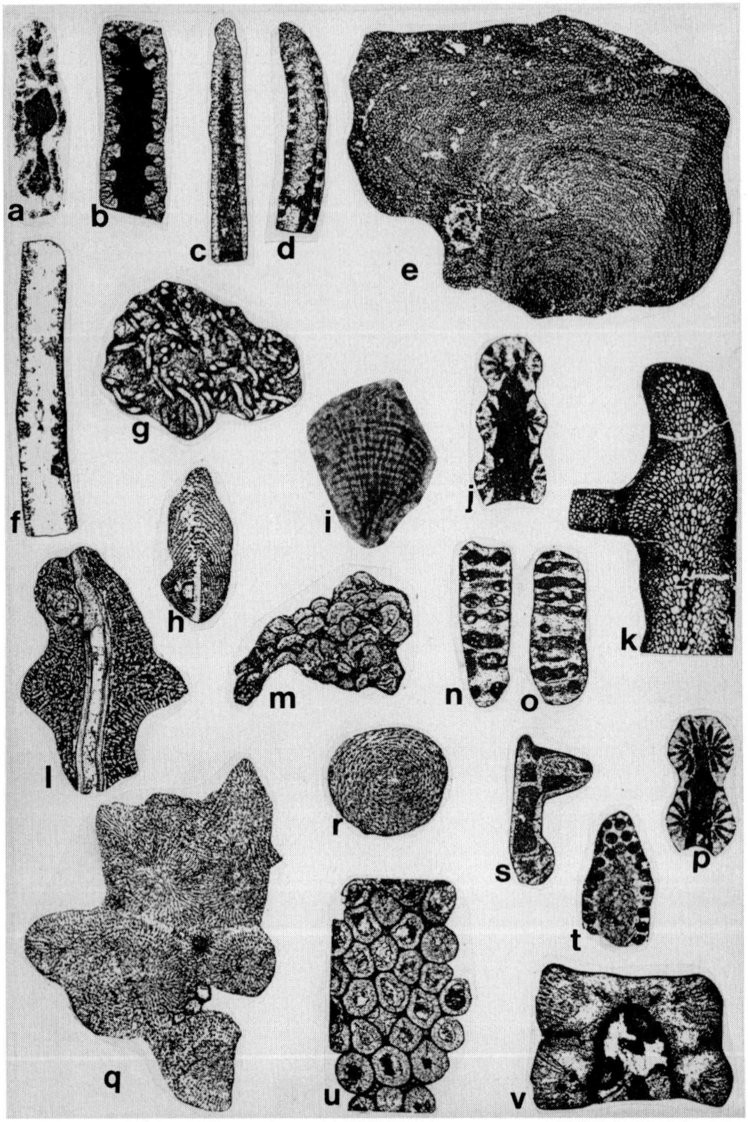

Plate III

8. *Aphralysia* Garwood 1914

Type of genus. *Aphralysia carbonaria* Garwood 1914. Late Tournaisian-Viséan, Great Britain.
Diagnosis. Thallus encrusting, nodular, maze of layered cystoid cells. In axial sections, cells are semi-elliptic with concave base attached to convex wall of preceding row. Wall micritic.
Other Carboniferous species.
Aphralysia capriorae Mamet and Roux 1975. Viséan-Namurian, Algeria, Belgium, France, Great Britain, Alaska, Tennessee.
Aphralysia ferreoli Mamet and Roux 1975. Silurian-Namurian, France, Australia, Arkansas, Québec.
Aphralysia garwoodi Hallett 1970. Late Viséan, Great Britain.
Aphralysia matthewsi Mamet and Roux 1975. Tournaisian-Namurian, Great Britain, Australia, Tennessee, Arkansas.
Suprageneric attributions. Incertae chlorophyte for Mamet and Roux (1975). Chlorophyte – Wetheredelleae according to Berchenko (1987).
Stratigraphic range and distribution. Known as early as Silurian and quite abundant in Eifelian of Belgium. Common in Tournaisian-Lower Namurian.

Plate III. **a-v** Illustrations of thalli. **a** *Anthracoporellopsis machaevi* Maslov 1956 (in Mamet and Pinard 1985, Peratrovich Island, Alaska, Zone 18, Early Namurian, x28). **b** *Exvotarisella index* (Ehrenberg 1854 *emend* von Möller 1879) (in Mamet and Roux 1974, Halecombe, Great Britain, Zone 13, Middle Viséan, x36). **c** *Nanopora* cf. *N. fragillissima* (Maslov 1939) (in Mamet and Rudloff 1972, Antigonish, Nova Scotia, Zone 17, Early Namurian, x29). **d** *Beresella* cf. *B. polyramosa* Kulik 1964 (in Mamet and Rudloff 1972, Ellesmere Island, mid-Moscovian, x29). **e** *Parachaetetes garwoodi* (Hinde in Garwood 1912) (in Mamet and Roux 1977, Wath, Great Britain, Zone 9, Upper Tournaisian, x12). **f** *Ivanovia tenuissima* Khvorova 1946 (in Mamet et al. 1987, van Hauen Pass, Ellesmere Island, Middle Carboniferous, x8). **g** *Girvanella problematica* Nicholson and Etheridge 1878 (in Mamet and Roux 1975b, Cabrières, France, Zone 16_{inf}, Late Viséan, x52). **h** *Fourstonella* sp. (in Mamet et al. 1987, Girty Creek, Ellesmere Island, mid-Moscovian or younger, x13). **i** *Masloviporidium delicata* (Berchenko 1982) (in Groves and Mamet 1985, Hale Mountain, Arkansas, Zone 20, Early Bashkirian, x17). **j** *Windsoporella radiata* Mamet and Rudloff 1972 (in Mamet and Rudloff 1972, Antigonish, Nova Scotia, Zone 17, Early Namurian, x29). **k** *Archaeolithophyllum* sp. (in Johnson 1963, Young Co, Texas, Pennsylvanian, x22). **l** *Epistacheoides connorensis* Mamet and Rudloff 1972 encrusting a brachiopod spine (in Mamet and Rudloff 1972, Connor Lakes, British Columbia, Zone 16_{sup}, Late Viséan, x12). **m** *Wetheredella silurica* Wood 1948 (in Mamet and Roux 1983, Jeremiah Hills, Australia, Early Tournaisian, x17). **n-o** *Epimastopora symetrica* (Johnson 1951) (in Mamet et al. 1987, Little Fish Creek, Yukon Territory, Asselian, x33). **p** *Eovelebitella robertsi* Mamet and Roux 1983 (in Mamet and Roux 1983, Bingleburra, New South Wales, Australia, Zone 9, Late Tournaisian, x33). **q** *Komia abundans* Korde 1951 (in Mamet and Rudloff 1972, McDame map area, British Columbia, Zone 18, Early Namurian, x13). **r** *Ungdarella uralica* Maslov 1956 (in Mamet et al. 1987, Girty Creek, Ellesmere Island, Bashkirian, x33). **s** *Kamaenella tenuis* von Möller 1879 (in Mamet and Roux 1980, Denbigh, Great Britain, Zone 15, Late Viséan, x24). **t** *Atractyliopsis weyanti* Mamet, Mortelmans and Roux 1979 (in Mamet and Roux 1983, Old Cannindah, Queensland, Australia, mid-Viséan, x16). **u** *Koninckopora tenuiramosa* Wood 1942 (in Mamet and Roux 1978, Fowler no 1 Traders, Scott Co, Tennessee, Zone 13, Middle Viséan, x33). **v** *Orthriosiphonoides tenuiramosa* Mamet and Rudloff 1972 (in Mamet and Rudloff 1972, Grotto Creek, Alberta, Zone 12, Middle Viséan, x12)

Algeria, France, Belgium, Great Britain, Germany, Donbass, Australia, Alaska (Peratrovich fm), Alberta, British Columbia, Arkansas, Tennessee.

"*Aphrostoma*" Gürich 1906

Type of genus. Aphrostoma tenerum Gürich 1906. Viséan, Belgium.
Diagnosis. Algal-sponge biscuit composed of regular semiplanar layers of, possibly, dissolved *Sphaerocodium*, irregularly disposed birdseyes, pelletoidal tissue with dissolved *Bevocastria* and sponges encrusted by *Bevocastria*.
Suprageneric attributions. Spongiostromidae Gürich 1906. Needs revision.
Stratigraphic range and distribution. Scarce. Viséan, Belgium, Poland, American Cordillera.

"*Aranea*" Hance 1983

Type of genus. Aranea eburei Hance 1983, early Viséan, Belgium.
Diagnosis. Spider-web of irregular polygonal cells, rounded cystoid cells and dichotomous tubes.
Suprageneric attributions. Rhodophyte (Solenoporaceae) for Hance (1983). Problematic nodule formed by different organisms. *Bevocastria* present in the holotype.
Stratigraphic range and distribution. Same as type of genus.

9. *Archaeolithophyllum* Johnson 1956

Type of genus. Archaeolithophyllum missouriense Johnson 1956. Upper Carboniferous, Missouri.
Diagnosis. Crustose multifoliate sheets of differentiated cells. Hypothallus prominent, coaxial, with large polygonal regular cells. Perithallic cells poorly developed, much smaller, rectangular or square. Conceptacles in the perithallus (see Shuysky 1987, for illustrations and also Plate 3k).
Other Carboniferous species.
Archaeolithophyllum delicatum Johnson 1956. Middle and Late Carboniferous, American midcontinent and Cordillera.
Archaeolithophyllum donbassicum Kossenko 1972. Viséan-Moscovian, Algeria, Donbass, Germany, Alaska, American Midcontinent (type of the genus *Principia* Brenckle 1982, an *Archaeolithophyllum* with poorly developed perithallus).
Archaeolithophyllum johnsoni Rácz 1965. Bashkirian-Moscovian, Algeria, Spain, Canadian Arctic.
Archaeolithophyllum lamellosum Wray 1964. Upper Carboniferous, Kansas.
Archaeolithophyllum vaillani Mamet and Roux 1977. Late Viséan-Early Namurian, France, Spain, Donbass.
Suprageneric attributions. Ancestral coralline for most authors (see Mamet et al. (1987) for discussion). Type of new family Archaeolithophyllaceae Chuvashov 1987.
Stratigraphic range and distribution. Quite common. Oldest reference in Early Viséan of Australia (Mamet and Roux 1983). Forms the "*Archaeolitho-*

phyllum community" (Flügel 1977) in association with stromatolites and *Tubiphytes* from Middle Carboniferous to Early Permian. Frame-builder in organic build-ups (in association with *Neoanchicodium*). Also forms "*Archaeolithophyllum* oncoid community". Algeria, Morocco, Spain, France, Germany, Carnic Alps, Donbass, Urals, Russian Platform, Kazakhstan, Siberia, China, Australia, Arctic Canada, Alaska, American Cordillera, American midcontinent.

Ardengostella Vachard in Perret and Vachard (1977)

See *Dvinella*.

"*Artophycus*" Johnson 1940

Type of genus. Artophycus columnaris Johnson 1940. Upper Carboniferous, Colorado.
Diagnosis. Described by Johnson as a spongiostromid made of "gently tapering cones of pillars built of thin, gently arched laminae". Needs revision.
Suprageneric attribution. Needs revision.
Stratigraphic range and distribution. See type of genus.

10. *Asphaltina* Mamet in Petryk and Mamet (1972)

Type of genus. Asphaltina cordillerensis Mamet in Petryk and Mamet (1972). Viséan, Alberta.
Diagnosis. Thallus(?) encrusting, composed of large, flexuous, interlaced cylindrical tubes with oblique partitions. Wall double-layered, a thin external micritic coating and a very thick pseudo-fibrous inner layer.
Suprageneric attributions. An incertae sedis reported to the tribe Wetheredellae by Berchenko in Shuysky (1987).
Stratigraphic range and distribution. Recorded from Devonian to Middle Carboniferous. Very common in North American Cordillera from which it was originally described in high energy carbonates. Great Britain, France, Siberia (Kusbass, Kolyma), Canadian Arctic, Alaska, including Peratrovich fm, Yukon Territory, Alberta, British Columbia, Idaho, Montana, Nevada, Utah, Oklahoma, Kansas, Iowa, Arkansas, Tennessee, Alabama.

11. *Asphaltinella* Mamet and Roux 1978

Type of genus. Asphaltinella horowitzi Mamet and Roux 1978. Tennessee, Early Namurian.
Diagnosis. Thallus (?) encrusting, composed of interwoven layers of cylindrical tubes as in *Asphaltina* (hence the name). Wall single layered.
Other Carboniferous species.
Asphaltinella? *bangorensis* Mamet and Roux 1978. Frasnian-Namurian. Morocco, American Midcontinent, Tennessee.
Asphaltinella peratrovichensis Mamet and Pinard 1983. Early Bashkirian, Peratrovich fm, Alaska.

Suprageneric attribution. Algae(?) incertae sedis.
Stratigraphic range and distribution. Common in North America in high energy carbonates. Scarce elsewhere.
Frasnian-Bashkirian. Morocco, Spain, Belgium, Afghanistan, Australia, Canadian Arctic, Alaska, Yukon Territory, Alberta, Wyoming, Nevada, Utah, Arizona, New Mexico, Arkansas, Kansas, Oklahoma, Tennessee, Alabama.

12. *Atractyliopsis* (Pia 1937) Accordi 1956

Type of genus. *Atractyliopsis lastensis* Accordi 1956. Permian, Italy.
Diagnosis. Thallus acicular or subcylindrical, undivided. Stem undivided, prominent. Incompletely calcified cortex containing a single row of spherical cysts? (endospores?).
Other Carboniferous species.
Atractyliopsis cumberlandensis Rich 1974. Upper Tournaisian-Namurian, Western Tethys (Morocco, Belgium, France), North America (Tennessee, Alabama, Alberta).
Atractyliopsis forezi Mamet and Roux 1975. Viséan, France, Belgium, Australia.
Atractyliopsis minima Mamet and Roux 1978. Viséan, Tennessee, Alaska.
Atractyliopsis weyanti Mamet, Mortelmans and Roux 1979. Viséan, France, Belgium, Australia (see Plate 3t).
Suprageneric attributions. Aciculelleae Bassoullet et al. 1979 *sensu* Berchenko and Saltovskaja (1987) (see Elliott 1971, for hypothetical ontogeny). Tribe emended by Deloffre (1988).
Stratigraphic range and distribution. Widespread, but fragile and often broken. Usually associated with red algae (*Archaeolithophyllum*) or phylloid algae (*Neoanchicodium*). Upper Tournaisian – Upper Permian. Eurasia (Belgium, France, Hungary, Italy, Yugoslavia, Afghanistan, China, Japan, Sumatra), North Africa (Libya), Australia, North America (Tennessee, Alabama).

"*Baculella*" Conil and Dreesen 1985

Type of genus. *Baculella gemina* Conil and Dreesen 1985. Devonian-Carboniferous transition, Belgium.
Diagnosis. Chain of pyriform chambers ending by narrow neck. Wall fibrous radial on a micritic layer, perforated by thin undivided pores.
Suprageneric attribution. Microproblematic. Morphologically similar to the incertae sedis *Saccamminopsis*, but with a wall structure similar to the alga *Issinella*.
Stratigraphic range and distribution. See type of genus. Reported to form "algal bindstones".

13. *Beresella* (Makhaev? 1939) Maslov and Kulik 1956

Type of genus. *Beresella erecta* Maslov and Kulik 1956. Moscovian, Russian Platform (see Plate 2e).
Diagnosis. Calcareous cylindrical tube, rarely dichotomous. Cortex perforated at regular intervals by short, thin, blind cortical branches, micritized and

forming "dark bands". They are separated by "clear bands" of non-perforated cortex. Thallus possibly formerly surrounded by mucilaginous coating, now preserved as a continuous cement layer.
Other Carboniferous species.
Beresella bilgutayae Güvenç 1965 (probably a *Beresella polyramosa*). Bashkirian, Turkey.
Beresella herminae Rácz 1965. Bashkirian, Spain, Algeria.
Beresella ishimica Kulik 1964. Middle Carboniferous, Russian Platform, Canadian Arctic.
Beresella polyramosa Kulik 1964. Bashkirian-Sakmarian, Eurasia, Canadian Arctic, North America (see Plate 3d).
Beresella translucea Kulik 1964. Middle Carboniferous. Russian Platform, Central Asia.
Suprageneric attributions. Bereselleae Maslov and Kulik 1956.
Beresellaceae (Maslov and Kulik 1956) *trans.* Shuysky 1985 (and Deloffre 1988) Concept of Beresellidae as a sponge (Termier, Termier and Vachard 1977) is untenable.
Stratigraphic range and distribution. One of the most important "reef" builders in Carboniferous time (bafflestone). First report in Serpukhovian Zone 18. Common to very abundant in Bashkirian-Moscovian and still widespread up to the Early Permian. Cosmopolitan: Eurasia, North Africa, Arctic, Nevada, Utah, Idaho.

14. *Berestovia* Berchenko 1982

Type of genus. *Berestovia filaris* Berchenko 1982. Serpukhovian, Donbass.
Diagnosis. Thallus crustose composed of continuous rows of highly calcified hemispherical to spherical cells. Rows asymmetrical, flattened at base and wavy at top. Wall yellowish. Morphology of pores, if any, doubtful.
Suprageneric attribution. Linked to the Donezelleae by Shuysky (1987), but presence of pores is debatable and attribution is unclear.
Stratigraphic range and distribution. Reported only once by Berchenko, but a common producer of oncoids in the Wahoo Limestone of Alaska (Zone 21). Also common in Oklahoma, Arkansas, Alabama.

15. *Bevocastria* Garwood 1931

Type of genus. *Bevocastria conglobata* Garwood 1931. Late Tournaisian, Great Britain.
Diagnosis. Thallus nodular, formed by meshwork of numerous anastomosed, branching, micritized tubes. Tubes have constrictions at regular intervals.
Other Carboniferous species.
Bevocastria amplefurcata Pia 1932. Devonian-Upper Carboniferous. Poland, Russian Platform, Iran.
Bevocastria fragilis Berchenko 1981 is synonymous with *Bevocastria conglobata* Garwood 1931.
Bevocastria kuzbassi Maslov 1956. Famennian-Viséan, Kuzbass.

Bevocastria hubbardi Mamet and Roux 1975. Lower Carboniferous, Western Europe.
Bevocastria uralica Kulik 1973. Serpukhovian, Urals.
Suprageneric attributions. Controversial alga. Codiaceae for Mamet and Roux 1975, Stignomenales for Ischenko 1985, Garwoodiaceae for Luchinina 1987.
Stratigraphic range and distribution. Cambrian (as "*Dictyophycus*") to Sakmarian. Although rarely recorded, omnipresent in Devonian-Carboniferous lagoons. Forms nodules ("*Glomus bevocastriosus*" Radionova 1976), spongiostromids, oncoids and composite nodules with ctenostomates and sponges. Eurasia, Canadian Arctic, North America.

16. *Biumbella* Mamet 1970

Type of genus. Biumbella braznikhovae Mamet 1970. Devonian-Carboniferous transition, Donbass.
Diagnosis. Utricular (?) cavity irregularly spherical. Apical opening closed by conical lid on a neck. Wall two-layered, with a prominent pseudo-fibrous yellowish layer.
Synonyms. Pseudoumbella Berchenko 1971. *Quasiumbelloides* Berchenko 1971.
Other Carboniferous species.
Biumbella donbassica (Berchenko 1971). Tournaisian, Donbass.
Biumbella tenuis (Berchenko 1971). Tournaisian, Donbass.
Biumbella venusta (Berchenko 1971). Tournaisian, Donbass.
Suprageneric attributions. Biumbellidae Berchenko 1974. Umbellaceae Fursenko 1959 for Berchenko in Shaikin (1987).
Stratigraphic range and distribution. Scarce. Famennian-Tournaisian. Endemic to the Donbass.

17. *Borisovella* Ivanova 1988

Type of genus. Borisovella turbinata Ivanova 1988. Serpukhovian, Urals.
Diagnosis. Thallus cylindrical, with continuous (?) stem. Cortex perforated by simple branches with very oblique insertion on medulla. Branches numerous, pedonculated, rapidly expanding and ampuliform.
Suprageneric attribution. Cyclocrinae Pia 1927.
Stratigraphic range and distribution. See type of genus.

18. *Brazhnikovia* Berchenko 1981

Type of genus. Brazhnikovia undata Berchenko 1981. Devonian-Carboniferous boundary, Donbass.
Diagnosis. A slightly tapering kamaenid with short irregular partitions. Morphology of branches is unclear.
Suprageneric attributions. Originally linked to the Palaeobereselleae by Berchenko (1981), then attributed to the Anthracoporellopsiae Shuysky 1985 by Shuysky (1987). Genus probably valid, but restudy necessary as nature of the branches is poorly known.
Stratigraphic range and distribution. See type of genus.

"*Buzgulella*" Korde 1951

Type of genus. Buzgulella serrata Korde 1951. Middle Carboniferous, Urals
Diagnosis, attribution and stratigraphy. Need revision.

19. *Cabrieropora* Mamet and Roux 1975

Type of genus. Cabrieropora pokornyi Mamet and Roux 1975. Late Viséan (Zone 16_i), France.
Diagnosis. Thallus long, cylindrical, composed or regular barrel-shaped articles. Expansion of medulla forms important spherical vestibule. Tufts of six divergent branches, each thick and phloiophore.
Suprageneric attributions. Velebitelleae (Vachard 1977) Mamet and Roux 1983. Incorrectly attributed to Linoporallinae Pia 1927 by Shuysky (1987).
Stratigraphic range and distribution. Scarce. Upper Viséan (Zones 14 to 16) France, Spain.

20. *Calcifolium* Schvetsov and Birina 1935

Type of genus. Calcifolium okense Schvetsov and Birina 1935. Russian Platform.
Diagnosis. Thallus composed of a succession of tubular siphons, rapidly expanding into cups or petals. Filaments occupy whole thickness of petals and branch dichotomously (see Plate 2j).
Other Carboniferous species.
Calcifolium bruntonense Johnson 1958 (identical to *Calcifolium okense*).
Calcifolium punctatum Maslov 1956. Latest Viséan (Zones 16_i and 16_s), Great Britain, Poland, Russian Platform, Donbass, Urals.
Suprageneric attributions. Codiaceae (Mamet and Roux 1977; Skompski 1981). Type of new tribe Calcifoliae Shuysky 1987, placed among Anchicodiaceae Shuysky 1987. Attribution to sponges by Termier et al. (1977) (as Calcifoliida) is unwarranted.
Stratigraphic range and distribution. Calcifolium is often referred to "phylloids", although there is no proof of a leaflike habit. A good stratigraphic marker for the Viséan-Namurian boundary. Occidental part of the Tethys: Algeria, Morocco, Spain, Great Britain, France, Poland, Russian Platform, Donbass, Urals, Tadzhikhistan.

21. *Calcisphaera* Williamson 1881

Type of genus. Calcisphaera laevis Williamson 1881. Late Viséan, Great Britain. (Type designation by Andrews 1955). Subsequent designation of *Calcisphaera cancellata* Williamson 1881 by Vachard (1977) is questionable.
Diagnosis. Microcrystalline, hollow, calcite spherule. Wall one-layered. Pores (?) scarcely noticeable. No known functional aperture.
Synonymy. Granulosphaera Derville 1931 OBJ, *Pachysphaera* Rauzer-Chernoussova 1948, OBJ, *Pachysphaera* Conil and Lys 1964 OBJ, *Pachysphaerina* Conil and Lys 1967 OBJ.
Other Carboniferous species.

Calcisphaera pachysphaerica (Pronina 1963). Viséan-Baschkirian, cosmopolitan.
Suprageneric attributions. Controversial since its origin. Kazmierczak (1976) and Vachard and Tellez-Giron (1986) have proposed volvocale affinity for the Calcisphaeridae. Most Russian authors still attribute genus to foraminiferal family Calcisphaeridae Reitlinger 1960 OBJ. The debate exists since Williamson who thought (1881, p. 525) that they were either "reproductive capsules of some marine form of vegetation" or "some type of extinct Protozoa". Recent authors have shown their resemblance with dasycladacean reproductive cysts (Wray 1977a). Mamet 1973 has illustrated an in situ crown of *Calcisphaera* refuting a protozoan origin. Loeblich and Tappan (1988) list them among the taxa erroneously regarded as foraminifers.
Stratigraphic range and distribution. Middle Devonian-Middle Carboniferous. Extremely abundant in Tournaisian-Viséan lagoons. Easily dispersed throughout entire carbonate platforms. "Calcispherites" are quite similar to the Late Carboniferous-Permian "algen sporen assemblages" of Flügel (1977). Cosmopolitan. Eurasia, North Africa, Australia, Arctic, North America.

Chabakovia Vologdin 1939

See *Renalcis*.

"*Chantonia*" Termier, Termier and Vachard 1977

Type of genus. Ungdarella maslovi Chanton-Güvenç 1965. Late Viséan, Algeria.
Diagnosis. Needs revision as original material is poorly preserved. Seems to be a mixture of *Ungdarella*, *Chuvashovia* and *Epistacheoides*.
Suprageneric attributions. The type described as a red alga, was transferred to the "sponges" (Aoujgaliida, Termier et al. 1977).
Stratigraphic range and distribution. Needs revision.

"*Chondrostoma*" Gürich 1906

Type of genus. Chondrostoma problematicum Gürich 1906. Viséan, Belgium.
Diagnosis. Needs revision. The four species described by Gürich (the type *Chondrostoma problematicum*, *Chondrostoma globuliferum*, *Chondrostoma intermixtum* and *Chondrostoma vermiculiferum*) show various cyanophyte mats, more or less bioturbated, some in situ, some reworked, some with mud cracks and algal chips. Faecal pellets, hard lumps, algal pellets and in situ sponge spicules are present.
Suprageneric attribution. Spongiostromidae Gürich 1906.
Stratigraphic range and distribution. Needs revision.

"*Claracrusta*" Vachard in Vachard and Montenat (1981)

Type of genus. Girvanella catenoides Homann 1972. Permian, Carnic Alps.

Diagnosis, suprageneric attribution, stratigraphic range and distribution. Needs revision. Most Middle Carboniferous forms are probably *Berestovia.*

22. *Clavaporella* Kochansky-Devidé and Herak 1960

Type of genus. Clavaporella caliciformis Kochansky-Devidé and Herak 1960. Permian, Yugoslavia.
Diagnosis. Thallus cylindrical, unramified, a succession of calyx-shaped asymmetrical euspondyl units. Composite vestibule followed by tufts of phloiophore branches. Stem isodiametrical, cylindrical, unsegmented. Three to four rows of simple pores form verticils.
Other Carboniferous species.
Clavaporella reinae Rácz 1966. Moscovian-Upper Carboniferous, Spain, Urals, Canadian Arctic, (see Plate 2a,b,c).
Suprageneric attributions. Albertaporelleae (Güvenç 1979) Mamet and Roux 1981 for Deloffre 1988. Triploporellinae (Pia 1920) Bassoullet et al. 1979 according to Shuysky (1987).
Stratigraphic range and distribution. Moscovian (scarce) to Permian (abundant). Spain, Yugoslavia, Turkey, Caucasus, Urals, Afghanistan, China, Canadian Arctic.

Coactilum Maslov 1956

See *Sphaerocodium.*

"*Codonophycus*" Fenton and Fenton 1939

Type of genus. Codonophycus austinii Fenton and Fenton 1939. Tournaisian, Wyoming.
Diagnosis, suprageneric attribution and stratigraphic range. Needs revision (algal biscuit).

23. *Coelosporella* Wood 1940

Type of genus. Coelosporella wetheredii Wood 1940. Late Viséan, Great Britain.
Diagnosis. Thallus cylindrical, with prominent unsegmented medulla. Poorly calcified cortex with one or two rows of lemon-shaped cavities ("sporangial" cavities) open to the outside.
Other Carboniferous species.
Coelosporella jonesii Wood 1940. Upper Viséan, Great Britain, Belgium, Donbass, Urals, Tennessee.
Suprageneric attributions. Aciculelleae Bassoullet et al. 1979 for Deloffre (1988) and Berchenko and Saltovskaja (1987). See Elliott (1971) for hypothetical ontogeny and function of rod.
Stratigraphic range and distribution. Quite scarce and fragile. Late Tournaisian to Serpukovian. Spain, Great Britain, Belgium, Germany, France, Donbass, Ural, Nepal, American Cordillera and midcontinent, Tennessee, Alabama, Newfoundland.

24. *Columbiapora* (Mamet 1974) Mamet and Roux 1981b

Type of genus. Columbiapora johnsoni Mamet 1974. Middle Tournaisian, British Columbia (see Plate 1a).
Diagnosis. Thallus, a succession of subspherical, verticillated units. Medulla subcylindrical, with some expansion at junction of verticils. Tufts of six fusiform branches diverge from a short composite vestibule, a cylindrical expansion of the medulla.
Suprageneric attributions. Incorrectly attributed to the Gyroporelleae Pal 1976 *emend* Bassoullet et al. 1979 by Shuysky (1987). Albertaporelleae (Güvenç 1979) Mamet and Roux 1981b for Deloffre (1988).
Stratigraphic range and distribution. Middle to Late Tournaisian. Endemic to the American Cordillera (*Columbiapora mameti* Vachard 1977 from the Serpukovian of France does not belong to the genus).

Conilalia Vachard in Massa and Vachard (1979)

See *Cuneiphycus.*

25. *Connexia* Kochansky-Devidé 1970

Type of genus. Connexia fragilis Kochansky-Devidé 1970. Moscovian, Yugoslavia.
Diagnosis. Thallus a succession of claviform units. Central medulla cylindrical, unsegmented. Each unit is stellate with five to six bifid projections ("horns"). Branches with second and third order dichotomy.
Suprageneric attribution. Triploporelleae (Pia 1920) Bassoullet et al. 1979.
Stratigraphic range and distribution. Both Deloffre (1988) and Shuysky (1987) restrict the genus to the Permian. However, the type is Moscovian and the correct range is Moscovian-Early Permian as reported by Milanović (1982). Yugoslavia, Central Asia.

Contortoporidium Maslov 1973 NOM. NUD.

See *Masloviporidium.*

"*Corydopodium*" Derville 1931

Type of genus. Corydopodium pruvosti Derville 1931. Viséan, France.
Diagnosis. Flattened algal biscuit composed of birdseyes (sponge canals?) interspersed with *Bevocastria* tubes and algal peloids of unknown affinity.
Suprageneric attribution. Spongiostromidae.
Stratigraphic range. Scarce. Lower Carboniferous, Great Britain, France, Belgium, Germany.

Costatumbella Berchenko 1974

See *Eoumbella* Platonov 1974

26. *Cribrokamaena* Brenckle 1985

Type of genus. Cribrokamaena citrosa. Brenckle 1985. Viséan, Nevada.
Diagnosis. Thallus (?) slender, cylindrical, branching. Medulla divided by regular cribrated pseudo-septa. Wall hyaline pseudo-fibrous, perforated by thin pores.
Note. Part of the original material may belong to *Issinella.* Cribrated partitions are unreported among Palaeosiphonocladales.
Other Carboniferous species.
Cribrokamaena furcillata Brenckle 1985. Viséan, Nevada, Missouri.
Suprageneric attribution. At origin, Palaeoberesellaceae? Mamet and Roux 1974.
Stratigraphic range and distribution. Endemic to North America. Common in American Cordillera (Alberta, British Columbia, Wyoming, Montana, Utah, Nevada, Colorado), American midcontinent (Iowa, Missouri, Illinois) and Tennessee-Alabama.

Cribroporidium Maslov 1973 NOM. NUD.

See *Masloviporidium* Groves and Mamet 1985.

27. *Cribrosphaeroides* Reitlinger 1959

Type of genus. Cribrosphaera simplex Reitlinger 1954. Frasnian, Russian Platform.
Diagnosis. Thallus? ovoid, irregular. Wall micritized, perforated by coarse, regular straight "pores". No aperture.
Carboniferous species.
Cribrosphaeroides ovalis Poyarkov 1961. Frasnian to Devonian-Carboniferous transition. Donbass, Urals, Siberian Platform, Tian-Shan, Turkey.
Cribrosphaeroides robusta (Mikluko-Maklai 1965). Givetian to Devonian-Carboniferous transition. Urals, Siberian Platform, Tian-Shan, Turkmenistan, Omolon Massif.
Cribrosphaeroides simplex donica Brazhnikova and Rostovceva in Aizenberg and Brazhnikova 1966. Devonian-Carboniferous transition, Donbass.
Suprageneric attributions. Controversial. *Cribrosphaeroides* (and *Auroria*) could be foraminifers? (Cribrosphaerinae Zadorozhnyi and Yuferev 1984), but they lack aperture and their "pores" normally indicate dasyclad branches.
Stratigraphic range and distribution. Eifelian to Tournaisian. Belgium, Poland, France, Germany, Donbass, Russian Platform, Urals, Kazakhstan, Tadhzikhistan, Turkmenia, Tian-Sian, Siberian Platform, Kuzbass, China, Omolon Massif.

28. *Cummingsella* Mamet and Roux 1980

Type of genus. Cummingsella lyoncrossi Mamet and Roux 1980. Lower Namurian, Scotland (see Plate 2b).

Diagnosis. Thallus, succession of barrel-shaped articles. Medulla cylindrical, unsegmented. Branches (?) or cortical tubes (?) numerous, as random "verticils" (?)
Dichotomy of first to second order depending on position of branch.
Other Carboniferous species.
Cummingsella bingleburrae Mamet and Roux 1983. Late Tournaisian, Australia.
Suprageneric attributions. Undetermined tribe among the Dasycladales for Deloffre (1988). Dasycladacean? or codiacean? for Mamet and Roux (1980). Could be a Udoteacean?
Stratigraphic range and distribution. Very scarce. Only two Lower Carboniferous occurrences.

29. *Cuneiphycus* Johnson 1960

Type of genus. Cuneiphycus texana Johnson 1960. Middle Carboniferous, Texas.
Diagnosis. Thallus tapering, more or less encrusting, without hypo- and perithallic differentiation. Rows of regular, thin, wedge-shaped cells. Thallus often difficult to separate from *Fourstonella*. Concept needs revision.
Synonym. Eflugelia, Conilalia.
Other Carboniferous species.
Cuneiphycus aliquantulus Johnson 1960. Viséan-Moscovian, Algeria, Spain, Belgium, Donbass, Siberia, Canadian Arctic, New Mexico, Texas.
Cuneiphycus johnsoni Flügel 1966. Viséan-Permian, Carnic Alps, Urals, Canadian Arctic, Idaho, Kansas.
Cuneiphycus texana Johnson 1960. Viséan-Upper Carboniferous, Western Tethys, (Morocco, Algeria, Great Britain, Russian Platform, Donbass), North America (Alaska, Texas, Oklahoma).
Suprageneric attributions. Undetermined rhodophycophyte family for Mamet et al. (1987). New tribe Mametelleae Chuvashov 1987. Attributed to the sponges by Termier et al. (1977) (as Ptychocladiidae Ellias 1950).
Stratigraphic range and distribution. Abundant. Long ranging from Viséan to Permian. Libya, Algeria, Morocco, Spain, Great Britain, France, Donbass, Siberia, Urals, Iran, Afghanistan, China, Alaska, Idaho, American midcontinent, Texas, New Mexico.

30. *Cylindrofolia* Brenckle and Groves 1987.

Type of genus. Cylindrofolia glenisteri Brenckle and Groves 1987. Tournaisian, Iowa.
Diagnosis. Thallus unsegmented, with continuous cylindrical medulla. Cortex poorly calcified, perforated by horizontal, wavy, irregular, frond-like appendages.
Suprageneric attribution. Incertae sedis. As noted by Brenckle and Groves, there are no known horizontal processes among bona fide dasycladales. This

puzzling genus has some similarities with *Menselina*, another Carboniferous microproblematicum.
Stratigraphic range and distribution. See type of genus.

31. *Dasycladophycus* Elliott 1982

Type of genus. Dasycladophycus ensomi Elliott 1982. Mid-Viséan, Great Britain.
Diagnosis. Non-calcified central stem with verticils of four branches. Each branch divides into four short branchlets.
Suprageneric attribution. Uncalcified Dasycladales for Elliott (1982).
Stratigraphic range and distribution. See type of genus.

Dasyporella Stolley 1893

Genus restricted to the Lower Palaeozoic, but often mentioned in the Viséan-Serpukhovian by Russian authors (e.g. *Dasyporella kulikae* Muromtseva 1979, *Dasyporella maslovi* Kulik 1973, *Dasyporella popovae* Kulik 1973, etc.). These taxa ought to be transferred to genuine Carboniferous Dasyporelleae and Uraloporelleae (*Anthracoporella*, *Issinella*, *Nanopora*, etc.).

32. *Dokutchaevskella* Berchenko 1981

Type of genus. Dokutchaevskella inaequalis Berchenko, Devonian-Carboniferous transition, Donbass.
Diagnosis. Thallus long, slender, cylindrical, unsegmented. Medulla divided by uneven projections of the cortex, forming irregularly-shaped cells. Branches thin, straight, perpendicular or slightly oblique to the medulla. Needs further study. Seems quite similar to *Pseudonanopora* Mamet and Roux 1975.
Suprageneric attributions. Dasyporelleae Pia 1927 for Berchenko (1981). Exvotariselleae Shuysky 1985 for Shuysky (1987).
Stratigraphic range and distribution. Devonian-Carboniferous transition, Donbass and Southern Urals. Needs revision.

33. *Donezella* Maslov 1929 non sensu Vachard in Meissami et al. 1978

Type of genus. Donezella lutugini Maslov 1929. Middle Carboniferous, Donbass.
Diagnosis. Thallus cylindrical, slightly tapering, tubular, sinuous, constricted and ramified in all directions. Medulla divided by thin irregular projections of the cortex. Cells grossly rectangular; height greater than width. Cortex perforated by blind pores, thin, unramified, forming dark micritic bands from one partition to the other. During life, thallus may have been covered by thin mucilagenous coating, now preserved as clear cement (see Plate 2i).
Synonyms. Goksuella Güvenç 1965 and probably *Praedonezella* Kulik 1973.
Other Carboniferous species.

Donezella cespaeformis (Kulik 1973). Latest Viséan, basal Bashkirian. Spain, France, Donbass, Urals.
Donezella lunaensis Rácz 1964 is probably synonymous with *Donezella lutugini*.
Suprageneric attributions. Palaeosiphonocladales of the tribe Donezelleae as emended by Shuysky (1985). Maslov's original description was very sketchy and has caused much taxonomic controversy. As with many Palaeozoic siphonocladales, this genus has been attributed to foraminifers, sponges, and pseudosponges.
Stratigraphic range and distribution. Very abundant in the Middle Carboniferous. Forms *Donezella* bafflestones (Mamet et al. 1979). Latest Viséan-Namurian (scarce), Bashkirian-Moscovian (very abundant). Late Carboniferous-Early Permian (scarce). Cosmopolitan in the Northern Hemisphere. Forms bafflestones and mud-mounds in Eurasia, North Africa and North America.

Dromastacheoides Vachard in Perret and Vachard (1977)

See *Epistacheoides*.

34. *Dutroella* Mamet and Roux 1978

Type of genus. *Dutroella scotti* Mamet and Roux 1978. Late Viséan (Zone 16_i), Tennessee.
Diagnosis. Thallus, a succession of thin barrel-shaped units. Medulla thick. Cortex perforated by a single row of perpendicular, simple, straight, phloiophore branches. Needs further study.
Suprageneric attribution. Unconvincingly placed with the Mizziinae Bassoullet et al. 1979 by Deloffre (1988). Should be placed among the Diploporaceae.
Stratigraphic range and distribution. See type of genus.

35. *Dvinella* Khvorova 1949

Type of genus. *Dvinella comata* Khvorova 1949. Middle Carboniferous, Russian Platform (see Plate 2f).
Diagnosis. Thallus, a long, cylindrical calcareous tube. Cortex perforated by blind pores, oblique and dichotomous, forming a triangular "dark ring", separated by cement filled "clear rings". During life, thallus may have been covered by a thin mucilagenous coating, now preserved as clear cement.
Three subgenera have been proposed, but they are often difficult to separate:
Dvinella (*Dvinella*) Khvorova 1949 has branches with single dichotomy.
Dvinella (Trinodella) Maslov and Kulik 1956 has pores with double dichotomy.
The subgenus *Dvinella* (*Ardengostella*) Vachard in Perret and Vachard 1977 is a *Dvinella* in Khvorova's sense.
Other Carboniferous species.
Dvinella (*Trinodella*) *aequalis* Kulik 1964. Moscovian, Russian Platform.
Dvinella (*Trinodella*) *bifurcata* Maslov and Kulik 1956. Bashkirian-Lower

Permian, Spain, Russian Platform, Canadian Arctic (Plate 2g).
Dvinella (*Dvinella*) *crassitheca* Kulik 1964. Moscovian, Hungary, Yugoslavia, Russian Platform, Canadian Arctic.
Dvinella cuvillieri Güvenç 1965. Moscovian, Turkey.
Dvinella (*Dvinella*) *distorta* Kulik 1964. Moscovian, Russian Platform.
Dvinella (*Dvinella*) *gracilis* Kulik 1964. Moscovian, Russian Platform, Canadian Arctic.
Dvinella (*Trinodella*) *varialonga* Kulik 1964. Moscovian, Russian Platform.
Suprageneric attributions. Bereselleae Maslov and Kulik 1956.
Beresellaceae (Maslov and Kulik 1956) *trans.* Shuysky 1985 (and by Deloffre 1987).
Stratigraphic range and distribution. Abundant and very important for carbonate production. Forms "bafflestones" in mud-mounds and in "reefs". Namurian-Bashkirian up to Early Permian. Cosmopolitan in the Northern Hemisphere. Eurasia, North Africa, Canadian Arctic, North America.

Eflugelia Vachard in Massa and Vachard (1979)

See *Cuneiphycus*. See also *Fourstonella*.

36. *Einorella* Saltovskaja 1984

Type of genus. Einorella globosa Saltovskaja 1984. Middle Carboniferous, Tadhzikistan.
Diagnosis. Thallus, a slender calcareous tube, rarely dichotomous, with regular constrictions at each internal pseudoseptation. Cortex perforated by thin blind pores forming micritized "black rings". Unperforated parts form thin "clear bands". Thallus may originally have been covered by a mucilaginous coating, now preserved as a continuous cement layer. Needs further study.
Suprageneric attributions. Bereselleae Maslov and Kulik 1956. Beresellaceae (Maslov and Kulik 1956) *trans.* Shuysky 1985 (and Deloffre 1987).
Stratigraphic range and distribution. See type of genus.

37a. *Elenia* Poyarkov 1965

Type of genus. Umbella famena Bykova in Bykova and Polenova (1955). Famennian, Russian platform.
Diagnosis. Utricular (?) spherical cavity. Parallel longitudinal ribs. No basal projection. Apertural lid unknown. Wall two-layered.
Synonym. Spinumbella Platonov 1974
Carboniferous species.
Elenia costulata Berchenko 1974. Devonian-Carboniferous transition, Donbass.
Elenia formosa Berchenko 1974. Devonian-Carboniferous transition, Donbass.
Elenia spinosa (Conil and Lys 1964) Devonian-Carboniferous transition, Belgium.
Suprageneric attributions. Umbellaceae Fursenko 1959.

Stratigraphic range and distribution. Mainly Upper Devonian (Frasnian-Famennian) up to the Devonian-Carboniferous transition. Restricted to the western Tethys. Belgium, France, Poland, Russian Platform, Urals, Donbass.

37b. *Eoclypeina* sensu Poncet 1987 non Vachard 1985

Type of genus. Eoclypeina crassa Poncet 1987. Moscovian, Algeria.
Diagnosis. Thallus, a succession of glass- or basket-shaped segments. Primary branches in verticils, of even diameter, first oblique to the axial cavity, then horizontal at their distal part.
Suprageneric attributions. Originally described as Dasycladaceae. Clypeineae (Elliott 1968) Bassoullet et al. 1979 for Deloffre (1988).
Stratigraphic range and distribution. See type of genus.

38. *Eokoninckopora* Saltovskaja 1984

Type of genus. Eokoninckopora einori Saltovskaja 1984. Tournaisian, Tadzhikistan.
Diagnosis. Thallus (?) very irregular, vermiporellid-shaped. Cortex perforated by very thick polygonal to subpolygonal pores. Wall structure unclear.
Suprageneric attribution. Linked to the Cyclocrinae Pia 1927 by Shuysky (1987) although there appears to be no relation with *Koninckopora.* Needs further study. Could be a partially dissolved bryozoan.
Stratigraphic range and distribution. See type of genus.

39. *Eolithoporella* Johnson 1966

Type of genus. Eolithoporella dawsoni Johnson 1966, Lower Carboniferous, Alberta.
Diagnosis. Reported by Johnson as a thin crust of vertically elongated cells resembling *Litostroma.*
Other Carboniferous species.
Eolithoporella europaea (Kochansky-Devidé 1970) (as *Litostroma*), Moscovian, Yugoslavia.
Suprageneric attribution. Johnson thought it belonged to an undetermined family among the red algae. It could very well be a megafaunal fragment.
Stratigraphic range and distribution. Scarce. Lower to Middle Carboniferous. Alberta, Yugoslavia.

40. *Eoumbella* Platonov 1974

Type of genus. Umbella ollaria Bykova in Bykova and Polenova (1955). Upper Devonian, Russian Platform.
Diagnosis. Utricular(?) cavity spherical. Coarse external irregular ribs. Extensive basal projection as in *Umbellina.* Apertural lid unknown. Wall two-layered.
Carboniferous species.
Eoumbella striatella (Berchenko 1974). Devonian-Carboniferous transition, Donbass.

Suprageneric attribution. Umbellaceae Fursenko 1959.
Stratigraphic range and distribution. Mainly Upper Devonian, but ranges into the Devonian-Carboniferous transition. Russian Platform, Urals, Donbass, Siberian Platform.

41. *Eouraloporella* Berchenko 1981

Type of genus. Eouraloporella kordeae Berchenko 1981. Devonian-Carboniferous transition, Donbass.
Diagnosis. Thallus long, slender, cylindrical, unsegmented with central stem conspicuous and occasional pseudoseptations. Pores functional, thin, straight, aspondyl, at right angle to the medulla. Needs revision.
Suprageneric attribution. In spite of its name, and attribution to the Uraloporellaeae by Shuysky (1985), the taxon is not a beresellid as the pores are functional.
Stratigraphic range and distribution. See type of genus.

42. *Eovelebitella* (Vachard 1974) emend. Mamet and Roux 1983

Type of genus. Eovelebitella occitanica (Vachard 1974) *emend.* Mamet and Roux 1983. Late Viséan, Zone 16_i, France. (*Note*: The original material comprises two genera, *Eovelobitella sensu stricto* as redefined and *Japhetella* Mamet and Roux 1983) (see Plates 1c and 2m).
Diagnosis. Thallus metaspondyl, a succession of subspherical articles. Medullar segments biconvex, with medullar projections forming a thick vestibule. Branches phloiophore, in thick dense tufts (about 6–8 branches in a random thin section, probable around thirty in three dimensions).
Other Carboniferous species.
Eovelebitalla robertsi Mamet and Roux 1983, Tournaisian-Viséan boundary, Australia (Plate 3p).
Suprageneric attribution. Diploporeae Pia 1920, emend. Güvenç 1979, non sensu Bassoullet et al. (1979).
Stratigraphic range and distribution. Locally abundant. Tournaisian and Viséan. France, Spain, Australia.

43. *Epimastopora* ("Pia 1922") (emend. Korde 1951) sensu Roux 1979.

Type of genus. Epimastopora piai Korde 1951 (*non piae* Bilgütay 1960). Late Carboniferous, Urals (see Plate 2n).
Diagnosis. Thallus cylindrical, with very large stem. Cortex perforated by dumb-bell shaped branches. In equatorial sections, branches are subpolygonal to spherical.
See Roux (1979) for discussion of type and elucidation of the taxonomy.
Synonym. Globuliferoporella Chuvashov 1974.
Other Carboniferous species.
Epimastopora primaeva (Chuvashov and Anfimov 1988). Moscovian, Urals.
Epimastopora symetrica (Johnson 1951) (as *Gyroporella*). Upper Carboniferous-Permian. Carnic Alps, Urals, Canadian Arctic, Texas, see Plate 3n-o.

Suprageneric attributions. Mastoporeae (Pia 1920) emend. Deloffre 1988. Cyclocrineae Pia 1927, Mastoporinae Pia 1927 for Shuysky (1987).
Stratigraphic range and distribution. Important and abundant. Forms the "*Epimastopora* community" in association with *Gyroporella* and *Mizzia* in outer shelf facies. Upper Carboniferous-Permian. Spain, Carnic Alps, Yugoslavia, Turkey, Urals, Japan, Canadian Arctic, Yukon Territory, Texas.

44. *Epimastoporella* Roux 1979

Type of genus. *Epimastopora japonica* Endo 1951. Early Permian, Japan, Plate 2p.
Diagnosis. Thallus cylindrical, with large, important medulla. Cortex perforated by ovoid branches, perpendicular to the cavity. In equatorial section, branches are subpolygonal to spherical.
See Roux (1979) for discussion. (Partially *Epimastopora auctores* and *Pseudoepimastopora auctores*).
Other Carboniferous species.
Epismastoporella bodoniensis (Rácz 1965) (as *Epimastopora*). Moscovian, Spain.
Epismastoporella camasobresensis (Rácz 1966) (as *Epimastopora*). Moscovian, Spain.
Epismastoporella hunzaensis (Zanin Buri 1965) (as *Epimastopora*). Upper Carboniferous-Permian, Carnic Alps, Pakistan, Canadian Arctic.
Epismastoporella iwaizakensis (Endo 1953) (as *Epimastopora*). Upper Carboniferous-Permian, Carnic Alps, Turkey, Japan, Canadian Arctic.
Epimastoporella japonica (Endo 1951) (as *Epimastopora*). Upper Carboniferous-Permian, Carnic Alps, Yugoslavia, Turkey, Japan, Canadian Arctic.
Epimastoporella macropora (Maslov 1956) (as "*Koninckopora*"). Serpukhovian-Bashkirian, France, Donbass, Urals, Tian-Shan.
Epimastoporella micropora (Maslov 1956) (as "*Koninckopora*'). Serpukhovian-Bashkirian, Donbass, Urals.
Epimastoporella rolloensis (Rácz 1965). (as *Epimastopora*). Moscovian-Permian, Spain, Urals.
Suprageneric attribution. Mastoporeae (Pia 1920) *emend.* Deloffre 1988.
Stratigraphic range and distribution. First occurrence, Middle Carboniferous. Abundant, Upper Carboniferous and Permian. Tunisia, Spain, Carnic Alps, Turkey, Russian Platform, Urals, Pakistan, Malaya, Japan, Canadian Arctic.

45. *Epiphyton* Bornemann 1886

Type of genus. *Epiphyton flabellatum* Bornemann 1886. Cambrian, Sardinia.
Diagnosis. Thallus branching, composed of highly dichotomous tufts of stems. Original cellular microstructures controversial.
Carboniferous species.
Epiphyton buldyricum Antropov 1955 (synonym, *Epiphyton aegeensis* Güvenç 1981) Frasnian-Tournaisian. Turkey, Russian Platform, Urals, China.
Suprageneric attributions. Controversial. Attributed to red, green and blue-green algae (see Saltovskaja 1984, for discussion). Placed by Luchinina (1987)

as Epiphytaceae Korde 1959 among the Hormogonophyceae (Geitler) Elenkin 1934.
Stratigraphic range and distribution. Reported from the Proterozoic to the Carboniferous. Important in Cambrian archaeocyathid mounds (nearly one hundred "species" in the literature). Still abundant in Devonian reefs. Scattered at the Devonian-Carboniferous transition and in the Tournaisian where it becomes extinct.

46. *Epistacheoides* Petryk and Mamet 1972

Type of genus. Epistacheoides nephroformis Petryk and Mamet 1972. Viséan, Alberta. (*Synonym*: *Dromastacheoides wilsoni* Vachard in Perret and Vachard 1977) (see Plate 11).
Diagnosis. Thallus encrusting, crustose, irregular. Hypothallic tissue poorly calcified, vesicular, surrounded by two or three rows of strongly calcified perithallic rectangular cells. Wall hyaline, yellowish calcite.
Synonyms. Dromastacheoides Vachard in Perret and Vachard 1977 and *Roquesselsia* Termier et al. 1977.
Other Carboniferous species.
Epistacheoides chantoni Mamet and Roux 1977 (synonym: *Roquesselsia radians* Termier et al. 1977). Upper Viséan, Algeria, France, Ireland.
Epistacheoides connorensis Mamet and Rudloff 1972 (synonym; *Dromastacheoides topi* Vachard 1981). Viséan-Namurian (Asselian-Sakmarian in the Arctic). Algeria, Morocco, Great Britain, Belgium, France, Urals, Afghanistan, China, Alaska, Canadian Arctic, British Columbia, Tennessee, Newfoundland (see Plate 31).
Suprageneric attributions. Stacheiinae Loeblich and Tappan 1961. New tribe Pseudostacheoidaea Chuvashov 1987. Erroneously attributed to the sponges by Vachard (1977) (as Aoujgaliida).
Stratigraphic range and distribution. Very common in the Viséan-Namurian-Baschkirian. Algeria, Morocco, Spain, France, Belgium, Great Britain, Germany, Urals, Afghanistan, China, Canadian Arctic (where is reaches Early Permian), Yukon Territory, Alaska, British Columbia, Alberta, Tennessee, Newfoundland.

47. *Eugonophyllum* Konishi and Wray 1961

Type of genus. Eugonophyllum johnsonii Konishi and Wray 1961. Late Carboniferous, New Mexico.
Diagnosis. Thallus composed of large blades, standing erect on substratum. Medulla with dichotomous filaments (?) often obliterated by early cementation. Subcortex contains anastomosed crescent-shaped utricles (?) communicating through the cortex by fine perpendicular branchlets. (Synonym. Could be similar to *Paradella* Maslov 1956 which has priority. However Maslov's taxon is ill defined and *Eugonophyllum* has been extensively used since its erection).
Other Carboniferous species.

Eugonophyllum johnsonii uralicum Rauser-Chernoussova and Korolyuk 1981. Moscovian, Urals.
Eugonophyllum magnum (Endo 1951). Middle Carboniferous-Lower Permian, Yugoslavia, Turkey, Japan.
Eugonophyllum mulderi Rácz 1965, Moscovian, Spain, Yugoslavia.
Suprageneric attributions. "Phylloid algae" (vernacular). Anchicodiaceae Shuysky 1987, new tribe Ivanoviae Shuysky 1987.
Stratigraphic range and distribution. Important and widespread. For a reconstruction of a phylloid algal community, see Toomey and Babcock (1983). Forms mounds in Moscovian-Sakmarian. Spain, Carnic Alps, Turkey, Urals, Kazakhstan, Japan, Canadian Arctic, Kansas, Texas, New Mexico.

48. *Evlania* Bykova 1952

Type of genus. Evlania transversa Bykova 1952. Upper Devonian, Russian Platform.
Diagnosis. Thallus(?) attached, irregularly shaped. Internal cavity divided by pseudo-septa. Wall hyaline, yellowish, perforated by numerous dichotomously divided pores. Needs revision.
Suprageneric attributions. Controversial. Variously interpreted as foraminifer, sponge or alga.
Stratigraphic range and distribution. Givetian-Viséan. Morocco, France, Belgium, Czechoslovakia, Urals, Donbass, Russian Platform, Alberta, Iowa.

49. *Exvotarisella* Elliott 1970

Type of genus. Nodosinella index Ehrenberg sensu von Möller 1879 (synonym: *Exvotarisella maponi* Elliott 1970). Late Viséan, Russian Platform (see Plate 3b).
Diagnosis. Thallus, a long cylindrical tube. Medulla divided by irregular cavities formed by thick pseudosepta. Branchlets thin, numerous, aspondyl. Second to third order ramifications.
Other Carboniferous species.
Exvotarisella dili Vachard in Dil et al. 1977. Viséan. Turkey.
Suprageneric attributions. Tribe Exvotariselleae Shuysky 1985 according to Shuysky (1987). Palaeobereselleae Mamet and Roux 1974 for Deloffre (1988).
Stratigraphic range and distribution. Scarce in Devonian-Tournaisian. Abundant in Late Viséan-Namurian. In association with other palaeoberesellids, forms bafflestones in protected lagoons. Algeria, Morocco, Spain, Great Britain, France, Belgium, Germany, Poland, Turkey, Russian Platform, Urals, Donbass, Kuzbass, Arctic Siberia, Omolon, Australia, Alberta, British Columbia, Newfoundland.

50. *Fasciella* Ivanova 1973

Type of genus. Fasciella kizilia Ivanova 1973 (synonyms: *Shartymophycus fusus* Kulik 1973, *Fasciella ivanovae* Saltovskaja 1984, and *Fasciella ramosa* Saltovskaja 1984). Late Viséan-Serpukhovian, Urals.

Diagnosis. Thallus encrusting, composed of irregular concentric layers of elongated, undulating cells. Recrystallization often obscures shape of cells. Walls calcareous, yellowish, hyaline.
Suprageneric attributions. Incertae sedis. Could be a chlorophyte or a rhodophyte.
Stratigraphic range and distribution. Very common in high energy open marine grainstones. Viséan-Bashkirian. Algeria, Morocco, Spain, Ireland, Great Britain, France, Belgium, Germany, Poland, Russian Platform, Urals, Donbass, Tadzhikhistan, China, Alaska, Idaho, Quebec.

51. *Foliophycus* Johnson 1960

Type of genus. Foliophycus llanoensis Johnson 1960. Middle Carboniferous, Texas. Type extensively recrystallized.
Diagnosis. Thallus long with slender leafy fronds. They are composed of long threads of rectangular or barrel-shaped cells emerging from central axis and branching dichotomously. No perithallic-hypothallic differentiation. Needs revision.
Suprageneric attribution. Undetermined family among the rhodophytes. The closest known taxon is *Masloviporidium* Groves and Mamet 1985.
Stratigraphic range and distribution. Scarce. Namurian-Moscovian. Algeria, Canadian Arctic, American midcontinent and Texas.

52. *Fourstonella* Cummings 1955

Type of genus. Stacheia fusiformis Brady 1876. Viséan, Great Britain.
Diagnosis. Thallus fusiform, symmetrical, encrusting around a central support. Thin rows of minute, subparallel laminae, slightly overlapping. Insertion of rows oblique on the support. Cells quadratic, regular. Wall calcareous, yellowish, hyaline (see Plate 3g).
Other Carboniferous species.
Fourstonella irregularis Mamet and Roux 1977. Late Viséan. Algeria, Morocco, France, Siberia.
Suprageneric attribution. Originally considered a foraminifer, it has been transferred to the ungdarellids by Mamet and Rudloff (1972), a transfer based on the nature of the wall. Stacheinaceae Loeblich and Tappan 1961. New tribe Mametellae Chuvashov 1987.
Stratigraphic range and distribution. Viséan-Namurian. Reaches the Moscovian in the Canadian Arctic. Algeria, Morocco, Great Britain, France, Germany, Poland, Russian Platform, Siberia (Verkhoyansk), Canadian Arctic, American Midcontinent, Nova Scotia, Newfoundland.

53. *Frustulata* Saltovskaja 1984

Type of genus. Frustulata asiatica Saltovskaja 1984. Viséan, Tadzhikhistan.
Diagnosis. Thallus cylindrical, regularly branching. Medulla divided by irregular pseudosepta. Cortex perforated by thin, simple pores. Further infor-

mation is needed to differentiate it from *Kulikaella* Berchenko 1981.
Suprageneric attribution. Systematic position debatable (see Shuysky 1987).
Stratigraphic range and distribution. See type of genus.

54. *Garwoodella* Paul 1938

Type of genus. Garwoodella thallus Paul 1938. Viséan, Germany.
Diagnosis. Thallus nodular composed of closely-packed layers of micritized tubes. Two different types of tube: a) regular, with simple, low-angle dichotomy; b) flexuous, irregular, pinched, with erratic dichotomy. Needs revision. Could be an association of *Ortonella* and *Mitcheldeania*. Needs revision.
Suprageneric attribution. Originally Schizophyceae. Probably a nodular codiacean.
Stratigraphic range and distribution. Viséan. Germany, Belgium, American midcontinent.

55. *Garwoodia* Wood 1941

Type of genus. Mitcheldeania gregaria Nicholson 1888 emend. Wood 1941. Tournaisian, Great Britain.
Diagnosis. Thallus nodular, composed of radiating, dichotomous, micritized tubes. These tubes, of even diameter, bifurcate at right angles.
Other Carboniferous species.
Garwoodia zonata (Derville 1931). Viséan, France.
Suprageneric attributions. Nodular codiacean for Mamet and Roux (1975b) Hormogonophyceaea (Geitler) Elenkin 1934 for Luchinina (1987).
Stratigraphic range and distribution. Common in lagoons and semi-restricted waters. Ordovician to Carboniferous. Scarce in Permian. Great Britain, Belgium, France, Carnic Alps, Turkey, Podolia, Urals, Siberia, Japan, Canadian Arctic, American Cordillera and midcontinent.

56. *Girvanella* Nicholson and Etheridge 1878

Type of genus. Girvanella problematica Nicholson and Etheridge 1878 emend. Wood 1957. Ordovician, Great Britain, see Plate 3g.
Diagnosis. Micritized sheaths of random filaments. Filaments do not branch and have even diameter. They are continuous and without partitions.
Synonyms: *Nicholsonia* Korde 1973, *Batinevia* Korde 1973.
Danielli (1981) cites many Paleozoic *Girvanella* which should be transferred to *Ellesmerella*, *Sphaerocodium*, *Mitcheldeania*, *Obruchevella*, *Razumovskia*, *Malakhovella*, etc.
Carboniferous species.
Mamet and Roux (1975b) recognize four species based on the diameter of the filament. The first three species are cosmopolitan and very long ranging:
Girvanella staminea Garwood 1931.
Girvanella wetheredii Chapman 1908 (*Girvanella incrustans* Wethered 1890, *non* Borneman 1886).

Girvanella problematica Nicholson and Etheridge 1878 emend. Wood 1957 (= *Girvanella ducii* Wethered 1890).
Girvanella kasakiensis Maslov 1949 is known from France, Great Britain, Poland, Russian Platform, Urals, Kazakhstan, China, Australia.
Suprageneric attributions. Controversial, but consensus is that girvanellids are calcified cyanobacteria. Otherwise attributed to Chlorophycophyta, Porostromata, Oscillatoriales, Nostocales, Hormogoneae, etc. Type of the family Girvanellacea Luchinina 1975.
Stratigraphic range and distribution. Very common, not only in restricted lagoons, but through the whole range of the photic zone in carbonate shelves or ramps. Many mudstones-wackestones are formed by intertwined girvanellid tubes. Forms "*Girvanella* community" of Flügel (1977). Proterozoic to Cretaceous. Eurasia including Siberia, North Africa, Australia, Canadian Arctic, North American Cordillera and midcontinent, Maritime Provinces of Canada.

57. *Gissarella* Saltovskaja 1979

Type of genus. Gissarella elegantula Saltovskaja 1979. Middle Carboniferous, Tadzhikhistan.
Diagnosis. Needs revision. Hypothetical reconstruction of Saltovskaja indicates a euspondyl thallus with cylindrical axial stem and unramified branches.
Suprageneric attribution. Doubtfully reported to the Luliporeae Shuysky 1986 among the Acetabulariaceae.
Stratigraphic range and distribution. See type of genus.

Globuliferoporella Chuvashov 1974

See *Epimastopora.*

Goksuella Güvenç 1965

See *Donezella.*

58. *Gouldina* Johnson 1940

Type of genus. Gouldina carbonaria Johnson 1940. Middle Carboniferous, Colorado.
Diagnosis. Needs revision. A mammillated algal biscuit composed of undulating laminae (see also *Shermanophycus*).
Suprageneric attribution. Spongiostromidae.
Stratigraphic age and distribution. See type of genus.

Grozdilovella Chermnykh 1972

See *Protoumbella.*

59. *Gyroporella* (Gümbel 1872) Benecke 1876

Type of genus. Gyroporella vesiculifera Gümbel 1872. Trias, Alps.

Diagnosis. Thallus aspondyl, club-shaped. Medulla of similar shape. Cortex perforated by important phloiophore? branches.
Other Carboniferous species.
? *Gyroporella constricta* Kochansky-Devidé 1970. Moscovian, Yugoslavia.
Gyroporella intraseptata Kochansky-Devidé 1970. Moscovian, Yugoslavia.
? *Gyroporella likana* Kochansky-Devidé 1970. Moscovian, Yugoslavia.
Gyroporella nipponica Endo and Hashimoto 1955. Late Carboniferous – Late Permian. Spain, Yugoslavia, Urals, Kazakhstan, China, Japan, British Columbia, Canadian Arctic (see Plate 2q).
? *Gyroporella primitiva* Rauser-Chernoussova and Korolyuk 1981. Moscovian, Urals.
? *Gyroporella prisca* Kochansky-Devidé 1970. Moscovian, Yugoslavia.
? *Gyroporella shimanensis* Chuvashov and Anfimov 1988. Moscovian, Urals.
Suprageneric attributions. Tribe Gyroporelleae (Pal 1976) Bassoullet et al. 1979 for Deloffre (1988). Subtribe Gyroporellinae Berchenko 1987 for Shuysky (1987).
Stratigraphic range and distribution. Moscovian (scarce) to Jurassic. Most abundant in Permian-Triassic where it is cosmopolitan in the Northern Hemisphere.

60. *Halenopora* Hance 1983

Type of genus. Halenopora radiata Hance 1983. Tournaisian-Viséan boundary, Belgium.
Diagnosis. Thallus(?) cylindrical, composed of three to five poorly calcified central canals of even diameter from which emerge thin, oblique lateral canals which become progressively thicker and more perpendicular to the thallus.
Suprageneric attribution. The idealized reconstruction of Hance shows two series of fertile and non-fertile branches derived from the stem and suggests an euspondyl dasycladacean. Illustrations show a udoteacean, similar to *Dimorphosiphonoides*.
Stratigraphic range and distribution. See type of the genus.

61. *Hedstroemia* Rothpletz 1913

Type of genus. Hedstroemia halimedoidea Rothpletz 1913. Silurian, Gotland.
Diagnosis. Nodule composed of radially disposed cylindrical tubes which have multiple (triple to quadruple and higher) dichotomies.
Carboniferous species.
Hedstroemia corymbosa Pia 1937. Viséan, France, Great Britain.
Hedstroemia koninckoporoides Vachard 1988 is not an alga.
Suprageneric attributions. Nodular codiacean for Mamet and Roux (1975b), Porostromata for Poncet (1986), Garwoodiaceae for Luchinina (1987).
Stratigraphic range and distribution. A widespread Ordovician-Silurian-Devonian taxon that peters out in the Lower Carboniferous. Highest recorded occurrence is Namurian. Eurasia. Australia. North America.

62. *Herakella* Kochansky-Devidé 1970

Type of genus. Herakella paradoxa Kochansky-Devidé 1970. Moscovian, Yugoslavia.
Diagnosis. Thallus cylindrical, segmented, asymmetrically annulated, metaspondyl. Medulla cylindrical. Vestibule followed by tufts of at least ten phloiophore branches, with first to second order dichotomy.
Suprageneric attributions. Diploporeae ? (Pia 1920) Güvenç 1979 according to Deloffre (1988). Linoporalinae (Pia 1927) Bassoullet et al. 1979 for Shuysky (1987).
Stratigraphic range and distribution. Middle to Late Carboniferous. Yugoslavia, Central Asia.

Hikorocodium Endo 1951

Type of genus. Hikorocodium elegantula Endo 1951. Permian, Japan. Although reported twice by Endo from the Carboniferous of Japan, this *incertae sedis* is restricted to the Permo-Triassic (Endo 1951, 1957).

63. *Iberiaella* Rácz 1984

Type of genus. Iberiaella carbonica Rácz 1984, Middle Carboniferous, Spain.
Diagnosis. Thallus encrusting or crustose, composed of colony-forming, thread-like wavy thin tubes with constrictions. They branch randomly, forming an anastomosing pattern. Wall calcareous, yellowish, hyaline.
Suprageneric attribution. Undetermined algal family for Rácz (1984). Seems akin to *Berestovia* Berchenko 1983.
Stratigraphic range and distribution. See type of genus.

64. *Intextulella* Petryk in Petryk and Mamet 1972

Type of genus. Intextulella agglomerata Petryk in Petryk and Mamet 1972. Viséan, Alberta.
Description. Thallus encrusting, composed of regular strands of crustose, layered, rectangular cells weaving uncalcified intercellular tissue. Wall yellowish, clear, hyaline. Needs revision.
Suprageneric attribution. Undetermined family among the rhodophytes.
Stratigraphic range and distribution. See type of genus.

65. *Issinella* (Reitlinger 1954) Mamet and Roux 1981a

Type of genus. Issinella devonica Reitlinger 1954 as a Devonian siphonales, Russian Platform (see Plate 1g).
Description. Thallus cylindrical, aspondyl, erect. Medulla important, divided by septiform processes. Cortex fibro-radial, perforated by thin, randomly disposed, unramified branches. Conceptacles subcortical in regular rows.
Synonym. Luteotubulus Vachard in Termier et al. (1977).
Other Carboniferous species.

Issinella grandis Chuvashov 1965. Upper Famennian – Upper Tournaisian. France, Urals, Donbass, Siberia (Kolyma).
Issinella ilychensis Ivanova 1988. Viséan, Urals.
Issinella sainsii Mamet and Roux 1975. Upper Famennian-Viséan. France, Belgium, Siberia (Kolyma, Omolon), Alaska, Utah.
Suprageneric attributions. A green alga for Reitlinger (1954) and an "algosponge" for Termier, Termier and Vachard 1977. Placed among the Issinelleae Deloffre 1987 by Deloffre (1988). Erroneously considered a Dasyporelleae Pia 1920 by Shuysky (1987).
Stratigraphic range and distribution. Middle Devonian-Late Viséan. Very common. Forms bafflestones associated with other tubular algae, palaeoberesellids and calcispheres. Morocco, Spain, Belgium, France, Ireland, Turkey, Russian Platform, Urals, Donbass, Siberia (Kuzbass, Kolyma, Omolon), China, Australia, Alaska, British Columbia, Alberta, Utah, Montana, Tennessee, Nova Scotia.

66. *Ivanovia* Khvorova 1946

Type of genus. Ivanovia tenuissima Khvorova 1946. Late Carboniferous, Russian Platform (see Plates 1i and 3f).
Diagnosis. Thallus composed of large blades, standing erect on substratum. Central zone with dichotomous poorly calcified filaments(?), usually obliterated by cementation. Cortex contains rows of cylindrical utricles(?) with branches perpendicular to the exterior.
Suprageneric attribution. "Phylloid algae" (vernacular). New tribe Ivanoviae Shuysky 1987 among the Anchicodiaceae Shuysky (1987).
Stratigraphic range and distribution. First occurrence in Middle Carboniferous. Abundant from Moscovian to Early Permian. Often associated with *Eugonophyllum* and *Neoanchicodium* in phylloid mounds characterized by *Tubiphytes*. Algeria, Spain, Carnic Alps, Russian Platform, Urals, China, Canadian Arctic, Yukon, Colorado, New Mexico, Texas.

Izhella Antropov 1955

See *Renalcis*.

67. *Japhetella* Mamet and Roux 1983

Type of genus. Japhetella boehmi Mamet and Roux 1983. Late Viséan, France (see Plate 1d).
Diagnosis. Thallus cylindrical, annulated. Verticils separated by slight constrictions. Medulla cylindrical with poorly defined expansions forming vestibule. Tufts of phloiophore branches, slightly arched.
Suprageneric attribution. Diploporeae (Pia 1920) Güvenç 1979 according to Deloffre (1988).
Stratigraphic range and distribution. Late Viséan Zone 16_i and 16_s, France, Spain.

68. *Kamaena* Antropov 1967

Type of genus. Kamaena delicata Antropov 1967 (= *Kamaena simplex* Eickhoff 1968). Devonian-Carboniferous transition, Russian Platform.
Diagnosis. Thallus long, slender, cylindrical, with dichotomic ramifications. Medulla divided by regular pseudosepta, of even thickness and perpendicular to the cortex. Inner cells rectangular to subrectangular. Cortex perforated by thin, straight, undivided, randomly disposed branches.
Synonyms. Carboniferous "*Nodosinella*" of the literature, *Devonoscalae* Langer 1979, *Subkamaena* Berchenko 1981.
Other Carboniferous species.
Kamaena awirsi Mamet and Roux 1974. Devonian-Carboniferous transition to Viséan. Ireland, Great Britain, Belgium, Donbass, Siberian Platform, Kolyma.
Kamaena concaviuscula (Berchenko 1981). Devonian-Carboniferous transition, Donbass.
Kamaena itkillikensis Mamet in Petryk and Mamet (1972). Lower Carboniferous, Great Britain, Belgium, Donbass, Urals, Siberia, Alaska, Alberta.
Kameana lata Ivanova 1988. Viséan, Urals, Siberian Platform (Kolyma, Omolon).
Kameana maclareni Mamet and Rudloff 1972. Lower Carboniferous, Alberta, British Columbia, Alaska.
Kamaena magna Ivanova 1988. Famennian-Viséan. Morocco, Ireland, France, Urals
Kameana pirleti Mamet and Roux 1974. Lower Carboniferous, Great Britain, France, Belgium, Donbass, Urals.
Kameana razdolnica (Berchenko 1981). Devonian-Carboniferous transition, Donbass, Urals
Kameana sibirica Ivanova 1988. Tournaisian, Siberian Platform.
Suprageneric attributions. Palaeobereselleae Mamet and Roux 1974. Tribe Kamaeneae Shuysky 1985 among the Palaeosiphonocladales Shuysky 1985.
Stratigraphic range and distribution. Abundant in Carboniferous lagoons associated with *Palaeoberesella*. Middle Devonian – Lower Carboniferous. Morocco, Algeria, Spain, Ireland, Great Britain, France, Belgium, Germany, Poland, Russian Platform, Donbass, Urals, Central Asia, Afghanistan, Siberia, (Taimyr, Kolyma, Omolon), Australia, Alaska, Alberta, American Cordillera, Iowa, Newfoundland.

69. *Kamaenella* Mamet and Roux 1974

Type of genus. Kamaenella denbighi Mamet and Roux 1974. Late Viséan, Great Britain (see Plate 1h).
Diagnosis. Thallus cylindrical, with extensive twisted bifurcations. Medulla divided by regular, even pseudoseptations, perpendicular to the cortex. Branches straight, very thin. Medullar cavities have height greater than width.
Other Carboniferous species.
Kamaenella tenuis. (von Möller 1879). Viséan, Morocco, Spain, France, Great Britain, Russian Platform (see Plate 3s).

Suprageneric attributions. Palaeobereselleae Mamet and Roux 1974. Tribe Kamaeneae Shuysky 1985 (as a Palaeosiphonocladales Shuysky 1985).
Stratigraphic range and distribution. Lower Carboniferous. Morocco, Spain, Great Britain, France, Russian Platform, Donbass, Urals, Tadhzikhistan, Newfoundland.

70. *Komia* Korde 1951

Type of genus. *Komia abundans* Korde 1951, Middle Carboniferous, Urals. (= *Komia eganensis* Wilson and Waines 1963) (see Plate 3q).
Diagnosis. Thallus ramose, cylindrical, robust. Hypothallic tissue thin, a flaring bundle of regular, elongated, rectangular cells. Perithallic tissue much more conspicuous, composed of quadrangular cells perpendicular to the central axis. Wall yellowish, hyaline. Conceptacles in cortical cells. Close to *Ungdarella* and hard to differentiate unless the section shows the hypothallic tissue.
Suprageneric attributions. Controversial with unfounded attributions to echinoderms, stromatoporoids, hydrozoans and calcisponges. Korde's original description is correct and with its companion *Ungdarella*, the genus belongs to the Ungdarellaceae Maslov 1962.
Stratigraphic range and distribution. Very abundant. Forms the "*Komia* community" of Flügel (1977) (shelf margin community of Wilson 1975). First probable occurrence in Namurian (Zone 18 of the Nizi Formation, British Columbia). Last occurrence, Early Permian. Cosmopolitan in the Northern Hemisphere. Morocco, Spain, Great Britain, Urals, Donbass, Kazakhstan, Iran, Japan, Alaska, Canadian Arctic, American Cordillera and midcontinent.

71. *Koninckopora* Lee 1912

Type of genus. *Calamopora inflata* de Koninck 1842. Viséan, Belgium.
Diagnosis. Thallus large, arched to cylindrical with an conspicuous medulla. Cortex, by comparison, thin, coarsely perforated by subpolygonal straight branches. Wall two-layered: an inner micritic layer and a yellowish pseudofibrous layer.
Synonym: *Coeloceratoides* Derville 1931, *Uragiella* Maslov 1935, *Koninckoporoides* Rich 1974.
Other Carboniferous species.
Koninckopora minuta Weyer 1968 (= *Koninckopora sahariensis* Chanton 1964 *pars*).
Koninckopora mortelmansi Mamet 1973 (= *Koninckoporoides monteaglensis* Rich 1974).
Koninckopora pruvosti Güvenç 1966 (= *Koninckopora sahariensis* Chanton 1964, *pars*).
Koninckopora tenuiramosa Wood 1942 (see Plate 3u).
All these species are Viséan and, except for *pruvosti*, are cosmopolitan in the Northern Hemisphere and Australia. *Koninckopora pruvosti* is known from Morocco, Spain, France, Belgium, Germany, Turkey, Iran and Australia.
Suprageneric attributions. Controversial with attributions to the tabulates,

bryozoans, cephalopods. Belongs to the cyclocrinitid Seletonellaceae either as Mastoporeae (Pia 1920) (Deloffre 1988) or as Cyclocrineae Pia 1929 (Shuysky 1987).
Stratigraphic range and distribution. Very common in high-energy shoals and banks. Fragile and usually observed as fragments. Cosmopolitan from Zone 11 to 16 (Early to Late Viséan) in Eurasia, including Siberian Arctic (Taymyr, Omolon), North Africa, Australia, Canadian Arctic and elsewhere in North America. Extends up to Zone 18 in Southern China.

Koninckoporoides Rich 1974

See *Koninckopora.*

72. *Kulikia* Golubtsov 1961

Type of genus. Kulikia sphaerica Golubtsov 1961. Upper Viséan, Byelorussia.
Diagnosis. Thallus metaspondyl, a succession of spherical units. Central medulla conspicuous, cylindrical, continuous but with constrictions. Verticils with first order ramifications. Branches(?) nearly spherical, communicating by round pores. Needs to be partially revised. Original description is confusing and the emendation of Skompski (1984) combines two taxa. The best published photograph of the genus is in Shuysky (1987, Pl. XIII, Fig. 5). Similar to, but distinct from, the aciculellid *Sphinctoporella.*
Suprageneric attribution. Aciculelleae Bassoullet et al. 1979 *emend.* Deloffre 1988.
Stratigraphic range and distribution. Scarce. Late Viséan. Belgium, Poland, Russian Platform, Ukraine.

73. *Litostroma* Mamay 1959

Type of genus. Litostroma oklahomense Mamay 1959, Middle Carboniferous, Oklahoma.
Diagnosis. Thallus, encrusting, flat platelets, one cell thick. Cells rather irregular, subpolygonal in axial section.
Suprageneric attribution. Undetermined family among the rhodophytes?
Stratigraphic range and distribution. Very scarce. Lower and Middle Carboniferous. Oklahoma, Newfoundland.

Luteotubulus Vachard in Vachard et al. (1977)

See *Issinella.*

74. *Macroporella* Pia 1912

Type of genus. Macroporella dinarica Pia 1912. Trias, Yugoslavia.
Diagnosis. Thallus cylindrical, unsegmented, with conspicuous medulla. Phloiophore primary branches in irregular oblique whorls.
Other Carboniferous species.

Macroporella ginkeli Rácz 1965. Bashkirian-Moscovian, Spain, Urals.
Macroporella ortashensis Rauser-Chernoussova and Korolyuk 1981. Moscovian, Urals.
Suprageneric attributions. Macroporellinae (Pia 1920) *emend.* Bassoullet et al. 1979 *trans.* Berchenko and Saltovskaja 1987.
Stratigraphic range and distribution. Bashkirian to Jurassic. Abundant from Permian to Triassic. Eurasia. North America.

"*Malacostroma*" Gürich 1906

Type of genus. Malacostroma concentricum Gürich 1906, Viséan, Belgium, France.
Diagnosis. Regularly layered algal biscuit, composed of loose peloidal meshwork of *Bevocastria* filaments and darker, heavily micritized layers with *Girvanella*. The two other species, "*Malacostroma*" *plumosum* Gürich 1906 and "*M*". *undulosum* Gürich 1906 have similar, but more irregular layering.
Suprageneric attributions. Spongiostromidae.
Stratigraphic range and distribution. Viséan of Belgium. Questionable in the Viséan of the American midcontinent.

75. *Malakhovella* Mamet and Roux 1977

Type of genus. Malakhovella malakhovae Mamet and Roux 1977, Viséan, Urals.
Diagnosis. Bundles of intertwined micritized sheaths of filaments forming pseudo-segments.
Other Carboniferous species.
Malakhovella sparsa Saltovskaja 1984. Viséan. Tadzhikhistan.
Suprageneric attributions. Incertae sedis. Cyanophyte? Girvanellaceae Luchinina 1975 according to Luchinina (1987).
Stratigraphic range and distribution. Lower Carboniferous, Belgium, Urals, Tadzhikhistan.

76. *Mametella* Brenckle 1977

Type of genus. Mametella chautauquae Brenckle 1977. Viséan, midcontinent (= *Stacheia? skimoensis* Mamet and Rudloff 1972).
Diagnosis. Thallus encrusting, fusiform or lemon-shaped. Cells quadratic, in regular concentric rows. Wall, a yellowish hyaline calcite.
Suprageneric attribution. New tribe Mametellae Chuvashov 1987. A Stacheiinae.
Stratigraphic range and distribution. Quite common in North America (Alaska, Cordillera, midcontinent) and scarce in Europe (Ireland, Great Britain and Siberia (Kuzbass, Verkoyansk). Tournaisian – Lower Namurian.

77. *Masloviporella* Kulik 1973

Type of genus. Masloviporella calixoidea Kulik 1973. Serpukhovian, Urals.
Diagnosis. Thallus cylindrical, a succession of verticillated, protruding, regular crowns. Branches euspondyl, simple. Insertion on stem is unclear.

Other Carboniferous species.
Masloviporella conili Vachard in Perret and Vachard 1977. Serpukhovian, France.
Suprageneric attribution. Clypeineae (Elliott 1968) Bassoullet et al. 1979, according to Deloffre (1988) and Shuysky (1987).
Stratigraphic range and distribution. Serpukhovian. France, Urals.

78. *Masloviporidium* Groves and Mamet 1985

Type of genus. Donezella delicata Berchenko 1982. Serpukovian-Bashkirian boundary, Donbass (see Plate 3i).
Diagnosis. Thallus sheet-like, foliate, branching, composed of arcuate rows of wedge-shaped cells. Horizontal partitions heavily calcified. Vertical partitions thinner, irregular. Pores connect continuous cells of adjacent rows. Wall calcareous, hyaline (see Plate 2d).
Suprageneric attribution. Incertae familia among the rhodophytes.
Stratigraphic range and distribution. The only known Carboniferous red alga which has precise stratigraphic connotations. Morrowan-Atokan (Zones 20–21) in North America, uppermost Serpukhovian-basal Bashkirian (Zones 20–21) in Eurasia. Algeria, Donetz, Central Asia, China, Alaska, Idaho, Utah, Arkansas, Oklahoma, Kansas, Texas.

79. *Mellporella* Rácz 1965

Type of genus. Mellporella anthracoporellaformis Rácz 1965. Moscovian, Spain.
Diagnosis. Thallus cylindrical, unbranched. Medulla very important. Cortex thin, perforated by numerous random, short, phloiophore pores which taper out before bifurcation.
Suprageneric attributions. Dasyporelleae (Pia 1920) Bassoullet et al. 1979, according to Deloffre (1988).
Stratigraphic range and distribution. Scarce. Late Bashkirian-Moscovian, Spain, Algeria.

80. *Menselina* Antropov 1967

Type of genus. Menselina clathrata Antropov 1967. Devonian-Carboniferous transition, Russian Platform.
Diagnosis. Thallus(?) conical. Wavy horizontal calcified partitions which could correspond to fronds? Vertical partitions poorly developed. No central stem.
Other Carboniferous species.
Menselina rotunda Berchenko 1981. Devonian-Carboniferous transition, Donbass, Siberia (Altai, Omolon).
Menselina triangula Berchenko 1981. Devonian-Carboniferous transition, Donbass.
Suprageneric attributions. Microproblematic. "Problematic clear cones" of Brazhnikova and Rostovceva (1966). "Problematic algae" of Bogush et al. (1970).
Stratigraphic range and distribution. Devonian-Carboniferous transition.

Belgium, France, Russian Platform, Donbass, Urals, Siberia (Altai, Omolon), Afghanistan.

81. *Microcodium* Glück 1912

Type of genus. Microcodium elegans Glück 1912. Miocene, Germany.
Diagnosis. A cob-shaped thallus(?) with a central micritized cavity surrounded by peripheral radial plates. Each plate contains a central, thin, cylindrical tube. The "corn-grain" plates often corrode the surrounding carbonates (see Plate 1e).
Suprageneric attributions. Highly controversial. Originally placed among the codiaceans. Maslov erected the family Microcodiaceae (1960). See Klappa (1978, table 1) for attributions ranging from fungi to vascular plant rootlets. Often considered as indicative of pedogenesis. Erroneously considered as Cenozoic infiltrations (Bodergat 1975). Refer to Mamet and Roux (1982) for discussion of Palaeozoic occurrences. See also the thesis of Plaziat (1984).
Stratigraphic range and distribution. Oldest documented occurrence in the Emma Fjord Formation of the Canadian Arctic (Viséan). Very common in Middle-Carboniferous and Permian shallowing-up sequences of the Canadian Arctic. Cosmopolitan in the Mesozoic and Cenozoic of the Northern Hemisphere.

82. *Mitcheldeania* (Wethered 1886) Mamet and Roux 1975b.

Type of genus. Mitcheldeania nicholsoni Wethered 1886 *pars.* Lower Carboniferous, Great Britain.
Diagnosis. Thallus nodular, an aggregate of micritized, highly dichotomous sheaths. Tubes are flexuous and pinched at point of bifurcation.
Other Carboniferous species.
Mitcheldeania distans (Conil and Lys 1964). Viséan-Serpukhovian, Belgium, France, Czechoslovakia, Nova Scotia.
Suprageneric attribution. Nodular codiacean for Mamet and Roux (1975b). Garwoodiaceae for Luchinina (1987).
Stratigraphic range and distribution. Common in Dinantian-Namurian lagoons, often associated with *Garwoodia* and *Ortonella*. Cosmopolitan, Eurasia, North Africa, North America.

83. *Nanopora* Wood 1964

Type of genus. Nanopora anglica Wood 1964. Viséan, Great Britain.
Diagnosis. Thallus minute, subcylindrical. Branches simple, euspondyl, perpendicular to the cylindrical, continuous, unconstricted medulla. In equatorial sections, branches are elliptical, oval and alternate.
Other Carboniferous species.
Nanopora fragilissima (Maslov 1939). Lower Carboniferous, Kazakhstan, Taimyr, Kolyma, Omolon (see Plate 3c).
Suprageneric attributions. Tribe Uraloporellae Shuysky 1985 for Shuysky (1987). Salpingoporellinae Bassoullet et al. 1979, according to Deloffre (1988).

Stratigraphic range and distribution. Viséan-Early Namurian. Cosmopolitan, Algeria, Morocco, Belgium, Great Britain, France, Germany, Donbass, Poland, Kazakhstan, Taimyr, Kolyma, Omolon, Alaska, Nova Scotia, Newfoundland.

84. *Nansenella* Mamet and Roux 1987

Type of genus. Nansenella multifurcata Mamet and Roux 1987. Bashkirian, Ellesmere Island (see Plate 1f).
Diagnosis. Thallus nodular, composed of micritized sheaths. Tubes very irregular, sinuous, dichotomous in all direction, sometimes cylindrical and continuous, then constricted ampuliform. Simple, double or triple ramifications.
Suprageneric attribution. Nodular codiacean (Mamet and Roux 1987). Cyanobacteria?
Stratigraphic range and distribution. Bashkirian to Sakmarian. Canadian Arctic.

85. *Neoanchicodium* Endo in Endo and Kanumai 1954

Type of genus. Neoanchicodium catenoides Endo in Endo and Kanumai 1954. Lower Permian, Japan (see Plate 1j).
Diagnosis. Thallus composed of large blades, erect on substratum. Blades end in cylindrical "fingers". Central zone with dichotomous poorly calcified tubes, contains one row of ellipsoidal "utricles"(?) with internal and external openings. The reconstruction of Shuysky (1987), inspired by Chuvashov (1974), indicates two rows of "utricles", which are not observed in the type material.
Suprageneric attributions. "Phylloid algae" (vernacular). New family Anchicodiaceae Shuysky 1987.
Stratigraphic range and distribution. Forms the "*Neoanchicodium* community" of Flügel (1977). Not as widespread as *Eugonophyllum* or *Ivanovia*, but important in formation of phylloid banks. Moscovian – Lower Permian. Yugoslavia, Carnic Alps, Urals, China, Japan, Canadian Arctic.

Nigriporella Rigby 1958

See *Tubiphytes*.

86. *Nostocites* Maslov 1929 *sensu* Maslov 1956

Type of genus. Nostocites vesiculosa Maslov 1929, Middle Carboniferous, Donbass.
Diagnosis. Thallus composed of cells serving as substratum to a palisade of elongate cylindrical cells containing an axial micritized filament.
Note. The original drawings of Maslov (1929) are sketchy and have led to numerous misinterpretations. The type-material is preserved in Moscow and Maslov's (1956) reinterpretation is correct.
Suprageneric attributions. Hormogonales for Maslov in Orlov (1963). Pleurocarpales? (Chroococcidiaceans) for Mamet and Roux (1983).
Stratigraphic range and distribution. Occurs as Early as the Middle Devonian

(Eifelian of Belgium). Quite common throughout the Carboniferous. Spain, Poland, Great Britain, Belgium, Donbass, Russian Platform, China, Australia, Canadian Arctic, Alaska, New Mexico, Arkansas, Oklahoma, Newfoundland.

87. *Orthriosiphon* Johnson and Konishi 1956

Type of genus. Orthriosiphon saskatchewanensis Johnson and Konishi 1956. Tournaisian-Viséan boundary, Saskatchewan.
Diagnosis. Thallus very large and thick. Cortex club-shaped, poorly calcified, fragile, often abraded. Cortical tubes dichotomous, diverging slowly a number of times and terminating is small funnels at the outer periphery. Conceptacles prominent, subspherical, irregularly arranged, subcortical.
Note. The taxon shows affinities with *Cummingsella* and *Orthriosiphonoides*.
Other Carboniferous species.
Othriosiphon turnhouti Mamet, Mortelmans and Roux 1979. Viséan, Belgium.
Suprageneric attributions. At the creation of the genus, thought to belong to the Codiaceae? Attributed to the Triploporelleae (Pia 1920) Bassoullet et al. 1979 according to Shuysky (1987) (cortical tubes interpreted as dasycladacean "branches"). Udoteaceae for Bassoullet et al. (1983).
Stratigraphic range and distribution. Late Tournaisian-Late Viséan. In the Canadian Arctic extends up to the Bashkirian. Belgium, Alaska, Canadian Arctic. Very prolific in the American Cordillera (Alberta, Wyoming, Idaho, Montana), Saskatchewan, Tennessee, Alabama. The type *Orthriosiphon saskatchewansis* is endemic to North America.

88. *Orthriosiphonoides* Petryk in Petryk and Mamet 1972, *non* Vachard 1981

Type of genus. Orthriosiphonoides salterensis Petryk in Petryk and Mamet 1972. Viséan, Alberta (see Plate 2l).
Diagnosis. Thallus large, club-shaped, segmented, "annulated". Large central stem. Segments outwardly rounded. Sutures deep. Cortical tubes thin, numerous, straight to slightly curved, of constant diameter, dichotomous. Conceptacles spherical to ellipsoidal, randomly distributed.
Other Carboniferous species.
Orthriosiphonoides tenuiramosa Mamet and Rudloff 1972. Viséan, Alberta, British Columbia (see Plate 3v).
Suprageneric attributions. Udoteaceae for Bassoullet et al. 1983 (see also critique of Vachard 1981). Attributed to the Triploporelleae (Pia 1920) Bassoullet et al. 1979 by Shuysky (1987) (cortical tubes interpreted as dasycladacean "branches").
Stratigraphic range and distribution. Endemic to North America (Eurasiatic reports of Vachard (1981) may be due to erroneous taxonomy). Viséan to Early Permian. Can be very common in high energy channelized grainstones. Canadian Arctic, American Cordillera (Alaska, Yukon Territory, British Columbia, Alberta, Idaho, Wyoming, Montana, Utah), Tennessee.

89. *Ortonella* (Garwood 1914) Mamet and Roux 1981

Type of genus. Ortonella furcata Garwood 1914. Viséan, Great Britain.
Diagnosis. Thallus nodular, composed of closely-packet radiating, dichotomous micritized sheaths. Even-diameter tubes branch at relatively low angle (10° to 45°).
Other Carboniferous species.
Ortonella balbinia (Poncet 1974). Siegenian-Viséan, France, Morocco.
Ortonella capnostyloides (Derville 1931). Viséan, France
Ortonella coloradoensis Johnson 1945. Lower to basal Middle Carboniferous, Urals, Alaska, Alberta, Saskatchewan, Colorado.
Ortonella gracilis Johnson 1951. Late Carboniferous-Permian, Canadian Arctic, Texas. (Probable in Germany and Russian Platform as *aff.*)
Ortonella kershopensis Garwood 1931. Lower Carboniferous, Ireland, Great Britain, France, Donbass, Kuznetsk, Alberta, Colorado.
Ortonella maksimovae Maslov 1956. Tournaisian, Donbass.
Ortonella moscovica Maslov 1956. Lower Carboniferous, Belgium, Russian Platform.
Ortonella myrae Rácz 1964. Moscovian, Spain.
Ortonella tenuissima Garwood 1931. Famennian – Lower Carboniferous, Great Britain, Donbass, Australia.
Ortonella tyrrelensis Mamet and Rudloff 1972. Tournaisian-Viséan, Belgium, Alberta, Nova Scotia.
Ortonella upensis Birina 1948. Devonian-Carboniferous transition to Tournaisian, Russian Platform, Urals, Kazakhstan.
Suprageneric attributions. Nodular codiacean (Mamet and Roux 1981). Garwoodiaceae (Johnson) Shuysky 1973 according to Luchinina (1987).
Stratigraphic range and distribution. Common in lagoons (see *Garwoodia*). Cosmopolitan in Northern Hemisphere. Very long ranging from Ordovician to Mesozoic.

"*Osagia*" Twenhofel 1919

Type of the "form-genus". Osagia incrustata Twenhofel 1919. Lower Permian, Kansas.
Diagnosis. Oncoid formed by the intimate intergrowth of algal filaments, sessile foraminifers and other organisms such as sponges or bryozoans. Carboniferous "*Osagia*" often produced by *Girvanella* sheaths, encrusting apterrinellids and sponges. The association of *Girvanella*, *Archaeolithophyllum*, phylloid blades and attached tuberitinids is common in the Moscovian. In the Bashkirian, "*Osagia*" is formed by *Berestovia*-cyanophytes, by *Iberiaella-Girvanella-Donezella* or by *Bevocastria*-sponges. As "*Osagia*" is formed by associations of different organisms, the generic and specific concepts have little taxonomic value.
Suprageneric attributions. See diagnosis.
Stratigraphic range and distribution. The "*Osagia*" community of Flügel (1977) is composed of *Girvanelle-Hedraites*-apterrinellid oncoids formed in shallow

near-coast or in off-shore environments. Doubtful reports in the Riphean (where it was probably formed by cyanobacteria). "*Osagia*" is scarce in the Lower Carboniferous and has a sudden outburst in the Bashkirian. Common from the Middle Carboniferous to the Permian. Cosmopolitan in the Northern Hemisphere.

"*Ottonosia*" Twenhofel 1919

Type of "form-genus". *Ottonosia laminata* Twenhofel 1919. Lower Permian, Kansas.
Diagnosis. Asymmetrical encrustations on the upper surface of a substratum. Microflora and microfauna are the same as observed in "*Osagia*". Needs revision.
Suprageneric attributions and stratigraphic range. See "*Osagia*".

Pachysphaera Conil and Lys 1964 OBJ

see *Calcisphaera*.

Pachysphaerina Conil and Lys 1967 OBJ

see *Calcisphaera*.

90. *Palaeoberesella* Mamet and Roux 1974

Type of genus. *Nodosinella lahuseni* von Möller 1879. Viséan, Russian Platform.
Diagnosis. Thallus cylindrical, long, with rare ramifications. Medulla subdivided by coarse conical projections forming irregular medullary cavities; they appear spherical in the centre of the thallus and angular in a section near the cortex. Branches straight, simple, rare, perpendicular to the cortex. Conceptacles spherical, cortical.
Synonym. *Stylaella* Berchenko 1981.
Suprageneric attributions. Palaeoberesellaea Mamet and Roux 1974.
Family Palaeoberesellaceae *trans.* Shuysky 1985 for Shuysky (1987).
Stratigraphic range and distribution. Originally described from the Tournaisian to the Namurian. Known to occur as early as the Middle Devonian. Very abundant in restricted platforms where it forms bafflestones. Cosmopolitan. Morocco, Algeria, Spain, France, Great Britain, Belgium, Germany, Russian Platform, Donbass, Urals, Siberia, Kuzbass, Kolyma, Omolon, China, Australia, Alaska, American Cordillera (Alberta, Wyoming, Montana, Idaho), American midcontinent and Newfoundland.

91. *Palaeomicrocodium* Mamet and Roux 1983

Type of genus. *Palaeomicrocodium devonicum* Mamet and Roux 1983. Frasnian, Australia.
Diagnosis. Thallus(?) composed of an aggregate of spheres of radial plates. The formal emendum of Mu and Zhang (1988) claiming that the plates are composed of radiate fibrous calcite is incorrect.

Suprageneric attributions. An algal incertae sedis belonging to the Microcodiaceae Maslov 1960. Related to *Microcodium* and *Baccanella*.
Stratigraphic range and distribution. Eurasia, Australia, Canadian Arctic. Doubtful as early as Late Silurian (Gotland). Middle Devonian (Eifelian) to Middle Carboniferous. Same ecological indicator as *Microcodium* and *Baccanella*.

92. *Parachaetetes* Deninger 1906 *emend* Mamet and Roux 1977

Type of genus. Parachaetetes tornquisti Deninger 1906. Bathonian, Sardinia.
Diagnosis. Thallus nodular, composed of closely packed, elongated filaments, subpentagonal in equatorial section, subdivided by horizontal, continuous, slightly concave partitions. Filaments also divided by longitudinal "septiform processes". Wall micritic.
Synonym. Tomilithon Maslov 1962 (as subgenus).
Carboniferous species.
Parachaetetes concentrica (Maslov 1956). Devonian-Lower Carboniferous, Germany, Russian Platform.
Parachaetetes donbassica Berchenko 1981. Devonian-Carboniferous transition, Donbass.
Parachaetetes garwoodi (Hinde in Garwood 1912). Tournaisian-Viséan, Western Europe, Arctic Siberia, British Columbia, Alberta (see Plate 3e).
Parachaetetes glenwoodensis (Johnson 1945). Tournaisian, Saskatchewan, Colorado.
Parachaetetes (Tomilithon) johnsoni Maslov 1962. Late Devonian-Late Tournaisian, Belgium, Great Britain, Poland, Donbass, Urals, Arctic Siberia, Kuzbass, Australia, Colorado.
Parachaetetes paleozoicus (Maslov 1935). (= *Solenopora nexa* Maslov 1956). Lower Carboniferous. France, Donbass, Urals, Arctic Siberia, Alaska, Kansas, Québec.
Parachaetetes regularis Konishi 1958. (= *Parachaetetes intermedius* Konishi 1958). Upper Devonian-Upper Tournaisian. Belgium, Poland, Donbass, Australia, Alaska, Alberta.
Parachaetetes russiensis (Maslov 1956) (= *Solenopora koivense* Chuvashov 1965). Devonian to Devonian-Carboniferous transition, Russian Platform, Urals, Donbass.
Parachaetetes velbertiana (Paul 1938). Tournaisian, Germany.
Suprageneric attribution. Solenoporaceae Pia 1927.
Stratigraphic range and distribution. Cosmopolitan. Very long ranging from Ordovician to Mesozoic.

"*Paradella*" Maslov 1956

Type of genus. Paradella recta Maslov 1956. Carboniferous, Kazakhstan.
Diagnosis. Needs revision. The type and the associated *Paradella adunca* Maslov 1956 and *Paradella fistulosa* Maslov 1956 are probably Carboniferous phylloid algae belonging to the Anchicodiaceae. A fourth Carboniferous

species, *Paradella arcuata* Maslov 1956, is a coelosporellid. A fifth species, *Paradella alveata* Maslov 1956, is a dissolved thallus of an udoteacean.
In 1981, Kulik published photographs of *Paradella recta* which are unusable. They are probably weathered blades of *Eugonophyllum*.
Paradella adunca re-illustrated by Kulik (1981) (not the type of the genus) can be differentiated form *Eugonophyllum* while the re-illustration of *Paradella fistulosa* bears no resemblance to Maslov's original drawing.
Paradella arcuata in the concept of Rauser-Chernoussova and Korolyuk (1981) also differs considerably from Maslov's illustration. The addition of a new Devonian udoteacean *Paradella crassa* by Shuysky (1973) and the illustration of a Devonian *Paradella*? by Deloffre (1988) (after Kochansky-Devidé and Güsic 1971) render the genus unusable without formal emendation.
Suprageneric attributions. Maslov 1956 erected the tribe Paradellae. Kulik (1981) includes *Paradella* and *Eugonophyllum* in that tribe. Shuysky (1987) attributes *Eugonophyllum* to his new tribe Ivanoviae while restricting *Paradella* to the Paradellae. Deloffre (1988) questionably places *Paradella* among the Aciculelleae, but with a Devonian age, which is implausible. Unless the generic status of *Paradella* is clarified, its taxonomy remains open to discussion.
Stratigraphic range. See above.

93. *Paraepimastopora* Roux 1979 (*Epimastopora auct*)

Type of genus. *Epimastopora kansasensis* Johnson 1946. Late Carboniferous, Kansas (see Plate 2o).
Diagnosis. Thallus cylindrical, large, unsegmented, with a thick stem and a slender poorly calcified cortex. Cortex with projections, perforated on both sides by numerous simple, isodiametric branches. In tangential sections these are subcylindrical to subpolygonal (see Roux in press).
Other Carboniferous species.
Paraepimastopora kanumai Endo *in* Endo and Kanumai 1954. Upper Carboniferous – Permian, Japan, China.
Paraepimastopora urtazymensis (Chuvaskov and Anfimov 1988). Moscovian, Urals.
Suprageneric attributions. Mastoporae (Pia 1920) *trans*. Deloffre 1988.
Stratigraphic range and distribution. Bashkirian to Permian. Carnic Alps, Afghanistan, Urals, Japan, China, Yukon Territory, Canadian Arctic, American midcontinent, New Mexico.

94. *Parakamaena* Mamet and Roux 1974

Type of genus. *Kamaena*(?) *tenuisepta* Mamet and Rudloff 1972. Viséan, Alberta.
Diagnosis. Thallus elongate, cylindrical, rarely ramified by simple bifurcation. Medulla divided by thin, straight partitions, perpendicular to the cortex. Cells rectangular, height greater than width. Branches thin, few, simple.
Other Carboniferous species.
Parakamaena exilis Ivanova 1988. Siberian Platform (Omolon Massif).

Parakamaena irregularis Berchenko 1981. Tournaisian, Donbass, Siberian Platform.
Suprageneric attributions. Palaeobereselleae Mamet and Roux 1974 according to Deloffre (1988). Unconvincingly reported to the Anthracoporellopsiae by Shuysky (1985).
Stratigraphic range and distribution. Scarce. Middle Devonian to Viséan. Belgium, Great Britain, Donbass, Siberian Platform, Omolon Massif, Alaska, Alberta. The type species is endemic to North America.

95. *Parastacheia* Mamet and Roux 1977

Type of genus. Parastacheia iglii. Mamet and Roux 1977. Late Viséan, Algeria.
Diagnosis. Thallus fusiform, encrusted on a central support. Rows of subparallel laminae with oblique insertion on support. Cells subrounded. Wall calcareous, yellowish, hyaline.
Suprageneric attributions. Stacheiinae Loeblich and Tappan 1961. New tribe "Aoujgaliae Chuvashov 1987".
Stratigraphic range and distribution. Scarce. See type of genus.

96. *Pekiskopora* Mamet 1974

Type of genus. Pekiskopora macqueeni Mamet 1974. Middle Tournaisian, British Columbia.
Diagnosis. Thallus, a succession of spherical units. Thick stem and metaspondyl verticils sharply protruding from the cortex. Medulla expands in numerous composite vestibules followed by crowns of sixteen branches. Branches expand slowly in diameter.
Suprageneric attributions. Unconvincingly reported to the Triploporellinae Pia 1920 *emend* Bassoullet et al. 1979 by Shuysky (1987). Albertaporelleae (Güvenç 1979) for Deloffre (1988).
Stratigraphic range and distribution. Locally abudant in Tournaisian-Viséan. Endemic to North America. British Columbia, Alberta, Tennessee, Newfoundland.

97. *Peristacheia* Mamet and Roux 1983

Type of genus. Peristacheia jonesi Mamet and Roux 1983. Late Tournaisian, Australia.
Diagnosis. Lemon-shaped stacheiin with differentiated thallus. Hypothallus with rows of quadratic cells. Perithallus composed of ovoid cells. Wall calcareous, yellowish hyaline.
Suprageneric attributions. Stacheinaceae Loeblich and Tappan 1961.
Stratigraphic range and distribution. Scarce. Late Tournaisian-Early Viséan, Australia.

98. *Petschoria* Korde 1951

Type of genus. Petschoria elegans Korde 1951. Middle Carboniferous, Urals.

Diagnosis. Thallus ramified, cylindrical, unsegmented. Hypothallus prominent, composed of long longitudinal filaments formed by rectangular cells. Perithallus poorly developed but of even thickness, composed of very fine filaments of rectangular cells, perpendicular to the hypothallic tissue.
Other Carboniferous species.
The genus is restricted to *Petschoria elegans. Petschoria antiqua* Berchenko 1981 and *Petschoria*(?) *tenuicrustata* Shuysky 1973 do not belong to the genus.
Suprageneric attribution. New tribe Petschoriae Chuvashov 1987. Ungdarellaceae Maslov 1962.
Stratigraphic range and distribution. Moscovian, Urals, Yukon Territory, New Mexico. Other reports are incorrect (France, Algeria, Morocco).

Pokorninella Vachard in Perret and Vachard (1977)

See *Proninella*?

"*Polymorphocodium*" Derville 1931

Type of genus. Polymorphocodium lapparenti Derville 1931. Viséan, France.
Diagnosis. Thallus composed of encrusting asymmetrical, cystoid tubes disposed in more or less regular rows and separated by micritized layers displaying sponge canals, *Bevocastria* and birdseyes. Needs revision. Very similar to "*Stylocodium*".
Suprageneric attributions. A codiacean at origin. Pia (1937) thought it was a consortium. Spongiostromidae Gürich 1906.
Stratigraphic range and distribution. Middle Viséan-Basal Bashkirian. France, Belgium, China, Alberta, Wyoming.

Praedonezella Kulik 1973

See *Donezella*.

Principia Brenckle 1982

See *Archaolithophyllum*.

99. *Proninella* Reitlinger in Menner and Reitlinger 1971

Type of genus. Proninella tamarae Reitlinger in Menner and Reitlinger 1971. Givetian, Siberia.
Diagnosis. Thallus encrusting, sinuous, irregular, branching. Tube compressed or inflated, subcylindrical, divided by irregular partitions. Wall yellowish, clear, with thin pores (?)
Carboniferous species.
Proninella gracilis (Vachard in Meissami et al. 1978) (= *Proninella enigmatica* Mamet and Roux 1978). Viséan, Morocco, France, Tennessee, Alabama.
Proninella strigosa (Vachard in Perret and Vachard 1977). Viséan-Early Bashkirian, Morocco, Spain, France, Iran, Alaska, Oklahoma, Kansas, Alabama, Tennessee.

Suprageneric attributions. A foraminifer at origin. Transferred to the sponges by Vachard in Perret and Vachard (1977). An algal incertae sedis for Groves (1983).
Stratigraphic range and distribution. Occurs in lagoonal conditions associated with Palaeosiphonocladales and calcispheres. Givetian-Bashkirian. Morocco, Spain, France, Iran, Afghanistan, Urals, Siberia (Verkoyansk), Alaska, Alberta, British Columbia, Oklahoma, Kansas, Alabama, Tennessee.

100. *Protoumbella* Mamet 1970

Type of genus. Umbella saccaminiformis Bykova in Bykova and Polenova 1955. Famennian, Russian Platform.
Diagnosis. Utricle (?) spherical. Exterior smooth, without ornamentation. No basal fibrous projection. Opening at end of small neck. No known apical lid. Wall two-layered. A micritic layer and a thick yellowish prismatic layer.
Synonym. Grozdilovella Chermnykh 1972.
Other Carboniferous species.
Protoumbella decurta (Reitlinger 1966). Devonian-Carboniferous transition, Armenia.
Protoumbella elliptica (Reitlinger 1966). Devonian-Carboniferous transition – Viséan, Armenia, Urals, American midcontinent.
Suprageneric attribution. Umbellaceae Fursenko 1959.
Stratigraphic range and distribution. Famennian-Viséan, Russian Platform, Urals, Armenia, Kazakhstan, American midcontinent. This is the youngest known representative of the umbellids.

101. *Pseudochaetetes* Haug 1883, *emend.* Mamet and Roux 1977

Type of genus. Pseudochaetetes champagnensis Peterhans 1929, Jurassic, France.
Diagnosis. Thallus nodular, composed or closely packed, elongated filaments, subpentagonal or subhexagonal in equatorial section, sparsely subdivided by horizontal partitions. Wall micritic.
Carboniferous species.
Pseudochaetetes elliotti (Bilan and Golonka 1973). Lower Carboniferous, Poland, China.
Pseudochaetetes similis (Paul 1938). Tournaisian, Germany.
Suprageneric attribution. Solenoporaceae Pia 1927.
Stratigraphic range and distribution. Very long ranging. Ordovician (?) – Silurian-Jurassic. Cosmopolitan in Northern Hemisphere.

Pseudoepimastopora Endo 1960

See *Epimastopa pars* and *Epimastoporella pars.*

102. *Pseudogyroporella* Endo 1959

Type of genus. Pseudogyroporella mizziaeformis Endo 1959. Permian, Japan.
Diagnosis. Thallus moniliform. Stem important. Cortex perforated by very

large vesiculiferous primary branches. They have a trapezoidal form toward the exterior. Needs revision.
Carboniferous species.
Pseudogyroporella mizziaeformis is present in the Upper Carboniferous of the Canadian Arctic archipelago.
Suprageneric attribution. Coniporelleae Bassoullet et al. 1979 according to Shuysky (1987) and to Deloffre (1988).
Stratigraphic range and distribution. Upper Carboniferous-Lower Permian. Yugoslavia, China, Japan, Canadian Arctic.

103. *Pseudohedstroemia* Mamet and Roux 1978

Type of genus. Pseudohedstroemia polyfurcata. Mamet and Roux 1978. Viséan, Tennessee.
Diagnosis. Nodule formed by radially disposed cylindrical, micritized sheaths of two types. First group of variable tubes is ramified by small angle dichotomies. It contains thin wavy tubes with double or triple branching.
Suprageneric attribution. Nodular codiacean? Cyanobacteria?
Stratigraphic range and distribution. Tournaisian-Namurian. Alaska, Alberta, Utah, Tennessee.

104. *Pseudoissinella* Mamet and Rudloff 1972

Type of genus. Pseudoissinella alaskaensis Mamet and Rudloff 1972. Viséan, Alaska.
Diagnosis. Thallus cylindrical, erect, with multiple branchings at right angle. Cortex, in fibro-radial yellowish calcite, perforated by scarce thin unramified branches.
Suprageneric attribution. Tribe Issinelleae Deloffre 1987 according to Deloffre (1988).
Stratigraphic range and distribution. Scarce in Devonian. Tournaisian to Lower Namurian. Belgium, Donbass, Alaska, Alberta, British Columbia.

105. *Pseudokamaena* Mamet in Petryk and Mamet 1972

Type of genus. Pseudokamaena armstrongi Mamet in Petryk and Mamet 1972. Viséan, Alaska.
Diagnosis. Thallus cylindrical, dichotomous, with faint outer constrictions. Stem divided by short transverse partitions. Cells ovoid to spherical. Wall calcareous, yellowish, hyaline, perforated by thin, unramified straight pores.
Other Carboniferous species.
Pseudokamaena boulderensis Mamet in Petryk and Mamet 1972. Viséan, Alberta.
Suprageneric attributions. Palaeoberesselleae Mamet and Roux 1974. Exvotariselleae Shuysky 1985 according to Shuysky (1987).
Stratigraphic range and distribution. Mostly in Lower Carboniferous lagoons associated with Palaeosiphonocladales and calcispheres. Scarce in Devonian (Eifelian-Frasnian). Mostly Tournaisian-Lower Bashkirian. Belgium, France,

Donbass, Urals, Siberia, Afghanistan, Kolyma, Omolon Massif, Australia, Canadian Arctic, Alaska, Alberta, Montana, Wyoming.

106. *Pseudokomia* Rácz 1965

Type of genus. Pseudokomia cansecoensis Rácz 1965. Moscovian, Spain.
Diagnosis. Thallus cylindrical, highly calcified. Hypothallic cells rather regular. Perithallic cells perpendicular to hypothallus, heavily calcified and very regular. The genus forms the morphological link between *Epistacheoides* and *Petschoria*.
Suprageneric attribution. Ungdarellaceae Maslov 1956. New tribe Petschoriae for Chuvashov 1987.
Stratigraphic range and distribution. Moscovian, Spain, Canadian Arctic, New Mexico.

107. *Pseudonanopora* Mamet and Roux 1975

Type of genus. Pseudonanopora stockmansi Mamet and Roux 1975. Middle Viséan, Belgium.
Diagnosis. Thallus tubular, cylindrical, with slight constrictions. Stem divided by rudimentary pseudo-septa. Cortex perforated by randomly disposed, straight, simple branches.
Suprageneric attribution. Anthracoporellopsiae Shuysky 1985 according to Shuysky (1987).
Stratigraphic age and distribution. Devonian-Carboniferous transition to Serpukhovian. Algeria, Morocco, Belgium, Donbass, Siberia (Kolyma).

108. *Pseudosolenopora* Mamet and Roux 1977

Type of genus. Solenopora owodenkoi Chanton-Güvenç 1972. Late Viséan, Morocco.
Diagnosis. Thallus nodular, composed of closely packed elongated, continuous, filaments, subpentagonal to subhexagonal in equatorial section. Horizontal partitions absent. Wall micritic.
Suprageneric attributions. Solenoporaceae Pia 1927. New tribe Pseudosolenoporaceae Chuvashov 1987 (ill-founded).
Stratigraphic range and distribution. Late Devonian-Lower Namurian. Morocco, Belgium, Poland, Germany, Afghanistan, Australia, Alberta, British Columbia.

109. *Pseudostacheoides* Petryk and Mamet 1972

Type of genus. Pseudostacheoides loomisi Petryk and Mamet 1972. Late Viséan, Alberta.
Diagnosis. Thallus encrusting, irregular, composed of closely-packed thick, porous, undulating, horizontal elements and of thin and scarce vertical elements. Cells elongated, flat to undulating. Wall calcareous, yellowish, perforated by thin unramified pores. Central part of thallus usually poorly calcified and micritized.

Suprageneric attribution. Stacheiinae Loeblich and Tappan 1961 *emend.* New tribe Pseudostacheoideae Chuvashov 1987.
Stratigraphic range and distribution. Early Viséan-Lower Namurian. "Cosmopolitan". Morocco, Algeria, Great Britain, Belgium, Afghanistan, Alaska, Alberta, British Columbia, Alabama.

Pseudoumbella Berchenko 1971

See *Biumbella* Mamet 1970.

110. *Pseudovelebitella* Mamet and Roux 1978

Type of genus. Velebitella simplex Kochansky-Devidé 1964. Early Permian, Yugoslavia.
Diagnosis. Thallus composed of a succession of barrel-shaped units, with a central cylindrical stem. Medulla expands into short vestibules. Tufts of five to six primary phloiophore branches are localized in the central part of the units.
Carboniferous species.
Pseudovelebitella pareyni (Mamet and Roux 1975). Viséan, Algeria, Great Britain, Russian Platform.
Pseudovelebitella simplicissima (Vachard in Massa and Vachard 1979). Moscovian, Algeria.
Suprageneric attribution. Velebitelleae (Vachard 1977) Mamet and Roux 1983.
Stratigraphic range and distribution. Viséan-Early Permian. Algeria, Great Britain, Yugoslavia, Russian Platform.

"Pycnostroma" Gürich 1906

Type of genus. Pycnostroma densius Gürich 1906. Viséan, Belgium.
Diagnosis. Oncolite? formed by a spongiostromid encrusting a central uncalcified support (now filled by cement or by sediment). Tissue composed of micritized and dissolved sheaths and a loosely interwoven meshwork of at least two types of algal tubes. Needs redescription.
Other Carboniferous species. The micritized part of *Pycnostroma spongilliferum* Gürich 1906 has in situ sponge spicules.
Suprageneric attribution. Spongiostromidae. An association of cyanobacteria and sponges.
Stratigraphic range and distribution. Known from Ordovician to Carboniferous (Namurian). Morocco, Belgium, France, Alberta, Nevada, Tennessee.

111. *Quasiumbella* Poyarkov 1965

Type of genus. Umbella rotunda Bykova in Bykova and Polenova (1955). Famennian, Russian Platform.
Diagnosis. Utricle(?) nearly spherical. External wall smooth, without ornamentation. No basal projections. Apical lid present.
Carboniferous species.

Quasiumbella cutis (Conil and Lys 1964). Famennian, Devonian-Carboniferous transition, Belgium.
Quasiumbella fragilis Berchenko 1981. Devonian-Carboniferous transition, Donbass.
Quasiumbella gravis (Conil and Lys 1964). Famennian, Devonian-Carboniferous transition, Belgium, Donbass.
Suprageneric attribution. Umbellaceae Fursenko 1959.
Stratigraphic range and distribution. As is the case for most umbellids, common in the Late Devonian and peters out in the Tournaisian. Belgium, Poland, Russian Platform, Donbass, Urals, Armenia, Iran, Siberia, Kolyma, Australia, Alberta.

Quasiumbelloides Berchenko 1971

see *Biumbella*.

112. *Queenslandella* Mamet and Roux 1983

Type of genus. Queenslandella jenkinsi Mamet and Roux 1983. Viséan, Australia.
Diagnosis. Metaspondyl dasyclad composed of a succession of annular units. Central stem subcylindrical with expansions at the junction of each unit. Composite vestibule. First order ramifications. Branches numerous, slightly arched and phloiophore (see Plate 1b).
Suprageneric attribution. Albertaporelleae (Güvenç 1979) Mamet and Roux 1981 according to Deloffre (1988).
Stratigraphic range and distribution. Middle to Late Viséan. Endemic to Australia.

"*Radiosphaera*" Reitlinger 1957
(or "radiospherid calcispheres" Stanton 1967)

Type of genus. Radiosphaera basilica Reitlinger 1957. Devonian-Carboniferous transition, Russian Platform.
Diagnosis. When disarticulated, observed as calcareous hollow spheres. Wall two-layered, a thin internal micritic layer and a thicker pseudo-fibrous wall, with coarse tapering prisms. No aperture. In situ, forms grape-like clusters of loosely-packed calcispheres around a noncalcified axis (see Mamet and Rudloff 1972). At origin (1957) considered as "undetermined spheres" without clear reference to botanical or zoological origin. (Most Russian authors consider the taxon as a foraminifer and the genus is thus a homonym of the Protozoan *Radiosphaera* Jorgensen 1905). A diagenetic stage of Tuberitinidae for Vachard and Tellez-Giron (1986).
Carboniferous species. There are many extant "species" in the literature which appear as ill-defined as the genus. Needs revision.
Suprageneric attributions. See *Calcisphaera* for discussion. The original *Calcisphaera sol* Williamson 1881 is a radiosphaere.

Stratigraphic range and distribution. Givetian-Viséan. Very common in restricted environments. Cosmopolitan in Northern Hemisphere.

113. *Rectangulina* Antropov 1959

Type of genus. Syniella tortuosa Antropov 1950. Frasnian, Russian Platform.
Diagnosis. Algal clast composed of bundles of subparallel, angular, unsegmented, straight tubes. Wall(?) micritic.
Suprageneric attributions. Controversial. Foraminifer (Chuvashov 1965), coprolite (Conil et al. 1969), alga (Antropov 1959, Mamet et al. 1987, Luchinina 1987), incertae sedis (Vachard 1981).
Carboniferous species.
Rectangulina geniculosa (Bykova 1961). Frasnian-Early Namurian. Great Britain, Belgium, Russian Platform, Urals, Siberia, Tennessee, Arkansas.
Stratigraphic range and distribution. Ordovician? – Carboniferous. Quite scarce. Libya, Morocco, Algeria, France, Belgium, Great Britain, Germany, Poland, Russian Platform, Urals, Kazakhstan, Afghanistan, Siberia, Australia, Canadian Arctic, Arkansas, Alabama, Tennessee, Georgia, Nova Scotia.

114. *Renalcis* Vologdin 1932 *emend.* Mamet and Roux 1983

Type of genus. Renalcis granosus Vologdin 1932. Cambrian, Siberia.
Diagnosis. Thallus an aggregate of hollow, inflated cells. Forms simple, straight, rectilinear colonies to complex botryodal aggregates. "Wall" micritic, but not directly secreted, dark, often fissured by retraction.
Synonyms. Chabakovia Vologdin 1939, *Nubecularites* Maslov 1937 (*pars*), *Eniseiella* Maslov in Malakhova 1959, *Izhella* Antropov 1955, *Nephelostroma* Dangeard and Doré 1957 and *Shuguria* Antropov 1959.
Suprageneric attributions. Controversial. Foraminifer (family Renalcidae Riding and Brasier 1975), red alga (solenoporid with "quadrangular cells" of Korde 1959), cyanophyte (as chroococcales, Chabakoviaceae Korde 1969), chroococcophyceae for Luchinina (1987). Consensus is usually to classify *Renalcis* among cyanobacteria (see Saltovskaya 1984 for discussion).
Stratigraphic range and distribution. Proterozoic. Very abundant from Cambrian to Devonian. Peters out in Lower Carboniferous. Cosmopolitan, but plays no important sedimentological role after the Frasnian.

115. *Retephycus* Johnson and Konishi 1956

Type of genus. Retephycus globosus Johnson and Konishi 1956. Late Tournaisian, Saskatchewan.
Diagnosis. Thallus(?) composed of closely packed algal filaments. Tubes are interwoven, appear to branch and are reported to have an irregular outline in transverse section. Needs revision.
Suprageneric attribution. Originally placed in the chlorophytes.
Stratigraphic range and distribution. See type of genus.

116. *Richella* Mamet and Roux 1987

Type of genus. Richella incrustata Mamet and Roux 1987. Mid-Moscovian, Canadian Arctic (Ellesmere) (see Plate 1k).
Diagnosis. Thallus encrusting, composed of semi-cylindrical, erratically interwoven, heavily calcified tubes, attached at their base, and perforated by straight, coarse, undivided branches.
Suprageneric attribution. Algal incertae sedis. In growth form, similar to *Wetheredella*.
Stratigraphic range and distribution. Namurian-Moscovian. Tennessee, Canadian Arctic.

Roquesselsia Termier, Termier and Vachard 1977

See *Epistacheoides*.

Samarella Maslov and Kulik 1956

See *Uraloporella*.

117. *Sandoella* Mamet and Roux 1978

Type of genus. Sandoella fowleri Mamet and Roux 1978. Late Viséan, Tennessee.
Diagnosis. Thallus subcylindrical, verticillated, a succession of irregularly shaped units. Medulla important. Cortex with ovoid cavities followed by bundles of straight branches.
Suprageneric attribution. Does not fit any extant tribe of Diploporaceae.
Stratigraphic range and distribution. Viséan, Alberta, Tennessee.

Shamovella Rauser-Chernoussova 1951

See *Tubiphytes*.

Shartymophycus Kulik 1973

See *Fasciella*.

"*Shermanophycus*" Johnson 1940

Type of genus. Shermanophycus gouldi Johnson 1940. Upper Carboniferous, Colorado.
Diagnosis. An algal biscuit. Needs revision. See *Stylophycus*, *Gouldina*.
Stratigraphic range and distribution. See type of genus.

Shuguria Antropov 1950

See *Renalcis*.

118. *Sinarella* Chuvashov in Chuvashov and Anfimov (1988)

Type of genus. Sinarella delicata Chuvashov and Anfimov 1988. Moscovian, Urals.
Diagnosis. Thallus important, cylindrical with poorly calcified thin cortex and big stem. Wall perforated by triangular pores, protruding from the outside periphery.
Suprageneric attribution. Cyclocrinaea Pia 1920.
Stratigraphic range and distribution. See type of genus.

Sinustacheoides Termier, Termier and Vachard 1977

See *Stacheoides.*

119. *Solenopora* Dybowski 1877 *emend.* Mamet and Roux 1977

Type of genus. Solenopora spongioides Dybowski 1877. Ordovician erratic block, Baltic.
Diagnosis. Thallus nodular, composed of closely packed elongated filaments subangular in equatorial section, subdivided by septiform processes. Wall micritic. Conceptacles ovoid to spherical.
Carboniferous species.
Solenopora dionantina Pia 1937. Viséan. Great Britain, Poland, Alaska. Most Carboniferous species of the literature ought to be transferred to *Parachaetetes.*
Suprageneric attribution. Solenoporaceae Pia 1927.
Stratigraphic range and distribution. Very long ranging. Ordovician-Mesozoic. Cosmopolitan. Forms the "*Solenopora* community" of Flügel (1977).

120. *Sphaerinvia* Vachard 1981

Type of genus. Sphaerinvia piai Vachard 1981. Devonian-Carboniferous transition, Afghanistan.
Diagnosis. Hollow calcitic sphere, with functional opening (basal or apical?) attached to a calcified tube. Wall two-layered, an inner micritic layer with radiating parietal expansions, and a segmented yellowish lamellar outer layer.
Suprageneric attribution. Incertae sedis. Vachard claims that the genus has no opening, but it is conspicuous in the material published by Bogush (1984). Similar material was considered as belonging to the Charophytes by Pia (1937) and a reasonable case for the same attribution has been made by Bogush (1984). However, *Sphaerinvia's* wall is different from that of bona fide Palaeozoic charophytes.
Stratigraphic range and distribution. Devonian-Carboniferous transition – Early Viséan, Belgium, Great Britain, Urals, Afghanistan, Kazakhstan, Siberia, Omolon. Very common in Alaska, Yukon Territory, British Columbia, Alberta, New Mexico, Arizona, Utah.

121. *Sphaerocodium* Rothpletz 1890 *emend* Wray 1967
(= *Rothpletzella* Wood 1948 OBJ.)

Type of genus. Sphaerocodium bornemanni Rothpletz, Trias, Austria (redefined as *Sphaerocodium gotlandicum* Rothpletz 1908).
Diagnosis. Thallus composed of palisades of highly dichotomous cylindrical tubes. Bifurcation is at low angle, always in the same plane and the tubes remain welded. In axial section, tubes are ovoid and their succession forms a characteristic "rosary". Wall micritic.
Synonyms. Rothpletzella Wood 1948, *Coactilum* Maslov 1956.
Suprageneric attributions. Algal incertae sedis. Wray attributed it to the Porostromata Pia 1927. Could be linked to the Proauloporales Luchinina 1975. Uncorrectly attributed to the Girvanellaceae by Luchinina (1987) (as *Rothpletzella*).
Stratigraphic range and distribution. Very common in Lower and Middle Palaeozoic. Forms bindstones in association with megafauna. Also forms bafflestones in mud-mounds. Last common occurrence is Frasnian and the genus is uncommon above that level. (Illustrated in Mamet and Roux 1975b from the Viséan of Germany).

122. *Sphaeroporella* Antropov 1967

Type of genus. Sphaeroporella aksubaica Antropov 1967. Devonian-Carboniferous transition, Russian Platform.
Diagnosis. Thallus(?) encrusting, an aggregate of ovoid or reniform cystoid cells. Wall two-layered, outside wall micritic with an inner pseudo-fibrous perforated layer.
Other Carboniferous species.
Sphaeroporella conchatiformis (Malakhova and Chuvashov 1965). Devonian-Carboniferous boundary, Urals, Siberia (Kuzbass), China.
Sphaeroporella tchussovensis (Malakhova and Chuvashov 1965). Devonian-Carboniferous boundary, Urals, China.
Suprageneric attribution. Algal? incertae sedis. Wetheredellidae Vachard in Dil et al. (1977), *trans. nov.*, Wetheredellaceae, *emend.* Berchenko in Shuysky (1987). Attributed to the new tribe Wetheredelleae Berchenko 1987. Akin to *Asphaltina* Mamet in Petryk and Mamet 1972, *Cateniphycus* Maslov 1960 and *Wetheredella* Wood 1948. Needs revision.
Stratigraphic range and distribution. Uncommon. Described from Early Devonian to Viséan. Ireland, Great Britain, France, Belgium, Siberia (Kuzbass), China, Australia, Alberta, Tennessee.

123. *Sphinctoporella* Mamet and Rudloff 1972

Type of genus. Sphinctoporella lisburnensis. Mamet and Rudloff 1972. Middle Viséan, Alaska.
Diagnosis. Thallus elongated, subpentagonal in cross-section with slight outer constrictions and much more accentuated divisions of the stem. Poorly calcified cortex contains two or three rows of spherical cavities ("sporangial cavities", "endospores").
Suprageneric attribution. Aciculelleae Bassoullet et al. 1979 for Shuysky (1987) and Deloffre (1988).

Stratigraphic range and distribution. Endemic to Alaska where it is restricted to the Viséan.

Spinumbella Platonov 1974

See *Elenia*.

"*Spongiostroma*" Gürich 1906

Type of genus. Spongiostroma maeandrinum Gürich 1906. Viséan, Belgium.
Diagnosis. Algal biscuit with dissolved tissue of *Bevocastria*, leaving a pelletoidal meshwork counterprint. Needs revision.
Other Carboniferous "species".
"*Spongiostroma*" *bacilliferum*, "*S*", *ovuliferum* and "*S*", *granuliferum* (all by Gürich 1960) are former exposed algal mats, heavily burrowed and cracked.
Suprageneric attributions. Spongiostromid Gürich 1906. Tentatively attributed to the Rivulariaceae by Ischenko (1985).
Stratigraphic range and distribution. Needs revision.

124. *Stacheia* Brady 1876

Type of genus. Stacheia marginuloides Brady 1876 (= *Stacheia pupoides* Brady 1876). Viséan, Great Britain.
Diagnosis. Thallus cylindrical to fusiform, encrusting a central support, and formed by a succession of overlapping arched rows of subquadratic cells. Wall hyaline, yellowish, unperforated.
Other Carboniferous species.
Stacheia acervalis Brady 1876. Viséan, Great Britain.
Suprageneric attributions. Originally described as a foraminifer. Incorrectly transferred to the aoujgaliid sponges by Vachard in Perret and Vachard (1977). Stacheiinae for Mamet and Roux (1977). Reported to the new tribe Mametellae by Chuvashov (1987).
Stratigraphic range and distribution. Common from Viséan to Bashkirian. Reaches the Moscovian in the Canadian Arctic. Morocco, Great Britain, France, Belgium, Germany, Poland, Donbass, Alaska, Canadian Arctic, Alberta, Arkansas.

125. *Stacheoidella* Mamet and Roux 1987

Type of genus. Stacheoides(?) *spissa* Petryk and Mamet 1972. Viséan, Alberta.
Diagnosis. Thallus encrusting on support. Perithallic and hypothallic cells well differentiated. Hypothallus thin, composed of one or two rows of subspherical cells. Perithallus, continuous rows of irregular cells. Wall yellowish, hyaline.
Suprageneric attribution. Stacheiinae Loeblich and Tappan 1961 *emend.*
Stratigraphic range and distribution. Viséan to Moscovian. Endemic to North America. Canadian Arctic, Alaska, American Cordillera, American Midcontinent.

126. *Stacheoides* Cummings 1955 *emend.* Petryk and Mamet 1972

Type of genus. Stacheia polytrematoides Brady 1876. Viséan, Great Britain.
Diagnosis. Thallus encrusting, very irregular, with protuberances. Grossly concentric arrangement of irregular rows of cells. Cells also irregular with walls of uneven thickness. Wall yellowish, hyaline, unperforated.
Other Carboniferous species
Stacheoides cannindahensis Mamet and Roux 1983. Late Viséan, Australia.
Stacheoides meandriformis Mamet and Rudloff 1972. Abundant from Viséan to Moscovian. Extends to Permian in Canadian Arctic. Spain, Siberian Platform, Kuzbass, Alaska, Yukon Territory, British Columbia, Alberta, Idaho, Montana, Utah, American midcontinent.
Stacheoides papillata Cummings 1955. Viséan. Morocco, France, Great Britain.
Stacheoides tenuis Petryk and Mamet 1972. Viséan-Namurian. Algeria, Morocco, Spain, Russian Platform, Donbass, Urals, Siberia (Kuzbass, Verkoyansk), China, Alaska, Alberta, British Columbia, Idaho, Utah, American midcontinent, Tennessee. Extends up to the Permian in the Canadian Arctic.
Suprageneric attributions. Originally described as a foraminifer. Incorrectly transferred to the aoujgaliid sponges by Vachard in Perret and Vachard (1977). Belongs to the Stacheiinae for Mamet and Roux (1977) and to the new tribe Aoujgaliae for Chuvashov (1987).
Stratigraphic range and distribution. Doubtful in Late Tournaisian? Abundant from basal Viséan (Zone II) to the Bashkirian. Still observed in the Early Permian of the Canadian Arctic Archipelago. Algeria, Morocco, Spain, Great Britain, France, Belgium, Russian Platform, Urals, Donbass, Afghanistan, China, Australia, Canadian Arctic, Alaska, Yukon Territory, British Columbia, Alberta, American Cordillera, American midcontinent, Tennessee.

"*Stipulella*" Maslov 1956

Type of genus. Stipulella fascicularis Maslov 1956. Lower Carboniferous, Russian Platform.
Diagnosis. Algal clast composed of bundles of straight, unsegmented, calcified, cylindrical sheats. Each bundle has an angular outline. Similar to *Rectangulina* but the tubes here are cylindrical.
Suprageneric attribution. Algal incertae sedis. Ranged among the cyanophytes by Luchinina (1987).
Stratigraphic range and distribution. Scarce, Late Tournaisian to Early Namurian. Great Britain, Belgium, France, Russian Platform, Australia.

Stylaella Berchenko 1981

See *Palaeoberesella.*

"*Stylocodium*" Derville 1931

Type of genus. Stylocodium rhopaloides Derville 1931. Viséan, France.
Diagnosis. Thallus columnar, composed of concentric rows of cystoid tubes

(*Aphralysia*? or *Sphaeroporella*?). Needs revision.
Suprageneric attribution. Wetheredelleae Berchenko in Shuysky (1987).
Stratigraphic range and distribution. Scarce. Ireland, France, Belgium, Alaska. Viséan. Needs revision.

"*Stylophycus*" Johnson 1940

Type of genus. Stylophycus carbonarius Johnson 1940. Upper Carboniferous, Colorado.
Diagnosis. Algal biscuit of uncertain affinity. Needs revision.
Other Carboniferous species.
Stylophycus calcarius Johnson 1940. Upper Carboniferous, Colorado.
Suprageneric attribution. See *Shermanophycus* and *Gouldina*.
Stratigraphic range and distribution. See type of genus. Needs revision.

Subkamaena Berchenko 1981

See *Kamaena*

127. *Sycidium* Sandberger 1849

Type of genus. Sycidium reticulatum Sandberger 1849. Devonian, Germany.
Diagnosis. Gyrogonite oblate spheroidal, non-coronulate, with meridional ribs and rectangular to hexagonal polygons. Small basal and important apical openings. Wall foliated, protruding at each rib.
Carboniferous species
Sycidium faveatum Peck 1934. Basal Carboniferous, Iowa, Missouri.
Also doubtful reports of *Sycidium melo* Sandberger 1880 and *Sycidium panderi* (Ehrenberg 1858) at the Devonian-Carboniferous transition in the Donbass (Berchenko 1981).
Stratigraphic range and distribution. Abundant in the Middle-Late Devonian of Eurasia and North America. Still present at the base of the Carboniferous in the same regions.

128. *Taldykites* Chuvashov 1988

Type of genus. Taldykites mugodjaricus Chuvashov and Anfimov 1988. Moscovian, Urals.
Diagnosis. Needs further study. The type material does not show the composition of the central tubes. Lateral tubes disposed as radiating fans. Tubes bifurcate at various angles and show some constrictions. They terminate as thin peripheral branches. Needs revision.
Suprageneric attribution. According to original description, belongs to the Dimorphosiphoneae Shuysky 1985. The description is that of a udoteacean.
Stratigraphic range and distribution. See type of genus.

Tomilithon Maslov 1962

See *Parachaetetes* (as subgenus).

Trinodella Maslov and Kulik 1956

See *Dvinella* (as subgenus).

129. *Tubiphytes* Maslov 1956

Type of genus. Tubiphytes obscurus Maslov 1956. Lower Permian, Urals.
Diagnosis. Thallus(?) massive, cylindrical, branching, encrusting on central support, now preserved as cement-filled cavity. Tissue heavily micritized, with darker and clearer concentric "zones". Although filaments have often been reported the tissue is probably composed of rectangular elongated cells.
Synonyms. Shamovella Rauser-Chernoussova 1951, NOM. NUD, *Nigriporella* Rigby 1958, "Hikorocodium" *sensu* Flügel *non* Endo 1951.
Suprageneric attributions. Controversial. Hydrozoan for Rigby 1958. Incertae alga for Maslov 1956. Cyanobacteria (Girvanellaceae Luchinina 1975) according to Luchinina (1987). Red alga for Kochansky-Devidé (1970). Stromatolite for Chuvashov and Dyupina (1973). An algal incertae sedis for Mamet et al. (1987). See also Roux (1985).
Stratigraphic range and distribution. Forms the "*Tubiphytes* community" of Flügel (1977). Often associated with phylloid algae in organic build-ups. Common on open shelf. Known from Middle Carboniferous to Triassic. Carboniferous (to Early Permian) occurrences are in Spain, Yugoslavia, Turkey, Urals, Kazakhstan, Iran, Afghanistan, China, Japan, Siberia, Canadian Arctic, American Cordillera, American midcontinent, Mexico.

130. *Ungdarella* Korde 1951 *non* Maslov "1950" or 1956

Type of genus. Ungdarella conservata Korde 1951 (non "*Ungdarella uralica* Maslov 1950", *nec* "*Ungdarella uralica* Maslov 1956"). Upper Carboniferous, Urals.
Diagnosis. Thallus elongated, ramified. Hypothallic tissue composed of a row of rectangular cells. Perithallic tissue, much more conspicuous, composed of numerous, thin, long rows of subquadratic cells, oblique to the hypothallus. Wall calcareous, yellowish, hyaline.
Other Carboniferous species.
Ungdarella(?) *americana* Toomey and Johnson 1968. Namurian-Bashkirian, Algeria, Russian Platform, New Mexico.
Ungdarella gracilissima Kulik 1973. Middle Carboniferous, Urals.
Ungdarella peratrovichensis Mamet and Rudloff 1972. Namurian-Early Permian, Alaska, British Columbia. Canadian Arctic.
Ungdarella uralica Maslov 1956 (= *Ungdarella deceanglorum* Elliott 1970). Common from Late Viséan-Moscovian. Scarce up to the Permian. Libya, Morocco, Algeria, Spain, France, Great Britain, Austria, Turkey, Irak, Russian Platform, Urals, Donbass, Afghanistan, Kazakhstan, Australia, China, Canadian Arctic Archipelago, Alaska, British Columbia, New Mexico, Arizona (see Plate 3r).
Suprageneric attributions. Ungdarellaceae Maslov 1956 or new tribe Ungdarelleae Chuvashov 1987.

Stratigraphic range and distribution. Very common. Forms boundstones in the Middle Carboniferous in a manner similar to *Komia*. Latest Viséan to Moscovian. Scarce up to the Permian. Libya, Algeria, Morocco, Spain, France, Great Britain, Austria, Turkey, Irak, Russian Platform, Donbass, Urals, Kazakhstan, Afghanistan, Australia, China, Canadian Arctic, Alaska, British Columbia, New Mexico.

131. *Unjaella* Korde 1951

Type of genus. Unjaella ovata Korde 1951. Late Carboniferous, Northern Urals.
Diagnosis. Thallus spindle-shaped, with thin cortex and important medulla. Pores subhexagonal in transverse section with honey-comb arrangement and regular interpores.
Suprageneric attribution. Tribe Cyclocrineae Pia 1927, subtribe Mastoporinae Pia 1927 according to Berchenko and Saltovskaja in Shuysky (1987).
Stratigraphic range and distribution. Only known by the type.

132. *Uraloporella* Korde 1950

Type of genus. Uraloporella variabilis Korde 1950. Middle Carboniferous, Northern Urals (see Plate 2h).
Diagnosis. Thallus elongated, cylindrical, unsegmented. Diaphragms present. Cortex perforated by numerous short, thin, "blind" cortical branches. The thallus was originally surrounded by a mucilaginous coating, now preserved as a continuous cement layer.
Synonym. Samarella Maslov and Kulik 1956.
Other Carboniferous species.
Uraloporella sieswerdai (pars) Rácz 1965. Moscovian, Spain. Part of Rácz's original material belongs to *Uraloporella* and part to *Zidella* Saltovskaya 1984.
Suprageneric attributions. Unconvincingly classified as hypercalcified sponge by Termier et al. (1977), as foraminifer (Riding and Jansa 1976) or as microproblematicum. (Wright 1982). Belongs to the Palaeosiphonocladales of Shuysky (1985). Placed among the Beresellaceae (Maslov and Kulik 1956) Shuysky 1985 by Shuysky (1987) and by Deloffre (1988). The tribe Uraloporelleae was also proposed by Shuysky (1985).
Stratigraphic range and distribution. Very common and often associated with *Beresella* or *Dvinella*. Total range Namurian to Early Permian. Very abundant from Late Bashkirian to Moscovian. Devonian and Early Carboniferous references are to be discarded. Spain, Turkey, Russian Platform, Urals, Donbass, Kazakhstan, Siberia, China, Canadian Arctic, Alaska.

Velebitella Kochansky-Devidé 1964

Type of genus. Velebitella triplicata Kochansky-Devidé 1964. Middle Permian, Yugoslavia. Although this genus is sometimes mentioned in the Carboniferous, it is restricted to the Permian. Carboniferous taxa such as "*Velebitella*" *simplex* Kochansky-Devidé 1964 ought to be transferred to *Pseudovelebitella*.

133. *Vermiporella* Stolley 1893 *sensu* Pia 1927

Type of genus. *Vermiporella fragilis* Stolley 1893. Ordovician erratic, Germany.
Diagnosis. Thallus cylindrical, irregularly ramified. Stem unconstricted. Cortex aspondyl, perforated by numerous straight cylindrical branches.
Carboniferous species.
Vermiporella(?) *shartymensis* Kulik 1973. Middle Carboniferous, Urals.
Suprageneric attributions. New tribe Vermiporelleae Saltovskaja in Shuysky 1987. Dasyporellinae (Pia 1920) according to Deloffre (1988).
Stratigraphic range and distribution. Abundant in Ordovician-Silurian-Devonian. Very scarce in the Carboniferous. Doubtful in the Viséan of Spain, Bashkirian of the Urals and in the Moscovian of the Canadian Arctic.

134. *Wetheredella* Wood 1948

Type of genus. *Wetheredella silurica* Wood 1948. Wenlock, Great Britain (see Plate 3m).
Diagnosis. Thallus(?) encrusting, an aggregate of intertwined, flexuous, undivided regular tubes. Perforated two-layered wall, external layer micritic, internal layer pseudo-fibro-radial.
Carboniferous species.
Wetheredella starobeshevense Berchenko 1983. Serpukhovian-Bashkirian, Donbass.
Suprageneric attributions. The original family Wetheredellaceae Vachard 1976 is a composite association of foraminifers, algae and incertae sedis. It has been emended by Berchenko (in Shuysky 1987) to fit the Dasycladales. Doubtful alga for Riding 1977, cyanophyte for Copper 1976, cyanobacterium for Ischenko (1985).
Stratigraphic range and distribution. Abundant in Silurian-Devonian in carbonate build-ups. Very scarce in the Carboniferous. Isolated occurrences in the Viséan (Great Britain and Turkey) and in the Middle Carboniferous (Donbass).

135. *Windsoporella* Mamet and Rudloff 1972

Type of genus. *Windsoporella radiata* Mamet and Rudloff 1972, Namurian Zone 17, Maritime Provinces of Canada (see Plate 3j).
Diagnosis. Thallus cylindrical, composed of a succession of units separated by constrictions. Annulated stem. Projections of medulla in the centre of each unit form around a vestibule from which expand tufts of coarse, simple, phloiophore branches.
Suprageneric attribution. Velebitelleae (Vachard 1977) Mamet and Roux 1983 according to Deloffre (1988). Unconvincingly reported to the Diploporeae (Pia 1920) Bassoullet et al. 1979 by Shuysky (1987).
Stratigraphic range and distribution. Endemic to the latest Viséan-earliest Namurian of the Maritime Provinces of Canada (Nova Scotia, Newfoundland shelf, Magdalen Islands of Quebec).

136. *Yukonella* Mamet and Rudloff 1972

Type of genus. Yukonella bamberi Mamet and Rudloff 1972. Viséan, Yukon Territory.
Diagnosis. Calcareous irregular tube with some external and internal constrictions. Wall micritic, perforated by numerous erraticaly disposed pores.
Suprageneric attribution. Algal *incertae sedis* (Shuysky 1987).
Stratigraphic range and distribution. Early Carboniferous. Endemic to the Yukon Territory, Alaska and the Omolon Massif (Siberia).

137. *Zaporella* Rácz 1965

Type of tenus. Zaporella cantabriensis Rácz 1965. Moscovian, Spain.
Diagnosis. Thallus club-shaped. Stem cylindrical, undivided. Cortex heavily calcified, perforated by randomly disposed cylindrical pores perpendicular to the exterior. Secondary branches grow out of the ultimate ends of the primary branches. Dichotomy is variable.
Remark. Deloffre (1988) and Shuysky (1987) put *Zaporella* in synonymy with *Anthracoporella*. Both genera are recognized here as the branching of the pores is different.
Suprageneric attribution. Dasyporellinae *ex* Dasyporelleae (Pia 1920) Bassoullet et al. 1979.
Stratigraphic range and distribution. Moscovian of Spain. Could occur in the Early Permian of China (*Zaporella baxoensis* Mu 1982).

138. *Zidella* Saltovskaya 1984

Type of genus. Zidella maxima Saltovskaya 1984. Viséan, Tadzhikhistan.
Diagnosis. Thallus, a long cylindrical tube, dichotomous, with slight internal and external constrictions. Cortex perforated by numerous "blind", thin, cortical branches. The thallus may have been surrounded by a mucilaginous coating, now preserved as a continuous cement layer. The genus is close to *Uraloporella* Korde 1950.
Other Carboniferous species.
Zidella minor Ivanova 1988. Viséan, Urals.
Suprageneric attribution. Uraloporelleae Shuysky 1985. Erroneously Beresselleae for Ivanova (1988).
Stratigraphic range and distribution. Viséan-Moscovian, Spain, Urals, Tadzhikhistan.

5 Stratigraphic Distribution

Figures 2 and 3 record the stratigraphic range of the microflora. As previously mentioned, all the simple cyanobacteria or Garwoodiaceae are very long ranging, some already known in the Precambrian and often observed throughout the whole Palaeozoic. As they do not evolve appreciably they can be proposed as reliable indicators for environmental comparison through the Palaeozoic. The same applies to the solenoporids.

Other algal groups are more restricted in time.

The umbellinids are mostly Middle-Late Devonian. However a few genera reach the "Strunian" and the Tournaisian (*Biumbella*, *Braznikovia*, *Eoumbella*) and *Protoumbella* is the youngest known representative in Viséan time.

Anchicodiaceae do not appear before the Moscovian, where they suddenly become important mound-builders.

The most precise time index is *Calcifolium* which spans the latest Viséan-earliest Namurian and is endemic to the Western Tethys.

Most of the Dasycladales are long ranging, but some are restricted to a stage. They are certainly not as precise as their Mesozoic counterparts and are too facies sensitive to be used in routine correlation. However, a number of characteristic floras are recognizable, some endemic to a continental block.

The Tournaisian of the American Cordillera is characterized by *Columbiapora*. The Viséan-Early Namurian in the Western Tethys has *Cabrieropora-Japhetella-Kulikia*; in the Eastern Tethys, *Queenslandella*; in North America, *Dutroella*, *Sandoella*, *Spinctoporella*, *Windsoporella*; *Koninckopora* is known from all the continental blocks.

The Bashkirian Dasycladaceae are not diagnostic and the stage is recognizable on the red alga *Masloviporidium* and the first abundance of Bereselleae (*Beresella*, *Dvinella*).

The Moscovian witnesses the first occurrence of *Clavaporella*, *Connexia*, *Epimastopora*, *Herakella*, *Sinarella*, *Zaporella*. They are associated with *Eugonophyllum* and *Ivanovia*.

The latest Carboniferous and Early Permian flora are difficult to separate. Palaeosiphonocladales are declining although *Beresella* is still common. *Epimastopora-Epimastoporella* are abundant associated with *Pseudogyroporella-Gyroporella*.

While many widespread groups have identical stratigraphic distributions from one basin to another, there are a few cases of strong heterochrony among

Fig. 2. Stratigraphic distribution of microflora. 1 *Albertaporella*; 2 *Amorfia*; 3 *Anatolipora*; 4 *Anchicodium*; 5 *Anthracoporella*; 6 *Anthracoporellopsis*; 7 *Aoujgalia*; 8 *Aphralysia*; 9 *Archaeolithophyllum*; 10 *Asphaltina*; 11 *Asphaltinella*; 12 *Atractyliopsis*; 13 *Beresella*; 14 *Berestovia*; 15 *Bevocastria*; 16 *Biumbella*; 17 *Borisovella*; 18 *Brazhnikovia*; 19 *Cabrieropora*; 20 *Calcifolium*; 21 *Calcisphaera*; 22 *Clavaporella*; 23 *Coelosporella*; 24 *Columbiapora*; 25 *Connexia*; 26 *Cribrokamaena*; 27 *Cribrosphaeroides*; 28 *Cummingsella*; 29 *Cuneiphycus*; 30 *Cylindrofolia*; 31 *Dasycladophycus*; 32 *Dokutchaevskella*; 33 *Donezella*; 34 *Dutroella*; 35 *Dvinella*; 36 *Einorella*; 37 *Elenia*; 38 *Eokoninckopora*; 39 *Eolithoporella*; 40 *Eoumbella*; 41 *Eouraloporella*; 42 *Eovelebitella*; 43 *Epimastopora*; 44 *Epimastoporella*; 45 *Epiphyton*; 46 *Epistacheoides*; 47 *Eugonophyllum*; 48 *Evlania*; 49 *Exvotarisella*; 50 *Fasciella*; 51 *Foliophycus*; 52 *Fourstonella*; 53 *Frustulata*; 54 *Garwoodella*; 55 *Garwoodia*; 56 *Girvanella*; 57 *Gissarella*; 58 *Gouldina*; 59 *Gyroporella*; 60 *Halenopora*; 61 *Hedstroemia*; 62 *Herakella*; 63 *Iberiaella*; 64 *Intextulella*; 65 *Issinella*; 66 *Ivanovia*; 67 *Japhetella*; 68 *Kamaena*; 69 *Kamaenella* (Illustration see page 438)

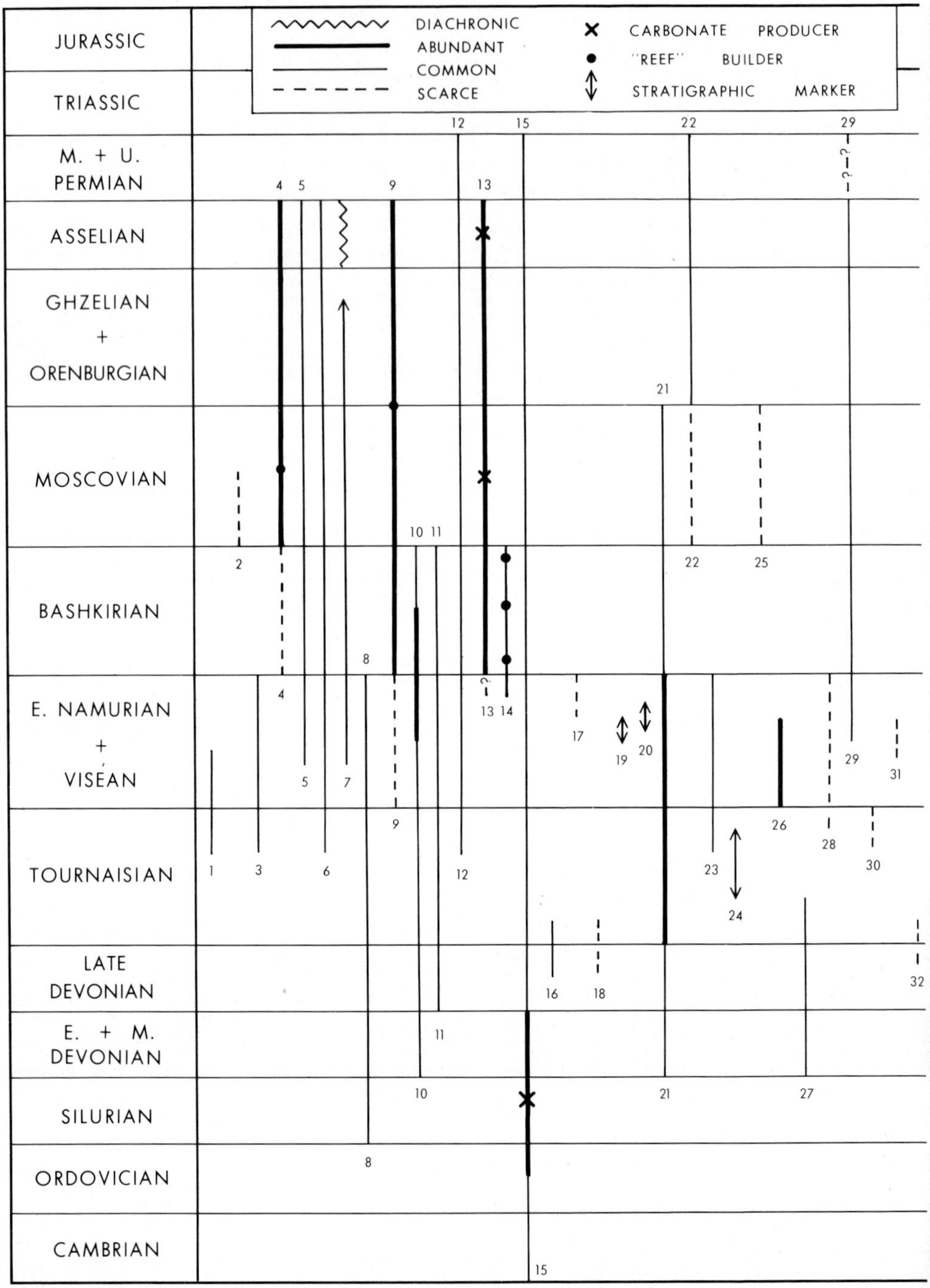

Fig. 2. (see caption page 437)

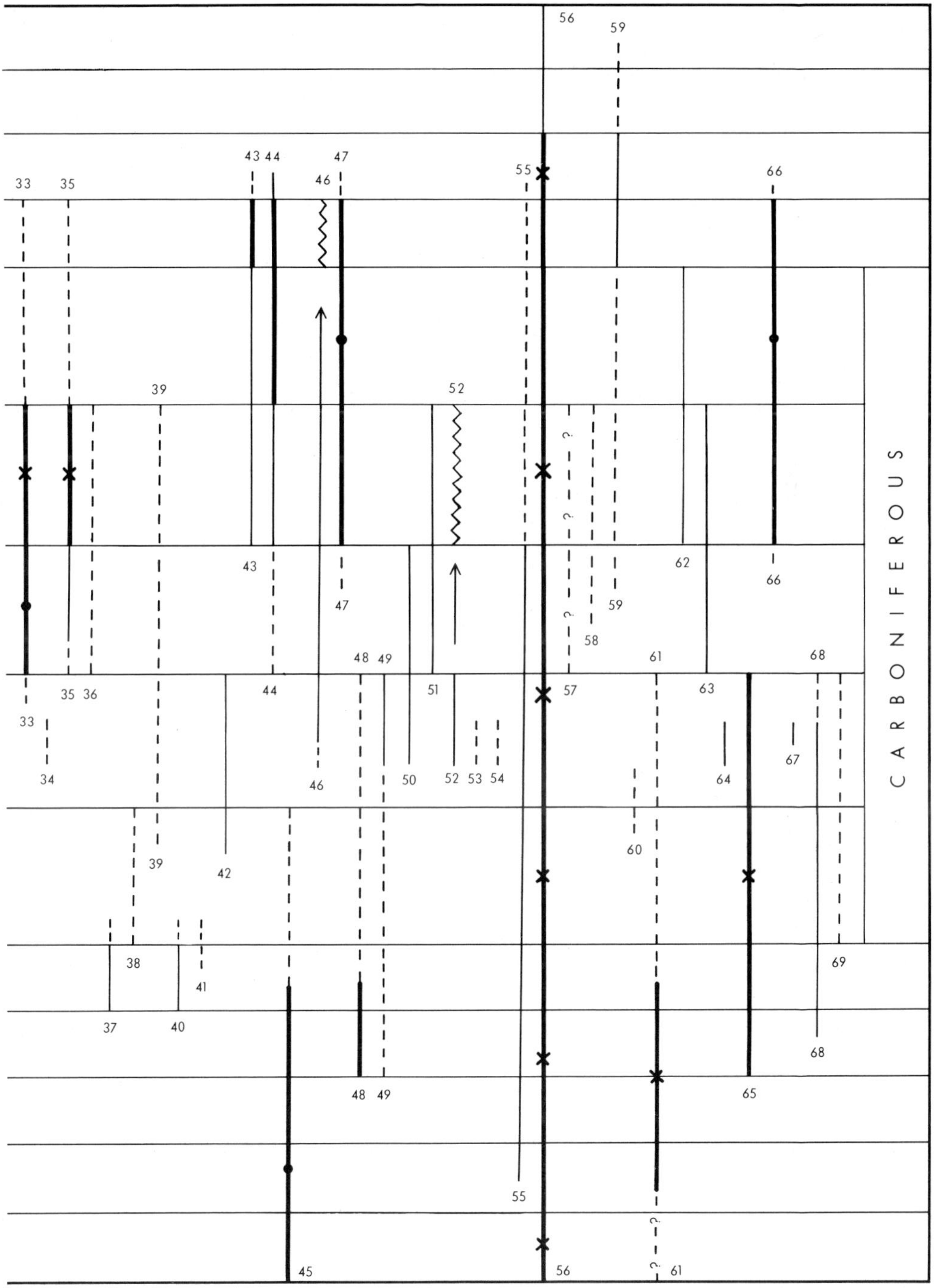
CARBONIFEROUS
33
34
35
36
37
38
39
40
41
42
43
44
45
46
47
48
49
50
51
52
53
54
55
56
57
58
59
60
61
62
63
64
65
66
67
68
69
?

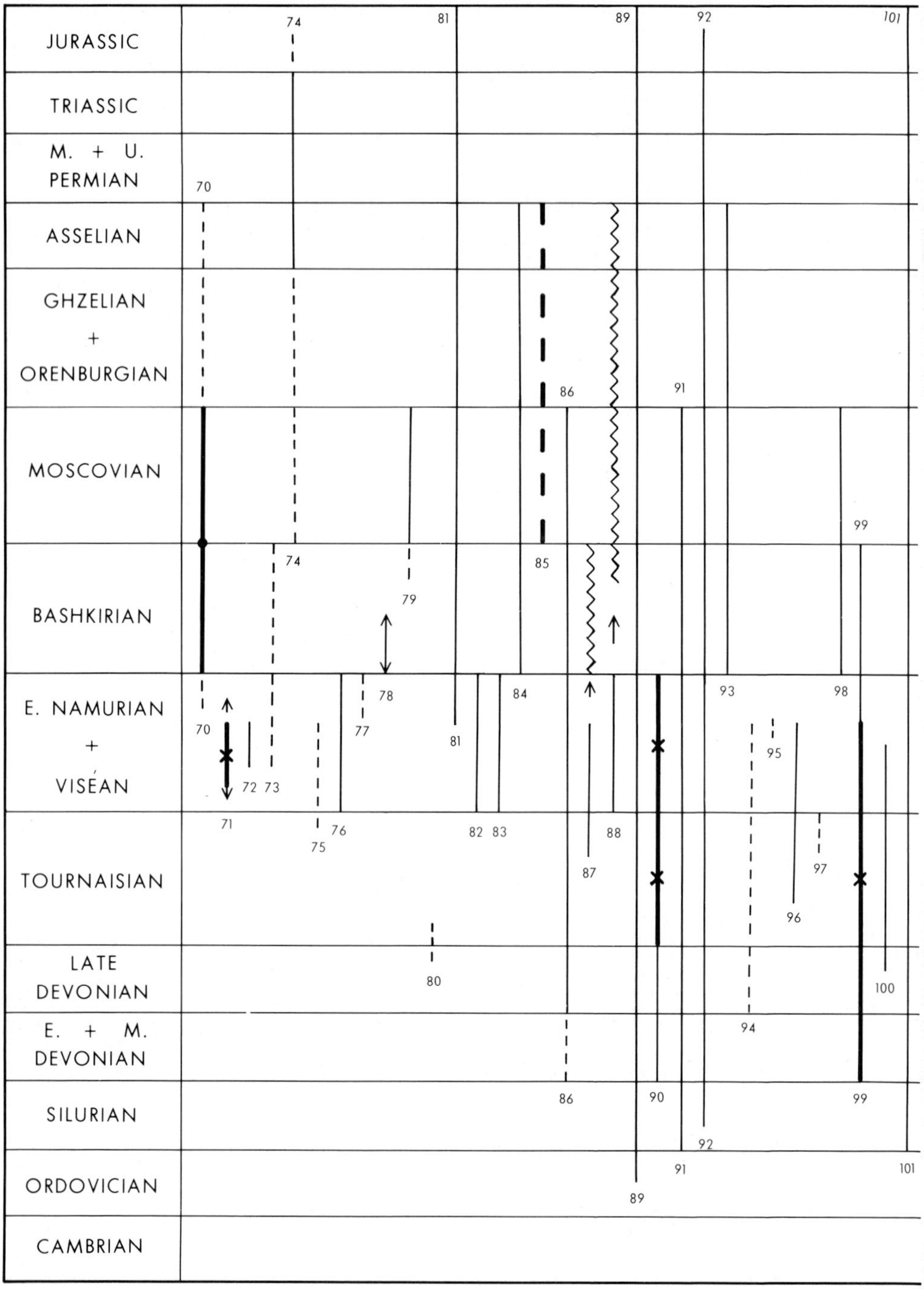

Fig. 3. (see caption page 442)

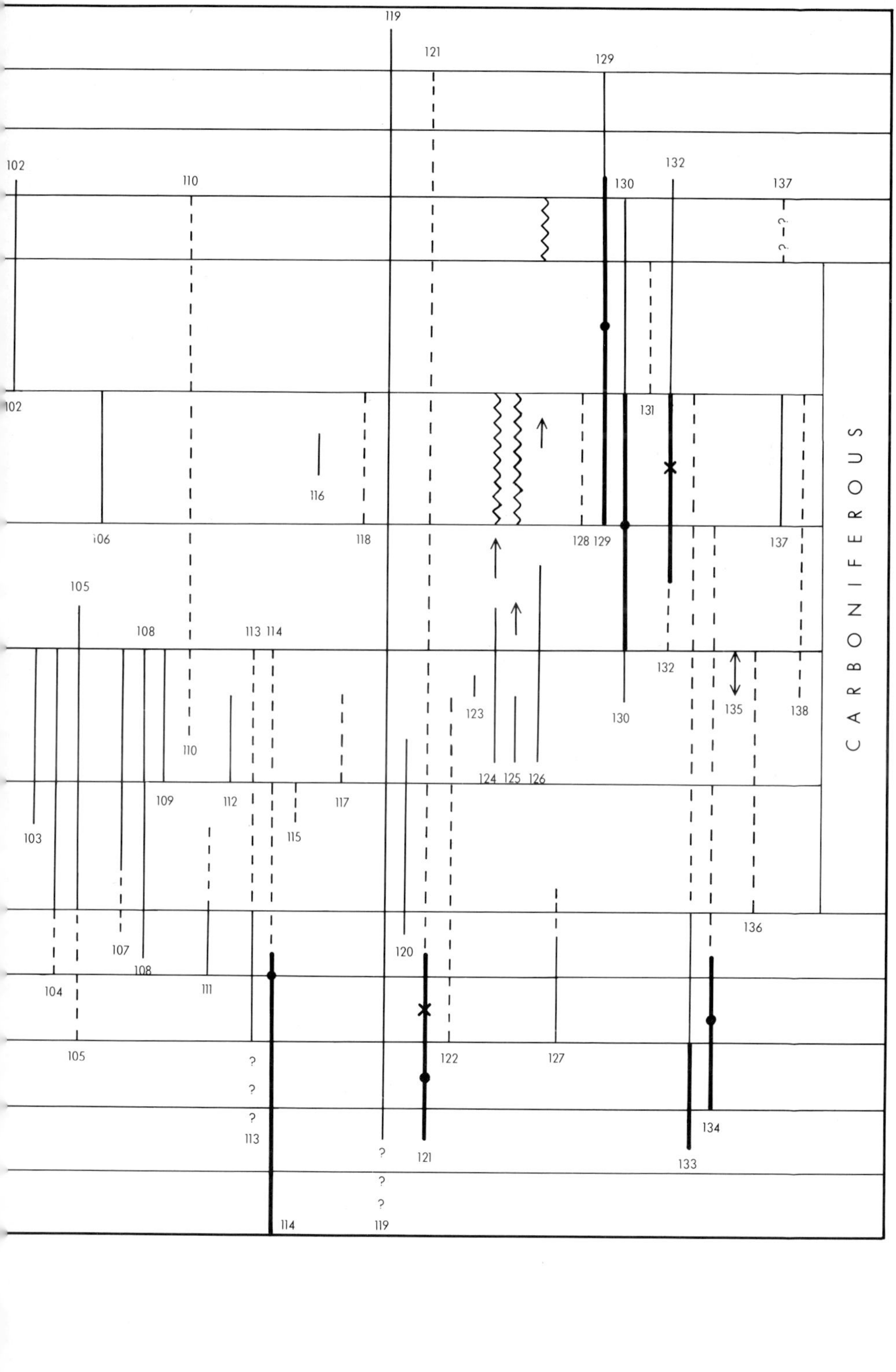

CARBONIFEROUS
102
110
119
121
129
130
132
137
116
118
106
105
108
113 114
123
124 125 126
128 129
131
135
136
138
103
104
107
109
111
112
115
117
120
122
127
133
134

Fig. 3. Stratigraphic distribution of microflora. 70 *Komia*; 71 *Koninckopora*; 72 *Kulikia*; 73 *Lithostroma*; 74 *Macroporella*; 75 *Malakhovella*; 76 *Mametella*; 77 *Masloviporella*; 78 *Maslovipo-ridium*; 79 *Mellporella*; 80 *Menselina*; 81 *Microcodium*; 82 *Mitcheldeania*; 83 *Nanopora*; 84 *Nansenella*; 85 *Neoanchicodium*; 86 *Nosticites*; 87 *Orthriosiphon*; 88 *Orthriosiphonoides*; 89 *Or-tonella*; 90 *Palaeoberesella*; 91 *Palaeomicrocodium*; 92 *Parachaetetes*; 93 *Paraepimastopora*; 94 *Parakamaena*; 95 *Parastacheia*; 96 *Pekiskopora*; 97 *Peristacheia*; 98 *Petschoria*; 99 *Proninella*; 100 *Protoumbella*; 101 *Pseudochaetetes*; 102 *Pseudogyroporella*; 103 *Pseudohedstroemia*; 104 *Pseudoissinella*; 105 *Pseudokamaena*; 106 *Pseudokomia*; 107 *Pseudonanopora*; 108 *Pseudosolen-opora*; 109 *Pseudostacheoides*; 110 *Pseudovelebitella*; 111 *Quasiumbella*; 112 *Queenslandella*; 113 *Rectangulina*; 114 *Renalcis*; 115 *Retephycus*; 116 *Richella*; 117 *Sandoella*; 118 *Sinarella*; 119 *Solenopora*; 120 *Sphaerinvia*; 121 *Sphaerocodium*; 122 *Sphaeroporella*; 123 *Sphinctoporella*; 124 *Stacheia*; 125 *Stacheoidella*; 126 *Stacheoides*; 127 *Sycidium*; 128 *Taldykites*; 129 *Tubiphytes*; 130 *Ungdarella*; 131 *Unjaella*; 132 *Uraloporella*; 133 *Vermiporella*; 134 *Wetheredella*; 135 *Windsoporella*; 136 *Yukonella*; 137 *Zaporella*; 138 *Zidella*

possible udoteaceans and stacheiins when one compares the Tethyan and North American blocks with the Arctic.

All known *Orthriosiphon-Orthriosiphonoides* are Lower Carboniferous with the exception of the Arctic where they are Middle Carboniferous and even reach the Lower Permian. The same heterochronism is observed for stacheiins such as *Stacheia*, *Epistacheoides*, and *Stacheoidella-Stacheoides* which are "normally" Viséan-Namurian but are Moscovian to Early Permian in the Arctic. This distribution suggests that the Tethyan, and to a lesser degree American, seas were the breeding ground from which the microflora repeatedly invaded the Arctic block during periods of warming.

References

Accordi B (1956) Calcareous algae from the Upper Permian of the Dolomites (Italy) with stratigraphy of the "*Bellerophon*-Zone". Palaeontol Soc India J 1:75–84

Aizenberg DE, Brazhnikova EV (1966) On the C_1t_a zone of the Donetz Basin. In: The Lower Tournaisian fauna (C_1t_a) of the Donetz Basin. Akad Nauk Ukr SSR, Inst Geol Nauk. Naukova Dumka, Kiev, pp 1–128

Andrews HN (1955) Index of generic names of fossil plants. US Geol Surv Bull 1013:1–262

Antropov IA (1950) New species of Foraminifera in the Upper Devonian of certain areas in the eastern part of the Russian Platform. Akad Nauk SSSR Kanzanskii Fil, Kazan, Izv Ser Geol 1:21–33

Antropov IA (1955) Blue-green algae from the Devonian of the central part of the Russian Platform. Uchenye Zapiski Kazan Gosud Univ, Tr Ova 115, 8:41–53

Antropov IA (1959) Devonian Foraminifera of Tatary. Akad Kauk SSSR Kazanskii Fil, Izv Ser Geol 7:11–33

Antropov IA (1967) Devonian and Carboniferous (Tournaisian) algae of the central part of the eastern Russian Platform. In: Zhuze AP (ed) Fossil algae of the USSR. Akad Nauk SSSR, Sibirsk Otdel, Inst Geol Geofiz, Nauka, Moscow, pp 118–125

Bassoullet JP, Bernier P, Deloffre R, Genot P, Jaffrezo M, Vachard D (1979) Essai de classification des Dasycladales en tribus. Bull Centre Rech Explor Prod Elf-Aquitaine 3, 2:429–442

Bassoullet JP, Bernier P, Deloffre R, Génot P, Poncet J, Roux A (1983) Les algues udotéacées du Paléozoique au Cénozoique. Bull Centre Rech Explor Prod Elf-Aquitaine 7, 2:449–621
Berchenko OI (1971) Charophytes. In: Aizenberg DE (ed) Atlas of the Tournaisian fauna in the Donetz Basin. Akad Nauk Ukr SSR, Inst Geol Nauk. Naukova Dumka, Kiev, pp 115–122
Berchenko OI (1974) New data on the charophyte algae (Umbellaceae) in the Donbass and Dnieper-Donetz Basin. Geol Zh 34, 2:104–116
Berchenko OI (1981) Tournaisian calcareous algae of the Donbass. Akad Nauk Ukr USSR, Naukova Dumka, Kiev, pp 1–71
Berchenko OI (1982) New species of calcareous algae of the Upper Serpukhovian deposits of the Donbass depression. In: Teslenko Yu V (ed) Systematics and evolution of fossil plants from Ukraine. Akad Nauk Ukr SSR, Naukova Dumka, Kiev, pp 51–55
Berchenko OI (1983) Calcareous algae. In: Upper Serpukhovian substage in the Donetz Basin. Akad Nauk Ukr SSR, Naukova Dumka, Kiev, pp 123–131
Berchenko OI (1987) Wetheredellaceae. In: Dubatolov VN (ed) Fossil calcareous algae. Akad Nauk SSSR, Sibirsk Otdel, Tr Inst Geol Geofiz 674:76–77
Berchenko OI, Saltovskaja VD (1987) Cyclocrinaceae. In: Dubatolov VN (ed) Fossil calcareous algae. Akad Nauk SSSR, Sibirsk Otdel, Tr Inst Geol Geofiz 674:65–76
Bilan W, Golonka J (1973) Upper Devonian/Lower Carboniferous calcareous algae from the Fore-Carpathian Depression. Acta Geol Pol 23, 1:149–169
Bilgütay U (1960) Some Permian calcareous algae from the vicinity of Ankara. Mineral Res Explor Inst Turk Bull 54:52–65
Birina LM (1948) New species of calcareous algae and foraminifers from the transitional Devonian-Carboniferous beds of the Moscow Basin. Sov Geol 28:154–159
Bodergat AM (1975) Les Microcodiums; milieux et modes de développement. Doc Lab Géol Fac Sci Lyon 62:1–235
Bogush OI (1984) New data on the genus *Sphaerinvia* Vachard 1980. In: Sokolov BI (ed) Paleozoic and mesozoic problematics. Akad Nauk SSSR, Sibirsk Otdel, Tr Inst Geol Geofiz 597:101–105
Bogush OI, Bushmina LS, Domnikova EI (1970) On the microfauna of the Devonian-Carboniferous transition beds of the Eltsov Synclinorium. Akad Nauk SSSR, Sibirsk Otdel, Tr Inst Geol Geofiz 71:49–59
Bornemann JG (1886) Die Versteinerungen des Cambrischen Schichtensystems der Insel Sardinien. Acad Caes Leopold-Carol N Acta 51:1–147
Brady AB (1876) A monograph of Carboniferous and Permian Foraminifera (the genus *Fusulina* excepted). Palaeontogr Soc Publ London 30:1–166
Brazhnikova NE, Rostovceva LP (1966) Foraminifera. In: The Lower Tournaisian fauna (C_1t_a) of the Donetz Basin. Akad Nauk Ukr SSR, Inst Geol Nauk. Naukova Dumka, Kiev, pp 1–128
Brenckle P (1977) *Mametella*, a new genus of calcareous red Algae (?) of Mississippian age in North America. J Palaeontol 51, 2:250–255
Brenckle P (1985) *Cribrokamaena* and *Crassikamaena*, new genera of Late Paleozoic algae from the United States. Micropaleontology 31, 1:55–67
Brenckle P, Groves JR (1987) Calcareous Foraminifers from the Humboldt Oolite of Iowa; Key to Early Osagean (Mississippian) Correlations between Eastern and Western North America. Palaios 1:561–581
Brenckle P, Marshall FC, Waller SF, Wilhem MH (1982) Calcareous Microfossils from the Mississippian Keokuk Limestone and Adjacent Formations, Upper Mississippi River Valley: Their Meaning for North American and Intercontinental Correlations. Geol Palaeontol 15:47–88
Bykova EV (1952) Foraminifers from the Devonian of the Russian Platform and Pre-Urals. Mikrofauna SSSR 5, Vses Neft Issl Geol Razved Inst (VNIGRI) Tr NS 60:5–64
Bykova EV (1955) Devonian foraminifers and radiolarians of the Volga-Urals region and the Devonian central field and their signification to stratigraphy. In: Bykova EV, Polenova EI (eds) Foraminifers, radiolarians and ostracods of the Devonian of the Volga-Urals region. Vses Neft Issl Geol Razved Inst (VNIGRI) Tr NS 87:1–141
Bykova EV (1961) Foraminifers of the Caradocian Stage of Eastern Kazakhstan. Akad Nauk Kazakh SSR, Inst Geol Nauk Tr no number:1–70

Chanton N (1964) Sur la présence d'Algues calcaires dans le Carbonifère de différents bassins sahariens (Colomb-Béchar, Reggane, Polignac). Bull Soc Géol Fr 7, 6:564–572

Chanton-Güvenç N (1972) Présence d'Algues calcaires dans le Carbonifère marocain. Bull Soc Géol Fr 7, 13:187–194

Chapman F (1908) On the relationship of the genus *Girvanella* and its occurrence in the Silurian limestones of Victoria. Australas Assoc Adv Sci Rep 2:377–386

Chermnykh VA (1972) New genera of Early Carboniferous foraminifers from the northern and pre-polar Urals. Akad Nauk SSSR, Inst Geol Komi Fil, Syktyvkar, Ejegodnik 1971 no number: 35–39

Chuvashov BI (1965) Foraminifera and algae from the Upper Devonian sediments in the western slope of the central and southern Urals. Akad Nauk SSSR, Ural Fil, Inst Geol Tr 74:3–93

Chuvashov BI (1974) Permian calcareous algae of the Urals. Akad Nauk SSSR, Ural Nauchn Centr, Inst Geol Geochem Trudy 109:3–76

Chuvashov BI (1987) Red algae (Rhodophytes). In: Dubatolov VI (ed) Fossil calcareous algae. Akad Nauk SSSR, Sibirsk Otdel, Inst Geol Geofiz Tr 674:109–140

Chuvashov BI, Anfimov AL (1988) New Late Carboniferous-Early Permian calcareous algae of the Urals and Pre-Urals. In: Dubatolov VI (ed) Calcareous algae and Stromatolites. Akad Nauk SSSR, Sibirsk Otdel, Inst Geol Geofiz. Nauka, Novosibirsk, pp 54–69

Chuvashov BI, Dyupina GV (1973) Palaeozoic terrigenous deposits of the western slope of the Middle Urals. Akad Nauk SSSR, Ural Nauchn Centr, Inst Geol Geochem Tr 105:3–208

Conil R, Lys M (1964) Matériaux pour l'étude micropaléontologique du Dinantien de la Belgique et de la France (Avesnois), pt 1: Algues et Foraminifères. Mém Inst Geol Univ Louvain 23:1–335

Conil R, Lys M (1967) Aperçu sur les associations de Foraminifères endothyroïdes du Dinantien de la Belgique. Ann Soc Géol Belg 90:395–412

Copper P (1976) The cyanophyte *Wetheredella* in Ordovician reefs and off-reef sediments. Lethaia 9:273–281

Cummings RH (1955) New genera of Foraminifera from the British Lower Carboniferous. Washington Acad Sci J 45:1–8

Dangeard L, Dorè F (1957) Observations nouvelles sur les Algues et les Stromatolithes du Cambrien de Carteret (Manche). Bull Soc Géol Fr 6, 7:1069–1075

Danielli HM (1981) The fossil alga *Girvanella* Nicholson and Etheridge. Brit Mus Nat Hist Geol Bull 35, 2:79–107

Deloffre R (1987) Nouvelle classification des Algues dasycladales fossiles. CR Acad Sci Paris Ser 2 305:1017–1020

Deloffre R (1988) Nouvelle taxonomie des Algues dasycladales. Bull Centre Rech Explor Prod Elf-Aquitaine 12, 1:165–217

Delvolvé JJ, Perret MF (1987) Foraminifères, algues et conodontes bashkiriens du Culm des Pyrénées béarnaises. Géobios 20, 2:193–213

Deninger K (1906) Einige neue Tabulaten und Hydrozöen aus mesozöischen Ablagerungen. N Jahrb Mineral 1:61–70

Derville H (1931) Les Marbres du Calcaire Carbonifère en Bas-Boulonnais. Boehm, Strasbourg, pp 1–322

Dil N, Termier G, Termier H, Vachard D (1977) Contribution à l'étude stratigraphique et paléontologique du Viséen Supérieur et du Namurien Inférieur du bassin houiller de Zonguldak. Ann Soc Géol Belg 99, 2:401–440

Dreesen R, Bless MJ, Conil R, Flajs G, Laschet C (1985) Depositional environment, paleoecology and diagenetic history of the "Marbre rouge à crinoïdes de Baelen" (Late Devonian, Verviers Synclinorium, Eastern Belgium). Ann Soc Géol Belg 108:311–359

Dubatolov VN (1981) Palaeozoic of the occidental Siberian Lowlands and their mountain belt. Akad Nauk SSSR, Sibirsk Otdel, Inst Geol Geofiz Tr 482:1–144

Dybowski WN (1877) Die Chaetetiden der ostbaltischen Silur-Formation. PhD Thesis, Kaiserlichen Univ Dorpat, pp 1–134

Ehrenberg CG (1858) Über andere massenhafte mikroskopische Lebensformen der ältesten Silurischen Grauwacken-Thone bei Petersburg. Preuss Akad Wiss Monatsber 1858: 295–311:324–337

Eickhoff G (1968) Moravamminidae (Foraminifera) aus dem Unterkarbon von Hermannsholz bei Frankenberg an der Eder. N Jahrb Geol Paläontol Monatsh 3:129–142
Elias MK (1950) Palaeozoic *Ptychocladia* and related Foraminifera. J Palaeontol 24, 3:287–306
Elliott GF (1968) Permian to Paleocene calcareous algae (Dasycladaceae) of the Middle East. Brit Mus Nat Hist Bull Geol 4:1–111
Elliott GF (1970) Calcareous algae new to the British Carboniferous. Palaeontology 13, 3:443–450
Elliott GF (1971) The nature of *Acicullela* Pia (calcareous algae). Palaeontology 14, 4:629–636
Elliott GF (1982) A possible non-calcified dasycladacean alga from the Carboniferous of England. Brit Mus Nat Hist Bull Geol 36, 2:105–107
Endo R (1951) Stratigraphical and Paleontological studies of the Later Paleozoic calcareous algae in Japan I. Trans Proc Palaeontol Soc Jpn NS 4:121–129
Endo R (1952) Stratigraphical and Paleontological studies of the Later Paleozoic calcareous algae in Japan II. Trans Proc Palaeontol Soc Jpn NS 5:139–144
Endo R (1953) Stratigraphical and Paleontological Studies of the Late Paleozoic calcareous algae in Japan V. Jpn J Geol Geogr Trans 23:117–126
Endo R (1957) Fossil algae from the Taishaku district, Hiroshima-ken and Kitami-no-kuni, Hokkaido. Sci Rep Saitama Univ Ser B 2, 3:279–305
Endo R, Hashimoto W (1955) Unquestionably Paleozoic (Permian) fossils found in Hokkaido, Japan. Proc Jpn Acad 31, 10:704–708
Endo R, Kanumai M (1954) Geology of the Mino Mountain Land and southern part of Hida Plateau, with the description of the algal remains found in those districts. Sci Rep Saitama Univ Ser B 1, 3:177–208
Fenton CL, Fenton MA (1939) Precambrian and Paleozoic Algae. Geol Soc Am Bull 50:89–126
Flügel E (1966) Algen aus dem Perm Karnischen Alpen. Carinthia 2, 25:3–76
Flügel E (ed) (1977) Environmental models for Upper Paleozoic benthic calcareous algal communities. In: Fossil algae. Springer Berlin Heidelberg New York, pp 314–343
Garwood EJ (1912) The Lower Carboniferous succession in the north-west of England. Geol Soc London Q J 68:449–586
Garwood EJ (1914) Some new rock-building organisms from the Lower Carboniferous beds of Westmoreland. Geol Mag NS 6, 1:265–271
Garwood EJ (1931) The Tuedian Beds of Northern Cumberland and Roxburghshire, east of the Liddel Water. Geol Soc London Q J 87:97–157
Glück H (1912) Eine neue gesteinbildende Siphonee (Codiacee) aus dem marinen Tertiär von Süddeutschland. Mitt Bad Geol Landesanst 7:3–24
Golubtsov VK (1961) *Kulikia* – new genus of Viséan calcareous algae. Akad Nauk Biel SSR, Inst Geol Nauk, Paleontol Stratigr Biel SSR, 3:348–353
Groves JR (1983) Calcareous foraminifers and algae from the type Morrowan (Lower Pennsylvanian) region of Northeastern Oklahoma and Northwestern Arkansas. Okl Geol Surv Bull 133:65
Groves JR, Mamet BL (1985) *Masloviporidium*, a cosmopolitan Middle Carboniferous Red Alga. In: Toomey DF, Nitecki MH (eds) Paleoalgology: contemporary research and applications. Springer, Berlin Heidelberg New York Tokyo, pp 85–90
Gümbel CW (1982) Über die dactyloporenähnlichen Fossilien der Trias. Verh Geol Reichsanst Wien 1872:91
Gürich G (1906) Les Spongiostromides du Viséen de la province de Namur. Mem Mus R Hist Nat Belg 3:1–55
Güvenç T (1965) Représentants des Bereselleae (Algues calcaires) dans le Carbonifère de Turquie et description d'un nouveau genre: *Goksuella* n.g. Bull Soc Géol Fr 7, 7:843–850
Güvenç T (1966) Description de quelques espèces d'algues calcaires (Gymnocodiacées et Dasycladacées) du Carbonifère et du Permien des Taurus occidentaux (Turquie). Rev Micropaleontol 9, 2:94–103
Güvenç T (1970) Description de deux espèces d'Algues calcaires dans le Permien des Taurus (Turquie). Bull Soc Géol Fr 7, 11:447–451
Güvenç T (1979) Dasycladacées métaspondyles du Paléozoïque supérieur et du Trias. Bull Centre Rech Explor Prod Elf-Aquitaine 3, 2:625–637

Güvenç T (1981) Epiphytonaceae (Algues calcaires marines) du Carbonifère et leurs représentants dans la région egeenne. Rev Micropaleontol 24, 3:157–167

Hallett D (1970) Foraminifera and algae from the Yoredale "Series" (Viséan-Namurian) of northern England. 6 ème Congr Inter Strat Géol Carbonifère, CR, Sheffield, 1967, pp 873–900

Hance L (1983) Micropaléontologie du Moliniacien belge. Trois genres nouveaux, *Halenopora* (Chlorophyceae), *Aranea* (Rhodophyceae) et *Globochernella* (Foraminiferida, Tournayellidae). Ann Soc Géol Belg 106:111–119

Haug E (1883) Über sogenannte *Chaetetes* aus mesozoischen Ablagerungen. N Jahrb Mineral 1:171–179

Homann W (1972) Unter- und tief-mittelpermische Kalkalgen aus den Ratendorfer Schichten, dem Trogkofel-Kalk und dem Tressdorfer-Kalk der karnischen Alpen (Österreich). Senckenbergiana Leth 53:133–313

Ischenko AA (1985) Silurian algae from Podolia. Akad Nauk Ukr SSR, Inst Geol Nauk. Naukova Dumka, Kiev, pp 1–113

Ivanova RM (1973) On the middle and upper Viséan stratigraphy of the eastern slope of southern Urals. Akad Nauk SSSR, Ural Nauchn Centr, Tr Inst Geol Geofiz 82:18–86

Ivanova RM (1988) Calcareous algae of the Viséan Stage in the Urals. In: Chuvashov BI (ed) Biostratigraphy and lithology of the Upper Paleozoic of the Urals. Akad Nauk SSSR, Uralsk Otdel, Sverdlovsk, pp 4–23

Johnson GAL (1958) Biostromes in the Namurian Great Limestone of Northern England. Palaeontology 1, 2:147–157

Johnson JH (1940) Lime-secreting algae and algal limestones from the Pennsylvanian of central Colorado. Geol Soc Am Bull 51:571–596

Johnson JH (1945) Calcareous algae of the Upper Leadville Limestone near Glenwood Springs, Colorado. Geol Soc Am Bull 56:829–848

Johnson JH (1946) Lime-secreting algae from the Pennsylvanian and Permian of Kansas. Geol Soc Am Bull 57:1087–1120

Johnson JH (1951) Permian calcareous algae from the Apache Mountains, Texas. J Paleontol 25, 1:21–30

Johnson JH (1956) *Archaeolithophyllum*, a new genus of Paleozoic coraline algae. J Paleontol 30, 1:53–55

Johnson JH (1960) Paleozoic Solenoporaceae and related red algae. Col Sch Mines Q 55, 3:1–77

Johnson JH (1966) New Mississippian algae from Alberta. J Paleontol 40, 6:1385–1389

Johnson JH (1966b) Pennsylvanian and Permian algae. Q Colorado School Mines 58, 3:211 pp

Johnson JH, Konishi K (1956) Studies of Mississippian algae. Mississippian algae from the Western Canada Basin and Montana. Col Sch Mines Q 51, 4:85–107

Kazmierczak J (1976) Volvocacean nature of some Paleozoic non-radiosphaerid Calcispheres and Parathuramminid "Foraminifera". Acta Palaeontol Pol 21, 3:245–258

Khvorova IV (1946) On a new genus of algae from the Middle Carboniferous deposits of the Moscow Basin. Akad Nauk SSSR Dokl 53, 8:737–739

Klappa CF (1978) Biolithogenesis of *Microcodium*: elucidation. Sedimentology 25:489–518

Kochansky-Devidé V (1964) Die Microfossilien des jugoslawischen Perms. Paläontol Z 38, 3/4:180–188

Kochansky-Devidé V (1970) Die Kalkalgen des Karbons vom Velebit-Gebirge (Moskovien und Kasimovien). Palaeontol Jug 10:1–32

Kochansky-Devidé V, Herak M (1960) On the Carboniferous and Permian Dasycladaceae of Yugoslavia. Geol Vjesn, Glasilo Zavoda Geol 13:65–95

Koninck LG de (1842) Description des animaux fossiles qui se trouvent dans le terrain carbonifère de Belgique, vol 3. Dessain, Liège, pp 1–650

Konishi K (1956) *Anatolipora*, a new dasycladacean genus and its algal associates from the Lower Carboniferous of Japan. Col Sch Mines Q 51, 4:111–127

Konishi K (1958) Devonian calcareous algae from Alberta. In: Studies of Devonian algae. Col Sch Mines Q 53, 2:89–109

Konishi K, Wray JL (1961) *Eugonophyllum*, a new Pennsylvanian and Permian algal genus. J Paleontol 35, 4:659–666

Korde KB (1950) On the morphology of Carboniferous verticilated Siphonales of northern Urals. Akad Nauk SSSR Dokl 73, 3:569–571
Korde KB (1951) New calcareous algal species and genera from the Carboniferous of northern Urals. Moskovskogo Obshestva Ispitatelei Prirody. Tr Otdel Geol 1:175–182
Korde KB (1959) Morphology and systematic position of the genus *Epiphyton*. Akad Nauk SSSR Dokl 126, 5:1087–1089
Korde KB (1973) Cambrian algae. Akad Nauk SSSR Tr Paleontol Inst 139:1–347
Kossenko ZA (1972) New calcareous algae from Ukraine and occidental Siberia. In: Zanina IE (ed) New species of fossil plants and invertebrates of the USSR. Akad Nauk SSSR Nauchn Sov Probl. Nauka Moskva, pp 293–295
Kulik YL (1964) Beresellids from the Carboniferous of the Russian Platform. Paleontol Zh 2:99–114
Kulik EL (1973) Cyanophyte. In: Stratigraphy and fauna of the Carboniferous succession of the Shartym River (Southern Urals). Uralsk Geol Upravl, Kievskii Gossud Univ, Izdat Universitet Lvov, pp 1–175
Kulik EL (1981) Systematic position of *Paradella* Maslov 1956. Akad Nauk SSSR Voprosy Mikropaleontol 24:152–156
Langer W (1979) Neue karbonatische *Microproblematica* aus dem westdeutschen Devon. N Jahrb Geol Paläontol Monatsh 12:723–733
Lee GW (1912) The British Carboniferous Trepostomata. Great Br Geol Surv Mem 1, 3:135–195
Loeblich AR, Tappan H (1961) Suprageneric classification of the Rhizopoda. J Paleontol 35, 2:245–330
Loeblich AR, Tappan H (1964) Treatise on invertebrate paleontology, pt C. In: Moore RC (ed) Protista. Geol Soc Am; and Univ Kansas Press, Lawrence, Kansas, pp 1–900
Loeblich AR, Tappan H (1988) Foraminiferal genera and their classification. Van Nostrand Reinhold, New York, 990 pp
Luchinina VA (1975) Palaeoalgological characteristics of the Early Cambrian of the Siberian Platform. Akad Nauk SSSR, Sibirsk Otdel, Inst Geol Geofiz Tr 216:1–99
Luchinina VA (1987) Blue-green algae (Cyanobacteria). In: Dubatolov VI (ed) Fossil calcareous algae. Akad Nauk SSSR, Sibirsk Otdel, Inst Geol Geofiz Tr 674:12–38
Mamay SH (1959) *Litostroma*, a new genus of problematical algae from the Pennsylvanian of Oklahoma. Am J Bot 46, 4:283–292
Mamet B (1970) Sur les Umbellaceae. Can J Earth Sci 7, 4:1164–1171
Mamet B (1973) Microfaciès viséens du Boulonnais (Nord, France). Rev Micropaleontol 16, 2:101–124
Mamet B (1974) Sur deux Dasycladacées carbonifères des Cordillères nord-américaines. Rev Micropaleontol 17, 1:38–44
Mamet B (1976) An atlas of microfacies in Carboniferous carbonates of the Canadian Cordillera. Geol Surv Can Bull 255:1–131
Mamet B, Pinard S (1985) Carboniferous algae from the Peratrovich Formation, southeastern Alaska. In: Toomey DF, Nitecki MH (eds) Paleoalgology: contemporary research and applications. Springer, Berlin Heidelberg New York Tokyo, pp 91–100
Mamet B, Roux A (1974) Sur quelques Algues tubulaires scalariformes de la Téthys Paléozoïque. Rev Micropaleontol 17, 3:134–156
Mamet B, Roux A (1975a) Dasycladacées dévoniennes et carbonifères de la Téthys occidentale. Rev Esp Micropaleontol 7, 2:245–295
Mamet B, Roux A (1975b) Algues dévoniennes et carbonifères de la Téthys occidentale (troisième partie). Rev Micropaleontol 18, 3:134–187
Mamet B, Roux A (1977) Algues rouges dévoniennes et carbonifères de la Téthys occidentale. Rev Micropaleontol 19, 4:215–266
Mamet B, Roux A (1978) Algues viséennes et namuriennes du Tennessee (Etats-Unis). Rev Micropaleontol 21, 2:68–97
Mamet B, Roux A (1980) *Cummingsella lyoncrossi*, n.gen.,n.sp., Algue nouvelle du Namurien d'Ecosse. Géobios 13, 5:787–793
Mamet B, Roux A (1981a) Note sur le genre *Issinella* (Algue verte paléozoique). Rev Micropaleontol 23, 3/4:151–158

Mamet B, Roux A (1981b) Note sur les Albertaporellinae (Dasycladaceae). Rev Micropaleontol 23, 3/4:159–168

Mamet B, Roux A (1982) Sur la présence de *Microcodium* (Algue?, Incertae sedis?) dans le Paléozoïque supérieur de l'Arctique canadien. Can J Earth Sci 19, 2:357–363

Mamet B, Roux A (1983) Algues dévono-carbonifères de l'Australie. Rev Micropaleontol 26, 2:63–130

Mamet B, Rudloff B (1972) Algues carbonifères de la partie septentrionale de l'Amérique du Nord. Rev Micropaleontol 15, 2:75–114

Mamet B, Mortelmans G, Roux A (1979a) Algues viséennes du sondage de Turnhout (Campine, Belgique). Ann Soc Geol Belg 101:351–383

Mamet B, Roux A, Nassichuk W (1979b) Algues et stratigraphie du Paléozoïque supérieur de l'Arctique canadien. Bull Centre Rech Explor Prod Elf-Aquitaine 3, 2:669–683

Mamet B, Roux A, Nassichuk W (1987) Algues carbonifères et permiennes de l'Arctique canadien. Geol Surv Can Bull 342:1–143

Maslov VP (1929) Microscopic algae from Carboniferous limestones of the Donetz Basin. Com Geol Bull 48, 10:115–138

Massa D, Vachard D (1979) Le Carbonifère de la Libye occidentale. Rev Inst Fr Pétrole 34, 1:3–64

Mu XN (1982) Some calcareous algae from Xizang. In: The series of the scientific expedition to the Qinghai-Xizang Plateau. Paleontology of Xizang, vol 5. Science Press, Beijing, pp 205–240

Mu XN, Zhang Y (1988) Calcareous algae and microproblematica. In: Yu CM (ed) Devonian-Carboniferous Boundary at Nanbiancum, Guilin, China. Science Press, Beijing, pp 149–158

Muromtseva AA (1979) Six species of calcareous algae from the Carboniferous of the Lvov-Volhynian Basin. Paleontol Sb 16; Lvov Gosud Univ Izdat, Lvov, pp 89–94

Nicholson HA, Etheridge R (1878) A monograph of the Silurian fossils of the Girvan district, Ayrshire, with special reference to those contained in the "Gray Collection", I, I. Blackwood, Edinburgh London, pp 1–341

Orlov Yu E (1963)(ed) Osnovy Paleontologii. In: Maslov VP, Korde KB, Kulik YL (eds) Algae, vol 14. Izdat Akad Nauk SSSR, Moscow, pp 13–267

Paul H (1938) Unterkarbonische Kalkalgen und Calcisphaeren Deutschlands. Preuss Geol Landesanst Abh Jahrb 58, A:276–281

Peck R (1934) The North American Trochiliscids. J Paleontol 8, 2:83–119

Perret MF, Vachard D (1977) Algues et pseudo-algues des calcaires serpoukhoviens d'Ardengost (Hautes Pyrénées). Ann Paléontol Invertébrés 63, 2:85–156

Petryk AA, Mamet B (1972) Lower Carboniferous algal microflora, southwestern Alberta. Can J Earth Sci 9, 7:767–802

Peterhans E (1929) Algues de la famille des Solénoporacées dans le Malm du Jura bâlois et soleurois. Soc Paléontol Suisse Mem 49:1–15

Pia J von (1912) Neue Studien über die triadischen Siphonaeae verticillatae, Beitr Paläontol Geol Öst Ung 25:25–85

Pia J von (1920) Die Siphoneae verticillatae vom Karbon bis zur Kreide. Verh Zool Bot Ges Wien 11, 2:1–267

Pia J von (1927) Thallophyta. In: Hirmer M (ed) Handbuch der Paläobotanik, vol 1: Thallophyta–Bryophyta–Pteridophyta. Oldenbourg, Munich, pp 36–99

Pia J von (1932) Algenkalknollen aus dem Russischen Devon. Acad Sci URSS Bull 1932:1345–1360

Pia J von (1937) Die wichtigsten Kalkalgen des Jungpaläozoikums und ihre geologische Bedeutung. In: 2nd Congr Av Strat Géol Carbonifère, vol 2, Abstr, pp 765–856

Platanov VA (1974) On the systematics of Umbellids (Charophyte algae). Paleontol Zh 1974, 1:101–110

Plaziat JC (1984) Le Domaine Pyrénéen de la fin du Crétacé à la fin de l'Eocène. Thèse Univ Paris-Sud, 1362 pp

Poncet J (1965) Sur une Dasycladacée nouvelle du Dévonien inférieur: *Diplopora constantini* sp.n. Bull Soc Géol Fr 7, 7:879–880

Poncet J (1974) Description de quelques Algues calcaires éodévoniennes du Nord-Est du Massif Armoricain. Bull Soc Géol Fr 7, 16:225–229

Poncet J (1986) Les Algues calcaires du Carbonifère moyen du bassin de Béchar (Sahara algérien). Rev Micropaleontol 29, 3:187–197

Poncet J (1987) *Eoclypeina*, un nouveau genre d'Algue verte calcaire (Dasycladaceae) dans le Carbonifère supérieur du Bassin de Béchar (Sahara algérien). Rev Micropaleontol 30, 2:117–121

Poyarkov BV (1961) On the Famennian and Lower Tournaisian stratigraphy of the occidental part of the Tian-Shian. Akad Nauk Kirgiz SSR, Izv Ser Est Tek Nauk 2, 9:23–48

Poyarkov BV (1965) On the systematic position of the Umbellids. Akad Nauk SSSR Dokl 163, 3:728–730

Poyarkov BV (1966) Devonian charophytes of the Tian-Shan. In: Maslov VP (ed) Fossil charophyta of the USSR. Akad Nauk SSSR, Geol Inst Tr 143:161–200

Rácz L (1965) Carboniferous calcareous Algae and their associations in the San Emiliano and Lois-Ciguera formations (Prov Leon, NW Spain). Leidse Geol Meded 31:1–112

Rácz L (1966) Late Paleozoic calcareous Algae in the Pisuerga Basin (North Palencia, Spain). Leidse Geol Meded 31:241–260

Rácz L (1984) *Iberiaella*, a new fossil alga from the Middle Carboniferous of NW Spain. Geol Mijnb 63:333–336

Rauser-Chernoussova DM (1948) Stratigraphy of the Viséan stage of the south wing of the Moscow Basin, after the foraminiferal fauna. Akad Nauk SSSR, Inst Geol Nauk Tr Ser Geol 19:3–40

Rauser-Chernoussova DM (1951) Carboniferous and artinskian facies in the Sterlitamak-Ishimbaevo region (pre-Urals), based on fusulinid studies. Akad Nauk SSSR Tr Ser Geol 119:1–108

Rauser-Chernoussova DM, Korolyuk IK (1981) On the morphology and systematic of Late Moscovian siphonal algae in the South Urals and their significance in rock formation. Akad Nauk SSSR, Vopros Mikropaleont 24:157–170

Reitlinger EA (1954) Devonian Foraminifera from some sections of the eastern part of the Russian Platform. VNIGNI, Leningrad, Paleontol Sb 1:52–81

Reitlinger EA (1957) Devonian spheres of the Russian Platform. Akad Nauk SSSR Dokl 115, 4:774–776

Reitlinger EA (1960) Characteristics of the Ozerko-Khovansk Formations on microscopic organic remains. Akad Nauk SSSR Tr Geol Inst 14:135–179

Reitlinger EA (1966) On the Umbellids in the European part of the USSR and in Trans-caucasia. In: Maslov VP (ed) Fossil Charophytes of the USSR. Akad Nauk SSSR, Tr Geol Inst 143:213–219

Riding R (1977) Problems of affinity,in Palaeozoic calcareous algae. In: Flügel E (ed) Fossil algae, recent results and developments. Springer, Berlin Heidelberg New York, pp 202–211

Riding R, Brasier M (1975) Earliest calcareous foraminifera. Nature (London) 257, 5523:208–210

Riding R, Jansa L (1976) Devonian occurrence of *Uraloporella* (?Foraminifer) in the Canning Basin, western Australia. J Paleontol 50, 5:805–807

Riding R, Voronova L (1985) Morphological Groups and Series in Cambrian Calcareous Algae. In: Toomey DF, Nitecki MH (eds) Paleoalgology: contemporary research and applications. Springer, Berlin Heidelberg New York Tokyo, pp 56–78

Rigby JK (1958) Two new Paleozoic hydrozoans. J Paleontol 32, 3:538–586

Rothpletz A (1890) Uber *Sphaerocodium Bornemanni*, eine neue fossile Kalkalge aus den Raibler Schichten der Ostalpen. Bot Centralbl 52:9

Rothpletz A (1890) Über Algen und Hydrozoen im Silur von Gotland und Oesel. Kungl Svensk Vetensk Handl 43, 5:1–23

Rothpletz A (1913) Über die Kalkalgen, Spongiostromen und einige andere Fossilien aus dem Obersilur Gotlands. Sver Geol Undersökning 10:1–54

Roux A (1979) Révision du genre *Epimastopora* "Pia, 1922" (Dasycladaceae). Bull Centre Rech Explor Prod Elf-Aquitaine 3, 2:803–810

Roux A (1985) Introduction à l'étude des Algues fossiles paléozoïques (de la Bactérie à la tectonique de plaques). Bull Centre Rech Explor Prod Elf-Aquitaine 9, 2:465–699

Roux A (in press) Study of *Paraepimastopora kansasensis* (dasyclad alga) topotypical material from the Upper Pennsylvanian of Kansas, USA. Rev Paleobiol

Saltovskaja VD (1979) New genus of calcareous algae from the Middle Carboniferous of Tadzikhistan. Akad Nauk Tadzik SSR, Izv Otdel Biol Nauk 74, 1:88–93

Saltovskaja VD (1984) On some new calcareous paleozoic algae of Tadzikhistan. Akad Nauk Tadzik SSR, Inst Geol, Donich, Duzhambe, pp 141–160

Sandberger G (1849) Eine neue Polypengattung *Sycidium* aus der Eifel. N Jahrb Mineral Geognos, Geol Petrefaktenk 1849:671–672
Schvetsov MS, Birina LM (1935) On the petrology and origin of the Oka Limestone in the Mikhailov-Aleksin region. Tr Moskovsk Geol Trest 10:1–86
Shaikin IM (1987) Charophyte algae. In: Dubatolov VN (ed) Fossil calcareous algae. Morphology systematics and study methods. Akad Nauk SSSR, Sibirsk Otdel, Trudy Inst Geol Geofiz 674:140–160
Shuysky VP (1973) Calcareous reef-building algae of the Lower Devonian of the Urals. Akad Nauk SSSR, Ural Nauchn Centr, Uralsk Nauchn Inst Geol Geochem Nauka, pp 1–156
Shuysky VP (1985) On the position of the Paleoberesellids and other segmented algae from the Siphonophyceae. In: New data on the geology, biostratigraphy and paleontology of the Urals. Akad Nauk SSSR, Ural Nauchn Centr, Inst Geol Geochem Akad, Zavaritski, pp 86–95
Shuysky VP (1986) New Dasyclad algae from the Lower Devonian of the Urals. Akad Nauk SSSR, Paleontol Zh 1986, 2:118–122
Shuysky VP (1987) Green algae (Chlorophyta). In: Dubatolov VI (ed) Fossil calcareous algae. Akad Nauk SSSR, Sibirsk Otdel, Tr Inst Geol Geofiz 674:38–109
Shuysky VP, Schirschova DI (1985) On a revision of the genus *Lancicula* (Maslov). In: New data on the geology, biostratigraphy and paleontology of the Urals. Akad Nauk SSSR, Uralsk Nauchn Centr, Inst Geol Geochem Akad Zavaritski, pp 95–104
Skompski S (1981) Morphology and systematic position of the Carboniferous algal genus *Calcifolium*. N Jahrb Geol Paläontol Monatsh 3:165–179
Skompski S (1984) The functional morphology of the Carboniferous dasycladacean genus *Kulikia*. N Jahrb Geol Paläontol Monatsh 7:427–436
Stolley E (1893) Über Silurische Siphoneen. N Jahrb Mineral Geol Paläontol 2:135–146
Termier G, Termier H (1950) Foraminifères, Spongiaires et Coelentérés. In: Paléontologie Maroccaine. Not Mém Serv Mines Maroc 73:30–40, 112–115
Termier H, Termier G, Vachard D (1975) Recherches micropaléontologiques dans le Paléozoïque supérieur du Maroc central. Cah Micropaléontol CRNS 4:1–99
Termier G, Termier H, Vachard D (1977) On Moravamminida and Aoujgaliida (Porifera, Ischyrospongia): Upper Paleozoic "Pseudo Algae". In: Flügel E (ed) Fossil algae. Springer, Berlin Heidelberg New York, pp 215–219
Toomey DF, Babcock JA (1983) Precambrian and Paleozoic algal carbonates, West Texas-southern New Mexico; field guide to selected localities of late Proterozoic, Ordovician, Pennsylvanian and Permian ages, including the Permian reef complex. Col Sch Mines Prof Contrib 11:1–345
Toomey DF, Johnson JH (1968) *Ungdarella americana*, a new red alga from the Pennsylvanian of southeastern New Mexico. J Paleontol 42, 2:556–560
Twenhofel WH (1919) Precambrian and Carboniferous algal deposits. Am J Sci 4, 43:339–352
Vachard D (1974) Sur les Dasycladacées métaspondyles "vestibulaires" à propos d'un de leurs représentants viséens: *Eovelebitella occitanica* n.gen., n.sp. CR Acad Sci Paris 279:1855–1858
Vachard D (1977) Etude stratigraphique et micropaléontologique (Algues et Foraminifères) du Viséen de la Montagne Noire (Hérault, France). Mém Inst Géol Univ Louvain 29:111–195
Vachard D (1981) Téthys et Gondwana au Paléozoïque supérieur. Les données afghanes. Doc Trav Inst Géol Lapparent 2:1–463
Vachard D (1988) Calcareous microfossils (Algae, Pseudo-algae and Foraminifera) from La Serre, Montagne Noire, France. Cour Forsch Inst Senkenberg 100:139–147
Vachard D, Massa D (1984) The Carboniferous of Western Libya. 9ème Congr Inter Strat Géol, Carbonifère, Washington 2:165–174
Vachard D, Montenat Ch (1981) Biostratigraphie, micropaléontologie, et paléogéographie du Permien de la région de Tezak (Montagnes Centrales d'Afghanistan). Palaeontographica 178, B:1–88
Vachard D, Tellez-Giron C (1986) El genero *Polyderma* y nuevas soluciones a la sistematica de las calcisferas, microfosiles problematicos del Paleozoico. Rev Inst Mex Petrol 18, 3:6–44
Vachard D, Termier G, Termier H (1977) La transgression viséenne au Jebel Goulib (Maroc Central). Bull Soc Linn Lyon 46, 8:267–281
Vologdin AG (1932) Archaeocyathinae from Siberia. Nauka, Moscow Leningrad, pp 1–106

Vologdin AG (1939) Middle Cambrian Archaeocyatha and algae from the southern Urals. Paleont Lab Moskov. Gosudarst Univ SSSR, Probl Paleontol 5:209–276

Wethered E (1886) On the structure and organisms of the Lower Limestone Shales, Carboniferous Limestone and Upper Limestones of the Forest of Dean. Geol Mag 3, 3:529–540

Wethered E (1890) The occurrence of the genus *Girvanella* in oolitic rocks and remarks on oolitic structure. Geol Soc London QJ 46:270–283

Weyer D (1968) Koninckopora Lee 1912 (Dasycladaceae) aus dem Unterkarbon der Insel Rügen. Geologie 17, 2:176–190

Williamson WC (1881) On the organization of the fossil plants of the Coal Measures. (Part X). Including an examination of the supposed Radiolarians from the Carboniferous rocks. Philos Trans R Soc London 171:493–539

Wilson JL (1975) Carbonate facies in geologic history. Springer, Berlin Heidelberg New York, 471 pp

Wilson EC, Waines RH (1963) A new species of *Komia* Korde and the systematic position of the genus. Palaeontology 6, 2:246–253

Wood A (1940) Two new calcareous algae of the family Dasycladaceae from the Carboniferous Limestone. Liverpool Geol Soc Proc 18:14–18

Wood A (1941) The Lower Carboniferous calcareous algae *Mitcheldeania* Wethered and *Garwoodia* gen. nov. Geol Assoc Proc, London 52, 3:216–226

Wood A (1942) The algal nature of the genus *Koninckopora* Lee and its occurrence in Canada and western Europe. Geol Soc London Q J 98, 3–4:205–222

Wood A (1948) *Sphaerocodium*, a misinterpreted fossil from the Wenlock Limestone. Geol Assoc Proc 59, 1:9–22

Wood A (1957) The type-species of the genus *Girvanella* (Calcareous algae). Palaeontology 1, 1:22–28

Wood A (1964) A new Dasycladacean algae, *Nanopora*, from the Lower Carboniferous of England and Kazakhstan. Palaeontology 7, 2:181–185

Wray JL (1964) *Archaeolithophyllum*, an abundant calcareous algae in limestones of the Lansing Group (Pennsylvanian), in southeastern Kansas. State Geol Surv Kansas Bull 170, 1:3–13

Wray JL (1967) Upper Devonian calcareous algae from the Canning Basin, Western Australia. Col Sch Mines Prof Contrib 3:1–76

Wray JL (1977a) Calcareous algae. Developments in Palaeontology and Stratigraphy, 4. Elsevier, Amsterdam, pp 1–185

Wray JL (1977b) Late Paleozoic calcareous red algae. In: Flügel E (ed) Fossil algae. Springer, Berlin Heidelberg New York, pp 167–176

Wright VP (1982) *Uraloporella* Korde from the Lower Carboniferous of South Wales. Br Mus Nat Hist Geol Bull 36, 2:151–155

Zadorozhnyi VM, Juferev OV (1984) Protozoa-Sarcodina-Foraminiferida. In: Kanygin AV (ed) Paleozoic of the southeast and western Siberian Plate. Akad Nauk SSSR, Sibirsk Otdel, Inst Geol Geofiz 568:70–113

Zanin Buri C (1965) A new Permian *Epimastopora* (calcareous alga) from Hunza Valley (Western Karakorum). Ital Exped Karakorum (K 2) and Hindu Kush Sci Rep IV. Paleont Zool Bot 1, 1:79–88

Chapter 20

Permian Marine Calcareous Algae

R. Riding and L. Guo[1]

Abstract

The Permian is characterized by development of distinctive gymnocodiacean non-reef floras and of the problematic reefal genera *Archaeolithoporella* and *Tubiphytes*. Lower Permian calcareous algae are dominated by dasycladaleans and phylloid algae. The importance of phylloids declines towards the middle Permian, as does that of problematic groups common in the Carboniferous such as *Stacheia, Ungdarella* and similar genera. Calcified cyanobacteria and genera traditionally referred to the Solenoporaceae are, overall, of minor importance during the period. Key locations for the study of Permian algal floras in New Mexico and Texas, the Eastern Alps and Yugoslavia, the central Urals, the Caucasus, northern Iran and Afghanistan, south China, and Japan, reflect their originally broadly equatorial distribution.

1 Introduction

An aura of mystery surrounds much that we know concerning Permian calcareous algae. There is a residue of the Carboniferous flora in the Early Permian, where phylloid algae are prominent in reef formation, continuing a role already well-established in the Pennsylvanian. This is not to deny that phylloid algae present problems of identity (some are probably green algae, others red), but their decline during the Lower Permian coincides with the rise in importance of the even more intriguing reef building genera *Archaeolithoporella* (red alga?) and *Tubiphytes* (cyanobacterium?, hydrozoan?, sponge?), together with the very successful, and also enigmatic, Gymnocodiaceae. The latter is the single most distinctive group of Permian algae, and it combines morphological features of both rhodophytes and chlorophytes. Gymnocodiaceans, together with dasycladaleans, are common constituents of Permian shallow marine non-reefal limestones.

Nevertheless, it is in the reefal facies, particularly in the middle Permian when *Archaeolithoporella* and *Tubiphytes* were most common, that problems are concentrated. Uncertainties concerning affinity make it difficult to determine whether these reefs are dominated by algae or sponges. Diagenetic problems make it difficult to say whether the reef fabric is altered skeleton or cement. These doubts concerning reef composition and structure are well observed in the history of research on the Capitan Reef Complex of New Mexico and west Texas, but there is a suite of other less famous reefs, particularly in the

[1] Department of Geology, University of Wales College of Cardiff, Cardiff CF1 3YE, UK

Mediterranean region, which further emphasizes the problems centred on middle Permian reefs. Only in the very latest Permian reefs, such as those in southern China, do *Archaeolithoporella* and *Tubiphytes* become subordinate to calcisponges, and a transition towards Triassic floras occurs. Latest Upper Permian (*Palaeofusulina* zone) reefs also occur in Skyros, Greece, where at least two reef types are present, characterized by the predominance of *Archaeolithoporella* and *Tubiphytes*, and by the predominance of calcisponges, respectively (Flügel, pers. comm. 1989).

Fundamental research on Permian calcareous algae is summarized by Pia (1937, pp. 813–837), Johnson (1963), Babcock (1979), and Flügel (1984). The principal features which are beginning to emerge are:

1. Decline during the Permian of Carboniferous groups, notably phylloid algae and algal microproblematica such as *Stacheia* Brady, *Ungdarella Maslov*, etc;
2. Middle Permian importance of dasycladaleans, but their decline in the Late Permian;
3. Stepwise progressive increase in the importance of gymnocodiceans during the Permian;
4. Importance, especially in the middle Permian, of the reef-associated problematica *Archaeolithoporella* Endo and *Tubiphytes* Maslov;
5. Broadly equatorial distribution, from North America to the Tethyan belt, although a northward trend in east Asia (particularly Japan) is anomalous and may indicate error in current continental reconstructions;
6. Two principal environmental associations: (a) reefal, dominated by phylloid algae, *Tubiphytes* and *Archaeolithoporella*, and (b) shallow normal marine non-reefal, dominated by Gymnocodiaceae and Dasycladales.

In addition, the evaporitic Zechstein deposits of north-west Europe contain a distinctive stromatolitic facies, although this is not dealt with here.

2 Algal Facies

2.1 Shelves

Fragments of dasycladaleans and gymnocodiaceans are conspicuous in both inner and outer platform environments throughout the Permian. In the Lower Permian (Asselian and Sakmarian) of the Carnic Alps, for example, *Girvanella* and *Efluegelia* occur in lagoonal environments whereas the inner platform is characterized by the dasycladaleans *Anthracoporella* Pia, *Epimastopora* Pia and *Mizzia* Schubert, plus the problematic *Ungdarella*, with a more diverse dasycladalean-dominated outer platform association (p. 573 and Fig. 2 in Flügel 1979). A similar distribution can be seen in rocks of Artinskian age in central Afghanistan where the dasycladalean *Atractyliopsis* Pia emend. Accordi occurs near the seaward margin of oncoid shoal deposits in association with the gymnocodiacean *Dzhulfanella* Korde and the phylloid red alga *Archaeolitho-*

phyllum Johnson, while the cyanobacterium *Zonotrichites* Bornemann occurs in more restricted environments shorewards of oolitic banks (Fig. 10 of Vachard and Montenat 1981). In the Upper Permian gymnocodiaceans generally become more abundant than dasycladaleans in both inner and outer platform environments (Elliott 1968).

2.2 Reefs

Wolfcampian (Early Permian) reefs, such as those of the southwestern United States which are characterized by calcareous sponges, phylloid algae, and bryozoans bound by *Archaeolithophyllum*, *Archaeolithoporella*, and *Tubiphytes*, may have their origins in late Pennsylvanian "thicket communities" on the margins of phylloid algal banks (Wahlman 1986). Nevertheless, phylloid algae, particularly *Eugonophyllum* Konishi and Wray and *Archaeolithophyllum*, continue as dominant organisms in reef formation in the Early Permian. Phylloid algal mounds with thicknesses of approximately 5–10 m form extensive complexes up to more than 120 m thick and several kilometres across in shelf environments in New Mexico and west Texas (Cys 1985). By Sakmarian and Artinskian time phylloids had declined and the *Tubiphytes-Archaeolithoporella* association dominated reefs such as those at Trogkofel in the Carnic Alps of southern Austria [Flügel 1966 (with *Archaeolithoporella* not named directly, but referred to as a stromatolite), Flügel 1981a]. However, the Guadalupian Capitan Reef Complex of New Mexico and Texas shows a more complicated zonation, with *Tubiphytes* Maslov, *Archaeolithoporella* Endo and phylloid algae all involved in reef construction, although phylloids remain generally subordinate.

In the late Permian reefs of Sichuan and Hubei, China, phylloid algae are not noticeable and even *Tubiphytes* and *Archaeolithoporella* are subordinate to sponges, "hydrozoans" (which, as in the Triassic, may also include sponges, Flügel 1981b, p. 307), tabulozoans and bryozoans.

3 Algal Groups

Despite the intricate problems of affinity in Permian calcareous algae, we attempt here to outline a preliminary classification based on existing information (Table 1). This provides a guide to uncertainties which need to be clarified by research, as well as to the emerging facts concerning calcareous algae at the end of the Palaeozoic.

3.1 Rhodophyta

Gymnocodiaceae, "ancestral corallines", genera attributed to the Solenoporaceae, and *Archaeolithoporella* may all be placed tentatively in the Rhodo-

Table 1. Classification of Permian marine calcareous algae and cyanobacteria, listing principal genera[a]

PROKARYOTES		
1 Cyanobacteria		*Garwoodia* Wood 1941 [H], *Girvanella* Nicholson and Etheridge 1878, *Hedstroemia* Rothpletz 1913, *Ortonella* Garwood 1914, *Zonotrichites* Bornemann 1887.
ALGAE		
1 Rhodophyta	a)	Gymnocodiaceae. *Gymnocodium* Pia 1920, *Nanjinophycus* Mu and Riding 1983, *Permocalculus* Elliott 1955 (= *Dzhulfanella* Korde 1965), *Succodium* Konishi 1954[H], *Tauridium* Güvenc 1966[H].
	b)	"Ancestral corallines". *Archaeolithophyllum* Johnson 1956[Ph], *Cuneiphycus* Johnson 1960 (probably not an alga), *Eflugelia* Vachard 1979, *Lysvaella* Chuvashov 1971.
	c)	"Solenoporaceans". *Parachaetetes* Deninger 1906, *Pycnoporidium* Yabe and Toyama 1928 [Ps], "*Solenopora*" Dybowski 1878[b].
2 Phylloid algae[RC]		*Anchicodium* Johnson 1946, *Eugonophyllum* Konishi and Wray 1961, *Hikorocodium* Endo 1951 (probably not an alga), *Ivanovia* Khvorova 1946.
3 Chlorophyta	a)	Dasycladales
		Albertaporella Johnson 1966
		Amicus Maslov 1956
		Anatolipora Konishi 1966
		Anthracoporella Pia 1920
		Archaeocladus Endo 1956
		Atractyliopsis Pia 1937
		Clavaphysoporella Endo 1958
		Clavaporella Kochansky and Herak 1960
		Clypeina Michelin 1845
		Diplopora Schafhautl 1863
		Dvinella Khvorova 1949
		Endoina Korde 1965
		Eogoniolina Endo 1953
		Eovelebitella Vachard 1974
		Epimastopora Pia 1922 (= *Globuliferoporella* Chuvashov 1974)
		Epimastoporella Roux 1979
		Goniolinopsis Milanović 1965
		Gyroporella Gümbel 1871
		Imperiella Elliott 1975
		Johnsonia Korde 1965
		Kochanskiella Milanović 1965
		Likanella Milanović 1966
		Macroporella Pia 1912
		Mizzia Schubert 1907
		Nanjinoporella Mu and Elliott 1982
		Nipponophysoporella Endo 1959
		Oligoporella Pia 1912
		Paraepimastopora Roux 1979
		Permopora Elias 1947
		Permoperplexella Elliott 1968
		Physoporella Steinmann emend Endo
		Pseudopimastopora Endo 1960

Table 1. *(Continued)*

ALGAE		
		Pseudogyroporella Endo 1959
		Pseudovermiporella Elliott 1958
		Pseudovelebitella Mamet and Roux 1978
		Salopekiella Milanović 1965
		Shigaporella Endo 1961
		Sinoporella Yabe 1949
		Teutloporella Pia 1912
		Thailandoporella Endo 1969 (= *Gyroporella* Gümbel)
		Triploporella Steinmann 1880
		Velebitella Kochansky-Devidé 1964
		Velomorpha Horiguchi 1961
		Vermiporella Stolley 1893
		Xainzanella Mu 1984
		Zaporella Rácz
	b)	Halimedaceae. *Aphroditicodium* Elliott, 1970, *Neoanchicodium* Endo 1954, *Orthriosiphonoides* Petryk 1972, *Siamporidium* Endo 1969
4 Problematica	a)	*Archaeolithoporella* Endo 1959[A]
	b)	Calcispheres[Ps]
	c)	*Collenella* Johnson 1942[Pb]
	d)	Receptaculitids [Ps]
	e)	*Stacheia* Brady 1876 group[Ps]
	f)	*Tubiphytes* Maslov 1956[Pb]
	g)	*Ungdarella* Maslov 1956 and *Komia* Korde 1951 group[Ps]

[a] A, Probably algal; H, may belong with the Halimedaceae; Ph, phylloid; Pb; probably not algal; Ps, possibly not algal; RC, probably include rhodophytes and chlorophytes; *Archaeolithophyllum* is also phylloid.

[b] Genuine *Solenopora* is restricted to the Ordovician.

phyta, but this necessitates careful discussion because very few of these can be regarded with certainty as red algae.

3.1.1 Gymnocodiaceae

This extinct group includes representatives that resemble the weakly calcified Recent chaetangiacean red alga *Galaxaura*. The family Gymnocodiaceae was erected by Elliott (1955) to include *Gymnocodium* Pia emend. Elliott (Fig. 1) and *Permocalculus* Elliott (Fig. 2). Mu and Riding (1983) referred *Nanjinophycus* to the Gymnocodiaceae. *Succodium* Konishi and *Dzhulfanella* Korde (see Korde 1965), as well as *Tauridium* Güvenc may also belong in the family (but also see Halimedaceae, below). *Dzhulfanella* is probably a junior synonym of *Permocalculus* (Xinan Mu, pers. comm. 1989).

These are cylindrical or sack-like fossils, typically occurring as fragments a few millimetres in size, with a filamentous internal medulla and a relatively thin cortex with pores. The plants may have been segmented and sporangia may be present. This latter detail is important for the affinities of the group: vegetatively

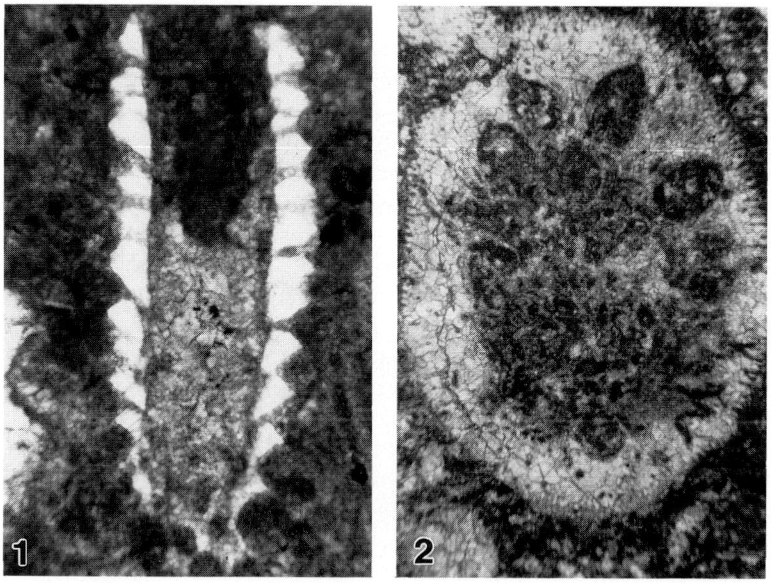

Fig. 1. *Gymnocodium*. Changxing Formation. Uppermost Permian, Sichuan, China. x50
Fig. 2. *Permocalculus*. Changxing Formation. Uppermost Permian, Sichuan, China. x20

the Gymnocodiaceae somewhat resemble halimedacean green algae, but they can also be compared with chaetangiacean red algae such as modern *Galaxaura* and, in addition, some gymnocodiaceans possess reproductive conceptacles similar to those of *Galaxaura* (see Mu and Riding 1983, pp. 270–274). At present the most likely affinity of the Gymnocodiaceae is with the red algae, which is the view arrived at by Pia (1937, pp. 830–831).

Gymnocodium and *Permocalculus* first appear in the Lower Permian. *Gymnocodium* is restricted to the Permian with the exception of a possible Eocene occurrence (Massieux 1966, pp. 119–121). *Permocalculus* is known to range to the Cretaceous, and has also recently been reported from the Lower Miocene of Papua New Guinea (M.D. Simmons, pers. comm. 1989). *Nanjinophycus* is, so far, only known from the Permian of China. Gymnocodiaceans are particularly abundant in the Upper Permian (Fig. 22), and *Gymnocodium* and *Permocalculus* are geographically widespread; occurring in North and Central America, southern Europe, Tunisia, Turkey, the Middle East, Pakistan, Thailand, Vietnam, China, and Japan.

Gymnocodiaceans typically occur as sand and gravel-size fragments in ramp and outer shelf environments, associated with dasycladaleans and *Pseudovermiporella*.

3.1.2 "Ancestral Corallines"

The origins of the coralline red algae are obscure. Several Upper Palaeozoic genera have been suggested to be ancestors of the Mesozoic Corallinaceae and Wray (1977, pp. 71–74) termed these ancestral corallines. In the Permian these include *Archaeolithophyllum, Cuneiphycus* Johnson, and *Lysvaella* Chuvashov. *Archaeolithophyllum* and *Lysvaella* are similar in internal structure, and possess hypothallial-perithalial tissue differentiation and conceptacle-like bodies, both of which suggest relationship to extant calcareous red algae.

Archaeolithophyllum has a thin crustose thallus similar to that of squamariacean (Peyssonneliaceae) red algae and to some phylloid algae (see below), whereas *Lysvaella* is cylindrical in form. *Cuneiphycus* has large cells arranged in anastomosing sheets and does not appear to have differentiated tissue. Its affinities are more problematic.

Lysvaella occurs in the Lower Permian of the Urals (Chuvashov 1971) and *Cuneiphycus*, best known from the Pennsylvanian, is present in the Lower Permian of the Carnic Alps (Flügel 1977, p. 320). *Archaeolithophyllum* is an important component of limestones, including reefs, in the Pennsylvanian and Permian of the central and south-western USA (Otte and Parks 1963; Wray 1964; Heckel and Cocke 1969; Mazzullo and Cys 1979; Toomey and Cys 1979). James et al. (1988, pp. 300–301) tentatively compare *Archaeolithophyllum* with squamariaceans (peyssonneliaceans).

3.1.3 Solenoporacean Genera

Massive skeletons composed of narrow closely packed cellular filaments which do not show tissue differentiation or conceptacles have traditionally been assigned to the Solenoporaceae. Genera grouped into this family have been reported to be geographically widespread in the Permian, although they do not appear to be very abundant.

Johnson (1942) reported *Solenopora* Dybowski from the Capitan Reef Complex and, later, *S. centurionis* and *S. texana* from the nearby Apache Mountains (Johnson 1951). Johnson (1951) and Rigby (1957) considered *S. texana* to occur as an intergrowth with the bryozoan *Acanthocladia guadalupiensis*. However, the bryozoan-algal consortium hypothesis in general, which originated with Condra and Elias (1944), was criticized by Duncan (1957, p. 789) who, according to Bain (1967, p. 213) maintained that "*Acanthocladia* represents bryozoan growth entirely". In line with this, Babcock (1974) suggested that *S. texana*, rather than being algal, may actually represent the outer surface of *Acanthocladia*, an observation supported by Reinhardt (1988, p. 256).

Babcock (1977, p. 19) noted that *Parachaetetes* Deninger "the only other solenoporacean red alga in the Capitan" is restricted to the Upper Capitan-massive, i.e. the shallower part of the reef complex. *Solenopora* and/or *Parachaetetes* have also been recorded from the Southern Alps (Fig. 18) (Flügel 1979, p. 575), southern Tunisia (Driggs 1977, p. 44), Saudi Arabia (Rezak 1959), the eastern slope of the Ural Mountains (Chuvashov 1969), south China (Mu 1981), and Japan (Konishi 1954; Endo 1959). Vachard 1980, p. 381) records *Pseudochaetetes* from the Upper Artinskian of Afghanistan. Fossils in which cross-partitions are often absent and which have been referred to *Solenopora* occur abundantly in the Zechstein limestones of the Holy Cross Mountains, Poland (Belka 1979).

Pycnoporidium Yabe and Toyama has been described from the Upper Permian of Japan (Endo 1956, 1961). Despite uncertainty concerning the affinity of Mesozoic examples of *Pycnoporidium*, Johnson (1963, p. 97) regarded the Permian specimens as being solenoporaceans.

3.1.4 Ungdarella Maslov, Komia Korde

These problematic genera show broad similarities. They are rodlike, branched fossils with a fibrous, apparently cellular, construction. They have commonly been regarded as rhodophytes (Korde and Maslov 1963, pp. 258–259) and both genera are present in the Permian. *Komia* is reported to range to the top of the Asselian, and *Ungdarella* throughout the period (Fig. 39 in Roux 1985), although it is most common in the Lower Permian.

3.1.5 Archaeolithoporella Endo

Archaeolithoporella (Figs. 15, 16) is a light-dark laminated encrusting organism which Endo (1959) originally described from the Permian of Japan. Babcock (1977, p. 15) verified Endo's report that cross-partitions are absent and discussed *Archaeolithoporella* in detail. The genus is, together with *Tubiphytes*, both abundant in the Permian and enigmatic. It makes crustose masses up to several millimetres thick in reefs in the Wolfcampian of Texas (Wahlman 1985), and Lower Artinskian of the Alps (Flügel 1981a), and also in Upper Permian reefs in Tunisia (Boyd et al. 1975), Sicily (Flügel 1984), south China (Fan et al. 1982), the northern Caucasus (Pisera and Zawidzka 1981), and the Capitan Reef Complex (Babcock 1977; Toomey and Cys 1977, pp. 136–138).

Although Endo (1959) failed to recognize any cross-partitions, the presence of conceptacle-like bodies led him to place *Archaeolithoporella* in the red algae. Johnson (1963, p. 98) agreed and went further by asserting that Endo's photograph does show partitions across the layers. He placed *Archaeolithoporella* in the Solenoporaceae and compared it with "masses of poorly preserved Tertiary *Lithocodium*". However, a cement origin has also been proposed for *Archaeolithoporella* (Dunham 1972). This was critically assessed by Babcock (1977, pp.16–17) who concluded in favour of an organic origin for *Archaeolithoporella*. Babcock drew attention to the regularity of the laminae, the palaeoecology of *Archaeolithoporella*, and the stratigraphic restriction of the genus, but also emphasized the need for further studies. Subsequently (Babcock 1986, pp. 17–19), he restated the difficulties of interpreting *Archaeolithoporella* but concluded that it is probably a red alga. On balance, at present it seems likely that *Archaeolithoporella* is organic and Johnson's (1963) broad comparison with corallinacean crusts remains valid, if tantalizing.

However, a likely complication in elucidating the nature of *Archaeolithoporella* is the fact that this name may have been applied too widely. A particular problem in this respect concerns confusion between *Archaeolithoporella* and stromatolites in earlier work on Permian reefs (see p. 52 and Fig. 9 in Flügel 1966). The important binding role of algae was also noted, for example, by Bain (1967, pp. 213–214, 226) in Leonardian (Artinskian-Kungurian) patch reefs of the Glass Mountains, Texas. However, apart from *Tubiphytes*, Bain did not identify the algae, which he described as "stromatolitic". These could be non-skeletal stromatolites, but it is also possible that they may be crusts of *Archaeolithophyllum*, other phylloids, or *Archaeolithoporella*, all of which are common encrusters in mid-Permian reefs. Vachard (1980, p. 344, Pl. 5, Fig. 2) mentions "*Archaeolithoporella* stromatolites" in the Upper Permian of Afghanistan. Laminated encrustations in the Zechstein of northern Europe resemble *Archaeolithoporella*, but may not necessarily be the real thing (Smith 1981, p. 178), although Peryt (1986) has suggested that *Archaeolithoporella* could be regarded as a junior synonym of *Stromaria* Walther (see also Flügel 1989, pp. 127–128). Caution is necessary!

4 Phylloid Algae

Pray and Wray (1963, p. 209) introduced the term phylloid algae to refer to a variety of leaflike genera which are especially common in the Pennsylvanian and Permian. Phylloid algae in the Permian include *Anchicodium* Johnson, *Eugonophyllum* Konishi and Wray (Fig. 14), *Ivanovia* Khvorova, and *Archaeolithophyllum* Johnson. The group is, however, morphological rather than biological and may include both red and green algae. However, the term phylloid algae is useful descriptively because the internal structure of these fossils is often lost during diagenesis, so that it can be impossible to distinguish them in many cases.

Early studies led to the view that genera such as *Anchicodium*, *Ivanovia* and *Eugonophyllum* were codiacean green algae (Konishi and Wray 1961; Johnson 1963, pp. 126–132). However, as work progressed, a rhodophyte affinity also had to be considered. The phylloid alga *Archaeolithophyllum* has a cellular internal structure which relates it to red algae (see "ancestral corallines", above).

Wray et al. (1975) compared Recent squamariaceans with Permian material and suggested that some phylloid algae are Squamariaceae. Flügel (1977, Table 2) listed *Anchicodium* and *Eugonophyllum* as possible ancestral squamariaceans. But *Eugonophyllum* has internal, or partly internal, conceptacle-like structures which suggest that it is not a squamariacean. *Ivanovia* has been mentioned as the phylloid genus which compares best with Squamariaceae (Wray pers. comm., in Riding 1977). However, it is also possible that some Palaeozoic phylloid algae, apart from *Archaeolithophyllum* are halimedacean green algae (see Johnson 1963) and James et al. (1988, p. 300) regard *Eugonophyllum* and *Ivanovia* as both belonging to this latter group.

Phylloid algae are best known from the southwestern United States. Phylloid algal mounds are the dominant type of reef in the Wolfcampian of New Mexico and Texas, and mound complexes are major hydrocarbon reservoirs in the region. *Ivanovia*, *Eugonophyllum* and *Archaeolithophyllum* are especially important genera here (Toomey 1976; Toomey and Cys 1979; Cys 1985; Wahlman 1985). Phylloids are also common until the upper Sakmarian in the Canadian Arctic (Mamet et al. 1979) and there is a similar stratigraphic distribution in the Southern Alps (Flügel 1977). *Anchicodium* occurs in the Lower Permian of Tibet (Mu 1982) and *Eugonophyllum* has been recorded in the Lower Permian of Guizhou, China (Zhang and Zhen 1985). In Japan *Anchicodium* occurs in the Lower Permian (Endo 1953).

Phylloid algae retain their importance from the Pennsylvanian into the Early Permian, but they suffered rapid decline during the Sakmarian. Phylloid genera continue to be present in the Trogkofel Limestone (Sakmarian and Lower Artinskian) of the Southern Alps (Edwards, pers. comm. 1988) and are locally important in the Capitan Reef Complex. In the latter, Babcock (1979) reported *Eugonophyllum* and *Archaeolithophyllum*. But phylloids are relatively rare in the Late Permian, although they are now known from the Norian (Late Triassic) of the Yukon (Reid 1986).

5 Chlorophyta

5.1 Dasycladales

Permian dasycladaleans (erect stick-like algae with relatively thick walls pierced by pores originally occupied by branches) are more diverse in the Lower Permian than in the Upper Carboniferous (Fig. 40 in Roux 1985). At least 47 genera (Table 1), which belong to at least seven tribes, have been described for the period as a whole. The most common genera are *Diplopora*, *Eogoniolina*, *Epimastopora* (Fig. 5), *Epimastoporella* (Fig. 8), *Gyroporella* (Fig. 10), *Mizzia* (Fig. 9), *Macroporella* (Fig. 4), *Paraepimastopora* (Fig. 6) and *Velebitella*. Genera especially common in the Lower Permian include *Anthracoporella* (Fig. 7), *Atractyliopsis* (Fig. 3), and *Globuliferoporella*. The latter genus is regarded as a junior synonym of *Epimastopora* by Roux (1979), whose taxonomy is followed here (see Figs. 5, 6 and 8). But for further details concerning *Epimastopora* also see Endo (1960) and Flügel and Flügel-Kahler (1980, pp. 138–142). Bebout and Coogan (1964) revised *Anthracoporella* using Wolfcampian material from Texas. *Pseudovermiporella* (Figs. 11, 12) is particularly common in the Upper Permian. There is no current agreement concerning the affinity of this latter genus, although Elliott (1968, p. 70) concluded that it is probably algal. Beresellids (recently shown by Skompski (1987) to be dasycladaleans) range from the Carboniferous up to the Asselian, as do the still problematic donezellids (Fig. 39 in Roux 1985).

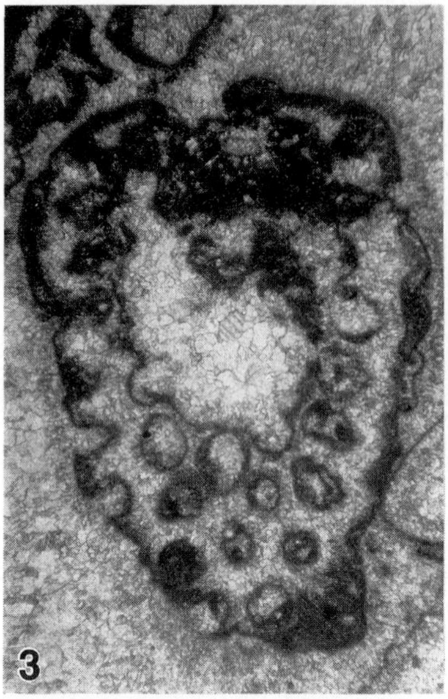

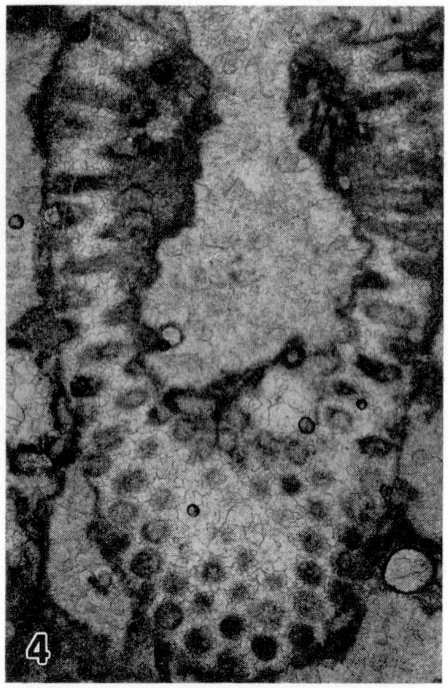

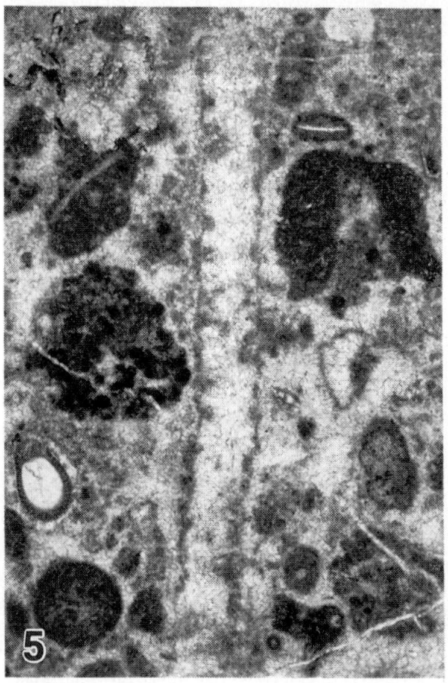

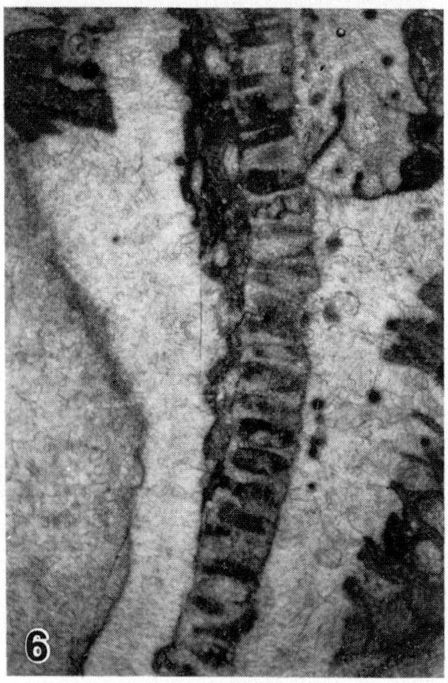

Fig. 3. *Atractyliopsis*. Trogkofel Formation. Lower Permian, Trogkofel, Carnic Alps, Austria. x20
Fig. 4. *Macroporella*. Trogkofel Formation. Lower Permian, Trogkofel, Carnic Alps, Austria. x20
Fig. 5. *Epimastopora*. Trogkofel Formation. Lower Permian, Trogkofel, Carnic Alps, Austria. x20
Fig. 6. *Paraepimastopora*. Trogkofel Formation. Lower Permian, Trogkofel, Carnic Alps, Austria. x20
Fig. 7. *Anthracoporella*. Trogkofel Formation. Lower Permian, Trogkofel, Carnic Alps, Austria. x7.2
Fig. 8. *Epimastoporella*. Upper Pseudoschwagerina Limestone. Lower Permian. Trogkofel, Carnic Alps, Austria. x20
Fig. 9. *Mizzia*. Capitan Formation. Upper Permian, Guadalupe Mountains, New Mexico. x21

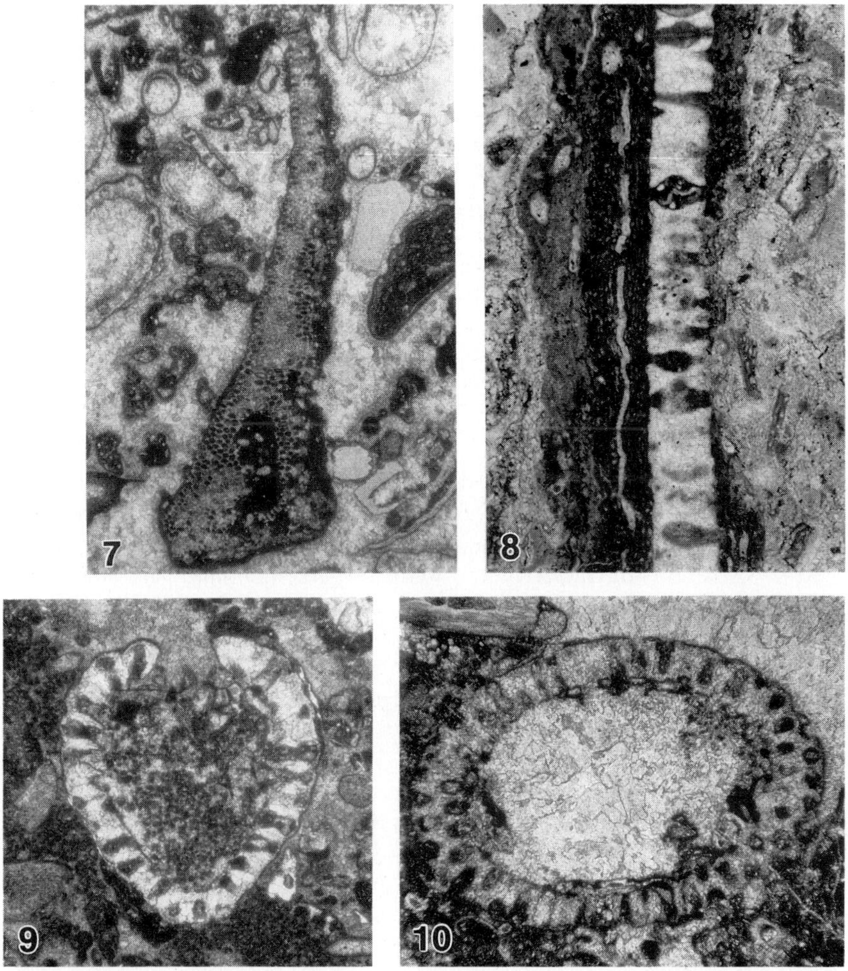

Fig. 10. *Gyroporella*. Trogkofel Formation. Lower Permian, Trogkofel, Carnic Alps. x21
Fig. 11. *Pseudovermiporella*. Changxing Formation. Uppermost Permian, Sichuan, China. x50
Fig. 12. *Pseudovermiporella sodalica* Elliott. Changxing Formation. Uppermost Permian, Sichuan, China. x63

Details of Permian dasycladalean associations are best known from Iraq (Elliott 1968), Japan (Endo 1956, 1958, 1959), the Carnic Alps (Flügel 1966, 1977, 1985; Homann 1972; Flügel and Flügel-Kahler 1980), Afghanistan (Vachard 1980, Vachard and Montenat 1981), Tibet (Mu 1982), southern China (Mu and Elliott 1982, Zhang and Zhen 1985), Tunisia (Glintzboeckel and Rabaté 1964, Vachard 1985), and New Mexico and Texas (Johnson 1942, 1951). Rezak (1959) reported a few genera from the Permian of Saudi Arabia.

Dasycladalean diversity during the period is greatest during the middle Permian (Fig. 2 in Flügel 1985), but declines sharply in the latest Permian (Xi-nan Mu, pers. comm. 1988) although individual genera continue to be abundant.

5.2 Halimedaceae

Apart from those phylloid algae (see above) which may be halimedaceans, this group of green algae, characterized by internal filamentous structure, is represented in the Permian by *Aphroditicodium* Elliott and *Orthriosiphonoides* Petryk. Neither is particularly abundant. *Neoanchicodium* Endo (Fig. 13) may be a halimedacean (Xinan Mu, pers. comm. 1989). Two additional genera, *Tauridium* Güvenc and the more common *Succodium* Konishi, could be halimedaceans, but may also belong to the Gymnocodiaceae (see above). *Succodium* has been reported from Japan (Konishi 1954a, Endo 1957), the Caucasus (Korde 1965), the Southern Alps (Homann 1972), Thailand (Endo 1969), and Vietnam (Nguyen 1970). Korde (1965) first suggested that *Succodium* could be a gymnocodiacean, and this has been followed by Vachard (1977). Mu and Riding (1983) placed *S. endoi* Nguyen in the gymnocodiacean genus *Nanjinophycus*.

6 Tubiphytes Maslov

Tubiphytes (Fig. 17) is commonly irregularly cylindrical or encrusting in form and has a characteristic dark, flocculent, internal structure, generally in a series of uneven bands. Internal tubes forming a central canal system are usually present. It occurs as small (a few millimetres) individual masses and also as thick

Fig. 13. *Neoanchicodium*. Trogkofel Formation. Lower Permian, Trogkofel, Carnic Alps, Austria. x18

Fig. 14. *Eugonophyllum*. Upper Pseudoschwagerina Limestone. Lower Permian, Trogkofel, Carnic Alps, Austria. x6

Fig. 15. *Archaeolithoporella*. Changxing Formation. Uppermost Permian, Sichuan, China. x20

Fig. 16. *Archaeolithoporella* encrusting *Tubiphytes*. Capitan Formation. Upper Permian, Guadalupe Mountains, New Mexico. x20

Fig. 17. *Tubiphytes*. Changxing Formation. Uppermost Permian, Sichuan China. x36

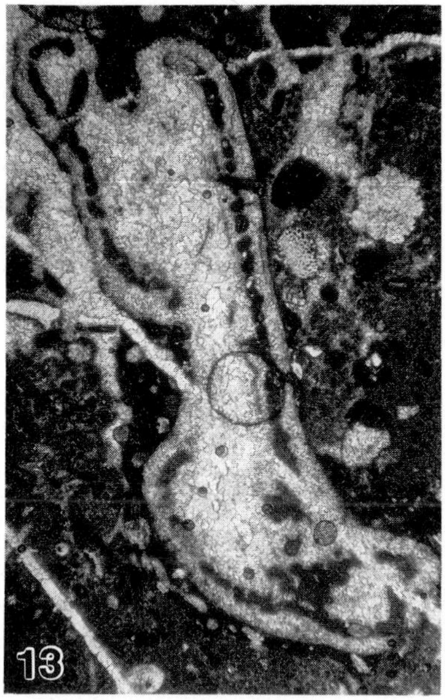

13

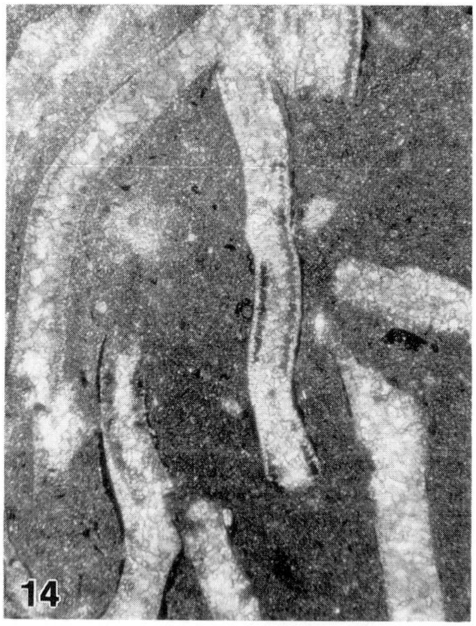

14

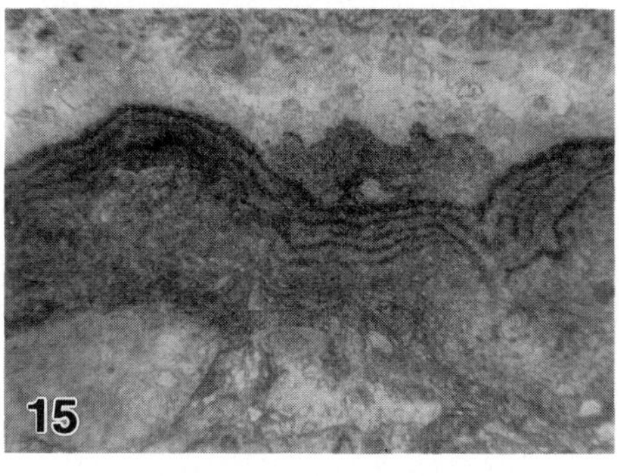

15

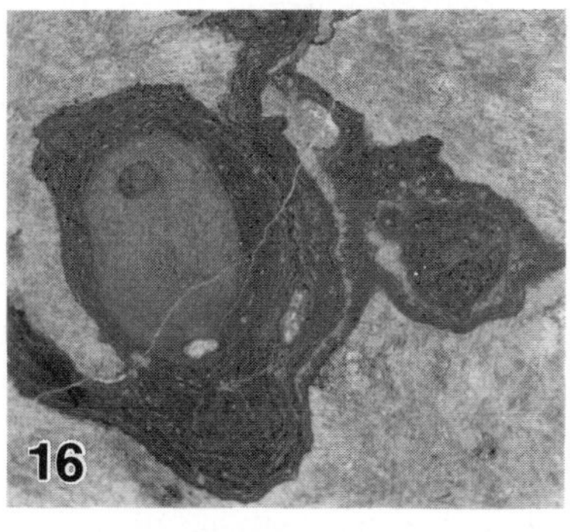

16

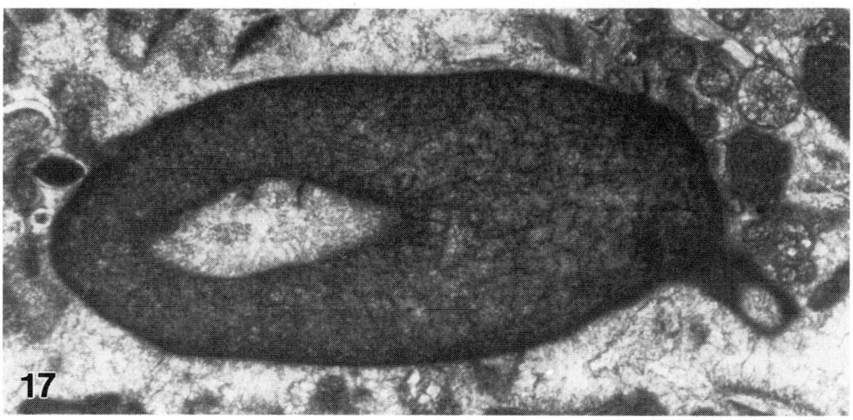

17

(up to several centimetres) encrustations on sponges and bryozoans. It ranges from Carboniferous to Cretaceous. It is especially abundant in the Permian, being a major element of reefs such as Capitan (Babcock 1977) and Trogkofel (Flügel 1981a).

Maslov (1956) described *Tubiphytes* as a possible cyanobacterium (Schizophyta) from the Kungurian of the central Urals. Previously, Rauser-Chernousova (1950) had named this fossil *Shamovella*, but without providing a description (see Maslov 1956, p. 21). Konishi (1959) recognized *Nigriporella* Rigby as a junior synonym of *Tubiphytes*. Babcock (1977, pp 17–18) briefly reviewed the affinities which had been proposed and concluded that neither a hydrozoan nor a cyanobacterial affinity for *Tubiphytes* seems likely. He has again (Babcock 1986, pp. 15–17) reviewed the genus in some detail. Ernst Ott (in Kraus and Ott 1968, p. 271) has suggested that *Tubiphytes* may be a sponge. A somewhat similar genus, *Plexoramea* Mello from the Triassic, has been compared with both chlorophyte algae and fungi (Flügel et al. 1988).

The question of the nature of *Tubiphytes* is made especially significant by the abundance of this genus in Permian reefs. It is a major component in the middle Permian and is also abundant in the Early and Late Permian. It now seems that earlier views, such as that of Rigby (1958), comparing *Tubiphytes* with hydrozoans are nearer the truth than suggestions of a cyanobacterial or algal affinity. Indeed, an invertebrate affinity seems likely, and if *Tubiphytes* should prove to belong to the calcisponges (the view favoured here) then this will only reinforce the already recognized importance of this group in middle to Upper Permian reefs generally.

7 Cyanobacteria

Girvanella Nicholson and Etheridge, *Hedstroemia* Rothpletz, and *Ortonella* Garwood, together with some other non-septate genera attributed to the Solenoporaceae, have been fairly widely reported from the Permian as a whole, but they only occassionally appear to be locally important. Flügel (1979, p. 573) notes *Girvanella* as a characteristic element of restricted lagoonal environments in the Lower Permian of the Carnic Alps.

Osagia Twenhofel and *Ottonosia* Twenhofel, oncoids which usually include *Girvanella* together with other encrusting organisms, have been reported from the Lower Permian (Twenhofel 1919; Johnson 1963, p. 134). Toomey et al. (1988, 1989) demonstrate that *Ottonosia* and *Osagia* nodules from the Lower Permian of Kansas and Oklahoma are dominated by "*Girvanella*, intimately associated with various calcareous and agglutinated encrusting tubiform foraminifers" (Toomey et al. 1989, p. 51), but note that *Girvanella* tends to be diagenetically degraded to micrite. They point out that the principal difference between *Ottonosia* and *Osagia* is one of size, with *Ottonosia* being larger and containing larger attached epizoans including worm tubes and bryozoans. Toomey et al. (1989, p. 61) propose that the formal names *Ottonosia* and *Osagia* be abandoned.

Two Permian genera which have been attributed to the cyanobacteria, *Collenella* Johnson and *Tubiphytes*, almost certainly do not belong in this group. *Collenella*, which is important in the upper part of the Capitan Reef Complex (Johnson 1942, 1963, pp. 136–137; Toomey and Cys 1977, p. 136) has, like *Tubiphytes* (see Sect. 6 above) been compared with metazoans such as stromatoporoids and bryozoans (see Babcock 1979, p. 424).

Hedstroemia, a cyanobacterium of rivulariacean type, is present quite commonly with *Collenella* in the upper Capitan-massive (Babcock 1977, Fig. 3b). However, other cyanobacterial genera are more rare. *Zonotrichites* Bornemann, a probable rivulariacean, is reported from the Lower Artinskian of eastern Afghanistan by Vachard (1980, p. 335). Vachard (1980, pp. 344–345) also reports *Bacinella* Radoičić from the Upper Permian of Afghanistan. This is probably the earliest record of this problematic meshwork-like fossil, which is best known from the Triassic-Cretaceous, and has variously been referred to chlorophytes and prokaryotes, as well as to invertebrates.

Renalcis Vologdin and *Tharama* Wray have been described from the Upper Permian of Tunisia (Vachard and Razgallah 1988), but these reports require clarification. One illustrated specimen (Vachard and Razgallah 1988, their Fig. 1) closely resembles *Izhella* Antropov, hitherto not known from rocks younger than Lower Carboniferous.

The relative lack of importance of calcified cyanobacteria in the Permian is in contrast to both the Lower Carboniferous and the Middle to Upper Triassic.

8 Receptaculitids

Receptaculitids are relatively large dish or vase-shaped fossils which have been compared with green algae, as well as with sponges. The group is most prominent in the middle Palaeozoic. Carboniferous and Permian examples are rare (Nitecki 1972, p. 306). Parona (1933) reported them from the Permian of Sicily and, if confirmed, this appears to be among the last known occurrences of this problematic group.

9 Regional Distribution

Low sea-level stands characterize the Permian but, nevertheless, extensive shallow marine sequences developed in some areas, notably western and Arctic North America, the Ural region, and the margins of Tethys including south China, and Japan (see Ross and Ross 1979). All these areas contain calcareous algae (Fig. 19). In addition, stromatolites are common in the evaporitic Zechstein area of northern Europe. There are at least seven key areas for the study of normal marine Permian calcareous algae worldwide, and these reflect a broadly equatorial distribution for these sequences during both the Lower and the Upper Permian (Figs. 20, 21).

Fig. 18. "*Solenopora*". Trogkofel Formation. Lower Permian, Trogkofel, Carnic Alps, Austria. x6

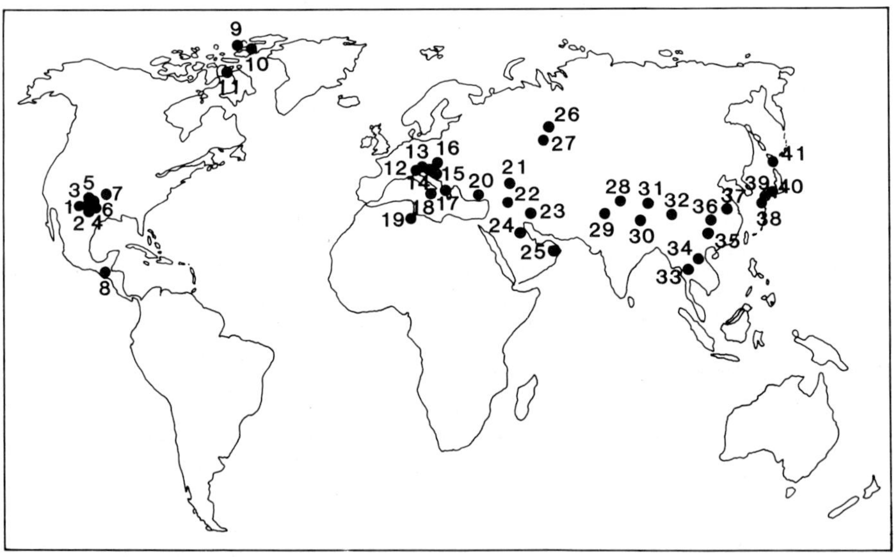

Fig. 19

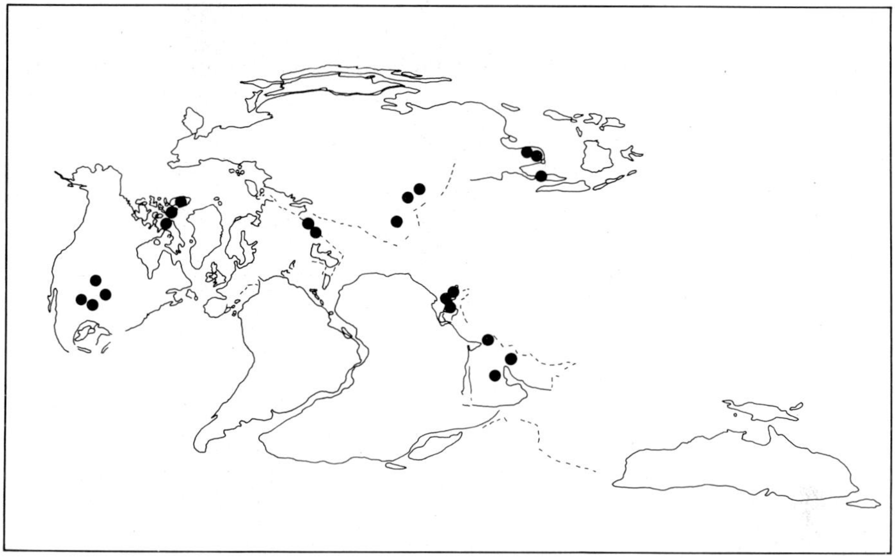

Fig. 20. Distribution of principal localities with marine Lower Permian calcareous algae. Base map from Smith et al. (1981)

Fig. 19. Main localities of described marine Permian calcareous algae, with principal references. *1* Capitan reef complex, Guadalupe Mountains, Texas and New Mexico (Babcock 1977); *2* southern Tatum Basin, southeastern New Mexico (Cys 1985); *3* northern Sacramento Mountains, south-central New Mexico (Toomey and Cys 1979); *4* Motley County, Texas (Wahlman 1985); *5* Apache Mountains, Texas (Johnson 1951); *6* Northern Midland Basin, Texas (Toomey 1985); *7* Kansas (Johnson 1946); *8* Gautemala (Johnson and Kaska 1965); *9* and *10* Sverdrup Basin, Canada (Davies et al. 1986); *11* Northern Franklin district, Canada (Mamet et al. 1979); *12* Sexten Dolomites, Italy (Flügel 1977, Fig. 1); *13* Carnic Alps (Flügel 1977, Fig. 1); *14* Sava Folds, Slovenia, Yugoslavia (Flügel 1977, Fig.1); *15* Velebit Mountains, Croatia, Yugoslavia (Fig. 1 in Flügel 1977); *16* Bükk Mountains, Hungary (Herak and Kochansky 1963); *17* Greece (Pia 1937); *18* Sicily (Flügel and Stanley 1984); *19* Djebel Tebaga, Tunisia (Elliott 1958; Emberger 1958; Glintzboeckel and Rabaté 1964; Vachard 1985); *20* Anatolia, Turkey (Bilgütay 1959); *21* Labariver Basin, northern Caucasus (Korde 1965; 1967; Pisera and Zawidzka 1981); *22* Mosul Liwa, northern Iraq (Elliott 1968, p. 85); *23* Isfahan, Iran; *24* Dammam, Saudi Arabia (Rezak 1959); *25* Oman (Elliott 1958); *26* Cherdyn, western Urals (Chuvashov 1983); *27* Sterlitamak, western Urals (Chuvashov 1983); *28* Bashkiria, USSR (Kulik 1978); *29* Salt Range, Pakistan (Rao and Varma 1953); *30* Pondu, Tibet (Mu 1982); *31* Karamulun, Tibet (Mu 1982); *32* Lago, Markam, Toba, Tibet (Mu 1982); *33* Khao Phlong Phrab, Thailand (Endo 1969); *34* Vietnam, Cambodia, Laos (Nguyen 1970); *35* Guizhou, southern China (Mu 1981); *36* Sichuan, southern China (Guo and Riding 1988); *37* Nanjing, China (Mu and Elliott 1982; Mu and Riding 1983); *38* Kyushu, Japan (Konishi 1954); *39* Hiroshima, Japan (Endo 1957); *40* Gifu-ken, Japan (Endo 1954); *41* Hokkaido, Japan (Endo 1957)

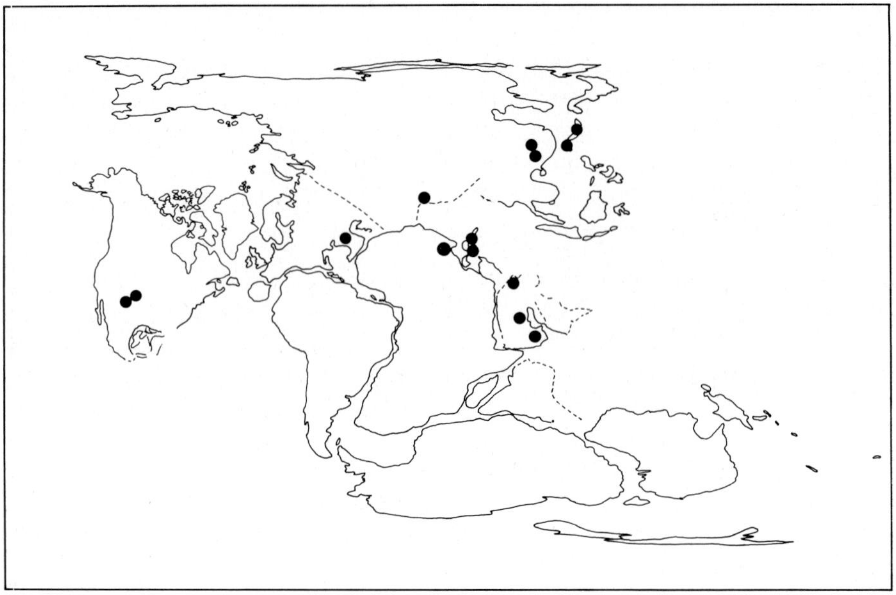

Fig. 21. Distribution of principal localities with marine Upper Permian calcareous algae. Base map after Smith et al. (1981)

9.1 New Mexico-Texas

Shallow carbonate shelves surrounding deep-water embayments in what is now the southwestern United States were sites of reefal sedimentation at several times during the Lower and Middle Permian (Fig. 19, localities 1–6; Figs. 20, 21), and phylloid algae, *Tubiphytes* and *Archaeolithoporella* are important additions to sponges in these reef environments. Phylloid algal mounds are well-developed in the Wolfcampian (Toomey and Cys 1979; Cys 1985). These Early Permian mounds contrast, in both morphology and algal flora, with the Upper Permian Guadalupian Capitan Reef Complex. The latter contains locally abundant *Archaeolithoporella*, *Eugonophyllium*, and *Tubiphytes*, distributed patchily mainly in the upper 50–60 m of the Capitan massive limestone (Babcock 1977, 1979). In addition, Babcock notes *Parachaetetes* and *Hedstroemia* in the uppermost parts of the reef, and the dasycladaleans *Macroporella* and *Mizzia* in the bedded back-reef facies. An important feature of the algal distribution at Capitan is the vertical succession, from bottom to top: (1) *Archaeolithoporella* and *Tubiphytes*, (2) phylloid algae, (3) *Parachaetetes* and *Hedstroemia*.

9.2 Eastern Alps and Yugoslavia (Fig. 19, Localities 12–15)

The Sakmarian to Artinskian Trogkofel Limestone of southern Austria and northern Italy is slightly younger in age than the Wolfcampian bioherms of the southwestern USA. Structural complications and patchy dolomitization obscure the morphology and setting of the Trogkofel reef limestones, but they contain abundant *Archaeolithoporella* and *Tubiphytes* in addition to *Archaeolithophyllum*, bryozoans, calcisponges, cyanobacteria, dasycladaleans and other algae (Flügel 1966, 1981a; Homann 1972; Flügel and Flügel-Kahler 1980). Edwards and Riding (1988) have suggested that sparite fabrics interpreted as marine cement at Trogkofel (and at Capitan) may be diagenetically altered algal skeletons. Calcareous algae are also present nearby in the Sexten Dolomites of northern Italy (Ogilvie-Gordon 1927; Pia 1937; Accordi 1956; Praturlon 1963). Here, and in adjacent parts of northern Yugoslavia, the Bellerophon Formation of Upper Permian age contains a non-reefal shelf flora dominated by Gymnocodiaceae and Dasycladales, with gymnocodiaceans more important in the more restricted inner shelf facies (Noé 1987). The Permian-Triassic boundary is located in the overlying Tesero Horizon and is marked by disappearance of the Permian algae and foraminifers due to climatic change (Noé 1987, p. 115).

The Permian calcareous flora, particularly the gymnocodiaceans and dasycladaleans, of Yugoslavia are known mainly through the work of Pia (1937), Herak and Kochansky (1960), and Kochansky and Herak (1960).

9.3 Central Ural Mountains

The Lower Permian of the central western Urals contains reefal facies with calcareous algae (Chuvashov 1983) (Fig. 19, localities 26, 27). In the Asselian and Sakmarian dasycladaleans (*Anthracoporella*, *Epimastopora*, *Globuliferoporella*, *Gyroporella*), phylloid algae (*Eugonophyllum*), and the possible halimedacean *Neoanchicodium* are common, and *Tubiphytes* is very abundant. However, floral diversity decreased during the Artinskian and Kungurian as evaporitic and non-marine conditions developed.

9.4 Caucasus

The Caucasus area (Fig. 19, locality 21) is comparable with both the Urals and with more thoroughly marine Tethyan areas. Korde (1965, 1967) described Guadalupian and Dzhulfian (partly equivalent to Kazanian, see Harland et al. 1982, p. 23) algae, which are dominated by dasycladaleans and gymnocodiaceans. These probably are from non-reef facies. Marine conditions persisted longer than in the Urals, but nevertheless from the late Guadalupian onwards there was reduced algal diversity, and by the mid-Dzhulfian no calcareous algae were present in the area.

9.5 Northern Iraq and Afghanistan

Shelf limestones of Permian age in northern Iraq (Fig. 19, locality 22) have been studied by Elliott (1968), who reported abundant *Mizzia*, *Gymnocodium* and *Permocalculus* throughout the 800 m thick Permian sequence. Dasycladaleans less common than *Mizzia* include *Anthracoporella*, *Atractyliopsis*, *Permoplexella*, *Pseudoepimastopora*, *Pseudovermiporella* (possibly not a dasycladalean), plus very scarce *Clypeina*. *Solenopora*, in contrast to the dasycladaleans and gymnocodiaceans, is uncommon (Elliott 1968, p. 93).

Vachard (1980) has described in detail Artinskian and Upper Permian calcareous algae from east-central Afghanistan. Common genera include *Tubiphytes*, numerous dasycladaleans (particularly *Mizzia*, *Clavaporella*, *Atractyliopsis*), *Orthriosiphonoides*, *Archaeolithophyllum*, and the gymnocodiacean *Permocalculus* (also reported as its junior synonym *Dzhulfanella*).

9.6 South China

The Permian of Tibet (Fig. 19, localities 30–32) and south China (Fig. 19, localities 35–37) contains shallow marine limestones at many levels throughout the sequence, which is relatively complete. Calcareous algae have been described from the Lower Permian of southern and eastern Tibet (Mu 1982, 1984) and the Lower Permian of Nanjing (Mu and Elliott 1982; Mu and Riding 1983). But the best known flora is from the Lower and Upper Permian of southern Guizhou (Kweichow) in southwest China (Mu 1981; Zhang and Zhen 1985). In Tibet, Nanjing, and Guizhou the sequences are essentially in non-reefal platform facies and the flora, as is that of Iraq, is dominated by Dasycladales and Gymnocodiaceae. An interesting feature of the Chinese Permian is the presence in eastern Sichuan and western Hubei of very late Permian (Changxing Stage) reefs. These contain abundant *Archaeolithoporella* and *Tubiphytes*, in addition to calcisponges and "hydrozoans", and are associated with dasycladaleans, gymnocodiaceans, and *Pseudovermiporella* in the off-reef facies (Fan et al. 1982; Reinhardt 1988; Guo and Riding 1988).

9.7 Japan (Fig. 19, Localities 38–41)

In numerous papers, mainly published between 1951 and 1961, Riuji Endo described calcareous algae from the Permian of Japan (see for example Endo 1959 and references therein and in Johnson 1963). Most of the algae are dasycladaleans and they appear to be in bedded limestones associated with fusuline foraminifers. Endo erected *Archaeolithoporella*, *Neoanchicodium*, *Hikorocodium* (probably not an alga), and the dasycladaleans *Archaeocladus*, *Clavaphysoporella*, *Eogoniolina*, *Nipponophysoporella*, *Pseudogyroporella*, and *Shigaporella*. In addition he reported occurrences of *Epimastopora*, *Girvanella*,

Gymnocodium, *Macroporella*, *Oligoporella*, *Physoporella*, *Solenopora*, *Teutloporella*, and *Vermiporella* (actually *Pseudovermiporella*).

Kenji Konishi also contributed significantly to description of Japanese Permian algae (see references in Johnson 1963), including erection of *Succodium* from the late Permian (Konishi 1954).

10 Conclusions

The most abundant and widespread fossils referred to the calcareous cyanobacteria and algae in the Permian are *Archaeolithoporella* and *Tubiphytes* (although it is likely that the latter is neither a cyanobacterium nor an alga), gymnocodiaceans, dasycladaleans, and phylloid algae (Table 1). In addition, the red alga *Archaeolithophyllum* (also phylloid in form), and the problematic genera *Cuneiphycus*, Stacheia and *Ungdarella* are present. Calcified cyanobacteria such as *Girvanella*, *Hedstroemia* and *Ortonella* occur, although rarely abundantly, throughout the period. Beresellids and donezellids which, like stacheiids and members of the *Ungdarella* group, are common in the Upper Carboniferous, have also been reported to persist into the Asselian (Fig. 39 in Roux 1985). Genera referred to the Solenoporaceae occur, but are not common, and receptaculitids make a final, very rare, appearance.

Although detailed information is still lacking from many areas, it is clear that there are marked changes in the importance of major groups of calcareous algae during the period (Fig. 22). The Early Permian is dominated by the persistence of Carboniferous groups, notably phylloid algae and problematical algae (Roux 1985, p. 670). Decline of these groups is counterbalanced from the Artinskian onwards by the rapid, stepwise expansion of the Gymnocodiaceae, which is dominant in non-reef shelf limestones in the Upper Permian, and by dramatic increase in importance in the reef-associated problematica *Archaeolithoporella* and *Tubiphytes*. Dasycladaleans are also common until at least the middle Permian, and probably longer. The result of these changes is a marked contrast between Early and Late Permian algal floras, with those of the middle Permian being intermediate in composition. These developments are seen just as clearly in non-reefal as in reefal facies. Thus, (1) Early Permian reefs are dominated by phylloid algae with dasycladaleans in bedded shelf limestones; (2) Trogkofel, Capitan, and other reefs of the middle Permian are characterized by *Archaeolithoporella* and *Tubiphytes*, with subordinate phylloids; while (3) Late Permian reefs in China have fewer of these organisms and more sponges, with gymnocodiaceans abundant in associated bedded limestones.

Permian calcareous algae are best known from reef environments. The Capitan Reef Complex, of Guadalupian age, is a good example both of vertical and lateral variations in algal distribution, and of problems of ecological and diagenetic interpretation of the algae. In the same area of the southwestern United States, Wolfcampian phylloid algal mounds are well developed. Per-

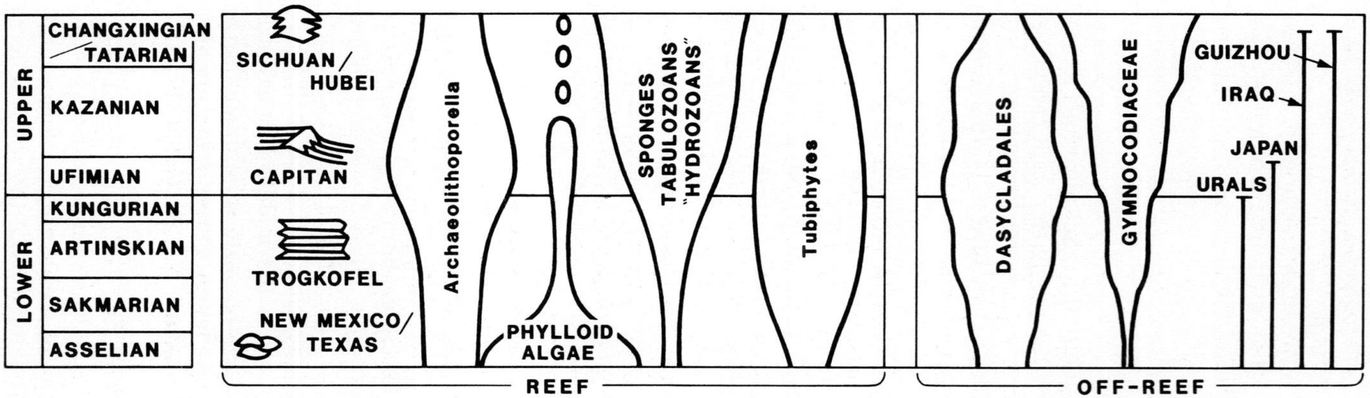

Fig. 22. Abundance estimates for major groups of marine calcareous algae and possible algae, plus sponges, tabulozoans and "hydrozoans". Stratigraphic positions and ranges of areas with notable reef, and off-reef sequences are shown on the *left* and *right* respectively. The Urals region contains both reef and off-reef facies. Time scale based on Harland et al. (1982, chart 2.7)

mian reefs are also seen in the Tethyan belt. In the Mediterranean region there are notable examples in Austria, Greece, Italy, Tunisia and Yugoslavia (Flügel and Stanley 1984) of which possibly the best known are in the Artinskian of the Carnic Alps (e.g. Trogkofel) close to the Austrian-Italian-Yugoslav border. In China, Late Permian (Changxing) reefs (e.g. in Hubei and Sichuan) are important for understanding developments immediately prior to the end of the period. More needs to be known concerning the algal floras of Lower Permian reefs in Arctic regions, such as Canada, Svalbard and the Barents Sea, and the central Ural Mountains.

Shallow normal marine non-reefal limestones with calcareous algae, notably dasycladaleans and gymnocodiaceans, are best known in northern Iraq, the Caucasus Mountains, southern China and Japan. The lithofacies of these deposits are, however, less well documented than those of the reef limestones.

The Zechstein stromatolitic deposits of northern Europe represent a quite different facies, related to developments also seen in the Urals, in which evaporitic conditions progressively eliminated normal marine algae, as well as invertebrates.

Permian calcareous algae present contrasts, as well as some similarities, with those of both the Carboniferous and Triassic. Several groups which are important in the Pennsylvanian decline rapidly in the Lower Permian while groups, such as calcified cyanobacteria, which are common in the Triassic are not conspicuous during much of the Permian. Gymnocodiaceae, and the problematic genera *Archaeolithoporella* and *Tubiphytes*, dominate the period, and it is the uncertainties concerning the affinities of these taxa which are responsible for much of the confusion, as well as not a little of the interest, characterizing the study of Permian calcareous algae.

Acknowledgements. Li Guo's work has been supported by the South Western Petroleum College in Nanchong, Sichuan, China, and TOTAL, Paris. We are grateful for André Maurin's support and encouragement. David Edwards very kindly provided us with examples of algae from Trogkofel in the Carnic Alps. We are particularly indebted to Xi-nan Mu for filling in gaps in our knowledge. Comments on the manuscript by Jack Babcock, Erik Flügel, and Xi-nan Mu have been most helpful and we appreciate their willingness to share their expertise with us.

References

Accordi B (1956) Calcareous algae from the Upper Permian of the Dolomites (Italy) with stratigraphy of the "Bellerophonzone". J Palaeontol Soc India 1, 1:75–84

Babcock JA (1974) The role of algae in the formation of the Capitan Limestone (Permian, Guadalupian), Guadalupe Mountains, West Texas and New Mexico. PhD Diss, Univ Wisc, Madison, 241 pp

Babcock JA (1977) Calcareous algae, organic boundstones, and the genesis of the upper Capitan Limestone (Permian, Guadalupian), Guadalupe Mountains, West Texas and New Mexico. In: Hileman ME, Mazzullo SJ (eds) SEPM Permian Basin Sec Publ 77-16, pp 3–44

Babcock JA (1979) Calcareous algae and algal problematica of the Capitan reef (Permian), Guadalupe Mountains, West Texas and New Mexico, USA. Bull Cent Rech Explor Prod Elf-Aquitaine 3, 2:419–428

Babcock JA (1986) The puzzle of alga-like problematica, or rummaging around in the algal wastebasket. In: Hoffman A, Nitecki MH (eds) Problematic fossil taxa. Oxford Univ Press, New York, pp 12–26

Bain RJ (1967) Paleoecology of some Leonardian patch reefs in the Glass Mountains, Texas. Brigham Young Univ Geol Stud 14:196–236

Belka Z (1979) Shallow-water Solenoporaceae and their environmental adaptation, Upper Permian of the Holy Cross Mountains. Bull Cent Rech Explor Prod Elf-Aquitaine 3, 2:443–452

Bebout DG, Coogan AH (1964) Algal genus *Anthracoporella* Pia. J Paleontol 38:1093–1096, pl 169

Bilgütay U (1959) The Permian calcareous algae from south-eastern Anatolia. Bull Mineral Res Explor Inst Turkey 52:48–58

Boyd DW, Driggs A, Newell ND, Rigby JK, Stehli FG (1975) Permian reef complex, Tunisia. Geol Soc Am Abstr Prog 7, 7:1006

Chuvashov BI (1969) Algae from the Devonian, Carboniferous and Lower Permian of the middle and southern Urals; their ecology and stratigraphic value. In: Vozzhennikova TF, Glezer ZI, Zhuze AN, Saks VN, Sheshukova-Poretskaya VS (eds) Fossil algae of the USSR. Nat Lending Libr Sci Technol, Boston Spa, Engl, pp 206–214 (Transl of original Russian version publ by USSR Acad Sci, Siberian Sec, Inst Geol Geophys. Nauka, Moscow 1967)

Chuvashov BI (1971) A new genus of late Palaeozoic red algae. Palaeontol J 2:85–89 (in Russian)

Chuvashov BI (1974) Permian calcareous algae of the Urals. In Algae, brachiopods and miospores from Permian deposits of the west Urals. Acad Sci USSR, Ural Sci Cent, Trans Inst Geol Geochem Sverdlovsk, pp 3–76 (in Russian)

Chuvashov BI (1983) Permian reefs of the Urals. Facies 8:191–212

Condra GE, Elias MK (1944) Study and revision of *Archimedes* (Hall). Geol Soc Am Spec Pap 53: 243 pp

Cys JM (1985) Lower Permian phylloid algal mounds, southern Tatum basin, southeastern New Mexico, USA In: Toomey DF, Nitecki MH (eds) Paleoalgology. Springer, Berlin Heidelberg New York Tokyo, pp 179–187

Davies GR, Nassichuk WW (1986) Ancient reefs in the high Arctic. GEOS 15, 4:1–5

Driggs AF (1977) The petrology of three Upper Permian bioherms, southern Tunisia. Brigham Young Univ Geol Stud 24:37–53

Duncan H (1957) Bryozoans. In: Ladd HS (ed) Treatise on marine ecology and paleoecology, vol 2: Paleoecology. Geol Soc Am Mem 67:783–800

Dunham RJ (1972) Capitan reef, New Mexico and Texas: facts and questions to aid interpretation and group discussion. Soc Econ Paleontol Mineral Permian Basin Sec Publ 72–14

Edwards DC, Riding R (1988) Permian reefs: aragonite cement or neomorphosed algal skeleton? In: 9th IAS Reg Meet Sedimentology, Leuven, Abstr, pp 64–65

Elliott GF (1955) The Permian calcareous alga *Gymnocodium*. Micropalaeontology 1, 1:83–91

Elliott GF (1958) Fossil microproblematica from the Middle East. Micropaleontology 4, 4:419–428

Elliott GF (1968) Permian to Palaeocene calcareous algae (Dasycladaceae) of the Middle East. Bull Brit Mus Nat Hist Geol Suppl 4:1–111, 24 pl

Emberger J (1958) Note préliminaire sur le faciès à algues du Permien du Djebel Tebaga (Sud Tunisien). CR Somm SG Fra 3:49–51

Endo R (1953) Stratigraphical and palaeontological studies of the later Palaeozoic calcareous algae in Japan VI: Saitama Univ Sci Rept Ser B 1, 2:97–104

Endo R (1954) Stratigraphical and paleontological studies of the later Paleozoic calcareous algae in Japan VIII: several species from Kinsho-zan, Akasaka-machi, Gifu-ken. Saitama Univ Sci Rep Ser B 1, 3:209–216

Endo R (1956) Stratigraphical and palaeontological studies of the later Palaeozoic calcareous algae in Japan X: Fossil algae from the Kwanto and Kitakami mountains: Saitama Univ Sci Rep Ser B 2, 2:221–248

Endo R (1957) Stratigraphical and palaeontological studies of the later Palaeozoic calcareous algae in Japan. XI: Fossil algae from the Taishaku district, Hiro-shima-ken, and Kitamino-Kuni, Hokkaido. Saitama Univ Sci Rep Ser B 2, 3:279–305

Endo R (1958) Recent stratigraphical and palaeontological studies of the younger Palaeozoic formations in Japan. In: 20th Congr Geology Int, Mexico City, 1956, Sec 7, pp 335–344

Endo R (1959) Stratigraphical and palaeontological studies of the later Palaeozoic calcareous algae

in Japan. XIV: Fossil algae from the Nyugawa Valley in the Hida Massif. Saitama Univ Sci Rep Ser B 3, 2:177–207
Endo R (1960) Stratigraphical and palaeontological studies of the later Palaeozoic calcareous algae in Japan. XV: A restudy of the genus *Epimastopora*. Saitama Univ Sci Rep Ser B 3, 3:267–270
Endo R (1961) Stratigraphical and palaeontological studies of the later Palaeozoic calcareous algae in Japan. XVII: Fossil algae from the Akiyoshi limestone group. Saitama Univ Sci Rep Ser B, Endo Commemorative vol, pp 119–142
Endo R (1969) Contributions to the geology and palaeontology of Southeast Asia. Fossil algae from Khao Phlong Phrab district in Thailand. Geol Paleontol SE Asia 7:35–85
Fan J-S, Ma X, Zhang Y-B, Zhang W (1982) The Upper Permian reefs in west Hubei, China. Facies 6:1–14
Flügel E (1966) Algen aus dem Perm der Karnischen Alpen. Carinthia II, 25:76 pp, 12 pl
Flügel E (ed) (1977) Environmental models for Upper Paleozoic benthic calcareous algal communities. In: Fossil algae, recent results and developments. Springer, Berlin Heidelberg New York, pp 314–343
Flügel E (1979) Paleoecology and microfacies of Permian, Triassic and Jurassic algal communities of platform and reef carbonates from the Alps. Bull Cent Rech Explor Prod Elf-Aquitaine 3, 2:569–587
Flügel E (1981a) Lower Permian *Tubiphytes/Archaeolithoporella* buildups in the southern Alps (Austria and Italy). In: Toomey DF (ed) European fossil reef models. Soc Econ Paleontol Mineral Spec Publ 30:143–160
Flügel E (1981b) Paleoecology and facies of Upper Triassic reefs in the Northern Calcareous Alps. In: Toomey DF (ed) European fossil reef models. Soc Econ Paleontol Mineral Spec Publ 30:291–359
Flügel E (1984) Permian reefs: evolution, structure and palaeoecology. In: Geister J, Herb R (eds) Géologie et paléoécologie des récifs. Inst Geol Univ Berne, pp 10.1–10.20
Flügel E (1985) Diversity and environments of Permian and Triassic dasycladacean algae. In: Toomey DF, Nitecki MH (eds) Paleoalgology. Springer, Berlin Heidelberg New York Tokyo, pp 344–351
Flügel E (1989) "Algen/Zement"-Riffe. Arch Lagerstättenforsch Geol B A 10:125–131
Flügel E, Flügel-Kahler E (1980) Algen aus den Kalken der Trogkofel-Schichten der Karnischen Alpen. Carinthia II, 36:113–182
Flügel E, Stanley GD Jr (1984) Reorganization, development, and evolution of post-Permian reefs and reef organisms. Paleontogr Am 54:177–186
Flügel E, Riedel P, Senowbari-Daryan B (1988) *Plexoramea cerebriformis* Mello, ein häufiges Mikrofossil in triadischen Flachwasserkalken: Alge oder Pilz? Mitt Ges Geol Bergbaustud Österr 34/35: pp 263–277
Glintzboeckel Ch, Rabaté J (1964) Microfaunes et microfacies du Permo-Carbonifère du sud Tunisien. Int Sediment Petrogr Ser 7:45 pp, 58 pl
Guo L, Riding R (1988) Late Permian reefs, Sichuan, China. In: Int Assoc Sediment 9th Reg Meet, Leuven, Abstr, p 88
Harland WB, Cox AV, Llewellyn PG, Pickton CAG, Smith AG, Walters R (1982) A geologic time scale. Univ Press, Cambridge, 131 pp
Heckel PH, Cocke JM (1969) Phylloid algal mound complexes in outcropping Upper Pennsylvanian rocks of mid-continent. Am Assoc Petrol Geol Bull 53:1058–1074
Herak M, Kochansky V (1960) Gymnocodiacean calcareous algae in the Permian of Yugoslavia. Geol Vjesn Zagreb 13:185–195
Herak M, Kochansky V (1963) Jungpaläozoische Kalkalgen aus dem Bükk-Gebirge (Nordungarn). Geol Hung Ser Palaeontol 28:45–77
Homann W (1972) Unter-und tiefmittelpermische Kalkalgen aus den Rattendorfer Schichten, dem Trogkofelkalk und dem Tressdorfer Kalk der Karnischen Alpen (Österreich). Senckenbergiana Lethaea 53, 3/4:135–313
James NP, Wray JL, Ginsburg RN (1988) Calcification of encrusting aragonitic algae (Peyssonneliaceae): implications for the origin of Late Paleozoic reefs and cements. J Sediment Petrol 58:291–303

Johnson JH (1942) Permian lime-secreting algae from the Guadalupe Mountains, New Mexico. Geol Soc Am Bull 53:195–226

Johnson JH (1946) Lime-secreting algae from the Pennsylvanian and Permian of Kansas. Geol Soc Am Bull 57, 12:1087–1120

Johnson JH (1951) Permian calcareous algae from the Apache Mountains, Texas. J Paleontol 25:21–30

Johnson JH (1963) Pennsylvanian and Permian algae. Col Sch Mines Q 58, 3:1–211

Johnson JH, Kaska HV (1965) Fossil algae from Guatemala. Col Sch Mines Prof Contrib 1:152 pp

Kochansky V, Herak M (1960) On the Carboniferous and Permian Dasycladaceae of Yugoslavia. Geol Vjesn Zagreb 13:65–94

Konishi K (1954) *Succodium*, a new codiacean genus, and its algal associates in the Late Permian Kuma Formation of southern Kyushu, Japan. J Fac Sci Univ Tokyo Sec 11, 9, 11:225–240

Konishi K (1959) Identity of algal *Tubiphytes* Maslov, 1956, and hydrozoan genus *Nigriporella* Rigby, 1958. Trans Proc Palaeontol Soc Jpn NS 35:142

Konishi K, Wray JL (1961) *Eugonophyllum*, a new Pennsylvanian and Permian algal genus. J Paleontol 35:659–666, 1 pl

Korde KB (1965) Algae. In: Rucenzev VE, Saryceva TG (eds) The development and alternation of marine organisms at the boundary between the Palaeozoic and the Mesozoic. Palaeontol Inst USSR Tr 108: 268–284; 414–429 (in Russian)

Korde KB (1967) The geological history of the earliest algae and their stratigraphical assemblages. In: Fossil algae of the USSR. Acad Sci USSR Siberian Sec, Inst Geol Geophys. Nauka, Moscow, pp 1–5 (in Russian)

Korde KB, Maslov VP (1963) Rhodophyta. In: Orlov Yu A (ed) Fundamental of palaeontology, algae, etc. USSR Acad Sci, Moscow, pp 243–261 (in Russian)

Kraus O, Ott E (1968) Eine ladinische Riff-Fauna in Dobratsch-Gipfelkalk (Kärnten, Österreich) und Bemerkungen zum Faziesvergleich von Nordalpen und Drauzug. Mitt Bayer Staatsamml Paläontol Hist Geol 8:263–290

Kulik EL, Korolyuk IK, Rauser-Chernousova DM (1978) On the stratigraphic and facial importance of Asselian and Sakmarian calcareous algae (the example of the Shakhtau bioherm massif, Bashkiria). USSR Acad Sci Quest Micropalaeont 21:216–233

Mamet B, Nassichuk W, Roux A (1979) Algaes et stratigraphie du Paléozoïque supérieur de l'Arctique Canadien. Bull Cent Rech Explor Prod Elf-Aquitaine 3, 2:669–683

Maslov VP (1956) Fossil calcareous algae of the USSR. Trans Acad Sci USSR Geol Inst 160:1–301 (in Russian)

Massieux M (1966) Les algues du nummulitique égyptien et des terrains crétacés-éocènes du quelques régions mésogéennes. Texte de Juliette Pfender revu et complété par M. Massieux, pt 1. Rev Micropaleontol 9:111–132, 5 pl

Mazzullo SJ, Cys JM (1979) Marine aragonite sea floor growths and cements in Permian phylloid algae mounds, Sacramento Mountains, New Mexico. J Sediment Petrol 49:917–937

Mu X (1981) Upper Permian calcareous algae from western Guizhou. Acta Palaeontol Sin 20, 2:33–49, 4 pls (in Chinese with English summary)

Mu X (1982) Some calcareous algae from Xizang. Palaeontology of Xizang. Sci Exped to the Qinghai-Xizang Plateau Ser, Chin Acad Sci Publ 5:205–240

Mu X (1984) A new Permian Dasycladales (calcareous alga) from Xizang, with a discussion on the genus *Goniolinopsis*. Acta Micropalaeontol Sin 1, 1:61–71 (in Chinese with English Abstr)

Mu X, Elliott GF (1982) *Nanjinoporella*, a new Permian dasyclad (calcareous algae) from Nanjing, China. Bull Brit Mus Nat Hist Geol 36, 2:109–116

Mu X, Riding R (1983) Silicified gymnocodiacean algae from the Permian of Nanjing, China. Palaeontology 26, 2:261–276

Nguyen LT (1970) Some Permian algae from Vietnam, Cambodia and Laos. Arch Geol Viet-Nam 13, 2:1–42

Nitecki MH (1972) The paleogeographic significance of receptaculitids. In: 24th Int Geol Congr, Montreal, Sec 7, pp 303–309

Noé SU (1987) Facies and paleogeography of the marine Upper Permian and of the Permian-Triassic boundary in the southern Alps (Bellerophon Formation, Tesero Horizon). Facies 16:89–142

Ogilvie-Gordon MM (1927) Das Grodener-Fassa und Enneberggebiet in den Südtiroler Dolomiten. Abh Geol Bundesanst Wien 24, 2:3–89

Otte C, Parks JM (1963) Fabric studies of Virgil and Wolfcamp bioherms, New Mexico. J Geol 71:380–396

Parona CF (1933) Le spugne della fauna permiana di Palazzo Adriano (Bacino del Sosio) in Sicilia. Mem Soc Geol Ital 1:58 pp

Peryt TM (1986) Zechstein *Stromaria* (= *Archaeolithoporella*) – cement reefs in Thuringia. N Jahrb Geol Paläontol Monatsh 1986/5:307–316

Pia J von (1937) Die wichtigsten Kalkalgen des Jungpaläozoikums und ihre geologische Bedeutung. CR 2e Congrès Carbonifère, vol 2, Heerlen, pp 765–856

Pisera A, Zawidzka K (1981) *Archaeolithoporella* from the Upper Permian reef limestones of the northern Caucasus. Bull Acad Pol Sci 29, 3:233–238

Praturlon A (1963) Dasycladaceae from the Upper Permian of the Dolomites. Geol Rom 2:119–150

Pray LC, Wray JL (1963) Porous algal facies (Pennsylvanian), Honaker Trail, San Juan Canyon, Utah. In: Shelf carbonates of the Paradox Basin. In: 4 Corners Geol Soc Symp, 4th Field Conf, pp 204–234

Rao SRN, Varma CP (1953) Permian algae from the middle *Productus* Beds. Palaeobotanist 2:19–23

Rauser-Chernousova DM (1950) Facies of the Upper Carboniferous and Artinskian deposits of the Sterlitamak-Ishimbaevo region of the Pre-Urals. Acad Sci USSR Inst Geol Sci, Works 119, Geol Ser 43:108 pp (in Russian)

Reid RP (1986) Discovery of Triassic phylloid algae: possible links with the Paleozoic. Can J Earth Sci 23:2068–2071

Reinhardt JW (1988) Uppermost Permian reefs and Permo-Triassic sedimentary facies from the southeastern margin of Sichuan Basin, China. Facies 18:231–288

Rezak R (1959) Permian algae from Saudi Arabia. J Palaeontol 33, 4:531–538

Riding R (1977) Problems of affinity in Palaeozoic calcareous algae. In: Flügel E (ed) Fossil algae, recent results and developments. Springer, Berlin Heidelberg New York, pp 202–211

Rigby JK (1957) Relationships between *Acanthocladia guadalupensis* and *Solenopora texana* and the bryozoan-algal consortium hypothesis. J Paleontol 31:603–606

Rigby JK (1958) Two new Upper Paleozoic hydrozoans. J Paleontol 32:583–586

Ross CA, Ross JRP (1979) Permian. In: Robison RA, Teichert C (eds) Treatise on invertebrate paleontology, pt A: Introduction. Geol Soc Am; and Kansas Univ Press Kansas, pp A291–A350

Roux A (1979) Révision du genre *Epimastopora* Pia, 1922 (Dasycladaceae). Bull Cent Rech Explor Prod Elf-Aquitaine 3, 2:803–810

Roux A (1985) Introduction à l'étude des Algues fossile paléozoïques (de la Bactérie à la tectonique des plaques). Bull Cent Rech Explor Prod Elf-Aquitaine 9, 2:465–699

Skompski S (1987) The dasycladacean nature of Late Paleozoic palaeoberesellid algae. Acta Geol Pol 37, 1-2:21–31

Smith AG, Hurley AM, Briden JC (1981) Phanerozoic paleocontinental world maps. Univ Press, Cambridge

Smith DB (1981) The Magnesian Limestone (Upper Permian) reef complex of northeastern England. In: Toomey DF (ed) European fossil reef models. Soc Econ Paleontol Mineral Spec Publ 30:161–186

Toomey DF (1976) Paleosynecology of a Permian plant dominated marine community. N Jahrb Geol Paläontol Abh 152, 1:1–18

Toomey DF, Cys JM (1977) Rock/biotic relationships of the Permian Tansill-Capitan facies exposed on the north side of the entrance to Dark Canyon, Guadalupe Mountains, southeastern New Mexico. In: Hileman ME, Mazzullo SJ (eds) Upper Guadalupian facies, Permian Reef Complex, Guadalupe Mountains New Mexico and West Texas. Soc Econ Paleontol Mineral Permian Basin Sec Publ 77-16:133–150

Toomey DF, Cys JM (1979) Community succession in small bioherms of algae and sponges in the Lower Permian of New Mexico. Lethaia 12:65–74

Toomey DF, Mitchell RW, Lowenstein TK (1988) "Algal biscuits" from the Lower Permian Herington/Krider limestones of southern Kansas – northern Oklahoma: paleoecology and paleodepositional setting. *Palaios* 3:285–297

Toomey DF, Lowenstein TK, Mitchell RW (1989) Reexamination of laminated osagid grains from a Lower Permian midcontinent limestone. *Palaios* 4:51–62

Twenhofel WH (1919) Precambrian and Carboniferous algal deposits. Am J Sci 4th Ser 48:339–352

Vachard D (1977) Etude stratigraphique et micropaléontologique (Alges et Foraminifères) du Viséan de la Montagne Noire (Hérault, France). Mem Inst Géol Univ Louvain 29:111–195

Vachard D (1980) Téthys et Gondwana au Paléozoique supérieur: les données afghanes; biostratigraphie, micropaléontologie, paléogéographie. Doc Trav IGAL Paris 2:463 pp, 35 pl

Vachard D (1985) Remarques sur les Dasycladales (algues vertes) du Permien supérieur du Jebel Tebaga. In: Trans 1st Cong Natl Sciences de la Terre, vol 1, Tunis 1981, pp 271–281

Vachard D, Montenat C (1981) Biostratigraphie, micropaléontologie et paléogéographie du Permien de la région de Tezak (Montagnes Centrales d'Afghanistan). Palaeontographica Ser B 178:1–88, 15 pl

Vachard D, Razgallah S (1988) Survie des genres *Tharama* et *Renalcis* (Epiphytales, algues prolématiques) dans le Permien supérieur du Djebel Tebaga (Tunisie). CR Acad Sci Paris Ser 2 306:1137–1140

Wahlman GP (1985) Lower Permian (Wolfcampian) *Archaeolithoporella-Tubiphytes*-sponge boundstones from the subsurface of West Texas. In: Toomey DF, Nitecki MH (eds) Paleoalgology. Springer, Berlin Heidelberg New York Tokyo, pp 208–215

Wahlman GP (1986) Lower Permian (Wolfcampian) boundstone reefs and the evolution of the Permian reef community. In: Soc Econ Paleontol Mineral Midcontin Sec, 4th Annu Meet, Abstr

Wray JL (1964) *Archaeolithophyllum*, an abundant calcareous alga in limestones of the Lansing Group (Pennsylvanian) southeastern Kansas. Kansas Geol Surv Bull 170:1–13

Wray JL (1977) Calcareous algae. Developments in paleontology and stratigraphy, vol 4. Elsevier, Amsterdam, 185 pp

Wray JL, James NP, Ginsburg RN (1975) The puzzling Paleozoic phylloid algae: a Holocene answer in squamariacean calcareous red algae. In: Am Assoc Petrol Geol, vol 3, Annu Mtg Abstr, pp 82–83

Zhang G, Zhen C (1985) Some calcareous algae from the Lower Permian carbonate platform of Guizhou and their distribution characteristics. In: Fossil Algae Committ Palaeontol Soc China (ed) Select Pap 1st Nat Fossil algae Symp. Geology Press, Beijing, pp 41–50 (in Chinese)

Chapter 21

Triassic and Jurassic Marine Calcareous Algae: A Critical Review

E. FLÜGEL[1]

Abstract

The information on benthic Triassic and Jurassic marine algae is scattered in about 1600 articles, most of which have been published during the past two decades. The Triassic algal floras yield porostromate "blue-green algae", udoteacean and dasycladacean green algae as well as solenoporacean and gymnocodiacean red algae. These groups also occur in the Jurassic, except for the gymnocodiaceans; in addition the first corallinaceans are recorded from the Upper Jurassic. About 50 genera and 200 species have been described from the Triassic, and about 60 genera and 130 species from the Jurassic.

Lower Triassic algae (except spongiostromates) are lacking on a global scale. The extinction rate near the Permian/Triassic boundary is 56% for dasycladacean genera and 53% for all algal genera. Because of an absence of originations and high extinction rates at the end of the Middle Permian, Upper Permian algal floras are characterized by low-diverse assemblages consisting predominantly of gymnocodiaceans and some dasycladaceans. Most dasycladaceans became extinct prior to gymnocodiaceans (as demonstrated by Late Permian floras from the Southern Alps and China), probably because gymnocodiaceans were better adapted to fluctuating salinity conditions during periods of regression.

The first Triassic skeletal algae are recorded from the early Middle Anisian of the western Tethys. Reef associations composed of encrusting spongiostromates, porostromates and solenoporaceans can be compared with Upper Permian algal associations described from "algal/cement reefs". Anisian lagoonal and platform environments are characterized by high-diverse dasycladacean floras which indicate a rapid post-Permian diversification. Diversity decreases from the Anisian to the Upper Triassic. Taxonomic turnover is high during Middle Triassic and Late Triassic but low at the Ladinian/Carnian boundary. Decreasing diversity is also evident within the four dasycladacean assemblages characteristic for Anisian to Rhaetian time intervals.

Liassic and early Middle Jurassic dasycladacean floras appear to be low-diverse subsequent to the Hettangiian-Sinemurian, but the information about Upper Liassic and Middle Jurassic algae is scanty. A significant increase in diversity during the Upper Jurassic reflects diversification during the Kimmeridgian and Portlandian (Tithonian).

Most Triassic and Jurassic time intervals are characterized by high origination rates at species level. Extinction rates are high during the Triassic and Liassic, low during the Middle Jurassic and increase significantly during the Upper Jurassic. Major extinction events, noted for animal biota, can be recognized also on generic and partly on specific level for the dasycladacean algae.

Dasycladaceans and udoteaceans exhibit a Tethyan, predominantly pan-tropical dispersal pattern since the Middle Triassic. Isocrymal distributional patterns have been proved for Jurassic algae but might have also existed in the Triassic (indicated by Middle Triassic non-Tethyan dasycladacean floras and perhaps also by the first Triassic dasycladaceans found in western North America).

Utilization of Triassic and Jurassic algae in biostratigraphy is restricted to dasycladaceans and strongly affected by environmentally controlled distribution patterns. Algal assemblages, rather than the time-range of species, enable the recognition of time intervals which might correspond to several ammonite or conodont zones in the Triassic. Jurassic platform and platform-edge car-

[1] Institut für Paläontologie, Universität Erlangen, Loewenichstraße 28, 8520 Erlangen, FRG

bonates can be subdivided into facies-controlled time units ("coenozones") which only partly might be regarded as synchronous within a global context. The stratigraphical range of Upper Jurassic dasycladaceans is relatively short for more than 50% of the species, especially during the Kimmeridgian and Portlandian.

The study of Triassic and Jurassic algae offers new insights into algae/sediment interactions, such as accumulation of bioclastic carbonate sands by halimediform green algae, evolution of spongiostromate and porostromate oncoids, together with the as yet not well understood significance of microbial crusts for the formation of carbonate buildups.

1 Introduction

This paper provides a summary of the current state of research in the field of marine Triassic and Jurassic calcareous algae, together with an assessment of future avenues of investigation. Not included are non-benthic as well as fresh-water algae, and the controversial information about evolutionary trends, e.g. of dasycladaceans (see Kamptner 1958; Herak et al. 1977). The review starts with a discussion of the extent and significance of the data base (number of taxa, taxonomic problems), followed by a discussion of taxonomic diversity which might be used to elucidate extinction and origination rates of Triassic and Jurassic algae. Despite the continued expansion of our knowledge on systematics and morphology there are still obstacles with regard to the relations between Late Permian and Triassic algae. The development of Upper Permian algal floras, therefore, is also considered. Speculations about the causes of changes in the composition of algal floras during time are only peripherally discussed. The groups dealt with here comprise "blue-green algae" (or cyanobacteria) including spongiostromate and porostromate taxa (which might include green algae as well, see Roux 1985); green algae (udoteaceans, dasycladaceans) and red algae (solenoporaceans, gymnocodiaceans, corallinaceans). Mesozoic phylloid algae (squamariaceans?) are only known from a unique occurrence in the Late Triassic of Northwestern Canada (Reid 1983).

The information on Triassic and Jurassic calcareous algae is scattered through about 1600 papers, published between 1820 and today. Sixty percent of the articles on Triassic algae and about 80% of the information about Jurassic (and Cretaceous) algae have been published during the last three decades. This situation is comparable with the strong increase in the knowledge of many plant protist groups (Tappan 1980). The increase in the number of "new" species generally depends on the efforts of only a few scientists as indicated by the distinct increase in the number of Triassic algal taxa in the years between 1910 and 1940, predominantly due to the fundamental work of Julius von Pia, and by the peak between 1960 and 1980, caused by Jan Bystricky's excellent contributions, and by the work of Milan Herak, Ernst Ott and new investigations of Triassic reef carbonates (Fig. 1). The striking expansion of our knowledge of Jurassic and Cretaceous algae since 1960 (Fig. 2) originates from the enormous work of the "Groupe française d'étude des algues fossiles" and from the contributions by Harlan Johnson and Graham F. Elliott. The common use of Jurassic and Triassic algae in microfacies studies and paleoecological analysis is

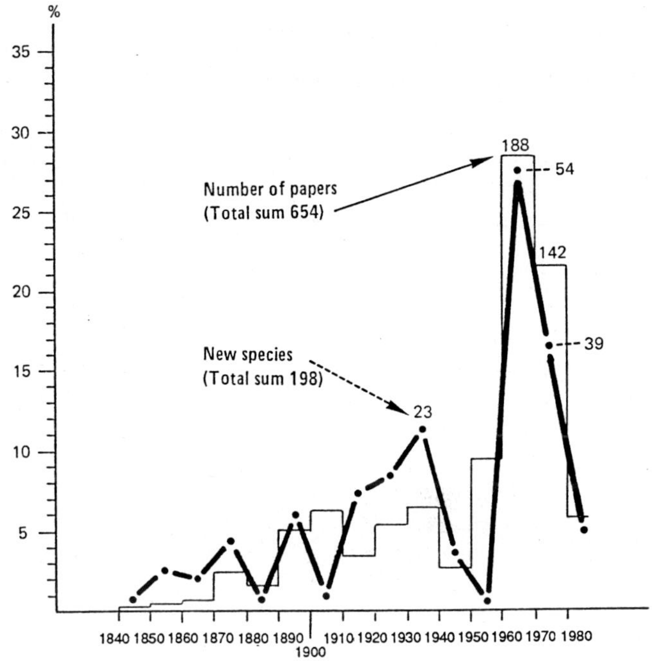

Fig. 1. Interest in the study of Triassic algae expressed by the number of relevant papers and by the number of species proposed within a decade. Source: Emberger (1979) and additions

another reason for the increase in algal literature. Excellent bibliographies are available for Triassic algae (Emberger 1979), Jurassic and Cretaceous dasycladaceans (Bassoullet et al. 1978) and udoteacean green algae (Bassoullet et al. 1983). Much of the following review is based on the data gathered in these critical compilations.

2 Data Base

The number of genera and species proposed for Triassic and Jurassic algae is tabulated in Table 1. A taxonomic census is always difficult because of the different state of the taxonomic concepts used and also because of the possibility that important papers have been overlooked. Table 1 is based on probably more than 90% of the relevant literature and, therefore, should receive a high degree of confidence with regard to the completeness of the data. Problematical or poor taxonomic concepts, however, will strongly influence the conclusive results of the census. Taxonomy and classification of Mesozoic solenoporacean red algae and especially of porostromate algae are highly artificial. Porostromate classification depends on the possibility of recognizing inter- and intraspecific morphological variability (Dragastan 1985), of differentiating various fos-

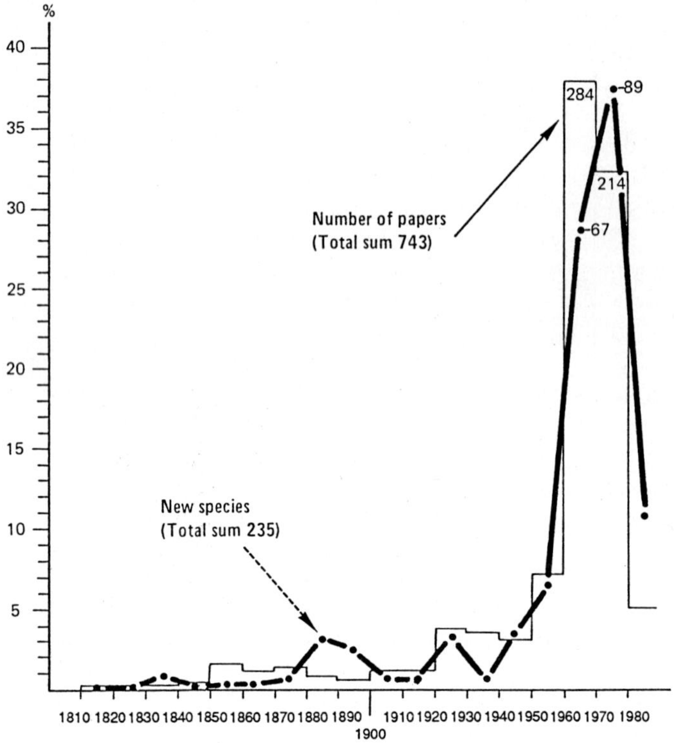

Fig. 2. Interest in the study of Jurassic and Cretaceous dasycladacean algae as shown by the number of relevant papers and by the number of new species proposed per decade. Source: Bassoullet et al. (1978) and additions

Table 1. Number of genera and species attributed to the major algal groups occurring in the Triassic and Jurassic[a]

		Triassic	Jurassic
	Genera	12	8
Porostromata	Species	36	24
	Genera	4	4
Udoteaceae	Species	4	7
	Genera	25	44
Dasycladaceae	Species	136	82
	Genera	5	4
Solonoporaceae	Species	19	12
	Genera	3	–
Gymnocodiaceae	Species	3	–
	Genera	–	2
Corallinaceae	Species	–	2

[a] The table is based on probably more than 90% of the relevant literature. Because of taxonomic problems the numbers of the porostromate algae are less confident than the data of the other algal groups.

silization patterns (Leinfelder 1985), and of finding Recent counterparts (Schäfer and Senowbari-Daryan 1983). Difficulties arising from these obstacles as well as from varying calcification patterns may even influence the taxonomy of fossil algae having Recent descendants such as dasycladacean and udoteacean green algae (Bassoullet et al. 1975, 1983, 1984; Deloffre 1988). The species concept of fossil green algae is still a matter of discussion. A biometrical approach, as exemplified for some Triassic and Jurassic dasycladaceans (Hurka 1967; Remane 1969; Zorn 1974, 1977; Bystricky 1984), might assist us to overcome some of these difficulties.

There is considerable discussion concerning whether counting and tabulating palaeontological taxa will provide sufficient information for recognizing origination and diversification phases (see Valentine 1985), because of the discrepancies between true biological species and "species" created on minor morphological differences or even on different preservation or ontogenetic stages of one and the same biological species. A census of Triassic and Jurassic dasycladacean species, therefore, can only reflect very rough trends of the overall morphological variability within the whole group. Because a modern revision of Triassic and Jurassic dasycladacean algae is lacking, the recognition of these trends (e.g. times of pronounced extinctions) is certainly based on a mixture of well-defined morphological species and questionable "species" which cannot be used for taxonomic comparisons because of insufficient definitions. The elimination of those species from the census, however, does not strictly influence the results shown in Figs. 5 and 6, probably because most of these questionable "species" have been described from the Middle Triassic and Upper Jurassic substages which, even after the subtraction of these taxa (about 20% of the total number of species, both in the Middle Triassic and the Upper Jurassic), represent high-diverse time units.

Most Phanerozoic diversity patterns have been described in terms of generic diversity or higher taxa diversity. Doing this with dasycladacean algae would result in a distinctly different pattern for diversity and origination as compared with the pattern indicated by species number. The substages of the Middle and Upper Triassic are characterized by corresponding numbers of genera, the number of new dasycladacean genera per substage remaining more or less constant. The number of genera increases during the Lower Jurassic, seems to be lower in the Middle Jurassic, and has a distinct peak in the Upper Jurassic (31 genera!). The origination rates are low during the Triassic (about 20 to 30%) and Liassic (31%) but high for the Middle Jurassic and Upper Jurassic (more than 80%). One reason for the difference between generic and specific diversity is the more intensive study of Upper Jurassic dasycladacean algae as compared with Triassic algae. The existence of time-dependent species groups (associations, Ott 1972b), characterized by morphologically related species, points to a rather broad generic concept applied to Triassic dasycladaceans.

The distribution of the major algal groups in time is shown in Fig. 3. All Triassic algal groups already existed in the Late Paleozoic. Jurassic groups continue into the Cretaceous. Dasycladaceans exhibit a high taxonomic coincidence of uppermost Jurassic and lowermost Cretaceous species; about 44% of the species known from the Tithonian continue to the Berriasian. A similar

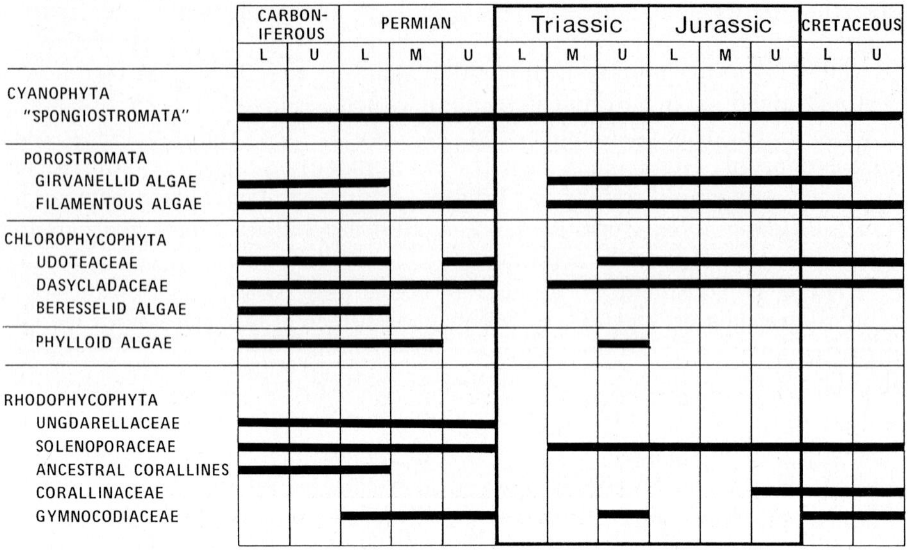

Fig. 3. Stratigraphical range of algal groups occurring in the Triassic and Jurassic. Note the gap in the paleontological record in the Lower Triassic, only non-skeletal algae are known from this interval. Phylloid algae are represented by a unique occurrence only

pattern can be recognized within the porostromate algae, if one accept the present classification of this group. Of special interest are the gaps in the record of the gymnocodiacean red algae; rare Triassic species have been described recently (H.W. Flügel 1971; Senowbari-Daryan and Schäfer 1980; Schäfer and Senowbari-Daryan 1983).

2.1 Relations Between Permian and Triassic Algae

A logical outcome of the data summarized in Fig. 3 is the extent to which Late Permian potential precursors of Triassic algae have been affected by the biotic crisis near the Permian/Triassic boundary. Recent discussions of the Phanerozoic diversity patterns do not consider calcareous algae (Valentine 1985). For invertebrates and protozoans a Late Permian extinction rate of 83% on generic level and on 94 to 96% on specific level is estimated (Sepkoski 1986). The extinction rates for calcareous algae seem to be different.

A survey of Late Permian algal floras from the southern Alps, Transcaucasia, Central Afghanistan, and China (Fig. 4) indicates a distinct change in diversity and taxonomic composition during time (Table 2). Middle Permian dasycladacean associations are characterized by high generic and specific diversity. Twenty-four genera are known from the Middle Permian *Neoschwagerina* zone, but only 13 genera from the Upper Permian. About 58% of the Middle Permian genera became extinct prior to the Upper Permian. Nearly all

Late Permian dasycladacean genera are already known from the Middle Permian. There is a significant lack of taxonomic originations. Algal limestones from the Late Permian *Palaeofusulina* zone yield 9 genera and about 14 species of dasycladaceans. At least 5 genera should have crossed the Permian/Triassic boundary (*Diplopora*, *Epimastopora*, *Gyroporella*, *Macroporella*, *Physoporella*) because they occur in Anisian limestones again. These recurring taxa suffered a distinct reduction in species diversity and population size during the late Permian. This is indicated by the development of algal floras within Late Permian sections in the Southern Alps (Bellerophon and Tesero Formations, Noë 1987) and in China (Wuchiapingian and Changhsiangian; Mu 1981; Rui et al. 1984; Reinhardt 1987): (1) both in the southern Alps and in China algal diversity decreases near the top of the sections; (2) dasycladaceans (with the exception of *Mizzia*) disappear prior to gymnocodiacean red algae which predominate in most Late Permian algal floras; (3) the algal floras are generally low-diversity; (4) *Gymnocodium bellerophontis*, *Mizzia* and *Atractyliopsis* are euryhaline taxa (Noe 1987) in contrast to *Permocalculus tenellus*, *Macroporella*, *Vermiporella* and *Solenopora* which seem to be adapted to normal or only slightly changing salinities.

The distribution of calcareous algae at the end of the Permian period, therefore, might have been strongly controlled by salinity fluctuations which, in the Southern Alps and probably also in Guizhou caused a significant reduction of dasycladacean diversity. Salinity control is evident for Uppermost Permian algae of Sichuan (Reinhardt 1987).

A strong salinity control of dasycladacean distribution might also be inferred from a comparison of algal associations described from Middle to Upper Permian sections in the Transcaucasian region (Korde 1965): from the Gnishinskian to the Kashikian and to the Dzhulfian (see Fig. 4) species diversity of algae decreases from 25 to 11 to 5, no algae are known from the Dorashamian of this region. Dasycladaceans disappear prior to gymnocodiaceans which are common in the Gnishikian part of the section (about 54% of the algal species) but rare in the Kashikian (27%) and Dzhulfian (20%). Low-diversity algal assemblages occurring in alternation with higher diverse floras, as well as the abundance of gymnocodiaceans within the low-diversity assemblages, might be explained by salinity changes.

The extinction rate at the Permian/Triassic boundary is 56% with regard to dasycladacean genera, and about 53% considering all algal genera known from the uppermost Permian. Algae living in reef environments seem to have been influenced by extinction events less than lagoonal and platform associations. Encrusting algae (*Archaeolithoporella*, *Tubiphytes*) are effective in building "algal/cement reefs" during the Late Permian (Pisera and Zawidzka 1981; Peryt 1986) as well as during the Triassic (Flügel et al. 1984; Flügel 1989; Martin and Braga 1987).

The problem of survivorship of Permian algae and of the post-Permian diversification, indicated by the extremely high species diversity of Anisian algae, becomes even more complicated if possible phylogenetic lines are taken into consideration. Kochansky-Devidé and Gusić (1971) have recognized that

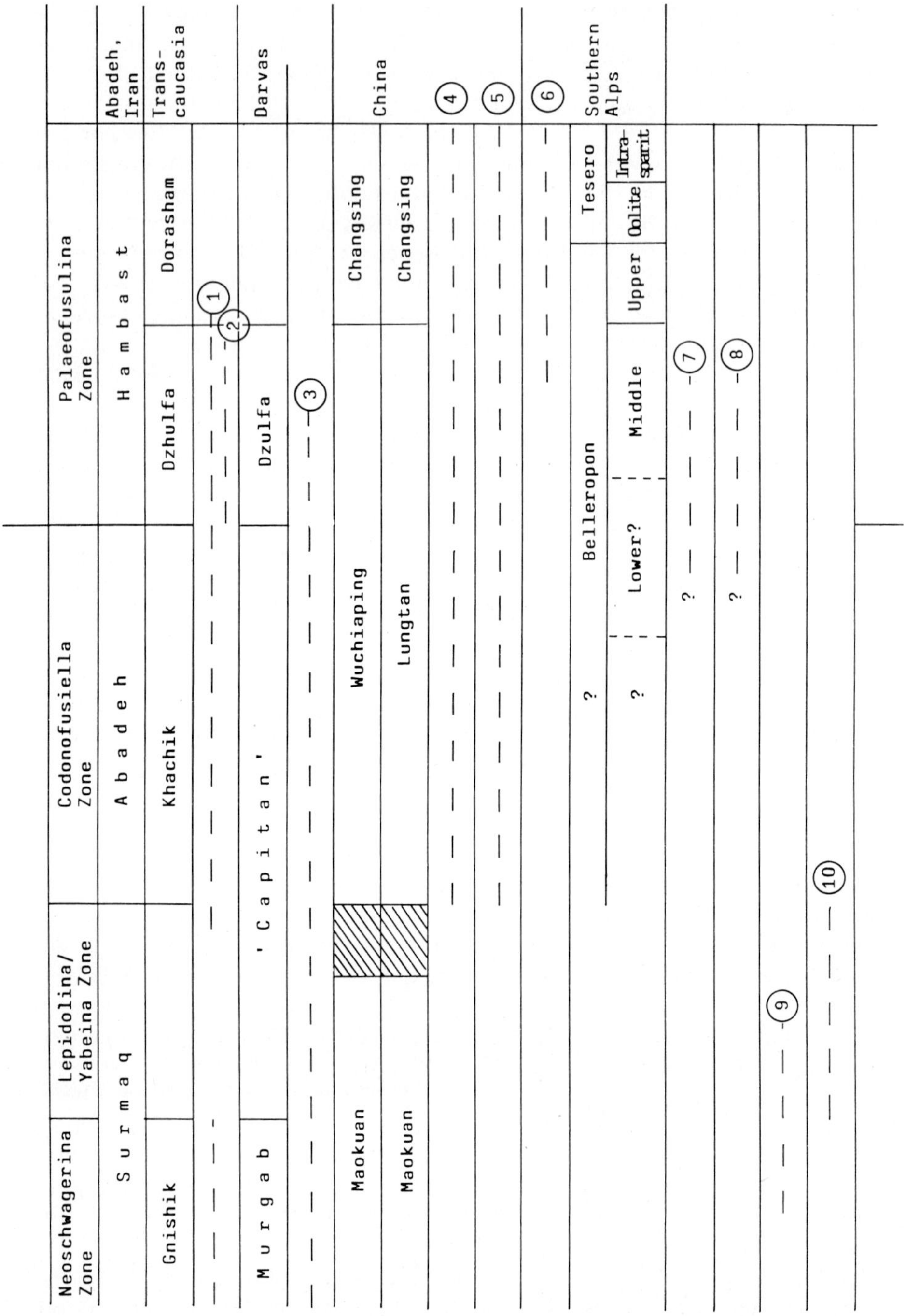
Neoschwagerina Zone
Lepidolina/ Yabeina Zone
Codonofusiella Zone
Palaeofusulina Zone
Abadeh, Iran
S u r m a q
A b a d e h
H a m b a s t
Trans-caucasia
Gnishik
Khachik
Dzhulfa
Dorasham
1
2
Darvas
M u r g a b
' C a p i t a n '
Dzulfa
3
China
Maokuan
Wuchiaping
Changsing
Maokuan
Lungtan
Changsing
4
5
6
Southern Alps
Bellerophon
Tesero
?
Lower?
Middle
Upper
Oolite
Intra-sparit
7
8
9
10

Table 2. Stratigraphical range of Middle Permian and Upper Permian dasycladacean algal genera[a]

	Middle Permian Neoschwagerina Zone	Upper Permian Lepidolina/ Yabeina Zone	Codonofosiella Zone	Palaeofusulina Zone
Anthracoporella	×			
Atractyliopsis	×			×
Clavaphysoporella	×			
Clavaporella	×			
Diplopora	×			×
Endoina	×			
Eoclypeina	×	×		
Eogoniolina	×			
Eovelebitella	×		×	
Epimastopora	×		×	×
Gyroporella	×	×		×
Imperiella	×			
Johnsonia	×			
Kochanskiella	×			
Macroporella	×		×	×
Mizzia	×	×	×	×
Pseudoepimastopora	×	×		
Pseudogyroporella	×			
Pseudovermiporella			×	×
Salopekiella	×			
Uragiellopsis	×			
Velebitella	×	×		
Vermiporella	×			×
Physoporella	×			×
Zaporella	×			

[a] Note the difference in taxonomic diversity between Middle and Upper Permian algae. At least five genera known from the uppermost Permian (*Palaeofusulina* zone) re-appear in the Anisian.

Fig. 4. Distribution of Middle and Upper Permian algal floras used in the discussion of the relations between Permian and Triassic algae. Sources: *1* Korde (1965); *2* Pisera and Zawidzka (1981), northern Caucasus; *3* Vachard and Montenat (1981), central Afghanistan; *4* Mu (1981); southern Guiszou, China; *5* Zhao et al. (1981), Shangxi, China; *6* Noé (1987), Southern Alps; *7* Kochansky (1954), Montenegro; *8* Herak and Kochansky (1963), Bükk Mountains, Hungary; *9* Termier et al. (1977), Djebel Tebaga, Tunisia; *10* Johnson and Danner (1966), Marble Canyon Formation, North America most morphological innovations characteristic of Triassic morphotypes evolved during the Late Palaeozoic and not in the Triassic. This includes also the development of metaspondylity in dasycladaceans (Güvenç 1979).

Lower Triassic algae are lacking on a global scale, with the exception of non-skeletal spongiostromate algae which seem to be common in lowermost Triassic environments (Bissell 1970; Taraz et al. 1981; Sheng et al. 1984; Noé 1987). Boring algae responsible for the formation of coated grains in many Scythian calcarenites are also common (e.g. Yun et al. 1983). Spongiostromate algae might have been even effective in the buildup of Scythian reef structures (Nazarevya et al. 1986).

There is no clear explanation for this gap in the paleontological record of calcareous algae in the Triassic. The causes of this worldwide phenomenon have been discussed in terms of rapidly changing transgression/regression patterns, salinity crisis, and reductions in the size of shallow-marine shelf environments. A change in skeletal mineralogy or in the degree of calcification (as discussed for some invertebrate groups; Mistiaen 1984; Cuif et al. 1987) at the Permian/Triassic boundary might be another explanation. Interestingly enough, the first algal associations of Triassic age consist only of algae which used calcite or Mg-calcite. Aragonitic dasycladacean green algae seem to appear within the Anisian after a short time lag. A change in the calcification potential due to non-tropical settings of Lower Triassic algal biotopes would perhaps also explain the lack in the palaeontological record.

2.2 Diversity Patterns of Triassic and Jurassic Algae

Summary of the species diversity of Triassic and Jurassic dasycladacean algae (Fig. 5) is based on a total of 253 species, 138 of which are Triassic, and 115 Jurassic in age. "Subspecies", described especially from the Middle Triassic, are counted as species. In addition, the number of species occurring in consecutive time units is indicated. Several conclusions concerning evolutionary trends (as well as the different state of knowledge) can be derived from Fig. 5:

(1) Dasycladacean diversity is highest at the beginning of the Mesozoic, reflecting a strong post-Permian, probably Anisian, taxonomic diversification proceeding from only a few Permian survivors; (2) diversity decreases from the Anisian to the Upper Triassic, although four distinct algal floras can be distinguished in this time (Ott 1972b) which do not coincide with a change in sedimentary facies and might, therefore, represent evolutionary innovation which declined towards the end of Triassic: the Anisian *Physoporella-Oligoporella* assemblage comprises the largest number of species, followed by the uppermost Anisian and Ladinian *Diplopora annulata-Teutloporella nodosa* assemblage. Even lower is the diversity of the Carnian *Poikiloporella duplicata-Clypeina besici* assemblage and of the Norian and Rhaetian *Diplopora phanerospora-Heteroporella* assemblage; (3) the number of species continuing to subsequent time units is high in the Middle Triassic but low in the Upper

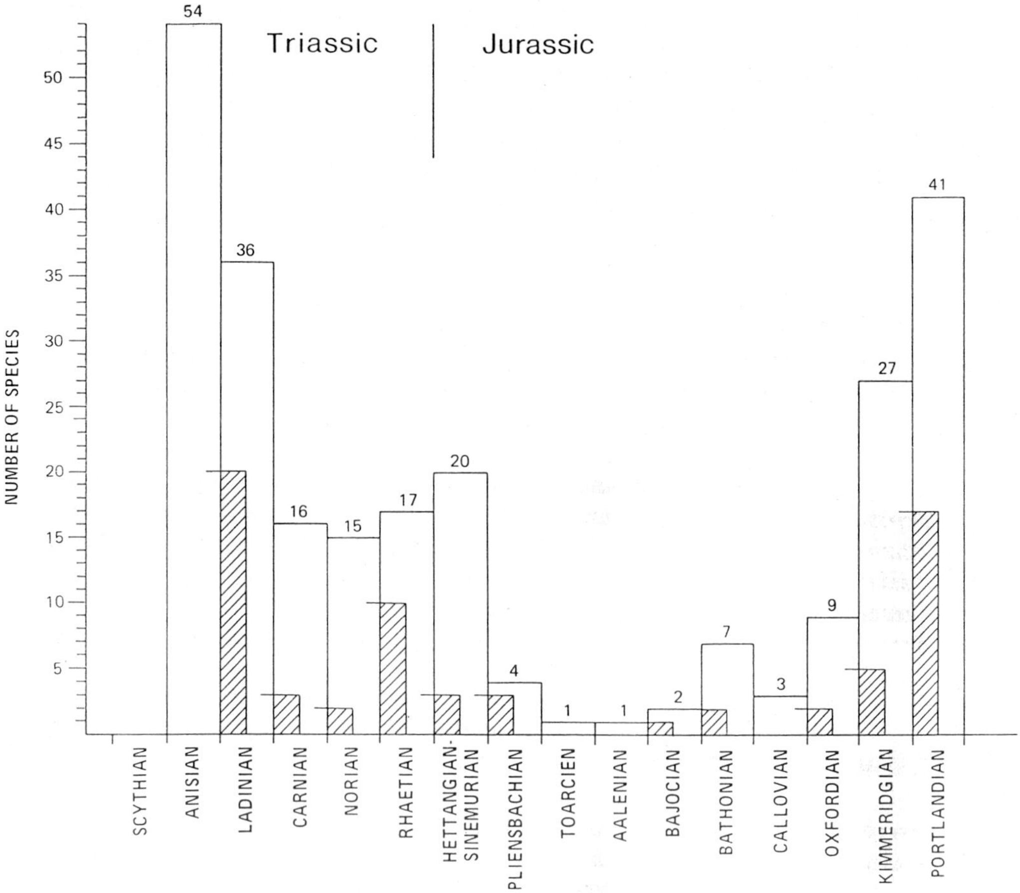

Fig. 5. Diversity of Triassic and Jurassic dasycladacean species. *Hatched* areas indicate the number of taxonomic turnovers. The figure reflects changing diversities but also gaps in our information (e.g. Lower and Middle Jurassic)

Triassic, except for the Norian-Rhaetian interval; (4) Liassic dasycladacean floras are characterized by a strong decrease in species diversity after the Hettanginian-Sinemurian. This pattern as well as the apparent low diversity of early Middle Jurassic floras should not be overrated because of the very poor information available for these time intervals. The very first dasycladaceans from the Toarcian have just been described (Cubaynes and Deloffre 1984); (5) Bathonian to Portlandian dasycladaceans are studied in much more detail. The increase in species diversity, recognizable for the Upper Jurassic, should, therefore, reflect new diversification phases during Kimmeridgian and Portlandian. This is also indicated by the low taxonomic turnover from the Oxfordian to the Kimmeridgian.

Diversity patterns of other algal groups are difficult to determine because of classification problems (porostromate algae) and because of the poor record (red algae). Mesozoic udoteacean green algae appear in the Upper Triassic (Flügel 1987, 1989b) but do not become highly diverse until the Lower Cretaceous (see Fig. 7). Both *Boueina* and *Arabicodium* are minor components of Upper Triassic and Jurassic algal assemblages.

2.3 Extinction and Origination Rates

Most Triassic and Jurassic time intervals exhibit high origination rates (Fig. 6). Low origination rates are only known for Rhaetian and Pliensbachian dasycladacean species, but the latter may be an artifact. The Bajocian and the Callovian origination rates are about 50%, all other origination rates are considerably higher.

The extinction rates are high in most parts of the Triassic and Liassic, low in the Middle Jurassic and increasing during the Upper Jurassic. Major generic extinction events recognized from animal biota (Sepkoski 1986) are also reflected by dasycladacean extinction rates: the Carnian and the Late Norian peak as well as the Pliensbachian/Toarcian peak are represented. The Late Tithonian peak (caused by the extinction of marine invertebrates as well as dinoflagellate algae) is even more significant for dasycladaceans than for other biota if the species level is considered. At generic level the Upper Jurassic origination rate for dasycladaceans is extremely high (approximately 80%).

2.4 Tethyan Dispersal Patterns

The first Triassic algal assemblages, similar in taxonomic composition to some Late Permian reef assemblages (see above), appear in the early Middle Anisian of the Southern Alps. Dasycladacean assemblages became common during the Anisian and flourished in the tropical settings of the western Tethyan region (Southern Spain, Alps, Carpathian Mountains, Yugoslavia, Hungary, Bulgaria, Romania, western Turkey) but some occurrences have also been reported outside the Tethys north of the Tethyan belt (e.g. Poland; Gadzicki and Kowalski 1974). The distribution of Ladinian algae corresponds more or less to that of Anisian algal assemblages. Carnian dasycladacean algae exhibit a somewhat extended distribution including occurrences in northern Tibet (Flügel and Mu 1982). Norian and Rhaetian algae known from the western Tethyan region as well from Asia. Recent discoveries have proved the existence of Upper Triassic dasycladaceans also in western Northern America (Flügel et al. 1989).

Elliott (1977, 1981, 1984) stressed the fact that aragonitic green algae (dasycladaceans and udoteaceans) have proliferated since Triassic times within the warm-water Tethyan sea, whereas subtropical tolerance within non-Tethyan settings has developed only for a minority of Triassic and Middle Jurassic

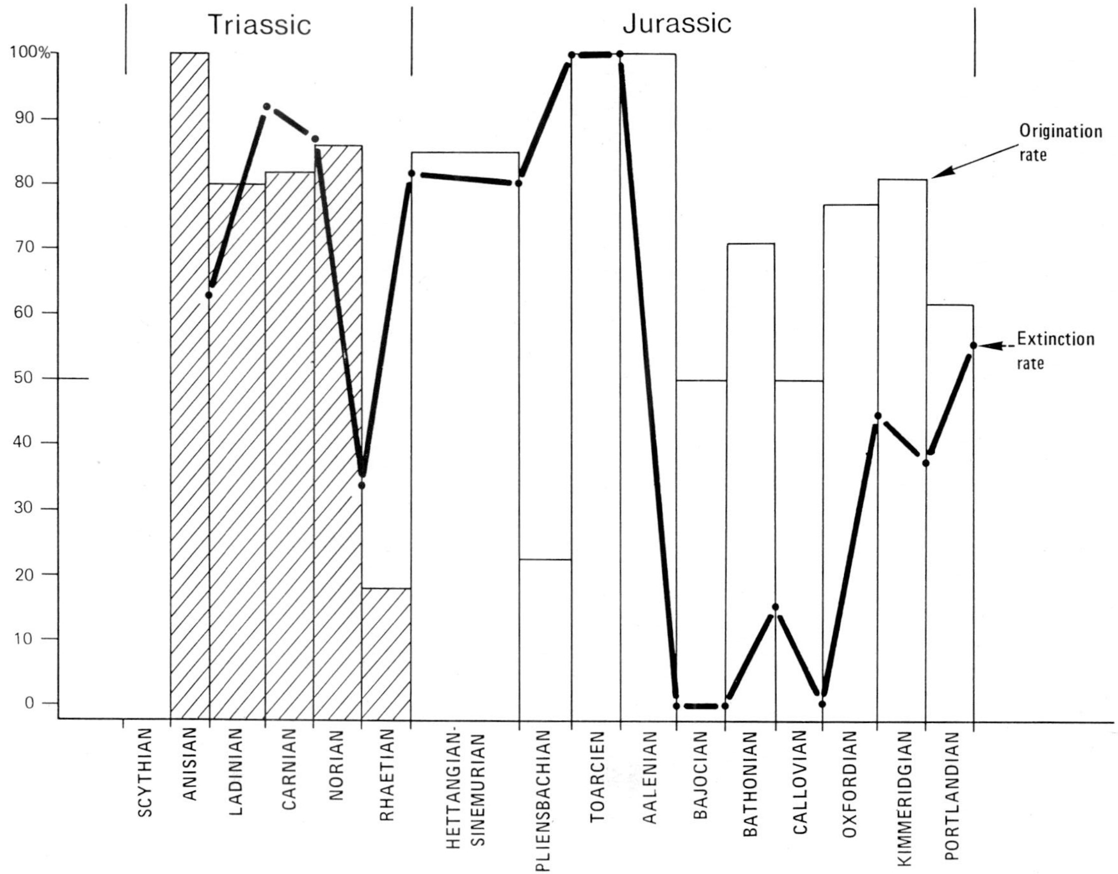

Fig. 6. Origination rates (percentage of species originating within a time interval) and extinction rates (percentages of species becoming extinct at boundaries between time units) of Triassic and Jurassic dasycladacean algae. The figure is based on a total of 253 species. High Liassic extinction rates may be due to insufficient information

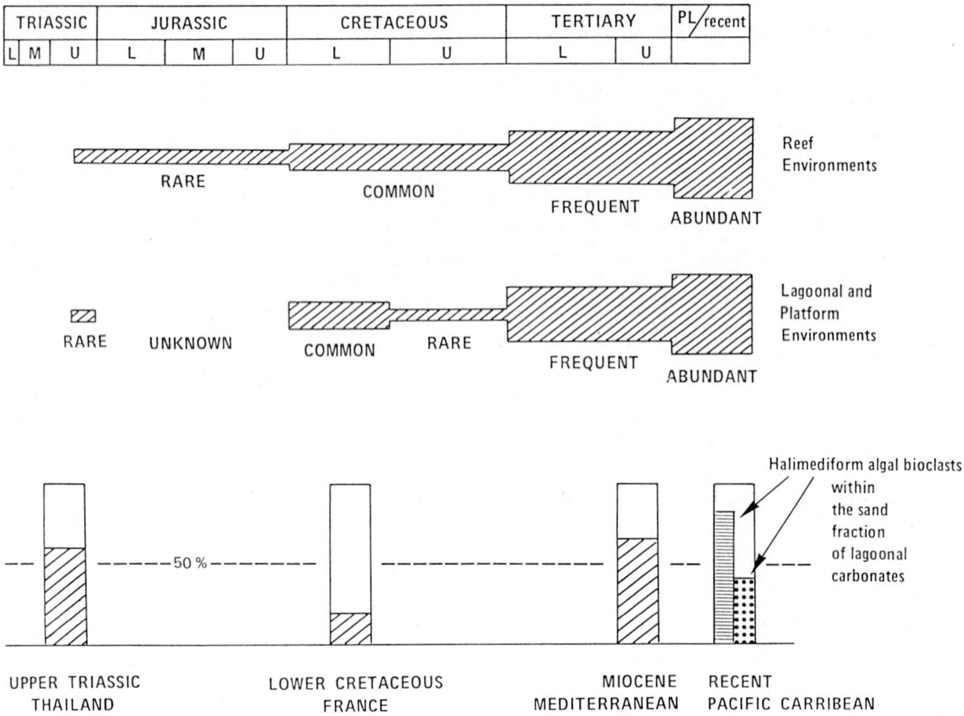

Fig. 7. Environmental distribution and sedimentary significance of halimediform udoteacean green algae. Various sources (see Flügel 1988)

taxa. There is a strong evidence for isocrymal distribution of dasycladacean algae in Europe, north Africa and southwestern Asia, at least since Liassic times (Elliott 1977). Water temperature and, probably, also substrate conditions (Flügel 1985) are regarded as principal factors governing the regional distribution of early and middle Mesozoic green algae.

3 Biostratigraphical Importance

Occurrence and distribution of ancient calcareous algae is strongly facies-controlled (Wray 1977). Utilization of Triassic and Jurassic algae in biostratigraphy is restricted to the Dasycladaceae. Pia had a very optimistic view concerning the value of dasycladaceans as Triassic index fossils (see Pia 1943). Although many of the algal samples studied by Pia lacked stratigraphical control, he recognized the general successional patterns of the dasycladacean assemblages fairly correctly. A better understanding of the influence of facies changes, and the study of algae from geological sections in the Alps, Carpathians and the Dinarides, led to some modifications concerning the time range of species (Bystricky 1964, Herak 1965). Ott (1972a,b) proposed a revised algal biochronology based on four dasycladacean assemblages occurring in vertical successions and separated by two major floral changes (within the uppermost Anisian and near the Ladinian/Carnian boundary). The absolute time-range of these units is rather long and embraces between two and about six ammonite zones, and two to five conodont zones, established for the Triassic. The practicability of this zonation, especially in lagoonal and backreef carbonates poor in macrofossils has been, however, confirmed by recent studies of Middle and Upper Triassic platform and reef carbonates (Fois 1979; Fois and Jadoul 1983; Dragastan 1981).

Many attempts have been made to subdivide Jurassic platform and platform-edge carbonates by the occurrence and/or dominance of particular calcareous algae (and benthic foraminifera). Starting with the fundamental papers by Sartoni and Crescenti (1962), Praturlon (1969), Farinacci and Radoičić (1964), Radoičić (1966), Crescenti (1971) and Ramalhao (1971), many studies of the zonation of Jurassic and Cretaceous platform carbonates have been published (e.g. Chiocchini et al. 1979; Peybernes 1979) most of which are critically summarized by Jaffrezo (1980), who provided a biozonation for the Jurassic and Lower Cretaceous of the Mesogean area. Because shallow-marine facies belts are easily affected by sea level fluctuations, which might change the position of the facies belts in time, the boundaries between the "coenozones" based on calcareous algae can hardly be regarded as synchronous time lines.

The stratigraphical range of Jurassic dasycladacean species varies, as exemplified by Upper Jurassic species; about 54% of the species seem to be rather shortlived (restricted to one substage, e.g. to the Kimmeridgian), about 20% of the species are known from two substages. The remaining species appear to be longlived, but this might be partly due to inconsistency of taxonomic determinations (e.g. *Salpingoporella annulata*, see Bassoullet et al. 1978).

The biostratigraphical value of Triassic and Jurassic calcareous algae is strongly restricted by the existence of environmentally controlled distributional patterns known from many Middle and Upper Triassic platform and reef carbonates (e.g. Ott 1967, 1972c; Flügel 1979; Senowbari-Daryan and Schäfer 1979) as well as from Jurassic carbonates (e.g. Flügel 1979). These patterns are defined by the occurrence of peculiar species within a distinct biotope, by the restriction of algal assemblages to special facies or microfacies types, or by the quantitative predominance of algal groups, adapted to changing environmental conditions (e.g. in lagoonal environments). Various parts of lagoonal and back-reef environments can be well distinguished, but a spatial subdivision of reef and fore-reef environments is also possible (using porostromate, solenoporacean and various encrusting algae in addition to green algae). An instructive example has been described by Schäfer and Senowbari-Daryan (1983) from the Upper Triassic Pantokrator limestone of Hydra Island, Greece. It must be stressed, however, that considerable changes in the settings of algal associations might have occurred during time: encrusting algae are common in reef environments during the Triassic, but are found also in open-shelf lagoonal environments in the Jurassic. *Tubiphytes* occurs in organic reefs in turbulent and foreslope positions during the Triassic, but in reefs and open platform setting in the Upper Jurassic. In contrast, filamentous blue-green algae (or green algae) seem to have been rather conservative with regard to their situation within Triassic and Jurassic environments.

4 Sedimentological Implications

Some of the many interrelationships between the formation of carbonate sediments and the activity of benthic calcareous algae are outlined below.

4.1 Green Algae and Bioclastic Sedimentation

Middle Triassic dasycladacean algae might have had a role in the rapid accumulation of sand-sized carbonate sediments similar to that of udoteacean green algae today. Grainstones and packstones of Ladinian lagoonal calcarenites in the Northern Alps may consist up to about 80% of dasycladacean fragments, but the proportion of limestones rich in dasycladaceans is often low (20 to 30%) as compared with other limestone types occurring in the sections (algal bindstones, bioclastic wackestones, mudstones). Upper Triassic (Norian and Rhaetian) shallow-marine carbonates are poor in mass-accumulations of dasycladaceans (generally not more than 15% of the rock volume). Liassic and Upper Jurassic shelf carbonates may be richer in dasycladacean remains (e.g. Liassic *Palaeodasycladus* limestone, up to 60%; Kimmeridgian and Tithonian *Clypeina* limestones, up to 80%).

A review of the paleontological record of udoteacean algae, related or identical to *Halimeda*, shows that these sedimentologically very important algae appeared during the Upper Triassic (Thailand, Northern Alps), are scarce during the Liassic and Middle Jurassic, and remained rare elements also within Upper Jurassic algal assemblages (Flügel 1987). Diversity increased during the Lower Cretaceous, both for *Boueina* and *Arabicodium*, and during the Tertiary for *Halimeda* which today embraces about 30 species (Fig. 8). The palaeoenvironmental setting of Mesozoic halimediform algae is partly comparable with that of recent *Halimeda*; lagoonal and reefal environments are known already in the Upper Triassic. The frequency, however, was low in the Upper Triassic and Jurassic as compared with other algal groups (porostromates, dasycladaceans). The high amount of halimediform algal bioclasts recorded from Upper Triassic occurrences (Fig. 7) should, therefore, not be overrated.

4.2 Oncoids

Oncoids with different microfabrics might reflect different paleoenvironments (Peryt 1981): Porostromate oncoids (skeletal oncoids, Riding 1977; cyanoids p.p., Riding 1983) are believed to have been formed from Cambrian to Jurassic

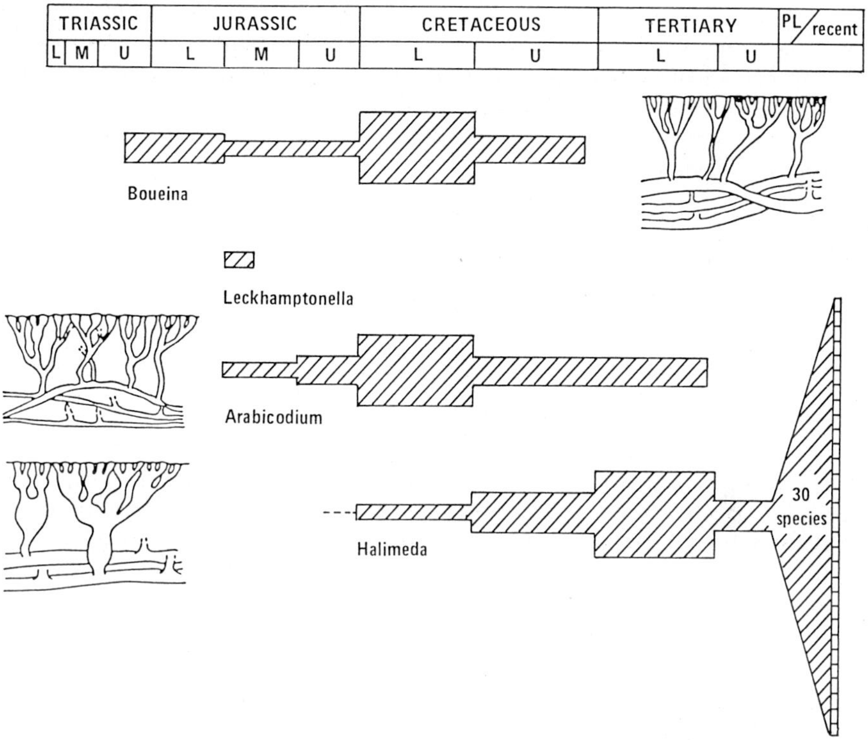

Fig. 8. Halimediform udoteacean green algae: Diversity and distribution in time. Height of the ashlar representing *Leckhamptonella* corresponds to one species. Pictures of the morphotypes after Elliott (1982)

times within marine, subtidal environments, generally with low-energy conditions, probably in somewhat deeper waters. Spongiostromate oncoids (non-skeletal oncoids, Riding 1977) occurred in lacustrine and, in association with stromatolites, in transitional marginal marine environments. Starting approximately during the Middle Jurassic, spongiostromate oncoids replaced porostromate oncoids, and became the only oncoid type in marine settings, but continued to flourish in lacustrine environments. During the Eocene spongiostromate marine oncoids were replaced by corallinacean rhodoids. Recent investigations of Triassic and Jurassic oncoids (e.g. Dahanayake 1977, 1978; Catalov 1983; Tichy 1983) seem to confirm this concept. Derivations might be explained by redeposition of oncoids (Gasiewicz 1983) or by obliteration of the primary microfabric by recrystallization (Peryt 1980).

4.3 Microbial Crusts

Micritic and pelmicritic or pelsparitic "crusts", sometimes with spongy, peloidal or fenestral microtextures, are common in Triassic and Jurassic reef carbonates.

The origin of these crusts is generally attributed to the binding and cementing effect of cyanobacteria, various "algae", or unknown microcommunities.

Of special interest are those microbial crusts which have been formed within subtidal environments and occur in close association with sessile invertebrates which may contribute to the formation of reefs. Several types can be distinguished in Triassic carbonates:

1. Micritic crusts, associated with encrusting and low-growing organisms (e.g. *Tubiphytes*, calcisponges);
2. Micritic and pelmicritic crusts, associated with a high-diverse fauna consisting of frame-building organisms (e.g. corals, calcisponges);
3. Thin encrusting micritic laminae, alternating with submarine carbonate cements which may be quantitatively more important than the crusts.

Type 1 is common in Middle Triassic, especially Anisian reef carbonates (e.g. Fois and Gaetani 1984). Type 2 seems to be restricted to Norian and Rhaetian reefs where spongiostromate crusts are important secondary framebuilders; no recent counterparts comparable with these crusts are known. The same seems to be true for type 3, which is common during the Permian and Middle Triassic (Flügel 1989a). An as yet undeciphered interaction between microbiota and unusually rapid synsedimentary cementation seems to be one of the main factors responsible for the formation of "algal/cement reefs".

Jurassic "cryptalgal crusts" (Dromart and Elmi 1986), "pellet crusts" (e.g. Brachert 1986) or "algal crusts" are other biogenic structures, probably caused by microbial activity. These crusts are common in the Upper Jurassic sponge reef facies (cf. Matyszkiewicz 1989). Indications for microbial activity get more and more strong. Photoacoustic studies of Oxfordian crusts from northern Franconia resulted in the recognition of anoxic bacteria which seem to have been responsible for the bacterially induced precipitation of carbonate and for a rapid submarine cementation (Krumbein, Oldenburg, pers. commun.). Comparable results have been obtained from bacteriological studies of Recent sea water and lagoonal carbonate mud (Morita 1980; Castanier et al. 1984).

5 Future Research

Despite increasing interest in Triassic and Jurassic algae during the last decades there is a strong need for an intensification of the investigations in several fields of research:

1. Data base: The information about the taxonomic inventory of Triassic and Jurassic calcareous algae is far from complete. Even better known groups, e.g. dasycladaceans, should yield a considerable number of as yet undescribed taxa judging from the gaps becoming visible in connection with higher taxonomic classifications (e.g. Bassoullet et al. 1979; Deloffre 1988).
2. Taxonomic revisions, as well as critical discussions, of classification concepts are necessary. The members of the "Groupe française d'étude des

algues fossiles" have made many very important contributions in this direction, but a re-study of Triassic dasycladaceans (based on type material) and a revision of porostromate algae (including comparison with Recent counterparts and considering changes in sea-water chemistry, cf. Pentecost and Riding 1986) are still lacking.

3. For some time intervals the information available is conspicuously poor (Liassic, early Middle Jurassic, Oxfordian) or completely lacking (Early Triassic). Discussions of evolutionary patterns as well as extinction and origination rates, can give just a very rough picture of main trends.
4. Studies dealing with the evolution of Triassic and Jurassic calcareous algae should be focussed on the change of morphological criteria through time. Those studies should also include investigations of the transition between Permian and Triassic algae, about the plasticity within algal populations (e.g. Hurka 1969) and about the possibility of a change in calcification patterns in time (Leadbeater and Riding 1986).
5. Non-Tethyan calcareous algae are of special interest because they might include offshoots of the Tethyan stock which could facilitate the discrimination of isocrymal distribution patterns.
6. Algal community analysis seems to provide a powerful tool in supporting paleoenvironmental reconstructions. A systematic investigation of algal floras, therefore, should be paralleled by microfacies studies in order to obtain better estimations of the importance of facies control for algal distribution patterns.
7. Because petrographical and microfacies criteria of "crusts" supposed to be of "algal" or microbial origin are controversial, biochemical and geochemical studies of crusts, formed in different paleoenvironments, are strongly needed.

Acknowledgements. This study is a part of the research program "Evolution of Reefs" supported by the Deutsche Forschungsgemeinschaft (project Fl 42/49-1, 2).

References

Bassoullet J-P, Bernier P, Deloffre R, Genot P, Jaffrezo M, Poignant A-F, Sengonzac G (1975) Classification Criteria of Fossil Dasycladales. In: Flügel E (ed) Fossil algae. Springer, Berlin Heidelberg New York, pp 154–166

Bassoullet J-P, Bernier P, Conrad M-A, Deloffre R, Jaffrezo M (1978) Les Algues Dasycladales du Jurassique et du Crétacé. Geobios Mem Spec 2: 330 pp, 40 pl

Bassoullet J-P, Bernier P, Deloffre R, Genot P, Jaffrezo M, Vachard D (1979) Essai de classification de Dasycladales en tribus. Bull Cent Rech Explor Prod Elf-Aquitaine 3:429–442

Bassoullet J-P, Bernier P, Deloffre R, Genot P, Poncet J, Roux A (1983) Les Algues Udoteacées du Paléozoique au Cénozoique. Bull Cent Rech Explor Prod Elf-Aquitaine 7:449–621

Bassoulet J-P, Bernier P, Deloffre R, Genot P, Poncet J, Roux A (1984) La notion de genre chez les Algues fossiles B. – Les Chlorophyceae calcaires. Bull Soc Geol Fr Ser 7, 26:605–607

Bissell HJ (1970) Petrology and petrography of Lower Triassic marine carbonates of Southern Nevada (USA). Int Sediment Petrogr Ser 14:27 pp

Brachert TC (1986) Kontinuierliche und diskontinuierliche Sedimentation im süddeutschen

Oberjura (unteres Kimmeridge; Ludwag/Oberfranken, Nördliche Frankenalb). Facies 15:233–284
Bystricky J (1964) Slovenský kras. Stratigrafia a Dasycladaceae mezozoika Slovenského krasu. Ustr Ustav Geol Bratislawa, 204 pp
Bystricky J (1984) *Physoporella croatica* Herak, 1958 of the Slovak Karst Anisian (Slovakia, The West Carpathian Mts.). Geol Zb Geol Carpath 35:507–533
Castanier S, Maurin A, Bianchi A (1984) Participation bactérienne à la précipitation du carbonate. C R Acad Sci Paris 299:1333–1336
Catalov GA (1983) Triassic Oncoids from Central Balkanides (Bulgaria). In: Peryt TM (ed) Coated grains. Springer, Berlin Heidelberg New York, pp 398–408
Chiocchini M, Manicelli A, Molinari-Paganelli V, Tilia-Zuccari A (1979) Répartition stratigraphique des algues dasycladales et codiacées dans les successions mésozoiques de plate-forme carbonatée du Lazio centre-méridional (Italie). Bull Cent Rech Explor Prod Elf-Aquitaine 3:525–535
Crescenti U (1966) Osservazioni sulla stratigrafia dell'Apennino meridionale alla luce delle recenti ricerche micropaleontologiche. Boll Soc Geol Ital 85:541–579
Crescenti U (1971) Biostratigraphic correlations in the Jurassic facies of central Italy by means of microfossils. Ann Inst Geol Publ Hung 54:209–213
Cubaynes R, Deloffre R (1984) *Tersella quercyensis* n. sp., Algue dasycladacée du Toarcien du Quercy (France). Bull Cent Rech Explor Prod Elf-Aquitaine 8:371–384
Cuif J-P, Dauphin Y, Denis A, Gautret P, Lawniczak A, Raguidaeu A (1987) Résultats récents concernant l'analyse des biocristaux carbonatés; implications biologiques et sédimentologiques. Bull Soc Geol Fr Ser 8, 3:269–288
Dahanayake K (1977) Classification of oncoids from the Upper Jurassic Carbonates of the French Jura. Sediment Geol 18:337–353
Dahanayake K (1978) Sequential Position and Environmental Significance of Different Types of Oncoids. Sediment Geol 20:301–316
Deloffre R (1988) Nouvelle taxonomie des algues dasycladales. Bull Centres Rech Explor Prod Elf-Aquitaine 12:165–217
Dragastan O (1981) Mesozoic Dasycladaceae from Romania: Distribution and Biostratigraphical Importance. Facies 4:165–196
Dragastan O (1985) Review of Tethyan Mesozoic Algae of Romania. In: Toomey DF, Nitecki MH (eds) Paleoalgology. Springer, Berlin Heidelberg New York, pp 101–161
Dromart G, Elmi S (1986) Développement de structures cryptalgaires en domaine pélagique au cours de l'ouverture des bassins jurassiques (Atlantique Central, Téthys occidentale). C R Acad Sci Paris 303:311–316
Elliott GF (1977) Inferred isocrymal distribution of Jurassic dasycladacean Algae in Europe, north Africa and southwestern Asia. J Geol Soc 133:363–373
Elliott GF (1981) The Tethyan Dispersal of some Chlorophyte Algae subsequent to the Paleozoic. Palaeogeogr Palaeoclimatol Palaeoecol 32:341–358
Elliott GF (1982) A new calcareous green alga from the Middle Jurassic of England: its relationships and evolutionary position. Palaeontology 25:431–437
Elliott GF (1984) Climatic tolerance in some aragonitic green algae of the Post-Palaeozoic. Palaeogeogr Palaeoclimatol Palaeoecol 48:163–169
Emberger J (1979) Les Algues (Euchlorophyceae, Prasinophyceae, Rhodophyceae) du Trias. Essai d'un inventaire bibliographique, géographique, stratigraphique. Bull Inst Geol Bassin d'Aquitaine Num Spec: 157 pp
Farinacci A, Radoičić R (1964) Correlazione fra serie gioresi et cretacee dell' Appenino centrale e delle Dinaridi esterne. Rendiconti 7, Ser 2:269–300
Flügel E (1979) Paleoecology and microfacies of Permian, Triassic and Jurassic algal communities of platform and reef carbonates from the Alps. Bull Cent Rech Explor Prod Elf-Aquitaine 3:569–587
Flügel E (1985) Diversity and environments of Permian and Triassic dasycladacean algae. In: Toomey DF, Nitecki MH (eds) Paleoalgology. Springer, Berlin Heidelberg New York, pp 344–351
Flügel E (1988) *Halimeda*: paleontological record and paleoenvironmental significance. Coral Reefs 6:123–130

Flügel E (1989) Algeu/Zement – Riffe Arch Lagerstättenforsch Geol Bundescenst Wien 10:125–131

Flügel E, Flügel-Kahler E, Martin JM, Martin-Algarra A (1984) Middle Triassic reefs from southern Spain. Facies 11:173–218

Flügel E, Mu X-N (1982) Upper Triassic Dasycladaceae from Eastern Tibet. Facies 6:59–74

Flügel E, Senowbari-Daryan B, Stanley GD Jr (1989) Late Triassic Dasycladacean Alga from northeastern Oregon: Significance of first reported occurrence in Western North America. J Paleont 63:374–381

Flügel HW (1971) *Oligoplagia* HERAK 1944, eine Gymnocodiaceae. N Jahrb Geol Paläontol Monatsh 1971:532–536

Fois E (1979) A new Dasycladacean (Calcareous Algae) assemblage from Triassic of M Popera (Belluno, Italy). Riv Ital Paleontol Stratigr 85:57–86

Fois E, Jadoul F (1983) La Dorsale Paleocarnica anisica di Pontebba. Riv Ital Paleontol Stratigr 89:3–30

Fois E, Gaetani M (1984) The recovery of reef-building communities and the role of cnidarians in carbonate sequences of the Middle Triassic (Anisian) in the Italian Dolomites. Palaeontogr Am 54:191–200

Gadzicki A, Kowalski WR (1974) Green Algae *Aciculella* Pia from the Muschelkalk of the Holy Cross Mts (Poland). Bull Acad Pol Sci Ser Sci Terre 22:27–32

Gasiewicz A (1983) Upper Oxfordian oncolites of SE-Poland. N Jahrb Geol Paläontol Monatsh 1983:205–212

Güvenç T (1979) Dasycladacées métaspondyles du Paléozoique supérieur et du Trias. Bull Centre Rech Explor Prod Elf-Aquitaine 3:625–637

Herak M (1965) Comparative Study of Some Triassic Dasycladaceae in Yugoslavia. Geol Vjesn Inst Geol Istrazivanja Zagreb 18:3–34

Herak M, Kochansky V (1963) Jungpaläozoische Kalkalgen aus dem Bükk-Gebirge (Nordungarn). Geol Hung Ser Palaeontol 28:45–77

Herak M, Kochansky-Devidé V, Gusic I (1977) The development of the Dasyclad Algae through the Ages. In: Flügel E (ed) Fossil algae. Springer, Berlin Heidelberg New York, pp 143–153

Hurka H (1967) Variationsstatistische Untersuchungen an anisischen Dasycladaceen aus den Pragser Dolomiten in Südtirol. N Jahrb Geol Paläontol Abh 128:41–100

Hurka H (1969) Umbildungstendenzen der Astformen in *Physoporella-Oligoporella*-Populationen (Dasycladaceen) aus dem Anis der Pragser Dolomiten (Italien). N Jahrb Geol Paläontol Monatsh 1969:104–120

Jaffrezo M (1980) Les formations carbonatées des Corbières (France) du Dogger a l'Aptien: micropaléontologie stratigraphique, biozonation, paléoecologie. Extension des resultats à la Mésogée. These Doct Etat, Univ Pierre et Marie Curie, Paris, 613 pp

Johnson HH, Danner WR (1966) Permian calcareous algae from northwestern Washington and southwestern British Columbia. J Paleontol 40:424–432

Kamptner E (1958) Über das System und die Stammesgeschichte der Dasycladaceen (Siphoneae verticillatae). Ann Naturhist Mus Wien 62:95–122

Kochansky-Devidé V (1954) Permske foraminifere i vapnenacke alge okolice Bara u Crnoj Gori. Geol Vjesn Zagreb 5–7:295–298

Kochansky-Devidé V, Gusić I (1971) Evolutions-Tendenzen der Dasycladaceen mit besonderer Berücksichtigung neuer Funde in Jugoslawien. Paläontol Z 45:82–91

Korde KB (1965) Les Algues Rhodophyces et Chlorophytes. Tr Inst Paleontol Akad Nauk SSR 108:268–284

Leadbeater BSC, Riding R (eds) (1986) Biomineralization in lower plants and animals. Syst Assoc Spec Vol 30:401 pp

Leinfelder RR (1985) Cyanophyte Calcification Morphotypes and Depositional Environments (Alenquer Oncolite, Upper Kimmeridgian? Portugal). Facies 12:253–274

Martin JM, Braga JC (1987) Aepujarride carbonate deposits (southern Spain) – marine sedimentation in a Triassic Atlantic. Paleogeogr Paleoclimatol Paleoecol 59:243–260

Matyszkiewicz J (1989) Sedimentation and diagenesis of the upper Oxfordian cyanobacterial-sponge limestones in Piekary near Krakow. Ann Soc Geol Poloniae 59:201–232

Mistiaen B (1984) Disparition des Stromatopores paléozoigues ou survie du groupe: hypothèse et discussion. Bull Soc Geol Fr Ser 7 26:1245–1250
Morita RJ (1980) Calcite precipitation by marine bacteria. Geomicrobiol 2:63–82
Mu Xi-nan (1981) Upper Permian Calcareous Algae from Western Guizhou. Acta Palaeontol Sin 20:33–49
Nazarevya BP, Nazarevya IA, Schiyabko NI (1986) Usloviya formirovaniya i osobennosti razmesheniya nizhnetriasikh iskopaemikh organogennikh postroek vostochnogo prekavkazya. Akad Nauk SSR, Otdel Geol Geofiz Geochim Gorn Nauk, pp 161–166
Noé S (1987) Facies and paleogeography of the marine upper Permian and the Permian triassic boundary in the Southern Alps (Bellerophon Formation, Tesero Horizon). Facies 16:89–142
Ott E (1967) Segmentierte Kalkschwämme (Sphinctozoa) aus der alpinen Mitteltrias und ihre Bedeutung als Riffbildner im Wettersteinkalk. Abh Bayer Akad Wiss München 131:1–96
Ott E (1972a) Zur Kalkalgen-Stratigraphie der Alpinen Trias. Mitt Ges Geol Bergbaustud Wien 21:455–464
Ott E (1972b) Die Kalkalgen-Chronologie der alpinen Mitteltrias in Angleichung an die Ammoniten-Chronologie. N Jahrb Geol Paläontol Abh 141:81–115
Ott E (1972c) Mitteltriadische Riffe der Nördlichen Kalkalpen und altersgleiche Bildungen auf Karaburun und Chios (Ägäis). Mitt Ges Geol Bergbaustud Wien 21:251–276
Ott E (1974) Algae (Dasycladaceae). Catal Foss Austriae, 17b:1–64
Pentecost A, Riding R (1986) Calcification in cyanobacteria. In: Leadbeater BSC, Riding R (eds) Biomineralization in lower plant and animals. Syst Assoc Spec vol 30:73–90
Peryt TM (1980) Structure of "Sphaerocodium kokeni WAGNER", a Girvanella-oncoid from the Upper Muschelkalk (Middle Triassic) of Württenberg (SW Germany). N Jahrb Geol Paläontol Monatsh 1980:293–302
Peryt TM (1981) Phanerozoic Oncoids – an overview. Facies 4:197–214
Peryt TM (1986) Zechstein *Stromaria* (= *Archaeolithoporella*) – cement reefs in Thuringia. Neues Jahrb Geol Paläont Monatshefte, 1986:307–316
Peybernes B (1979) Les Algues jurassique et du Crétace inférieur des Pyrenées franco-espagnoles. Interet biostratigraphique et paléoecologique. Bull Cent Rech Explor Prod Elf-Aquitaine 3:733–741
Pia J von (1943) Übersicht über die fossilen Kalkalgen und die geologischen Ergebnisse ihrer Untersuchung. Mitt Geol Ges Wien 33:11–34
Pisera A, Zawidzka K (1981) *Archaeolithoporella* from the Upper Permian Reef Limestones of the Northern Caucasus. Bull Acad Pol Sci Ser Sci Terre 29:233–238
Praturlon A (1966) Algal Assemblages from Lias to Paleocene. Boll Soc Geol Ital 85:167–194
Radoičić R (1966) Microfacies du Jurassique des Dinarides externes de la Yougoslavie. Geologija 9:377 pp
Ramalhao MM (1971) Contribution à l'étude micropaléontologique et stratigraphique du Jurassique supérieur et du Crétacé inférieur des environs de Lisbonne (Portugal). Serv Geol Portugal Mem 19:212 pp
Reid P (1983) The first occurrence of Triassic Phylloid Algae – possible links with the Paleozoic. In: 3rd Int Symp Fossil algae, Program Abstr, p 26
Reinhardt J (1987) Oberpermische Riffe und permo-triassische Sedimentfazies am Südostrand des Beckens von Sichuan, China. Dipl Thesis, Inst Paläontol Univ Erlangen, 176 pp
Remane J (1969) Morphological and statistical analysis of *Clypeina jurassica* (FAVRE) (Dasycladaceae) by means of serial sections. Geol Rom 8:85–116
Riding R (1977) Skeletal stromatolites. In: Flügel E (ed) Fossil algae. Springer, Berlin Heidelberg New York, pp 57–60
Riding R (1983) Cyanoliths (Cyanoids): Oncoids Formed by Calcified Cyanophytes. In: Peryt TM (ed) Coated grains. Springer, Berlin Heidelberg New York, pp 276–283
Roux A (1985) Introduction á l'étude des Algues fossiles paléozoiques (de la bactérie à la tectonique des plaques). Bull Cent Rech Explor Prod Elf-Aquitaine 9:465–699
Rui L, Zhao J-M, Mu-X-N, Wang K-L, Wang T-H (1984) Restudies on the Wujiaping Limestone from Liangshan of Hanzhong, Shaanxi. *Dicengxue Zazhi*, J Stratigr 8:179–193
Sartoni S, Crescenti U (1962) Recherche biostratigrafiche nel Mesozoico dell' Appennino meridionale. Giorn Geol 29:161–304

Schäfer P, Senowbari-Daryan B (1983) Die Kalkalgen aus der Obertrias von Hydra, Griechenland. Palaeontographica Abt B 185:83–142

Senowbari-Daryan B, Schäfer P (1979) Distributional patterns of calcareous algae within Upper Triassic patch reef structures of the northern calcareous Alps (Salzburg). Bull Centres Rech Explor Prod Elf-Aquitaine 3:811–820

Senowbari-Daryan B, Schäfer P (1980) *Abatea culleiformis* n.g., n.sp., eine neue Rotalge (Gymnocodiaceae) aus den "oberrhätischen" Riffkalken südlich von Salzburg (Nördliche Kalkalpen, Österreich). Verh Geol (Bundesanst Wien 1979:393–399

Sepkoski JJ (1986) Phanerozoic overview of mass extinction. In: Raup DM, Jablonski D (eds) Patterns and processes in the history of life. Dahlem Kon Life Sci Res Rep 36:277–296

Sheng Jin-Zhang, Chen Chu-Zhen, Wang Y-Gang, Rui Lin, Liao Zhuo-Ting, Yuji Bando, Ken-ichi Ishii, Keji Nakazawa, Koji Nakamura (1984) Permian-Triassic Boundary in Middle and Eastern Tethys. J Fac Sci Hokkaido Univ Ser 4, 21:133–181

Tappan H (1980) The paleobiology of plant protists. Freeman, San Francisco, 1028 pp

Taraz H, Golshani F, Nakazawa K, Shimizui D, Bando Y, Ishii, Ken-ichi, Murata M, Okimura Y, Sakagami S, Nakamura K, Tokuoka T (1981) The Permian and Lower Triassic Systems in Abadeh Region, Central Iran. Mem Fac Sci Kyoto Univ Ser Geol Min 47:61–133

Termier H, Termier G, Vachard D (1977) Monographie paléontologique des affleurements permiens du Djebel Tebaga (Sud Tunisien). Palaeontographica Abt A 156:1–109

Tichy G (1983) Carnian Oncolites of the Northern Calcareous Alps and the Drau Range (Austria). In: Peryt TM (ed) Coated Grains. Springer, Berlin Heidelberg New York, pp 386–389

Vachard D, Montenat C (1981) Biostratigraphie, micropaléontologie et paléogéographie du Permien de la Région de Tezak (Montagnes Centrales d'Afghanistan). Palaeontographica Abt B 178:1–88

Valentine JW (ed) (1985) Phanerozoic diversity patterns. Profiles in macroevolution. Princeton Univ Press, Princeton New Jersey, 441 pp

Wray JL (1977) Calcareous algae. Developments in paleontology and stratigraphy, vol 4. Elsevier, Amsterdam, 185 pp

Yun Fu Zeng, Nan Hao Lee, Yang Zhou Huang (1983) Sedimentary characteristics of oolitic carbonates from the Jialing-Jiang Formation (Lower Triassic T_1^2 J1)), South Sichuan Basin, China. In: Peryt TM (ed) Coated grains. Springer, Berlin Heidelberg New York, pp 176–187

Zhao Jinke, Sheng Jinzhang, Yao Zhaogi, Liang Xilou, Chen Chuzhen, Rui Lin, Liao Zhuoting (1981) The Changhsingian and Permian-Triassic boundary of South China. Bull Nanjing Inst Geol Paleont 2:1–85

Zorn H (1974) Variationsstatistische Untersuchungen an *Diplopora annulata* (SCHAFH. und *D. philosophi* (PIA), Kalkalgen der alpinen Mitteltrias. Eclog Geol Helv 67:1–30

Zorn H (1977) Der Einfluβ der Umwelt auf die Variation fossiler Arten. Nat Mus 107:1–5

Chapter 22

Mesozoic and Cenozoic Marine Benthic Calcareous Algae with Particular Regard to Mesozoic Dasycladaleans

F. Barattolo[1]

Abstract

During the Mesozoic, benthic calcareous marine green algae are represented principally by the Caulerpales and Dasycladales. Starting from this era Caulerpaleans consist of genera similar to extant ones. Dasycladaleans are represented by forms with primary branches only and by ones with higher orders of branches. Furthermore, the branch arrangement is mainly euspondyle and the reproductive organs are mainly cladosporous. On the basis of morphological characters and of the fluctuation in the number of genera and species it is possible to recognize the following stages: persistence of Late Palaeozoic forms into the Triassic, decline and renewal during the Lower-Middle Jurassic, Upper Jurassic-Lower Cretaceous acme, and Upper Cretaceous collapse.

During the Cenozoic era, calcified cyanobacteria undergo a numerical reduction. As for the red algae, gymnocodiaceans and solenoporaceans disappear, while there is an explosive diversification of the Corallinaceae, and Solieriaceae and Squamariaceae appear.

Among the dasycladaleans, thalli with branches of higher orders in an euspondyle arrangement and with choristoporous reproductive organs prevail. During the Palaeocene this order shows the greatest number of species.

1 Introduction

The marine benthic calcareous algae are mainly represented by the divisions Cyanophyta, Rhodophyta, and Chlorophyta.

The literature on calcareous algae has increased remarkably in the last 25 years. It deals mainly with specific subjects; syntheses are still rather poor and there all still far too few publications on the general significance of the calcareous algae.

Among the syntheses, the monographs of Johnson (1952, 1958, 1966), Johnson and Konishi (1956, 1959), Johnson and Høeg (1961), Roux (1985) all deal with Palaeozoic algae, and the monograph of Korde (1973) concerns Cambrian algae. Johnson (1964, 1969) deals with Jurassic-Lower Cretaceous algae, and Johnson (1960, 1962, 1963, 1965) with the corallinaceans and solenoporaceans. The syntheses of Bassoullet et al. (1978) and Deloffre and Génot (1982) cover post-Triassic fossil dasycladaleans, while that of Bassoullet et al. (1983) deals with the udoteaceans. Among the reviews on the calcareous algae, the monographs of Pia (1927), Johnson (1961), Korde et al. (1963), Maslov and Korde (1963), Maslov et al. (1963), Wray (1977b), Emberger (1968), Dragastan (1980), and Chuvashov et al. (1987), are especially important.

[1]Dipartimento di Paleontologia, Largo S. Marcellino 10, 80138 Napoli, Italy

The relatively recent bibliographical reviews of Emberger (1976, 1978, 1979) about Paleozoic and Triassic algae, and that of Emberger and Jaffrezo (1975) about the Jurassic-Lower Cretaceous dasycladaleans, are particularly useful.

For algal systematics, I refer to Bold and Winne (1978), but for the taxa above ordinal level the most commonly used suffixes have been utilized. As regards the geological time-scale, I refer to Odin (1982).

The diagrams shown in the text have tried to utilize, as far as possible, the data of the most recent literature.

2 Division Cyanophyta

2.1 Stromatolites

Stromatolitic structures (stromatolites and oncolites) are cosmopolitan and reached their maximum abundance in the Proterozoic and in the Lower Palaeozoic. They are also well represented in the Mesozoic, although we witness, in this period, their global gradual decrease. In the Mesozoic they are well distributed in the carbonate platform facies of the Alpine Chain, particularly in very shallow environments.

Triassic stromatolitic structures are described by Fischer (1964), D'Argenio (1966a), Catalano et al. (1974), and liassic stromatolites by, for example, Colacicchi et al. (1975) and D'Argenio (1966b).

Stromatolitic structures are less frequent after the Lias. Examples are known from the Palaeogene of the USSR (Maslov 1962), the Miocene of Saudi Arabia (Whybrow et al. 1987), and of Mazara del Vallo in Sicily (Esteban et al. 1982). Recent stromatolitic structures (laminated algal mats and algal pebbles) grow mainly in shallow waters and terrestrial high-humidity conditions of tropical and subtropical areas (Monty 1965, 1967; Gebelein 1969). Rare occurrences in temperate zones are also known (Passeri 1974). For further information see Monty (1977).

Stromatolitic structures, both stromatolites and oncolites, are also well known in the pelagic facies of the Mesozoic, particularly in the Jurassic Ammonitico Rosso of the Mediterranean area. Their presence here has modified interpretations of the sedimentary environment that had been considered deep water (e.g. Massari 1983; Massari and Dieni 1983; Clari et al. 1984).

2.2 Calcified Cyanobacteria

Among the most common of these in the Mesozoic are the nodular thalli made up of branched filaments which radiate from a restricted area of the substratum belonging to the genus *Cayeuxia* Frollo and, subordinately, to the genera *Garwoodia* Wood, *Bevocastria* Garwood, *Hedstroemia* Rothpletz and *Ortonella* Garwood.

This type of fossil has also been reported, even if rarely, after the Mesozoic: in the Miocene by Reis (1923), who attributed them to *Zonotrichites* Bornemann; in the Pleistocene by Richter et al. (1979), who attributed them to *Rivularia haematites* (De Candolle). Same of these genera were considered green algae by Johnson (1961), Elliott (1965), and Emberger (1968). Other authors ascribe these morphologically similar forms to cyanophytes and group them in the Garwoodiaceae (e.g. Chuvashov et al. 1987). Recently Dragastan (1985) has revised the systematics of this group which he assigns partly to cyanophyceans and partly to codiaceans; *Cayeuxia* Frollo in particular is considered a junior synonymous of *Rivularia* Roth.

In the back-reef facies of the Upper Cretaceous, the fragments of filamentous thalli of hormogonaleans can assume a certain local lithogenetic value (e.g. *Aeolisaccus kotori* Radoičić and *A. barattoloi* De Castro: Pl. I, 1–2; De Castro 1975, 1989).

3 Division Rhodophyta

The red algae (subclass Florideophycidae) have a wider stratigraphic and geographic distribution (Fig. 1). The orders and the families known as fossils are: Epiphytales (Epiphytaceae), Cryptonemiales (Ungdarellaceae, Solenoporaceae, Corallinaceae and Squamariaceae), Gigartinales (Solieriaceae), Nemalionales (Gymnocodiaceae).

3.1 Epiphytales

The systematic position of these algae is somewhat controversial (see Roux 1985). Chuvashov et al. (1987) ascribe these algae to Cyanophytes.

Epiphytaleans become less frequent after the Cambrian; Mesozoic *Epiphyton* has been pointed out by Masse (1979) from the Cretaceous of France.

Plate I. Cyanophyta and Incertae sedis. 1 Aeolisaccus barattoloi De Castro. The distribution of *A. barattoloi* in sediment often looks like that of cyanophycean sheath fragments inside the older parts of laminated algal mats from Andros Island. Thin section A.5238.3, Upper Cretaceous (Senonian) of Regia Piana near Cusano Mutri (Campania, Italy), ca. x120. *2 Aeolisaccus kotori* Radoičić. Fragments of sheaths similar to extant Scytonemataceae. Thin section A.1049.5, Upper Cretaceous (Senonian) of Mt. Tobenna near Giffoni V.P. (Campania, Italy), ca. x120. *3 Paronipora* (= *Microcodium*) *penicillata* Capeder. Inside the rock cavities both entire specimens and minute fragments of *Paronipora* can be seen. Thin section A.6990.2, Lower Palaeocene (Danian) of Colle di Medea near Manzano (Friuli, Italy), ca. x40. *4–5 Tubiphytes morronensis* Crescenti. Subtransversal (*4*) and longitudinal (*5*) sections. *4* Thin section A.6646.1, Uppermost Jurassic of Capri (Campania, Italy), ca. x30. *5* Thin section A.6402.1, Uppermost Jurassic-Lowermost Cretaceous of Capri (Campania, Italy), ca. x30

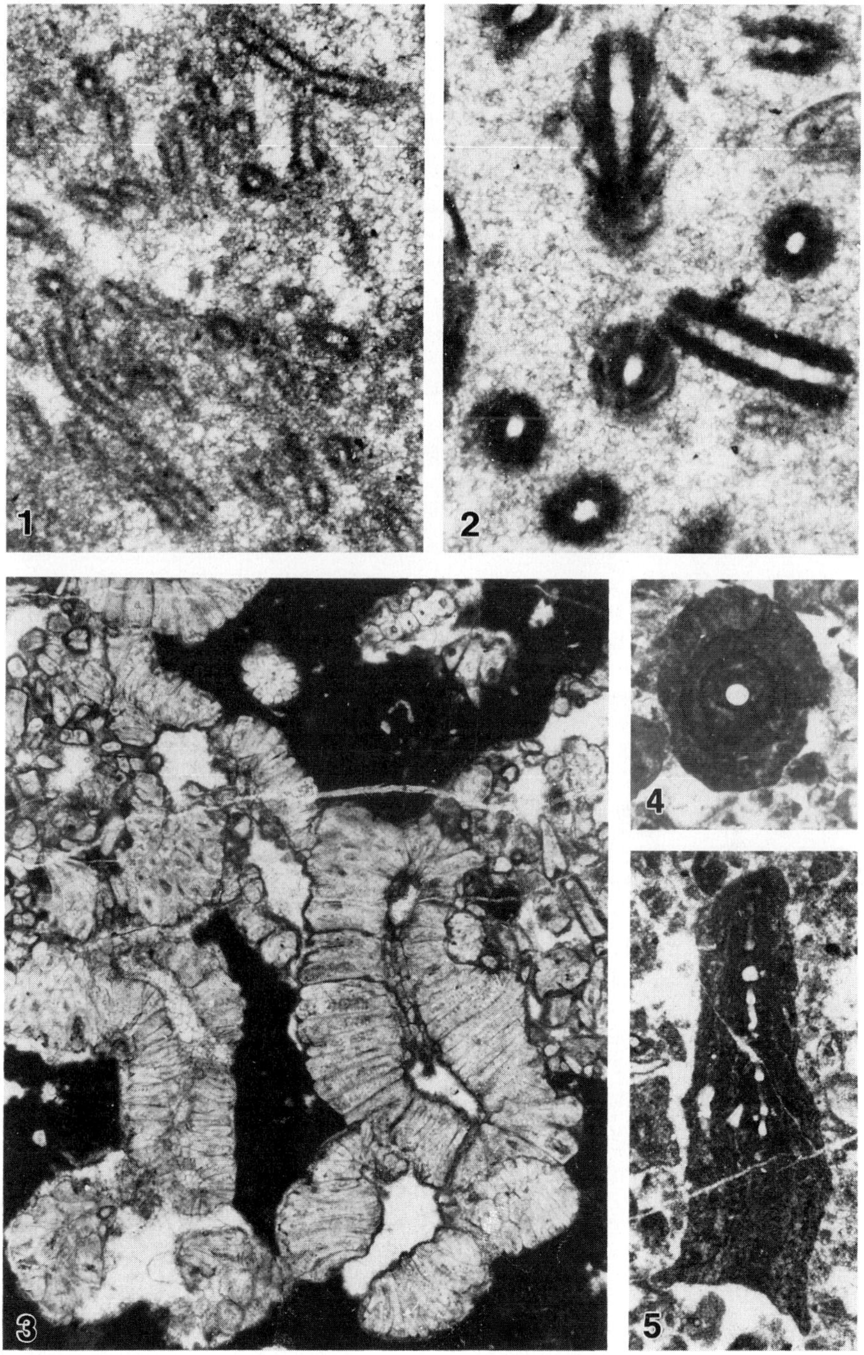

Plate I

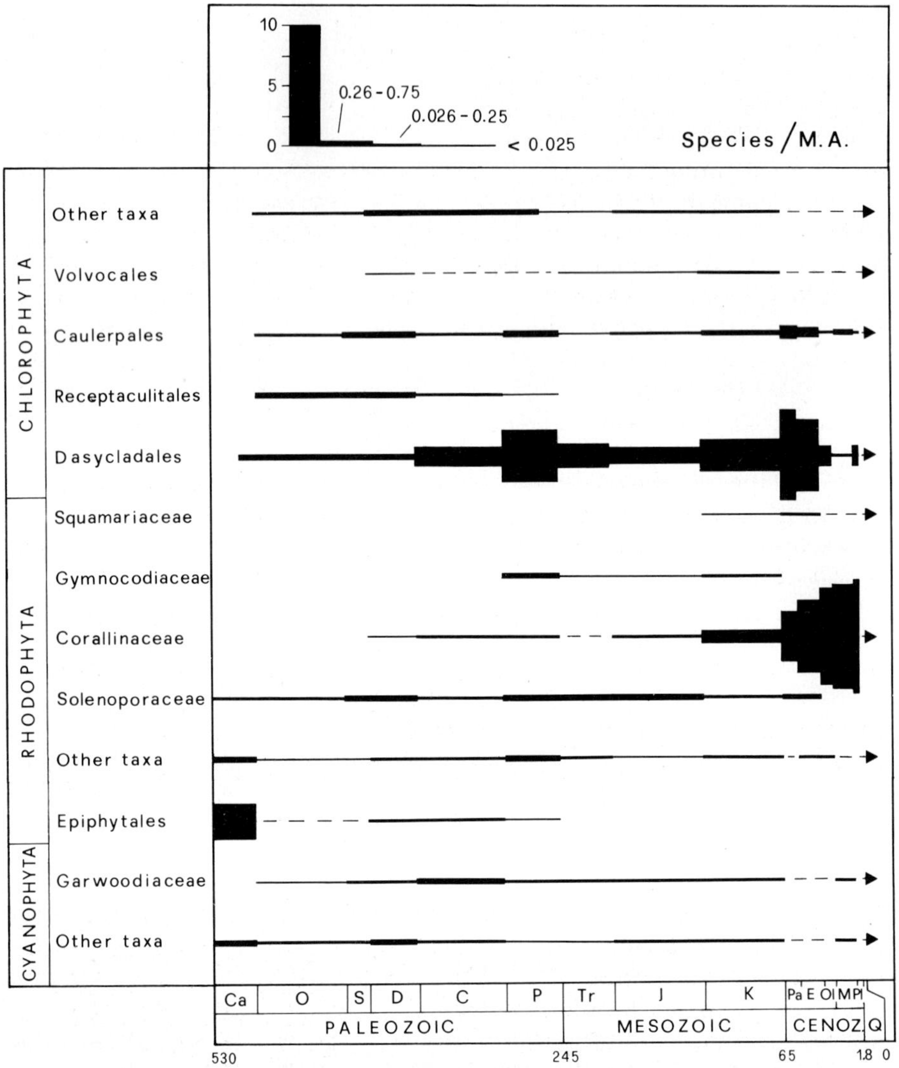

Fig. 1. Quantitative distribution of fossil marine benthic calcareous algae during the Phanerozoic. *Arrows* indicate extant categories; living species are not included. Time frequency of each group is represented by the number of species per Ma. Volvocales includes, in the chart, Volvocales, Thaumatoporellales and Tetrasporales. Time scale is from Odin (1982)

3.2 Solenoporaceae

In the Upper Cambrian the family Solenoporaceae appears, the more representative genera of which are *Solenopora* Dybowski and *Parachaetetes* Deninger. The group reached its maximum development in the Palaeozoic and disappeared during the Eocene. In the Mesozoic and the Cenozoic we find the genera *Solenopora*, *Parachaetetes* and *Pycnoporidium* Yabe and Toyama; this latter one, however, according to Emberger (1968) and Schäfer and Senowbari-Daryan (1983), should be referred to the green algae. *Guvencipora* Vachard et al. (1978) is, up to now, only known in the Trias; *Marinella* Pfender is mainly known from the Upper Jurassic to the Cretaceous (see Barattolo and Del Re 1984), and *Elianella* Pfender and Basse in the Cretaceous-Eocene (Hagn and Ott 1975; Poignant 1980).

Neosolenopora Mastrorilli, from the French and Italian Miocene, according to Tillier's studies (1975), is to be assigned to the Bryozoa. *Solenomeris* Douvillé, from the Upper Cretaceous-Palaeogene reef facies, according to some authors (Perrin 1987), is to be ascribed to the foraminifers (Acervulinidae).

Solenopores lived mainly in reef environments or on platforms characterized by rough waters.

3.3 Corallinaceae

The Corallinaceae, probably evolved from the Solenoporaceae, represents by far the most abundant group of fossil rhodophytes.

The family shows a progressive development from the Upper Cretaceous up to the present times especially with the genera *Archaeolithothamnium* Rothpletz, *Lithophyllum* Philippi, *Lithothamnium* Philippi and *Mesophyllum* Lemoine. The genera *Kymalithon* Lemoine and Emberger, *Paraphyllum* Lemoine and *Hemiphyllum* Lemoine occur exclusively in the Cretaceous. Apart from a Permian species (Endo 1969), the genus *Distichoplax* Dietrich is restricted to the Paleocene-Eocene. *Leptolithophyllum* Airoldi is an Oligocene genus, *Aethesolithon* Johnson and *Pseudoaethesolithon* Elliott are exclusively Miocene genera.

The number of species per million years is about 0.02–0.11 in the Middle-Upper Palaeozoic, and about 0 and 0.08 respectively in the Triassic and Jurassic. It increases in an explosive way from the Cretaceous (1.05 species per Ma.) to Pliocene (9.52 species per Ma.). In the present, corallinaceans densely populate numerous habitat among the supralittoral to circalittoral zones, and are spread from Arctic to tropical seas between latitudes 83° N and 73° S.

In the Mediterranean area they flourish spectacularly in the tidal zone with the "trottoirs", in the infralittoral zone with the buildups of the "coralligéne de plateau" and in the circalittoral zone with the conspicuous deposits of "maërl" (Pérès and Picard 1964; Pérès 1967); they also form banks of rhodoliths in the sublittoral areas with strong currents. Corallinaceans are mostly widespread in the "Lithothamnium facies" ("Nullipore facies" *auct.*) of the tropical reefs, between low-tide level and a depth of about 30 metres.

For further information about the ecology of coralline algae see: Adey and MacIntyre (1973), Littler and Doty (1974), Adey and Vassar (1975), van den Hoek et al. (1975), Adey et al. (1982), Bosence (1985) and Minnery et al. (1985).

In the geological past corallinaceans probably occupied the same environments; they are known mainly from high energy habitats. In central and southern Italy the Lower Miocene Cusano Formation, considered of circalittoral environment (Barbera et al. 1978), is represented by a massive limestone sequence up to 40 m in thickness, constituted essentially of rhodoliths, which can grow up to 25 cm in diameter, together with ostreids, pectinids, bryozoans and foraminifers (Ogniben 1958).

For the phylogeny, and a review of the main taxa of corallinaceans see, for example, Poignant (1977, 1979b), and for some palaeogeographical details see Poignant (1979a) and Poignant and Laville (1972).

3.4 Gymnocodiaceae, Squamariaceae and Solieriaceae

The Gymnocodiaceae, probably related to the living *Galaxaura*, are known as fossils mainly from *Gymnocodium* Pia (Permian) and *Permocalculus* Elliott (Permian-Palaeocene). The genera *Nanjinophycus* Mu and Riding and *Oligoplagia* Herak are restricted respectively to Permian and Triassic. According to Chuvashov et al. (1987), *Succodium* Konishi and *Dzhulfanella* Korde are also to be ascribed to this family; gymnocodiaceans are referred by these authors to green algae (order Siphonales). Gymnocodiaceans lived in platform environments associated with caulerpaleans and dasycladaleans. For further information see Elliott (1955, 1956, 1961), Mu and Riding (1983), and Emberger's (1976, 1979) bibliographic reviews.

The Squamariaceae are known, starting in the Mesozoic, from the genera *Ethelia* Weber Van Bossae (= *Pseudolithothamnium* Pfender and *Polystrata* Hydrich) (Cretaceous-Present) and *Peyssonnelia* Decaisne (Eocene-Present). For details of the synonymy of *Ethelia*, *Pseudolithothamnium* and *Polystrata* see Massieux and Denizot (1964), Denizot (1968) and Buchbinder and Halley (1985).

The extant family Solieriaceae is known in the Mesozoic from *Agardhiellopsis* Lemoine (1966), occurring in the Lower Cretaceous (Aptian-Albian) reef facies in France and Spain.

4 Division Chlorophyta

The Chlorophyceae, like the Rhodophyceae, has left most evidence of its existence through geologic time in the littoral environments. It is represented as fossils mainly by the following orders and families: Wetheredellaceae, Anchicodiaceae, Receptaculitales (Receptaculitidae), Caulerpales (Udoteaceae,

Codiaceae), Volvocales (Volvocaceae), Thaunatoporellales (Thaumatoporellaceae), Tetrasporales (Chlorangiaceae), Ulotrichales, Chaetophorales (Chaetophoraceae), Dasycladales (Seletonellaceae, Beresellaceae, Diploporaceae, Dasycladaceae, Acetabulariaceae).

4.1 Wetheredellaceae and Anchicodiaceae

Wetheredellaceae populated reef environments mainly from the Devonian to the Permian and subordinately from the Middle Jurassic to the Upper Cretaceous. For further information on this family see Cherchi and Schroeder (1985) and Chuvashov et al. (1987); the latter authors ascribe the Wetheredellaceae to the Dasycladales. During the Mesozoic the family is mainly represented by *Koskinobullina socialis* Cherchi and Schroeder (1979) which shows close similarities to *Aphralysia jurassica* Elliott (1982).

The Anchicodiaceae Shuysky (1987b) (= phylloid udoteaceans *auct.*) lived in Carboniferous-Permian platform environments, especially in the reef facies. This family has been assigned both to red algae (e.g. Wray 1977a) and to the green algae (e.g. Roux 1985). Shuysky (1987b) and Chuvashov et al. (1987) ascribe Anchicodiaceans to the green algae order Siphonales; however several recent classifications do not include the Siphonales anymore, which have been split into several orders.

4.2 Caulerpales

Udoteaceae

The Udoteaceae is represented in the Recent by fourteen genera; they are well known in geologic time from about twenty-five genera of which one is still living. Udoteaceans appeared in the upper Cambrian and became widespread in the Silurian-Devonian, in the Permian, and subsequently in the Palaeocene-Eocene. The more frequent Mesozoic-Cenozoic genera are *Boueina* Toula (Pl. II, Fig. 2), *Arabicodium* Elliott, *Halimeda* Lamouroux and *Ovulites* Lamarck (see Bassoullet et al. 1983). Udoteaceans lived in environments similar to those of the dasycladaleans, populating lagoon and also shelf margin facies (e.g. *Nipponophycus* Yabe and Toyama and *Juraella* Bernier 1984).

Codiaceae

About forty genera have been assigned to the Codiaceae but subsequently, with the improvement of the systematics, have been transferred to other families. Forms like *Halimeda* Lamouroux and *Boueina* Toula are generally placed in the udoteaceans. There is today a certain consensus in reassigning the Garwoodiaceae (= Mitcheldeaniaceae *auct.*) to cyanophyceans (Riding 1977; Schäfer and Senowbari-Daryan 1983; Chuvashov et al. 1987). The *Anchicodium*-like forms (= phylloid Udoteaceans *auct.*) are grouped together in

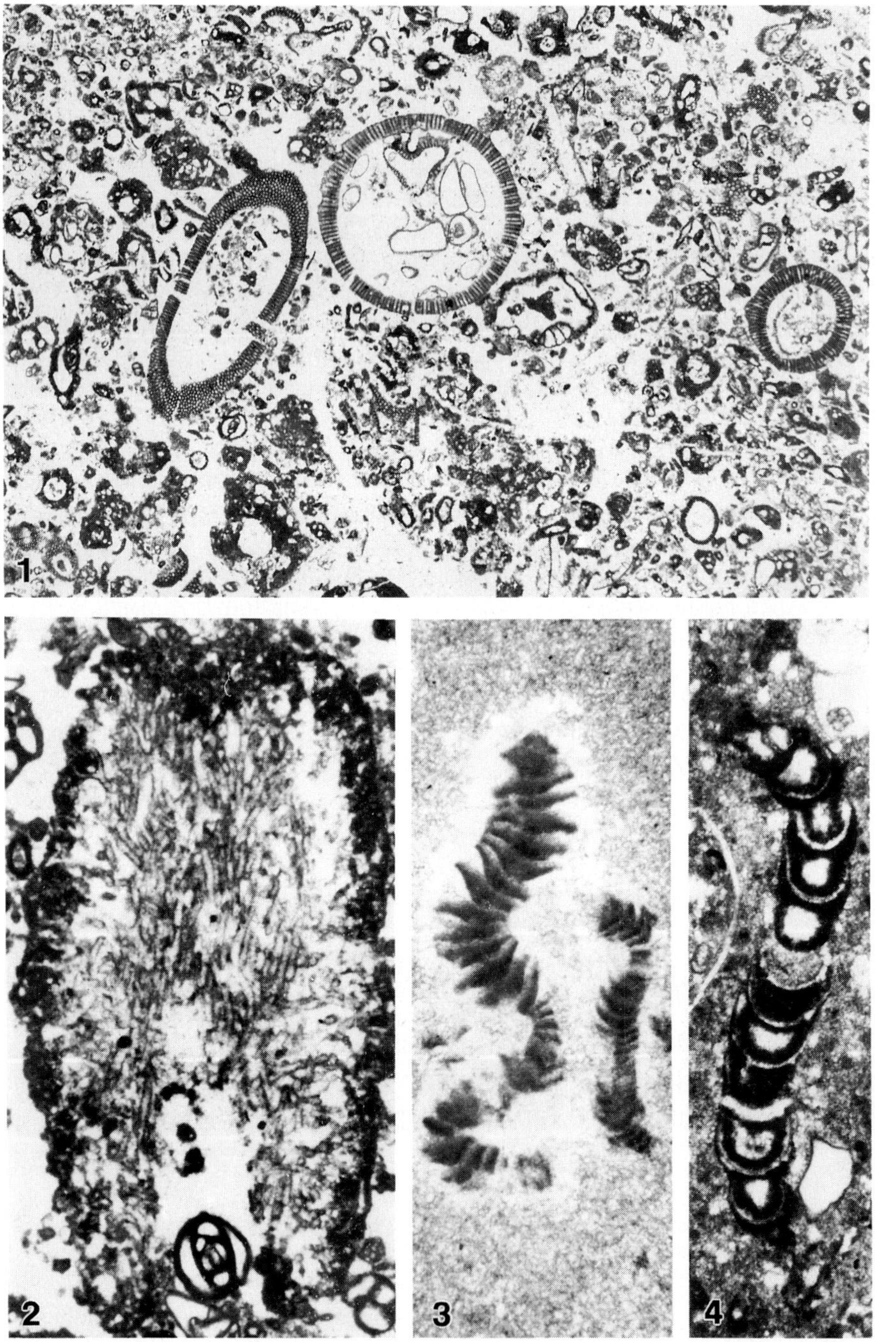

Plate II

the Anchicodiaceae (*Siphonales*) by Shuysky (1987b) and Chuvashov et al. (1987).

Recently Dragastan (1985) attributed some new genera established by himself from the Mesozoic of Romania to codiaceans, and also considered *Garwoodia* Wood and *Bevocastria* Garwood to be codiacean algae.

4.3 Volvocales, Thaumatoporellales and Tetrasporales

The order Volvocales is represented as fossils by mobile colonies (Volvocaceae) that lived particularly in shallow-water carbonate facies. According to Kazmierczak (1979, 1981) it is known in the Early Proterozoic with *Eosphaera* Barghoorn, according to Hofmann and Schopf (1983) *Eosphaera* is an incertae sedis genus (prokaryote?). Volvocaleans occur in the Devonian and Carboniferous with *Eovolvox* Kazmierczak and other taxa formerly attributed to non-radiosphaerid calcispheres and parathuramminid Foraminifera (Kazmierczak 1975, 1976, 1981).

The order Thaumatoporellales (Pl. II, 1) is wide-spread in the Mesozoic (from the Late Triassic onwards) carbonate platform facies. This order has been recently established by De Castro (1988a); it resembles volvocaleans in some characters and chladophoraleans in others.

The order Tetrasporales is represented as fossils by sessile dendroid forms (Chlorangiaceae) that flourished in shelf lagoons; they are well known in the Cretaceous with *Sgrossoella* De Castro (1969) (Pl. II, 4), related to the living genus *Ecballocystis* Yengar, and probably by the genus *Cretacicladus* Luperto Sinni (1979) (Pl. II, 3) related to the living *Prasinocladus* Kuckuck.

4.4 Ulotrichales and Chaetophorales

The Ulotrichales is mainly known from the genus *Vermiporella* Stolley. This genus, once attributed to dasycladaleans (Pia 1920), is referred by Kozlowski and Kazmierczak (1968) to this order (family uncertain) while it is considered a foraminifer by Vachard (1976). This Palaeozoic genus also doubtfully occurs in the Lower Cretaceous (Conrad 1970).

The genus *Heterotrichella* established by Schäfer and Senowbari-Daryan (1983), is tentatively referred by the same authors to the Chaetophorales.

Plate II. Chlorophyta (Caulerpales, Thaumatoporellales and Tetrasporales). *1* Assemblage of thaumathoporellaleans. Sections of cylindrical and globose thalli, sometimes with younger thalli inside (middle part of the figure), can be seen. Thin section A.5580.20, Uppermost Cretaceous of Pliačiaz (Mali Ston, Yugoslavia), ca. x10. *2 Boueina* sp. Longitudinal section. Thin section A.1974.27. Upper Cretaceous (Cenomanian) of Mt. Cerreto near Caserta (Campania, Italy), ca. x30. *3 Cretacicladus minervini* Luperto-Sinni. Longitudinal sections. Thin section A.1982.1. Cretaceous (Late Albian-Early Cenomanian) of Mt. Cerreto near Caserta (Campania, Italy), ca. x80. *4 Sgrossoella partenopeia* De Castro. Longitudinal section. Thin section A.3028.35, Upper Cretaceous (Senonian) of Cerreto Sannita (Campania, Italy), ca. x40 (De Castro 1969)

4.5 Dasycladales

The dasycladaleans are the most numerous Mesozoic fossil algae when compared to other groups. They are in fact represented by nearly 200 genera and little less than 900 species ranged from the upper Cambrian to the Recent. They attained high abundances in the Carboniferous-Permian, in the Upper Jurassic-Lower Cretaceous, and in the Palaeocene-Eocene. The genera have an average existence of about 53 Ma. The average number of species per genus is 4.7 (Fig. 2).

Understanding of this group of organisms has been made possible thanks mainly to the numerous contributions by Julius Pia, above all his "Neue studien über triadischen Siphoneae Verticillatae" (1912), the specifications given in Pia (1920), and finally the systematic scheme (Pia 1927) contained in Hirmer's Handbuch der Paläobotanik. Other important contributions include the pioneering work of Steinmann (1899, 1903), who was the first to show the importance of the position of reproductive organs for the systematics and the phylogeny of this order. Morellet and Morellet's studies (1913, 1922) on the Cenozoic dasycladaleans of the Paris Basin also hold particular interest.

For reference to numerous and often valuable contributions which have appeared in more recent times see Johnson and Konishi (1959), Johnson and Høeg (1961), Korde (1973) and Emberger's reviews (1976, 1978) for the Palaeozoic; Emberger (1979) for the Triassic and the synthesis of Bassoullet et al. (1978) for the Jurassic-Cretaceous and, finally, Deloffre and Génot (1982) and Génot (1987) for the Cenozoic. For recent classifications see Chuvashov et al. (1987), Deloffre (1987, 1988) and Shuysky (1987b).

General considerations of taxonomic criteria and concepts are supplied by Pia (1920), and Bassoullet et al. (1977, 1984). Statistical and biometrical methodologies have also been used to describe fossil dasyclad species (e.g. Remane 1969; Zorn 1974, 1977a; De Castro and De Rosa 1977; De Castro 1979; Barattolo 1978, 1980, 1981, 1982a,b,c, 1983a,b, 1984).

Aspects of reproduction in fossil dasycladaleans and its evolutionary significance are discussed by Steinmann (1903), Pia (1920), Elliott (1972), Barattolo (1980, 1983a) and Bucur (1985). Considerations of evolution, are supplied by Herak et al. (1977). The function of the branches within the framework of the algal construction is discussed by Pia (1920), Zorn (1972), Barattolo (1978), Elliott (1979b), Skompski (1984).

Palaeozoic Dasycladaleans

It is relevant here to briefly review the pre-Mesozoic history of dasycladaleans. Cambrian-Devonian dasycladaleans belong mainly to the Seletonellaceae, secondly to the Beresellaceae, occasionally to Diploporaceae and Dasycladaceae. In the Devonian, Acetabulariaceae seem also to be represented by the genera *Lulipora* Shuysky (1986) and *Lopsiella* Shuysky (1987a). These Lower Palaeozoic forms are mainly characterized by the following features: irregular shaped or simple thallus, the presence of only primary branches which

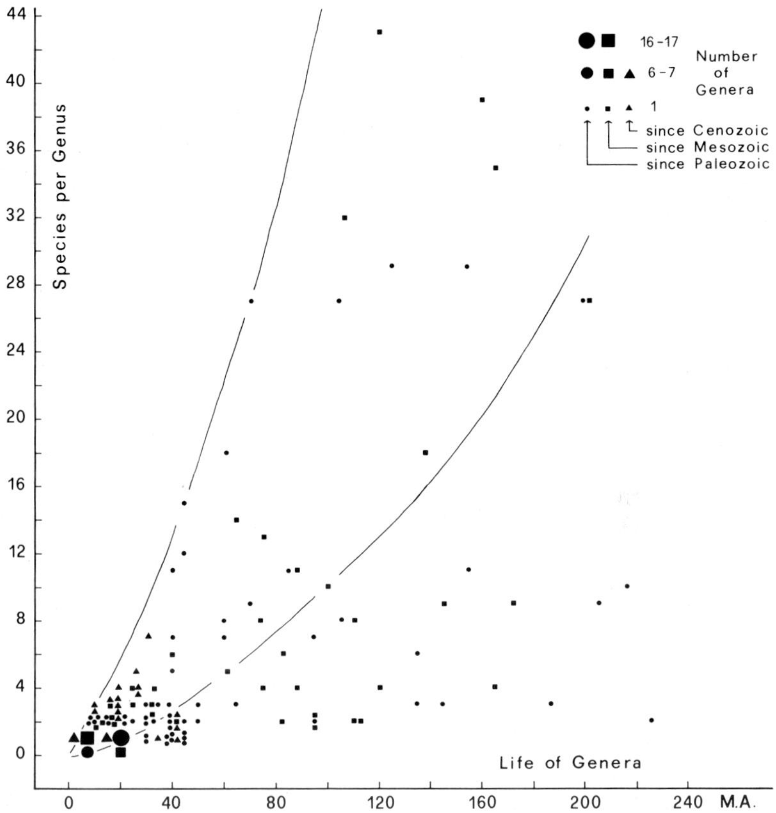

Fig. 2. Dasycladales. Number of species per genus plotted against the life of genera in Ma. *Circles, squares* and *triangles* represent genera which appeared in the Palaeozoic, Mesozoic and Cenozoic respectively; *small symbols* represent a single genus, *medium-sized ones* represent 6–7 genera and *large ones* represent 16–17 genera. The number of species per genus and the life of genera seem to be linked by an exponential law. The chart shows a high frequency of genera with few species and a low frequency of genera with many species. Outside the two curves, towards the right, relatively new and/or poorly known genera, together with others having doubtful species or an uncertain stratigraphical range fall

are in aspondyle arrangement, and reproductive organs of endosporate-type. Reproductive features can be only inferred for the majority of taxa with relatively thin branches because these ones were presumably not suited to contain reproductive organs. Fewer genera show twofold branches, euspondyle arrangement, and reproductive organs of cladosporate-type (Fig. 3). With the genus *Archaeobatophora* Nitecki (1976) in the Ordovician the presence of a choristosporous tribe (Batophoreae) has been noted. Nevertheless, this genus has no trace of choristosporate reproductive organs. During the Lower Palaeozoic and Devonian dasycladaleans are principally represented by a small number of genera (45), mainly characteristic of each period, and by ca. 80

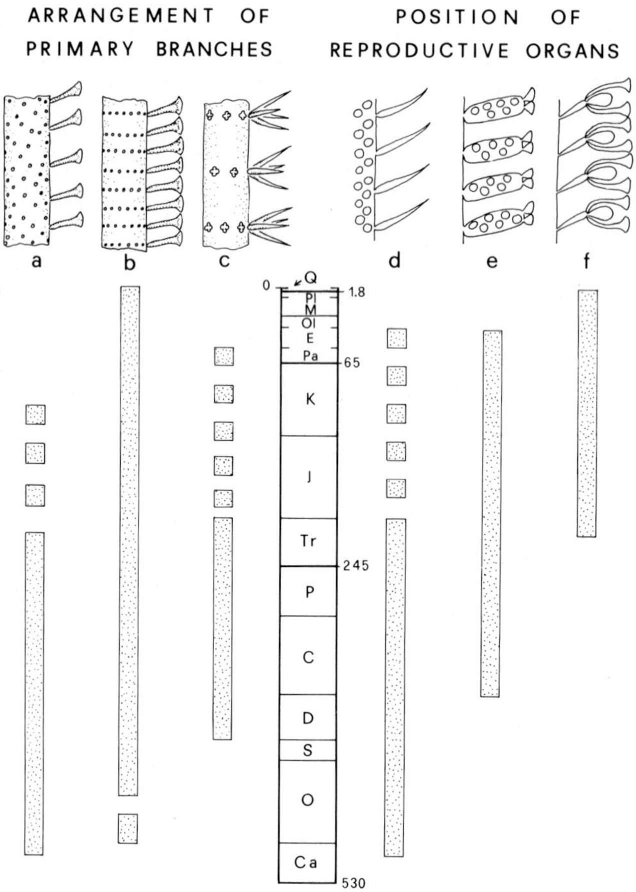

Fig. 3a-f. Dasycladales. Stratigraphic distribution of primary branch arrangement (**a-c**) and position of reproductive organs (**d-f**) during the Phanerozoic. **a** aspondyle; **b** euspondyle; **c** metaspondyle; **d** endosporate-type; **e** cladosporate-type; **f** choristosporate-type. A *broken line* indicates doubtful or very rare presence. Time scale is from Odin (1982)

species. The number of species per genus is also rather low: about 1.4 (Fig. 4).

The Carboniferous-Permian represents the period of maximum Palaeozoic expansion of dasycladaleans. They belong particularly to the Seletonellaceae, Beresellaceae, Diploporaceae and Dasycladaceae. The order is represented by about 80 genera and 300 species; the importance of the dasycladaleans in this interval is also shown by the relatively high number of species (2.6 species per Ma) related to the duration of the period (115 Ma).

The morphological styles which appeared in the preceding periods (coarsely branched thalli, aspondyle branch arrangement, endosporous re-

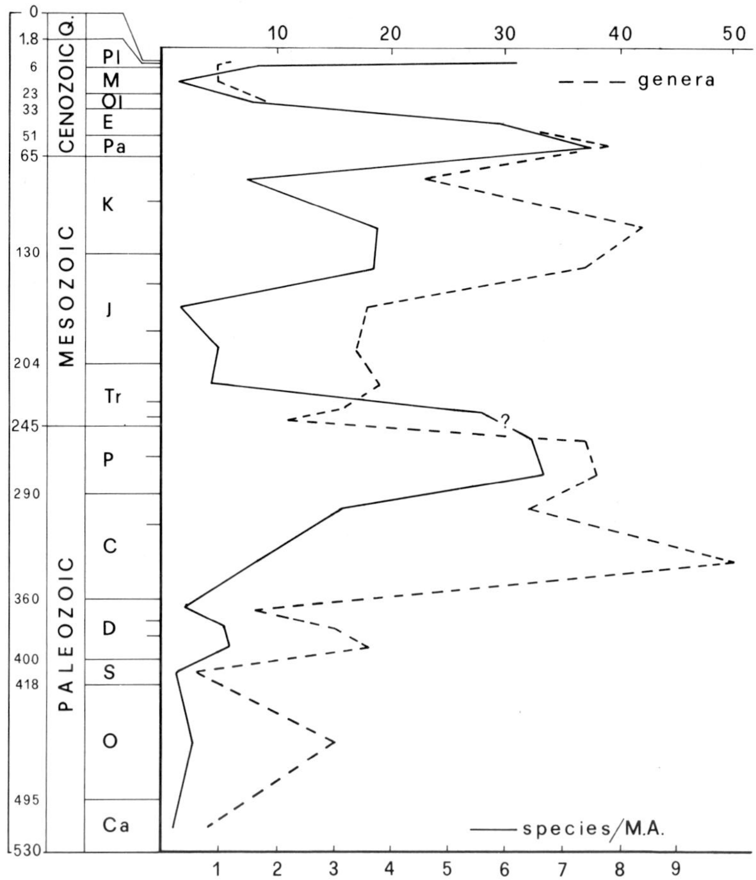

Fig. 4. Dasycladales. Number of genera per time interval and number of species per time interval per Ma during the Phanerozoic. No Lower Triassic species are known, the value supplied is merely speculative. Lower Triassic genera are the ones present both in the Palaeozoic and in the Middle Triassic. Only strongly calcified living species are taken into consideraton for the Quaternary values. Time scale is from Odin (1982)

productive organs) persisted in the Carboniferous-Permian, while the more "evolved" ones that were present, but not well represented in the Cambrian-Devonian, become important.

The principal latter characters of the thalli are:

1. A prevalently euspondyle distribution of the branches, even though aspondyle (e.g. *Antrachoporella* Pia, *Gyroporella* Gümbel) and metaspondyle arrangements are also well represented. Metaspondyle dasycladaleans are represented by as many "vestibuled" taxa (e.g. *Velebitella* Kochansky-Devidé and *Albertaporella* Johnson) as "non-vestibuled" ones (e.g. *Diplopora* Shafhäutl

and *Clavapora* Güvenc). For details of the shape and the systematic value of the vestibule in the metaspondyle forms, see Mamet and Roux (1981, 1983);
2. Reproductive organs are mainly of endosporate type and, secondly, of cladosporate type; the taxa with relatively swollen branches, that presumably could have contained reproductive organs, are probably to be assigned to the cladosporous forms: *Gyroporella* Gümbel, characterized by vesiculiferous branches, *Physoporella* Steinmann characterized by piriferous branches and *Kulikia* Golubtshov with swollen phloiophorous branches;
3. Prevalence of forms with only primary branches, even though twofold branched genera with dichotomic (e.g. *Antracoporella* Pia) and the tuft-like ramifications (e.g. *Dasycladophycus* Elliott) also occur. The branches are of akrophorous (e.g. *Queenslandella* Mamet and Roux), trichophorous (e.g. *Likanella* Milanovic and *Teutloporella* Pia) and phloiophorous (*Mizzia* Shubert) type; vesiculiferous (e.g. *Gyroporella* Gümbel) and piriferous (e.g. *Physoporella* Steinmann) morphological variations also occur.

Mesozoic Dasycladaleans

The Mesozoic dasycladaleans belong mainly to the Dasycladaceae, to a lesser extent to the Acetabulariaceae and Diploporaceae, occasionally to the Seletonellaceae. The Mesozoic is on the whole an era of great diversification of dasycladaleans; these undergo remarkable variations both in the vegetative parts and the reproductive organs. The vegetative parts show the disappearance of the metaspondyle and aspondyle forms, although the metaspondyle arrangement seems to reappear, almost like an accident, with the genus *Furcoporella* Pia in the Palaeocene. The endosporous forms disappear in the Mesozoic; while the cladosporous forms are widespread and the cladosporous reproductive organs exhibit the greatest degree of diversification. Choristosporous forms appear and attain a moderate importance in the Mesozoic.

The following phases can be identified during the era in the history of dasycladaleans: persistence of Late Palaeozoic forms into the Triassic, decline and renewal during the Lower-Middle Jurassic, Upper Jurassic-Lower Cretaceous acme, Upper Cretaceous collapse.

Persistence of Late Palaeozoic Forms into the Triassic. Contrary to what was once believed, the Triassic represents a period of decrease of dasycladaleans when compared to the Permian and especially to the Lower Permian, and in fact no dasycladaleans are known to occur in the Lower Triassic. In the period as a whole (41 Ma), altogether about 24 genera are present (16 in the Middle Trias and 19 in the Upper), and 80 species (56 in the Middle Trias and 26 in the Upper); the number of species per Ma is about 1.95.

Compared to the Permian and especially to the Upper Permian the number of species per genus is the same (4.5) while the number of genera decreases considerably (37 in the Upper Permian, 16 in the Middle Trias).

During the Triassic the characters of the thallus, and most of the taxa, are still those of the Carboniferous-Permian. In the Middle Triassic the genera

Diplopora Schafhäutl, *Teutloporella* Pia, *Oligoporella* Pia, *Physoporella* Steinmann and *Macroporella* Pia prevail and *Kantia* Pia, *Clavapora* Güvenc, *Anisoporella* Botteron, *Griphoporella* Pia and *Neoteutloporella* Bassoullet et al. appear.

Diplopora annulata Schafhäutl (Pl. III, 2) and *Oligoporella chia* Herak seem to be species adapted to live in the middle parts of the lagoons (Ott 1972); *Diplopora nodosa* Schafhäutl (Pl. III, 5) and *Teutloporella herculea* (Stoppani) seem to be adapted to live in the outer lagoon facies (Ott 1972, Zorn 1976, De Castro 1979); *Teutloporella echinata* Ott flourished in the reef facies (Zorn 1976).

Zorn (1976), in particular, has shown that the epicontinental platform facies that received a terrigenous sedimentary contribution, were populated by *Diplopora hexaster*-like forms and *Oligoporella* and *Physoporella* species with paired whorls. The epioceanic shelf lagoons were mainly populated by *Macroporella*, *Teutloporella* and *Oligoporella* or *Physoporella* species with simple whorls. Further information on the distribution of numerous dasycladaleans in the lagoon facies is given by Dragastan (1981).

In the Upper Triassic most of the preceding genera disappear, there is a rapid reduction of the dasycladaleans and at the same time same renewal. The more frequent genera are *Griphoporella* Pia, *Gyroporella* Gümbel and *Poikiloporella* Pia (= *Andrusoporella* Bystricky). *Heteroporella* Praturlon, *Pentaporella* Senowbari-Daryan and *Placklesia* Bilgutäy appear; the last two rather recently established genera do not seem to survive to the end of the period. With the appearance of *Heteroporella* the Upper Trias marks the appearance of choristosporous reproductive organs. However, this character, seen in *Heteroporella* and in other genera (*Cylindroporella* Johnson, *Neomeris* Lamouroux and *Cymopolia* Lamouroux), only gains importance in the Cretaceous.

In the lagoon facies, more or less close to the outer lagoon margin, *Griphoporella curvata* (Gümbel; Pl. III, 1), *Heteroporella zankli* Ott and *Heteroporella crosi* Ott dominate; in shelf margin facies *Diplopora tubispora* Ott and *Diplopora phanerospora* Pia flourished; in the shoal facies it is possible to find *Placklesia multipora* Bilgutäy, probably redeposited (Abate et al. 1977; Flügel 1979; Senowbari-Daryan and Schäfer 1979; Di Stefano and Senowbari-Daryan 1985).

Decline and Renewal During the Lower-Middle Jurassic. Compared with the Upper Triassic, the Lower-Middle Jurassic represents a period of decrease of dasycladaleans. The number of genera (about 17 in the Lias and 18 in the Dogger), when compared with the Upper Triassic (about 19 genera), remains more or less the same; similar behaviour is seen by the species/genera ratio (1.4 in the Upper Trias, 1.5 and 1.4 respectively in the Lias and in the Dogger). However, the number of species tends to decrease even compared with the duration of the time interval (species/Ma.).

Lower-Middle Jurassic dasycladaleans are almost exclusively euspondyle and the number of genera with only primary branches is approximately the same as that with more orders of branches. Reproduction was prevalently of cladosporate-type (e.g. *Dissocladella* Pia, *Fanesella* Cros and Lemoine, *Sarfa-*

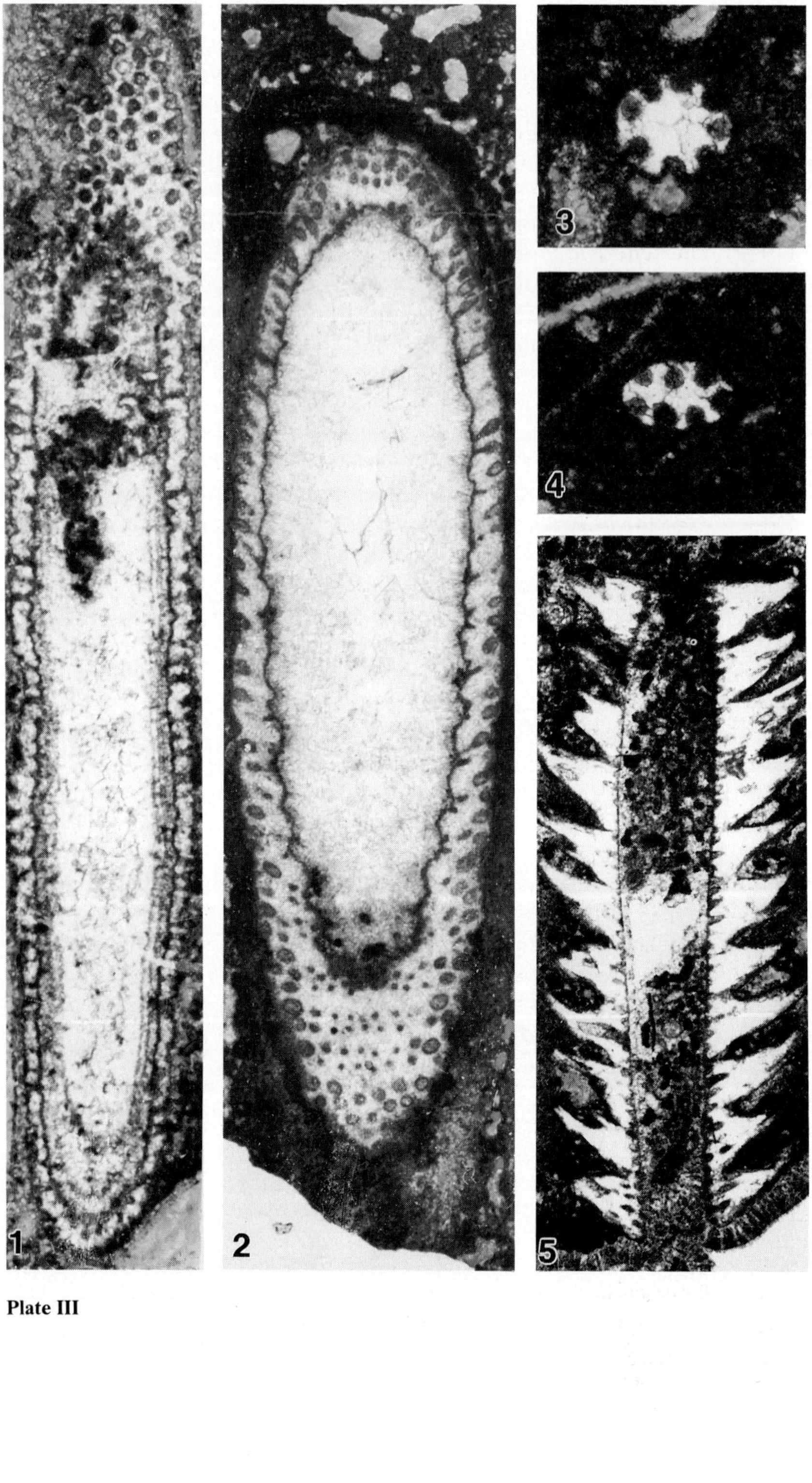

Plate III

tiella Conrad and Peybernés, *Stichoporella* Pia etc.), and maybe of choristosporate-type as well (*Eodasycladus* Cros and Lemoine). The branches are trichophorous (e.g. *Selliporella* Sartoni and Crescenti and *Neoteutloporella* Bassoullet et al.), akrophorous (e.g. *Fanesella* Cros and Lemoine, *Dobuniella* Elliott etc.), and phloiophorous (e.g. *Neomizzia* Levy, *Dinarella* Sokăć and Niklér, *Sarfatiella* Conrad and Peybernés and *Stichoporella* Pia).

Almost half of all Lower Jurassic genera appear during this time interval; the other half had already mainly appeared in the Upper Triassic. The more frequent Liassic genera are *Palaedasycladus* Pia and *Sestrosphaera* Pia; besides these two genera, *Dinarella* Sokăć and Niklér, *Tersella* Morellet and *Eodasycladus* appear and are also restricted to the Lias; whereas *Neomizzia*, *Dissocladella* and *Fanesella* are also known in younger rocks. *Palaedasycladus mediterraneus* Pia (Pl. IV, 1) and *P. barrabei* Lebouché and Lemoine seem to be present both in the lagoon facies and in those close to the shelf margin. For further detailed information see mainly Sartoni and Crescenti (1959), De Castro (1962), Chiocchini (1977), Chiocchini and Mancinelli (1977, 1978) and Peybernés (1979).

The genera *Sarfatiella* Conrad and Peybernés, *Coniporella* Fischer and Thierry, *Dobunniella* Elliott, *Stichoporella* Pia, *Selliporella* Sartoni and Crescenti appear in the Dogger; the last three are restricted to the Dogger. *Selliporella* has been recently redescribed by Barattolo, De Castro and Radoičić (1988).

In the lagoon facies *Selliporella donzellii* Sartoni and Crescenti (Pl. IV, 5) is dominant; *Sarfatiella dubari* Conrad and Peybernés dominates in the lagoon facies next to the shelf margin (De Castro 1962, 1987; Sartoni and Crescenti 1962; Chiocchini and Mancinelli 1977; Sokăć and Velić 1978a,b; Peybernés 1979).

According to Elliott (1977b, 1982, 1984) dasycladaleans gradually spread out, starting in the Triassic, from tropical towards temperate zones; in the Middle Jurassic in particular it is possible to find a differentiation due to climatic causes between the dasycladaleans of northern and southern Europe.

Upper Jurassic-Lower Cretaceous Acme. The Upper Jurassic-Lower Cretaceous represents a time interval of considerable increase of dasycladaleans; in fact we pass from 11 species in the Dogger to 73 species in the Malm and to 131 species in the Lower Cretaceous. The same goes for the behaviour of the number

Plate III. Chlorophyta (Dasycladales). *1 Griphoporella curvata* (Gümbel). Longitudinal-oblique section. Thin section A.7776.4, Upper Triassic (Rhetian), Erchie (Campania, Italy), ca. x15. *2 Diplopora annulata* Schafhäutl. Oblique section. Thin section A.6884.1, Middle Triassic, vicinity of Gračać (Velebit, Yugoslavia), ca. x10. *3–4 Terquemella* sp. Respectively equatorial and axial sections. These small lenticular bodies represent calcified reproductive organs of some dasycladaleans (e.g. Bornetelleae). Thin section A.2959.2, Lower Cretaceous (Aptian-Albian) of Piano di Pecore near Laviano (Campania, Italy), ca. x120. *5 Diplopora nodosa* Schafhäutl. Longitudinal section. Thin section A.3010.24, Middle Triassic of Il Castello near Marsico Nuovo (Basilicata, Italy), ca. x10

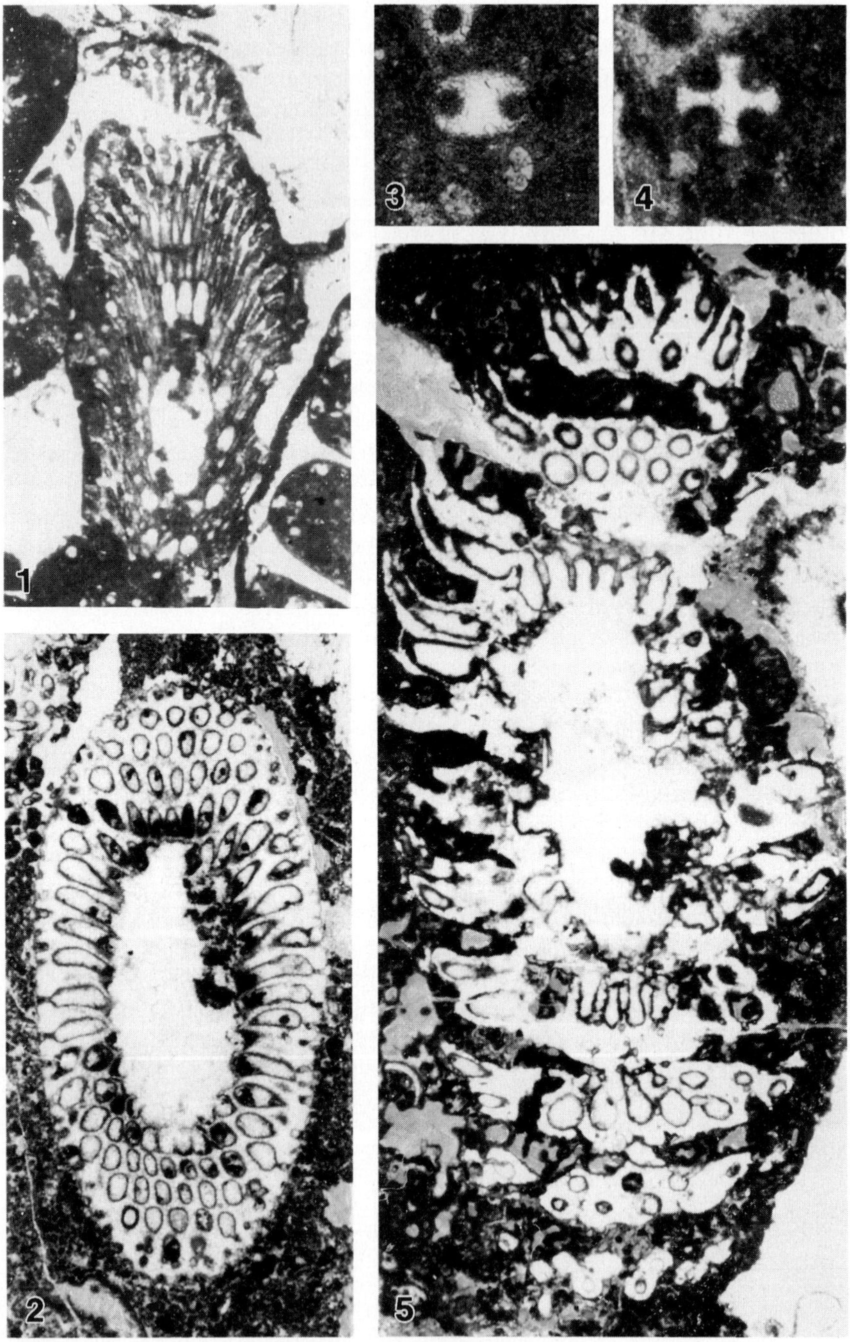

Plate IV

of genera, passing from 18 genera in the Dogger to 37 genera in the Malm and to 43 genera in the Lower Cretaceous; this last value represents the generic maximum during the Mesozoic and Cenozoic.

Particularly in the Malm there is a considerable generic renewal (about 60%), while the number of species per genus remains rather low; in the Lower Cretaceous the opposite happens: the percentage of new genera decreases while the number of species per genus increases considerably (Fig. 5). A similar phenomenon happens in the Palaeocene-Eocene.

As regards morphology, thalli have mainly euspondyle branches; aspondyle-branched forms (e.g. *Megaporella* Deloffre and *Gyroporella* Gümbel) only occasionally occur. The reproductive organs are mainly cladosporous (*Neogyroporella* Yabe and Toyama; *Sarfatiella* Conrad and Peybernés; *Praturlonella* Barattolo, Pl. VII, 1; *Triploporella* Steinmann; *Suppiluliumaella* Elliott; *Pseudotriploporella* Jaffrezo and Poisson; *Megaporella* Deloffre; *Gyroporella* Gümbel, etc.) even though the choristosporate-type is not uncommon (*Cylindroporella* Johnson, *Heteroporella* Praturlon, *Neomeris* Lamouroux, Pl VII, 2; *Actinoporella* Gümbel, etc.); doubtful endosporous forms also occur (*Linoporella* Steinmann, *Campbelliella* Radoičić, etc.).

About two-thirds of the genera included in this time interval have twofold branches.

In the Upper Jurassic-Lower Cretaceous *Salpingoporella* Pia, *Clypeina* Michelin, *Cylindroporella* Johnson, *Heteroporella* Praturlon, and *Triploporella* Steinmann (Pl. IV, 2) are dominant.

Isolated calcified reproductive organs, probably assignable to dasycladaceans (*Russoella* Barattolo and *Terquemella* Morellet and Morellet; Pl. III, 3–4 and Pl. IV, 3–4) and acetabulariaceans (*Acicularia* D'Archiac), also occur.

In the lagoon and/or inner margin facies *Salpingoporella grudii* Radoičić, *S. annulata* Carozzi (Pl. VI, 2), *Clypeina jurassica* Favre (Pl. V, 1) and *Campbelliella striata* (Carozzi) (Pl. V, 2) dominate. In the margin facies *Salpingoporella pygmaea* (Gümbel), *Neoteutloporella socialis* (Praturlon) (Pl. VI, 1) and *Petrascula bursiformis* (Etallon) (Praturlon 1963, 1966; Chiocchini and Mancinelli 1978; Flügel 1979; Peybernés 1979; Bernier 1984) dominate. In the Lower Cretaceous lagoon facies *Triploporella neocomiensis* Radoičić,

Plate IV. Chlorophyta (Dasycladales). *1 Palaeodasycladus mediterraneus* Pia. Oblique section. Primary, secondary and, less clearly, tertiary branches can be seen. Thin section 420.1, Lower Jurassic of L'Ascensione near Nocera Inferiore (Campania, Italy), ca. x15. *2 Triploporella marsicana* Praturlon. Oblique section. Notice the large, stout primary branches and the tufts of secondaries. Thin section A.5874.45, Lower Cretaceous of Serra Campo dell'Acero near Matese Lake (Campania, Italy), ca. x15. *3–4 Russoella radoiciciae* Barattolo. Axial (*3*) and equatorial (*4*) sections. *Russoella* represents calcified reproductive organs of some dasycladaleans (e.g. Triploporelleae and Bornetelleae). For both figures: thin section A.2168.4, Lower Cretaceous (Albian) of Mt. Calvi near Caserta (Campania, Italy), ca. x120. *5 Selliporella donzellii* Sartoni and Crescenti. Oblique section by De Castro (1962). Short and stout primary branches and tufts of two trichophorous secondary branches can be seen (see also Barattolo, De Castro and Radoičić 1988, Fig. 1). Thin section 269.1, Jurassic (Middle Dogger), Case Ciaramello near Nocera Inferiore (Campania, Italy), ca. x15

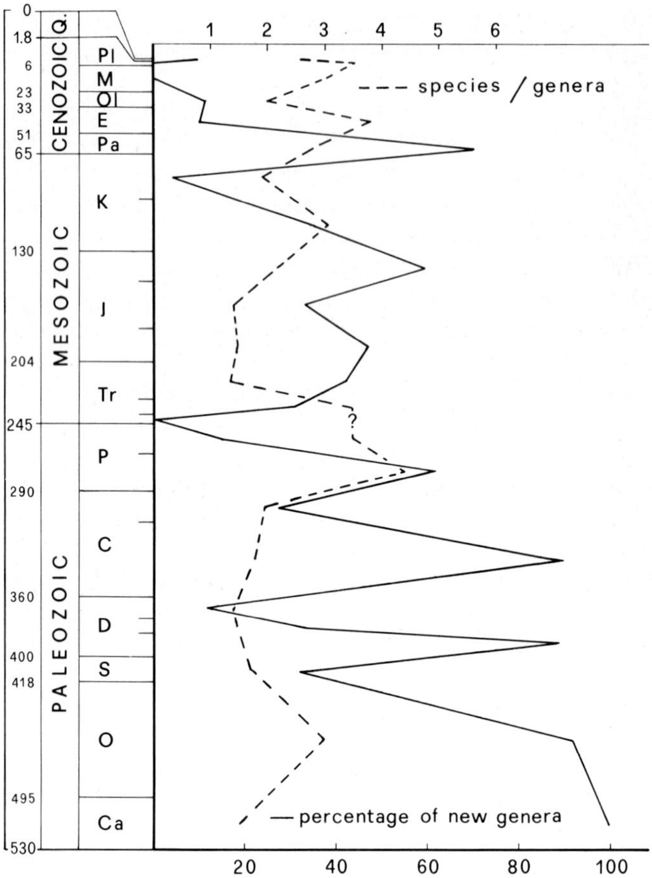

Fig. 5. Dasycladales. Species-genera ratio and percentage of new genera (first appeared genera-total genera ratio per time interval) during the Phanerozoic. The species-genera ratio in the Lower Triassic is merely speculative. Only strongly calcified living species are taken into consideration for the Quaternary values. Time scale is from Odin (1982)

Plate V. Chlorophyta (Dasycladales). *1 Clypeina jurassica* Favre. Variously sectioned verticils of disarticulated thalli. Thin section 421.1, Uppermost Jurassic of Colle di S, Pietro near Nocera Inferiore (Campania, Italy), ca. x10. *2 Campbelliella striata* (Carozzi). Variously sectioned verticils of disarticulated thalli. Thin section A.6834, Uppermost Jurassic-Lowermost Cretaceous, Tuoro di Sasso near Nola (Campania, Italy), ca. x10

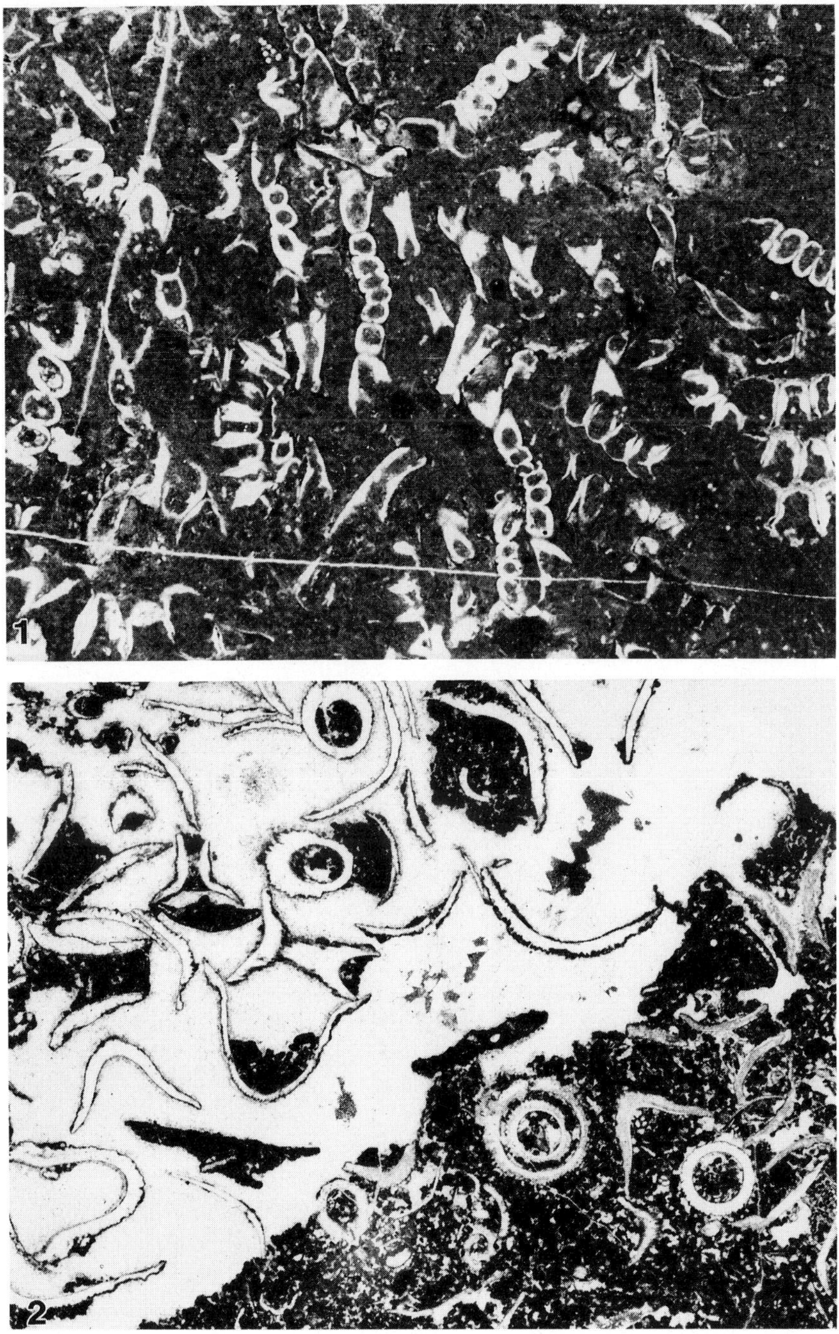

Plate V

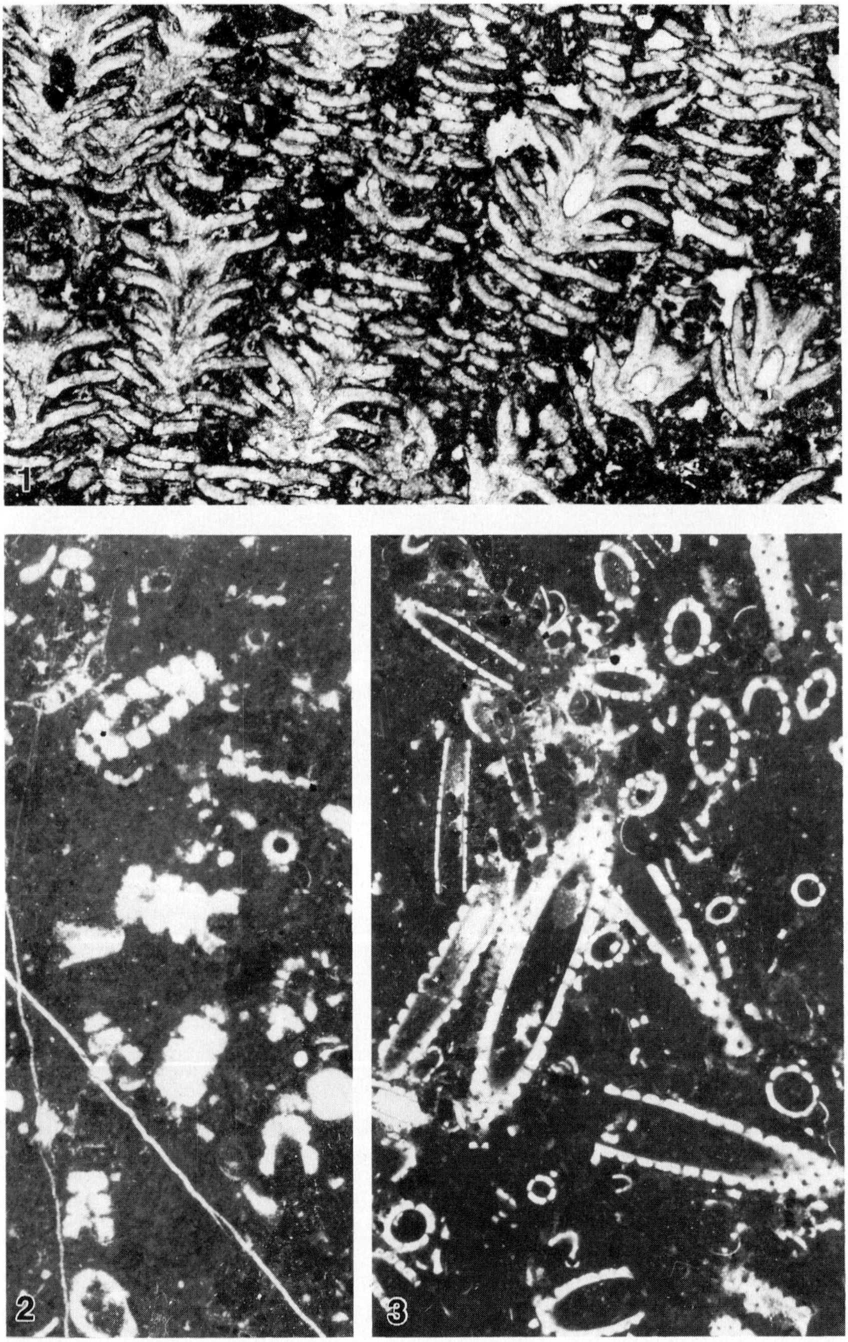

Plate VI

Macroporella embergeri Boroullec and Deloffre, *Likanella campanensis* Azema and Jaffrezo, *Clypeina radici* Sokăć and Velić and *Salpingoporella dinarica* Radoičić (Pl. VI, 3) dominate.

In the margin facies *Macroporella praturloni* Dragastan, "*Griphoporella*" *piae* Dragastan and *Salpingoporella genevensis* Conrad, *Acroporella radoiciciae* Praturlon etc. (e.g. Praturlon 1964, 1966; Chiocchini and Mancinelli 1977, 1978; Conrad 1977; Velić 1977; Sokăć and Velić 1978b) dominate.

Upper Cretaceous Collapse. In the Upper Cretaceous (duration about 30 Ma) we witness a rapid decrease in the number of both genera and species of dasycladaleans; the former decreases from 44 (Lower Cretaceous) to 23; the calculation takes into account the Permian genus *Salopekiella* Milanovic, the presence of which in the Maastrichtian (*S. inopinata* Gušić) is perplexing (Herak et al. 1979). The number of species decreases from 131 (Lower Cretaceous) to 45. The number of species per Ma is only 1.5.

All the Upper Cretaceous genera apart from *Harlanjohnsonella* Elliott were present earlier and mainly in the Lower Cretaceous; *Likanella* Milanović is known from the Permian too. The percentage of new genera is very low (4%); ever lower values can also be found in the Upper Devonian, Lower Triassic and Neogene.

As regards morphology, the thalli are exclusively euspondyle; the reproductive organs are mainly either cladosporous (*Triploporella* Steinmann, *Harlanjohnsonella* Elliott, *Suppiluliumaella* Elliott, *Dissocladella* Pia, etc.) or choristosporous (*Neomeris* Lamouroux, *Cymopolia* Lamouroux, *Cylindroporella* Johnson, etc.), even if the presence of endosporous species cannot be excluded (*Acroporella* Praturlon).

Most genera show twofold branches. The most common genera are *Trinocladus* Raineri, *Cymopolia* Lamouroux (Pl. VII, 4), *Neomeris* Lamouroux, *Salpingoporella* Pia, *Clypeina* Michelin, *Cylindroporella* Johnson (Pl. VII, 3) and *Heteroporella* Praturlon.

Cenozoic Dasycladaleans

Cenozoic dasycladaleans belong mainly to the Dasycladaceae and Acetabulariaceae. Other families are either not known, or their presence is questionable.

The beginning of the Cenozoic marks a renewal in the development of dasycladaleans. In the Palaeocene (duration about 14 Ma) the number of genera

◄

Plate VI. Chlorophyta (Dasycladales). *1 Neoteutloporella socialis* (Praturlon). Longitudinal-oblique sections of closely packed thalli. Thin section 697. Uppermost Jurassic of Mt. Corno (Abruzzi, Italy), ca. x10. *2 Salpingoporella annulata* Carozzi. Variously sectioned specimens. Thin section A.4498.1. Lowermost Cretaceous (Neocomian) of Mt. Monaco di Gioia near Faicchio (Campania, Italy), ca. x25. *3 Salpingoporella dinarica* (Radoičić). Variously sectional specimens. Thin section A.5940.4, Lower Cretaceous (Aptian), Suio Terme (Latium, Italy), ca. x25

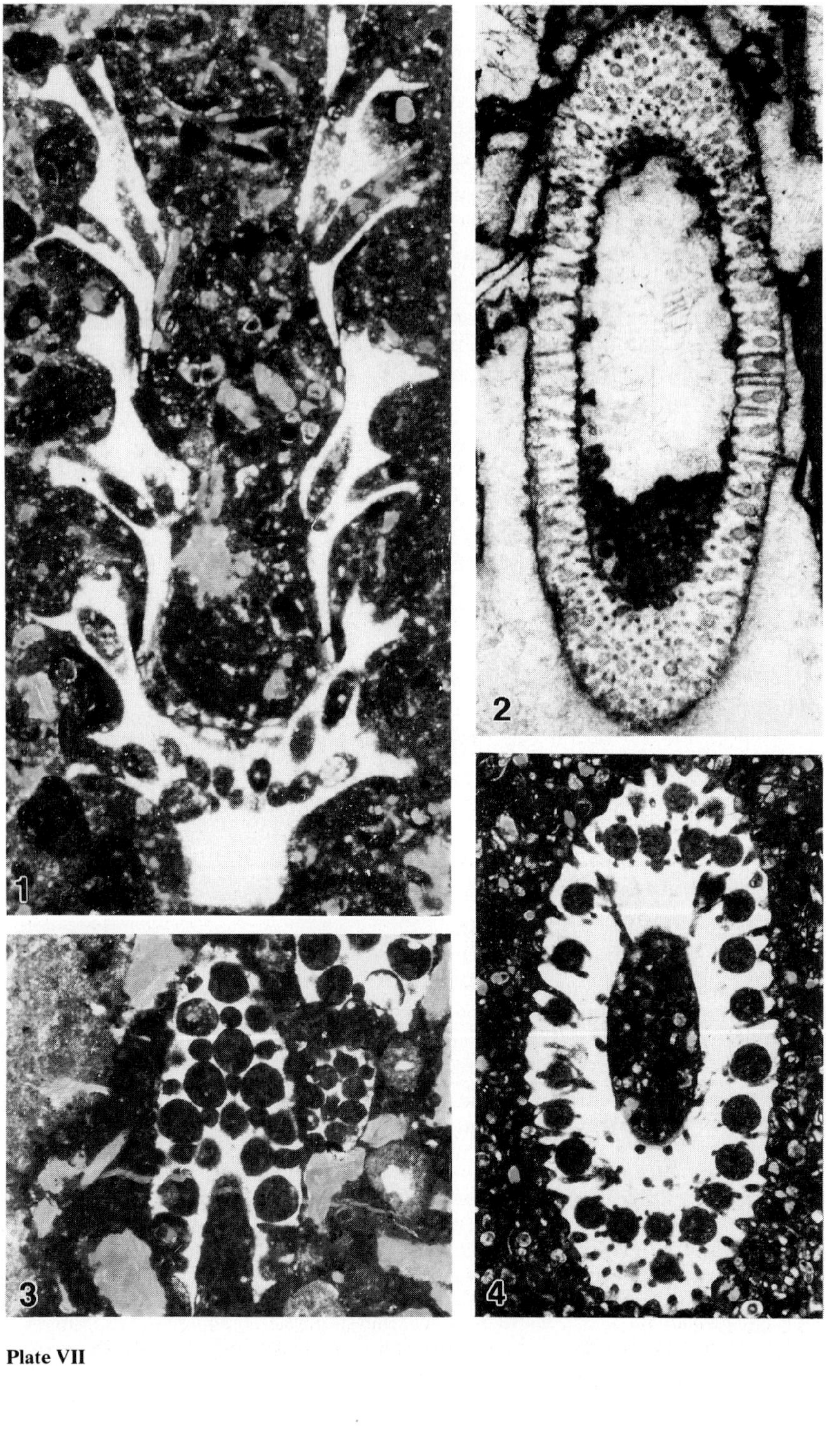

Plate VII

rises from 23 (Upper Cretaceous) to 39; the number of species rises from 45 (Upper Cretaceous) to 106. The number of species per Ma is 7.5; when compared to those of other geologic intervals this value represents the absolute maximum (Fig. 4). The percentage of new genera and the number of species per genus (Fig. 5) show a similar behaviour to the Upper Jurassic-Lower Cretaceous interval. After the Eocene the group gradually decreases in importance up to present.

In Figs. 4 and 5, in order to make the Recent data comparable with the fossil ones, only extant genera and species with significant calcification have been taken into consideration.

In the Recent the order is represented only by the two families, Acetabulariaceae and Dasycladaceae, both living in environments comparable to those of the past, i.e. mainly in the shelf lagoon or inner platform and subordinately in shelf margin habitats (Valet 1979). They mainly populate the tropical and subtropical zones, with few species in the Mediterranean: *Acetabularia mediterranea* Lamouroux, *A. calyculus* Quoy and Gaimard, *A. parvula* Solms-Laubach and *Dasycladus vermicularis* (Scopoli). Only the first of these species calcifies significantly.

The family Acetabulariaceae is at present represented by two living genera and 20 species. The family Dasycladaceae is represented by 6 genera and 19 species. The forms that live in reef environments belong mainly to the genera *Bornetella* Munier-Chalmas, *Neomeris* Lamouroux and *Acetabularia* Lamouroux (section *Polyphysa*).

For the stratigraphical distribution of Mesozoic and Cenozoic dasycladaleans see Fig. 6.

5 Incertae Sedis

Among the benthic algae a certain number of taxa is present which, although assignable to the algae on the basis of their vegetative simplicity, are not closely comparable with known taxa of thallophytes. They are not taken into consideration in Fig. 1. These *incertae sedis* forms are numerous; some of them have already been dealt with and assigned to higher taxa as proposed by some authors.

Plate VII. Chlorophyta (Dasycladales). *1 Praturlonella salernitana* Barattolo. Longitudinal section. Thin section A.3519.a.76, Lower-Middle Eocene of Mt. Vesole near Trentinara (Campania, Italy), ca. x30 (Barattolo 1978). *2 Neomeris cretacea* Steinmann. Oblique section. Note the ellipsoidal ampullae and the slender secondary branches, primary branches are uncalcified. Thin section n.9, Lower Cretaceous (Albian) of Cerro Escamela near Orizaba (Varacruz, Mexico), ca. x20. *3 Cylindroporella kochanskyae* Radoičić. Oblique-tangential section. Note the sterile (secondary) branches and reproductive organs (ampullae) alternated in the same verticil. Thin section A.4156.1, Upper Cretaceous (Early Turonian) of Milanovăć plantăć (Okovan, Yugoslavia), ca. x40. *4 Cymopolia* cf. *paronai* Raineri. Oblique section. Primary branches, ampullae and tufts of phloiophorous secondary branches can be seen. Thin section A.6750.1, Upper Palaeocene (Thanetian) of Colle di Medea near Manzano (Friuli, Italy), ca. x40

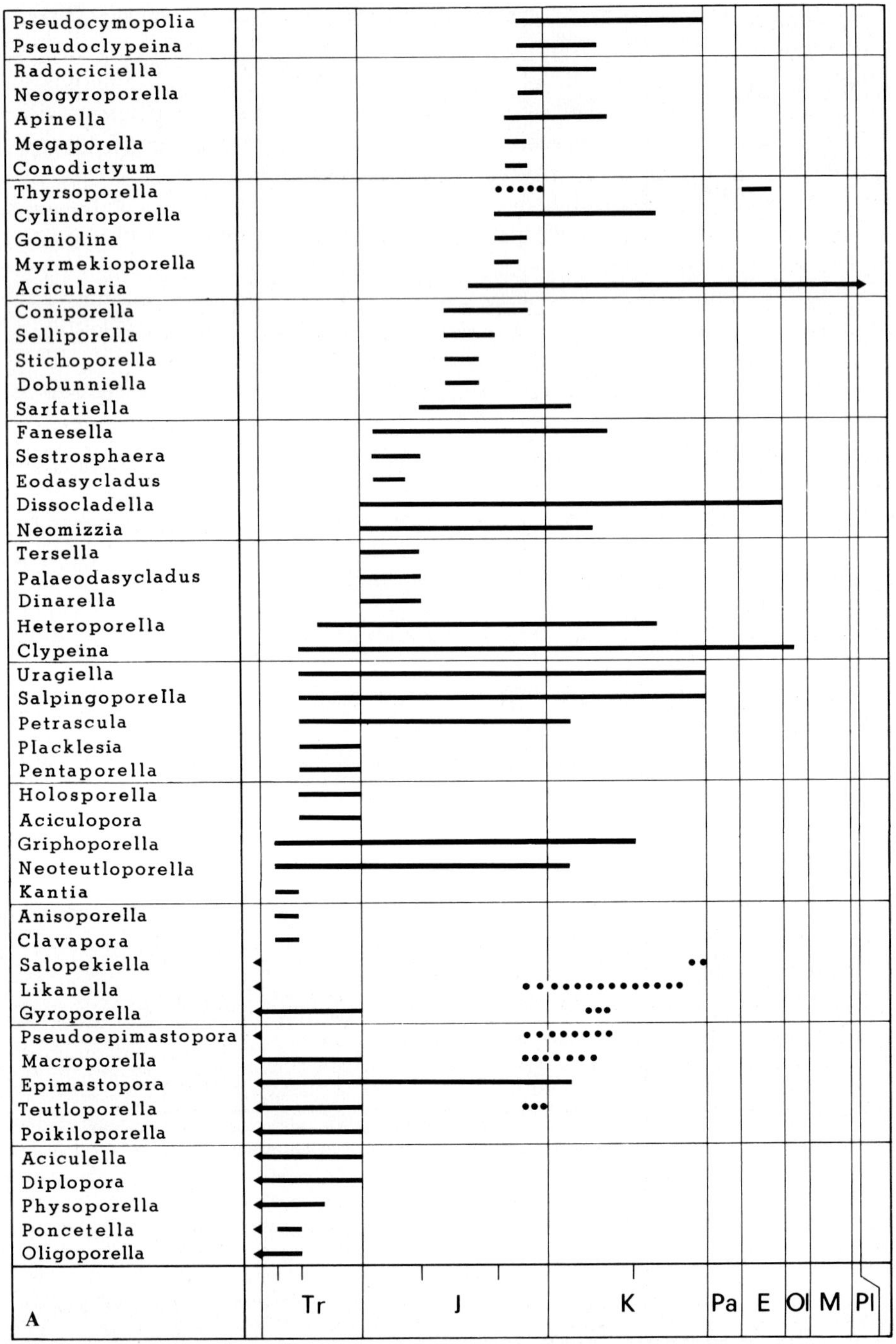

Fig. 6A,B. Dasycladales. Stratigraphic distribution of Mesozoic and Cenozoic genera. ◆ Genus known also in the Palaeozoic; ➧ extant genus; ----: genus with uncertain stratigraphical range; ••••: uncertain occurrence of the genus. Time scale is from Odin (1982)

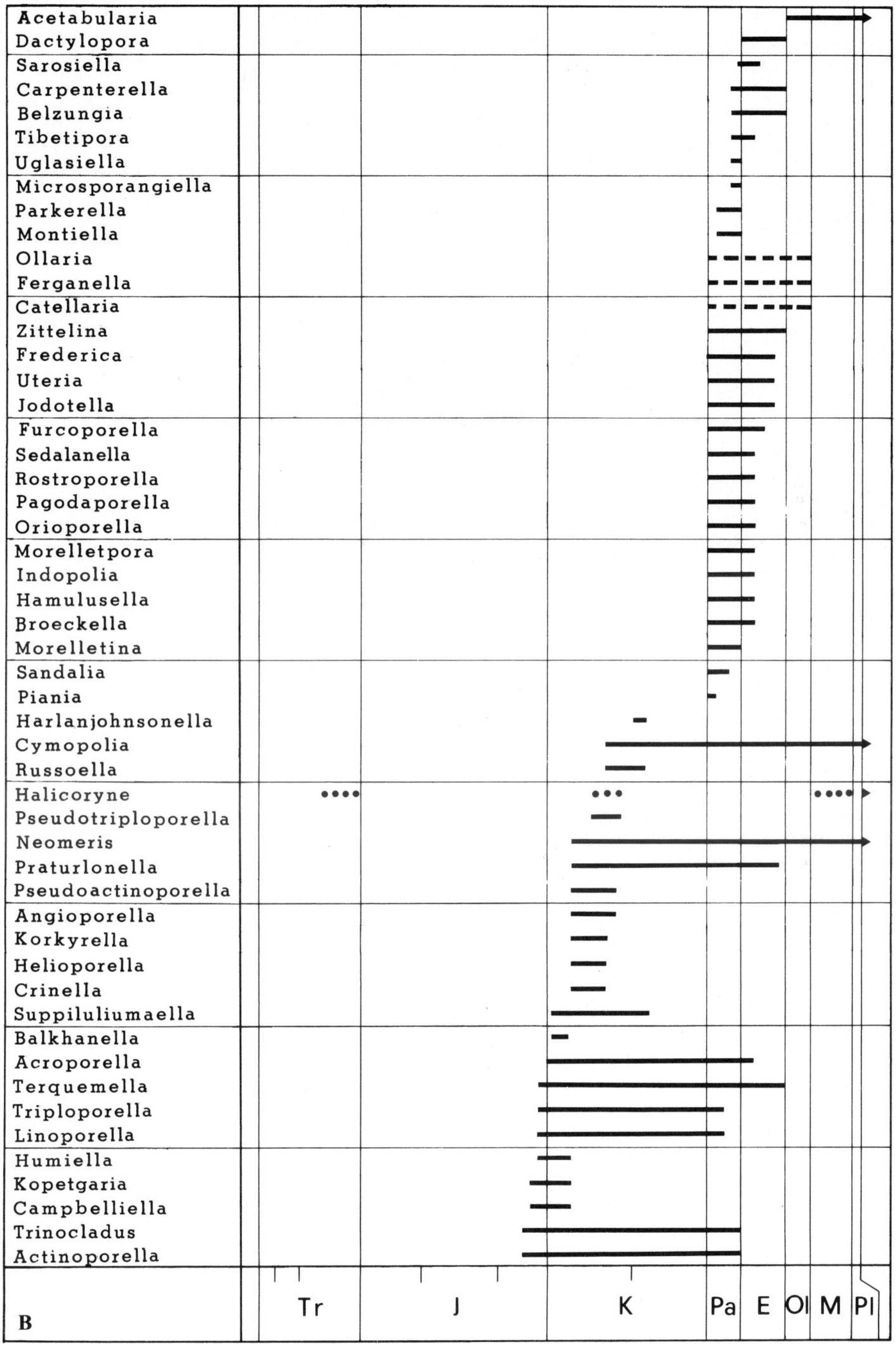
Acetabularia
Dactylopora
Sarosiella
Carpenterella
Belzungia
Tibetipora
Uglasiella
Microsporangiella
Parkerella
Montiella
Ollaria
Ferganella
Catellaria
Zittelina
Frederica
Uteria
Jodotella
Furcoporella
Sedalanella
Rostroporella
Pagodaporella
Orioporella
Morelletpora
Indopolia
Hamulusella
Broeckella
Morelletina
Sandalia
Piania
Harlanjohnsonella
Cymopolia
Russoella
Halicoryne
Pseudotriploporella
Neomeris
Praturlonella
Pseudoactinoporella
Angioporella
Korkyrella
Helioporella
Crinella
Suppiluliumaella
Balkhanella
Acroporella
Terquemella
Triploporella
Linoporella
Humiella
Kopetgaria
Campbelliella
Trinocladus
Actinoporella
B
Tr
J
K
Pa
E
Ol
M
Pl

Some other *incertae sedis* taxa mentioned mainly in the Mesozoic are *Paronipora* Capeder, *Bacinella* Radoičić, *Lithocodium* Elliott, *Tubiphytes* Maslov, and *Baccanella* Pantić.

The taxonomic and stratigraphic problems of *Paronipora* Capeder 1904 (= *Microcodium* Gluck 1912) (Pl. I, 3) have been reviewed recently by Cherchi and Schroeder (1988). This fossil shows similarities with the mineral productions caused by the action of filamentous boring bacteria; these dissolve the carbonate of the substrate that then precipitates around the bacterial filaments (Lucas and Montenat 1967).

The genus, which occurs from the Carboniferous to the Pliocene, is mainly present in the Palaeogene. According to Lucas and Montenat (1967) and Smit (1979) many literature reports should be checked and *Paronipora*, which occurs prior to the end of the Cretaceous, would probably not be actually recognized up to the Recent, nor would the colonies have developed in a marine environment. The capacity that it has for penetrating underground, through joints, cracks and diaclases, could explain the occurrences in pre-Cenozoic rocks (Smit 1979) and also reflects its terrestrial origin.

Palaeomicrocodium Mamet and Roux, from the Devonian of Australia and Canada is, according to Mamet and Roux (1983) and Mamet et al. (1987), a marine genus similar to *Paronipora*. According to Cherchi and Schroeder (1988) *Palaeomicrocodium* is a junior synonym of *Baccanella* Pantić of the Alpine Trias reef facies.

Bacinella Radoičić and *Lithocodium* Elliott, according to some authors (Segonzac and Marin 1972), represent two different parts of the same thallus; respectively the older and the younger.

The two genera range from Upper Trias to the Cretaceous and flourished mainly in the external parts of the lagoons and in the shelf margin facies (Conrad 1977; Flügel 1979; Peybernés 1979; Senowbari-Daryan and Schäfer 1979). The majority of authors consider these two genera as *incertae sedis*; but Elliott (1955) and some other authors (e.g. Johnson 1969 and Dragastan 1980) have assigned *Lithocodium* to the Codiaceae; *Bacinella* has been referred to cyanophyceans by Johnson (1969).

The genus *Tubiphytes* Maslov is known from Carboniferous to Lower Cretaceous margin facies: see Maslov (1956a), Crescenti (1969), Dragastan (1969, Micro-oncholithes), Flügel (1981), Bernier and Gaillard (1974) and Barattolo and Pugliese (1987). According to Maslov (1956a) *Tubiphytes* could be attributed to the incrustations of foreign bodies by thin calcified cyanophyceans. Elliott (1962) also noted that *Tubiphytes* develops around other bodies. In the Mesozoic of the Apennines, as Crescenti (1969) has pointed out, *Tubiphytes* (Pl. I, 4–5) does not develop around foreign material, but contains a tubular cavity with periodical narrowings. Similar observations had been made by Dragastan (1969) for the "Micro-oncholithes sous type Nubecularioïdal"; according to this author it is possible that these structures originate from the interrelations among *Nubecularia*-like foraminifers, cyanophyceans and sediment. According to Bernier (1984), *Tubiphytes morronensis* Crescenti, and probably also the genus *Tubiphytes*, is a foraminifer (Nubeculariidae) in which the thick outer wall was an integral part of the porcellanaceous test and therefore was not due to an oncoidal envelope.

The genus *Tubiphytes*, as *Nigriporella* Rigby, was considered a hydrozoan by Rigby (1958). For further information see Crescenti (1969), Bernier (1984) and Roux (1985).

References

Abate B, Catalano R, D'Argenio B, Di Stefano P, Riccobono R (1977) Relationships of algae with depositional environments and faunal assemblages of the Panormide Carbonate Platform, Upper Triassic, northwestern Sicily. In: Flügel E (ed) Fossil algae. Springer, Berlin Heidelberg New York, pp 301–313

Adey WH, MacIntyre IG (1973) Crustose coralline algae: a revaluation in the geological sciences. Geol Soc Am Bull 84:883–904

Adey WH, Vassar JM (1975) Colonization, succession, and growth rates of tropical crustose coralline algae (Rhodophyta, Criptonemiales). Phycologia 14:55–69

Adey WH, Townsend RA, Boykins WT (1982) The crustose coralline algae (Rhodophyta: Corallinaceae) of the Hawaiian Islands. Smithsonian Contrib Mar Sci 15:1–74

Barattolo F (1978) Su di una nuova dasicladacee (alghe verdi) nel Paleocene dell'Appennino meridionale. Boll Soc Nat Napoli 87:83–157

Barattolo F (1980) Su alcune nuove Triploporelle (alghe verdi, dasicladacea) del Cretacico inferiore dell'Appennino campano. Boll Soc Nat Napoli 89:97–167

Barattolo F (1981) Osservazioni su *Triploporella fraasi* Steinmann 1880 (alghe verdi, Dasicladali). Atti Accad Pontaniana 30:41 pp

Barattolo F (1982a) Osservazioni su *Triploporella steinmannii* n. sp. (alghe verdi, Dasicladali) del Cretacico del Messico. Boll Soc Nat Napoli 91:89–123

Barattolo F (1982b) Osservazioni su *Triploporella apenninica* Baretti 1922 (alghe verdi, Dasicladali) del Paleocene dell'Abruzzo aquilano. Geol Rom 21:29–59

Barattolo F (1982c) Osservazioni su *Triploporella praturlonii* n. sp. (alghe verdi, Dasicladali) del Cretacico inferiore dell'Appennino centrale. Geol Rom 21:825–847

Barattolo F (1983a) Su alcuni organi riproduttori isolati di Dasicladacee nel Cretacico dell'Appennino. Boll Soc Nat Napoli 92:17–49

Barattolo F (1983b) Osservazioni su *Suppiluliumaella schroederi* n.sp. (alghe verdi, Dasicladali) del Cenomaniano del Matese (Appennino meridionale, Italia). Boll Soc Nat Napoli 92:415–461

Barattolo F (1984) New data on tribe *Bornetelleae* (Chlorophyta, Dasycladales). Boll Soc Nat Napoli 93:143–201

Barattolo F, Del Re MC (1984) Alghe rosse Solenoporacee del Cenomaniano dei dintorni di Caserta (Appennino Meridionale, Italia). Boll Soc Nat Napoli 93:203–255

Barattolo F, Pugliese A (1987) Il Mesozoico dell'Isola di Capri. Q Accad Pontaniana 8:1–37

Barattolo F, De Castro P, Radoičić R (1988) Notes on the genus *Selliporella* Sartoni and Crescenti 1963 (Chlorophyta, Dasycladales). In: 74th Congr Soc Geol Ital Abstr, vol A, pp 35–38

Barbera C, Simone L, Carannante G (1978) Depositi circalitorali di piattaforma aperta nel Miocene Campano. Analisi sedimentologica e paleoecologica. Boll Soc Geol Ital 97:821–834

Bassoullet JP, Bernier P, Conrad MA, Deloffre R, Jaffrezo M, Poignant AF, Segonzac G (1977) Classification criteria of fossil Dasycladales. In: Flügel E (ed) Fossil algae. Springer, Berlin Heidelberg New York, pp 154–156

Bassoullet JP, Bernier P, Conrad MA, Deloffre R, Jaffrezo M (1978) Les Algues Dasycladales du Jurassique et du Crétacé. Geobios Mem Spec 2:330

Bassoullet JP, Bernier P, Deloffre R, Génot P, Jaffrezo M, Vachard D (1979) Essai de classification des Dasycladales en Tribus. Bull Cent Rech Explor Prod Elf-Aquitaine Ser 2, 3:429–442

Bassoullet JP, Bernier P, Conrad MA, Deloffre R, Poncet J, Roux A (1983) Les algues Udoteacées du Paléozoïque au Cénozoïque. Bull Cent Rech Explor Prod Elf-Aquitaine 2, 7:449–621

Bassoullet JP, Bernier P, Deloffre R, Génot P, Poncet J, Roux A (1984) La notion de genre chez les

Algues fossiles. B. – Les Chlorophyceae calcaires. Bull Soc Geol Fr Ser 7, 26:605–607
Bernier P (1984) Les formations carbonatées du Kimmeridgien et du Portlandien dans le Giura meridional. Stratigraphie, micropaléontologie, sedimentologie. Doc Lab Geol Lyon Ser 2, 92:1–803
Bernier P, Gaillard C (1974) Ultrastructure observées dans des encroûtements calcaires de type stromatolitique. C R Acad Sci Paris Ser D 278:557–559
Bold HC, Wynne MJ (1978) Introduction to the Algae. Structure and reproduction. Prentice-Hall, Englewood Cliffs
Bosence DWJ (1985) The "Coralligène" of the Mediterranean, a recent analog for tertiary coralline algal limestones. In: Toomey DF, Nitecki MH (eds) Paleoalgology: contemporary research and applications. Springer, Berlin Heidelberg New York Tokyo, pp 216–225
Buchbinder B, Halley RB (1985) Occurrence and preservation of Eocene squamariacean and coralline rhodoliths: Eua, Tonga. In: Toomey DF, Nitecki MH (eds) Paleoalgology: contemporary research and applications. Springer, Berlin Heidelberg New York Tokyo, pp 248–256
Bucur I (1985) Sur la présence des cystes reproducteurs dans les ramifications de *Macroporella praturloni* et quelques considérations concernant l'évolution de la cladosporie chez les Dasycladales fossiles. Evol Adapt 2:101–109
Catalano R, D'Argenio B, Lo Cicero G (1974) I ciclotemi triassici di Capo Rama (Monti di Palermo) Studi sulle piattaforme carbonatiche siciliane. Geol Rom 13:125–145
Cherchi A, Schroeder R (1979) *Koskinobullina* n. gen., micro-organisme en colonie incertae sedis (algues? du Jurassique-Crètacè de la règion méditerranéenne. Note prèliminaire. Bull Cent Rech Explor Prod Elf-Aquitaine Ser 2, 3:519–523
Cherchi A, Schroeder R (1985) *Koskinobullina socialis* Cherchi and Schroeder 1979: a colonial microfossil incertae sedis (algae?) from Jurassic-Cretaceous of the Mediterranean region. Boll Soc Paleontol Ital 23:361–374
Cherchi A, Schroeder R (1988) Osservazioni sui microproblematica *Paronipora* Capeder, *Microcodium* Gluck, *Baccanella* Pantic e *Palaeomicrodium* Mamet and Roux. Boll Soc Paleontol Ital 27, 1:79–81
Chiocchini M (1977) Ricerche geologiche sul Gran Sasso d'Italia (Abruzzo). I. Facies di margine della Piattaforma carbonatica durante il Lias medio e superiore nei pressi di S. Stefano di Sessanio (F. 140 Teramo). Stud Geol Camerti 3:165–173
Chiocchini M, Mancinelli A (1977) Microbiostratigrafia del Mesozoico in facies di piattaforma carbonatica dei Monti Aurunci (Lazio meridionale). Stud Geol Camerti 3:109–152
Chiocchini M, Mancinelli A (1978) Ricerche geologiche sul Mesozoico del Gran Sasso d'Italia (Abruzzo). III. Correlazioni microbiostratigrafiche tra facies di margine della piattaforma carbonatica e facies pelagiche del Giurassico e Cretacico inferiore. Stud Geol Camerti 4:19–36
Chuvashov B, Riding R (1984) Principal floras of Palaeozoic marine calcareous algae. Palaeontology Ser 3, 27:487–500
Chuvashov B, Luchinina VA, Shuysky VP, Berchenko OI, Istechenko AA, Saltovskaya VA, Schirschova DI (1987) Iskopaemye izvestkovye vodorosli. Morfologija, sistematika, metody izuchenija. Izdat Nauka, Sibirsk Otdel, Novosibirsk
Clari PA, Marini P, Pastorini M, Pavia G (1984) Il Rosso Ammonitico Inferiore (Baiociano-Calloviano) nei Monti Lessini settentrionali (Verona). Riv Ital Paleontol Stratigr Ser 1, 90:15–86
Colacicchi R, Passeri L, Pialli P (1975) Evidences of tidal environment deposition in the calcare Massiccio Formation (Central Apennines, Lower Lias). In: Ginsburg RN (ed) Tidal deposits. Springer, Berlin Heidelberg New York, pp 345–354
Conrad MA (1970) *Vermiporella ?tenuipora*, n. sp., une algue incertae sedis des calcaires urgoniens de la region genevoise. C R Seances Soc Phys Hist Nat Genève N S 5:37–42
Conrad MA (1977) The Lower Cretaceous calcareous algae in the area surrounding Geneva (Switzerland): biostratigraphy and depositional environments. In: Flügel E (ed) Fossil algae. Springer, Berlin Heidelberg New York, pp 295–300
Crescenti U (1969) Biostratigrafia delle facies mesozoiche dell'Appennino centrale: correlazioni. Geol Rom 8:15–40
D'Argenio B (1966a) Stromatoliti triassiche della Calabria settentrionale. Boll Soc Nat Napoli 75:433–457

D'Argenio B (1966b) Le facies littorali mesozoiche nell'Appennino meridionale. Boll Soc Nat Napoli 75:497–552

De Castro P (1962) Il Giura-Lias dei Monti Lattari e dei rilievi ad ovest della Valle dell'Irno e della Piana di Montoro. Boll Soc Nat Napoli 71:73 pp

De Castro P (1969) Su alcune tallofite del Mesozoico in Campania. Stratigrafia e Paleontologia. Boll Soc Nat Napoli 78:87–167

De Castro P (1975) Osservazioni su *Aeolisaccus kotori* Radoicic 1959, Cyanoschizophyta. Boll Soc Nat Napoli 84:44 pp

De Castro P (1979) Osservazioni su *Diplopora nodosa* Schafhäutl 1863 (alga verde Dasicladacea) delle scogliere triassiche della Lucania (Appennino meridionale). Boll Soc Nat Napoli 88:101–191

De Castro P (1987) Le facies di piattaforma carbonatica del Giurassico italiano. Diffusione areale e lineamenti biostratigrafici. Boll Soc Paleontol Ital 26:309–325

De Castro P (1988) Observations on Thaumatoporellaceans. In: 74th Congr Soc Geol Ital, Abstr, vol A, pp 245–250

De Castro P (1989) On some Foraminifera and Algae in Apennine Upper Cretaceous and Paleocene. Mem Soc Geol Ital 40, 10 pp

De Castro P, De Rosa C (1977) Osservazioni su Salpingoporella adriatica (Gusic 1966). Boll Soc Nat Napoli 86:169–207

Deloffre R (1987) Nouvelle classification des Algues Dasycladales fossiles. C R Acad Sci Paris Ser 2, 305:1017–1020

Deloffre R (1988) Nouvelle taxonomie des Algues Dasycladales. Bull Cent Rech Explor Prod Elf-Aquitaine 12, 1:165–217

Deloffre R, Génot P (1982) Les Algues Dasycladales du Cénozoique. Bull Cent Rech Explor Prod Elf-Aquitaine Mem 4:1–248

Denizot M (1968) Les algues floridées encroûtantes (à l'exclusion des corallinacées) Lab Cryptogam Mus Nat Hist Nat Paris, pp 1–280

Di Stefano P, Senowbari-Daryan B (1985) Upper Triassic Dasycladales from the Palermo Mountains (Sicily, Italy). Geol Rom 24:189–220

Dragastan O (1969) "Micro-oncolithes" dans le Jurassique supérier et du Crétacé inférieur de Romanie. Rev Micropaleontol 12:53–62

Dragastan O (1980) Alge calcaroase din Mezozoicul si Tertiarul României. Acad RS Rom, Bucaresti

Dragastan O (1981) Mesozoic Dasycladaceae from Romania: distribution and biostratigraphical importance. Facies 4:165–196

Dragastan O (1985) Review of tethyan mesozoic algae of Romania. In: Toomey DF, Nitecki MH (eds) Paleoalgology: contemporary research and applications. Springer, Berlin Heidelberg New York Tokyo, pp 101–161

Elliott GF (1955) The permian calcareous alga Gymnocodium. Micropaleontology Ser 1, 1:83–90

Elliott GF (1956) Galaxaura (calcareous algae) and similar fossil genera. J Washington Acad Sci Ser 11, 46:341–343

Elliott GF (1961) The sexual organization of Cretaceous Permocalculus (calcareous algae). Palaeontology Ser 1, 4:82–84

Elliott GF (1962) More microproblematica from the Middle East. Micropaleontology 8:29–44

Elliott GF (1964) Tertiary Solenoporaceae Algae and the reproductive structures of the Solenoporaceae. Palaeontology Ser 4, 7:695–702

Elliott GF (1965) The interrelationships of some Cretaceous Codiaceae (Calcareous Algae). Palaeontology Ser 2, 8:199–203

Elliott GF (1972) Lower Paleozoic green algae from Southern Scotland, and their evolutionary significance. Bull Brit Mus Nat Hist Geol Ser 4, 22:357–377

Elliott GF (1977a) A consideration of the tribe Thyrsoporelleae, dasyclad algae. Palaeontology Ser 3, 20:705–714

Elliott GF (1977b) Inferred isocrymal distribution of Jurassic dasycladacean Algae in Europe, north Africa and southwestern Asia. Jl Geol Lond 133:363–373

Elliott GF (1978) Ecologic significance of post-Palaeozoic green calcareous algae. Geol Mag Ser 6, 115:437–442
Elliott GF (1979a) Influence of organic reefs on the evolution of post-Palaeozoic algae. Geol Mag Ser 5, 116:375–383
Elliott GF (1979b) Taxonomy and opercular function of the Jurassic alga Stichoporella. Palaeontology 22:407–412
Elliott GF (1982) Calcareous algae and Middle Jurassic facies of southern England. Geol Mag Ser 3, 119:309–313
Elliott GF (1984) Climatic tolerance in some aragonitic green algae of the post-palaeozoic. Palaeogeogr Palaeoclimatol Palaeoecol 48:163–169
Emberger J (1976) Les Algues (Chlorophyceae, Prasinophyceae, Rhodophyceae) du Carbonifére et du Permien. Essai d'un inventaire bibliographique, géographique, stratigraphique. Bull Inst Geol Bassin Aquitaine Num Spec, pp 1–168
Emberger J (1978) Les Algues (Chlorophyceae, Prasinophyceae, Rhodophyceae) du Devonien. Essai d'un inventaire bibliographique, géographique, stratigraphique. Bull Inst Geol Bassin Aquitaine Num Spec, pp 1–94
Emberger J (1979) Les Algues (Euchlorophyceae, Prasinophyceae, Rhodophyceae) du Trias. Essai d'un inventaire bibliographique, géographique, stratigraphique. Bull Inst Geol Bassin Aquitaine Num Spec, pp 1–157
Emberger J, Jaffrezo M (1975) Complement à un essai d'inventaire des Algues Dasycladacèes du Jurassique et du Crètacé inférieur. Bull Inst Géol Bassin Aquitaine 18:59–132
Emberger L (1968) Les plantes fossiles dans leurs rapports avec les végétaux vivants, 2nd edn. Masson, Paris
Endo R (1969) Contributions to the geology and paleontology of Southeast Asia, 74. Fossil algae from the Khao Phrab district in Thailand. Geol Paleontol SE Asia 7:5–85
Esteban M, Calandra D, Catalano R, Di Stefano E (1982) La scogliera messiniana di Mazara del Vallo. In: Catalano R, D'Argenio B (eds) Guida alla geologia della Sicilia occidentale. Mem Soc Geol It 24, Suppl A:146–151
Fischer AG (1964) The Lofer cyclothems of the Alpine Triassic. Geol Surv Kansas Bull 169:149–170
Flügel E (1979) Palaecology and microfacies of Permian, Triassic and Jurassic algae communities of Platform and Reef Carbonates from the Alps. Bull Cent Rech Explor Prod Elf-Aquitaine Ser 2, 3:569–587
Flügel E (1981) Lower Permian Tubiphytes/Archaeolithoporella buildups in the Southern Alps (Austria and Italy). SEPM Spec Publ 30:143–160
Flügel E (1985) Diversity and environments of Permian and Triassic dasycladacean algae. In: Toomey DF, Nitecki MH (eds) Paleoalgology: contemporary research and applications. Springer, Berlin Heidelberg New York Tokyo, pp 343–351
Gebelein CD (1969) Distribution morphology and accretion rate of recent subtidal algal stromatolites, Bermuda. J Sediment Petrol 39:49–69
Génot P (1987) Les Chlorophycées calcaires du Paleocene d'Europe Nord-Occidentale (Bassin de Paris, Cotentin, Bassin de Mons). These Doct d'Etat, Univ Nantes, Fac Sci Tech, vol 1:500 pp, vol 2:49 pp
Güvenç T (1979) Dasycladacées mètaspondyles du Paléozoique superieur et du Trias. Bull Cent Rech Explor Prod Elf-Aquitaine Ser 2, 3:625–637
Hagn H, Ott E (1975) Ein Geröll mit *Elianella elegans* Pfender und Basse (Paleozán, Kalkalpin) aus der subalpinen Molasse N. Salzburg. Mitt Bayer Staat Paláontol Hist Geol 15:119–129
Herak M, Kochansky-Devidé V, Gušić I (1977) The development of dasiclad algae through the ages. In: Flügel E (ed) Fossil algae. Springer, Berlin Heidelberg New York, pp 143–153
Hofmann HJ, Schopf JW (1983) Early Proterozoic Microfossils. In: Schopf JW (ed) Earth's earliest biosphere. Its origin and evolution. Univ Press, Princeton, pp 214–239
Johnson JH (1952) Ordovician rock-building algae. Col Sch Mines Q Ser 2, 47:29–56
Johnson JH (1958) A review of Devonian algae. In: Studies of Devonian algae. Col Sch Mines Q Ser 2, 53:1–84
Johnson JH (1960) Paleozoic Solenoporaceae and related red algae. Col Sch Mines Q Ser 3, 55:1–77

Johnson JH (1961) Limestone-building algae and algal limestone. Col Sch Mines Golden edn: 1–297
Johnson JH (1962) The algal genus *Lithothamnium* and its fossil representatives. Col Sch Mines Q 57, 1:1–111
Johnson JH (1963) The algal genus *Archaeolithothamnium* and its fossil representatives. J Paleontol Ser 1, 37:175–211
Johnson JH (1964) The Jurassic algae. Col School Mines Q 59:1–129
Johnson JH (1965) The genus *Lithophyllum* and its fossil representatives. Col School Mines Q 60, 2:1–65
Johnson JH (1966) Pennsylvanian and Permian calcareous algae. Col School Mines Q Ser 3, 58:1–211
Johnson JH (1969) A review of the Lower Cretaceous algae. Prof Contrib Col School Mines 6:1–180
Johnson JH, Høeg OA (1961) Studies of Ordovician Algae. Col Sch Mines Q Ser 2, 56:1–120
Johnson JH, Konishi K (1956) A review of Mississipian algae. In: studies of Mississipian algae. Col School Mines Q Ser 4, 51:1–84
Johnson JH, Konishi K (1959) A review of Silurian (Gothlandian) algae. Col School Mines Q Ser 1, 54, 1: pp 144
Kazmierczak J (1975) Colonial Volvocales (Chlorophyta) from the Upper Devonian of Poland and their palaeoenvironmental significance. Acta Palaeontol Pol 20, 1:75–85
Kazmierczak J (1976) Volvocacean nature of some Palaeozoic non-Radiosphaerid Calcispheres and Parathuramminid "Foraminifera". Acta Palaeontol Pol 21, 3:245–258
Kazmierczak J (1979) The eukaryotic nature of Eosphaera-like ferriferous structures from the Precambrian Gunflint Iron Formation, Canada: a comparative study. Precambrian Res 9:1–22
Kazmierczak J (1981) The biology and evolutionary significance of Devonian Volvocaceans and their Precambrian relatives. Acta Palaeontol Pol 26, 3/4:299–337
Korde KB (1973) Vodorosli kembrija. Tr Paleontol Inst Akad Nauk SSSR 139:1–349
Korde KB, Maslov VP, Krylov IN (1963) Embranchement des Cyanophyta (Schizophyceae). Algues bleues. Osnovy Paleontol SSSR 14:29–54 (French Transl 4845 BRGM Paris)
Kozlowski R, Kazmierczak J (1968) On two Ordovician calcareous algae. Acta Palaeontol Pol Ser 3, 13:325–346
Lemoine P (1966) Un nouveau genre d'Algue du Crètacè inférieur: *Agardhiellopsis* nov. gen. Rev Micropalaeont 4:203–210
Littler M, Doty MS (1974) Ecological components structuring the seaward edges of tropical Pacific reefs: the distribution, communities and productivity-ecology of Porolithon. J Ecol 63:117–129
Lucas G, Montenat C (1967) Observations sur les structures internes et developpement des Microcodium. Bull Soc Geol Fr Ser 7, 9, 6:909–918
Luperto Sinni E (1979) *Cretacicladus minervini* n. gen. n. sp., nuova alga (Chlorophyta) del Cenomaniano delle Murge. Atti Ist Geol Paleontol Univ Bari 8:30 pp
Mamet B, Roux A (1981) Note sur les Albertaporellinae (Dasycladaceae). Rev Micropalaeontol Ser 3–4, 23:159–168
Mamet B, Roux A (1983) Algues Devono-Carboniferes de l'Australie. Rev Micropaleontol Ser 2, 26:63–131
Mamet B, Roux A, Nassichuk WW (1987) Algues Carboniferes et Permiennes de l'Arctique Canadien. Bull Geol Surv Can 342:1–77
Maslov VP (1956a) Les Algues calcaires fossiles de l'URSS. Akad Nauk SSSR, Tr Geol Inst 160:1–301 (French Transl BRGM Paris 3517)
Maslov VP (1956b) Nouvelle famille d'Algues fossiles pourpes et deux nouveaux genres d'Algues fossiles Cyanophycées du Carbonifère. Dokl Akad Nauk SSSR Ser 1, 107:151–154 (French Transl BRGM Paris 1381)
Maslov VP (1962) Stromatolites du Paléogène de la crête de Gissar. Dokl Akad Nauk SSSR 142:690–691 (French Transl BRGM Paris 3682)
Maslov VP, Korde KB (1963) Embranchement des Rhodophyta. Algues rouges o pourpres. Osnov Paleontol SSSR 14:243–260 (French Transl BRGM Paris 4934)

Maslov VP, Naumova SM, Korde KB (1963) Embranchement des Chlorophyta. Algues vertes. Osnov Paleontol SSSR 14:198–223 (French Transl BRGM Paris 4933)
Massari F (1983) Oncoids and stromatolites in the Rosso Ammonitico sequences (Middle-Upper Jurassic) of the Venetian Alps, Italy. In: Peryt TM (ed) Coated grains. Springer, Berlin Heidelberg New York, pp 358–366
Massari F, Dieni I (1983) Pelagic oncoids and ooids in the Middle-Upper Jurassic of Eastern Sardinia. In: Peryt TM (ed) Coated grains. Springer, Berlin Heidelberg New York, pp 367–376
Masse J-P (1979) Schizophytoides du Crétacé inférieur. Caractéristiques et signification écologique. Bull Cent Rech Explor Prod Elf-Aquitaine Ser 2, 3:685–703
Massieux M, Denizot M (1964) Rapprochement du genre *Pseudolithothamnium* Pfender avec le genre actuel *Ethelia* van Bosse (algue Florideae, Squamariaceae). Rev Micropaleontol 7:31–42
Minnery GA, Rezak R, Bright TJ (1985) Depth Zonation and Growth Form of Crustose Coralline Algae: Flower Garden Banks, Northwestern Gulf of Mexico. In: Toomey DF, Nitecki MH (eds) Paleoalgology: contemporary research and applications. Springer, Berlin Heidelberg New York, Tokyo, pp 237–246
Monty CLV (1965) Recent algal stromatolites in the Windword Lagoon, Andros Island, Bahamas. Ann Soc Geol Belg Bull Ser 6, 88 B:269–276
Monty CLV (1967) Distribution and structure of Recent stromatolitic algal mats, Eastern Andros Id., Bahamas. Ann Soc Geol Belg Bull 90:55–100
Monty CLV (1977) Evolving concepts on the nature and ecological significance of stromatolites: a review. In: Flügel E (ed) Fossil algae. Springer, Berlin Heidelberg New York, pp 15–35
Morellet L, Morellet J (1913) Les Dasycladacées du Tertiaire parisien. Mem Soc Geol Fr 21 Mem 47:1–43
Morellet L, Morellet J (1922) Nouvelle contribution a l'ètude des Dasycladacées Tertiaires. Mem Soc Geol Fr 25 mem 58:1–35
Mu X-N, Riding R (1983) Silicified Gymnocodiacean Alga from the Permian of Nanjing, China. Palaeontology Ser 2, 26:261–276
Nitecki MH (1976) Ordovician Batophoreae (Dasycladales) from Michigan. Fieldiana Geol Ser 4, 35:29–40
Odin GS (1982) The Phanerozoic time scale revisited. Episodes Ser 3:3–9
Ogniben L (1958) Melobesie basso-elveziane di Caiazzo (Caserta). Palaeontogr Ital 53:49–73
Ott E (1972) Mitteltriadische Riffe der Nördlichen Kalkalpen und altersgleiche Bildungen auf Karaburun und Chios (Ägäis). Mitt Ges Geol Bergbaustudwien 21:251–276
Passeri L (1974) Sedimentazione carbonatica attuale e diagenesi precoce nella laguna di Porto Cesareo (Penisola Salentina). Boll Soc Geol Ital Suppl 92:3–40
Pérès JM (1967) The Mediterranean benthos. Oceanogr Mar Biol Annu Rev 5:449–533
Pérès JM, Picard J (1964) Manuel de Bionomie benthique de la Mer Mèditerranée. Recl Trav Stn Mar Endoume Bull Ser 47, 31:1–161
Perrin C (1987) *Solenomeris* — un foraminifére Acervulinidae constructeur de récifs. Rev Micropaleontol Ser 3, 30:197–206
Peybernés B (1979) Les algues du Jurassique et du Crétacé inferieur des Pirénées Franco-Espagnoles. Intérêt biostratigraphique et paléoécologique. Bull Cent Rech Explor Prod Elf-Aquitaine Ser 2, 3:733–741
Pia J von (1912) Neue Studien über die triadischen Siphoneae Verticillatae. Beitr Geol Paläontol Öster Ung Orients 25:25–81
Pia J von (1920) Die Siphoneae Verticillatae vom Karbon bis zur Kreide. Abh Zool Bot Ges Wien Ser 2, 11:1–262
Pia J von (1927) Tallophyta. In: Hirmer M (ed) Handbuch der Paläobotanik, vol 1. Oldenbourg, Munich, pp 31–136
Poignant AF (1977) Les Algues fossiles, point de vue du gèologue. Bull Soc Phycal Fr 22:87–98
Poignant AF (1979a) Les Algues rouges crétacées. Relations mer Borèale. Tèthys. Aspekte der Kreide Europas. IUGS Ser A 6:273–278
Poignant AF (1979b) Dètermination générique des Corallinacées Mésozoiques et Cénozoiques. Bull Cent Rech Explor Prod Elf-Aquitaine Ser 2, 3:757–765
Poignant AF (1980) Sur deux algues calcaires du Crétacé supèrieur du nord de l'Espagne. Reflexions sur le genre *Elianella*. In: 106th Congr Natl Soc Savantes, Caen 1980. Sciences 2:323–328

Poignant AF, Laville P (1972) Un exemple de migration algaire. Consèguences stratigraphiques. CR Acad Sci Paris 274:2563–2565

Praturlon A (1963) Una nuova Teutloporella (alga calcarea) nel Giurese superiore di M. Corno (parco Nazionale d'Abruzzo). Geol Rom 2:199–206

Praturlon A (1964) Calcareous algae from Jurassic-Cretaceous limestones of Central Apennines (Southern Latium-Abruzzi). Geol Rom 3:171–202

Praturlon A (1966) Algal assemblages from Lias to Palaeocene in Southern Latium-Abruzzi: a review. Boll Soc Geol Ital 85:167–194

Radoičić R (1980) Contribution to the reconsideration of a group of cylindroporelliform dasyclads. Boll Acad Serb Sci Arts Cl Sci Nat Mat Ser 20, 72:109–115

Reis MO (1923) Kalkalgen und Seesinterkalke aus dem rheinpfälzischen Tertiär. Geogn Jahresh 36:103–130

Remane J (1969) Morphological and statistical analysis of *Clypeina jurassica* (Favre) (Dasycladaceae) by means of serial sections. Geol Rom 8:85–116

Richter KD, Heyforth A, Ott E (1979) Pleistozäne, brackische Blaugrunalgenriffe mit *Rivularia haematites* auf der Perachara Halbinsel bei Korinth (Griechenland). N Jahrb Geol Palaeontol Abh Ser 1, 159:14–40

Riding R (1977) Problems of affinity in Palaeozoic calcareous algae. In: Flügel E (ed) Fossil algae. Springer, Berlin Heidelberg New York, pp 202–211

Rigby JK (1958) Two new Upper Palaeozoic Hydrozoans. J Paleontol 32:583–586

Roux A (1985) Introduction à l'étude des Algues fossiles paléozoiques (de la Bactérie à la tectonique des plaques). Bull Cent Rech Explor Prod Elf-Aquitaine Ser 2, 9:465–699

Sartoni S, Crescenti U (1959) La zona a *Palaeodasycladus mediterraneus* (Pia) nel Lias dell'Appennino meridionale. Giorn Geol Ser 2, 27:115–139

Sartoni S, Crescenti U (1962) Ricerche biostratigrafiche nel Mesozoico dell'Appennino meridionale. Giorn Geol Ser 2, 29:161–302

Schäfer P, Senowbari-Daryan B (1983) Die Kalkalgen aus der Obertrias von Hydra, Griechenland. Paleontographica Abt B 185:83–142

Segonzac G, Marin P (1972) *Lithocodium aggregatum* Elliott et *Bacinella irregularis* Radoicic de l'Aptien de Teruel (Espagne): deux stades de croissance d'un seul et meme organisme incertae sedis. Boll Soc Geol Fr Ser 7, 14:331–335

Senowbari-Daryan B, Schäfer P (1979) Distributional pattern of calcareous algae within Upper Triassic patch reef structures of the Northern Calcareous Alps (Salzburg). Bull Cent Rech Explor Prod Elf-Aquitaine Ser 2, 3:569–587

Shuysky VP (1986) Novyi predstavitel dazikladieryx vodoroslej iz niznego devona urala. Paleontol Zh, Otdel' Ottisk 2:118–122

Shuysky VP (1987a) *Lopsiella mansica* – novyj predstavitel' acetabuljarievyx vodoroslej iz zivets kix otlozenij urala. Paleontol Zh Otdel Ottisk 3:67–75

Shuysky VP (1987b) Sistema nadrodovyx taksonov iskopaemyx iz vestkovix vodoroslej. Akad Nauk SSSR, Geol Paleontol Urala, pp 33–40

Skompski S (1984) The functional morphology of the Carboniferous dasycladacean genus *Kulikia*. N Jahrb Geol Paläontol Monatsh 7:427–436

Smit J (1979) Microcodium, its earliest occurrence and other considerations. Rev Micropaleontol 22:44–50

Sokâč B (1987) On some controversial Dasyclad genera and species and their stratigraphic position in the Lower Cretaceous deposits of the Dinarides. Geol Vjesn 40:9–38

Sokâč B, Velić I (1978a) Redescription of the genus Selliporella (Calcareous algae, Dasycladaceae). Geol Vjesn 30:225–242

Sokâč B, Velić I (1978b) Biostratigrafska istrazivanja donje Krede vanjskih Dinarida (I). Neokom zapadne Istre. Geol Vjesn 30:243–250

Steinmann G (1899) Über fossile Dasycladaceen von Cerro Escamela (Mexico). Bot Z 57:137–154

Steinmann G (1903) Tetraploporella remesi, eine neue Dasycladacea aus dem Thiton von Stramberg. Beitr Paläontol Geol Öster Ung Orients 15:45–54

Tillier S (1975) Précision morphologiques, structurales et systématiques sur *Reptomulticava parviporosa* Canu et Lecointre (Bryozoa, Cyclostomata). Geobis Ser 1, 8:75–78

Vachard D (1976) Problématique des micro-organismes du Paléozoique inferieur. In: 4. C R Reun Sciences de la terre, Paris, pp 385

Vachard D, Termier G, Termier H (1978) Sur l'appartenance systematique des Algues Solenoporacées. CR Acad Sci Paris Ser D 286:1865–1868

Valet G (1979) Approche palèoècologique du monde des Dasycladales à partir de l'Ecologie des formes actuelles. Bull Cent Rech Explor Prod Elf-Aquitaine Ser 2, 3:859–866

van den Hoek C, Cortel-Breeman AM, Wanders JB (1975) Algal zonation in the fringing coral reefs of Curaçao, Netherlands Antilles, in relation to the zonation of corals and gorgonians. Aquat Bot 1:269–308

Velić I (1977) Jurassic and Lower Cretaceous assemblage-zones in Mt. Velika Kapela, Central Croatia. Acta Geol 9:37 pp

Whybrow PJ, Mc Clure HA, Elliott GF (1987) Miocene stratigraphy, geology and flora (Algae) of eastern Saudi Arabia and the Ad Dabtiyah vertebrate locality. Bull Brit Mus Nat Hist Geol Ser 4, 41:371–382

Wray JL (1977a) Late Paleozoic calcareous red algae. In: Flügel E (ed) Fossil algae. Springer, Berlin Heidelberg New York, pp 167–176

Wray JL (1977b) Calcareous algae. Elsevier, Amsterdam

Zorn H (1972) Bau und Funktion der Wirteläste einiger triassischer Dasycladaceen (Kalkalgen). N Jahrb Geol Palaontol Mouatsh Ser 12, 1972:727–740

Zorn H (1974) Variationstatistische Untersuchungen an *Diplopora annulata* (Schafh.) und *D. philosophi* (Pia), Kalkagen der alpinen Mitteltrias. Eclog Geol Helv 67:1–30

Zorn H (1976) Über den Lebensraum fossiler Wirtelalgen in der Trias der Alpen. Naturwissenschaften 63:426–429

Zorn H (1977a) Statistical analysis of fossil algae. In: Flügel E (ed) Fossil algae. Springer, Berlin Heidelberg New York, pp 220–224

Zorn H (1977b) Der Einfluss der Umwelt auf die Variation fossiler Arten. Nat Mus Ser 1, 107:1–5

Chapter 23

Modern Stromatolites: A Review

S. Golubic[1]

Abstract

Modern stromatolites comprise a wide spectrum of various, laminated and nonlaminated organo-sedimentary structures that form in various marine and freshwater habitats. Some stromatolites form under extremely harsh ecological conditions, such as those found in hypersaline and ephemeral ponds, or in thermal springs, others form in environments with normal marine salinity or in freshwater. Microbial communities of different species composition participate in their formation, lithification and destruction, often leaving marks of their particular activities.

1 Introduction

Historically, the main driving force for studies of modern stromatolites was the need for interpretation of ancient stromatolites, particularly those found abundantly in pre-Phanerozoic strata. The discovery of Precambrian microbial fossils, and their association with stromatolitic structures stimulated the interest in modern stromatolitic microbiota.

Understandably, there was no shortage of pertinent questions. Are stromatolites shaped by environmental forces, or by the genetically determined properties of the organisms that built them? What is the significance of their lamination; does it record solar or lunar cycles, or perhaps less regularly recurring events such as storms? Can they serve as a measure of ancient astronomic events, or palaeolatitudes? Can they be used in biostratigraphy? How do stromatolites trap and bind sediment, and how are they lithified and preserved? Do they form in intertidal or subtidal environments? What limits their growth and accretion? Which forces or organisms contribute to stromatolite construction, and which to their destruction?

The hope was that many of these questions could be answered by studying modern stromatolites, but some disappointment about the slowness of the progress in this research area has been recently expressed (Burne 1986). An important, often overlooked reason, that the research on modern stromatolites did not fulfil expectations is that simple answers were expected regarding exceedingly complex situations. The euphoria accompanying early discoveries sometimes led to overgeneralizations and simplistic models that often persisted and hindered further analytical work.

[1]Department of Biology, Boston University, 2 Cummington St., Boston, MA 02215, USA

Modern stromatolite studies concentrated on extreme environments in which microbial ecosystems dominate today, more or less unimpaired and unmasked by the presence of macroscopic plants and animals. They provided valuable information on microbial ecology. However, modern counterparts for most of the ancient stromatolitic situations have not been found. Conversely, a number of modern stromatolite settings lack fossil counterparts. Are these stromatolites relatively new, or are they old ones that do not preserve? How relevant are they then for the interpretation of fossil stromatolites?

Over the past three decades, modern stromatolites and microbial mats were discovered in various environments, ranging from hypersaline to freshwater, and from tropical to polar. They attracted the attention of researchers from different disciplines, such as sedimentology, geochemistry, limnology and microbial ecology, and became the focal point of an interdisciplinary research effort. Awramik (1976) and co-workers (Awramik et al. 1976, 1979) compiled an extensive list of references (which, incidentally, needs another update).

As a result of these studies, many overgeneralizations could be corrected, definitions refined, and the scope of stromatolite-related phenomena and environments within which they occur significantly widened. The insight into structural, and particularly, functional complexity of modern stromatolites highlighted the importance of the role of microbes in promotion and regulation of biogeochemical cycling of elements in modern ecosystems. The principles learned in the process are probably the best contribution which can be applied to our understanding of fossil stromatolites and their environments.

When different scientific disciplines meet to form an interdisciplinary approach to a problem, each discipline carries its own separate traditions, concerns, and historic biases. Thus, misunderstandings and differences in the use of terminology are to be expected. It is surprising, however, how deeply rooted some traditional practices have become in the course of establishment and development of various schools of thought, and how difficult it is to find enough openmindedness and cooperative spirit to reconcile these differences and arrive to reasonable new conventions. The field of stromatolite research is no exception to this experience (see Monty 1982 vs Krumbein 1982, 1983).

2 Definitions and Terminology

The term 'stromatolites' entered the literature with full recognition of their organo-sedimentary nature, together with an interpretation of their genesis (Kalkowsky 1908). This view is shared by most researchers today, although some prefer to use the term in a strictly descriptive sense, meaning a laminated rock. The latter approach encounters problems when dealing with unrelated laminated structures that are only superficially similar to stromatolites (e.g. geyserite, calcrete, varved lake sediments, Liesegang rings). To study a modern or living stromatolite means to study a stromatolite in the process of its formation. Stromatolite lithification occurs as a part of stromatolite genesis or diagenesis. As with other fossilized structures, lithification may help preservation, but it is neither a requirement nor a warranty for it. Therefore, the term 'potential stromatolites' when used for mats and non-lithified stromatolites (Krumbein 1983), may be misleading, because it raises expectations that may not be fulfilled. Preservation potential (see, for example, Park 1977), is a probabilistic measure analogous to life expectancy calculated by insurance companies. Early lithification increases the expectancy of preservation.

The portion of the stromatolite in which the living community of stromatolite-building microorganisms operates, is often spread as a thin, soft mat over the structure's surface. Microbial activity in this mat may result in accretion of the underlying sediment or rock. The stromatolitic structure then becomes a cumulative record of an upwardly displaced microbial mat over time (Hofmann 1969). The lower portion of the mat may be 'dead', relating to the living community as the wood tissue relates to the cambium (or the foliage). Thus the term 'Recent stromatolites' includes both living and fossil stromatolites of the Holocene age (see Monty 1982).

When first found, fossil stromatolites were treated as organisms and named according to biosystematic practice using Linnean binomial nomenclature: e.g. *Cryptozoon proliferum* Hall 1884 and *Collenia undosa* Walcott 1914. This traditional classification practice provides reference points, which can be hierarchically organized into a system, or used in stratigraphic and facies correlations. Since there are no skeletal fossils in pre-Phanerozoic rocks, a stratigraphic subdivision of these strata based on stromatolites has been introduced (Krylov 1963; Raaben 1969; Semikhatov 1976). For this purpose, Russian geologists developed an elaborate methodology to study three-dimensional morphology of fossil stromatolites, described a number of form-genera and form-species, and named them according to Linnean nomenclature. This approach was followed by Australian, French and, with some reluctance, also by American researchers.

There was no problem in using Linnean nomenclature for shapes, as long as modern and fossil stromatolites were treated separately. Krylov (1976) argued convincingly that the Linnean binomial nomenclature has been historically applied to different unrelated items such as parts of the human body, chemical compounds and drugs, plant extractions, and entire communities, so why not for stromatolites? However, once modern stromatolites were approached by a combined team of palaeontologists and microbiologists, there was difficulty in reconciling the usage of Linnean nomenclature for both the structure *and* the organism that built it (Walter et al. 1976). In this particular example from the Yellowstone thermal springs, the cyanobacterium *Phormidium tenue* var. *granuliferum* was found to be the main constructive element of the stromatolite structure *Conophyton weedii*, whereas *Phormidium truncatum* var. *thermale* built the flat-topped stromatolite *Vacerrilla walcottii*. Such confusion in dual naming could be resolved by new conventions, e.g. by restricting the use of italics to organismal names only.

3 What Controls Stromatolite Shape?

The use of stromatolites in biostratigraphy is based on the assumption that biological, genetically controlled properties of (mostly unknown) ancient stromatolite-builders determined the morphological features recognized in fossil stromatolites. Consequently, evolutionary changes and/or extinctions of taxa would be reflected in the record of changing stromatolite morphologies. The theoretical basis of this approach was explained by Maslov (1960).

This view was challenged by Logan et al. (1964). According to their view, various stromatolite shapes are produced by a 'generic' microbial mat, which is shaped differently by environmental forces, such as desiccation, waves and currents. Their assumption was that stromatolites are built by entire microbial communities, thus properties of individual organisms could not be expressed in the shape of the resulting structure.

Successful application of stromatolites in stratigraphic and interbasinal correlations (Walter 1972) supported the 'biological' view, while determination of palaeocurrents in Precambrian shallow water facies (Hoffman 1967) supported the 'environmental' view.

A conclusive answer has been sought in studying modern stromatolites. The evidence that has accumulated from these studies over several decades, shows that a combination of biological and environmental factors is responsible for stromatolite shaping. The relative importance of these factors for the resultant shape, however, is proportioned differently from case to case.

Distribution of organisms in a community, microbial communities notwithstanding, is not even, but rather log-normal. This observation has been formulated by MacArthur (1960) as the 'broken stick' theorem. Accordingly, every community is dominated by only few species which leave also the highest impact in the community's structure and function. As one approaches extreme environments, the total number of taxa decreases, and the discrepancy between the dominant and subdominant species increases (further breaking the 'stick'). Lowering the number of species, and discrepancy in their distribution, both express a decrease in diversity. Thus, it is explained why a microbial community may be shaped by the 'biological' properties of its dominant members. Moreover, some modern stromatolitic structures are shaped by virtually monospecific populations. Good examples are *Phormidium hendersonii* (Golubic and Focke 1978), *Entophysalis major* (Golubic 1976a) and *Rivularia haematites* (Golubic 1976b; Schneider 1977).

On the other hand, in high energy environments it becomes evident that wave scouring and currents contribute directly to the trimming and shaping of stromatolites (Plate II, 11; Logan 1961). Less apparent is the fact that each taxon has different, genetically determined environmental requirements, and responds accordingly to environmental changes. Under extreme conditions of the intertidal zone, for example, this response is reflected in zonal distribution of taxa. In the case of the Yellowstone stromatolite study (Walter et al. 1976) it was the environmental change that caused the shift in microbial composition and dominance from one to the other species of *Phormidium*, whereas the inherently different growth habits of these two taxa resulted in the change of the stromatolite shape.

Gebelein (1974) expressed the fact that stromatolites are mostly built by microscopic organisms by formulating the generalization, that the stromatolites micromorphology is largely 'biological', while their macromorphology is 'environmental' in origin. The studies of the microenvironments within which the stromatolite growth takes place (e.g. Cohen et al. 1977; Krumbein et al. 1977) provided further refinement, indicating that complex interactions between organisms and environmental determinants take place at the microscopic as well as macroscopic scale.

We may conclude (1) that genetically determined properties of dominant micro-organisms, including their specific growth habits and behaviour (e.g phototropism and phototaxis) are often revealed in stromatolite morphology. A good example is provided in the genesis of conical stromatolite shapes, where analogous relationships between micro- and macromorphology was found in modern and ancient stromatolites (Awramik and Vanyo 1986); (2) that because of specifically different environmental requirements of microorganisms, their distribution and dominance depends on environmental conditions and restraints, and (3) within these restraints, the environmental forces can influence stromatolite morphology directly, by trimming and shaping form analogous to the way trees and shrubs are trimmed by the wind in the mountains and along the coasts.

4 What Is Lamination?

The word stromatolite implies layering. Yet, as we learn more about modern and fossil stromatolites, we find (1) that lamination is a property which is differently expressed, ranging from fine and regular to coarse, to hardly noticeable or non-existent, and (2) that following the genesis of a stromatolite we need to distinguish between community stratification and structure lamination, both resulting in distinct layers.

The living stromatolite-building microbial community is often stratified. Layers dominated by different organisms that arrange themselves in strata of optimal growth, follow a vertical gradient of light, O_2, Eh, pH and concentrations of dissolved minerals. Such a gradient, or chemocline, becomes established within the interstitial waters as a result of microbial metabolic activities in a homeostatic fashion. Thus, the mat is microenvironmentally and biologically differentiated into strata, analogous to lake stratification, but with each stratum only a fraction of a millimetre in thickness (e.g. Guerrero and Mas 1989).

The primary producers of this community, usually cyanobacteria and algae, form the top layer. Analogous to the plankton in the epilimnion of lakes, they may show an inhibition of photosynthetic output at the very top of the mat due to excessive light. The organisms may respond either by elevated intracellular carotenoid production, as protection against photo-oxidation, or by production of extracellular shielding pigments within their polysaccharide sheaths, or both.

The highest output of organic matter production by oxygenic photosynthesis is observed in the subsurface layer, which is also supersaturated with respect to oxygen, and usually blue-green or green in colour. The light inhibition is reduced, or eliminated, in deeper water or in shaded places (caves).

The density of the surface layers acts as a light filter, so that self-shading becomes effective deeper within the mat. The initial reaction of cyanobacteria to light attenuation is a compensative increase in pigment production, particularly of the red light-gathering pigment phycoerythrin (Jorgensen 1989). An abrupt drop in primary production follows, accompanied by intensive aerobic decomposition of organic matter (primary product) which, in turn, removes the ambient oxygen, forming an anaerobic layer beneath the green portion of the mat. Release of H_2S by bacterial protein decomposition and, more importantly, by SO_4 reduction through the activity of anaerobic respirers, fuels the non-oxygenic bacterial photosynthesis. This photosynthesis uses light of longer wavelengths (700–900 nm) that filters through the cyanobacterial layers. Some cyanobacteria show the ability to switch between oxygenic and non-oxygenic mode of photosynthesis (see Cohen 1989). A layer of purple sulfur bacteria usually follows the blue-green one. Filamentous chlorobacteria of the *Chloroflexus*-type are often intermingled, providing mechanical support and modifying the colouration of this zone. The layers beneath are in darkness, dominated by fermenting decomposers, and possibly methanogens. Excessive H_2S is bound in a black FeS precipitate, which may diagenetically recrystallize into pyrite.

The conditions inside microbial mats vary diurnally and seasonally, in fact the chemocline often shifts up and down in the same rhythm. Microorganisms may respond either by vertical migration (Golubic 1973a) or by physiological adjustments (de Wit 1989).

The described stratification is typical of many water-logged intertidal mats. Such mats line the channels, ponds and mud flats of Abu-Dhabi, (UAE Plate II, 9, 10), of protected embayments in Shark Bay, Australia, and of lagoons of Baja California, Mexico, and can be considered cosmopolitan in distribution. Intertidal sediments with better drainage are covered with a thinner mat in which oxygen depletion occurs transiently, restricting the development of anaerobic microorganisms. Decomposition by the more efficient aerobic respiration is usually able to recycle all of the organic matter produced, which keeps the mat thin. Change in drainage regime may trigger a microorganismal succession, promoting colonization by heliotropic cyanobacteria (Awramik and Vanyo 1986). In Abu-Dhabi flats this change results in formation of upright pinnacles formed by the cyanobacterium *Schizothrix splendida* (Plate II, 10; Golubic 1973a,b).

In perpendicular sections through a mat, the microbial stratification is conspicuous as coloured bands, however, this layering is not the predecessor of lithified stromatolite laminae. As the mat accretes due to sediment trapment (Plate I, 1), or mineral precipitation (Plate I, 2), members of the community escape by either growing or moving upward. An upward displacement usually starts with the primary producers of the community, and is then followed by a cohort of consumers, degraders, and anaerobic photobacteria. Their activities, in turn, causes an upward displacement of the entire chemocline. If this progression proceeds at an even rate, the resulting stromatolitic structure, i.e. the cumulative record, will also appear uniform and smooth. Only recurring disturbances in the process of stromatolite accretion are recorded as laminae. The underlying causes behind a sequence of alternating laminae, typical of most stromatolites may be quite different. In the case of *Phormidium hendersonii*, diurnal alternation in light was sufficient to change filament orientation and cause lamination (Monty 1965; Golubic and Focke 1978).

The thickness of organic-rich laminae may depend on the timing of organic production within the cycle, but may also reflect the relationship between organic production and decomposition. Observations on intertidal mats of Abu Dhabi show that decomposition rates are more sensitive to drainage patterns than production rates are. Water shortage, on the other hand, is likely to affect production rates directly, while decomposition continues for a while unimpaired protected by the overlying mat. The thickness of organic-poor stromatolitic laminae may depend on sediment supply. Stromatolites that trap sediment particles carried by tidal currents (Gebelein and Hoffman 1968), may record lunar cyclicity. Storm-associated sediment transport is often reflected in organic-poor laminae of varying thickness. If storms cluster in a seasonal pattern, such laminae may record a solar cyclicity.

With so many interacting sources of lamination, is it reasonable to expect that palaeoastronomic interpretations could be reliably made from stromatolite lamination patterns (Pannella et al. 1968; Awramik and Vanyo 1986)? This may be so, providing there is some limitation in the number of irregular disturbances in the palaeoenvironment. Subtidal stromatolites growing continuously in relatively protected and calm environments should qualify as the best potential models for such studies. However, this hypothesis is difficult to test, as no such model has been found in modern environments (*Phormidium hendersonii* has a short season of about 2 months).

5 Preservation and Lithification of Modern Stromatolites

Living microbial mats are metabolically active ecosystems which effectively recycle nutrients. Aerobic decomposition of organic matter is almost complete, leaving very little contribution for the fossil record. Under anaerobic conditions (e.g. in water-logged mats) complete decomposition of organic matter is possible (Doemel and Brock 1977), but it involves a succession of several metabolically different groups of microbes. Organic matter is more likely to be preserved and fossilized anaerobically, because the required multimembered, slower-growing set of anaerobic degraders is, statistically speaking, less likely to congregate, diminishing the potential for complete removal. Sedimentary components of the mat, unless consolidated, may easily fall victim to erosion or pressure distortion. The best chance for stromatolite preservation (by no means a requirement) is achieved through early lithification. Most commonly, this

Plate I

involves cementation by calcium carbonate. However, early lithification is not conducive to preservation of microbial fossils, which are best preserved in silicified sediments (see Knoll 1989). (Embedding in silica, iron or manganese occurs under special conditions and will not be discussed here.)

Biogenic destruction of lithified stromatolite surfaces by endolithic microorganisms that leave specific traces (microborings) has been demonstrated in marine (Golubic 1983) and freshwater (Anagnostidis and Pantazidou 1988; Winsborough et al. 1990) environments (Plate I, 7, 8). Microbial endoliths are known from pre-Phanerozoic oolites (Campbell 1982; Knoll et al. 1986), as well as stromatolites (Zhang and Golubic 1987).

6 Abiotic and Biogenic Factors in Calcium Carbonate Deposition

Carbonate deposition is often an integral part of stromatolite formation. The role of microbial activity in this process is not easily determined and remains controversial. In the process of calcification the carbonate ion passes through several phases, starting from an aqueous solution and ending as calcium carbonate within a limestone rock. The contribution of microorganisms in each of these phases needs to be assessed separately.

In its initial phase, the process of carbonate precipitation belongs to the realm of solution chemistry. The solubility of different chemical species of the calcium carbonate buffer system in natural waters is under constraints of temperature, pH and overall ionic composition. The solution can reach states of saturation and supersaturation, which may lead to precipitation of $CaCO_3$ from solution. This process can be influenced by abiotic factors such as evaporation. It may also be influenced by biotic factors such as photosynthetic removal of CO_2 and HCO_3^-, or heterotrophic ammonia release and sulfate reduction, all of these being associated with pH increase.

Following chemical precipitation from solution, carbonate molecules organize, following a crystal lattice. This process belongs to the realm of mineral

Plate I. *1* Aragonite particles trapped on the surface of the polysaccharide sheath of the marine cyanobacterium *Microcoleus chthonoplastes*. Florida Keys. Nomarski Interference contrast; scale *bar* is 10 μm. *2* Calcite precipitated on the surface of the polysaccharide sheath of the freshwater cyanobacterium *Scytonema julianum*. Nomarski Interference contrast; same scale as in *1*. *3* Species-specific calcite skeleton formed by the desmid *Oocardium stratum*. The entire cell population formed a single calcite monocrystal. The cells are removed in this preparation using sodium hypochlorite (Clorox). SEM; scale *bar* is 100 μm. *4* Detail from *3*; scale *bar* is 20 μm. *5* Calcitic tubules formed within the sheath of the cyanobacterium *Geitleria calcarea*. SEM; scale *bar* is 20 μm. *6* Detail from Fig. 5 showing the monocrystalline structure and orientation of calcite; scale *bar* is 1 μm. *7* Endolithic filaments of the calcium carbonate penetrating freshwater cyanobacterium in stromatolites of the El Mojarral, Cuatro Cienegas, Mexico, 3 m depth. *Arrow* indicates a smaller endolith *Plectonema* sp. Nomarski interference contrast; scale same as in *8*. *8* Resin casts of endoliths microborings produced by the cyanobacterium *Plectonema* sp. Same sample as in *7*. SEM; scale *bar* is 10 μm formation and is commonly referred to as 'nucleation'. It may depend on inorganic factors such as temperature or the presence of other ions or colloids, that serve as potential nucleation sites that compete for carbonate molecules. Biogenic influences on the nucleation process may involve structural macromolecules (e.g. polysaccharides and polypeptides) or metabolic products that come into contact with the carbonate ions. They may promote or retard nucleation. Nucleation and crystal growth occurs either in the water column, forming a suspension (sometimes referred to as a 'whiting'), or at the sediment-water interface.

Ultimately, all the precipitated carbonate settles and contributes to the formation of sediments. The rules that are followed thereafter belong to the realm of sedimentology. This phase includes depositional and diagenetic processes in which prominent abiotic factors include current and wave energies, and pore water chemistry, respectively. Biogenic modifications include trapping and binding of particles, bioturbation, as well as release of metabolites.

The abiotic and biogenic influences may be qualitatively different in each of the sequential phases of carbonate deposition, and their relative importance may shift from phase to phase. Keeping them conceptually separate as 'chemical', 'mineralogical' and 'sedimentological', may help to resolve some controversies that persist in the literature merely because the disagreeing parties are referring to different parts of a multifaceted, complex process.

7 Degree of Biological Control

One can view the cell as a special type of microenvironment in which the processes are subject to tight biological control, principally through biocatalytic action of enzymes. By extension, in multicellular organisms, tissues and intercellular spaces comprise microenvironments where the combined activity of many cells exercises its influence. A decreased degree of biological control is expected outside the organism, within its immediate surrounding that comprises its microenvironment. The degree of biotic influence here will depend on the size of this microenvironment in relation to the size and activity of the organisms within it, as well as on the degree of seclusion of this microenvironment from the environment at large – the macroenvironment. It is increasingly becoming recognized that there is no environment on Earth that has not been modified by biotic processes throughout the geological past (e.g. Westbroek 1983). We have just grown accustomed to taking the present state of the environment as a given background, against which we measure the influences of organisms that are presently active.

Mineral deposition under close biological control also follows the organization of biological structures (templates) resulting in species-specific skeletons. A comprehensive comparative treatment of biomineralization processes has been recently published by Lowenstam and Weiner (1989). Eukaryotes have produced, and left in the fossil record, a wide variety of skeletons and skeletal fragments. Mineral deposits associated with prokaryotic organisms are usually not considered to be skeletal. However, clear distinction between skeletal and non-skeletal mineral deposits is not always possible, because the degree of biological control varies from phase to phase of the process, and from taxon to taxon in eukaryotes and prokaryotes alike. Similarly, the distinction between biologically induced and biologically controlled mineralization processes is a conceptual simplification, whereas in the real world there is a wide spectrum of possibilities with different degrees of biological vs environmental control (Lowenstam and Weiner 1989).

A common characteristic of living systems is their ability to act against chemical gradients. This is exemplified at the cellular level by the active transport of nutrients across the plasma membrane, and at the environmental level by the biogenic polarization of reduced and oxidized microenvironments in a mat. Thus, in assessing the biogenicity of mineral precipitation in a complex natural situation, it is good strategy to study situations with known environmental 'resistance' to the process, against which the biological impact is well contrasted. The opposite is true for situations where both abiotic and biotic conditions favour a process, and it is difficult to separate their respective contributions. This may explain some controversies accompanying the assessment of the organismal impact on exergonic oxidation processes (e.g. oxidation of iron and manganese) along redox interfaces or, for that matter, on the precipitation of calcium carbonate from heavily supersaturated waters.

Marine waters of normal salinity, although often chemically supersaturated with respect to calcium carbonate, show a strong resistance to spontaneous precipitation. Calcification is therefore largely under direct organismal control resulting in production of various skeletons. The prevalent crystallization form of $CaCO_3$ in sea water is aragonite, biogenic calcification, however, frequently takes the mineral form of calcite. (There is a possibility, however, that at different times of the Earth's history the overall ocean chemistry may have been different, and more conducive to spontaneous precipitation of calcite rather than aragonite). Low Mg-calcite precipitation is today prevalent in nonmarine waters of normal temperatures, however, rapid precipitation from hot springs is often aragonitic. The following examples illustrate different degrees of biological impact on the calcification process.

In eukaryotes, calcitic coccoliths of species-specific morphology are examples of marine intracellular calcification (Westbroek et al. 1984). Mg-calcite is deposited in the cell walls of crustose rhodophytes in a microenvironment that is technically extracellular (outside plasma membrane), but in intimate contact with cellular tissues. Open intracellular spaces in marine siphonaceous chlorophytes are filled with aragonitic encrustations, and show correspondingly less cellular control over the calcification process, although the initial precipitation seems to be guided by biogenic templates (Borowitzka 1984).

In prokaryotes, biogenic carbonate encrustations of species-specific morphology is found in two marine cyanobacteria *Rivularia mesenterica* and *R. polyotis*. It takes place in an environment in which spontaneous carbonate precipitation is not known to occur (Adriatic Sea). Calcium carbonate is deposited outside cells, but within an extracellular polysaccharide gel. The shapes of the concretion in the two species differ markedly, indicating specifically different modes of nucleation templating. The mineral form of the precipitate is aragonite in both cases, although the initial deposition appears amorphous (Golubic and Campbell 1981).

Spontaneous precipitation of calcium carbonate from supersaturated freshwater occurs readily, making determination of biogenic control of the process more difficult. Critical differences between calcifying and non-calcifying taxa, however, can be observed in waters close to equilibrium or slightly undersaturated with respect to calcium carbonate. Under such conditions, within the same microenvironment, the cyanobacterium *Phormidium incrustatum* is heavily calcified while *Ph. autumnale* is not (Golubic and Fischer 1975). The fact that both species are found calcified in highly supersaturated waters of the Plitvice waterfalls (Yugoslavia) masks this distinction. Similarly the cyanobacterium *Scytonema julianum* is consistently covered with a coat of characteristic $CaCO_3$ crystals (Plate I, 2; see Couté 1985), while *S. myochrous* remains completely uncalcified. Species-specific templating and imprinting of characteristic biogenic patterns over those determined by inorganic mineral lattices is evident in many freshwater stromatolite-building organisms, both prokaryotic and eukaryotic. Biologically modified crystals occur in the cyanobacteria *Rivularia haematites* (Schneider 1977), *Schizothrix fasciculata*, *Hydrocoleum homoeotrichum*, and *Geitleria calcarea* (Plate I, 5, 6; Friedmann 1955; Couté 1985), in the filamentous chlorophyte *Gongrosira incrustans* (Golubic and Fischer 1975), and desmid *Oocardium stratum* (Plate I, 3, 4). Carbonate deposition on gelatinous stalks of benthic diatoms also carries biologically specific marks. It takes place at a certain distance from the cells indicating that it is the nucleation of calcium carbonate crystals on the polysaccharide cell product, rather than metabolic activity (photosynthesis) of the cell, that directly affects calcification (Winsborough and Golubic 1987).

The crystals in all these taxa conform to some degree with the mineralogically expected rhombohedral form. However, they are deposited tightly around templating cyanobacterial or algal products and bear their species-specific imprints. In this sense they can be considered skeletal (Riding 1977).

Experimental evidence from cultures shows that heterotrophic bacteria engage in activities that result in an increase in pH (e.g. sulfate reduction or ammonia release), and can cause precipitation of carbonates in their vicinity. Bacterial precipitation of calcium carbonate in nature has also been reported (Krumbein 1974). The evidence is based on findings of carbonate-encased bacteria in the hypersaline environment of Solar Lake, Sinai. Under the circumstances of critical supersaturation, the actual role of these bacteria in the calcification process is not clear. Their role may have been metabolic, promoting selective nucleation of chemically precipitated minerals, or entirely passive in the sence of serving as incidental surfaces.

The experimental evidence that particular bacteria in pure culture have the capacity to initiate mineral precipitation does not necessarily explain the origin of the minerals found in natural environments, where processes may act synergistically or in opposition to each other. For example, under natural conditions, organic matter is decomposed via fermentative and respirative processes in which H_2, CO_2, and organic acids are released, generally lowering the environmental pH. Some processes, such as deamination and sulfate reduction act in the opposite direction, elevating pH. The balance between these processes and their microenvironmental distribution need to be known before the ultimate effect of bacteria on carbonate dissolution or precipitation in particular cases can be assessed.

8 Are Stromatolites Intertidal, Subtidal or Both?

Important information about ancient stromatolites, which is expected from the study of their modern counterparts, entails assessment of the conditions under which they form, in order to facilitate their use in reconstruction of palaeoenvironments. Again, we need to resort to particular case histories. The study of modern stromatolites of Shark Bay, Australia has provided most of our current knowledge, but has also spawned serious misconceptions, which now need to be sorted out. Logan (1961) described several important features of one type of intertidal stromatolite (which we now know to be formed by the cyanobacterium *Entophysalis major*, see Golubic 1973a, 1983; Golubic and Hofmann 1976). Logan established the wave exposure-dependence of their domal morphology, assessing the role of erosion and wave scouring on their shape and directional orientation. By extrapolation from these observations he concluded that all stromatolites, modern and fossil, must be intertidal in origin and are shaped by environmental forces prevalent there. The presence of modern subtidal stromatolites in Hamelin Pool were dismissed by Logan as relics of intertidal stromatolites of earlier times when water levels were lower than today.

On the other extreme, some 20 years later, Burne and James (1986) claim that subtidal ranges of the Hamelin Pool are the real home of stromatolitic structures. They explain today's intertidal stromatolites as being relics of earlier higher water levels, stranded by the relative fall in sea level, and recolonized by intertidal microbial mats. Which stromatolites are actively accreting today and which can be considered fossil? Marking of subtidal stromatolites with stainless steel nails and following the subsequent increments showed that their accretion is patchy and slow (Playford and Cockbain 1976). In the intertidal zone, the co-existence of growing *Entophysalis* mat and hard lithified surfaces, implies similar irregularity of the process.

In order to test the validity of either of these postulates, the following question needs to be answered for both subtidal and intertidal stromatolites: Are microbial mats that currently colonize stromatolite surface contributing to the accretion of the stromatolite upon which they grow, are they neutral to it, or perhaps even contributing to stromatolite destruction?

Subtidal stromatolites are colonized by a distinct (colloform) mat community composed of cyanobacteria and algae, and dominated by a coccoid cyanobacterium, new to science (Golubic 1985). The role of the eukaryotic component of this community is discussed by Awramik and Riding (1988).

Sediment particles are trapped and cemented by in situ carbonate precipitation. The same mineral elements that occur within the active mat are also found consolidated in the interior of the stromatolite. Thus, there is evidence for the contribution of the current microbiota to stromatolite accretion, and no evidence of any discontinuity in this process, that would separate the past, vestigial structures and processes from those occurring today.

Intertidal stromatolites are colonized by the mamillate (pustular) mat, an entirely different, much less diverse microbial community, which is dominated by the coccoid cyanobacterium *Entophysalis major*. For the most part of the year, this mat colonizes loose sediment and is shaped by wave action and erosion (Plate II, 11). At critical times at the peak of the austral summer, precipitous lithification occurs within the *Entophysalis* mat, destroying the microorganisms in the process. Thus, there is evidence that the soft *Entophysalis* mat turns periodically into hard stromatolite (Plate II, 11, arrow), contributing significantly to its accretion. In the intermittent times between two colonizations by *Entophysalis* the stromatolites are attacked by carbonate-boring endolithic microflora (Golubic 1983).

In addition to the two stromatolite types under discussion, several others, which are built by different microorganisms, show different degrees of lithification and occur in distinct zones. The shapes, densities and orientation of subtidal, lower intertidal, and upper intertidal stromatolites are all different (compare illustrations in Golubic 1985) and cannot be explained as 'inherited' from one another as Logan (1961) or Burne and James (1986) suggested. As compared with the predominantly large, round-based subtidal stromatolites (Golubic 1985, Pl. 1.1, Fig. 3), the lower intertidal ones are smaller and denser (Golubic 1985, Pl. 1.1, Fig. 6), whereas the upper intertidal ones are often elongated, following the prevailing wind direction and wave scouring (Golubic 1985, Pl. 1.2, Fig. 2), an activity that the subtidal stromatolites are hardly exposed to. We deal here with separate, but active ranges in which stromatolite formation and lithification takes place, each formed by different microbiota and lithified differently.

Among modern stromatolites some are intertidal, and others subtidal. In all probability, the same should be true for their fossil counterparts. The case

Plate II. *9* Heliotropic pinnacles built by the cyanobacterium *Schizothrix splendida* occupying slightly elevated points on top of a periodically waterlogged, *Microcoleus chthonoplastes* dominated mat. Intertidal flats east of Abu Dhabi, Arabian Gulf (Photo courtesy of R. Park). *10* Vertical section through one of the heliotropic pinnacles in Fig. 9 showing continuity of pinnacle differentiation in the process of mat accumulation (Photo courtesy of R. Park). *11* Domal stromatolites 'carved' by wave erosion from a contiguous sediment platform (in the *background*). Sediments are stabilized by the mamillate mat dominated by the coccoid cyanobacterium *Entophysalis major*. *Arrow* marks the lithified portion of the stromatolite. Hamelin Pool, Shark Bay, Western Australia. *12* Large subtidal stromatolites built by trapping and binding of oolitic sands, Exuma Island, Bahamas (Photo courtesy of S.M. Awramik). *13* Freshwater domal stromatolites in Pozo Azul, Cuatro Cienegas, Mexico

Plate II

history of Shark Bay teaches us, among other things, that simple sweeping generalizations, no matter how attractive, are unlikely to be useful as interpretation models.

9 Are Modern Stromatolites Restricted to Extreme Environments?

Diversity of forms, vast distribution ranges and absolute dominance of stromatolites in the pre-Phanerozoic have always intrigued researchers, particularly when contrasted with their relative paucity in the Phanerozoic. Awramik (1971) has shown the decline of stromatolites in the fossil record at the end of the Proterozoic to coincide with the appearance of skeleton bearing organisms. Garrett (1970) offered an explanation for this decline showing that in modern marine environments the formation of microbial mats, as precursors of stromatolites is effectively prevented by the grazing pressure of invertebrates. Monty (1973) later extended this view, by including competition with newly evolving reef-forming organisms. These may have caused displacement of stromatolites from normal marine environments into the less favorable intertidal zones and into freshwaters.

Intertidal ranges and ephemeral pools within them, with their fluctuating water supply and chemistry, particularly in arid climates certainly qualify to be called extreme environments. The studies of such environments (e.g. Kinsman and Park 1976; Park 1976, 1977; Horodyski et al. 1977; Krumbein et al. 1977; Campbell and Golubic 1985) show organismal distribution patterns that are consistent with Garrett's model, and confirm an almost absolute dominance of prokaryotes. Similarly, extreme conditions in the non-marine realm are found in thermal springs (Brock 1978; Castenholz 1973; Ward et al. 1989), and somewhat less extreme in the Antarctic lakes (Parker et al. 1981; Parker and Wharton 1985). However, some modern stromatolites persist in normal marine environments as well as in nonmarine environments with less extreme conditions (Moore et al. 1984; Dill et al. 1986). In addition, fossil Phanerozoic stromatolites are often found surrounded by normal marine floras and faunas (see Monty 1981). These findings testify that in many environments stromatolites did survive competition and accommodated animal grazing, in an apparent contradiction to the Garrett's dictum, and require a different explanation.

Although the research of the Exuma stromatolites (Plate II, 12; Dill et al. 1986) is still in its initial stages, several peculiar properties of this environment are known. They grow under a regime of strong tidal currents that carry loads of ooidal sands. The key to their accretion is the ability of stromatolite building microorganisms to trap the sediments and escape burial. The key to their persistence lies probably in rapid cementation and induration of these structures. Whether, and to what extent this process is facilitated by organisms remains to be established.

Carbonate deposition in non-marine setting occurs predominantly as tufa (cool water travertine), forming porous biohermal structures which show certain stromatolitic properties such as lamination. They are usually associated with intensive in situ precipitation of carbonate from springs and running waters that are supersaturated with respect to $CaCO_3$, although they also involve participation of micro- and macroorganisms. Their early lithification seems responsible for their preservation and accretion (Golubic 1967, 1973a). Benthic carbonate deposition in lakes has been known to result in formation of oncoids (Irion and Müller 1968), and calcareous crusts, which usually do not exceed a thickness of about 1 cm. Their accretion is counteracted by bacterial degradation and corrosion (Golubic 1962), destruction by endoliths (Schneider 1977), probably combined with animal grazing and wave and, in cooler regions, ice erosion (Kann 1941).

Lacustrine stromatolites in lakes with normal freshwater chemistry, are rare and known only from warmer climates. In habitats where such lacustrine stromatolites occur, they accommodate rich and diverse animals populations, which include grazers, such as snails and fish (Winsborough et al. 1990). Well-developed lacustrine stromatolites are found on the margins of limnocrene springs, locally called 'pozos' in Cuatro Cienegas basin, Coahuila, Mexico (Plate II, 13). They are domal in outline, with corrugated brain-like surface configuration. The surface of these stromatolites is soft, and loaded with trapped faecal pellets. Toward the interior, these stromatolites become gradually harder due to cementation. They grow best in shallow water, frequently coalescing into reef-like bioherms, that prograde from the shore toward the centre of the 'pozo' creating platforms with overhangs. With increasing depth, they show evidence of decreased accretion rate, accompanied by increase in hardness, and in deepest ranges they show signs of corrosion (Plate I, 7, 8).

10 Conclusions

Modern stromatolites offer an opportunity to study relationships between microbiota and sedimentary structures within their ecological context, and to follow their genesis. Actualistic models can be postulated and experimentally tested. As such they can provide useful interpretational tools for fossil stromatolites. However, a number of modern stromatolite settings are not known from the fossil record and, conversely, most of the known ancient stromatolites do not have close modern counterparts. Thus, the studies of modern stromatolites rarely allow direct comparisons. They provide, however, information of more general nature, about the principles and ecological constraints involved. We learn that the shapes of modern stromatolites may be biologically controlled by growth habits of dominant stromatolite-building microorganisms. Hence, there is a need for accurate taxonomic identification of these builders. The environment may act on two levels: by selecting for or against microorganisms with particular growth requirements, and by physically trimming the shape of the resulting structure. These interactions determine the distribution patterns of stromatolites. Like a coral reef, a stromatolite is a habitat as well as a growing structure. The role of organisms associated with it can be constructive, neutral, or destructive. The balance of combined constructive and destructive forces determines the stromatolite accretion rates and preservability.

Acknowledgements. Barbara and Philip Winsborough are thanked for field guidance in Cuatro Cienegas, and Jacob Seeler, Susan E. Campbell, and Stanley M. Awramik for critically reading the manuscript and offering valuable suggestions. The work has been partially supported by the NSF grants EAR 8306179, INT-8509365, and Boston University grant GRS 8021-BI.

References

Anagnostidis K, Pantazidou A (1988) *Hyella kalligrammos* sp. nov., *Hyella* maxima (Geitl.) comb. a nov., and other freshwater morphotypes of the genus *Hyella* Born. et Flah. (Chroococcales, Cyanophyceae). In: Anagnostidis K, Golubic S, Komárek J, Lhotsky O (eds) Cyanophyta (Cyanobacteria): morphology, taxonomy, ecology. Arch Hydrobiol Suppl 80 (Algol Stud 50/53):227–247

Awramik SM (1971) Precambrian columnar stromatolite diversity: reflection of metazoan appearance. Science 174:825–827

Awramik SM (1976) Selective subject index to the bibliography. In: Walter MR (ed) Stromatolites. Developments in sedimentology, vol 20. Elsevier, Amsterdam, pp 697–704

Awramik SM, Riding R (1988) Role of algal eukaryotes in subtidal columnar stromatolite formation. Proc Natl Acad Sci USA 85:1327–1329

Awramik SM, Vanyo JP (1986) Heliotropism in modern stromatolites. Science 231:1279–1281

Awramik SM, Hofmann HJ, Raaben ME (1976) Bibliography. In: Walter MR (ed) Stromatolites. Developments in sedimentology, vol 20. Elsevier, Amsterdam, pp 705–771 (2034 references, 433 on modern stromatolites)

Awramik SM, Haupt A, Hofmann HJ, Walter MR (1979) Stromatolite bibliography 2. Precambrian Res 9:105–166 (1238 references, 161 on modern stromatolites)

Borowitzka MA (1984) Calcification in aquatic plants. Plant Cell Environ 7:457–475

Brock TD (1978) Thermophilic microorganisms and life at high temperatures. Springer, Berlin Heidelberg New York, 465 pp

Burne DR (1986) Creation of an international collaboration programme: stromatolites. IGCP Project 261. Stromatolite Newslett 13:1–5

Burne RV, James N (1986) Subtidal origin of club-shaped stromatolites, Shark Bay. In: Sediments down-under, 12th Int Sedimentol Congr, Canberra, Aust, Abstr, p 49

Campbell SE (1982) Precambrian endoliths discovered. Nature (London) 299:429–431

Campbell SE, Golubic S (1985) Benthic cyanophytes (cyanobacteria) of Solar Lake (Sinai). In: Golubic S, Komárek J, Lhotsky O (eds) Cyanophyta (Cyanobacteria): morphology, taxonomy, ecology. Arch Hydrobiol Suppl 71 (Algol Stud 38/39):311–329

Castenholz RW (1973) Ecology of blue-green algae in hot springs. In: Carr N, Whitton BA (eds) The biology of blue-green algae. Blackwell, Oxford, pp 379–414

Cohen Y (1989) Photosynthesis in cyanobacterial mats and its relation to the sulfur cycle: a model for microbial sulfur interactions. In: Cohen Y, Rosenberg E (eds) Microbial mats. Am Soc Microbiol, Washington, pp 22–36

Cohen Y, Krumbein WE, Goldberg M, Shilo M (1977) Solar Lake (Sinai). 1. Physical and chemical limnology. Limnol Oceanogr 22:597–608

Couté A (1985) Essai préliminaire de comparaison de deux Cyanophycées cavernicoles calcifiées: *Geitleria calcarea* Friedmann et *Scytonema julianum* Meneghini. In: Golubic S, Komárek J, Lhotsky O (eds) Cyanophyta (Cyanobacteria): morphology, taxonomy, ecology. Arch Hydrobiol Suppl 71 (Algol Stud 38/39):91–98

de Wit K (1989) Interactions between phototrophic bacteria in marine sediments. PhD Thesis, University of Groningen, The Netherlands, 143 pp
Dill RF, Shinn EA, Jones AT, Kelly K, Steinen RP (1986) Giant subtidal stromatolites forming in normal salinity waters. Nature (London) 324:55–58
Doemel WN, Brock TD (1977) Structure, growth, and decomposition of laminated algal-bacterial mats in alkaline hot springs. Appl Environ Microbiol 34:433–452
Friedmann I (1955) *Geitleria calcarea* n.gen. et n.sp. A new atmophytic lime-encrusted blue-green alga. Bot Not 108:439–445
Garrett P (1970) Phanerozoic stromatolites: noncompetitive ecologic restriction by grazing and burrowing animals. Science 169:171–173
Gebelein CD (1974) Biologic control of stromatolite microstructure: implication for Precambrian time stratigraphy. Am J Sci 274:575–598
Gebelein CD, Hoffman P (1968) Intertidal stromatolites from Cape Sable, Florida. Geol Soc Am Spec Pap 121:109 (Abstract)
Golubic S (1962) Zur Kenntnis der Kalkinkrustation und Kalkkorrosion in Seelitoral. Schweiz Z Hydrol 24:229–243
Golubic S (1967) Algenvegetation der Felsen, eine ökologische Algenstudie im dinarischen Karstgebiet. Binnengewässer 23:1–183
Golubic S (1973a) The relationship between blue-green algae and carbonate deposits. In: Carr N, Whitton BA (eds) The biology of blue-green algae. Blackwell, Oxford, pp 434–472
Golubic S (1973b) Three new species of *Schizothrix* Kützing (Cyanophyta) from marine algal mats. Schweiz Z Hydrol 35:152–156
Golubic S (1976a) Organisms that build stromatolites. In: Walter MR (ed) Stromatolites. Developments in sedimentology, vol 20. Elsevier, Amsterdam, pp 113–126
Golubic S (1976b) Taxonomy of extant stromatolite building cyanophytes. In: Walter MR (ed) Stromatolites. Developments in sedimentology, vol 20. Elsevier, Amsterdam, pp 127–140
Golubic S (1983) Stromatolites, fossil and recent: a case history. In: Westbroek P, de Jong EW (eds) Biomineralization and biological metal accumulation. Reidel, Dordrecht, pp 313–326
Golubic S (1985) Microbial mats and modern stromatolites in Shark Bay, Western Australia. In: Caldwell DE, Brierley JA, Brierley CL (eds) Planetary ecology. Van Nostrand Reinhold, New York, pp 3–16
Golubic S, Campbell SE (1981) Biogenically formed aragonite concretions in marine *Rivularia*. In: Monty CLV (ed) Phanerozoic stromatolites. Springer, Berlin Heidelberg New York, pp 209–229
Golubic S, Fischer AG (1975) Ecology of calcareous nodules forming in Little Connestoga Creek near Lancaster, Pennsylvania. Verh Int Verein Limnol 19:2315–2323
Golubic S, Focke JW (1978) *Phormidium hendersonii* Howe: identity and significance of a modern stromatolite building microorganism. J Sediment Petrol 48:751–764
Golubic S, Hofmann HJ (1976) Comparison of modern and mid-Precambrian Entophysalidaceae (Cyanophyta) in stromatolitic algal mats: cell division and degradation. J Paleontol 50:1074–1082
Guerrero R, Mas J (1989) Multilayered microbial communities in aquatic ecosystems: growth and loss factors. In: Cohen Y, Rosenberg E (eds) Microbial mats. Am Soc Microbiol, Washington, pp 37–51
Hoffman PF (1967) Algal stromatolites: use in stratigraphic correlation and paleocurrent determination. Science 157:1043–1045
Hofmann HJ (1969) Attributes of stromatolites. Geol Surv Can Pap 69-39:1–58
Horodyski RJ, Bloeser B, von der Haar S (1977) Laminated algal mats from a coastal lagoon, Laguna Mormona, Baja California, Mexico. J Sediment Petrol 47:680–696
Irion G, Müller G (1968) Mineralogy, petrology, and chemical composition of some calcareous tuffa from the Schwäbische Alb, Germany. In: Müller G, Friedman GM (eds) Carbonate sedimentology in Central Europe. Springer, Berlin Heidelberg New York, pp 157–171
Jorgensen BB (1989) Light penetration, absorption, and action spectra in cyanobacterial mats. In: Cohen Y, Rosenberg E (eds) Microbial mats. Am Soc Microbiol, Washington, pp 123–137
Kalkowsky E (1908) Oolith and Stromatolith im norddeutschen Buntsandstein. Z Dtsch Geol Ges 60:68–125

Kann E (1941) Krustensteine in Seen. Arch Hydrobiol 37:504–532
Kinsman JJ, Park RK (1976) Algal belt and coastal sabkha evolution, Trucial Coast, Persian Gulf. In: Walter MR (ed) Stromatolites. Developments in sedimentology, vol 20. Elsevier, Amsterdam, pp 421–433
Knoll AH (1989) The paleomicrobiological information in Proterozoic rocks. In: Cohen Y, Rosenberg E (eds) Microbial mats. Am Soc Microbiol, Washington, pp 469–484
Knoll AH, Golubic S, Green J, Swett K (1986) Organically preserved microbial endoliths from the Late Proterozoic of East Greenland. Nature (London) 321:856–857
Krumbein WE (1974) On the precipitation of aragonite on the surface of marine bacteria. Naturwissenschaften 61:167
Krumbein WE (1982) Stromatolites – the challenge of a term through space and time. Biogeochem Newslett 1:26–28
Krumbein WE (1983) Stromatolites – the challenge of a term in space and time. Precambrian Res 20:493–531
Krumbein WE, Cohen Y, Shilo M (1977) Solar Lake (Sinai). 4. Stromatolitic cyanobacterial mats. Limnol Oceanogr 22:635–656
Krylov IN (1963) Columnar branching stromatolites of the Riphean deposits of the southern Urals and their importance for the stratigraphy of the Upper Precambrian. Tr Geol Inst Leningrad 69:1–133 (in Russian)
Krylov IN (1976) Approaches to the classification of stromatolites. In: Walter MR (ed) Stromatolites. Developments in sedimentology, vol 20. Elsevier, Amsterdam, pp 31–43
Logan BW (1961) Cryptozoan and associate stromatolites from the Recent of Shark Bay, Western Australia. J Geol 69:517–533
Logan BW, Rezak R, Ginsburg RN (1964) Classification and environmental significance of algal stromatolites. J Geol 72:68–83
Lowenstam HA, Weiner S (1989) On biomineralization. Oxford University Press, Oxford, 324 pp
MacArthur R (1960) On the relative abundance of species. Am Nat 94:25–36
Maslov VP (1960) Stromatolity. Tr Geol Inst Akad Nauk SSSR 41:3–188
Monty CLV (1965) Recent algal stromatolites in the Windward Lagoon, Andros Island, Bahamas. Ann Soc Geol Belg Bull 88:269–276
Monty CLV (1973) Precambrian background and Phanerozoic history of stromatolitic communities, an overview. Ann Soc Geol Belg Bull 96:585–624
Monty CLV (ed) (1981) Phanerozoic stromatolites. Springer, Berlin Heidelberg New York, 249 pp
Monty CLV (1982) Editorial. Stromatolite Newslett 9:1–11
Moore L, Knott B, Stanley N (1984) The stromatolites of lake Clifton, Western Australia. ANZAAS Search 14, pp 309–314
Pannella G, MacClintock C, Thompson MN (1968) Paleontological evidence of variations in length of synodic month since Late Cambrian. Science 162:792–796
Park RK (1976) A note on the significance of lamination in stromatolites. Sedimentology 23:379–393
Park RK (1977) The preservation potential of some recent stromatolites. Sedimentology 24:485–506
Parker BC, Wharton RA Jr (1985) Physiological ecology of bluegreen algal mats (modern stromatolites) in Antarctic oasis lakes. In: Golubic S, Komárek J, Lhotsky O (eds) Cyanophyta (Cyanobacteria): morphology, taxonomy, ecology. Arch Hydrobiol Suppl 71 (Algol Stud 38/39):331–348
Parker BC, Simmons GM Jr, Love FG, Seaburg KG, Wharton RA Jr (1981) Modern stromatolites in Antarctic Dry Valley lakes. BioScience 31:656–661
Playford PE, Cockbain AE (1976) Modern algal stromatolites at Hamelin Pool, a hypersaline barred basin in Shark Bay, Western Australia. In: Walter MR (ed) Stromatolites. Developments in sedimentology, vol 20. Elsevier, Amsterdam, pp 389–411
Raaben ME (1969) Columnar stromatolites and Late Precambrian stratigraphy. Am J Sci 267:1–18
Riding R (1977) Skeletal stromatolites. In: Flügel E (ed) Fossil algae, recent results and developments. Springer, Berlin Heidelberg New York, pp 57–60
Schneider J (1977) Carbonate construction and decomposition by epilithic and endolithic microorganisms in salt and freshwater. In: Flügel E (ed) Fossil algae, recent results and developments. Springer, Berlin Heidelberg New York, pp 248–260

Semikhatov M (1976) Experience in stromatolite studies in the USSR. In: Walter MR (ed) Stromatolites. Developments in sedimentology, vol 20. Elsevier, Amsterdam, pp 337–357

Walter MR (1972) Stromatolites and biostratigraphy of the Australian Precambrian and Cambrian. Paleontology Spec Pap 11, 256 pp

Walter MR, Bauld J, Brock TD (1976) Microbiology and morphogenesis of columnar stromatolites (Conophyton, Vacerrilla) from hot springs in Yellowstone National Park. In: Walter MR (ed) Stromatolites. Developments in sedimentology, vol 20. Elsevier, Amsterdam, pp 273–310

Ward DM, Weller R, Shiea J, Castenholz RW, Cohen Y (1989) Hot spring microbial mats: anoxygenic and oxygenic mat of possible evolutionary significance. In: Cohen Y, Rosenberg E (eds) Microbial mats. Am Soc Microbiol, Washington, pp 3–15

Westbroek P (1983) Biological metal accumulation and biomineralization in a geological perspective. In: Westbroek P, de Jong EW (eds) Biomineralization and biological metal accumulation. Reidel, Dordrecht, pp 1–11

Westbroek P, de Jong EW, van der Wal P, Borman AH, de Vring JPM, Kok D, de Bruijn WC, Parker SB (1984) Mechanisms of calcification in the marine alga *Emiliania huxleyi*. Philos Trans R Soc London Ser B 304:435–444

Winsborough B, Golubic S (1987) Stromatolitic structures of inland waters built by diatoms. J Phycol 23:195–201

Winsborough BM, Maguire B Jr, Golubic S, Folk RL (1990) Fresh-water oncolites, coated pebbles, bioherms, and microbial crusts from a spring-fed lake in northeastern Mexico. In: Monty CLV (ed) Phanerozoic stromatolites, 2. Springer, Berlin Heidelberg New York (in press)

Zhang Y, Golubic S (1987) Endolithic microfossils (cyanophyta) from early Proterozoic stromatolites, Hebei, China. Acta Micropaleontol Sin 4:1–12

Index